To save time and space on architectural drawings, some words are abbreviated. Abbreviated means written in a shortened manner, such as HW represents Hot Water. Most abbreviations are written in capital letters. Most metric abbreviations are lowercase. Usually, there is no period after the abbreviation unless it could be confused for a word such as CAT, which represents catalog.

Carpenters must be familiar with these so that they can read architects' drawings.

Length	LG
Length overall	LOA
Light	LT
Linear	LIN
Linen closet	L CL
Live load	LL
Living room	LR
Long	LG
Louver	LV
Lumber	LBR
Machine stress rated	MSR
Main	MN
Manhole	MH
Manual	MAN.
Material	MATL
Maximum	MAX
Medicine cabinet	MC
Membrane	MEMB
Metal	MET.
Meter (the instrument)	M
Meter (metric length)	m
Minimum	MIN
Minute	(') or MIN
Miscellaneous	MISC
Mixture	MIX.
Model	MOD
Modular	MOD
Moisture content	MC
Molding	MLDG
Motor	MOT
Natural	NAT
Newton	N
Nominal	NOM
North	N
Not to scale	NTS
Number	NO.
Obscure	OB
On center	OC
Opening	OPNG
Opposite	OPP
Overall	OA
Overhead	OVHD
Panel	PNL
Parallel	PAR.
Part	PT
Partition	PTN
Penny (nail size)	d
Permanent	PERM
Perpendicular	PERP
Piece	PC
Plaster	PL
Plate	PL
Plumbing	PLMB
Pound	LB
Precast	PRCST
Prefabricated	PREFAB
Preferred	PFD
Quality	QUAL
Quantity	QTY
Radiator	RAD
Radius	R
Random	RDM
Random lengths	RL
Range	R
Receptacle	RECP
Reference	REF
Refrigerate	REF
Refrigerator	REF
Register	REG
Reinforce	REINF
Required	REQD
Return	RET
Riser	R
Roof	RF
Room	RM
Rough	RGH
Round	RD
Safety	SAF
Sanitary	SAN
Scale	SC
Schedule	SCH
Second	(") or SEC
Section	SECT
Select	SEL
Service	SERV
Sewer	SEW.
Sheathing	SHTHG
Sheet	SH
Shower	SH
Side	S
Siding	SDG
Similar	SIM
Sink	S
Soil pipe	SP
South	S
Specification	SPEC
Square	SQ
Square meter	m^2
Stairs	ST
Standard	STD
Steam	ST
Steel	STL
Stock	STK
Storage	STG
Street	ST
Structural	STR
Supply	SUP
Surface	SUR
Switch	SW
Symmetrical	SYM
System	SYS
Tangent	TAN.
Tar and gravel	T & G
Tarpaulin	TARP
Tee	T
Telephone	TEL
Television	TV
Temperature	TEMP
Terra-cotta	TC
Terrazzo	TER
Thermostat	THERMO
Thick	THK
Thousand	M
Through	THRU
Timber	TBR
Toilet	T
Tongue and groove	T & G
Total	TOT.
Tread	TR
Tubing	TUB.
Typical	TYP
Unfinished	UNFIN
Urinal	UR
Valve	V
Vapor proof	VAP PRF
Vent pipe	VP
Ventilate	VENT.
Vertical	VERT
Vitreous	VIT
Volt	V
Volume	VOL
Washing machine	WM
Water closet	WC
Water heater	WH
Waterproofing	WP
Watt	W
Weather stripping	WS
Weatherproof	WP
Weep hole	WH
Weight	WT
West	W
Width	W
Window	WDW
With	W
Without	WO
Wood	WD
Wrought iron	WI
Yard	YD

GENERAL CARPENTRY

GENERAL CARPENTRY

William P. Spence

Construction Engineering Technology
Pittsburg State University
PITTSBURG, KANSAS

Prentice-Hall, Inc., Englewood Cliffs, New Jersey 07632

Library of Congress Cataloging in Publication Data

Spence, William
General carpentry.

Includes index.
1. Carpentry. I. Title.
TH5604.S63 694 81-15735
ISBN 0-13-349209-5 AACR2

Editorial/production supervision by *Barbara A. Cassel*
Interior design by *Jayne Conte*
Page makeup by *Diane Koromhas*
Cover design by *Jayne Conte*
Manufacturing buyer: *Joyce Levatino*

Printed in the United States of America
10 9 8 7 6 5 4 3 2 1

ISBN 0-13-349209-5

Prentice-Hall International, Inc., *London*
Prentice-Hall of Australia Pty. Limited, *Sydney*
Prentice-Hall of Canada, Ltd., *Toronto*
Prentice-Hall of India Private Limited, *New Delhi*
Prentice-Hall of Japan, Inc., *Tokyo*
Prentice-Hall of Southeast Asia Pte. Ltd., *Singapore*
Whitehall Books Limited, *Wellington, New Zealand*

CONTENTS

3 HAND TOOLS 14

4 POWER TOOLS 41

5 SAFETY ON THE CONSTRUCTION SITE 66

6 CONSTRUCTION DRAWINGS, PERMITS, AND CODES 77

7 WOOD—A BASIC CONSTRUCTION MATERIAL 95

8 LOCATING AND LEVELING A BUILDING 121

9 FOOTINGS AND FOUNDATIONS 131

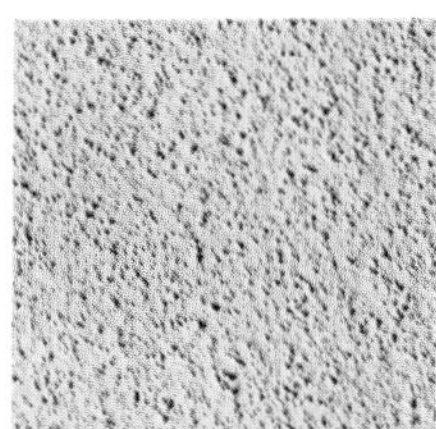

10 FLOOR CONSTRUCTION 166

11 WALL FRAMING 195

12 CEILING FRAMING 212

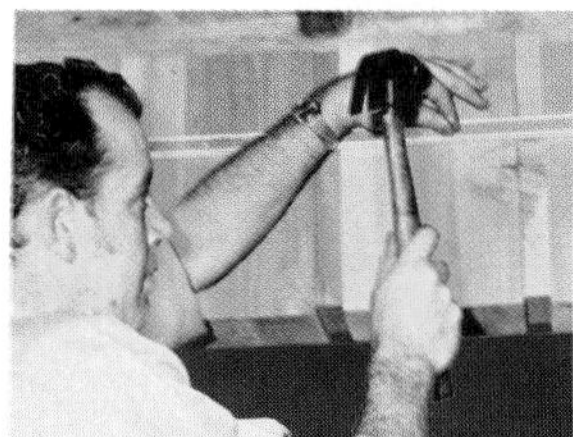

13

FRAMING THE ROOF 217

14

FINISHING THE ROOF 265

15

DOORS AND WINDOWS 291

16

FINISHING THE EXTERIOR 318

17

THERMAL INSULATION AND SOUND CONTROL 347

18 FINISHING THE INTERIOR 365

19 INSTALLING INTERIOR TRIM 404

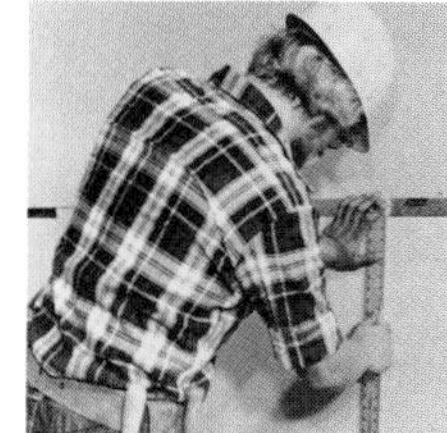

20 FINISH FLOORING 411

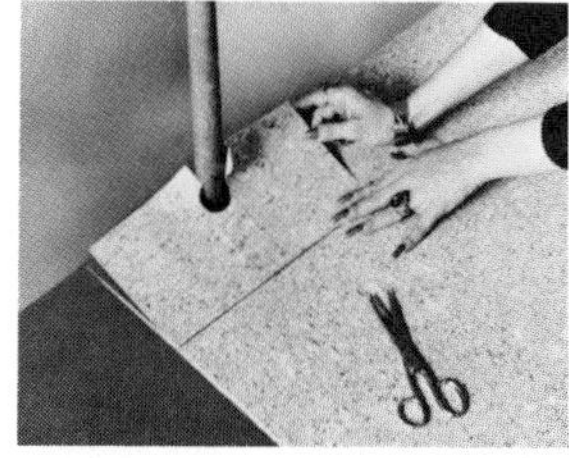

21 STAIRS 428

22 CABINETS AND OTHER BUILT-IN UNITS 449

PREFACE

General Carpentry has been designed to meet the needs of students in high school, vocational school, and community college carpentry and building construction programs. It emphasizes the basic fundamentals of carpentry. The student has the opportunity to learn the technical information and develop the skills necessary for success on the job. The text will be helpful for those enrolled in apprenticeship programs or individuals who desire to do their own construction and repair.

The text acquaints the reader with the career opportunities in carpentry and explains how a building is constructed. Considerable attention is given to the safe use of hand and power tools commonly used by carpenters. Construction is one of the more dangerous occupations. Therefore emphasis on construction safety is included. Every carpenter must thoroughly understand the safety practices expected on a job.

The carpenter is involved in assemblying the building so that it is level and true. Industrial practices to do this are stressed. In addition the carpenter must be able to work from the architect's drawings and specifications. The text explains how to read construction drawings and the use of typical specifications.

A carpenter works with a wide variety of materials. These are carefully examined and design data for frequently used items, such as beams, are included.

The text thoroughly covers the construction details beginning with forming the footing and foundation wall. Commonly used floor, wall, ceiling, and roof framings systems are covered in detail. Clear, carefully drawn details show every construction detail. Doors and windows are installed by carpenters. The various types available and installation details are included. Commonly used hardware installation steps are shown. There are many types of exterior finish. These are fully explained and illustrated.

An important part of a building is the finishing of the interior. The text covers the various types of interior wall finish and how they are installed. The final finishing step, the trim, is carefully explained and illustrated. Various finished flooring materials are detailed, along with installation instructions.

Energy conservation is becoming increasingly important. Thermal and sound insulation are covered in detail. Special items such as stair construction, cabinets, and built-in units are explained. This includes factory-built and custom-made units.

A brief report on factory manufactured building sections, panelized units, and precut members concludes the text. Many carpenters are employed in building these units and erecting them on the site.

The text contains hundreds of drawings and photographs to help the reader understand the techniques presented. An effort has been made to place the reading level at a point where most students can comprehend the written material. A second color has been used to help call attention to the important features on each illustration. All of these will help the learner grasp the meaning of the material.

The spelling of metric terms used in the United States has not been resolved at this time. In this text the two commonly used spellings, meter versus metre, millimeter versus millimetre, etc., are used interchangeably in the illustrations and the text so the reader can become accustomed to the use of both spellings.

This text would not be possible without the assistance of many companies and individuals who made contributions.

William P. Spence

GENERAL CARPENTRY

1 CARPENTRY AS A CAREER

The construction occupations are made up of skilled workers. They are the largest group of skilled workers in the U.S. work force.

Employment in the various construction areas is shown in Fig. 1-1. Several of the occupational areas have over 100,000 jobs. Of these the largest is that of the carpenters, with over 1 million jobs.

The construction occupations are divided into three major types: structural, finishing, and mechanical. *Structural* work includes the preparation for constructing a building and building the load-carrying frame and associated members

(Fig. 1-2). Carpenters are a part of the structural group.

The *finishing* occupations are listed in Fig. 1-3. Finishing work includes those parts of a building that are generally visible when finished and complete the structure (Fig. 1-4). Some of these workers, such as the roofer and the lather, use some of the same tools used by the carpenter (Fig. 1-5).

Mechanical work includes the installation of the plumbing, electrical, heating, air-conditioning, and ventilation systems (Fig. 1-6). These workers have to plan their work in cooperation with the structural work. The carpenter will have frequent contacts with those in the mechanical trades (Fig. 1-7).

WHERE DO CARPENTERS WORK?

Most carpenters are employed by general contractors. Some are self-employed and undertake work on their own. Sometimes they become experienced enough to form a small construction company of their own.

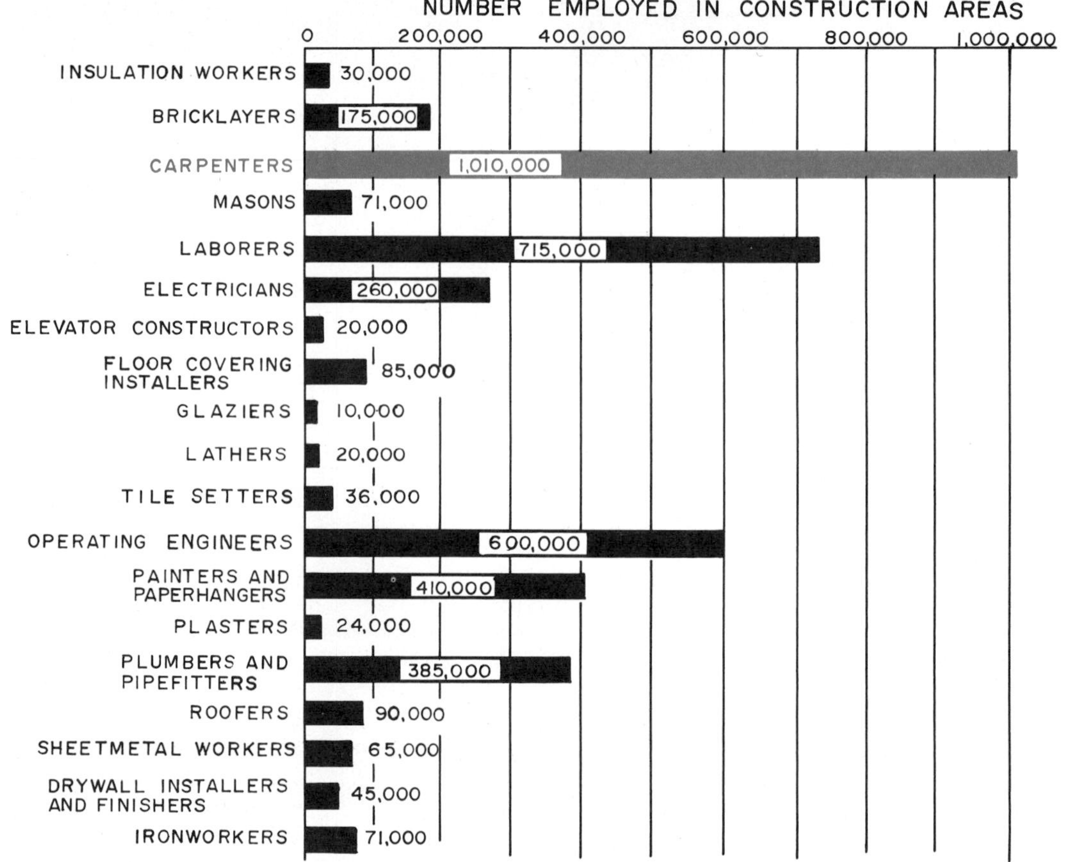

FIGURE 1-1

The demand for skilled persons is shown by the number of persons employed in the various construction areas.

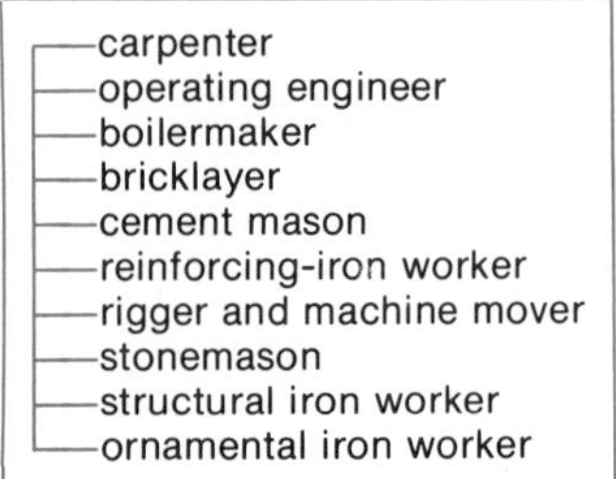

FIGURE 1-2

The structural trades.

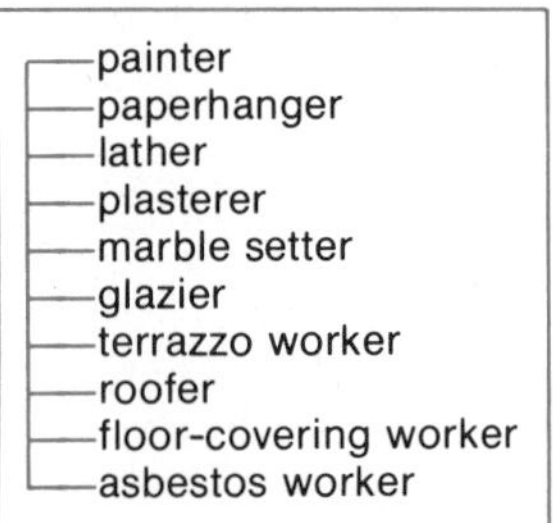

FIGURE 1-3

The finishing trades.

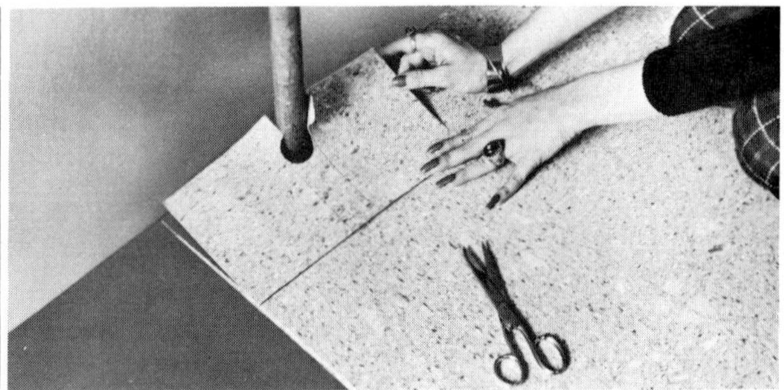

FIGURE 1-4

The installation of flooring is a finishing trade. *(Courtesy of Azrock Floor Products)*

FIGURE 1-5

A roofer uses some of the same tools as a carpenter. *(Courtesy of W. F. Whelan, Jr., CertainTeed Corporation, Valley Forge, PA.)*

- electrician
- plumber
- pipefitter
- sheet-metal worker
- elevator constructor
- millwright

FIGURE 1-6

The mechanical trades.

FIGURE 1-7

Carpenters have to cooperate with those working in the mechanical trades. *(Courtesy of Armstrong World Industries, Inc.)*

Most carpenters work on new construction. Remodeling is another common work area. Some carpenters work in factories where mobile homes or modular housing units are built.

Some carpenters work for large manufacturing companies as part of a physical plant maintenance force. They do the general carpentry work needed within the buildings operated by the company. Government agencies also employ carpenters. A few work in related industries, such as shipbuilding and mining.

WHAT ARE THE WORKING CONDITIONS?

Working conditions vary considerably. New construction requires that most of the time be spent out-of-doors. This subjects the carpenter to extremes of heat and cold. When carpenters are finishing interior work, they are completely or partially protected from the weather (Fig. 1-8).

Carpenters do considerable bending, stooping, and lifting (Fig. 1-9). They must be strong and have endurance. Some work is done on ladders and scaffolds. Commercial construction requires carpenters to work at considerable heights (Fig. 1-10).

Carpenters use a wide array of hand and power tools. Accidents are a constant hazard. Care must be exercised constantly in tool use. Attention to potential injury from electrical shock

FIGURE 1-8

Carpenters work outdoors in heat and cold. *(Courtesy of American Plywood Association)*

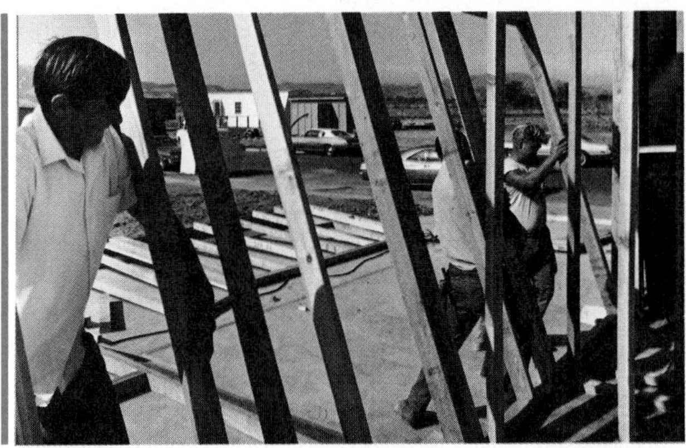

FIGURE 1-9

Carpenters do much lifting and heavy work. *(Courtesy of Western Wood Products Association)*

FIGURE 1-10

Carpenters work at considerable heights. They work on ladders and scaffolds. *(Courtesy of American Plywood Association)*

is mandatory. Carpenters must be alert to danger from falling materials and potential accidents from actions of others working on the job. Most jobs require the carpenter to travel to the job, which is sometimes at a considerable distance. Occasionally, they have to be away from home during the week to work on an isolated job, returning home on weekends.

WHAT WORK DO THEY PERFORM?

Carpenters perform a variety of tasks. They are often among the first of the skilled trades to be called to the construction site. They build the forms used for concrete footings and walls. Any concrete above ground level is also formed by carpenters. They construct the wood framework of buildings and install sheathing, wood siding, and roof decking. Window and door installation requires their skills.

Carpenters do a great deal of the interior trim. They cut trim around doors and windows and nail it in place (Fig. 1-11). They lay wood floors and baseboards. Cabinets are installed and in some cases built by carpenters (Fig. 1-12).

In some areas, carpenters install metal framework upon which interior wall finish is applied. They install metal fastening devices, exterior prefinished metal products such as cornices or louvers, and metal doors and windows. Sometimes, they install gypsum interior wall panels (Fig. 1-13).

In heavy construction, carpenters install heavy timbers in docks and trestles. They build forms for multistory concrete structures. Often, they have to build special-purpose scaffolding on the job.

FIGURE 1-11

Trim work must be carefully fitted. *(Courtesy of California Redwood Association)*

FIGURE 1-12

Carpenters do considerable interior work. *(Courtesy of American Plywood Association)*

FIGURE 1-13

Carpenters often install gypsum wallboard. *(Photo courtesy of Georgia-Pacific Corp.)*

In rural areas and small cities, carpenters tend to do all types of carpentry. In large cities, they tend to specialize: for example, some would build only concrete forms.

HOW TO YOU LEARN TO BE A CARPENTER?

A person planning to become a carpenter should complete a high school education. While in high school, a basic understanding of mathematics and general science should be gained. Skill in writing and speaking effectively is important. Industrial arts courses in general drafting, architectural drafting, electricity, woodworking, and general carpentry are valuable.

After graduation from high school a person could enter a vocational carpentry curriculum in a vocational school or community college. Some high schools offer a vocational carpentry program. In addition, courses in electric wiring, masonry, concrete, plumbing, and sheet-metal work are helpful. The carpenter constantly works in cooperation with professionals in these areas and should understand something about how they perform their work.

Another way to prepare for a career in carpentry is through an *apprenticeship*. An apprentice is a person who is learning a skilled trade by working on a job with an experienced worker. The skilled workers teach the beginner the things needed to do the job. This helps them develop manipulative skills. In addition, the apprentice attends night school for at least 144 hours per year. These classes teach the related theory the apprentice needs to know. For example, carpenter apprentices will learn about the tools and materials used in the trade. They also will study in areas such as blueprint reading and mathematics related to carpentry. These subjects supply technical knowledge that is not easily learned on the job.

Apprentices sign a formal written agreement which is filed with the state apprenticeship office or the Bureau of Apprenticeship and Training of the U.S. Department of Labor. The apprentice is guided by a local committee made up of people from management and the organized carpenters' labor organization. The local schools participate by offering night courses as needed.

Requirements for a formal apprenticeship in carpentry include graduation from high school, being 17 to 27 years of age, working 4 years as an apprentice, and passing courses in related knowledge.

The apprentice is paid a salary while learning. The salary is usually set by the local apprenticeship committee. It is usually about half of that paid to an experienced journeyman carpenter. As the apprentice gains experience, the salary is gradually increased.

When the apprentice completes all the requirements and passes the final examinations, a journeyman's certificate is awarded. This is recognized nationwide and is a record of considerable accomplishment.

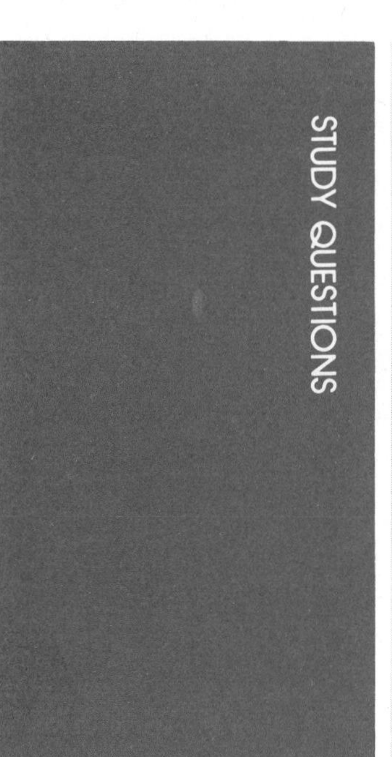

After reading the chapter, answer each of the following questions. If you do not know the answer, review the chapter.

1. How many carpenters are employed in the United States?
2. Into what major types are construction companies divided?
3. Where do carpenters work?
4. What kinds of tasks do carpenters perform?
5. How can you learn to be a carpenter?

IMPORTANT TECHNICAL TERMS

Following are technical terms that you should be able to use as part of your working vocabulary. Write a brief description of the meaning of each term.

Structural construction
Finishing construction
Mechanical construction
Apprenticeship

2 HOW A HOUSE IS BUILT

Carpenters are often among the first persons on the building site. They should understand the general schedule for constructing a building. A typical plan is discussed below, but there are many points at which it could be varied. Sometimes, several activities take place at the same time.

WHO DOES THE WORK?

Most buildings are built by general contractors. A *general contractor* is a company that has the contract for constructing the building. The company has agreed to deliver a completed building according to plans and specifications at a specific price. Their work includes securing the materials, equipment, and workers, and erecting the building.

Most general contractors have a variety of skilled workers on the payroll. However, many of the jobs to be performed are given to subcontractors.

A *subcontractor* is a company that specializes in one type of construction work. Following are jobs that a general contractor often gives to subcontractors: surveying; excavating; concrete and masonry work; plumbing; electrical; heating and air conditioning; sheet-metal work; painting; drywall finishing; plastering; laying tile, carpet, and hardwood floors; roofing; and grading and landscaping the site.

The general contractor reaches an agreement with the subcontractor for the work to be done. They agree on the price the subcontractor will charge the general contractor. This agreement is called a *subcontract.*

The subcontractor then comes to the job with the needed materials and skilled workers and performs that portion of the work.

PRELIMINARY WORK

Before construction can be started, complete plans and specifications are developed. Plans are drawings that show the design, size, and construction details. Specifications are a written document which lists details that are not shown on the plans. (See Chapter 6 on reading construction drawings.)

After the plans and specifications are finished, they are given to general contractors who have indicated an interest in constructing the building. These contractors estimate the cost of the building. They then give the owner a firm price. This is called *bidding.*

The owner of the new building decides which bid to accept. Sometimes, changes are made in the design or materials in an effort to reduce the finished cost.

The owner will have purchased a lot or acreage upon which the building is to be constructed. The land must be surveyed and its boundaries located.

The contractor will arrange with local utility companies for temporary electrical service (Figs. 2-1 and 2-2). Sometimes, temporary water can be provided. A building permit must be obtained from local authorities.

The contractor will also arrange for insurance on the job. This protects the owner and contractor from lawsuits due to accidents on the job.

A surveyor will be employed to locate the corners of the building on the lot. The grade will also be noted. (See Chapter 8 on locating a building on the site.) It is from these markings that the excavations for footings and foundations are made. The carpenter uses them to locate and construct the forms for the footing and foundation.

FIGURE 2-1

Temporary residential construction electric meter installation for single-phase, three-wire service using a permanently mounted transformer on a pad with temporary service on a pole.

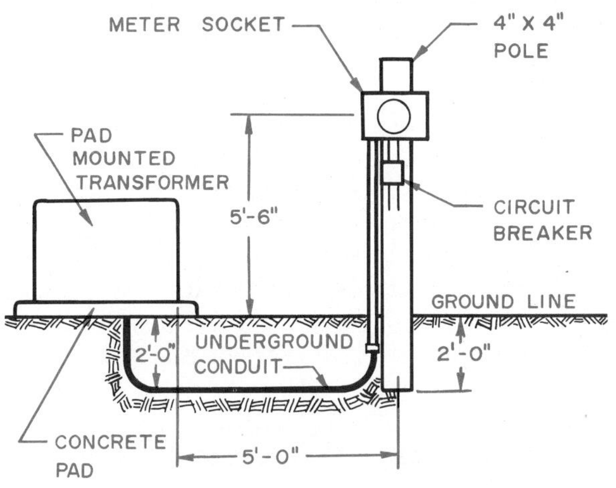

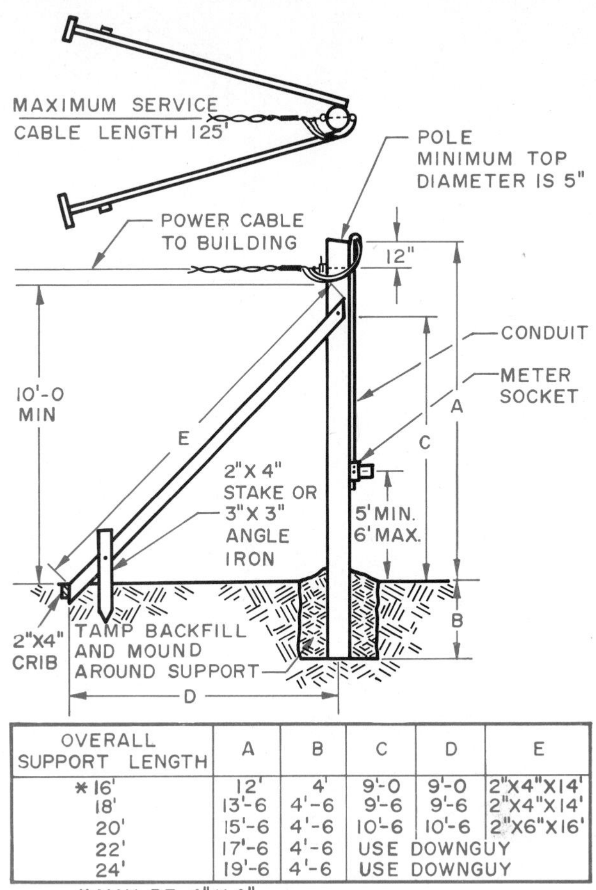

OVERALL SUPPORT LENGTH	A	B	C	D	E
*16'	12'	4'	9'-0	9'-0	2"X4"X14'
18'	13'-6	4'-6	9'-6	9'-6	2"X4"X14'
20'	15'-6	4'-6	10'-6	10'-6	2"X6"X16'
22'	17'-6	4'-6	USE DOWNGUY		
24'	19'-6	4'-6	USE DOWNGUY		

*MAY BE 4" X 4"

FIGURE 2-2

Temporary residential construction electrical meter installation for single-phase, three-wire service using a wood pole.

When the soil is removed for the footings and basement, the topsoil is removed first and piled separately from the other soil. It is saved for finishing the grading of the lot to its final contour or shape.

FOUNDATION CONSTRUCTION

The typical building rests on a concrete pad called a footing (Fig. 2-3). The carpenter builds the forms for the *footing*, which are then filled with concrete. Next, the carpenters set the forms for the *foundation*. The foundation is a masonry wall that rests on the footing. The forms are filled with concrete, forming the foundation (Fig. 2-4). Some foundations are built by laying up a concrete block wall. This is done by bricklayers.

Next, the columns and beam are installed. The columns have a footing formed by carpenters. The forms are filled with concrete.

The plumbing that is below grade is installed next. Sometimes, water and natural gas lines are laid at this point.

Before the excavated area around the basement is backfilled, the soil is treated for termites.

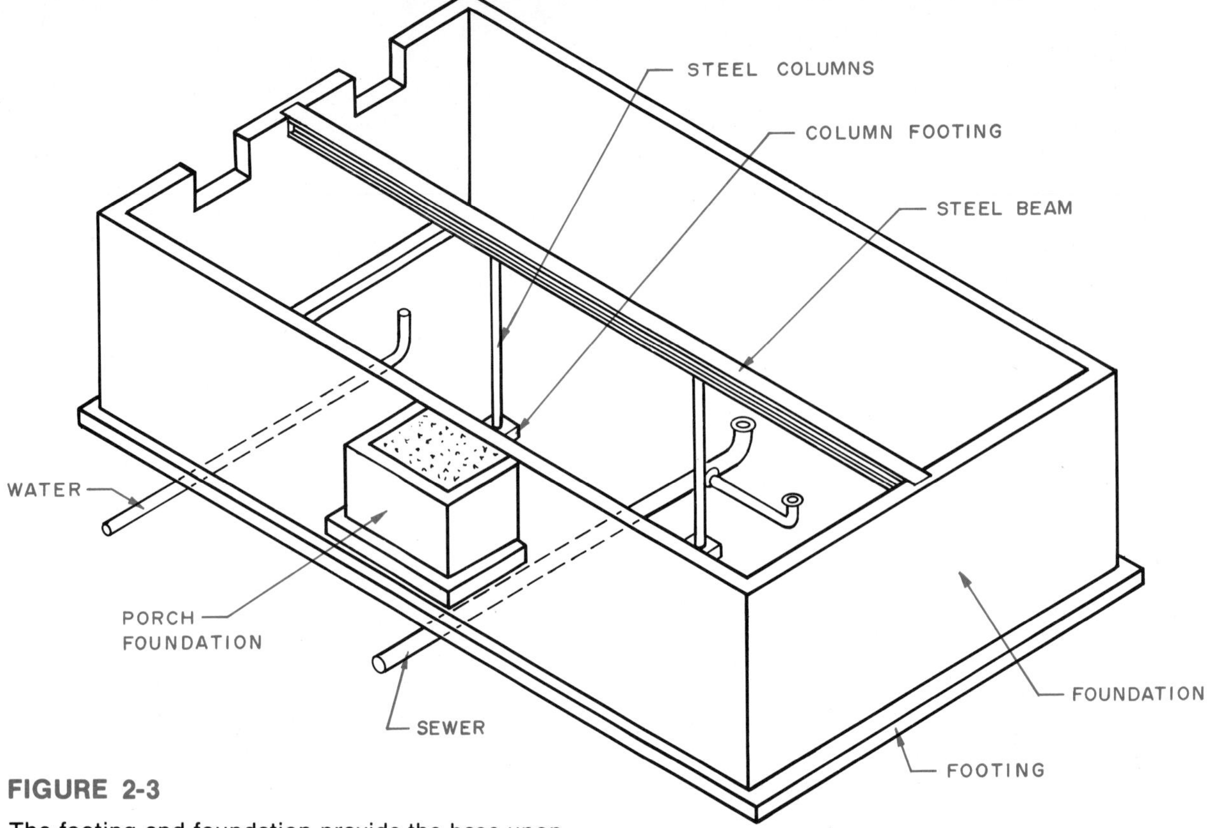

FIGURE 2-3

The footing and foundation provide the base upon which a building is built.

Backfilling involves pushing the soil in the void around the outside of the foundation.

Next, the foundation receives a waterproof coating. This extends down over the footing. Then drain tiles set in a gravel bed are placed around the footing (Fig. 2-5).

Usually, the foundation is not backfilled until the first-floor framing is in place. This is needed to reinforce the foundation wall. If the area is to be backfilled before the floor is nailed into place, the foundation walls will have to be braced.

FIGURE 2-4

The foundation is built using forms that are filled with concrete.

FIGURE 2-5

Subsurface water is carried away from the foundation with drain tiles.

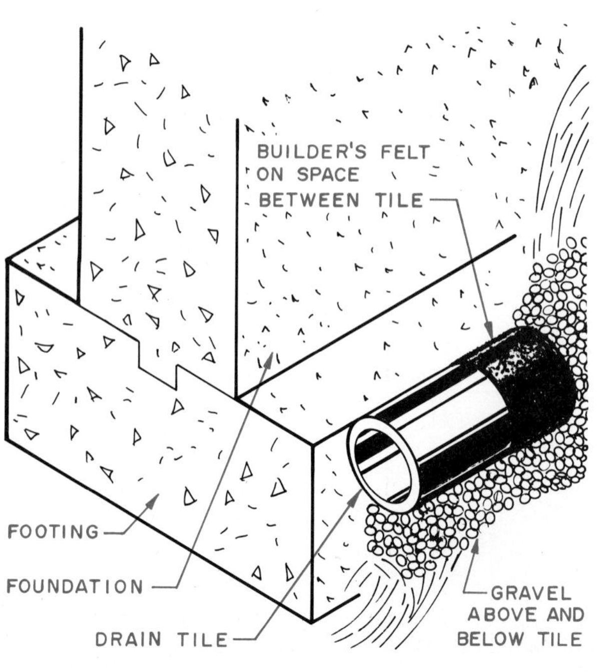

FRAMING THE FLOOR

After the foundation is finished, the first floor is framed (Fig. 2-6). There are several ways to frame the floor of a house. (See Chapter 10 on floor construction.)

A *sill sealer* is laid on the top of the foundation. It fills irregularities and stops air leakage. Sometimes a termite shield is also used (Fig. 2-7). A wood member called the *sill* is bolted over the sill sealer. The bolts are set in the concrete foundation before the concrete hardens (Fig. 2-8). The floor *joists* are set in place and nailed to the header. Joists are doubled around openings, such as for a stair. They are doubled under interior walls.

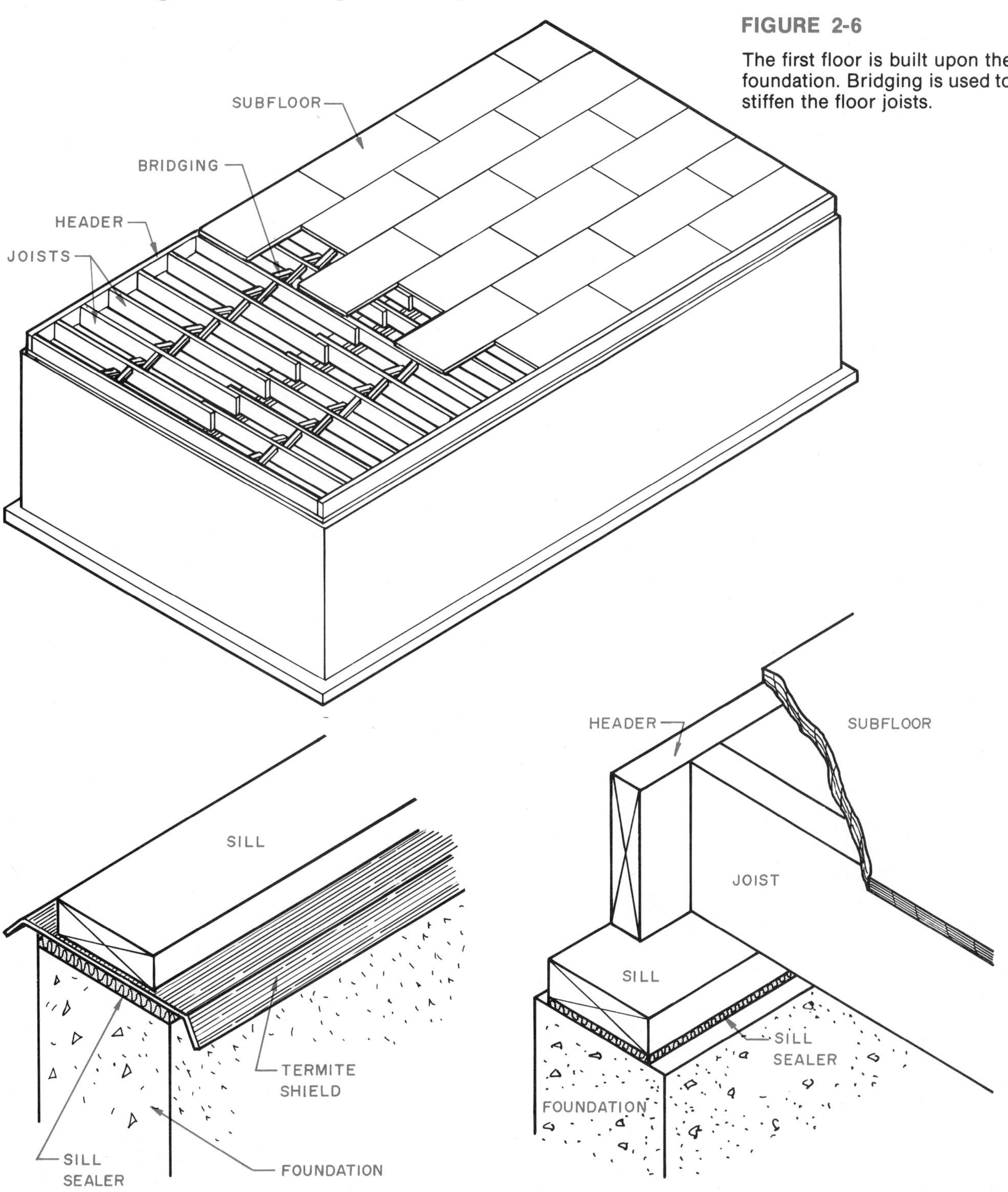

FIGURE 2-6

The first floor is built upon the foundation. Bridging is used to stiffen the floor joists.

FIGURE 2-7

A sill sealer is placed on top of the foundation. A termite shield is sometimes placed over it.

FIGURE 2-8

A wood sill is bolted to the foundation. The floor joists and header are placed on top of the sill.

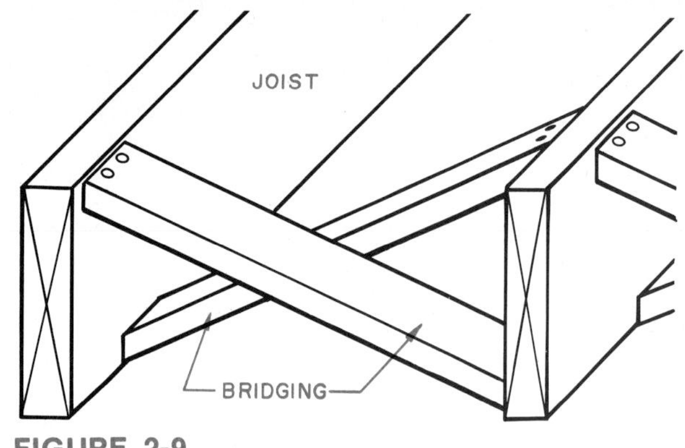

FIGURE 2-9

Bridging is nailed between the floor joists.

Bridging is nailed between the joists. It is nailed on the top with the bottom left loose (Fig. 2-9). Next, the *subfloor* is nailed to the joists. Then the bottom end of the bridging is nailed tight. The bridging stiffens the joists. Metal bridging is also used.

Now the void around the foundation can be safely backfilled.

FRAMING THE WALLS

After the subfloor is nailed in place, the interior and exterior walls are assembled and set in place (Fig. 2-10). (See Chapter 11 on wall framing.)

The exterior walls are framed first. The *studs* are laid out on the floor and nailed to the sole plate and the top plate (Fig. 2-11). The studs are doubled around all openings.

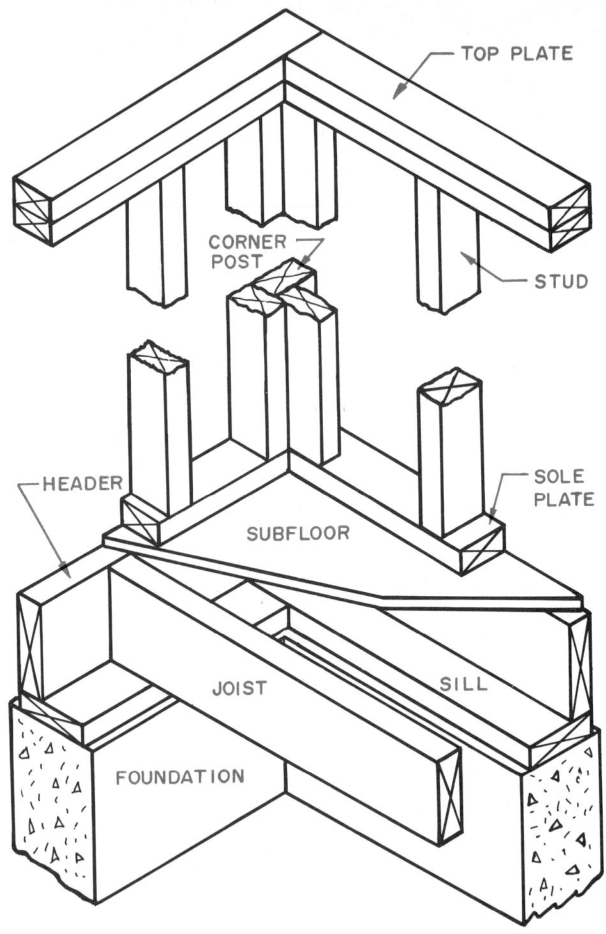

FIGURE 2-11

A typical way to frame exterior walls and form a corner.

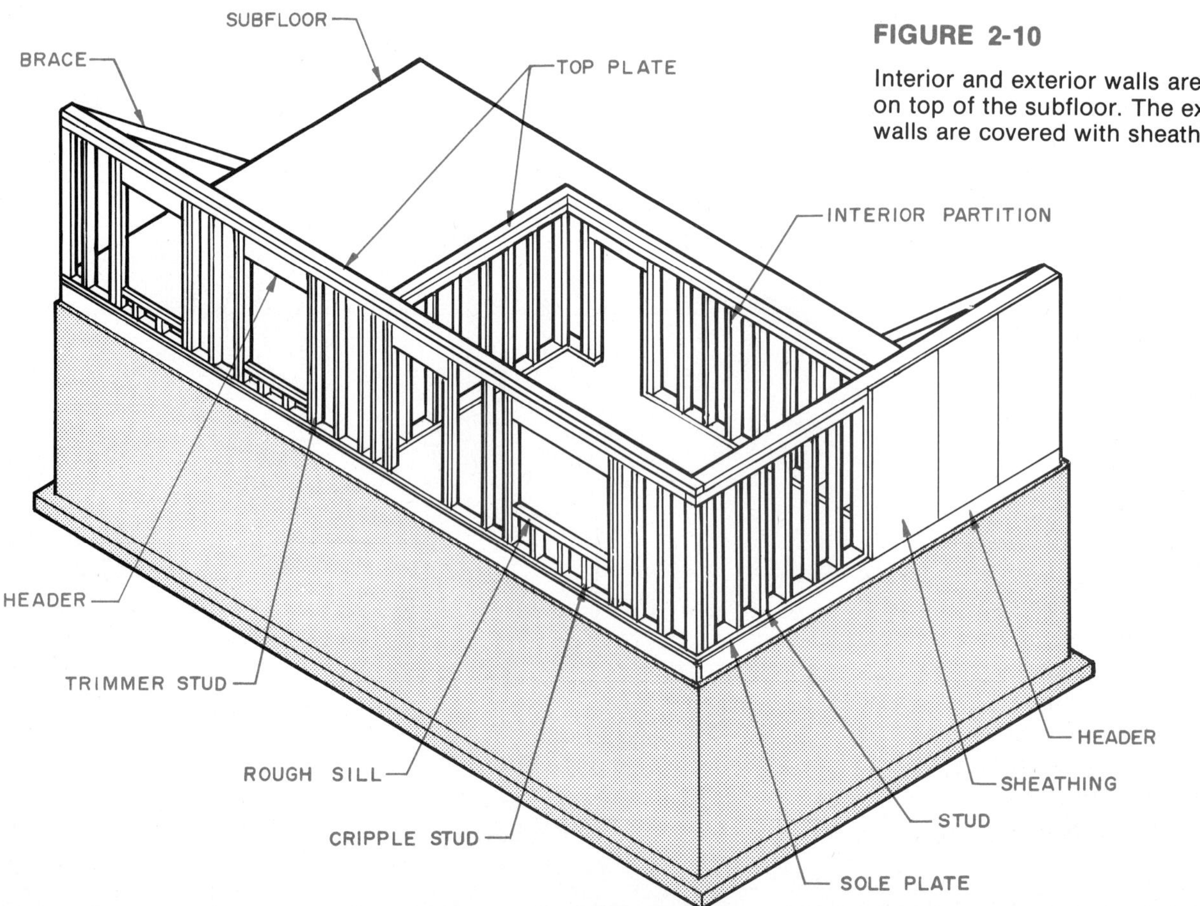

FIGURE 2-10

Interior and exterior walls are built on top of the subfloor. The exterior walls are covered with sheathing.

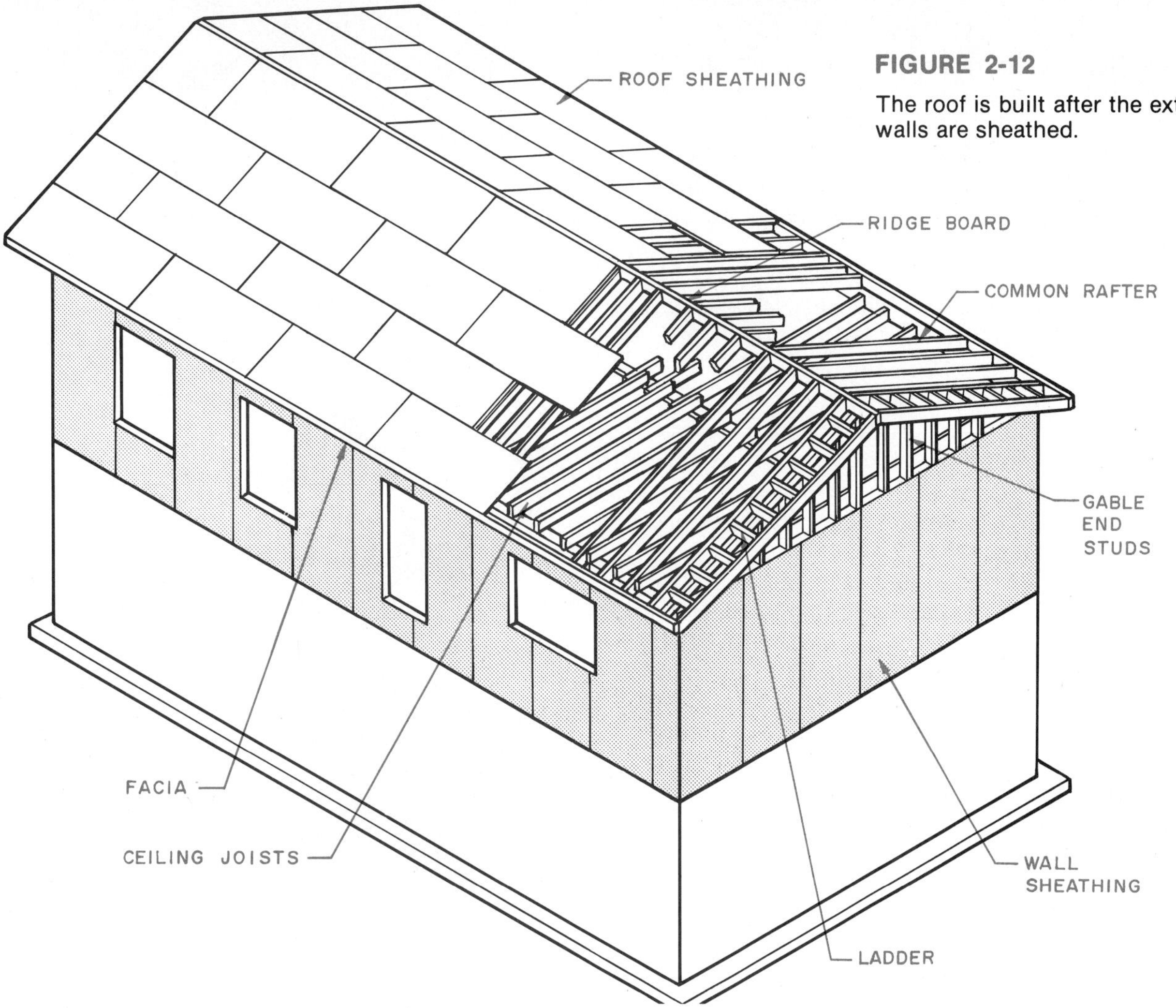

ROOF SHEATHING
RIDGE BOARD
COMMON RAFTER
GABLE END STUDS
FACIA
CEILING JOISTS
WALL SHEATHING
LADDER

FIGURE 2-12

The roof is built after the exterior walls are sheathed.

After the exterior walls are assembled, they are lifted to a vertical position and braced. The interior partitions are also assembled and set in place. As they begin to be joined to each other, a strong wall system is formed. If there is a brick chimney on the interior, it should be built to at least the ceiling height. (See Chapter 16 on finishing the exterior.)

Sheathing is applied to the exterior wall. Windows and door frames are nailed in place. Then the roof is built.

FRAMING THE ROOF

After the walls are assembled, the ceiling and roof are framed (Fig. 2-12). (See Chapters 12 on ceiling framing and 13 on roof framing.)

If trusses are to be used, they are lifted in place and nailed. If rafters are used, they are cut and laid against the exterior wall. The ridge board is set in place. The rafters are nailed to the ridge board and top plate.

The roof is then covered with sheathing. Finally, the shingles are applied (Fig. 2-13).

At this point the building is weather-tight.

FINISHING THE BUILDING

The exterior siding is nailed in place. The trim around the edges of the roof is installed and can be primed. (See Chapter 16 on finishing the exterior.)

FIGURE 2-13

After the roof is covered with sheathing, shingles are applied. *(Courtesy of American Plywood Association)*

Inside, the plumbers, electricians, and heating and air-conditioning technicians are at work. The electricians must attach to the rough framing the boxes for lights, switches, and outlets. They then must run wires to all of these. The light fixtures, switches, and outlets are not installed until the interior is finished.

The plumber runs the needed hot- and cold-water lines and sewer pipes. The bathtub is set in place. The other fixtures are not installed until later.

The heating and air-conditioning unit is set in place. The electrician runs the needed electric power to it. For a hot-air system, the air ducts are installed and insulated. For a hot-water system, the pipes are run. For electric heat, the wires are run to the unit locations.

The walls are insulated (Fig. 2-14). If the house is over a crawl space, the floors are insulated. The ceiling is insulated after the finished ceiling is set in place. (See Chapter 17 on insulation.)

Usually, the basement floor is poured before too much is done with interior wall finish. The concrete generates a lot of moisture, which could damage interior finish materials. The garage floor and porches can be poured anytime after the backfill is completed.

Next, the interior wall finish is applied (Fig. 2-15). This is commonly *gypsum drywall*, plaster, or paneling. Drywall requires that the seams and nails be taped (Fig. 2-16). This requires time for drying and for sanding between coats. Plaster generates considerable moisture and often requires 7 to 10 days to dry completely. (See Chapter 18 on finishing the interior.)

When the interior wall finish is completed, the trim around doors and windows is installed. The baseboard is then nailed in place. The shoe molding is cut and loosely nailed so that it can be removed by the floor-covering workers.

FIGURE 2-15

After the exterior walls are insulated, the interior finishing material is applied. This construction worker is applying gypsum wallboard. *(Courtesy of U.S. Gypsum)*

Now the floors can be finished. Underlayment is nailed down in areas where tile or carpet is to be laid. Hardwood floors are nailed into place.

Cabinets and built-in units are installed. The plumbing fixtures are set in place.

FIGURE 2-14

This construction worker is placing insulation in the ceiling. *(Courtesy of Owens-Corning Fiberglass Corporation)*

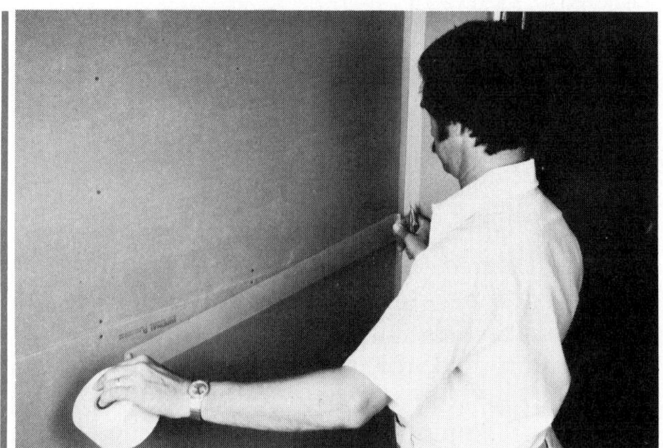

FIGURE 2-16

The seams in gypsum wallboard are taped and cement is applied to cover joints and nails. *(Courtesy of U.S. Gypsum)*

The painters can now paint the interior walls. The exterior painting and staining can be done at any time after the exterior trim is finished. Paperhangers can complete their work. Then the building must have a thorough cleaning.

Finally, the finished floor is laid. The hardwood floors are sanded and finished. Tile and carpet are laid (Fig. 2-17). (Read Chapter 20 on finish flooring.)

At about this time the electricians can install the light fixtures, switches, and outlets.

While the interior finishing is in progress, the yard could be graded, sidewalks formed and poured, the yard sodded or seeded, and shrubs planted.

FIGURE 2-17

Applying the finished flooring is usually the final step in interior finish. This worker is applying vinyl floor tile. *(Courtesy of Azrock Floor Products)*

STUDY QUESTIONS

After reading the chapter, answer each of the following questions. If you do not know the answer, review the chapter.

1. What is a general contracting company?
2. What is a subcontractor?
3. What is happening when bidding takes place?
4. List the steps to complete a foundation.
5. Explain how exterior and interior walls are framed and erected.
6. List the work to be done to finish the building after the roof is in place.

IMPORTANT TECHNICAL TERMS

Following are technical terms that you should be able to use as part of your working vocabulary. Write a brief description of the meaning of each term.

General contractor
Subcontractor
Subcontract
Bidding
Footing
Foundation
Backfill
Sill sealer
Sill
Joists
Bridging
Subfloor
Studs
Sheathing
Gypsum drywall

3 HAND TOOLS

A carpenter is expected to own and use a variety of hand tools. Although power tools are used for an ever-increasing number of tasks, there are many jobs for which hand tools are needed. A journeyman carpenter must be highly skilled in the use of hand tools.

Care and sharpening of tools is also important. Even when given good care, cutting tools become dull. Carpenters must know how to restore the cutting edge. Often, this must be done on the job under less-than-desirable conditions. The work is accomplished faster, easier, and better if the tools used are in good repair.

A carpenter should purchase only high-quality tools. These will last longer, are less likely to bend or break, and will stay sharp longer. Tools should be stored in a strong toolbox. Every tool should have its place so that it will be protected in storage. Also, the carpenter will know where to look for each tool.

Hand tools should be used properly so that they are not damaged. Probably the biggest cause of tool damage is that a tool was not used for the purpose for which it was designed. For example, if used as a chisel, a screwdriver will soon have a damaged blade and broken handle.

The following pages discuss the selection, care, and use of common hand tools.

MEASURING TOOLS

A carpenter must be able to measure distances accurately and to lay out work to be cut and assembled.

A *folding rule* is possibly the most used tool. The standard rule is 6 feet (ft) long (Fig. 3-1). Some have a metal piece on one end that

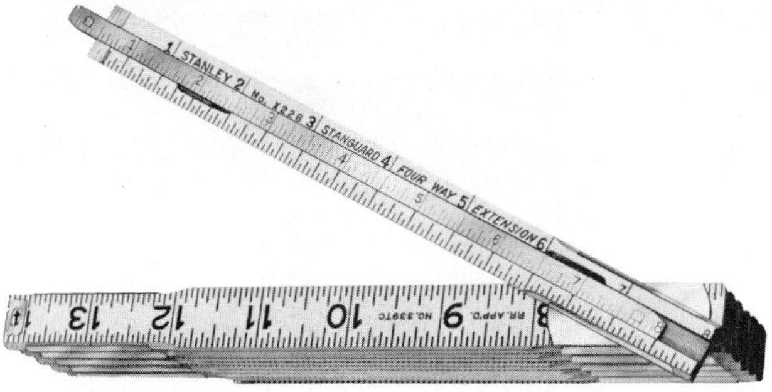

FIGURE 3-1

A folding rule. *(Courtesy of Stanley Tools)*

can be extended to help make inside measurements (Fig. 3-2).

A more accurate measurement can be made if the rule is placed on edge. This places the divisions next to the surface upon which the measurement is to be marked (Fig. 3-3).

Tape rules of various sizes are also essential. Common lengths for the small pocket-size tapes are 6, 8, 10, and 12 ft. Long tapes are available in 50- and 100-ft lengths (Fig. 3-4).

The three squares often used are the rafter square, the try square, and the combination square.

The *rafter square* is designed especially for carpentry work. The body of the square is 24 inches (in.) long. The tongue is 16 (in.) long (Fig. 3-5). The scales are 1/16, 1/8, and 1/2 in. The rafter square is designed especially for laying out rafters. (This is explained in Chapter 13 on roof framing.) It is also used to check parts that have to be at right angles and to lay out 90° lines. A metric square is also available. Its body is 600 millimeters (mm) long and 50 mm wide, and the tongue is 400 mm long and 40 mm wide.

The *try square* is smaller than the rafter square (Fig. 3-6). The blade is usually 6, 8, 10, or 12 in. long. It has a scale divided into 1/8-in. units.

FIGURE 3-2

The extension tip helps measure inside distances.

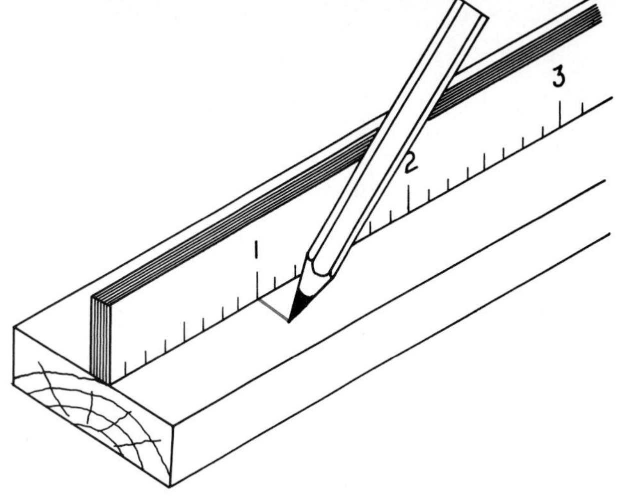

FIGURE 3-3

Stand the rule on edge to get accurate measurements.

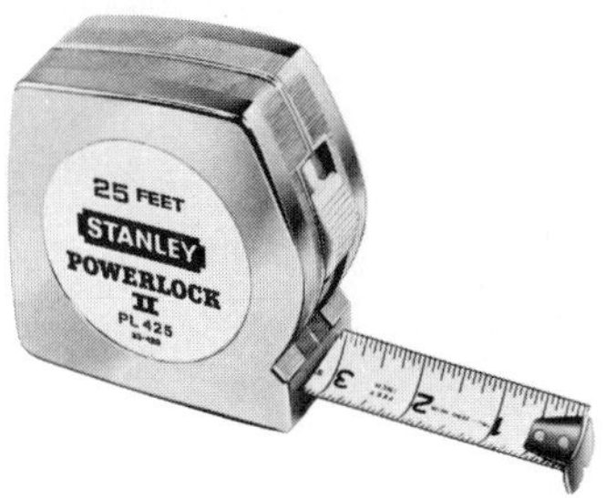

FIGURE 3-4

A pocket-size tape. *(Courtesy of Stanley Tools)*

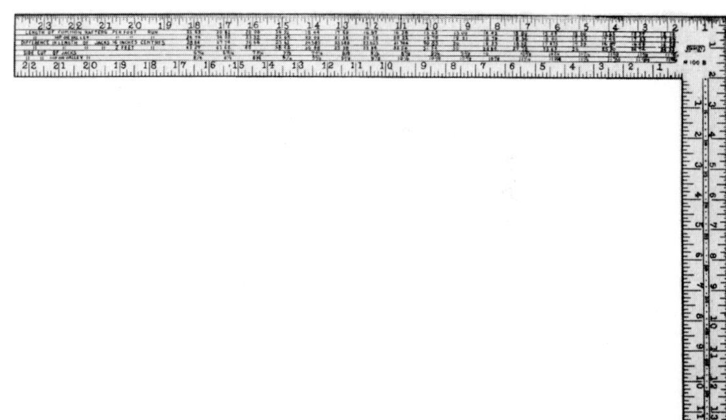

FIGURE 3-5

A framing square. *(Courtesy of Stanley Tools)*

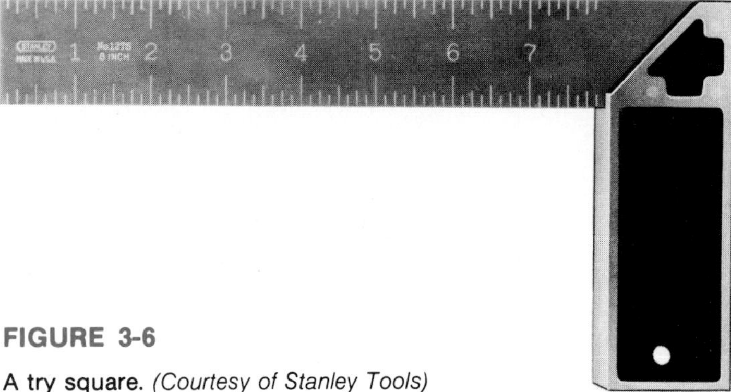

FIGURE 3-6

A try square. *(Courtesy of Stanley Tools)*

The handle is thick and the blade is set in the center of the handle. This makes it especially useful for laying out right angles (Fig. 3-7).

The *combination square* serves the same purpose as a try square (Fig. 3-8). It has an adjustable handle. This permits a length to be set and accurately marked many times (Fig. 3-9). The handle also has a surface set on a 45° angle. This permits rapid, accurate layout of this angle (Fig. 3-10).

The *T-bevel* has an adjustable blade (Fig. 3-11). It can be set on any angle desired and used to lay out that angle. It can be used to measure an angle on one piece and transfer it to another piece (Fig. 3-12). It is especially useful in laying out cuts on rafters.

FIGURE 3-7

Laying out right-angle cuts with a try square.

FIGURE 3-8

A combination square. *(Courtesy of Stanley Tools)*

FIGURE 3-9

Marking to width using a combination square.

FIGURE 3-10

Laying out a 45° angle with a combination square.

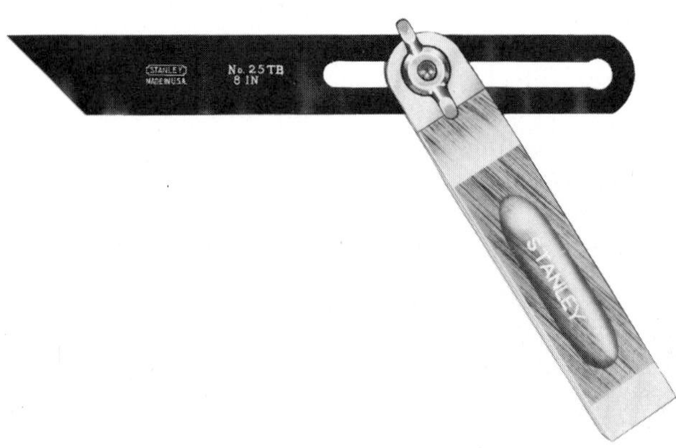

FIGURE 3-11

A T-bevel. *(Courtesy of Stanley Tools)*

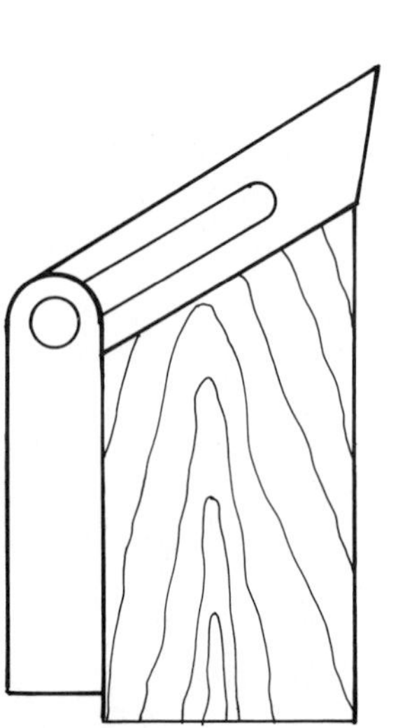

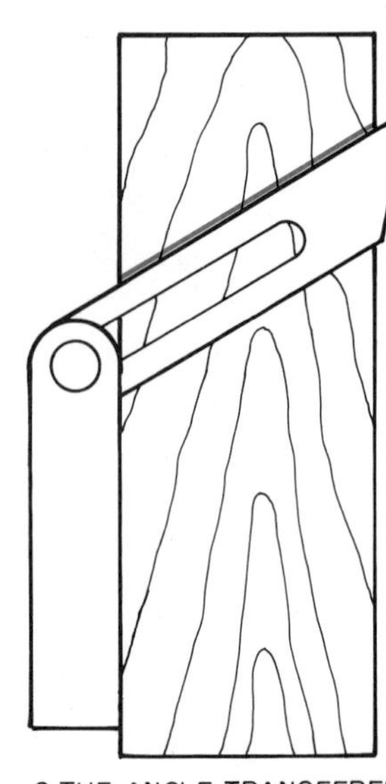

FIGURE 3-12

Transferring an angle with a T-bevel.

The squares and T-bevel must be handled carefully so that they are not damaged. Because they are layout tools they must remain accurate. Any abuse through misuse, such as improper storage, could damage them and reduce their usefulness.

Another layout tool is the *marking gauge* (Fig. 3-13). The tool is used to lay out lines parallel with the edge of a board. The bar has a pinpoint. The distance between the point and the head is set. The head slides along the edge of the board and the pin scratches a mark on the board (Fig. 3-14).

Dividers and trammel points are used to scribe circles (Fig. 3-15). The *dividers* will draw circles up to about 12 in. in diameter. The *trammel points* will scribe circles of any size. The points are fastened to a wood member. The longer the piece of wood, the larger the circle that can be scribed (Fig. 3-16). These tools are also used to transfer distances. The points can be set on a distance and used to mark it anywhere desired.

Levels are used to place members in a perfectly horizontal or vertical position. There are many sizes of levels (Fig. 3-17). For general carpentry work levels 24 to 48 in. in length are most commonly used. To ascertain if a member is horizontal, place the level on its surface. The bubble in the glass tube that parallels the surface should be centered on the mark (Fig. 3-18). Ver-

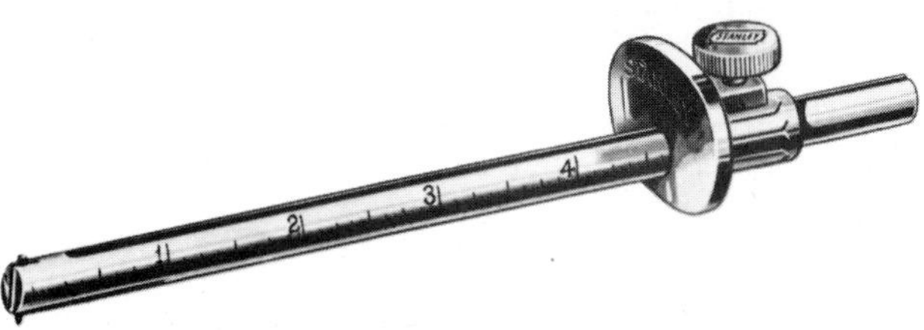

FIGURE 3-13

A marking gauge. *(Courtesy of Stanley Tools)*

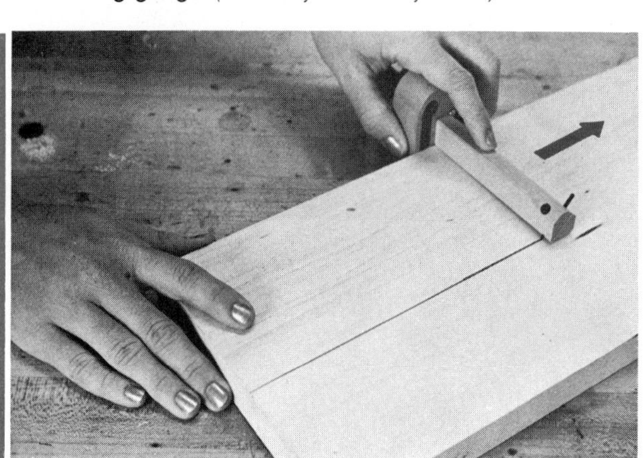

FIGURE 3-14

Marking to width using a marking gauge.

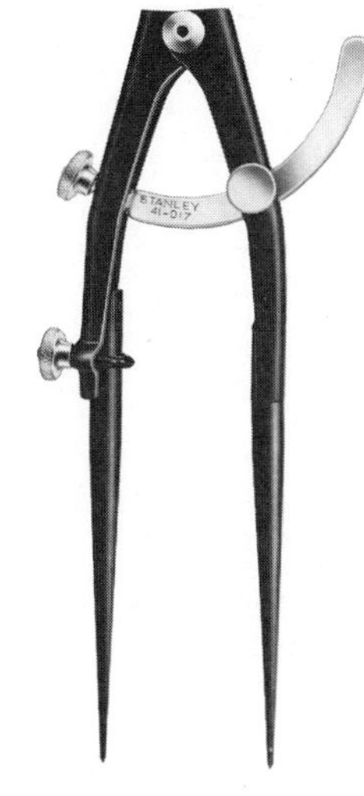

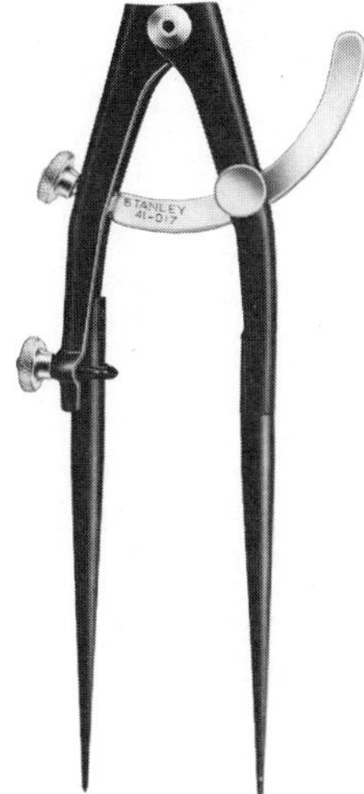

FIGURE 3-15

A divider. *(Courtesy of Stanley Tools)*

FIGURE 3-16

Trammel points. *(Courtesy of Stanley Tools)*

FIGURE 3-17

A carpenter's level. *(Courtesy of Stanley Tools)*

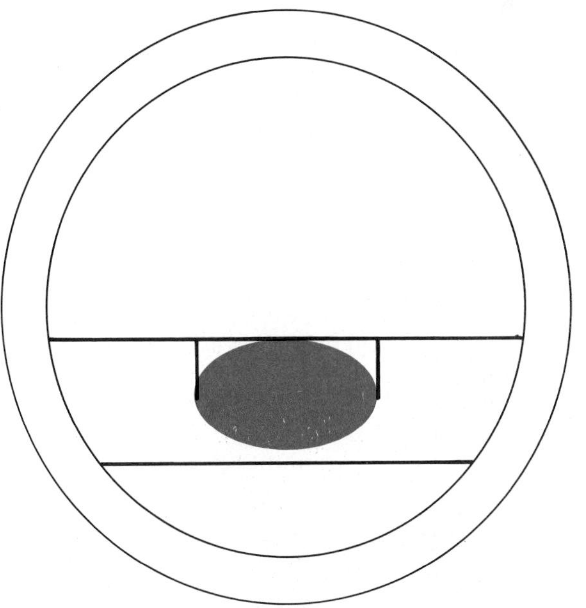

FIGURE 3-18

When the bubble on the level is centered in the glass tube, the board is level.

tical members can be checked with the other glass tubes on the instrument (Fig. 3-19). The glass tubes on levels are usually protected with a flat glass plate. When stored, these must be protected against breakage. When in use, they should be carefully handled.

Carpenters use *chalk line* to lay out footings and foundations. To check to see if the line is level, a *line level* is used (Fig. 3-20). It is very lightweight and hangs on the chalk line. It has a bubble in a glass tube, as do other levels.

A *plumb bob* is used to pull vertical chalk lines tight (Fig. 3-21). It is tied to the end and the line is hung in a vertical position. In addition, the plumb bob can be lined up with a point on the ground, thus forming a true vertical marker located above a predetermined point (Fig. 3-22).

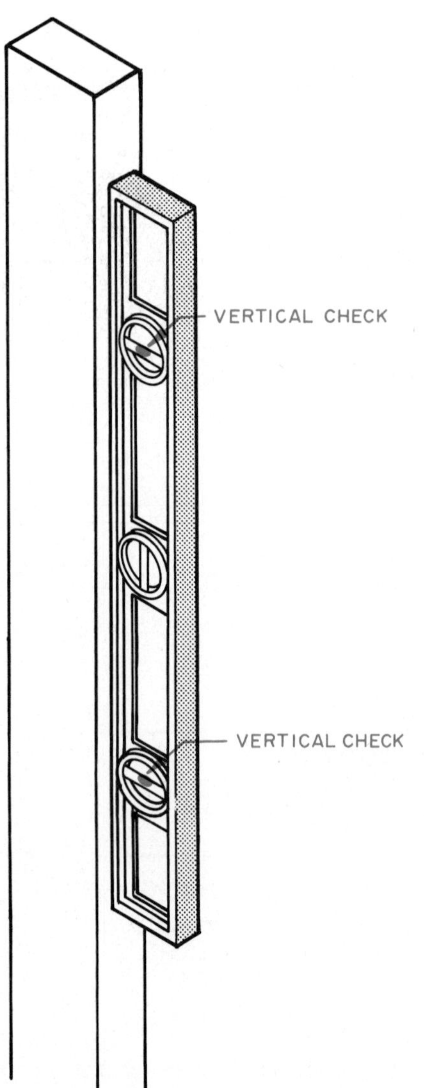

FIGURE 3-19

Using a level to check to see if a stud is vertical.

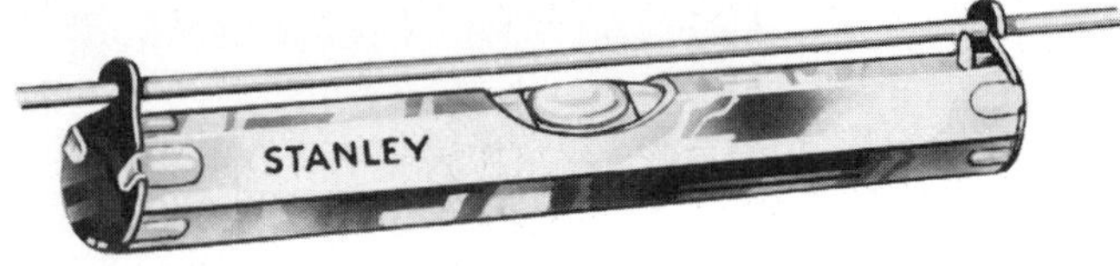

FIGURE 3-20

A line level. *(Courtesy of Stanley Tools)*

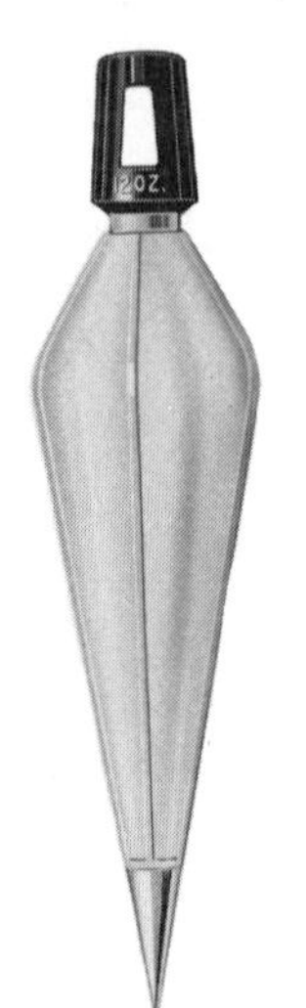

FIGURE 3-21

A plumb bob. *(Courtesy of Stanley Tools)*

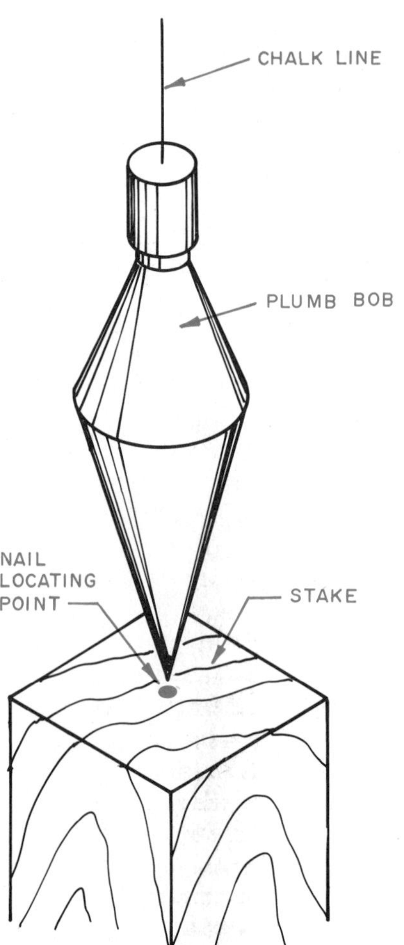

FIGURE 3-22

A plumb bob is used to locate vertical corners.

The major types of hand saws used by carpenters include the rip, crosscut, back, keyhole, and hack saws.

The rip and crosscut saws appear to be identical (Fig. 3-23). The major difference is the layout of the teeth. *Crosscut saws* are designed to cut across the grain of a board. They are usually 26 in. long and have either 8 or 11 cutting points per inch on the cutting edge. A saw has one more point per inch than it has teeth (Fig. 3-24). The teeth on a saw are *set* (bent) alternately to the right and left. A saw is set so that the slot it cuts is wider than the thickness of the blade. This prevents the saw from binding on cut surfaces (Fig. 3-25). Some saws have a taper ground from the teeth to the top of the blade. The taper reduces the amount of set required.

The sharp points of the teeth on a crosscut saw sever the wood fibers as the saw moves across them. This produces a kerf wider than the thickness of the saw blade. A *kerf* is the slot formed by the saw as it cuts the board. Notice in Fig. 3-25 that the teeth are slanted on the bottom. The sharp outside corner cuts the wood fiber and the slanted portion removes the wood behind this.

The crosscut saw tooth angles are shown in Fig. 3-24. Notice that the front of the tooth is 15° from perpendicular. The back of the tooth is 45°. The cutting edge is sharpened at an angle of about 65° to the side of the blade.

FIGURE 3-23

Crosscut and rip hand saws. *(Courtesy of Stanley Tools)*

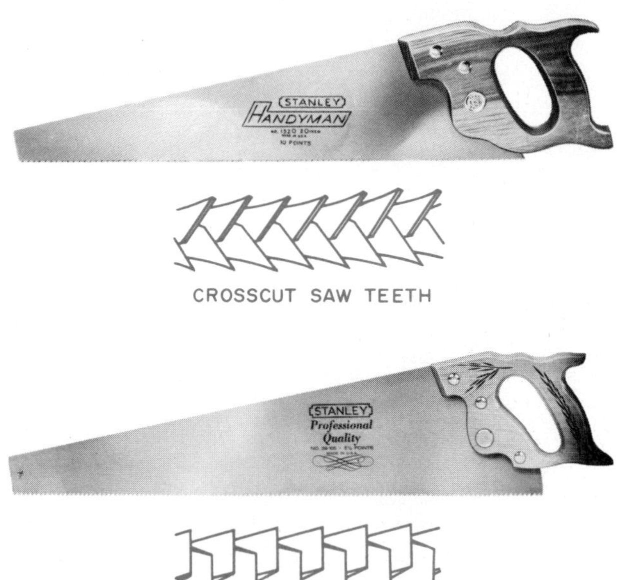

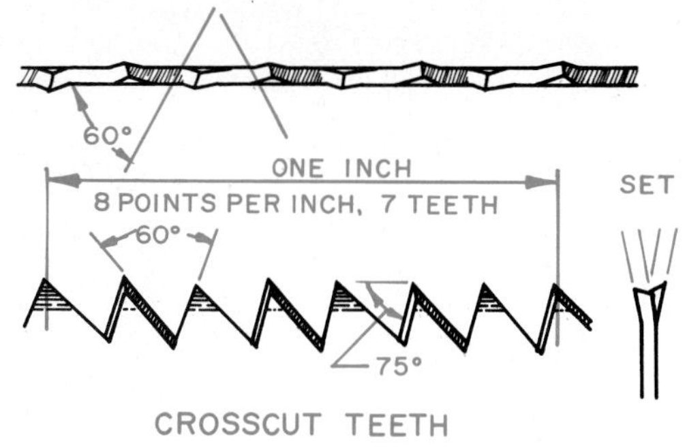

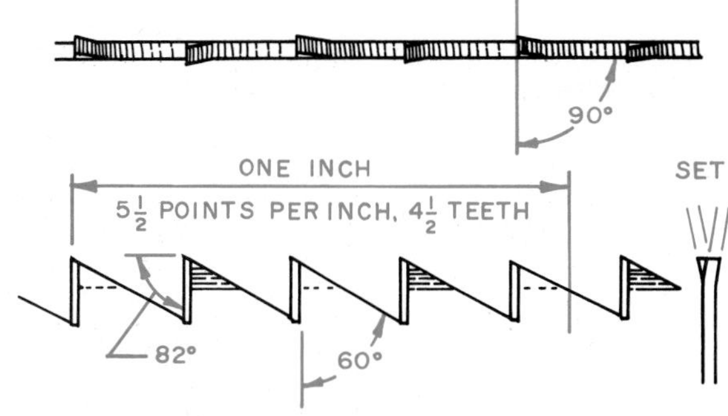

FIGURE 3-24

Crosscut and rip saw teeth.

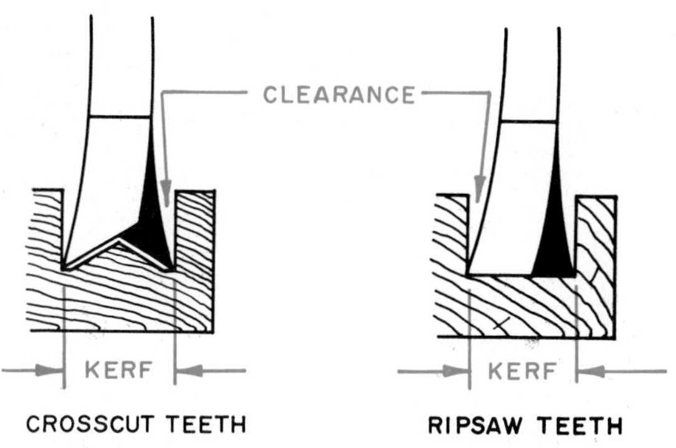

FIGURE 3-25

Saw teeth are set to provide a kerf wider than the thickness of the saw blade.

A *ripsaw* is used to cut with the grain of a board. The teeth are set differently from those on a crosscut saw. While the teeth have a set, they are also flat on the bottom. Actually, they are more like a series of small chisels which remove small chips of wood (Fig. 3-25). A common ripsaw size is 26 in. long with 5½ points per inch (Fig. 3-24). The tooth is designed so that the cutting face is on a 90° angle. The back of the tooth is on a 60° angle. When being sharpened, the front of the tooth is kept perpendicular to the side of the saw blade (Fig. 3-24).

Other saws used are the wallboard saw and the drywall saw (Fig. 3-26). These are made of extra-heavy-gauge metal. The teeth are de-

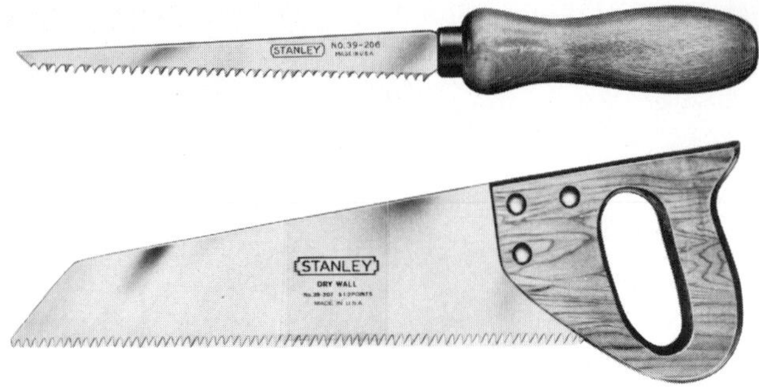

FIGURE 3-26

Drywall saws. *(Courtesy of Stanley Tools)*

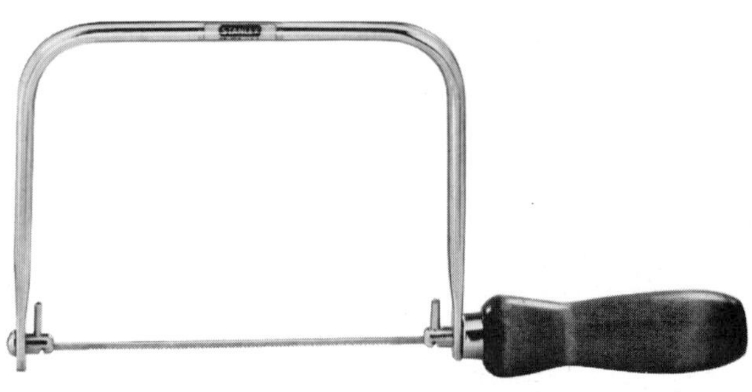

FIGURE 3-27

A coping saw. *(Courtesy of Stanley Tools)*

signed to prevent clogging. The *wallboard saw* is used to cut small openings in plasterboard, such as for an electrical outlet. The *drywall saw* is used for longer cuts.

A *coping saw* (Fig. 3-27) is used to cut small curves in thin wood. The blades have very fine teeth.

Sharpening Hand Saws

Saws are usually sharpened by persons specializing in this work. They use a saw filing machine which automatically files each tooth on the proper angle. Each tooth is then set the proper amount to give the blade clearance.

A carpenter can touch up the cutting edges on a saw. Following is a recommended procedure:

1. Clamp the saw in a vise or saw clamp (Fig. 3-28).
2. The saw is sharpened with a small triangular file. The file is held in two hands. Hold the file at the proper angle to the blade. For a crosscut saw, this is about 65°. For a ripsaw, it is 90°. Slant the file about 15° below horizontal.
3. Start filing the teeth from the toe of the saw. The toe is the end opposite the handle. File toward the heel. The heel is the end of the handle.

FIGURE 3-28

To sharpen a saw, clamp it in a saw clamp. File the teeth on the angles shown in Fig. 3-24.

4. Study the teeth so that you start on a tooth with the cutting edge slanting in the direction you are holding the file. Start the filing stroke with an even pressure. Raise the file and release the pressure as the stroke ends.
5. File every other tooth in this manner. You skip every other tooth because it has its cutting edge facing the other side of the saw.
6. After filing one alternate set of teeth, remove the saw from the vise, turn it around and file the series of teeth skipped in the first filing.
7. File each tooth the same number of strokes.

Setting Hand Saws

Filing reduces the set. After a saw is hand-filed a few times, the teeth will have to be reset.

Following is a procedure for resetting a saw:

1. Clamp the saw in a vise or saw clamp.
2. Make certain that each tooth is the same height. If it is not, the teeth must be jointed. This is done by pushing a flat, fine-tooth mill file along the length of the cutting edge. Repeat until all teeth are the same length. If it was necessary to joint the teeth heavily, they will need to be refiled to their proper shape.
3. Using a saw set, bend each tooth the desired amount (Fig. 3-29). A saw set has a plunger that strikes a tooth bending it. The amount of bend can be adjusted. Begin at the toe and set every other tooth. Be certain to start on the proper tooth. Then reverse the saw in the vise and set the other alternate teeth.

The amount of set needed varies with the type of wood to be sawed. Soft and wet woods require more set than dry or hard woods.

FIGURE 3-29

Setting the teeth on a hand saw.

Usually, hand saws are sharpened on a saw-sharpening machine. This is done by someone specializing in sharpening edge tools (Fig. 3-30).

Using Hand Saws The following is a suggested procedure for crosscutting.

1. Place the stock on a firm foundation. Sawhorses are commonly used. The board should be supported so that it will not fall after the cut is finished. The end to be removed should also be supported or be small enough for the carpenter to hold in one hand.
2. Use a try square or T-bevel to draw the line to be cut.
3. To begin the cut, a right-handed person usually places the left knee on the board (Fig. 3-31). The left hand grips the board. The saw is in the right hand.

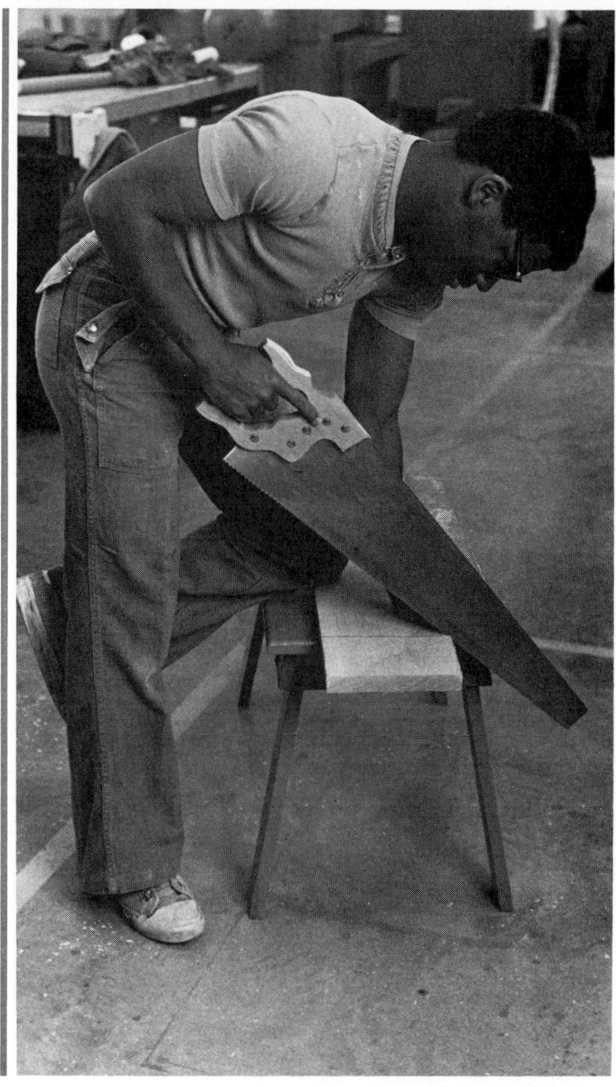

FIGURE 3-31

To crosscut, hold the board firmly with your knee and one hand.

FIGURE 3-30

A commercial saw filing machine.

4. Begin the cut by placing the cutting edge of the saw on the waste side of the line. Draw the saw back a time or two to start a kerf. Then proceed with a few short back-and-forth strokes. Once the saw is cutting easily, use full-length strokes. The saw should be at an angle of 45° with the board (Fig. 3-32).
5. After some practice a person can cut a straight line and produce a square cut. It helps to sight down the blade of the saw to see if it is running square.
6. As the cut finishes, grasp the waste piece being cut off. If it is allowed to drop, it will most likely splinter pieces off the good piece (Fig. 3-33).

FIGURE 3-32

To crosscut, hold the saw 45° to the board. To rip, hold the saw 60° to the board.

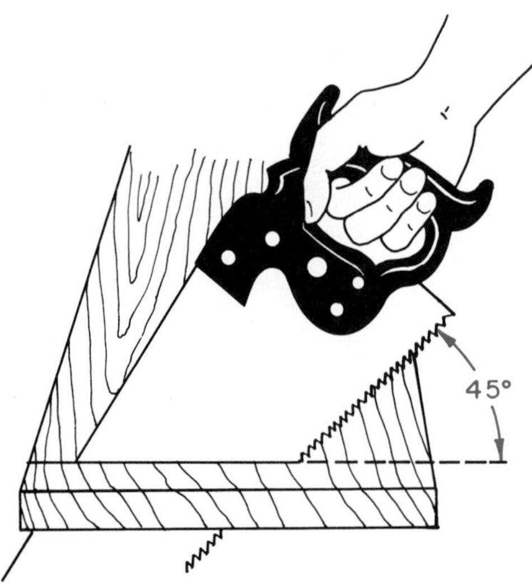

FIGURE 3-33

When finishing a cut, hold the piece to be removed so that it will not drop and split the edge of the board.

Remember to use long, full, steady strokes. Do not push down hard on the saw. This will crowd it and clog it. If the saw is sharp, it will cut with very little pressure.

The procedure for ripping is much like crosscutting. The biggest problem is getting the work properly supported. Sawhorses are commonly used (Fig. 3-34). If the kerf binds the saw, a wood wedge can be forced in the kerf. This spreads it so that the saw does not bind.

The ripsaw should be held so that the cutting edge has an angle of 60° with the cutting surface (Fig. 3-32). Be certain to cut on the waste side of the line.

Large sheets of plywood should be supported with several boards running between sawhorses (Fig. 3-35).

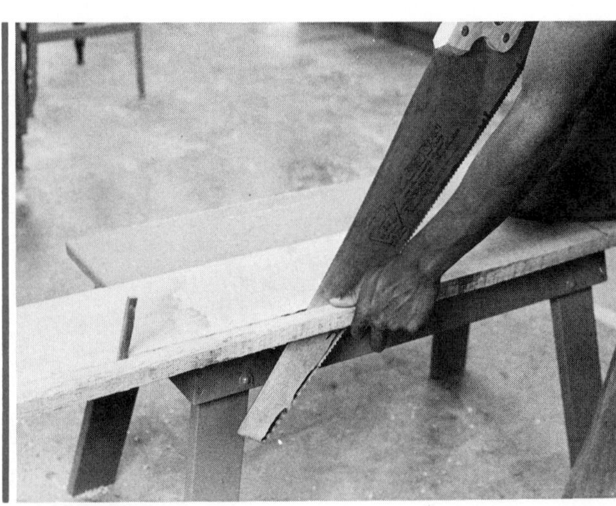

FIGURE 3-34

Use a wedge to hold the kerf open when ripping.

FIGURE 3-35

Support large sheets of plywood with several boards below.

Other Hand Saws The *compass saw* has a narrow, tapered blade (Fig. 3-36). It is used to cut in tight places or to remove material from the center of a panel. For example, a large opening could be cut in a piece of plywood by boring a hole and inserting the compass saw into the hole. It can then proceed to cut out the shape desired (Fig. 3-37). The teeth on a compass saw are similar to a ripsaw. A *keyhole saw* is like a compass saw but smaller.

The *backsaw* is used for fine cutting (Fig. 3-38). It has small fine teeth. A typical backsaw is about 14 in. long and will have 13 points per inch. The teeth are similar to those of a crosscut saw. The backsaw leaves a rather smooth edge and greatly reduces chipping on the edge cut.

The backsaw cuts with its teeth parallel with the wood. To start a cut, steady it with your thumb as you pull it across the board a few times (Fig. 3-39).

A large backsaw is made for use in a *miter*

FIGURE 3-36

A compass saw. *(Courtesy of Stanley Tools)*

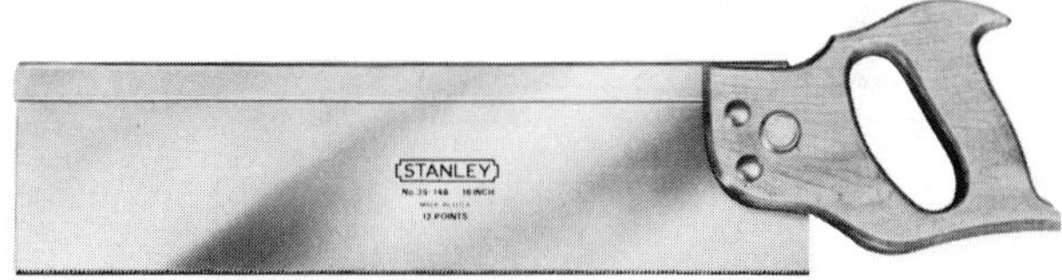

FIGURE 3-38

A backsaw. *(Courtesy of Stanley Tools)*

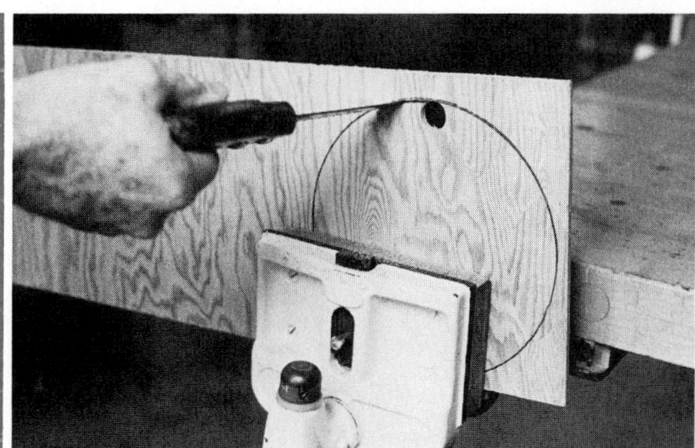

FIGURE 3-37

Cutting an opening with a compass saw.

FIGURE 3-39

Crosscutting with a backsaw.

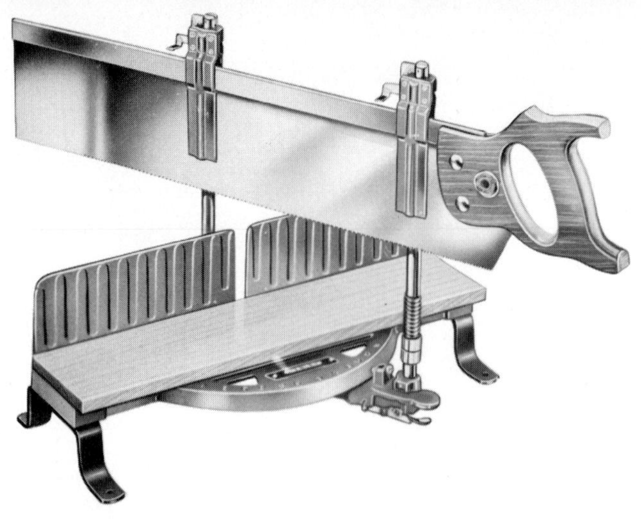

FIGURE 3-40

A miter box. *(Courtesy of Stanley Tools)*

FIGURE 3-41

Cutting on an angle using the miter box.

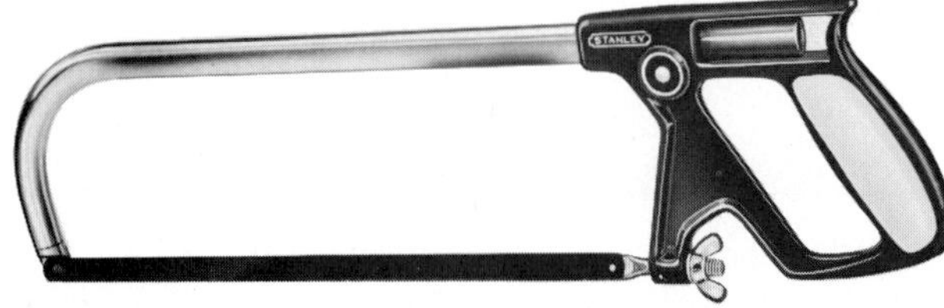

FIGURE 3-42

A hacksaw. *(Courtesy of Stanley Tools)*

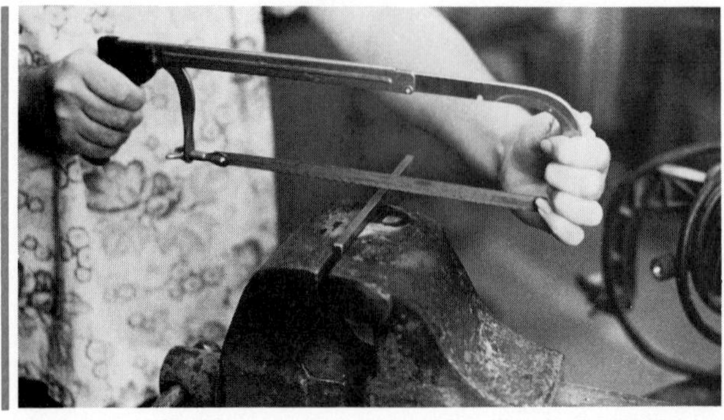

FIGURE 3-43

Cutting a metal bar with a hacksaw.

box (Fig. 3-40). The miter box is used to saw boards on various angles. It is very accurate (Fig. 3-41).

A *hacksaw* is used to cut metal (Fig. 3-42). The blade is of a hardened steel. The saw frame is made to hold blades of different lengths. The blades are bought separate from the frame. When they are worn out, they are discarded.

The hacksaw is held in a horizontal position. It cuts on the push stroke, so ease up on the pressure on the return stroke (Fig. 3-43).

PLANES AND EDGE TOOLS

There are a variety of planes available (Fig. 3-44). Of these the *bench*, *jack*, *fore*, and *jointer* are almost identical except for size. The bench plane is about 8 in., the jack about 14 in., the fore about 18 in., and the jointer about 22 in. long. The longer planes are used to smooth longer boards. Planes are used to remove a thin layer of wood from the surface of a board. This smooths the surface and gets it square. The plane uses a cutting action. The plane iron is the part that has a bevel cutting edge. It is set at an angle of 45° and the bevel on it is down (Fig. 3-45). The depth of cut is regulated with the adjusting nut.

FIGURE 3-44

Commonly used hand planes. *(Courtesy of Stanley Tools)*

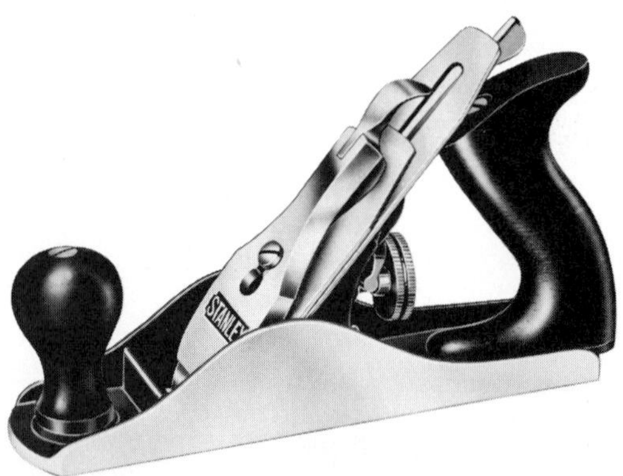

Bench Plane

Jack Plane

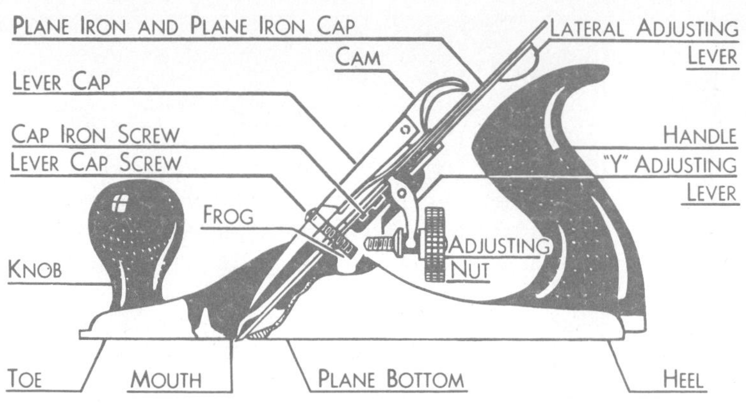

FIGURE 3-45

Parts of a hand plane. *(Courtesy of Stanley Tools)*

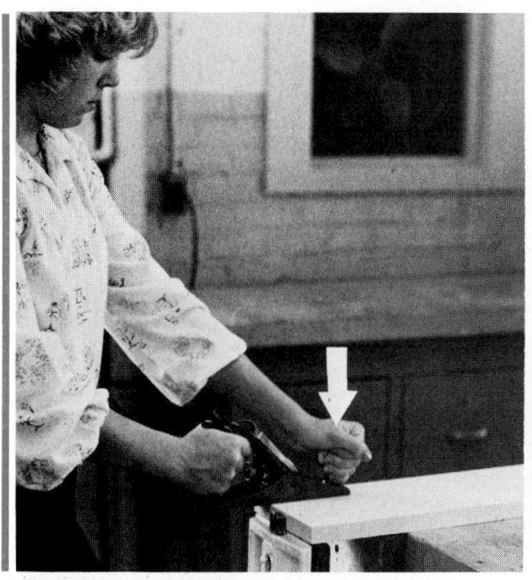

Start with Pressure on the Knob

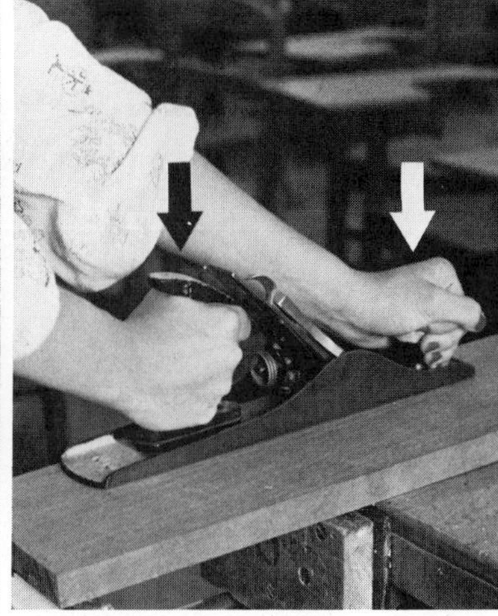

Uniform Pressure on Handle and Knob

Finish Stroke with Pressure on Handle

FIGURE 3-46

The three steps required to properly dress a board with a hand plane.

Following are suggestions for using the plane:

1. The board to be planed should be clamped down tightly.
2. Adjust the depth of cut of the plane. This is done by turning the adjusting nut. This pushes the plane iron out of an opening in the bottom of the plane. At first, set this cut very light. As you try to smooth the surface, the depth of cut can be increased until you get it the way you want it.
3. A right-handed person holds the handle in the right hand and the knob in the left. Place the front of the plane on the board. Apply pressure to the knob and begin the stroke. In the center of the board, apply uniform pressure to the handle and knob. At the end of the stroke, apply more pressure to the handle (Fig. 3-46).
4. Plane in the direction of the grain. This can usually be found by examining the board. The direction of the hard rings points in the direction of the grain (Fig. 3-47). If you are unable to ascertain the direction, set the plane to make a thin cut-and-try one direction. If it is the wrong way, the surface of the board will have rough places.
5. Use a try square to check a board for squareness. Edges and ends are checked with the blade and handle (Fig. 3-48). The surface can be checked with the blade. The surface

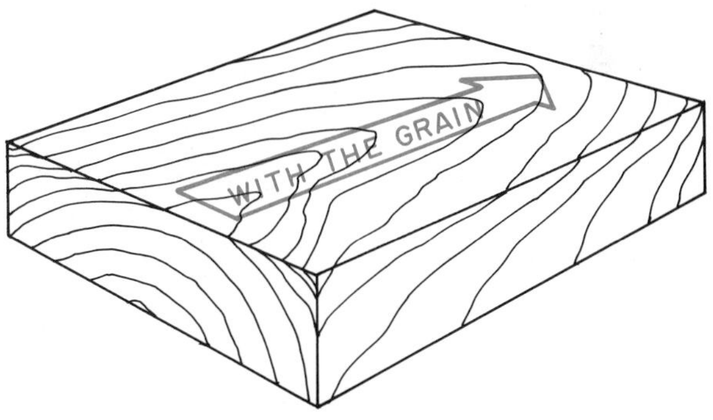

FIGURE 3-47

The grain of a board points in the direction you should plane.

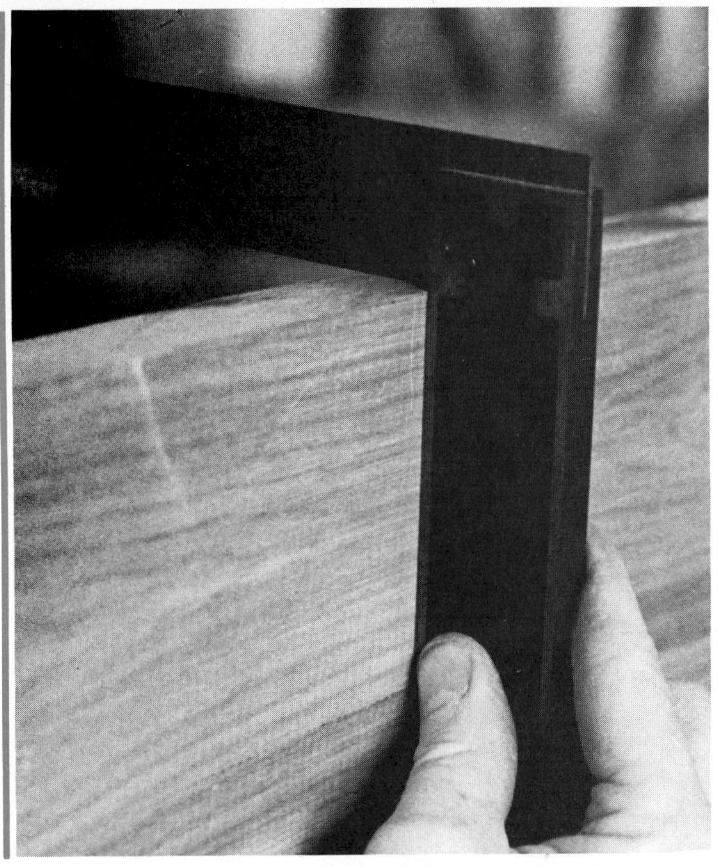

Checking the Edge for Squareness

Checking the End for Squareness

FIGURE 3-48

How to check an edge and end for squareness.

FIGURE 3-49

How to check a surface to see if it is flat.

is checked across the grain (Fig. 3-49), with the grain (Fig. 3-50), and on both diagonals (Fig. 3-51).

6. When planing end grain, either chamfer the edge of the board or back it up with waste stock (Fig. 3-52). If this is not done, the plane will splinter the edge at the end of the cut.

FIGURE 3-50

Use a framing square to check the length of the board for flatness.

FIGURE 3-51

Check a board across its diagonals to see if it is flat.

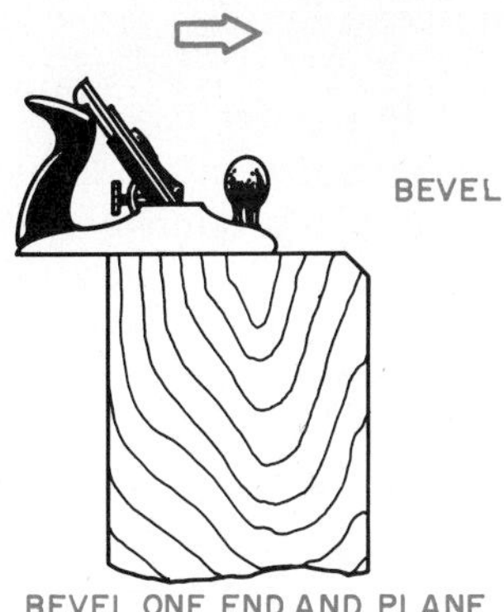
BEVEL
BEVEL ONE END AND PLANE TOWARD THE BEVEL.

STOP
PLANE TO THE CENTER FROM EACH END.

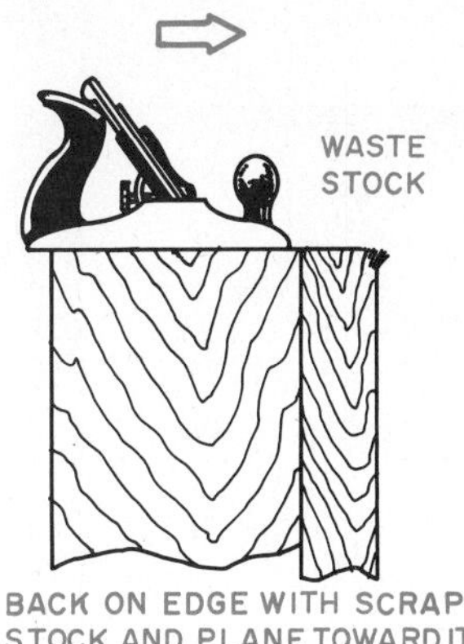
WASTE STOCK
BACK ON EDGE WITH SCRAP STOCK AND PLANE TOWARD IT.

FIGURE 3-52

Three ways to plane end grain without splitting the edge.

Another plane is the *block plane* (Fig. 3-53). It is about 6 in. long. It is held in one hand. Its main use is to smooth end grain and across grain. The plane iron is set at an angle of 20° and works with the bevel up (Fig. 3-54).

A *rabbet plane* is used to plane rabbets (Fig. 3-55). A rabbet is a rectangular notch cut from the edge of a board (Fig. 3-56).

A *router plane* is used to smooth dados and grooves (Fig. 3-57). A *dado* is a rectangular notch cut across the grain. A *groove* is a rectangular notch cut with the grain (Fig. 3-58).

FIGURE 3-53

A block plane. *(Courtesy of Stanley Tools)*

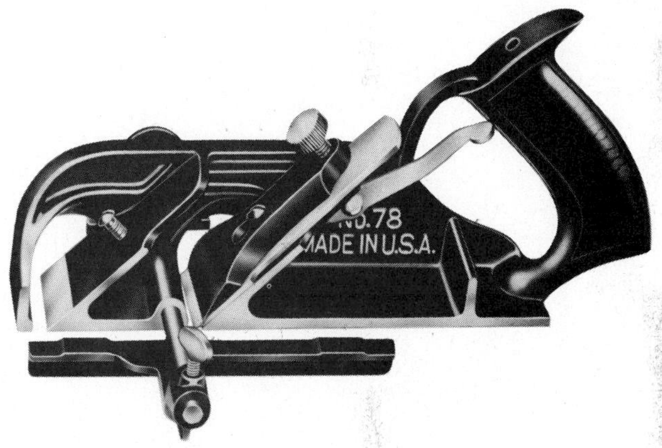
MADE IN U.S.A.

FIGURE 3-55

A rabbet plane. *(Courtesy of Stanley Tools)*

FIGURE 3-54

Planning end grain with a block plane.

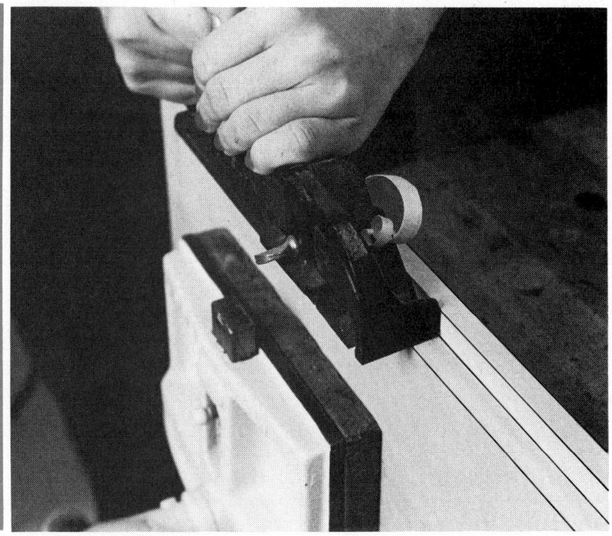

FIGURE 3-56

Using the rabbet plane to smooth a rabbet.

FIGURE 3-57

A router plane is used to smooth grooves and dadoes.

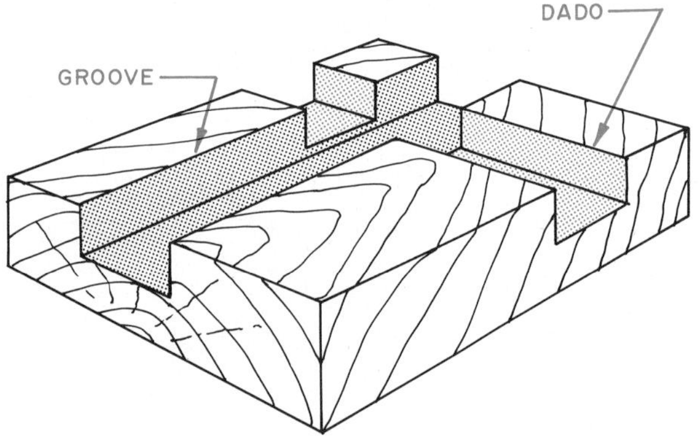

FIGURE 3-58

A groove and a dado.

Two other tools used to remove wood are the *multiblade forming tool* (Surform) and the *tungsten carbide-coated file.* They remove material rapidly but do not leave a finished smooth surface (Fig. 3-59).

Wood chisels are used to cut joints and recesses. They are made in a variety of sizes. A set will usually have chisels in the following widths: 1/4, 3/8, 1/2, 5/8, 3/4, 1, 1 1/4, 1 1/2, and 2 in. (Fig. 3-60). They are made to be driven with a wood- or plastic-headed mallet. A steel hammer is very damaging to the handle. Light cuts are made using only hand pressure. Always keep both hands behind the cutting edge (Fig. 3-61).

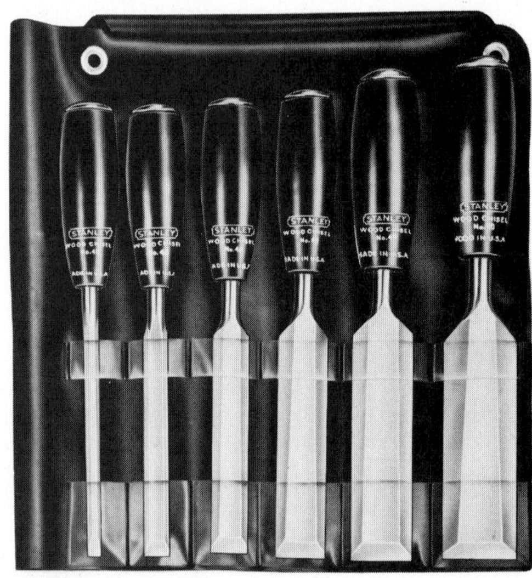

FIGURE 3-60

A typical set of wood chisels. *(Courtesy of Stanley Tools)*

FIGURE 3-59

Multiblade forming tools are available in several shapes and styles. *(Courtesy of Stanley Tools)*

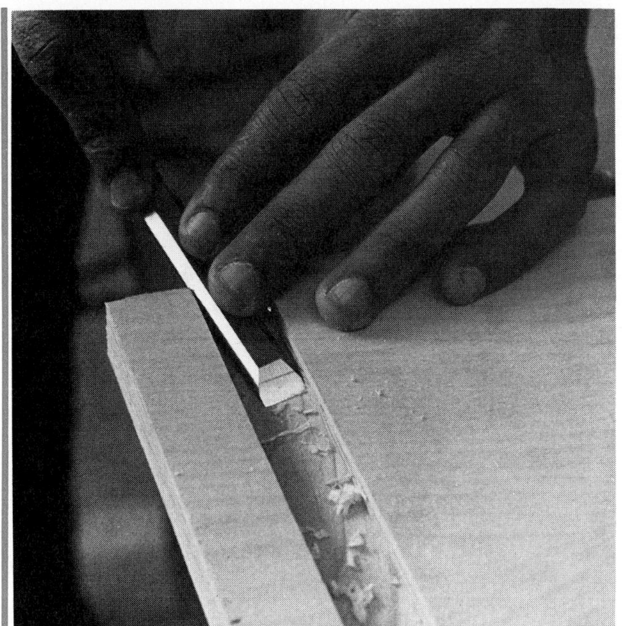

FIGURE 3-61

Smoothing a dado with a wood chisel. Notice that the bevel is up.

Sharpening Edge Tools When a plane iron becomes dull, it can be resharpened several times by honing it on an *oil stone*. Following are the steps needed to do this:

1. Put a light coating of kerosine or lightweight oil on the face of the stone.
2. Separate the plane iron and the plane iron cap. Place the plane iron on the stone with the bevel flat on the stone (Fig. 3-62).
3. Raise the back edge slightly so that the cutting edge rests on the stone.
4. Move the plane iron over the face of the stone in a circular motion.
5. After circulating it several times, test to see if it is sharp. If the edge has dull or flat spots, hone it again.
6. When sharp, place the plane iron flat on the oil stone with the bevel up. Move lightly until the wire edge caused by the original honing is removed (Fig. 3-63). Sometimes, the wire edge can be removed by running a sharp bevel across a piece of wood.
7. Some persons prefer to use a leather strap to give the plane iron a final honing.
8. A popular test for sharpness is to run the sharp edge against a piece of paper. If it slices into it easily, it is sharp.

After the plane iron has been honed several times, it will lose its concave surface. It needs to be reground (Fig. 3-64).

FIGURE 3-62

Edge tools are sharpened on an oil stone. This plane iron is being sharpened after it was reground to the correct bevel.

FIGURE 3-63

Place the plane iron flat on the oil stone to remove the wire edge.

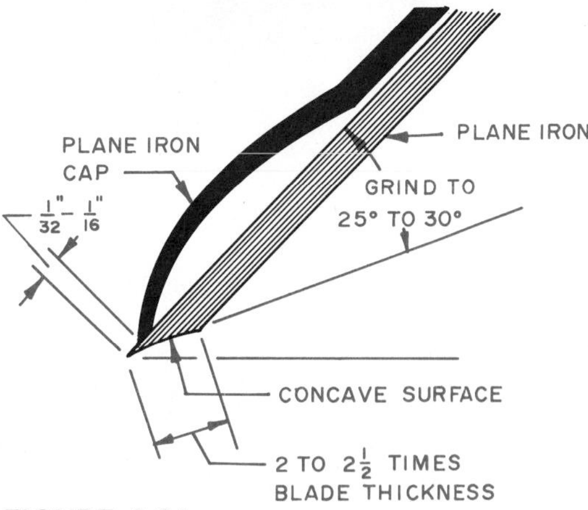

FIGURE 3-64

The correct angles for grinding a plane iron.

Following are the steps needed to grind a plane iron:

1. Remove the plane iron from the plane. Remove the plane iron cap.
2. Set the tool rest on the electric grinder so that the bevel of the plane is ground on an angle of 25 to 30°. When properly ground, the bevel will be 2 to 2½ times as long as the iron is thick (Fig. 3-64).
3. Start the grinder and move the plane iron into the wheel until it touches lightly. Then move it from side to side. Keep it square with the wheel. Do not press hard against the wheel. This will cause the plane iron to heat, turn blue, and ruin the metal (Fig. 3-65).

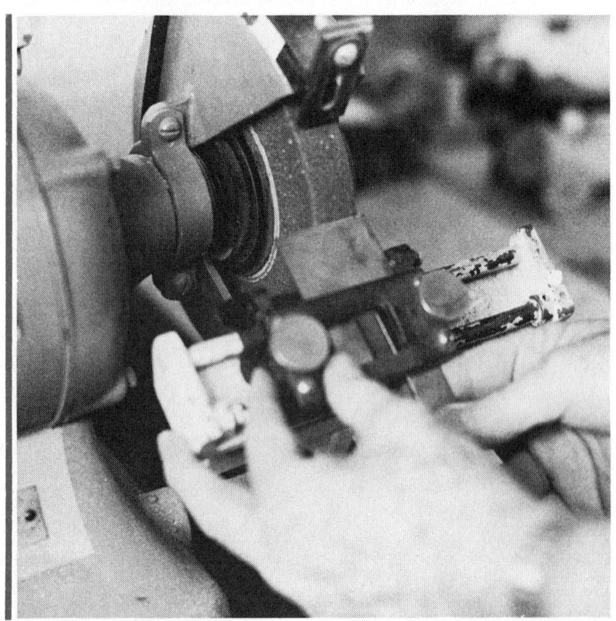

FIGURE 3-65

Grinding a plane iron with a grinding attachment on a bench grinder.

4. Continue grinding until the bevel is formed across the entire width of the iron. If it appears to be getting hot, remove it from the wheel and let it cool. It can be dipped in water to help cool it. A correctly ground plane iron is square with its edge and straight (Fig. 3-66).

Wood chisels are sharpened in the same manner as plane irons. They are also ground at an angle of 25 to 30°.

FIGURE 3-66

Check the cutting edge of the plane iron for squareness when it is being ground.

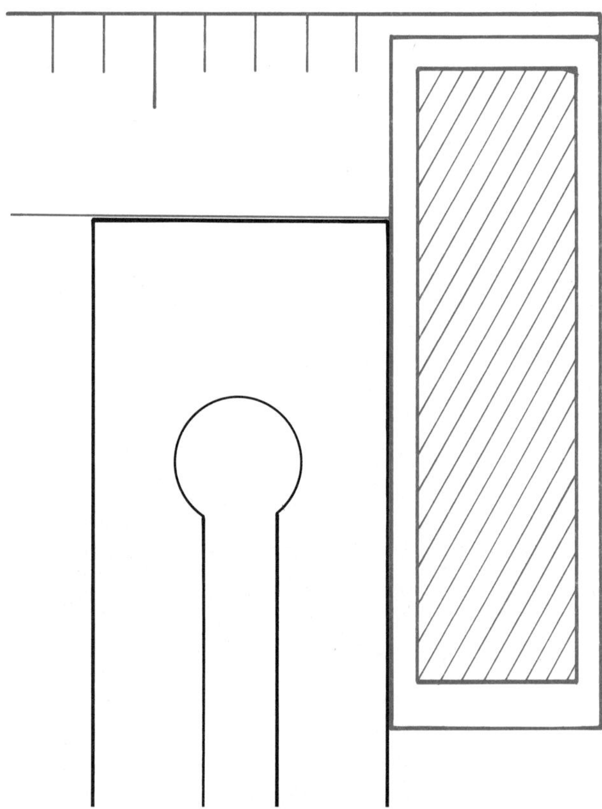

DRILLING AND BORING TOOLS

Holes up to ¼ in. in diameter are made with tools called *drills*. Forming these holes is called drilling. Holes larger than this are made with tools called *auger bits*. This is referred to as boring holes.

A *hand drill* and *straight-shank twist drills* are used for small-diameter holes (Fig. 3-67). These drills can be used on wood, metal, or plastics. The drills are sold in sets and usually range in size from 1/16 in. to ¼ in. by 1/16s. The hand drill comes in two sizes, ¼ in. and ⅜ in. This refers to the largest size drill the chuck will accept. The drill is placed in the chuck. The hand

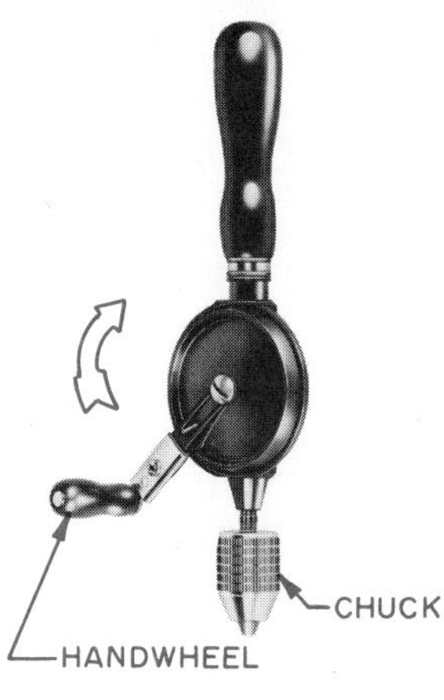

FIGURE 3-67

A hand drill. *(Courtesy of Stanley Tools)*

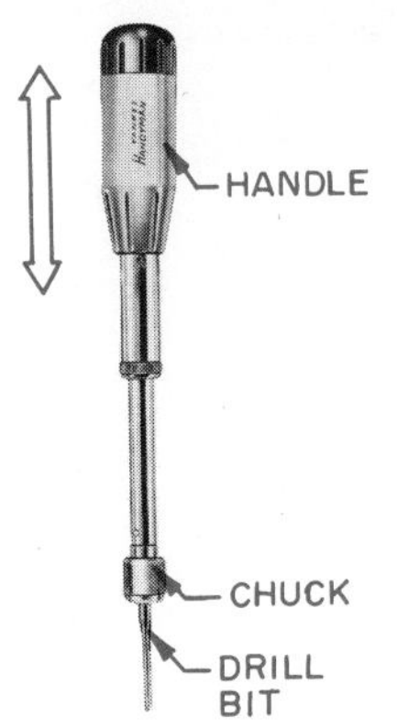

FIGURE 3-69

A push drill. *(Courtesy of Stanley Tools)*

wheel is turned in a clockwise driection to drill a hole (Fig. 3-68).

A popular drill for carpentry use is the *push drill* (Fig. 3-69). It uses a fluted bit. The bit is placed in the chuck. The hole is drilled by pushing down on the handle. This causes the chuck to rotate. Drill sizes are from 1/16 in. to 1/4 in. (Fig. 3-70).

FIGURE 3-68

Drilling a hole with a hand drill. Notice that it uses a straight-shank twist drill.

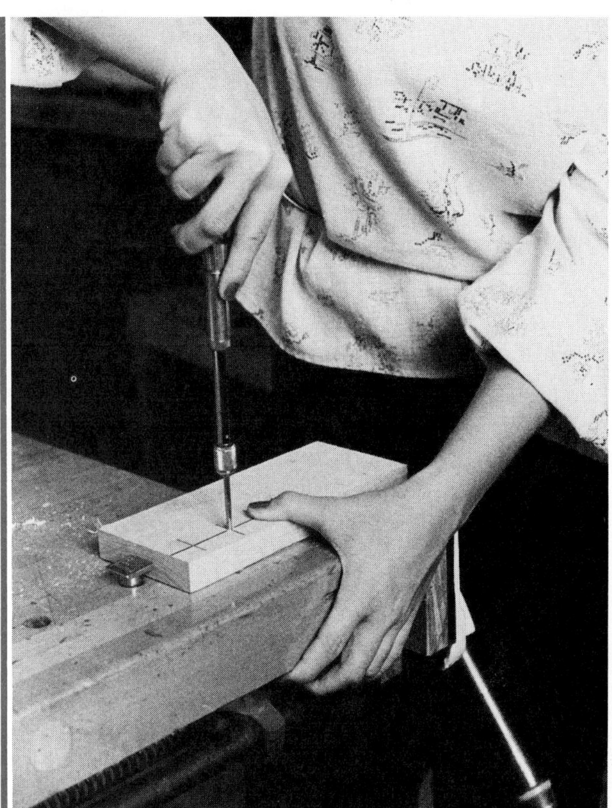

FIGURE 3-70

Drilling a hole with a push drill. Push up and down on the handle.

Auger bits are designed for boring large-diameter holes in wood. A typical set would run from 1/4 in. to 1 in. in diameter by 1/16 s. Larger sizes are available (Fig. 3-71). The size is shown by a number stamped on the tang. This refers to the number of 1/16 s in. in diameter. A No. 8 auger bit is 8/16 or 1/2 in. in diameter.

The parts of an auger bit are shown in Fig. 3-72. The feed screw pulls the bit into the wood. If it is ever damaged, it becomes very difficult to bore a hole. The screw must be protected at all times. The *spur* cuts into the wood, forming a circle. The lips act as chisels and scoop out the wood inside the circle.

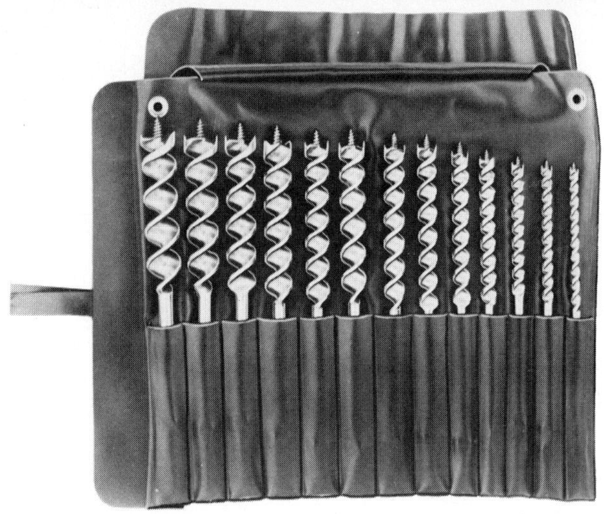

FIGURE 3-71

A set of auger bits. *(Courtesy of Stanley Tools)*

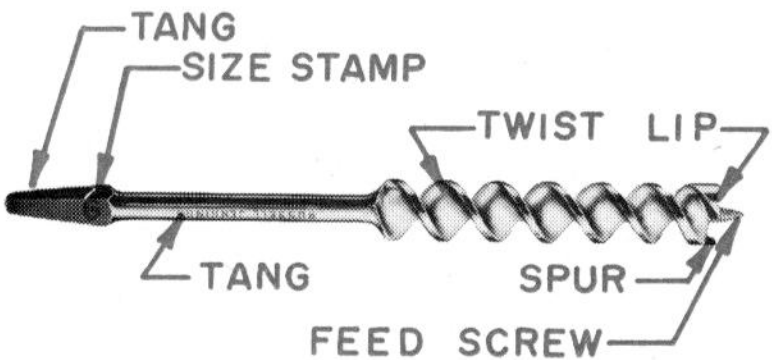

FIGURE 3-72

The parts of an auger bit. *(Courtesy of Russell Jennings)*

The auger has a square *tang*. It fits into the chuck of a brace (Fig. 3-73). The size of a brace is given by the diameter of the swing. This is the diameter of a circle scribed by the handle as it rotates. Typical sizes are 8, 10, 12, and 14 in.

To use the brace and auger bit, hold the head in your right hand and the handle in your left hand. Push down firmly on the head but not too hard. Rotate the handle in a clockwise direction (Fig. 3-74).

If an auger bit gets dull, it can be sharpened. This must be done with great care. If improperly done, it could ruin the auger.

The spur is sharpened with an auger bit file on the inside. Place the auger bit against the edge of a table and lightly sharpen the cutting edge of each spur (Fig. 3-75). Be certain to keep both spurs the same length. Do not hit the screw with the file.

Next, sharpen the lips. Place the auger point down on a table and file the lips from above. Keep each at the same angle. Do not overfile. File just enough to bring back the cutting edge (Fig. 3-76).

When boring holes with auger bits, do not bore all the way through. If this is done, the auger will split out the face of the board on the bottom. The best thing to do is bore until the screw comes through. Then remove the auger and bore from the other side to finish the hole.

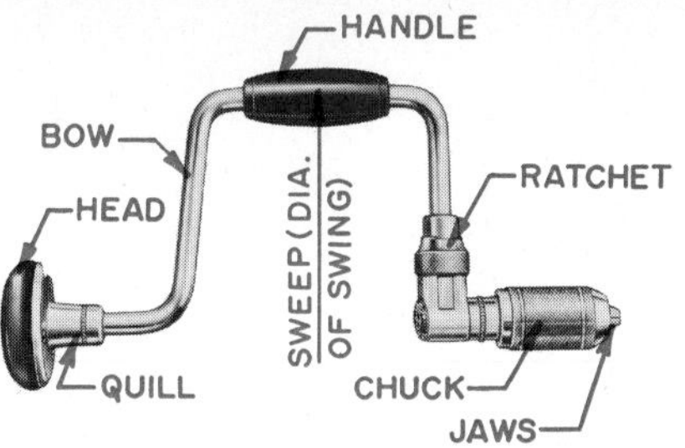

FIGURE 3-73

Parts of a brace. *(Courtesy of Stanley Tools)*

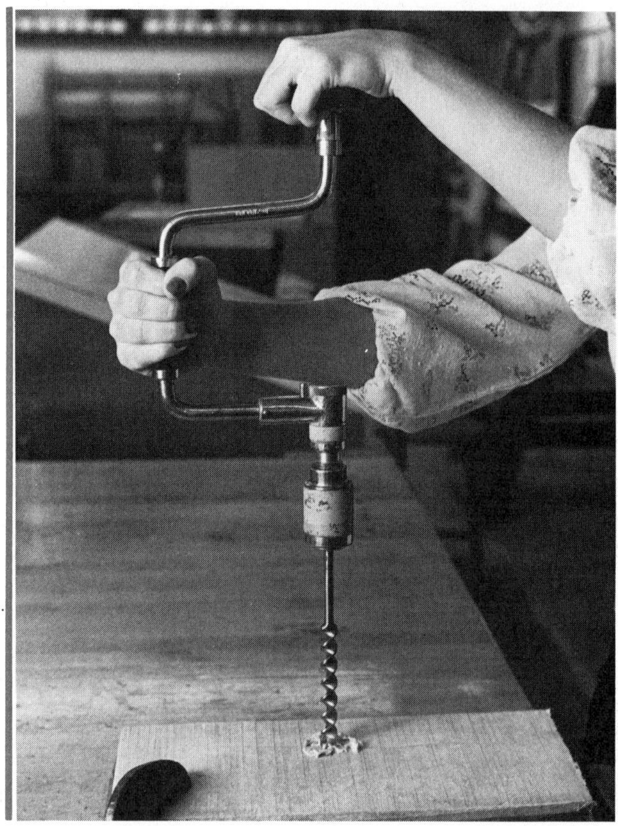

FIGURE 3-74

Boring a hole with a brace and an auger bit.

FIGURE 3-75

Sharpening the nibs of an auger bit with an auger bit file.

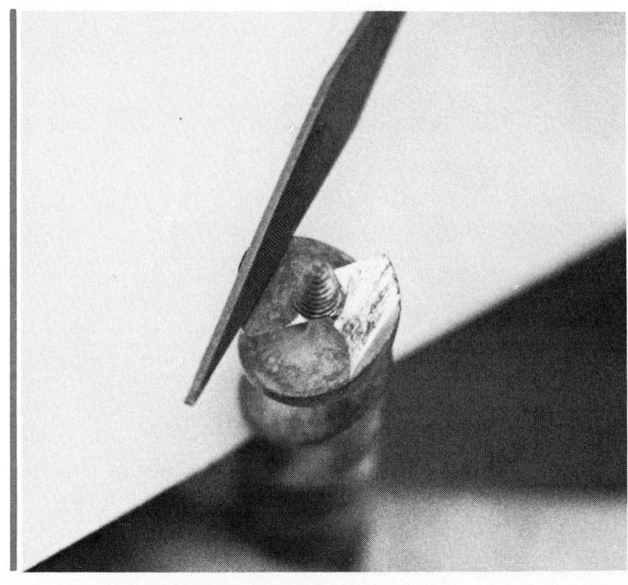

FIGURE 3-76

Sharpening the lips of an auger bit using a small triangular file.

Another way is to bore through into a piece of scrap wood (Fig. 3-77).

Holes can be bored to a predetermined depth using a *bit stop* (Fig. 3-78). The depth of hole is measured from the lip to the foot of the stop. The stop is then clamped to the bit.

An *expansive bit* is a type of auger used to bore large-diameter holes. It has one spur and lip. The distance the spur is from the screw can be adjusted. There are several sizes available. These include bits that will bore from ⅞ to 1½ in., 1½ to 2½ in., and 2½ to 3 in. (Fig. 3-79).

A *countersink* is a tool with a cone-shaped tip having cutting edges (Fig. 3-80). It is held in the chuck of a brace. After a hole is drilled or bored, it is inserted in the hole and revolved just like an auger. This produces a cone-shaped depression. This is used to set the heads of flathead screws and bolts flush with the surface (Fig. 3-81).

A *lockset bit* is used to bore the holes needed for installing door locks (Fig. 3-82). It is designed for use in a brace or electric drill. These bits are available in 1¾-, 2-, and 2⅛-in. diameters.

FIGURE 3-77

Two ways to bore through holes with an auger bit without splitting out the back of the wood.

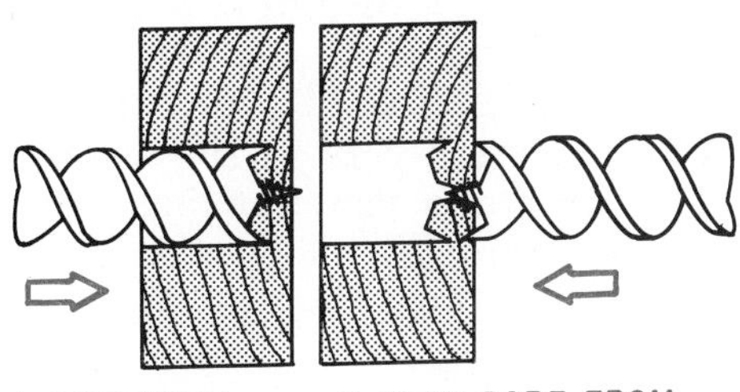
1. BORE UNTIL SCREW SHOWS.
2. THEN BORE FROM OTHER SIDE.

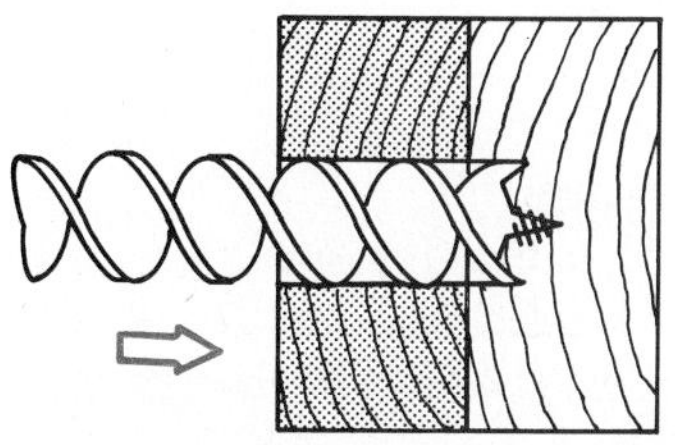
BORE THROUGH INTO A PIECE OF SCRAP.

FIGURE 3-78

Boring a hole to a predetermined depth with a bit stop.

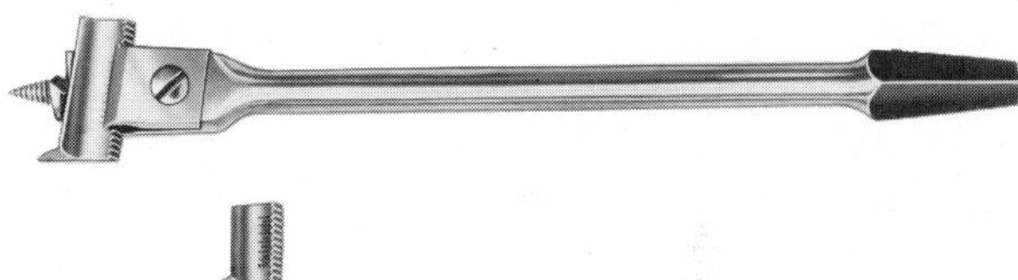

FIGURE 3-79

An expansive bit. *(Courtesy of Stanley Tools)*

FIGURE 3-80

A countersink with a tang for use in a brace. *(Courtesy of Stanley Tools)*

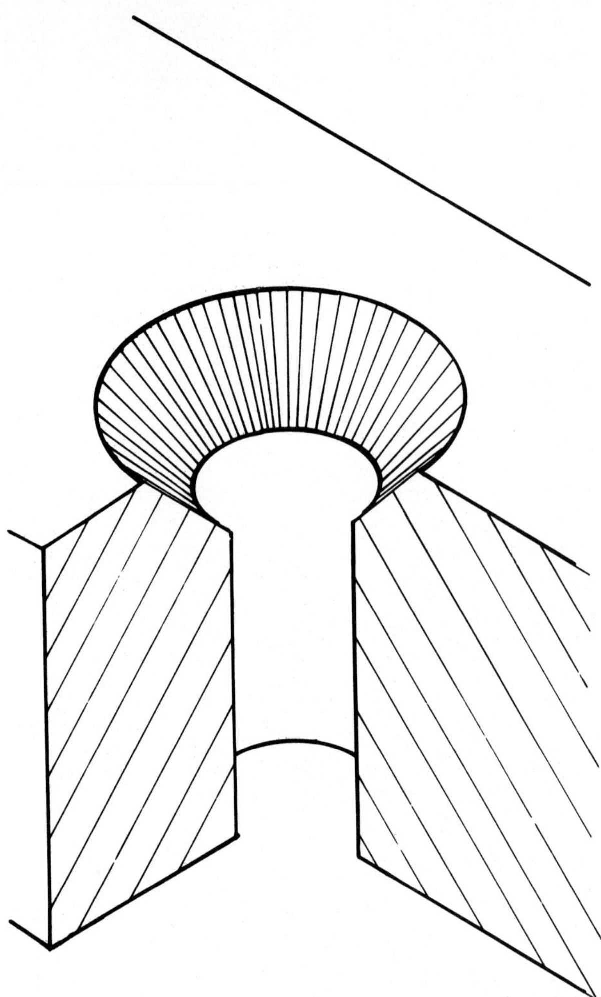

FIGURE 3-81

A countersunk hole is used to receive the head of flathead screws.

FIGURE 3-82

A lockset bit used to install door locks. *(Courtesy of Stanley Tools)*

JOINING AND FASTENING TOOLS

The basic joining activities performed by a carpenter include fastening with nails and screws. Sometimes bolts or special fasteners are used.

The *hammer* is probably the most used carpentry tool. Two types are used: the *curved claw* (Fig. 3-83), and the *ripping claw* (Fig. 3-84). The curved claw is used for most work. It is best for pulling nails. The ripping claw is designed to be driven between pieces nailed together to pry them apart.

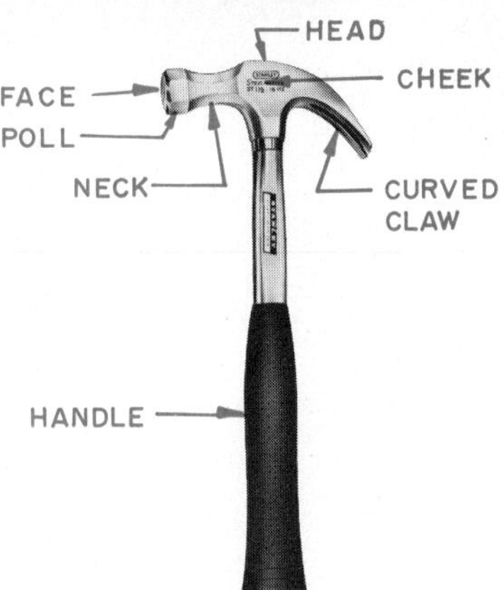

FIGURE 3-83

A curved claw hammer. *(Courtesy of Stanley Tools)*

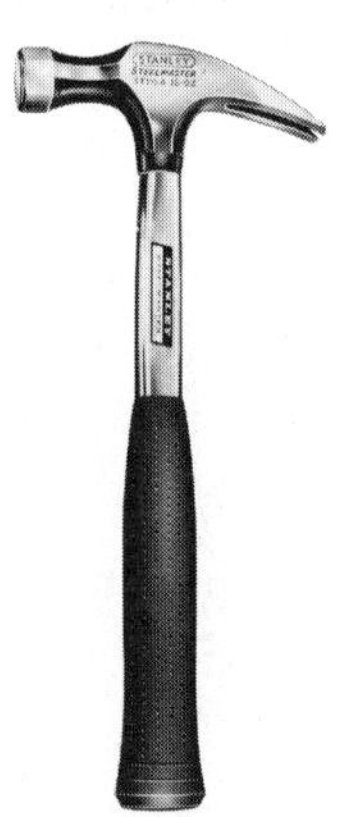

FIGURE 3-84

A ripping claw hammer. *(Courtesy of Stanley Tools)*

The parts of a hammer are shown in Fig. 3-83. The face can be either flat or slightly convex. The advantage to the convex face is that it permits driving a nail close to the surface without leaving a hammer mark in the wood. These dents around nails are often referred to as "sorry marks."

Hammer heads are made of high-quality steel. Some have steel handles and a rubber covering on the grip. Others have wood handles. Hammers are specified by the weight of the head. Common weights are 13, 16, and 20 ounces (oz). Most hammers are 13 to 16 in. long. The 13-oz size is generally used for most work, including interior trim. The heavier hammers are used for rough framing. Floor layers use a special 32-oz hammer.

A *hammer holster* is used to carry the hammer when not in use (Fig. 3-85). It is fastened on the carpenter's belt, making the hammer always within easy reach. This is especially useful when working on ladders, scaffolds, or in the structure above the ground.

With a little practice, driving nails becomes automatic. The nail is held in one hand and placed where it is to be driven. The hammer is

FIGURE 3-85

A hammer holster. *(Courtesy of Stanley Tools)*

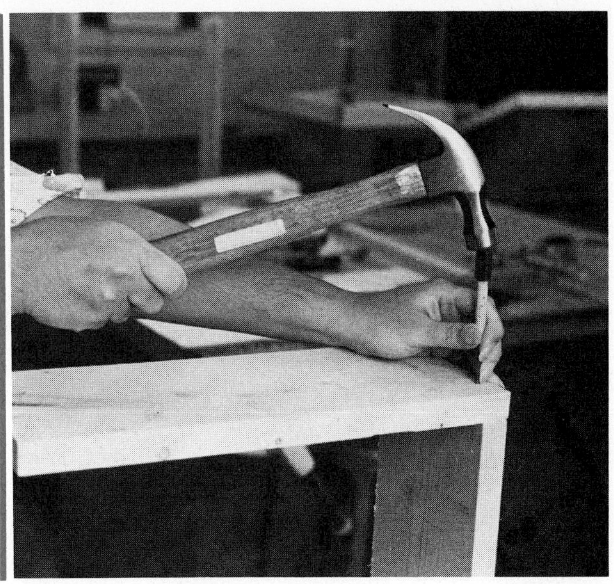

FIGURE 3-87

Nails are set below the surface with a nail set.

held in the other hand. Tap the nail lightly so that it penetrates the wood. The nail will then stand by itself and can be released. Strike the nail with the hammer until it is flush or nearly flush with the surface of the board. Large nails are driven using a movement including the entire arm—wrist, elbow, and shoulder. Small nails and brads are driven using only a wrist movement (Fig. 3-86).

If rough carpentry is being done the nail is driven flush with the surface of the board. The last blows can be a little lighter to prevent serious marring of the wood. In finish carpentry the nail is left sticking slightly above the surface. It is set below the surface with a *nail set* (Fig. 3-87). Nail sets are available with tips having diameters from 1/32 to 5/32 in. by 1/32s (Fig. 3-88).

FIGURE 3-88

A nail set. *(Courtesy of Stanley Tools)*

The hammer is held at the end of the handle. In this way it is in balance. Some persons prefer to grip the handle shorter when driving small nails.

If a nail bends, remove it and start a new one. To remove a nail, the curved claw hammer is best. Slide the claws under the nail head and pull back on the hammer. The hammer head will mar the surface. If this is to be prevented, slip a metal piece, such as a putty knife, or a thin piece of wood under the head (Fig. 3-89). Pull

FIGURE 3-86

Small nails and brads are driven using only a wrist movement.

FIGURE 3-89

When pulling nails, place a piece of wood or metal on the surface of the wood to prevent damage from the hammer head.

FIGURE 3-90

Use a thicker block to pull nails that are only partially driven.

the handle until it is in a vertical position. Then lean the handle forward and place a block of wood under it (Fig. 3-90). Pull back on the handle again. Add more wood pieces until the nail is removed.

Generally, nails should be driven at least ¾ in. from the edge of the board. If they are closer, the board is likely to split. To prevent this, drill a small hole where the nail is to be driven. Sometimes the grain of the board will cause the nail to drive crooked. To overcome this, drill a small hole through the board and drive the nail into the hole.

Carpenters use hammers with a plastic head to drive wood chisels (Fig. 3-91). The plastic faces can be replaced when worn or damaged. Sizes range from 1½-to 32-oz heads. The head diameters range from ⅝ to 1⅝ in.

A carpenter will often use a *ripping bar* to remove boards nailed together. This is especially needed on remodeling jobs where old material must be removed. They are available in a wide variety of types and sizes (Fig. 3-92).

Upon occasion a *hatchet* is needed. The one most useful to carpenters has a blade on one side and a face to drive nails on the other. It is a half hatchet (Fig. 3-93). They are used for rough work such as building and setting up concrete forms.

The *screwdriver* is another frequently used fastening tool. There are many sizes and shapes available. The basic types have either a cabinet, standard, clutch head or a Phillips tip (Fig. 3-94). The size of *cabinet and standard screwdrivers* is specified by giving the length of the blade from the tip to the handle. Common sizes include 3, 4, 6, 8, and 10 in. The size of the tip on

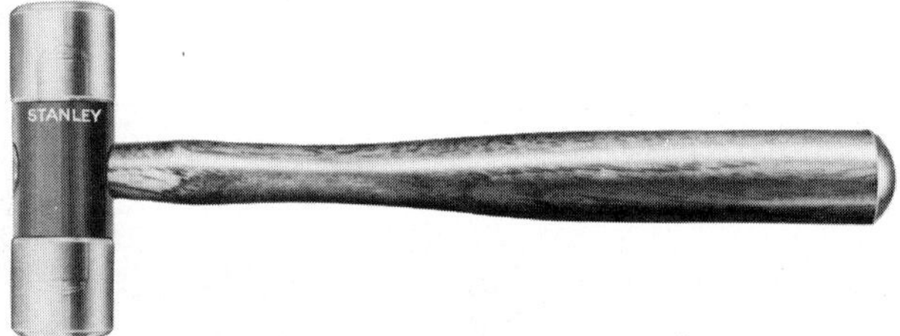

FIGURE 3-91

A plastic head mallet. *(Courtesy of Stanley Tools)*

FIGURE 3-92

A ripping bar. *(Courtesy of Stanley Tools)*

FIGURE 3-93

A carpenter's hatchet. Notice that it has a nailing face. *(Courtesy of Stanley Tools)*

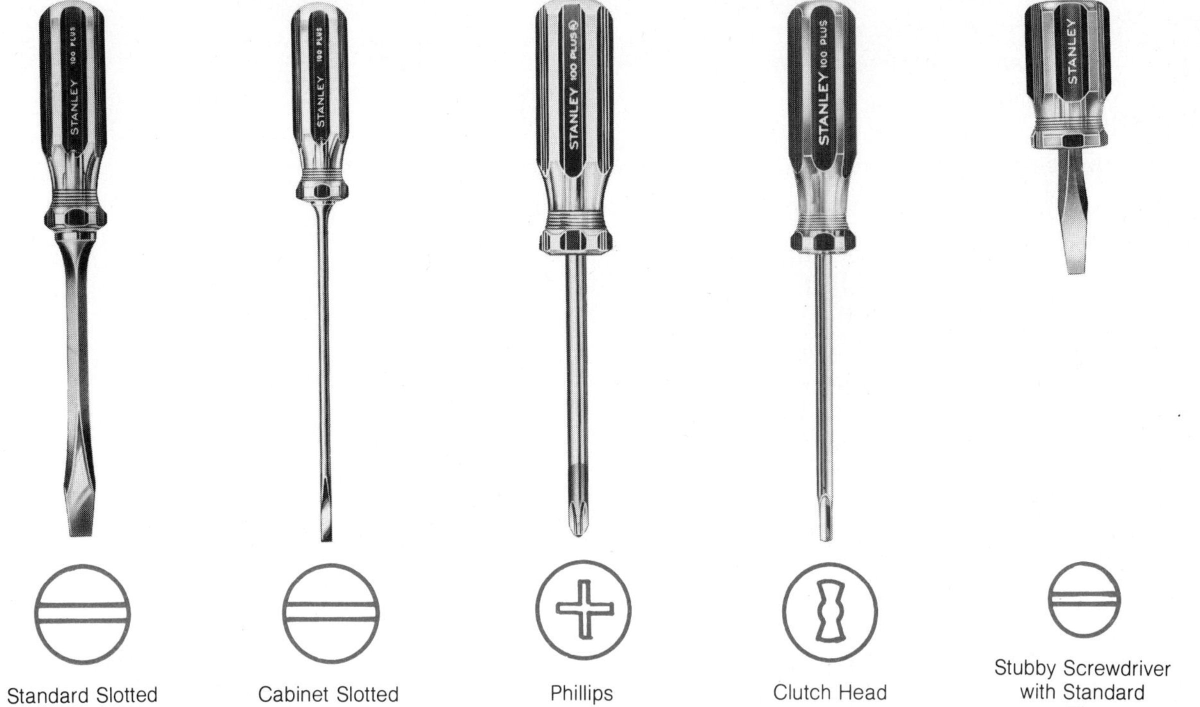

FIGURE 3-94

Common types of screwdrivers. *(Courtesy of Stanley Tools)*

standard and cabinet screwdrivers is related to their length. *Phillips screwdrivers* specify size by point number. The common point numbers are 0, 1, 2, 3, and 4. A 0 point is about ⅛ in. in diameter. A 4 point is ⅜ in. in diameter.

The proper screwdriver must be selected for the screw. The tip must fit securely in the screw slot. Cabinet and standard screwdriver blades should be about as wide as the diameter of the screw head (Fig. 3-95). The Phillips screwdriver should fit snugly in the recess in the screw head. It should fill it with no play.

The tip of cabinet and standard screwdrivers must be kept in good condition. It must be ground so that it is flat and square on the bottom (Fig. 3-95). If it becomes rounded or chipped, it can be ground to shape on a power tool grinder. Never use a screwdriver as a chisel. The blade is not prepared for such use. The handle also will eventually break.

A *spiral ratchet screwdriver* helps drive screws rapidly (Fig. 3-96). The handle is pushed down, which causes the chuck to rotate. Various size screwdriver bits can be used in the chuck. They range in size when extended from 18 to 25 in. Bits for standard and Phillips head screws are available.

A screwdriver bit is placed in a brace (Fig. 3-97). This combination permits rapid driving of screws. Considerable force can be exerted. If the holes are not properly prepared for a screw, this could twist the screw into two pieces.

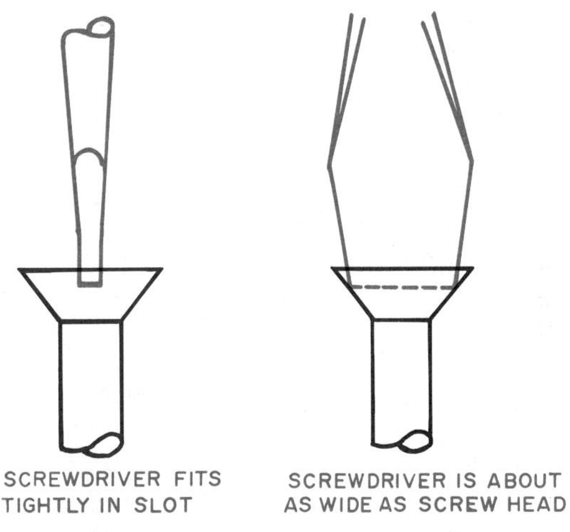

FIGURE 3-95

The screwdriver block must fit properly in the slot of the screw.

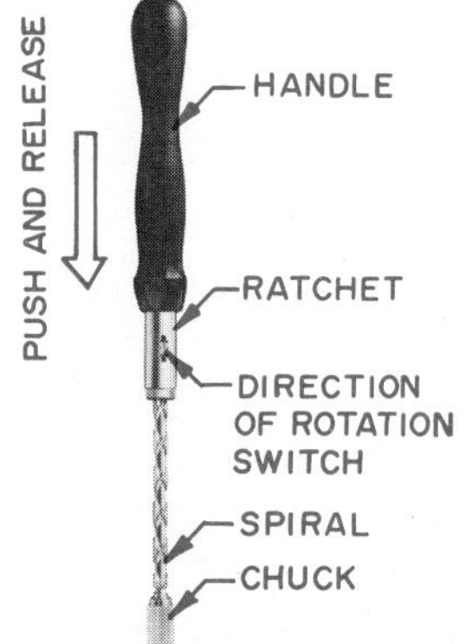

FIGURE 3-96

A spiral ratchet screwdriver. *(Courtesy of Stanley Tools)*

FIGURE 3-97

A screwdriver bit with a tang for use in a brace. *(Courtesy of Stanley Tools)*

How to Drive Screws To properly drive a screw, two holes must be drilled. The top board to be fastened has a hole drilled the size of the shank of the screw. The *shank* is the steel body below the head. The board into which the screw threads will cut requires a *pilot hole* (Fig. 3-98). The pilot hole should be about the same diameter as the core of the screw. The core is the steel center about which the threads are cut. The core fits in the pilot hole. The threads cut grooves in the wood at the sides of the hole.

Round-head screws are installed with the bottom of their heads on the surface of the board. Cone-shaped heads have the shank hole countersunk so that they are flush with the surface (Fig. 3-99). The tool used to prepare this hole is a *countersink* (Fig. 3-80).

Pliers are sometimes used for loosening or tightening bolts. Typical types are shown in Fig. 3-100. The *vice-grip pliers* are adjustable and can grip with great force. They can also serve as small clamps.

There are various types of nailing and *stapling* devices used in carpentry. Some are operated by hand whereas others use compressed air or electricity. Examples are shown in Figs. 3-101 and 3-102.

FIGURE 3-98

To properly install wood screws, drill a pilot hole and an anchor hole.

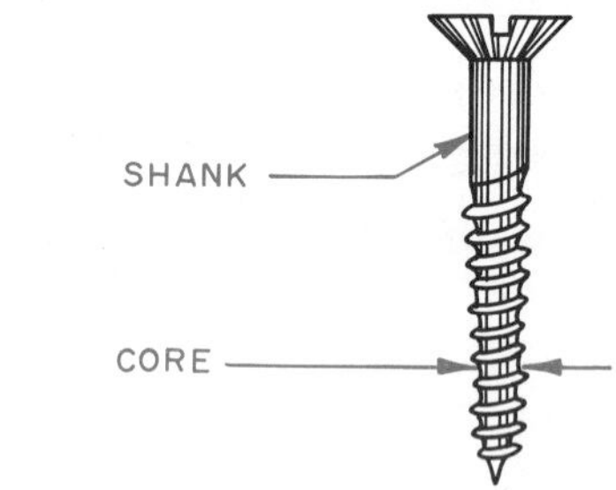

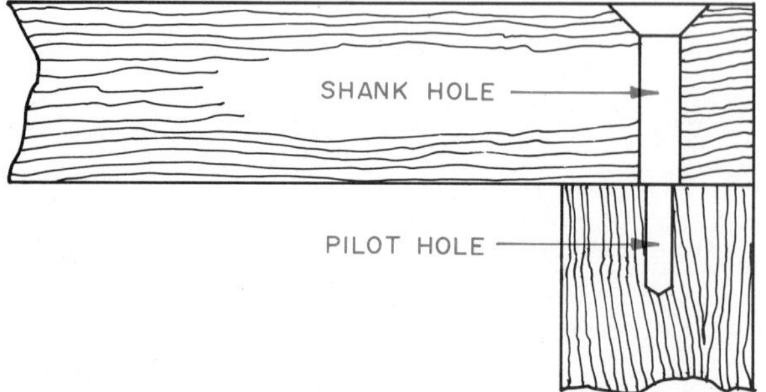

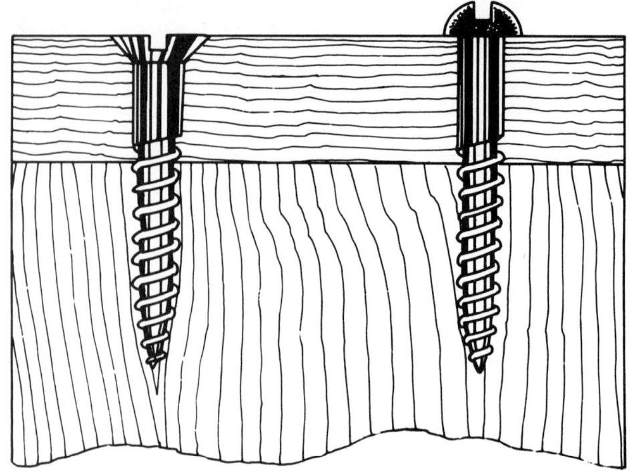

FIGURE 3-99

Flat-head screws are set in a countersink so that the head is flush with the wood. Round-head screws have the bottom of their heads on the surface of the board.

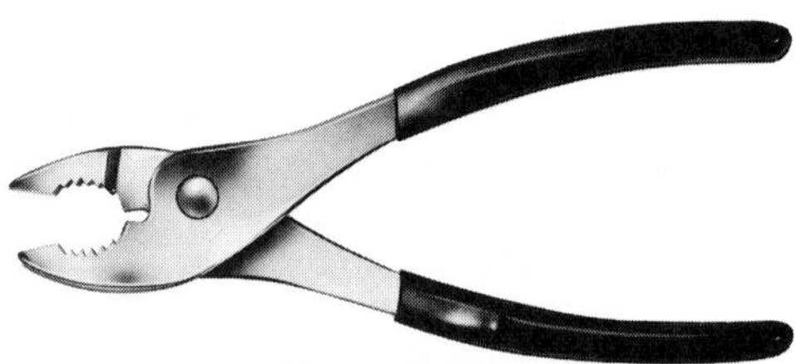

Regular Pliers

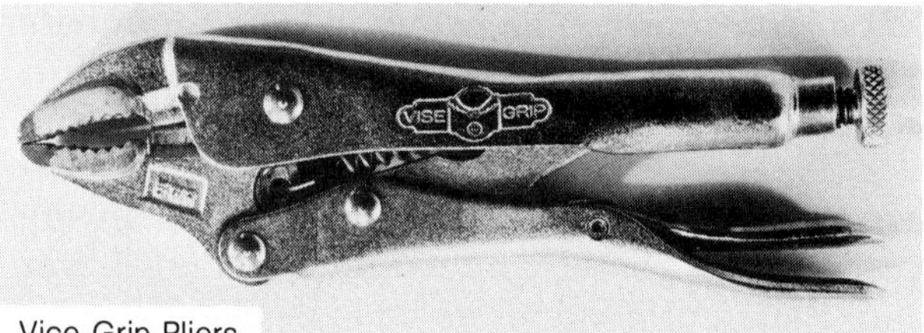

Vice-Grip Pliers

FIGURE 3-100

Common types of pliers. *(Courtesy of Stanley Tools)*

FIGURE 3-101

Using a hand-powered stapler to install insulation. *(Courtesy of Owens Corning Fiberglas)*

FIGURE 3-102

Fastening trim around a door with a compressed air-powered nailer. *(Courtesy of Bostitch Textron)*

CLAMPS

There are a wide variety of clamping tools. Probably the most common is the *C-clamp* (Fig. 3-103). Common sizes have openings up to 3 in. *Bar clamps* are quickly adjustable and can open to 6 ft or more (Fig. 3-104). *Spring clamps* hold by pressure of a spring in the handle. Sizes up to 3 in. are available (Fig. 3-105). *Hand screws* have jaws which give a wide flat surface. They can pull boards together and provide a uniform pressure over the entire area (Fig. 3-106).

A *lever wrench* is shown in Fig. 3-107. It is much like a C-clamp but is self-adjusting. Lift the lever-like handle to open the jaws. Place the upper jaw on the work and squeeze the lever closed. The clamp automatically adjusts the lower jaw to fit tightly over the work to be clamped.

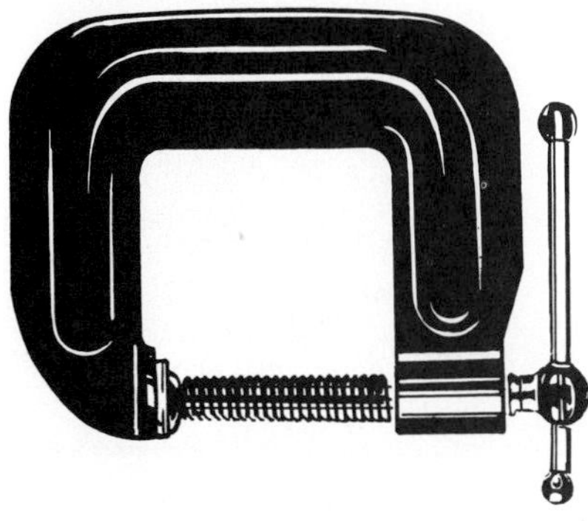

FIGURE 3-103

A C-clamp. *(Courtesy of Stanley Tools)*

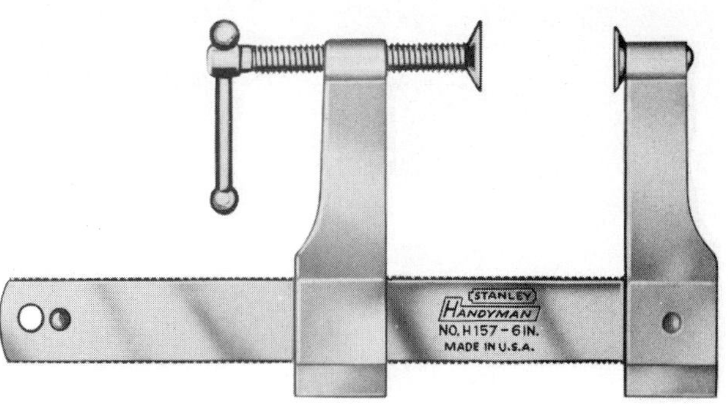

FIGURE 3-104

An adjustable bar clamp. *(Courtesy of Stanley Tools)*

FIGURE 3-105

A spring clamp. Notice the plastic-covered handles and jaw tips. *(Courtesy of Stanley Tools)*

FIGURE 3-106

Wood hand screws are useful for face-to-face clamping.

FIGURE 3-107

A lever wrench. *(Courtesy of Leverage Tools, Inc.)*

STUDY QUESTIONS

After reading the chapter, answer each of the following questions. If you do not know the answer, review the chapter.

1. What is the main reason hand tools are damaged?
2. How is a rule held to get the most accurate measurement?
3. What are the parts of a framing square?
4. For what purposes are levels used?
5. What are the major types of hand saws, and for what uses are they designed?
6. How does a crosscut saw differ from a ripsaw?
7. What are the different kinds of hand planes?
8. What are the commonly used sizes of wood chisels?
9. What tools are used to form small-diameter and large-diameter holes?
10. What are the different kinds of hammers?
11. What types of screwdrivers are in common use?
12. List the types of clamps often used by carpenters.

IMPORTANT TECHNICAL TERMS

Following are technical terms that you should be able to use as part of your working vocabulary. Write a brief description of the meaning of each term.

Folding rule
Tape rule
Rafter square
Try square
Combination square
T-bevel
Marking gauge
Dividers
Trammel points
Levels
Chalk line
Plumb bob
Crosscut saw
Saw teeth set
Ripsaw
Keyhole saw
Backsaw
Miterbox
Hacksaw
Kerf
Bench plane
Jack plane
Fore plane
Jointer plane
Block plane
Rabbet plane
Multiblade forming tool
Tungsten carbide file
Chisel
Oil stone
Straight-shank twist drill
Auger bit
Spur
Tang
Bit stop
Expansive bit
Countersink
Lockset bit
Curved claw hammer
Nail set
Ripping bar
Hatchet
Screwdriver
Pilot hole
Pliers
Staplers
C-clamp
Lever wrench

4 POWER TOOLS

A great deal of the on-site sawing, planing, and boring is done with power tools. These save time and energy and produce accurate work if used correctly. Every carpenter will have in addition to a complete set of hand tools a variety of electrically powered saws, sanders, planers, and drills.

Power tools will fall into two broad types: stationary and portable. The *stationary* tools are mounted on floor stands. The carpenter brings work to be done to the machine. *Portable* tools are carried about the building and used wherever the carpenter has to work.

When selecting such tools, several things should be considered:

1. Purchase only high-quality tools. They last longer, perform better, and are less likely to cause accidents.
2. Consider the weight. Stationary tools have to be carried from a truck into the building. They should be heavy enough to remain steady when in use yet light enough to move without undo problems. Since portable tools are carried by the carpenter onto the roof and other places, they should be as lightweight as possible without sacrificing quality.
3. Examine the tools for the quality of their electrical system. Danger from electrical shock is constant, especially when working on the ground.
4. The equipment should have guards that thoroughly protect the operator. Examine the guards carefully to see that they operate easily.
5. Make certain that repairs can be made on the tools. Examine the guarantee to see if the manufacturer backs up the equipment.

GENERAL SAFETY RULES

Workers on a construction site are constantly faced with potential for accidents. The use of power tools is one of these dangers. When using any power tool, consider the following:

1. Make certain that the tool is in good repair.
2. The tool should be sharp. Accidents are caused by dull tools.
3. The blades, cutters, belts, or drills should be firmly installed in the unit.
4. Be constantly on the alert. Someone may bump you. The tool may slip. Are you overly tired and is your attention lacking?
5. Make certain that the material to be worked is solidly in place. It must not move or slip as it is being worked upon.
6. Keep all guards in place.
7. Wear safety glasses.
8. Do not overcrowd the tool. Let it do its job at the normal pace.
9. After finishing an operation, let the tool stop before you move on to something else.
10. Check the electric cord. It must have a third wire for a ground. It must not be worn.
11. Be certain your source of electrical power is wired so that the third wire (ground) is available. The third wire on your tools is of no use if the source is not grounded.
12. Be certain that the switch on the tool is operating correctly. Many tools have switches that turn off the motor when the operator releases pressure on them. These should always be in working order.
13. Do not wear rings, bracelets, necklaces, and so on. They tend to get tangled up in moving machinery.

USING A STATIONARY CIRCULAR SAW

Stationary circular saws are most often used for ripping stock and cutting plywood panels. The table provides a steady surface for the work. The saw can be used for crosscutting, mitering, and cutting moldings.

Circular saws are specified by the diameter of the blade. For general carpentry an 8- or 10-in. blade is adequate. Also, saws of this size are light enough to move from job to job.

Safety

1. All adjustments should be made after the blade has stopped.
2. The blade should be the right one for the saw. The diameter and arbor hole should be as specified for the saw.
3. Keep the blade sharp. If it wobbles or is cracked, throw it away.
4. The blade should rise above the wood being cut no more than ⅛ to ¼ in.
5. Always cut with the guards in place.
6. Use *feather boards* and *push sticks* to protect your fingers. Never, ever get your fingers near the saw blade.
7. Never stand directly behind the blade when cutting. It often kicks back a board, which will injure anyone behind the saw.
8. Use the *miter gauge* for crosscutting. Never use the fence for crosscutting.
9. When ripping, keep the smooth straight edge of the board next to the fence.
10. Always keep the table free of scraps. Push them off with a wood push stick. Do not use your fingers.
11. Before operating make certain that everything—including the fence, miter gauge, and depth control—is locked tightly.
12. Warped boards will bind the blade and kick back. Do not cut warped boards.
13. The person who may be removing boards as they are cut should not pull on them. This might choke the saw and cause a kickback. They should only support the wood.
14. Never cut stock freehand. Always use the miter gauge for crosscutting and the fence for ripping.
15. Do not adjust the saw while it is running.

Blade Selection Select the proper blade. For saws to be used for some ripping and crosscutting, a combination blade will be best. It is designed for both operations. If the saw is used mostly for ripping, a rip blade would be needed. If it is used mostly for crosscutting, a crosscut blade should be used. A hollow-ground blade produces a smooth cut and is preferred for cabinet and finishing work (Fig. 4-1).

Parts of a Circular Saw A typical stationary circular saw is shown in Fig. 4-2. The table supports the work. The guard covers the blade. The fence is used to guide the edge of a board as it is ripped. It is adjusted by control knobs. The miter gauge is used to hold a board as it is crosscut. Below the table are handwheels, used to raise, lower, and tilt the blade. Notice that it has wheels, which help move it about the building under construction.

Rough Cut Combination.

Rip.

Crosscut.

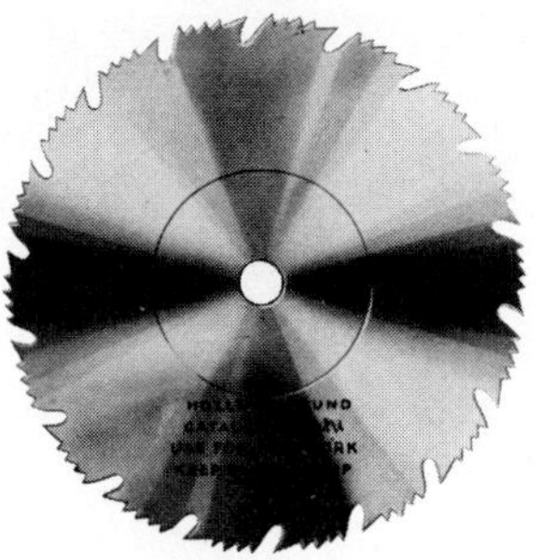
Hollow Ground Combination.

FIGURE 4-1

Commonly used types of circular saw blades. *(Courtesy of Rockwell International, Industrial Tool Division)*

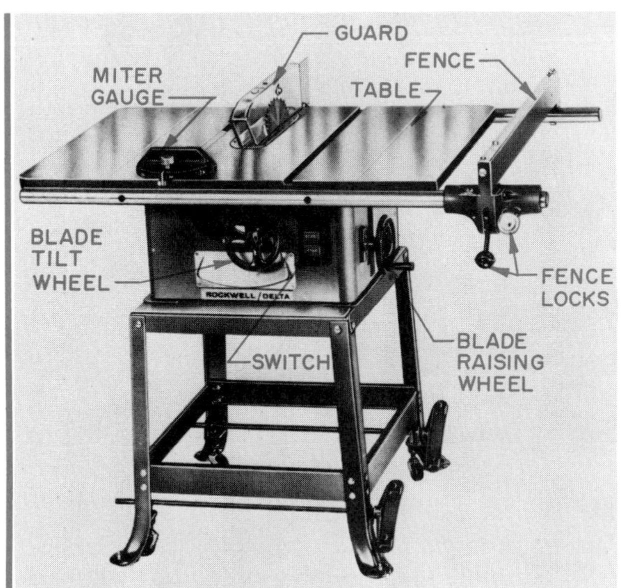

FIGURE 4-2

A 10-in. circular saw which is light enough to be moved around a construction site. The wheels make it easily movable. *(Courtesy of Rockwell International, Industrial Tool Division)*

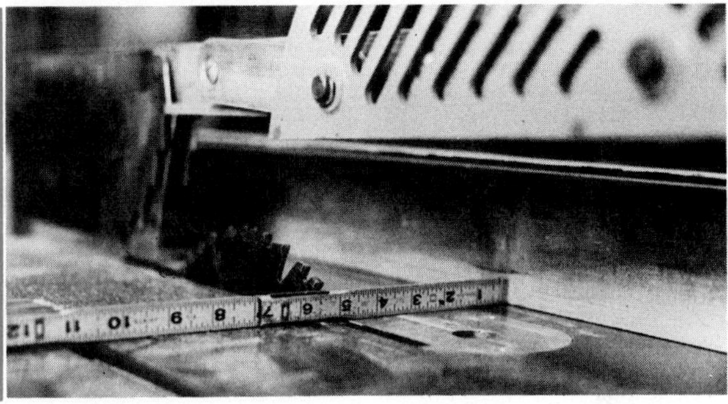

FIGURE 4-3

The folding rule can be used to set the distance between the fence and the blade.

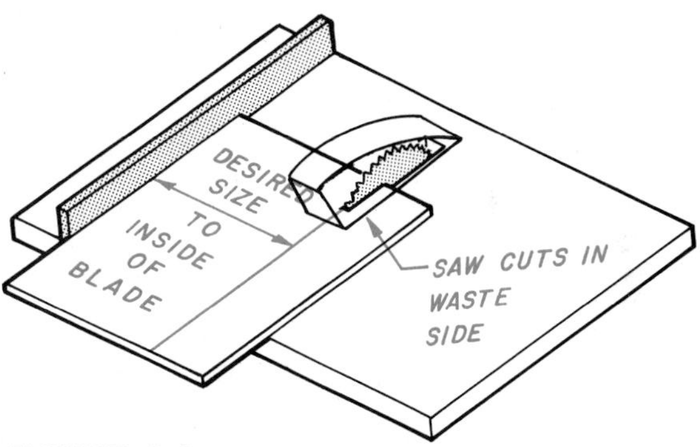

FIGURE 4-4

The saw blade should cut on the waste side of the line.

Ripping

1. Set the distance between the blade and the fence. If the saw is in adjustment, the scale on the guide bar can be used. If not, a ruler must be used (Fig. 4-3). Be certain that the saw cuts on the waste side of the board (Fig. 4-4).
2. Raise the blade so that it is slightly above the surface of the board.
3. Lower the guard over the blade.
4. Start the saw.
5. Press the smooth edge of the board against the fence. If the board does not have a smooth edge, run one edge on the jointer until it is straight and smooth.
6. Keep the board against the fence and slide it forward into the saw. As the cut begins, keep firm pressure on the board (Fig. 4-5).
7. If the board is long, have someone on the other side of the saw to support it.
8. As the end of the board nears the saw, use push sticks to push it between the saw and the fence (Fig. 4-6). If the board is wide, such as 12 in. or more, push sticks are not absolutely necessary. Still be careful to keep fingers away from the blade. A good technique is to let the hand run along the fence.

FIGURE 4-5

When ripping stock to width, use a push stick to keep the board against the fence.

FIGURE 4-6

Use push sticks to slide the end of a board between the fence and the saw blade. In this picture the guard has been removed to show the blade.

Crosscutting

1. Set the miter gauge on the angle desired for the cut. Place it in the groove in the table.
2. Raise the blade so that it is slightly above the surface of the board.
3. Lower the guard over the blade.
4. Move the fence so that it is completely out of the way. For some long boards it is necessary to remove it from the saw.
5. Place the board to be cut on the table in the miter gauge.
6. Start the saw.
7. Hold the board firmly to the miter gauge. Slide the gauge and board up to the saw. Line up the mark on the board with the edge of the rotating saw. Cut on the waste side of the line. This is best done by touching the saw to the board to see where it will cut. Then keep moving it closer to the line until it cuts on the waste edge of the line. This requires careful observation. Safety glasses are important to keep sawdust out of the eyes.
8. Push the board and miter gauge on past the saw (Fig. 4-7). Remove the board from the miter gauge.

Sawing Plywood Large plywood panels are cut using the fence as described for ripping

NOTE: The guard has been raised off the saw to show the position of the blade.

Using a miter gauge to crosscut a narrow board.

Turn the miter gauge around to crosscut a wide board.

FIGURE 4-7

Crosscutting is done with the miter gauge.

(Fig. 4-8). A combination or hollow-ground blade should be used. Sometimes the saw will cause the edge of the sheet to splinter. The surface next to the table is the one that splinters. Therefore, keep the good face of the sheet up.

Cutting Grooves and Dados Grooves and dados are cut by placing a special set of cutters on the saw in place of the blade. One type is shown in Fig. 4-9. The metal insert in the table with the slot through which the saw moves must be removed. An insert with a wide slot is inserted.

The operation of the saw is the same as for ripping and crosscutting. Figure 4-10 shows a

FIGURE 4-8

Large panels can be cut using the ripping fence as a guide.

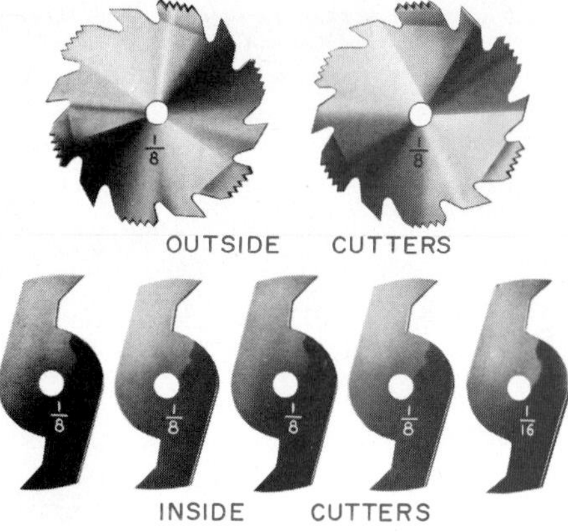

FIGURE 4-9

A dado head made of outside cutters and inside cutters. *(Courtesy of Rockwell International, Industrial Tool Division)*

FIGURE 4-10

Grooves can be cut with a dado head.

FIGURE 4-11

Cutting dados with a dado head. *(Courtesy of Rockwell International, Industrial Tool Division)*

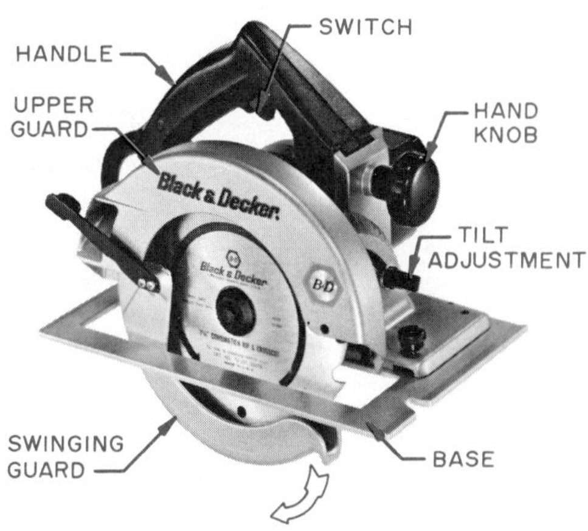

FIGURE 4-12

The parts of a portable circular saw. The top guard is fixed. The low guard swings back by the stock as the saw cuts into it.

groove being cut. A *groove* is a rectangular slot cut with the grain. A dado is shown in Fig. 4-11. A *dado* is a rectangular slot cut across the grain.

USING A PORTABLE CIRCULAR SAW

Most *portable electric circular saws* use 7- or 8-in. blades. The parts of a typical saw are in Fig. 4-12. The depth of cut is controlled by moving the saw. It can also be tilted to produce angle cuts.

The saw is held in two hands. Some cuts permit the use of one hand. The switch is in the handle. The lower guard swings down as it crosses a board. When the cut is finished, a spring pulls it back over the blade.

The basic safety rules are the same as described for a stationary circular saw. In addition, the following should be observed:

1. If the saw stalls in a cut, back the saw out with the motor running.
2. Be certain to adjust the depth of cut so that it is slightly greater than the thickness of the board.
3. Before changing a blade, remove the electric plug from the source of power.
4. Be certain that the teeth on the blade point in the direction of rotation. This saw cuts from the bottom of the board toward the top (Fig. 4-13).
5. Do not start a cut until the saw is running at full speed.
6. Always leave the guards in place.

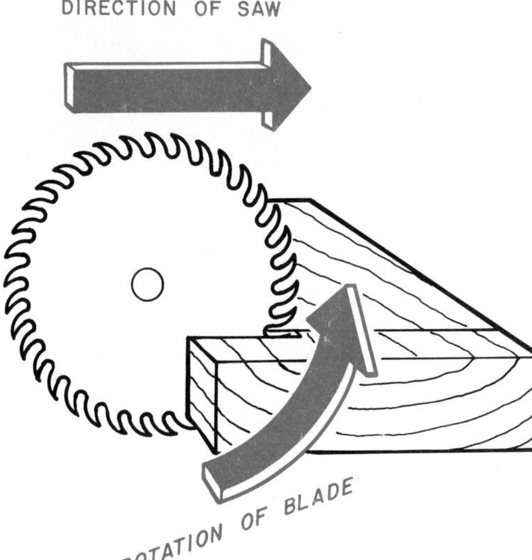

FIGURE 4-13

The portable circular saw cuts up from the bottom of the board.

FIGURE 4-14

Ripping with a portable circular saw. *(Courtesy of Rockwell International, Industrial Tool Division)*

FIGURE 4-15

Crosscutting with a portable circular saw. *(Courtesy of Rockwell International, Industrial Tool Division)*

7. Remember, the saw cuts under the board, so keep all fingers far away from the line to be cut.
8. Make certain that there is nothing below the board the blade will strike as the cut is made.
9. Never set a saw down until the blade has stopped running.

Operating Procedure

1. Mark where the board is to be cut. The saw will have a guide mark on the shoe which you will line up with this mark.
2. Place the board so that it is firmly seated and will not wiggle or fall.
3. Set the depth of cut.
4. Start the saw. When it is running full speed, place the shoe on the board. Line up the guide mark and push the saw into the board (Figs. 4-14 and 4-15).
5. Keep a firm pressure on the saw until the cut is completed. Grasp the piece of board to be cut off. This will prevent it from falling and binding the blade.

USING A RADIAL ARM SAW

The *radial arm saw* is a circular-type saw with the blade above the table. The wood to be cut remains in place and the saw moves out on an arm and cuts it (Fig. 4-16). The saw will crosscut, rip, groove, dado, miter, and cut some moldings.

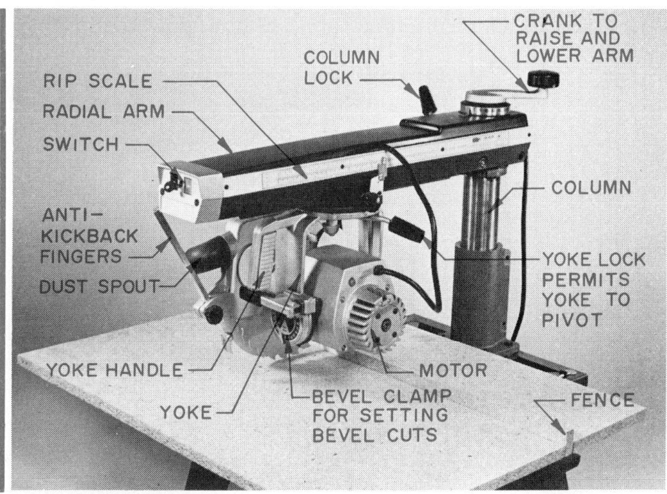

FIGURE 4-16

The parts of a small radial arm saw. This unit is light enough to be moved around a construction site. *(Courtesy of DeWalt)*

The arm moves vertically on a column. The depth of cut is adjusted by raising or lowering the arm on the column.

The motor is mounted in a yoke that slides horizontally on the arm. The blade is mounted to the motor shaft.

The blade is installed with the teeth pointing down toward the table. The saw rotates in a clockwise direction which puts the thrust down and to the rear (Fig. 4-17). This helps the stock stay tight to the fence.

The saw can be tilted to cut angles. The motor with the blade turns on the yoke (Fig. 4-16).

The arm can also be swung right or left so that boards can be crosscut on angles other than 90°. This saw is especially useful for cutting rafters. In addition to the normal cuts, it can easily cut compound meters (Fig. 4-18).

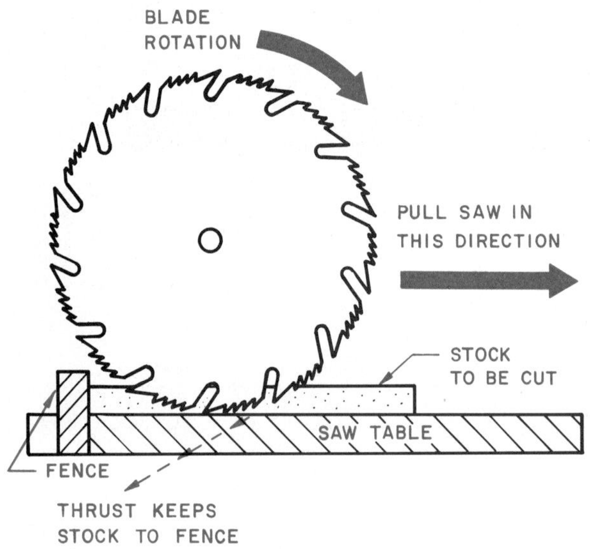

FIGURE 4-17

The saw blade on a radial arm saw down toward the table.

FIGURE 4-18

Cutting a compound miter with a radial arm saw. *(Courtesy of DeWalt)*

Safety

1. Keep all guards in place.
2. After adjusting and before starting, make certain that all locking handles are secure.
3. Use the *anti-kickback fingers*. Keep them about ⅛ in. above the top of the board.
4. Make certain that the saw is set so that it is lower at the back. This tilts the table and arm to the rear. This keeps the yoke from moving forward on its own. It helps to return the saw to the rear position when a cut is finished.
5. The stock must be pressed firmly against the table and fence. The stock must rest flat on the table. If stock is bowed, it will bind the saw.
6. After crosscutting a board, return the saw to its rear position.
7. Let the saw reach its full speed before starting to cut.
8. When ripping, be absolutely certain to feed the stock into the blade so that the teeth are turning upward and toward you.
9. Before starting a cut, be certain that your fingers are out of the path of the blade. This is a difficult saw to guard. Do not depend on the guards to protect your fingers.
10. Keep the table free of scraps.
11. When crosscutting the saw tends to feed itself into the work. As you move it toward you, control forward movement with the handle. It may be necessary to resist forward movement if the saw feeds too fast.
12. Do not attempt to rip stock unless it has one square edge to place against the fence.

Crosscutting and Mitering

1. Set the arm at the desired angle to the table. Lock it in place.
2. Lower the blade until it is 1/16 in. below the surface of the table. The table will have a kerf cut into it.
3. Adjust the anti-kickback fingers until they are about ⅛ in. above the surface of the board to be cut.
4. Place the board on the table and press it against the fence.
5. Line up the line to be cut with the blade of the saw. Be certain that the blade will cut on the waste side of the line.
6. With the blade in the rear position, start the motor. Allow the saw to get to full operating speed.

7. Firmly grasp the *yoke* handle in one hand. Hold the board firmly with the other hand. Be certain your fingers are out of the line of cut.
8. Pull the yoke handle forward and the saw will begin to cut the board. Pull the saw so that it cuts completely through the board (Fig. 4-19). A saw set to crosscut forming a miter is shown in Fig. 4-20. A saw set to crosscut forming a bevel cut is shown in Fig. 4-21.
9. Return the saw to the rear position.
10. Turn off the power and remove the boards from the table.

FIGURE 4-19

To cut stock to length, place it against the fence and pull the blade forward. *(Courtesy of Rockwell International, Industrial Tool Division)*

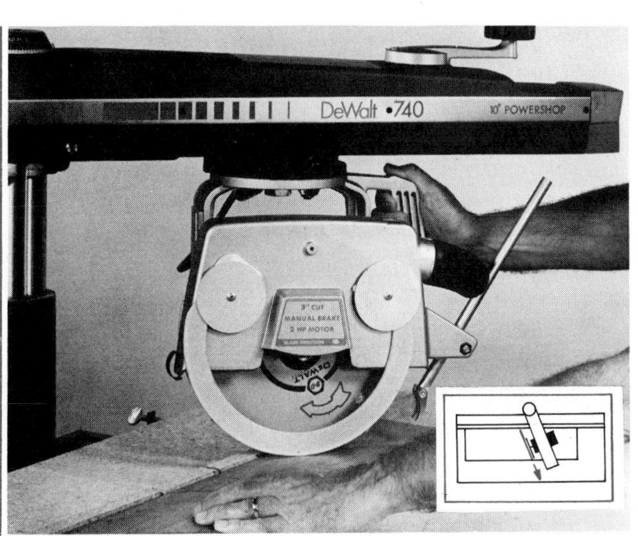

FIGURE 4-20

A flat miter is cut by swinging the arm on an angle. The stock is placed against the fence. The saw is pulled forward. *(Courtesy of DeWalt)*

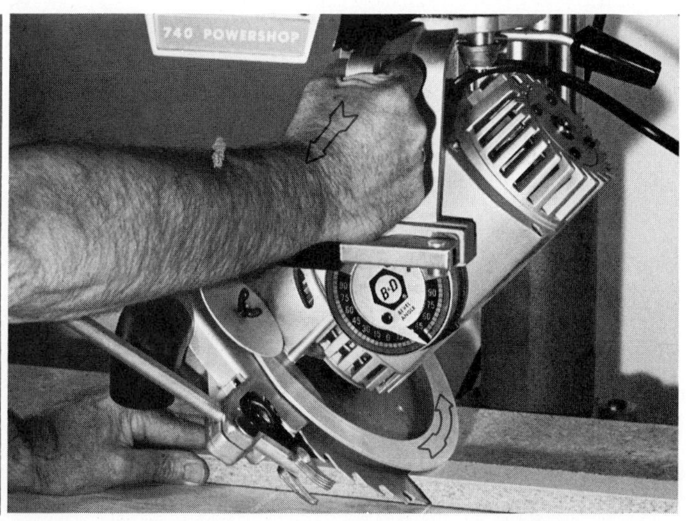

FIGURE 4-21

The blade is set on an angle to make bevel cuts. *(Courtesy of DeWalt)*

1. Move the yoke out to the end of the arm.
2. Remove the locating pin and revolve the yoke 90°. This will place the blade parallel with the fence.
3. Measure the desired width of the board to be cut. This is from the fence to the blade (Fig. 4-22). Lock the yoke to the arm.
4. Lower the saw until the blade touches the table surface.
5. Adjust the guards so that they clear the board to be cut. Set the anti-kickback fingers ⅛ in. above the surface.
6. After checking to see that everything is locked, start the saw. Be certain that it is rotating upward toward you (the side from which you will feed the board) (Fig. 4-23).
7. Put the board against the fence and feed it into the saw. If it is long, have someone on the other end to balance it. Use a push stick to push the end of the board between the fence and the blade (Fig. 4-24).
8. When the cut is finished, shut off the saw.

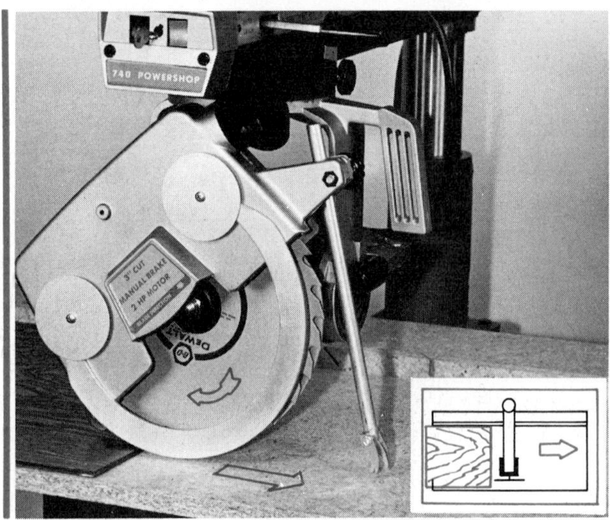

FIGURE 4-22

The saw blade is set parallel to the fence for ripping stock. The yoke is locked to the arm so that it does not move. The wood is pushed into the rotating blade. *(Courtesy of DeWalt)*

FIGURE 4-23

When feeding stock to be ripped, always feed into the blade from the side that is rotating up.

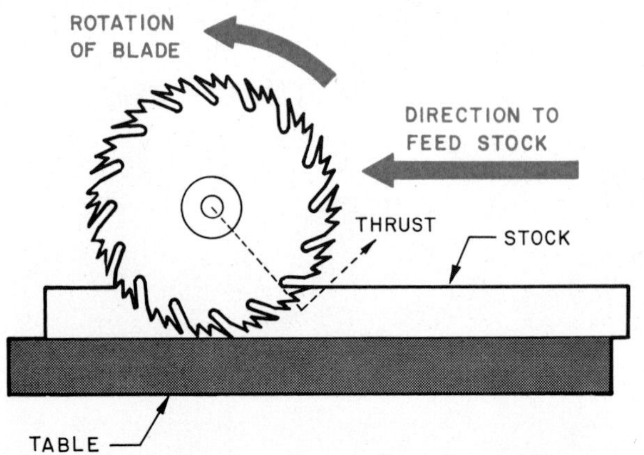

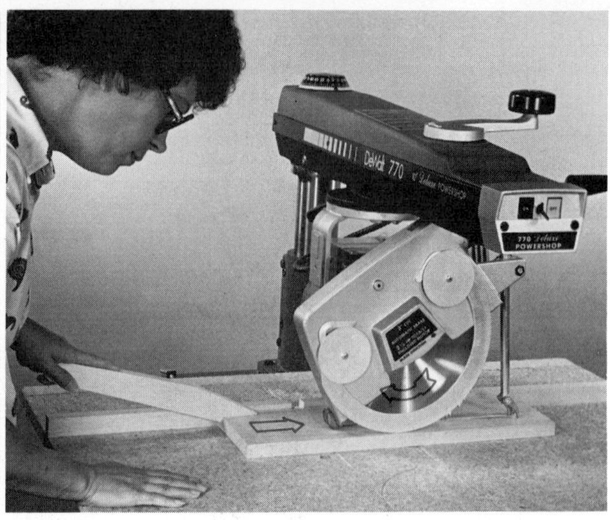

FIGURE 4-24

Use a push stick to complete the ripping operation. It helps to have someone supporting the stock on the other side of the saw. *(Courtesy of DeWalt)*

Special Cuts There are many special cuts that can be made on a radial arm saw. Two are frequently performed by carpenters. One is cutting a bird's-mouth in rafters. A *birdsmouth* is a notch cut in the rafter so that it sits squarely on the top plate of the wall (Fig. 4-25). This cut is made with a special cutter. See the setup in Fig. 4-26.

Another cut is forming dados and grooves. The saw blade is replaced with a dado head (Fig. 4-27). The machine is operated in the same manner as for crosscutting and ripping (Fig. 4-28).

FIGURE 4-25

A birdsmouth cut on a rafter.

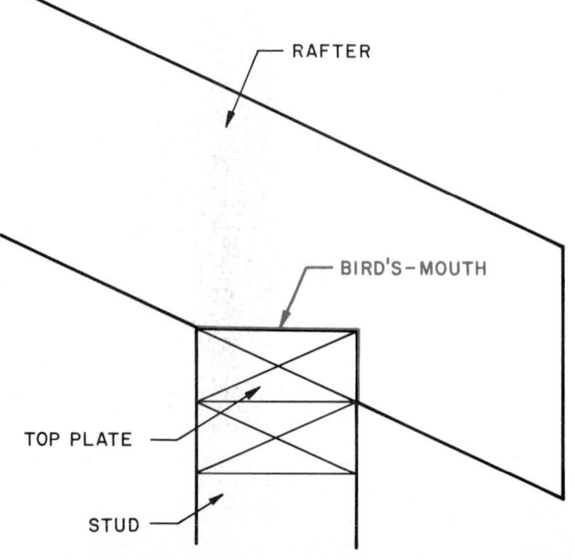

FIGURE 4-26

Bird's-mouth cuts on rafters are easily made with a radial arm saw. This setup is using a special cutter for bird's-mouth cuts rather than a saw blade. *(Courtesy of DeWalt)*

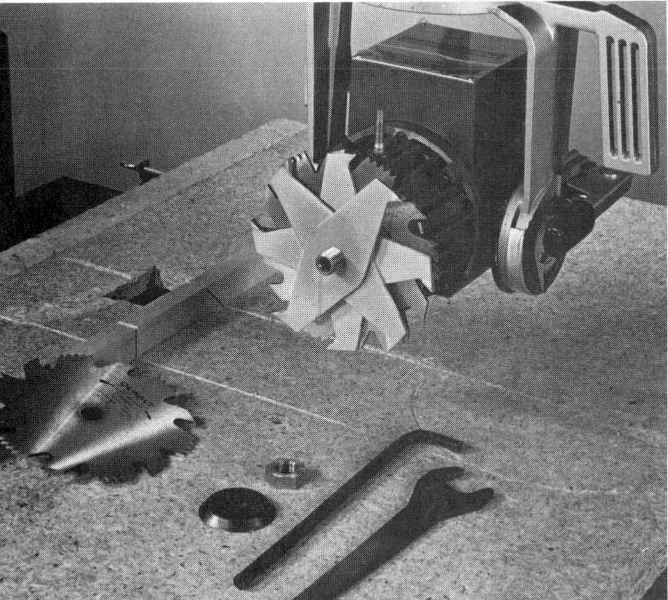

FIGURE 4-27

Grooves and dados are cut with a radial arm saw. Here a dado head is being installed. Notice that the chippers are evenly spaced inside the outside cutters. *(Courtesy of DeWalt)*

FIGURE 4-28

Dados are cut in the same manner as crosscutting. *(Courtesy of DeWalt)*

USING A POWER MITER BOX

The *power miter box* is used to accurately crosscut boards and cut miters. It is smaller than the radial arm saw. The difference is that the arm is on a pivot. The cut is made by lowering the arm with the revolving saw into the wood (Fig. 4-29).

The saw has a scale in degrees on the front. A handle in front triggers the release, which permits the saw to pivot to cut angles. It also locks the saw in place (Fig. 4-30).

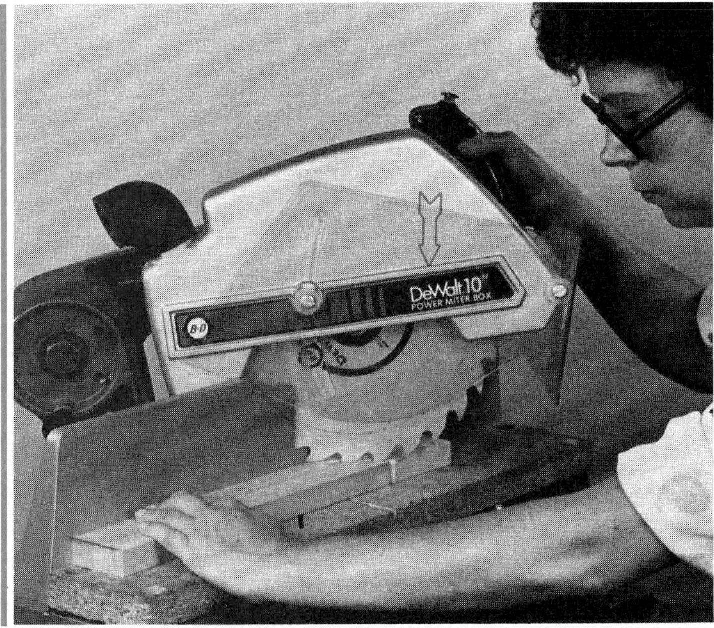

FIGURE 4-29

A power miter saw. *(Courtesy of DeWalt)*

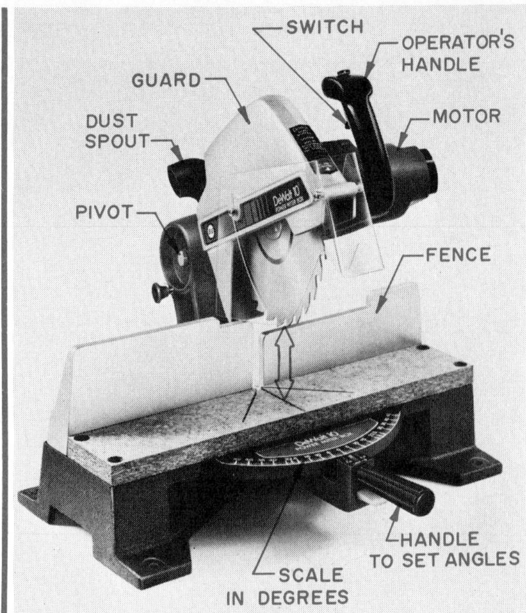

FIGURE 4-30

Parts of a power miter saw. *(Courtesy of DeWalt)*

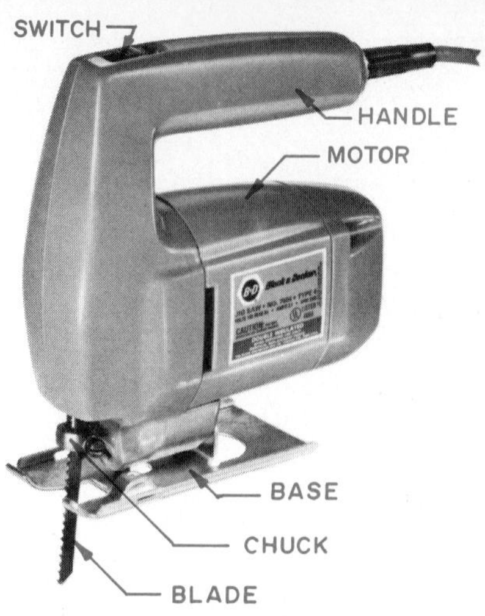

FIGURE 4-31

Parts of a saber saw. *(Courtesy of Black and Decker)*

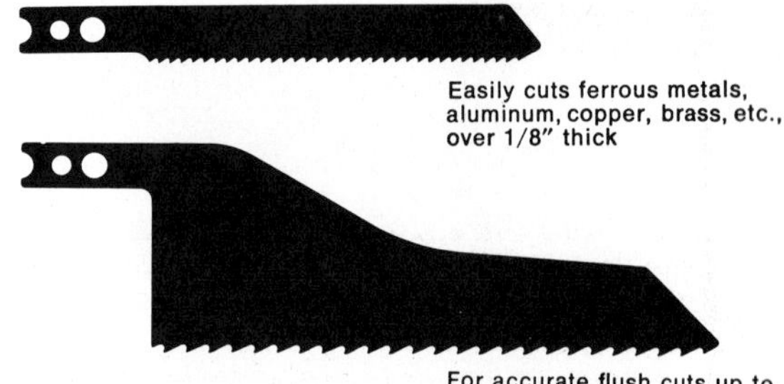

FIGURE 4-32

Typical blades for use in a saber saw. *(Courtesy of Rockwell International, Industrial Tool Division)*

Safety

1. Keep all guards in place.
2. Make sure that the saw is tightly locked at the angle it is to cut.
3. Keep the stock pressed firmly against the table and fence. Be especially careful when cutting moldings.
4. Be certain that hands are clear of the cutting area before swinging the saw down for a cut.
5. Wear safety glasses or a face shield to keep eyes free of sawdust.
6. Keep the saw blade sharp.

USING SABER AND RECIPROCATING SAWS

A *saber saw* is used for light cutting. It is especially useful for cutting curves and internal cuts. It uses a back-and-forth cutting action (Fig. 4-31). There are a variety of blades available (Fig. 4-32). For cutting thin stock a blade with 12 teeth per inch is used. For general cutting a blade with 10 teeth per inch is common. Heavy cuts, such as in 2-in. stock, require a blade with 6 teeth per inch. For best results at least two teeth should always be in contact with the wood.

Saber saws can cut 2-in. stock easily. They can crosscut and rip (Fig. 4-33) and cut bevels (Fig. 4-34) and curves (Fig. 4-35).

Internal cuts can be made by drilling a hole in the waste part of the board and placing the blade in the hole (Fig. 4-36).

FIGURE 4-33

Crosscutting with a saber saw.

FIGURE 4-34

Cutting bevels with a saber saw. *(Courtesy of Rockwell International, Industrial Tool Division)*

FIGURE 4-35

Circles and irregular curves are easily cut with a saber saw. *(Courtesy of Rockwell International, Industrial Tool Division)*

FIGURE 4-36

Internal cuts can be made by boring a hole in the waste area and cutting from this opening.

Heavy-duty saber saws can do this by *plunge cutting*. The saw is tilted on its base and the edge of the blade touches the surface of the board (Fig. 4-37). Start the saw. Gently press the blade against the wood. It will cut its way through. Then place the base flat on the panel and cut in the normal way.

A *reciprocating saw* uses a back-and-forth cutting action like the saber saw (Fig. 4-38). It might be described as a portable electric handsaw (Fig. 4-39). Its operation is much like a saber saw.

FIGURE 4-37

Internal cuts can be made with a saber saw by plunge cutting. *(Courtesy of Black and Decker)*

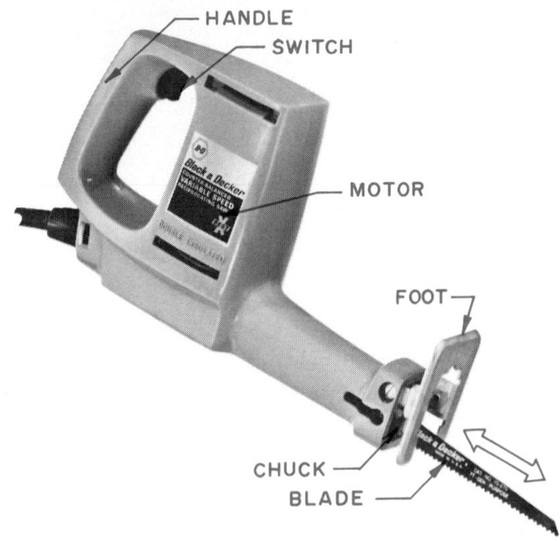

FIGURE 4-38

A reciprocating saw. *(Courtesy of Black and Decker)*

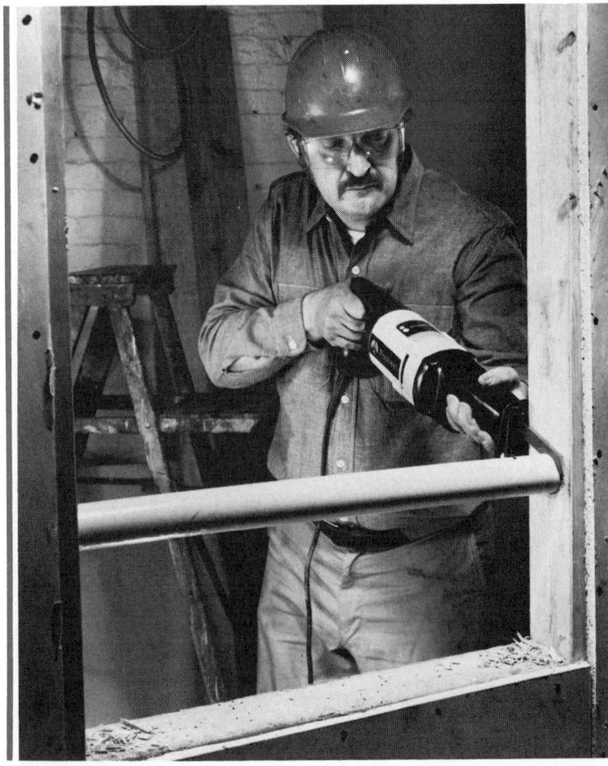

FIGURE 4-39

A reciprocating saw serves much the same purpose as a hand saw. *(Courtesy of Rockwell International, Industrial Tool Division)*

Safety

1. Use sharp blades. Worn or bent blades should be discarded.
2. Be certain that the blade is securely mounted in the chuck. Follow the manufacturer's directions.
3. Unplug the machine before changing blades.
4. Never use an ungrounded saw.
5. Always let the motor get to full operating speed before starting a cut.
6. Be certain that the space below the board being cut is clear.
7. Do not let your fingers hang below the board. Remember that the blade is below the board.
8. Before cutting, make certain that the work is properly supported.
9. Do not walk around with the saw running. Let it stop before moving.

USING A JOINTER

A *jointer* is a stationary power tool used to smooth edges and surfaces of stock. The main parts include an *infeed* and *outfeed table*, a fence, cutter head, and adjusting handles (Fig. 4-40). The size of jointers is specified by the length of the cutter head. For general carpentry work a 6-in. jointer is commonly used. It is light enough that two people can move it about a construction site.

The operating principle is that the infeed table is set lower than the outfeed table. This difference is the depth of cut wanted on the board. A ⅛-in. cut is common on edges and $^1/_{16}$ in. on faces. The cutter rotates so that the knives move toward the board (Fig. 4-41). The outfeed table must be kept set so that it is exactly the same height as the rotating knife. The outfeed table is usually reset when the knives are replaced with a new sharp set.

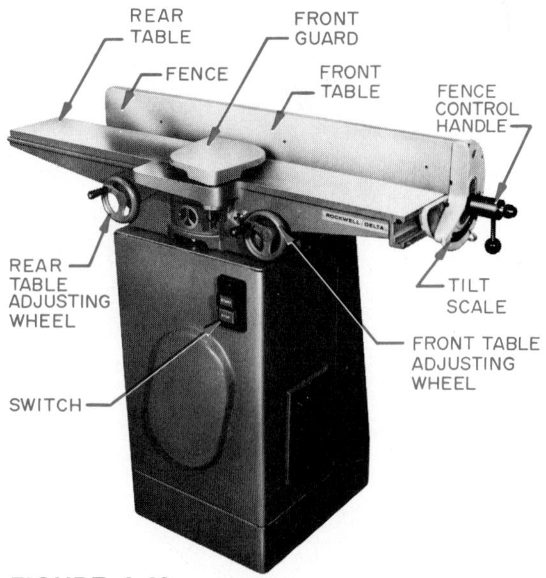

FIGURE 4-40

This small jointer can be moved about a construction site. *(Courtesy of Rockwell International, Industrial Tool Division)*

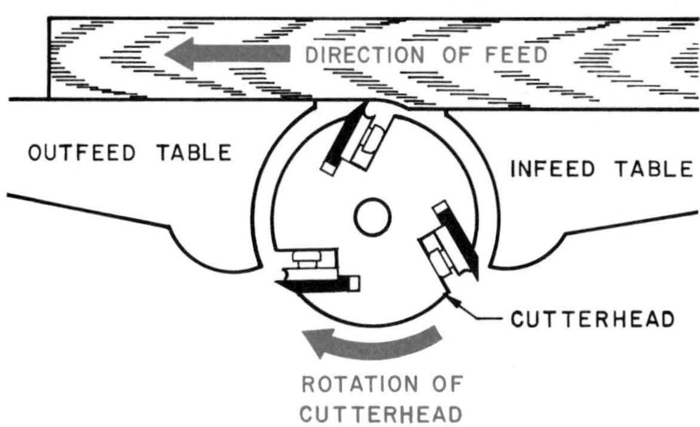

FIGURE 4-41

The jointer knives rotate into the stock as it is moved over them.

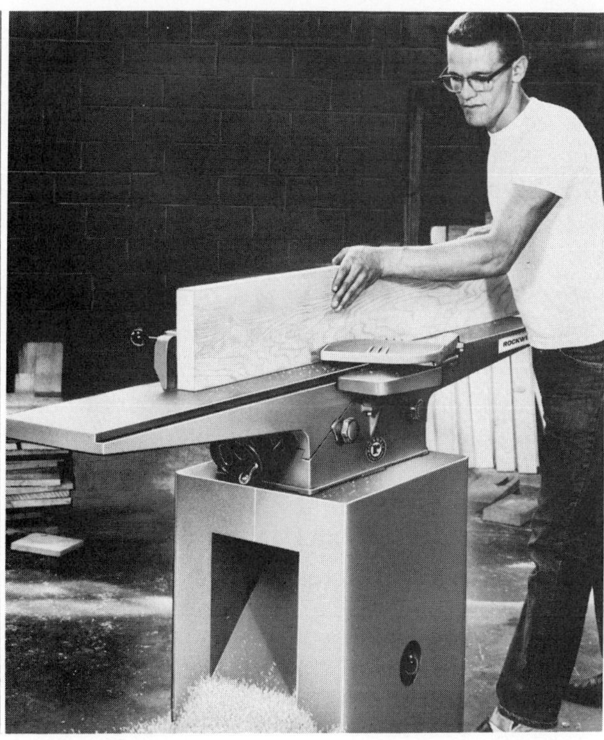

FIGURE 4-42

Jointing edges of a board. Keep both hands from above the cutter head. Never pass them over the cutter head. *(Courtesy of Rockwell International, Industrial Tool Division)*

Safety

1. Never make any adjustments when the machine is running.
2. Before starting, check the handles to make certain that everything is tight.
3. The knives should be replaced only by someone trained in doing this. If improperly done, they can fly out of the cutter head.
4. Always keep all guards in place.
5. Never pass your hands across the cutter head. Keep them clear on one side or the other (Fig. 4-42).
6. When finishing a cut on small stock, use wood push sticks or push blocks (Fig. 4-43).
7. Never run stock shorter than 12 in. or thinner than ⅜ in.
8. Never take too deep a cut. Follow the manufacturer's directions.
9. Do not joint stock with knots. They will chip the knives and often come loose and be thrown.
10. Always use sharp knives. Dull or improperly ground knives cause vibration and will kick the stock out of your hands.
11. Knives must be sharpened by someone having the proper equipment.

FIGURE 4-43

Use a push block when jointing face grain or small pieces.

Jointing an Edge

1. Set the fence at the proper angle with the table. This is 90° for a square edge or some lesser angle for a beveled edge.
2. Examine the board to find the direction of the grain. Always plane in the direction of the grain (Fig. 4-44).

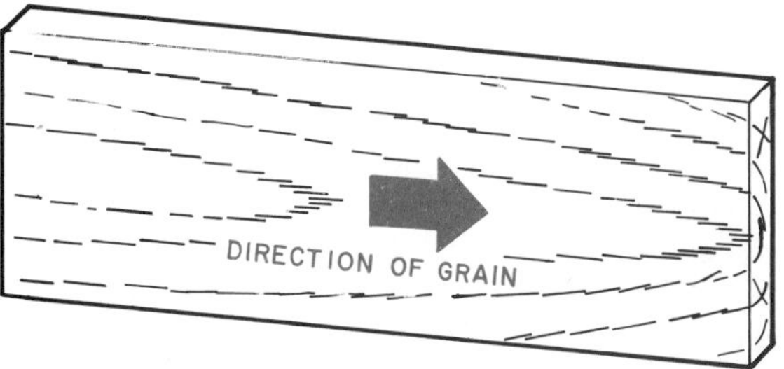

FIGURE 4-44

Always joint stock in the direction of the grain.

3. Set the depth of cut. The jointer will have a handle to unlock the infeed table. Another handle will raise or lower the table. A depth of cut scale will show the depth set. After setting the depth of cut, relock the infeed table.
4. Be certain that the guards are in place and operating.
5. Start the motor and let it reach full speed.
6. Place the stock on the table and press against the fence.
7. Push forward into the rotating cutter head.
8. Proceed to push across the cutter head at a slow uniform speed. Be certain to hold the stock firmly so that it will not chatter. Refer to Figs. 4-42 and 4-43.
9. After passing across the cutter head, turn off the power. Do not leave the machine until it has stopped.

Jointing a Surface The procedure is the same as for edges. A push block is absolutely essential. Take a much lighter cut than used for jointing edges (Fig. 4-43).

Jointing End Grain

1. Because of the structure of wood, the jointer will split the edge of the board unless special precautions are taken.
2. Use a light depth setting.
3. Do not joint stock narrower than 12 in.
4. First joint the end about 1 in. into the board. Then lift the board from the jointer and run from the other side. Stop when you reach the jointed 1-in. section (Fig. 4-45).

FIGURE 4-45

To joint end grain make a short cut from one side. Then turn the board around and finish the cut from the other side.

USING A PORTABLE ROUTER

A *portable router* is basically an electric motor with a chuck, a base, and handles (Fig. 4-46). The router is used to shape the edges of boards. It can also form grooves and dados. A wide variety of cutters are available (Fig. 4-47). Cutters with flat bottoms are used for grooves and dado. Those with curved cutters are used to shape edges.

The depth of cut is adjusted by moving the base up or down the sides of the motor. The depth lock secures the base to the motor shell.

Routers are usually specified by the horsepower of the motor. For general carpentry work a 1½- to 2½-horsepower (hp) motor is recommended.

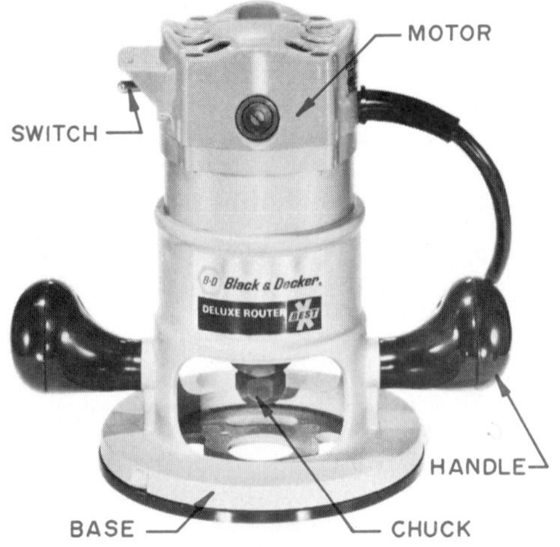

FIGURE 4-46

Parts of a portable router. *(Courtesy of Black and Decker)*

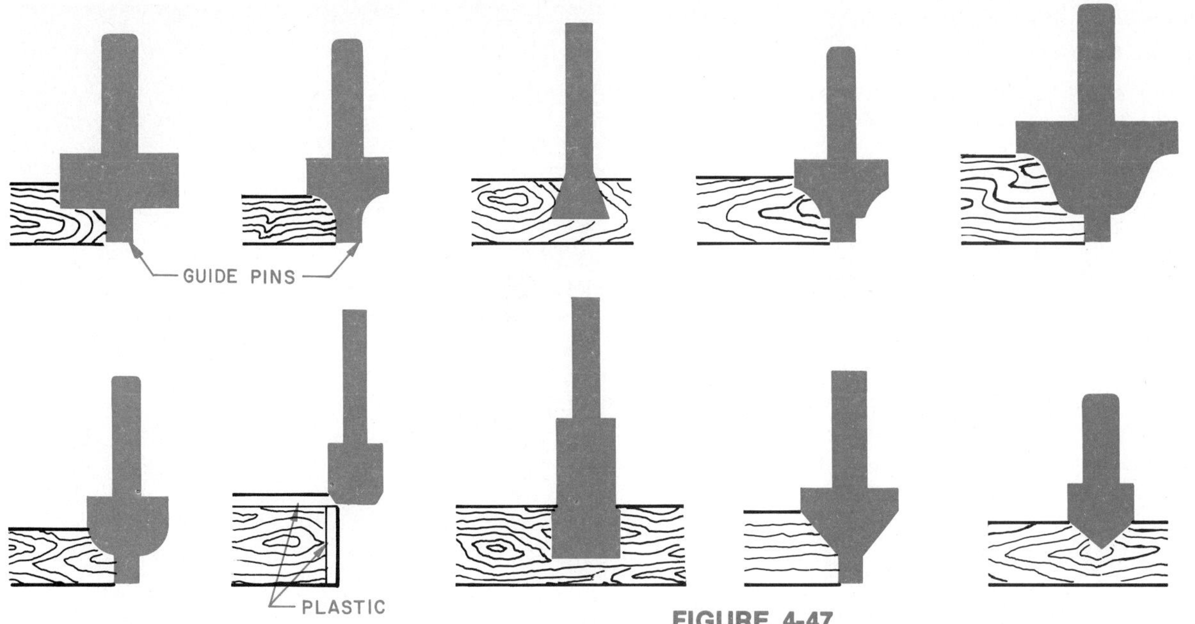

FIGURE 4-47

Typical shapes of router cutters.

Installing Cutters

1. Unplug the power plug.
2. Lock the shaft or hold it with a wrench.
3. Insert the shank of the cutter.
4. Tighten the chuck with a wrench.
5. Unlock the shaft.

Operating the Router

1. Install the cutter.
2. Plug in the power plug.
3. Adjust the depth of cut. This is done by raising or lowering the base.
4. Start the motor. Hold the router handles and place the base on the face of the board (Fig. 4-48). Move the router until the cutter cuts full depth. The guide pin on the cutter will be touching the edge of the board (Fig. 4-49).
5. The cutter revolves in a clockwise direction. Make cuts on straight edges from left to right. Make cuts on curved edges in a counterclockwise direction.

Safety

1. Always unplug the router before changing cutters.
2. Make certain that the cutter is tight in the chuck.
3. Wear eye shields.

FIGURE 4-48

Hold the router by both handles. Keep the base squarely on the surface to be routed. *(Courtesy of Black and Decker)*

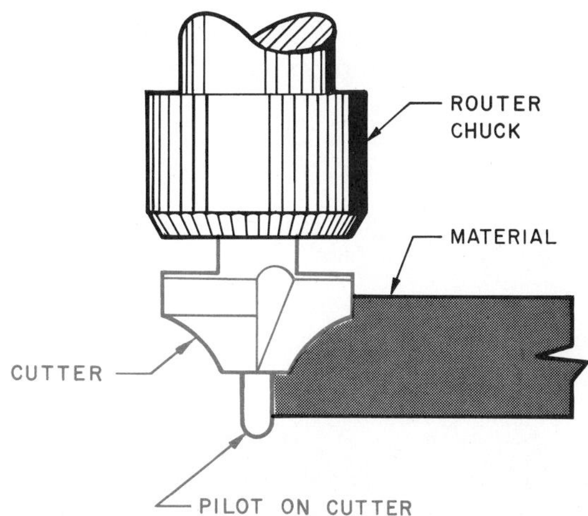

FIGURE 4-49

The guide pin regulates the depth of cut. It will burn the wood if it is not kept moving along the board.

4. Clamp the work to be routed so that it will not be thrown.
5. When starting the router, torque from the motor tends to spin it. Hold it tightly.
6. When operating, hold the router with both hands.

USING A PORTABLE ELECTRIC PLANE

A *portable electric plane* is used to smooth edges and surfaces of boards. It rotates the cutter at high speed. The depth of the cut is adjusted by raising or lowering the front shoe. The rear shoe is kept level with the outside diameter of the revolving cutter (Fig. 4-50).

Operating the Electric Plane

1. Set the depth of cut. If a surface is to be planed, the depth will be less than when planing an edge.
2. Position the fence. It can be set at 90° for a square edge or on an angle for a chamfer (Fig. 4-51).
3. Start the motor. Let it reach full speed.
4. Place the front shoe on the board. Holding the plane in two hands, move it across the board at a uniform speed.
5. When starting the cut, keep most of the pressure on the front shoe. When finishing, keep the pressure on the rear shoe.

The same general principles apply to the portable power block plane (Fig. 4-52). It is much like the larger plane but is smaller.

Safety

1. Always hold the plane in two hands.
2. Follow the manufacturer's directions for adjusting the plane.

FIGURE 4-50

Parts of an electric portable plane. *(Courtesy of Rockwell International, Industrial Tool Division)*

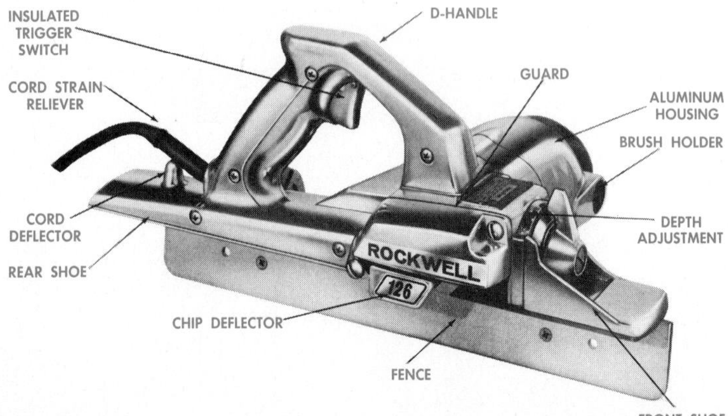

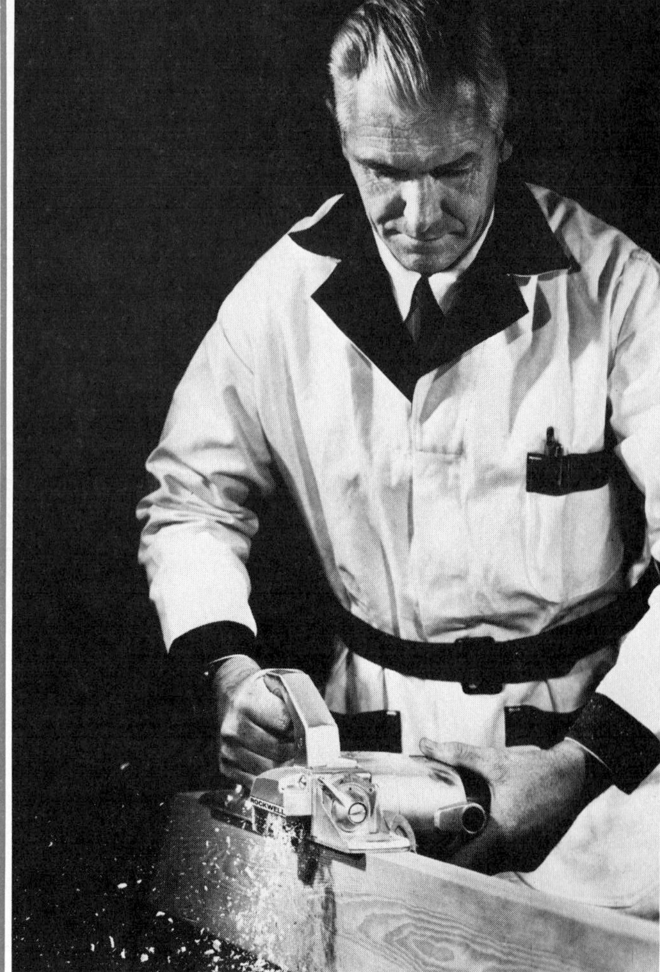

FIGURE 4-51

The shoes slide on the surface to be planed. *(Courtesy of Rockwell International, Industrial Tool Division)*

FIGURE 4-52

A power block plane. *(Courtesy of Rockwell International, Industrial Tool Division)*

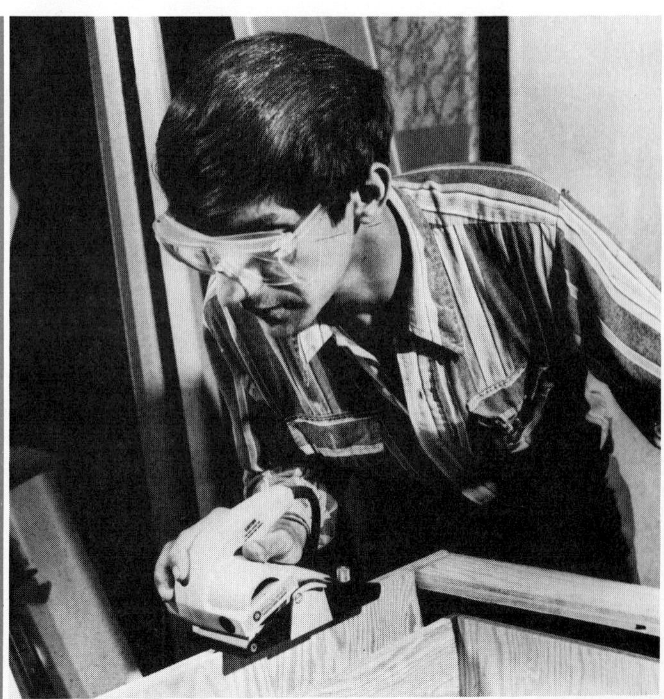

3. Clamp the work securely so that it will not kick back.
4. Disconnect the power plug before making adjustments or replacing cutters.

USING PORTABLE ELECTRIC DRILLS

The *portable electric drill* is used for drilling and boring holes in all kinds of materials. In addition, attachments are available for grinding, buffing, and other rotary-type operations.

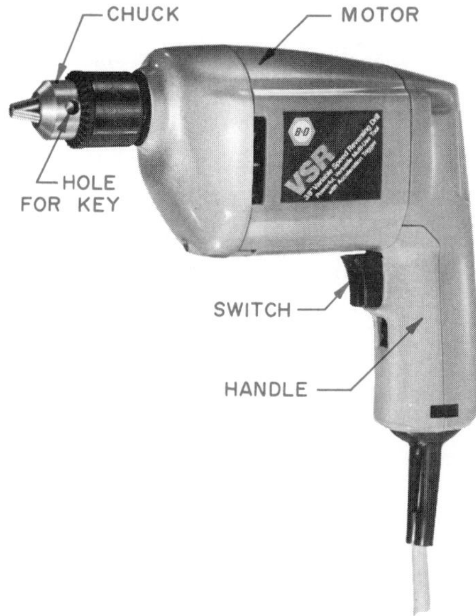

FIGURE 4-53

Parts of an electric portable drill. *(Courtesy of Black and Decker)*

FIGURE 4-54

Heavy-duty electric drills are used for large-diameter holes. *(Courtesy of Black and Decker)*

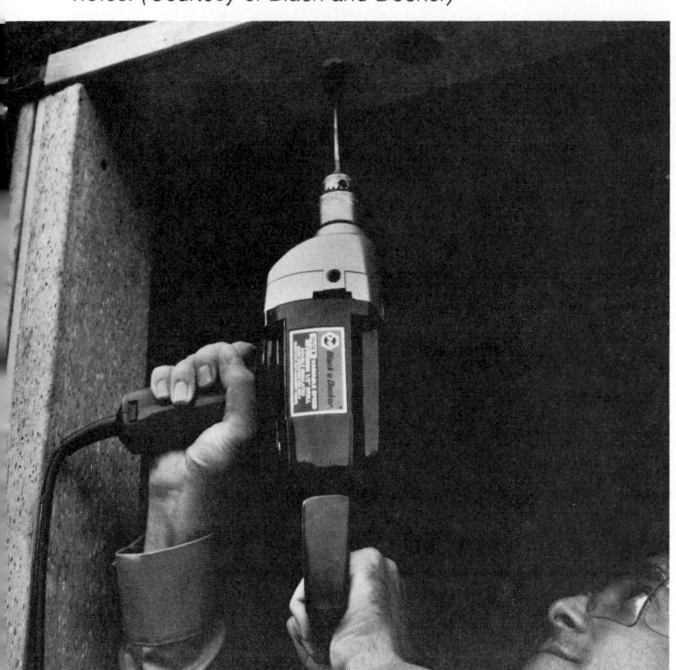

A drill is primarily an electric motor with a handle and a chuck (Fig. 4-53). The *chuck* holds the drills. Electric drills are specified by the chuck capacity and motor horsepower. The chuck capacities commonly used are 1/4, 3/8, and 1/2 in. This refers to the largest-diameter drill shank it will hold. A shank is the round part of a drill which goes in a chuck. Heavy-duty drills will have motors with more horsepower than light-duty drills, even though the chuck capacity may be the same (Fig. 4-54). Drills of different speeds are available. For carpentry work a speed of about 1000 rpm is good.

Available are cordless electric drills. They are helpful when working in wet places or above ground (Fig. 4-55).

The reversing-end-handle drill permits drilling holes in close spaces (Fig. 4-56).

FIGURE 4-55

A cordless electric drill is helpful when working in high places or on wet ground. *(Courtesy of Black and Decker)*

FIGURE 4-56

This is a reversing-end-handle drill. It permits drilling in tight places. *(Courtesy of Black and Decker)*

There are a wide variety of drills and bits available. The *twist drill* is used in wood, metal, and plastics (Fig. 4-57). A *spade drill* is used to bore large-diameter holes. The cutters are considerably larger than the shank (Fig. 4-58). A *hole saw* is used to bore even-larger-diameter holes. The hole saw has a drill in the center to start the hole. The drill should be set so that it sticks out about ¼ in. beyond the circular saw (Fig. 4-59). Combination drills have a drill section for making a hole and a countersink section following it. This enables a hole to be drilled and countersunk at the same time (Fig. 4-60).

Installing a Drill

1. Open the chuck.
2. Place the drill into the chuck. Twist drills should not have any of the spiral part in the chuck.
3. Tighten the chuck by hand.
4. Use the key to finish tightening the chuck (Fig. 4-61). Be certain to remove the key before starting the motor.

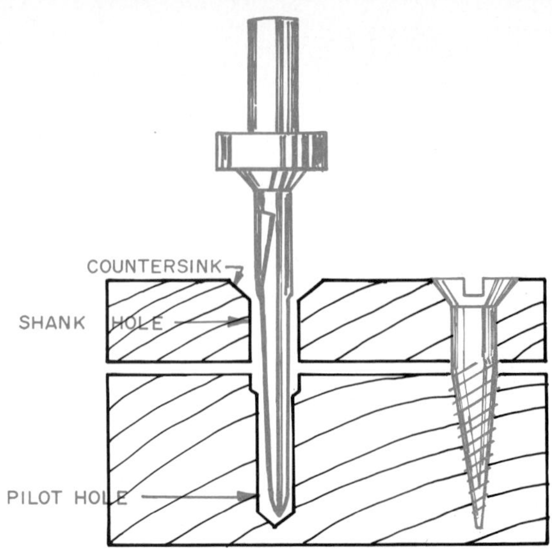

FIGURE 4-60

Various types of combination drills are used to drill and countersink holes.

FIGURE 4-57

A straight-shank twist drill.

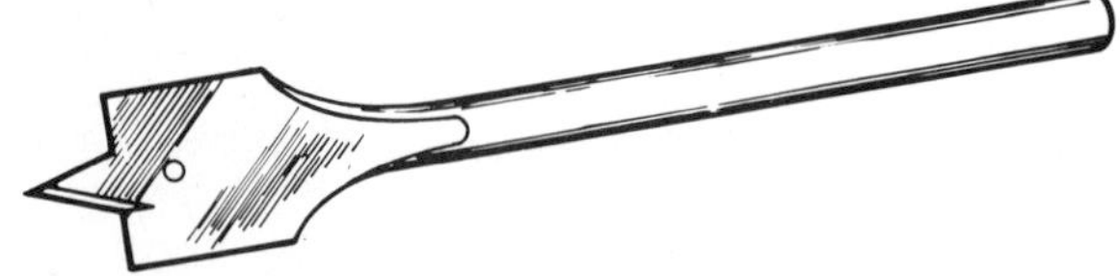

FIGURE 4-58

A spade bit is used to bore larger diameter holes in wood.

Safety

1. Use double-insulated drills or wireless drills in wet places.
2. Be certain that the drill is tight in the chuck. Remove the key before starting the motor.
3. Use eye protection.

Starting the Cut.

Finishing the Cut.

FIGURE 4-59

A hole saw cuts large-diameter holes using an electric drill as the power source.

FIGURE 4-61

A key is used to tighten the drill in the chuck.

4. When drilling do not force the drill beyond its normal cutting speed.
5. Use sharp drills.

USING PORTABLE SANDERS

Belt, finishing, and disc sanders are available. The portable *belt sander* uses an endless belt that runs over two wide pulleys (Fig. 4-62).

To use the sander, place it flat on the board. Let it rest under its own weight. Do not push down on it. Hold it in two hands (Fig. 4-63). Start the motor and move it forward and backward

FIGURE 4-62

Hold the portable sander in two hands. Let it sand using its own weight. *(Courtesy of Rockwell International, Industrial Tool Division)*

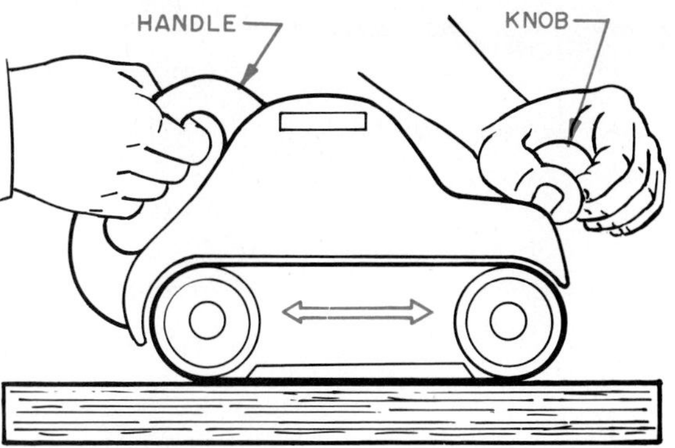

FIGURE 4-63

To start the portable belt sander, set it flat on the wood surface, resting under its own weight. Hold it tightly by the handle and the knob. Then turn on the power and move it across the board.

FIGURE 4-64

A finishing sander.

and at the same time move it from side to side. Keep it flat and moving or it will gouge the surface of the wood.

When starting the portable belt sander, be certain you have a firm grip on the handle and the knob. When it starts the belt will cause it to move, and you must be ready to guide and control it.

The *finishing sander* is shown in Fig. 4-64. It has the abrasive in sheet form. It is placed over the pad on the sander (Fig. 4-65). These sanders are available with three different actions. One is straight back-and-forth action. Another is a side-to-side vibrating action. The third is an orbital action (Fig. 4-66). The straight-line action gives the smoothest surface. The other two tend to cut across the grain, leaving fine scratches (Fig. 4-67).

The *disc sander* has the abrasive paper glued to a flat rotating disc. The work is placed on a table and moved into the disc (Fig. 4-68). Be certain to touch the disc on the side where it is rotating down toward the table (Fig. 4-69). If

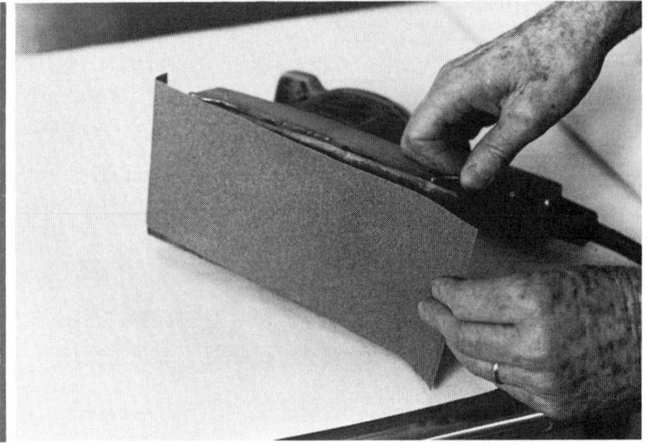

FIGURE 4-65

Finishing sanders use abrasives in sheet form.

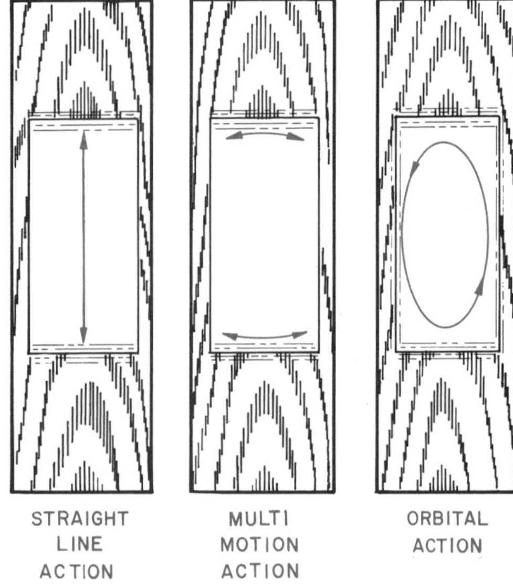

FIGURE 4-66

Finishing sanders are available with three types of motion. Straight-line action produces a surface with the fewest scratches across the grain.

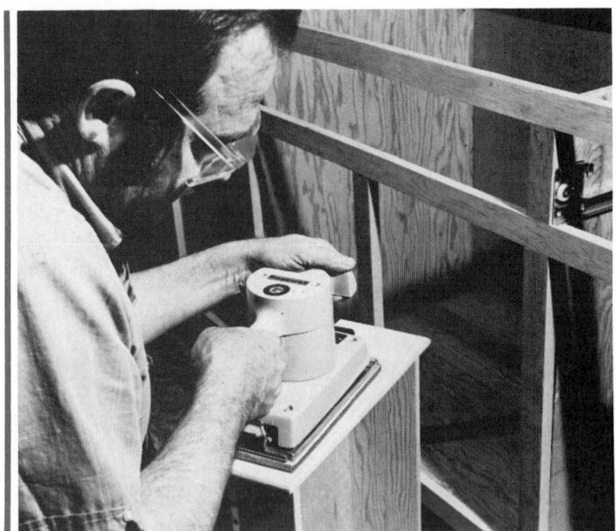

FIGURE 4-67

A finishing sander should be kept parallel with the grain. *(Courtesy of Rockwell International, Industrial Tool Division)*

FIGURE 4-68

A disc sander has abrasive sheets glued to a flat rotating metal disc. *(Courtesy of Rockwell International, Industrial Tool Division)*

FIGURE 4-69

Sand on the down-rotation side of the disc. *(Courtesy of Rockwell International, Industrial Tool Division)*

you sand on the other side, the wood may be thrown out of your hands.

A stationary belt sander is shown in Fig. 4-70. It can be operated in a vertical position to sand end grain. In this position it serves the same purpose as the disc sander. When laid in a horizontal position it can sand the surface of a board (Fig. 4-71). The fence holds the wood in place so that it is not thrown off the belt. Notice that the machine in Fig. 4-71 has both disc and belt sanders on one frame.

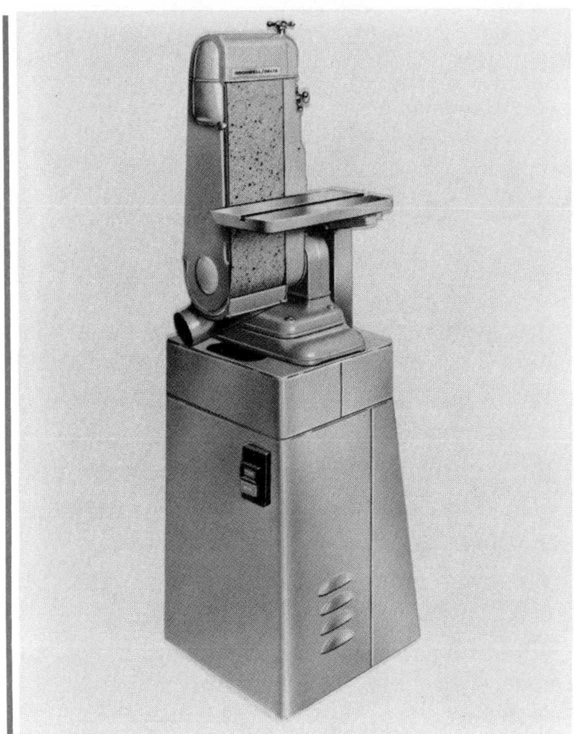

FIGURE 4-70

A stationary belt sander. The belt can be used in a vertical or horizontal position. This belt is in a vertical position. *(Courtesy of Rockwell International, Industrial Tool Division)*

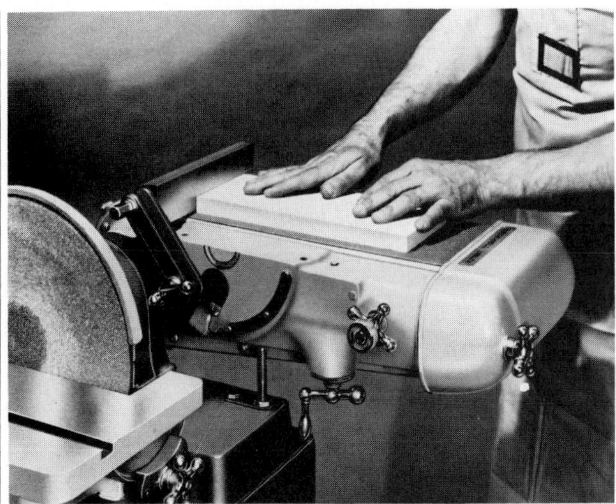

FIGURE 4-71

A stationary belt sander finishing face grain. It can be used for edges and end grain. The belt is in a horizontal position. *(Courtesy of Rockwell International, Industrial Tool Division)*

USING POWER NAILERS AND STAPLERS

There are many types of power nailers and staplers. In factories where buildings are mass-produced, many are mounted as stationary machines. Portable units are used by carpenters on the construction site.

Portable units are either electric or pneumatic (air) powered. *Portable nailers* have a magazine that holds the nails. A variety of nails can be driven (Fig. 4-72). They are fed into a nailing position. The portable nailer can be used in any position (Fig. 4-73). This is a pneumatic unit which drives nails 2½ to 3⅞ in. long. It weighs 10 pounds (lb) and operates on 60 to 100 lb per square inch of air pressure.

A *pneumatic stapler* is shown in Fig. 4-74. It drives staples with leg lengths up to 1½ in. having a crown (width between legs) of 1 in. It weighs 4½ lb and operates on 60 to 100 lb per square inch of air pressure.

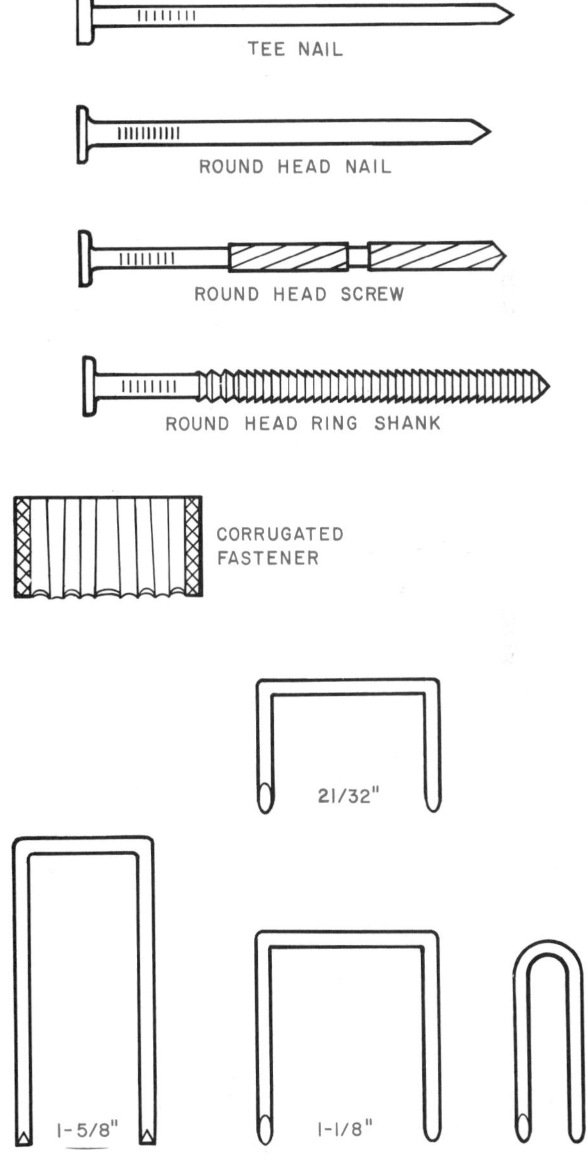

FIGURE 4-72

Typical fasteners for power nailing and stapling. *(Courtesy of Bostitch Textron)*

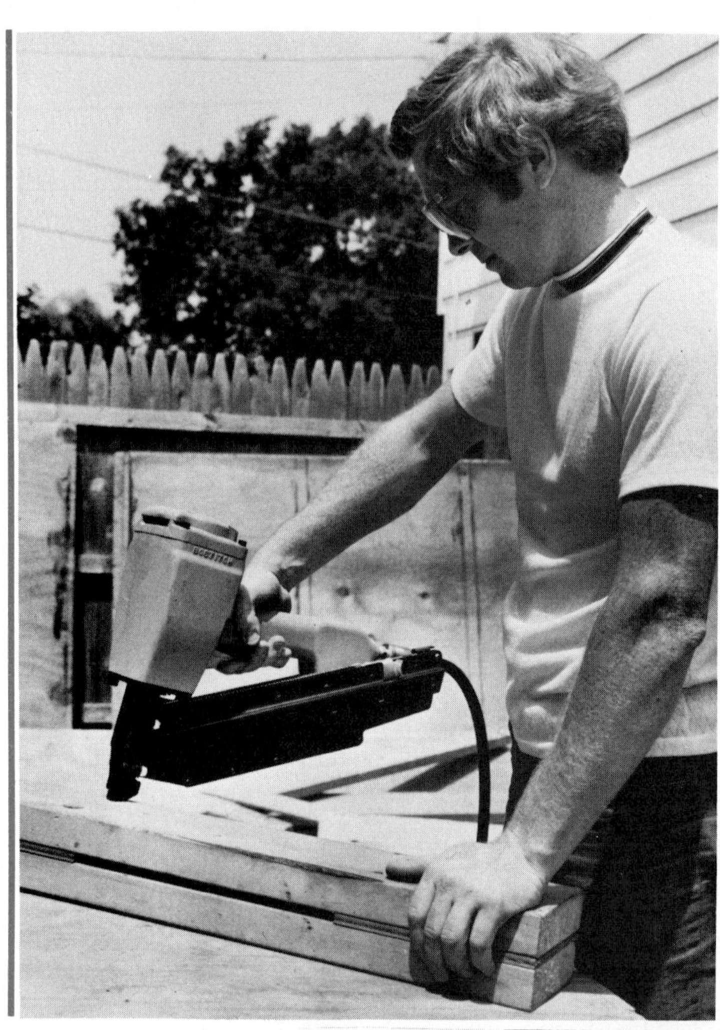

FIGURE 4-73

A power nailer joining 2-in.-thick material forming a header. *(Courtesy of Bostitch Textron)*

Safety

1. Since operation principles vary, study the manufacturer's operating manual.
2. Be certain to use the type of nail or staple recommended.
3. Be certain that electric units are grounded.
4. Use the recommended air pressure for pneumatic units.
5. Treat the machine as you would a gun. Do not point it at yourself or others.
6. When not in use, disconnect from the power source to prevent accidental releasing of fasteners.
7. Always keep the nailer or stapler tight against the surface to receive the staple or nail.

FIGURE 4-74

A power stapler fastening shingles to the roof sheathing. *(Courtesy of Bostitch Textron)*

STUDY QUESTIONS

After reading the chapter, answer each of the following questions. If you do not know the answer, review the chapter.

1. What factors should be considered when selecting power tools?
2. What are the general safety rules for power tool use?
3. What do you use to protect your fingers when cutting on a stationary circular saw?
4. List the operations that can be performed on a stationary circular saw.
5. How does a radial arm saw differ from a stationary circular saw?
6. What are the differences between a saber saw and a reciprocating saw?
7. What type of work is done with a jointer?
8. What operations are performed with a portable router?
9. Explain how to smooth an edge with a portable electric plane.
10. What types of portable sanders are in use?
11. How are nails held in portable power nailers?

IMPORTANT TECHNICAL TERMS

Following are technical terms that you should be able to use as part of your working vocabulary. Write a brief description of the meaning of each term.

Stationary power tools
Portable power tools
Ripping
Crosscutting
Miter gauge
Feather boards
Push sticks
Groove
Dado
Portable electric circular saw
Radial arm saw
Anti-kickback fingers
Mitering
Yoke
Bird's-mouth
Power miter box
Saber saw
Plunge cutting
Reciprocating saw
Jointer
Infeed table
Outfeed table
Portable router
Portable electric plane
Portable electric drill
Belt sander
Finishing sander
Disc sander
Power nailer
Power stapler

5 SAFETY ON THE CONSTRUCTION SITE

Safety is everyone's business. Even though a contractor may have a safety supervisor on the construction site, this does not guarantee safe conditions. All those working on the job must understand the basic principles of safe behavior and act accordingly. Only then can accidents on a construction site be reduced. Everyone should be safety conscious. Safety consciousness is the awareness of good safety practices. This includes knowledge, attitudes, obedience, and concern.

In an attempt to improve the working conditions for industrial workers, the U.S. government passed the Occupational Safety and Health Administration Act of 1970.

OCCUPATIONAL SAFETY AND HEALTH ADMINISTRATION

OSHA was formed by the Occupational Safety and Health Administration Act of 1970. The act provides minimum safety and health standards for working conditions. OSHA is part of the U.S. Department of Labor. Its main duties are to:

1. Encourage employers and employees to reduce hazards in their workplaces.
2. Establish responsibilities and rights of employers and employees.
3. Encourage new safety and health programs.
4. Establish record-keeping procedures to keep track of injuries and illnesses that happen on or because of the job.

5. Develop health and safety standards and enforce them.
6. Encourage the states to establish safety and health programs.

There are two types of standards: those that apply to all industries and those that apply to one industry, such as construction.

GENERAL HOUSEKEEPING RULES

1. Keep walkways, runways, stairs, aisles, and work areas free from debris.
2. Store tools and materials in a safe manner.
3. Remove all nails from used lumber.
4. Keep oily rags and other combustible material in nonflammable containers.
5. Clean up oil, grease, and other liquids that have been spilled on walkways.
6. Remove trash and oily materials.
7. When dropping materials more than 20 ft, use an enclosed chute. When dropping materials short distances through holes on the inside, barricade off the area where they will land.

PERSONAL SAFETY

1. Always wear proper head protection. The *hard hat* must meet safety standards set by the American National Standards Association (Fig. 5-1).
2. The hard hat must fit properly. The lining must be adjusted so that the hat sits squarely upon the head.

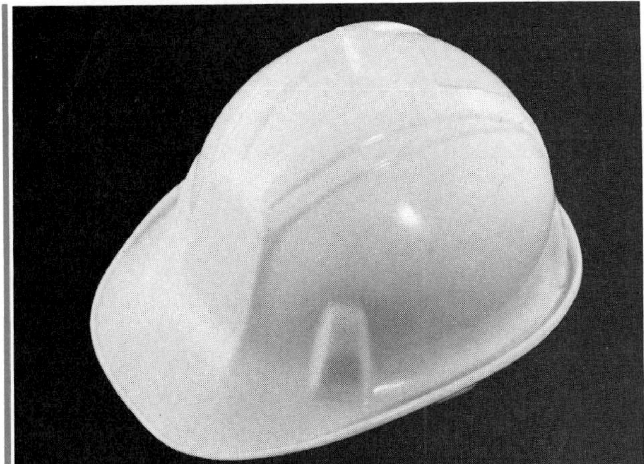

FIGURE 5-1

A hard hat that meets safety standards set by the American National Standards Association. *(Courtesy of Apex Safety Products)*

3. Never wear a hard hat over another hat.
4. Wear eye protection on the construction site at all times. *Safety glasses* with safety lenses, rims, and side shields are good for general construction (Fig. 5-2). Safety goggles can be used over regular eye glasses or as the primary means of eye protection (Fig. 5-3).
5. Eye protection should fit properly and meet safety standards. Lenses should be kept clean.
6. If there is danger to the face, a *full-face shield* should be worn (Fig. 5-4).

FIGURE 5-2

These safety glasses have side shields. *(Courtesy of American Optical Corporation)*

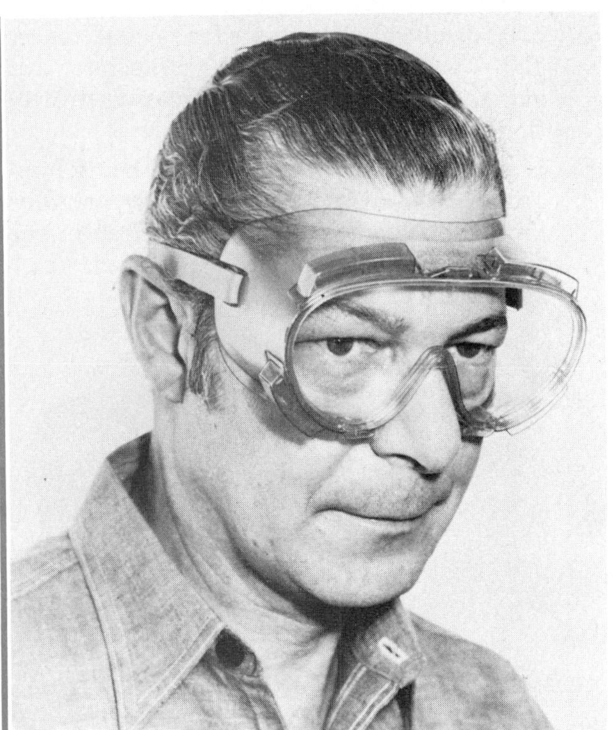

FIGURE 5-3

These safety goggles provide total eye protection. *(Courtesy of American Allsafe Co., Inc.)*

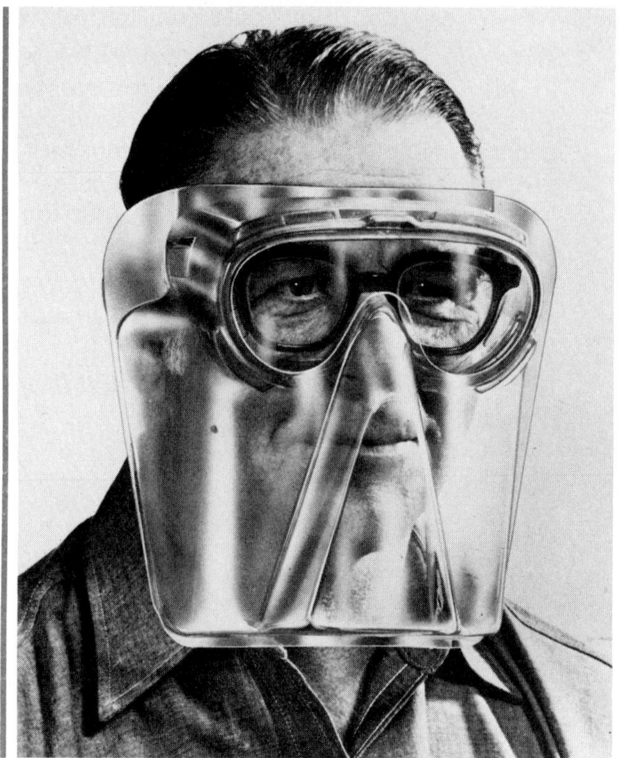

FIGURE 5-4

This full-face shield can be worn over eyeglasses. *(Courtesy of American Allsafe Co., Inc.)*

FIGURE 5-6

This particle mask type of respirator provides respiratory protection against dust. *(Courtesy of 3M Co.)*

7. Hearing-protection devices must be worn if working in areas subjected to noise levels of 90 decibels (db) or more for several hours. *Ear inserts* or *muff-type protectors* are satisfactory (Fig. 5-5). Keep hearing-protection devices clean.
8. If working in areas where a breathing hazard is present, some type of *respiration-protection device* is needed. Mechanical filter types protect against nontoxic dust such as sawdust (Fig. 5-6).

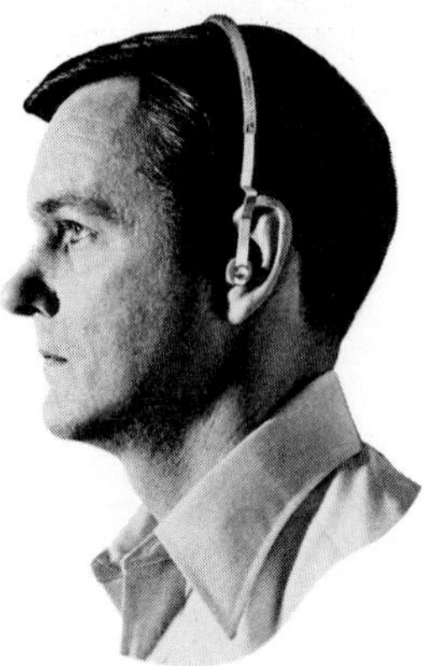

FIGURE 5-5

Ear inserts or muffs are needed to protect against noise levels beyond 90 decibels. *(Courtesy of Willson Products Division of ESB, Inc.)*

9. Keep the respiration device clean. Change filters often.
10. Proper foot protection is needed at all times. Safety-toe shoes with a steel toe covering are recommended.
11. If working in wet conditions, rubber boots are needed.
12. Always wear a shirt and pants with long legs. Clothes should be loose enough to permit easy bending. They should not be too big because they can get caught.
13. Never wear jewelry of any kind. Rings, necklaces, bracelets, watches, ties, and other such articles must be removed.

FIRST AID

A carpenter must understand basic first-aid techniques. Red Cross first-aid manuals and courses are available. A small first-aid kit could be carried by a carpenter. A large contractor will also have extensive first-aid equipment.

Some of the more common situations a carpenter should be able to handle are skin abrasions; sharp, deep cuts; puncture wounds; shock due to the accident or electrical shock, heat exhaustion; heat stroke; broken bones; burns; and frostbite.

TOOL SAFETY

Chapter 4 gives extensive safety procedures when using power tools. Following are some general safety hints for tool safety.

1. All tools should be kept in good repair. Broken handles should be replaced. Dull saws and edge tools should be sharpened. A dull tool is a dangerous tool.
2. Keep tools clean.
3. Never remove guards from tools.
4. Use the right tool for the job. A chisel is a poor screwdriver. A screwdriver is a poor pry bar.
5. Read the manufacturer's instructions before using a tool. Obey them.
6. Store edge tools so that the cutting edges are covered.
7. When using wrenches, select the proper size. Do not overtighten threaded fasteners.
8. Use only a hammer for hammering. Pliers and other tools make poor hammers.
9. When sawing boards, be certain that they are held securely.
10. When using portable electric tools, be certain that they have a grounding wire and that it is connected to an approved ground.
11. Use only high-quality extension cords. A 100-ft cord of No. 12 wire will safely carry 20 amperes (A). A No. 8 wire will carry 35 A. A No. 4 wire will carry 60 A.
12. Examine electric extension cords for cuts and breaks.
13. Unplug any electric tool before you adjust or replace the cutting device.
14. Use only the attachments recommended for the tool by the manufacturer.
15. If working near welders, do not look at the arc.
16. Stay clear of all heavy equipment in operation. Stay alert for overhead and crane work as well as on-the-ground operations.
17. If demolition work is under way, observe the orders to seek safe quarters prior to the explosion.

LADDER AND SCAFFOLD SAFETY

Carpenters make extensive use of *ladders* and *scaffolding*. Much of their work is above the ground. They have to carry materials and tools up on the structure, store them there for use, cut and fit them, and fasten them into place.

Ladder Safety The three basic ladders used are the *straight*, *extension*, and *step* (Fig. 5-7).

1. Be certain that ladders are in good repair. Any cracked, broken, or missing parts should be replaced.
2. Be careful where ladders are placed. If they are in front of a door or drive, they could be accidentally bumped and knocked down.
3. The ladder should stand on a firm base. A square-leg ladder on a wood floor is certain to slip. It should be cleated (Fig. 5-8). Most ladders have nonskid feet. Even with these, special precautions against sliding are advisable.

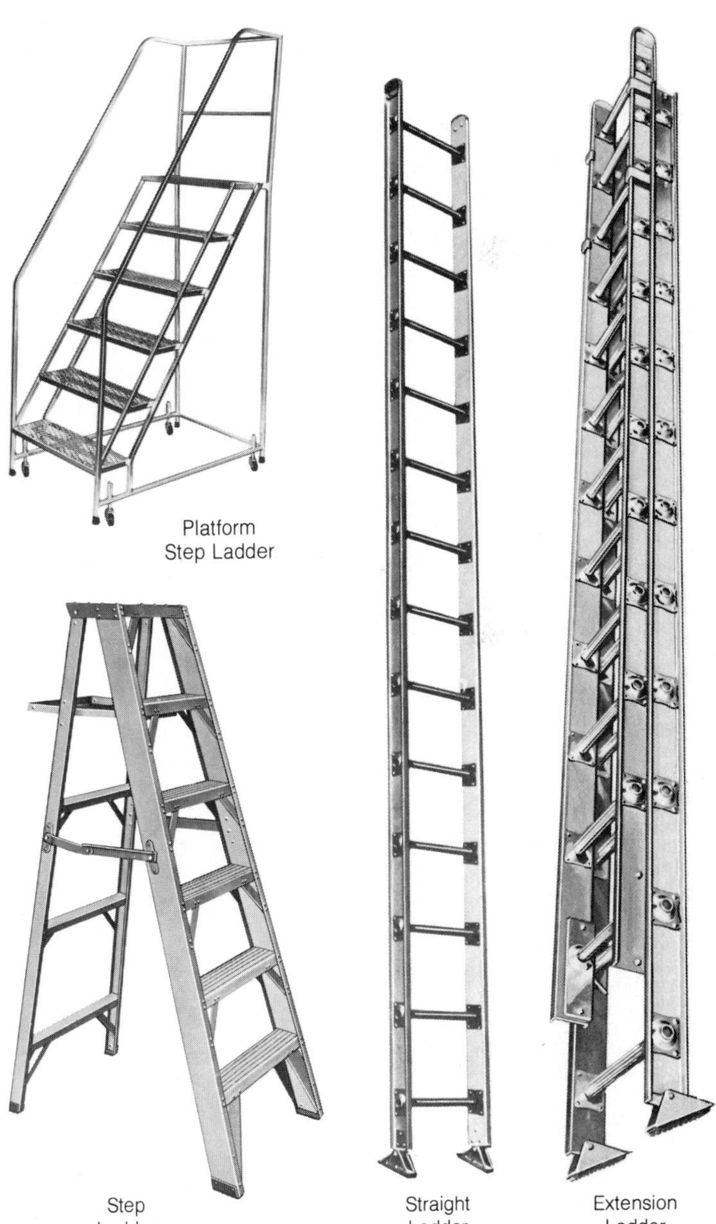

FIGURE 5-7

The basic types of ladders are straight, extension, step, and platform step.

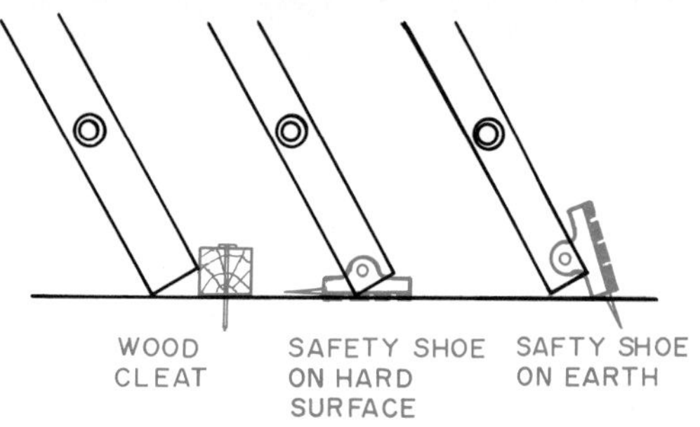

FIGURE 5-8

Ladders should have some type of safety foot to protect against sliding.

4. Keep the area at the foot of the ladder clear of debris.
5. To raise a ladder, place the feet against something to keep them from sliding. Lay it away from the house. Raise the top end and raise it by pushing up on the rungs. Move your hands from rung to rung as it raises (Fig. 5-9).
6. Use a ladder the correct length (Fig. 5-10). A ladder that is too short or too long is unsafe. The top of the ladder should extend 3 ft above the top of the wall.
7. When properly in place, the foot of the ladder will be away from the wall a distance of one-fourth the working distance of the ladder (Fig. 5-11).
8. Never use aluminum ladders around electrical wiring.
9. Face the ladder when climbing up or down. Keep both hands on the rails.
10. It is best to move materials above the ground using ropes. It is dangerous to try to carry them up a ladder.
11. Do not overload a ladder.

FIGURE 5-9

To raise a ladder, place the ladder feet against something to keep them from sliding.

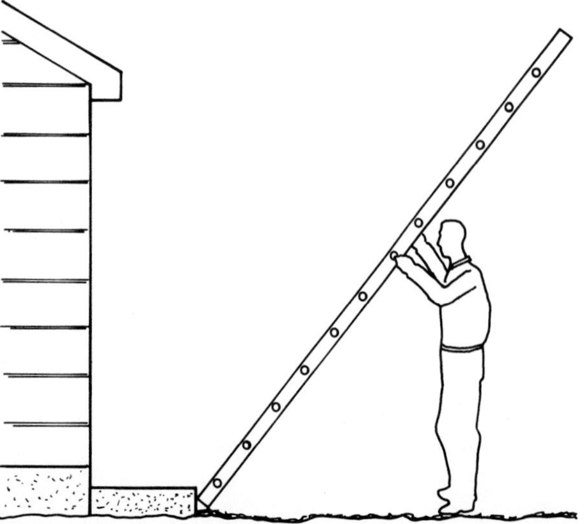

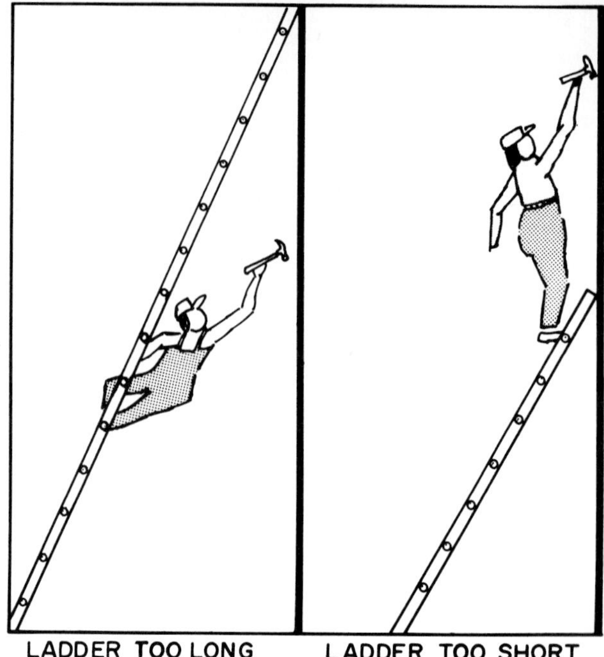

FIGURE 5-10

Always use a ladder that is the proper length.

FIGURE 5-11

Set the feet of the ladder the proper distance from the wall. Make certain that it extends 36 in. above the top of the roof.

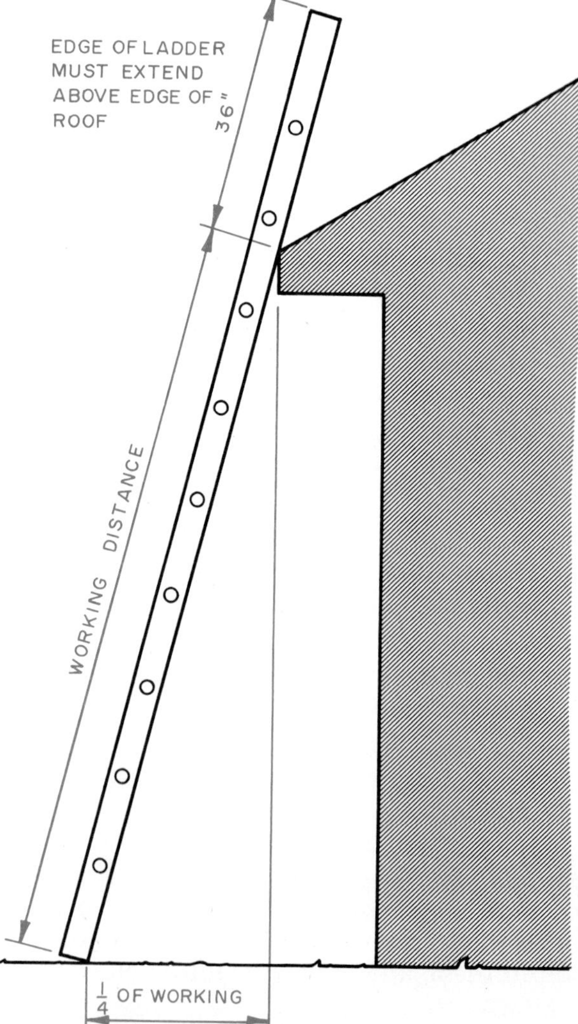

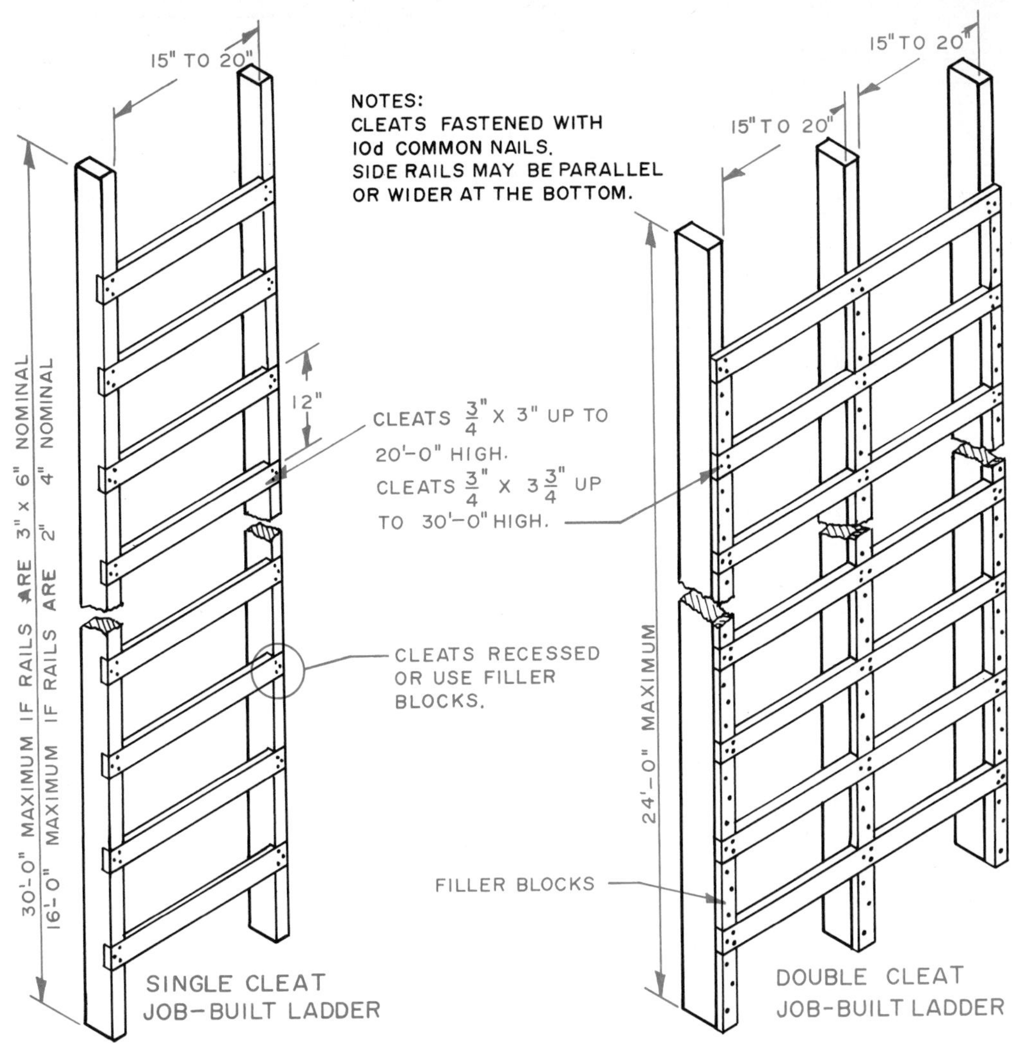

FIGURE 5-12

Typical job-built ladders.

12. Do not use ladders in a horizontal position on a scaffold. They are not designed to carry these loads.
13. Do not paint wood ladders. This hides cracks and defects. Seal them with a clear finish.
14. Store ladders in a horizontal position.
15. Do not try to reach out to the right or left from a ladder. Get down and move it over.
16. Do not climb ladders when you are wearing wet or muddy shoes.
17. Never stand on the top three rungs. They are needed for a hand hold. Get a longer ladder.
18. Extension ladders should have adequate overlap between sections. A 36-ft ladder should have at least a 3-ft overlap. A 48-ft ladder should have a 4-ft overlap.
19. Job-built ladders must be made to OSHA standards. A single-cleat ladder provides one-way traffic. A double-cleat ladder provides two-way traffic (Fig. 5-12).
20. Single-cleat, job-built ladders cannot exceed 30 ft in length. Double-cleat ladders cannot exceed 24 ft in length.

Stepladder Safety

1. The stepladder should be fully open and the legs locked in place.
2. Do not stand on the top two steps.
3. Be certain that all four legs are set on a firm surface. The surface should be even.
4. Do not leave tools on the top of a step ladder unless some type of toolholder is provided.

Scaffold Safety There are several kinds of scaffolds in common use. One is built on the site by the carpenters (Figs. 5-13 and 5-14). Another is a manufactured scaffold made from metal parts which are assembled (Fig. 5-15). A platform ladder as shown in Fig. 5-7 also serves as a type of scaffold. The following are some things that must be considered when using scaffolding.

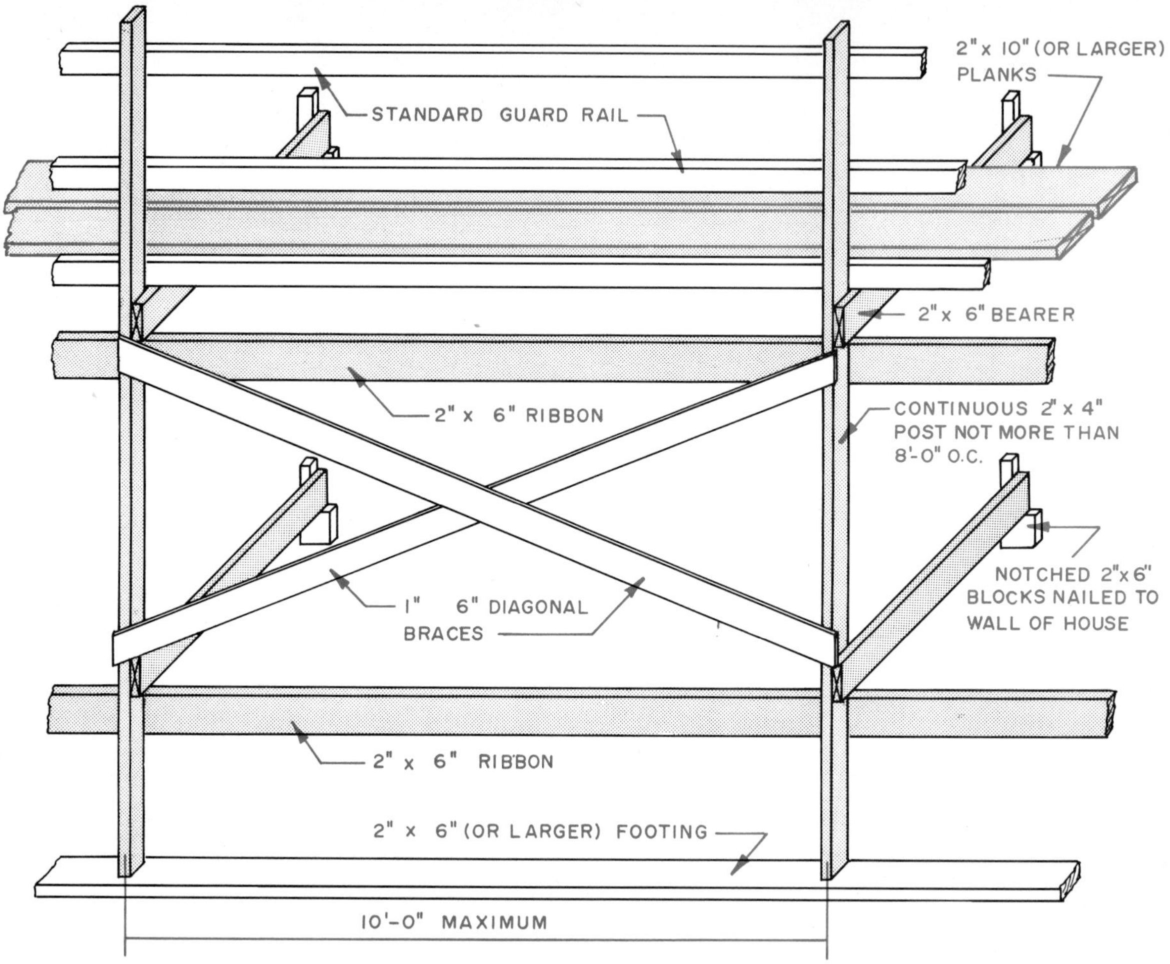

FIGURE 5-13 *(above)*

A job-built scaffold fastened to the wall of the building.

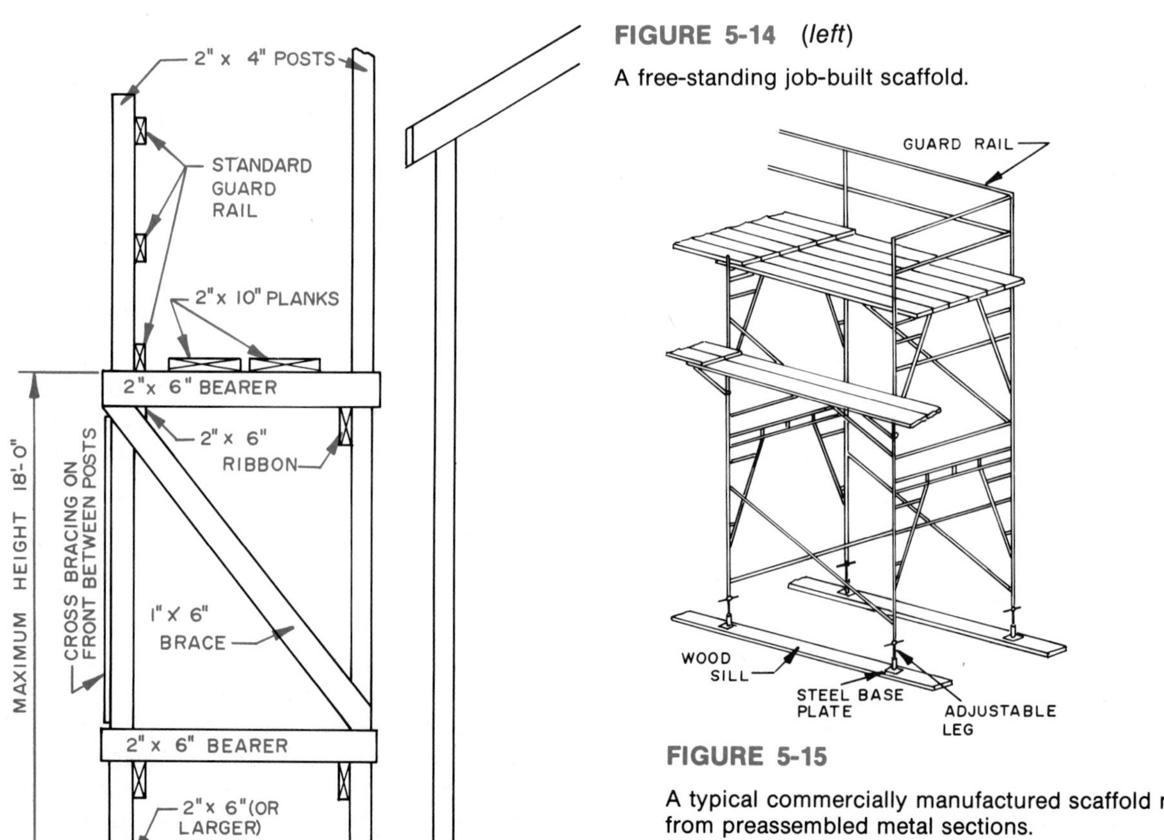

FIGURE 5-14 *(left)*

A free-standing job-built scaffold.

FIGURE 5-15

A typical commercially manufactured scaffold made from preassembled metal sections.

1. A scaffold must be able to hold four times the load it is expected to carry.
2. The footing for a scaffold must be solid and must not have motion when weight is applied. The scaffold must be level and plumb.
3. Scaffolds must be erected by experienced carpenters.
4. The lumber used must be free of defects.
5. Wood scaffolding should be designed to meet OSHA standards.
6. Use duplex nails on wood scaffolding to help when taking it apart (Fig. 5-16).
7. Scaffolds from 4 to 10 ft high must have guard rails. Those over 10 ft high must have a toeboard and a guard rail (Fig. 5-17).
8. If people must walk under scaffolds as they work, the space between the toeboard and guard rail should be covered with screen. This prevents materials from falling on them.
9. Always use a ladder to climb up to a scaffold. Never climb on the scaffold frame.
10. If work is going on above the scaffold, overhead protection is needed for those working on the scaffold.
11. Do not put more material on the scaffold than is needed for the immediate job. Keep it clear of debris.
12. The boards forming the scaffold platform should not have spaces between them. The minimum-size board is 2 × 10 in.
13. Posts for scaffolding up to 20 ft long are 2 × 4 in. members. Use 4 × 4 in. members up to 60 ft high.
14. When using manufactured scaffolding, observe the maximum design load.

FIGURE 5-16

A duplex nail.

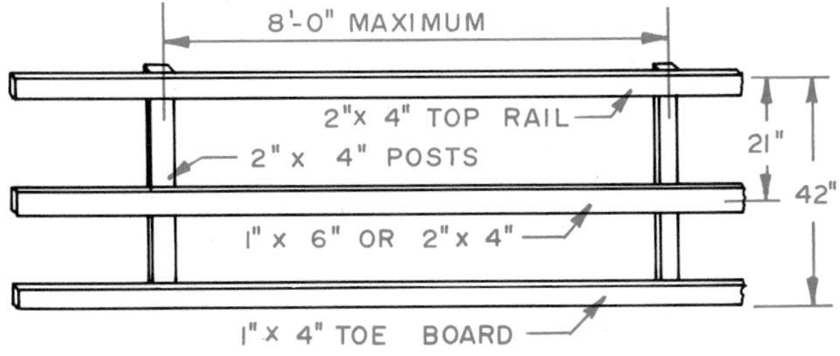

FIGURE 5-17

A standard wood guard railing.

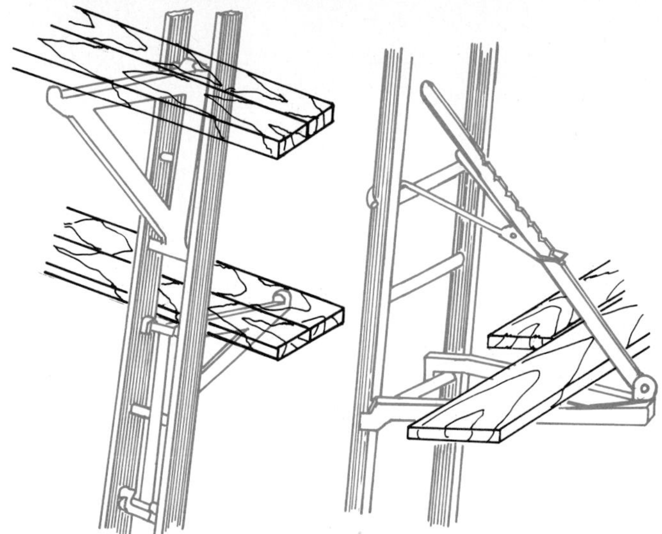

FIGURE 5-18

Two typical types of ladder jacks.

15. Follow the manufacturer's instructions for assembling manufactured scaffolding.
16. Be certain that all hand screws and wing nuts are tight.
17. If the unit is on wheels, be certain the wheels are locked or blocked so that they do not roll.

Ladder Jack Safety *Ladder jacks* are metal frames fastened to ladders to support scaffold boards (Fig. 5-18). There are a variety of types available.

1. Use ladder jacks for light duty. Use only on manufactured ladders.
2. Use two 2 × 10 in. boards as a platform. Do not space ladder jacks over 8 ft apart.
3. The 8-ft space will support a maximum of two people. One person per span is preferred.

Roofing Bracket Safety A *roofing bracket* is a metal frame that goes over the roof ridge or is nailed in place on the roof. It has space for a wood plank (Fig. 5-19). They can be adjusted to fit the pitch of the roof. When working with a roofing bracket, the worker should wear a safety belt.

WALL BRACKETS

A *wall bracket* is a metal frame that is either nailed to the building frame or is hooked around a stud (Fig. 5-20). These brackets must be installed carefully. Each nail should penetrate solid wood. Use 16d or 20d common nails. Nails that bend or have damaged heads should be replaced.

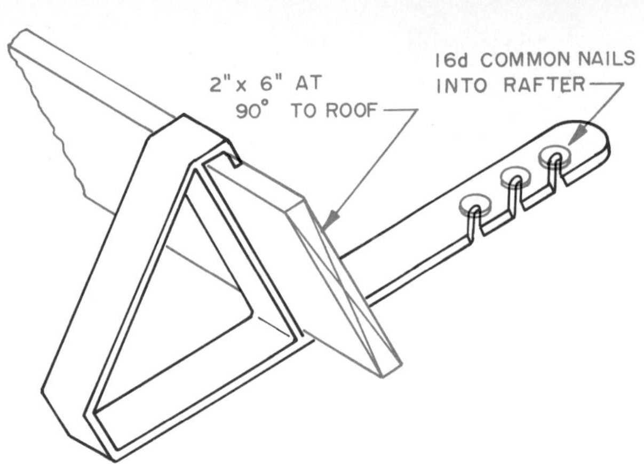

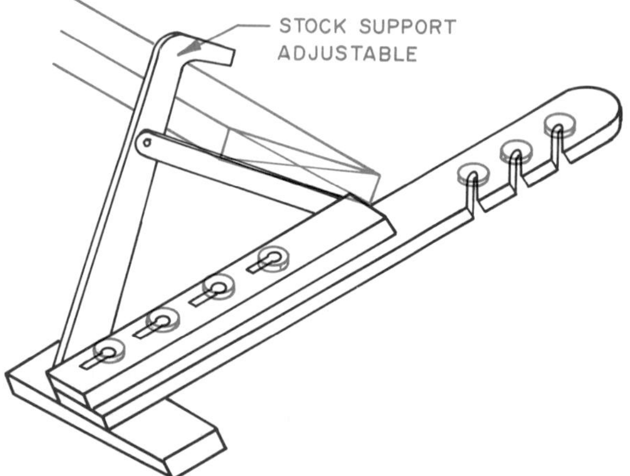

FIGURE 5-19

Two typical types of roofing brackets.

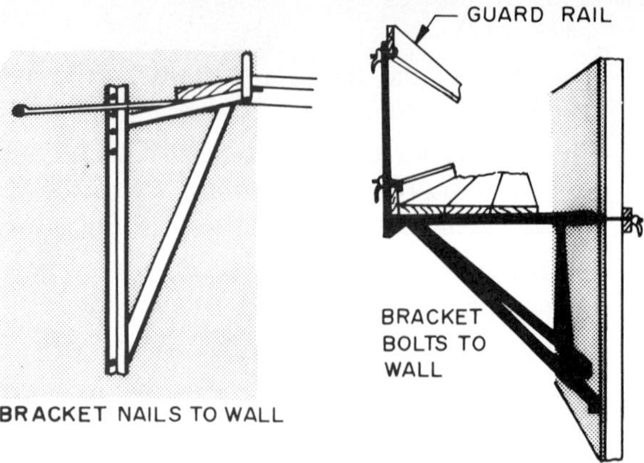

FIGURE 5-20

Wall brackets are nailed or bolted to the building wall.

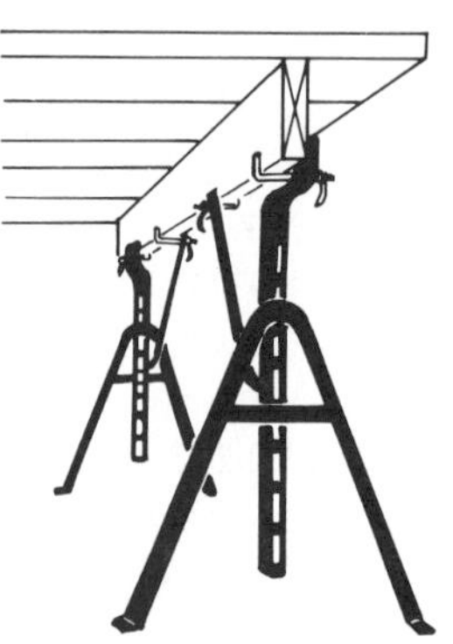

FIGURE 5-21

Trestle jacks rest on the ground or the floor. They are used to build a low scaffold.

TRESTLE JACKS

Trestle jacks are used to support scaffolding boards for low work. They are most often used for interior jobs such as applying drywall sheets to a ceiling. Their height is adjustable (Fig. 5-21). The boards must be 2-in. material and free from defects that would weaken them. The jacks are spaced to provide a firm support for the boards.

FIGURE 5-22

Elevated platforms provide a safe working area off the ground. *(Courtesy of Reynolds Televator Corporation)*

ELEVATED PLATFORMS

Elevated platforms are widely used on construction sites. They are hydraulically operated. The work platform has guard rails and a toe board as required by OSHA safety regulations.

These platforms lower into a compact unit for easy moving (Fig. 5-22). They are controlled by the workers on the platform. Capacities vary, but the unit in Fig. 5-23 will lift 800 lb to a maximum of 35 ft. Notice the outriggers used to stabilize the base unit.

FLOOR AND WALL OPENINGS

A *floor hole* is more than 1 in. in its smallest size and less than 12 in. in its largest size. It can permit debris to fall on those below, but people cannot fall through. A floor hole over 12 in. in its smallest dimension will permit a person to fall through. A *wall opening* 18 × 30 in. is considered large enough for a person to fall through.

FIGURE 5-23

Elevated platforms are operated by the worker on the platform. *(Courtesy of Reynolds Televator Corporation)*

1. Floor holes must be guarded with a standard railing or cover. A railing is shown in Fig. 5-24. If a cover is used, it must be able to carry twice the load that may pass over it. The cover should not be able to be accidentally moved out of place.
2. If the opening is a stairway or ladderway, it must have a swinging gate. The gate must be able to be locked.
3. Hatchways and chutes must have a standard railing or cover.

FIGURE 5-24

A recommended guard rail for enclosing floor openings.

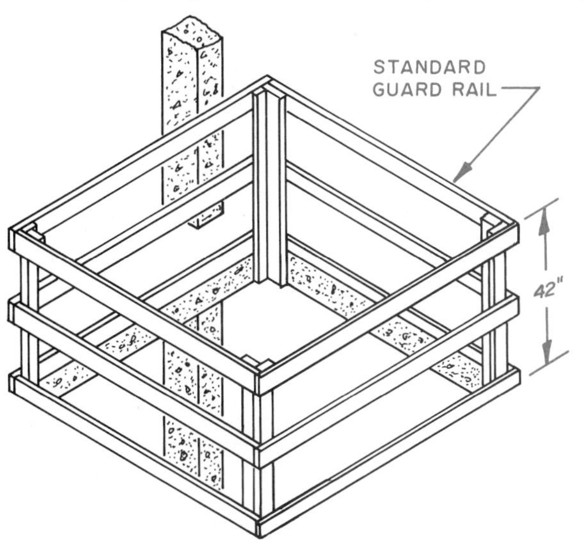

4. Wall openings with a drop of over 4 ft must have a standard railing.
5. Stairs should have a handrail.
6. Stairs should be kept clean.
7. Runways and ramps should have a standard rail if they are over 4 ft above the ground or floor.
8. Ramps should be securely fastened on both ends to prevent them from moving when used.

MATERIALS HANDLING SAFETY

1. Do not stack lumber and other materials too high. A maximum of 20 ft is normal for dressed lumber. When stacking, be certain to keep piles level and well supported.
2. Remove nails before handling used lumber.
3. When removing lumber from a stack, remove it so that the pile is kept level as it goes down.
4. When lifting materials, keep your back straight and lift with your legs (Fig. 5-25).
5. Stay out from under material being moved by a crane or forklift.
6. If a load is too heavy to lift alone, get some help.
7. Use mechanical aids to lift and move heavy objects. These include such things as wheeled carts, chain hoists, and forklifts.

FIGURE 5-25

Commonsense rules for lifting heavy materials.

LIFTING BY HAND

1. Size up the load.
2. Check the route.
3. Get good footing.
4. Bend knees, keep back straight.
5. Get a good grip on a solid part.
6. Lift with legs.
7. Keep balance, do not twist.
8. Face in the direction you are moving.
9. Lower the load by bending knees, not back.
10. Watch out for fingers and feet.

STUDY QUESTIONS

After reading the chapter, answer each of the following questions. If you do not know the answer, review the chapter.

1. What is mean by safety consciousness?
2. What are the provisions of the OSHA Act of 1970?
3. What are the general housekeeping rules to follow on a construction site?
4. What organization sets safety standards for hard hats?
5. List the safety suggestions for eye protection.
6. How do carpenters protect their feet?
7. What are the requirements for extension cords used with portable electric tools?
8. What are the three types of ladders?
9. How do you place a ladder so that it meets safety recommendations?
10. How much overlap should be left when using extension ladders?
11. What is the recommended safe loading requirement for scaffolding?
12. What are the requirements for railings and toeboards for scaffolding?
13. What is the difference between a ladder jack and a roofing bracket?
14. When do carpenters use trestle jacks?
15. What are the regulations for protecting floor and wall openings?

IMPORTANT TECHNICAL TERMS

Following are technical terms that you should be able to use as part of your working vocabulary. Write a brief description of the meaning of each term.

OSHA
Hard hat
Safety glasses
Face shield
Ear inserts
Muff protectors
Respiration-protection device
Ladder
Scaffold
Ladder jack
Roofing bracket
Wall bracket
Trestle jack
Elevated platform

6 CONSTRUCTION DRAWINGS, PERMITS, AND CODES

A carpenter must be able to read construction drawings and specifications. Together the drawings and specifications tell everything that is needed to construct a building.

A typical set of *construction drawings* will include a floor plan, foundation plan, plot plan, elevations of all sides of the building, sections, and special details. Also shown are electrical, plumbing, and heating and air-conditioning systems.

Specifications provide a word description of things about a building that cannot be easily shown on drawings. For example, information about the concrete mix wanted or the type of flooring desired are described in the specifications.

METHODS OF MEASUREMENT*

The customary (English) system of linear measurement uses inches, feet, yards, and miles as the units. The metric system uses millimeters, centimeters, meters, and kilometers.

The *meter* is the basic metric unit of length. It is slightly longer than the yard (Fig. 6-1). An inch is equal to 25.4 millimeters (Fig. 6-2).

The metric system of *linear measurement* is based on units of 10. A meter is divided into 10 decimeters (dm), 100 centimeters (cm), or 1000 millimeters (mm). Since the system is based on 10, it is possible to change from one metric linear unit to another by only moving the decimal point. For example, 2.575 meters equals 25.75 decimeters, 257.5 centimeters, or 2575 millimeters.

The metric units of linear measure, symbols used to show them on drawings, and conversion factors are shown in Fig. 6-3.

*The spellings of metric terms such as meter or metre, millimetre or millimeter, are used interchangeably in the illustrations and text so that the reader can become accustomed to the use of both spellings.

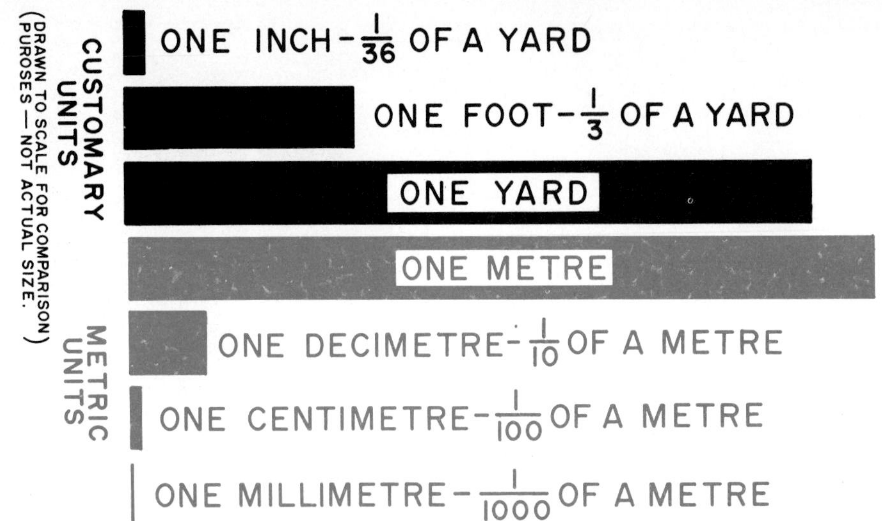

FIGURE 6-1

A comparison of the U.S. Customary and metric units of linear measure.

FIGURE 6-2

A comparison of an inch ruler and a millimeter scale.

1 meter (m) = 10 decimeters (dm) 1 decimeter (dm) = 10 centimeters (cm) 1 centimeter (cm) = 10 millimeters (mm) (A)
inches to millimeters—multiply inches by 25.4 millimeters to inches—multiply millimeters by 0.0394 inches to centimeters—multiply inches by 2.54 centimeters to inches—multiply centimeters by 0.394 inches to meters—multiply inches by 0.0254 feet to meters—multiply feet by 0.3048 (B)

FIGURE 6-3

(A) Metric measures of linear distances. (B) Linear measure conversion factors.

1 kilogram (kg) = 1000 grams (g) 1 hectogram (hg) = 100 grams (g) 1 dekogram (dag) = 10 grams (g) 1 decigram (dg) = one-tenth of a gram (g) 1 centigram (cg) = one-hundredth of a gram (g) 1 milligram (mg) = one-thousandth of a gram (g) (A)
pounds to kilograms—multiply pounds by 0.454 kilograms to pounds—multiply kilograms by 2.205 (B)

FIGURE 6-4

(A) Metric measures of mass. (B) Mass conversion factors.

The metric unit of *mass* is the *kilogram* (km). The kilogram is the mass of standard cylinder stored at the International Bureau of Weights and Measures in France. Although mass and weight are not exactly the same thing, for practical purposes mass is used as a measure of weight.

The metric units of mass, the symbols used to show them on drawings, and conversion factors are shown in Fig. 6-4.

The basic metric unit of *volume* is the *cubic meter* (m^3). A cubic meter is a cube measuring 1 meter on each side (Fig. 6-5).

FIGURE 6-5

The metric units for measuring volume.

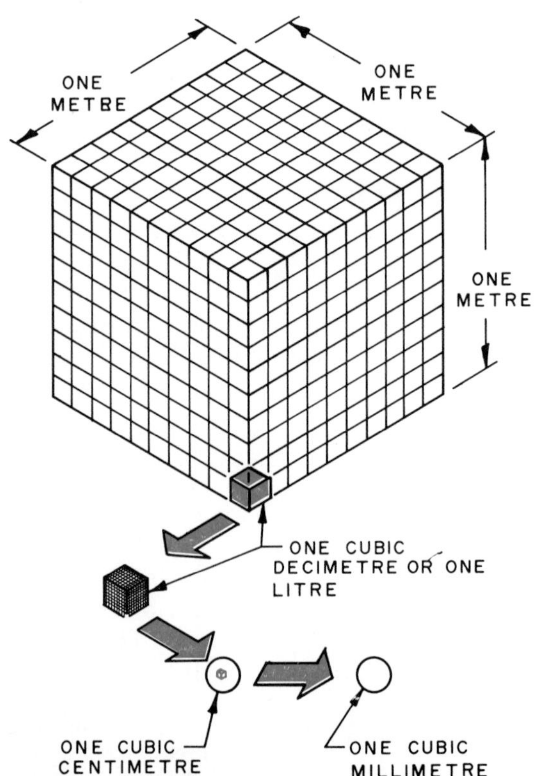

1000 cubic millimeters (mm^3) = 1 cubic centimeter (cm^3)
1000 cubic centimeters (cm^3) = 1 cubic decimeter (dm^3)
1000 cubic decimeters (dm^3) = 1 cubic meter (m^3)

(A)

cubic yards to cubic meters (m^3)—multiply cubic yards by 0.764
cubic feet to cubic meters (m^3)—multiply cubic feet by 0.0028
gallons to cubic meters (m^3)—multiply gallons by 0.0038
quarts to cubic meters (m^3)—multiply quarts by 0.00095
ounces to cubic meters (m^3)—multiply ounces by 0.000029

(B)

FIGURE 6-6

(A) Metric measures of volume. (B) Conversion factors for measures of volume.

The metric units of volume, the symbols used to show them on drawings, and conversion factors are shown in Fig. 6-6.

The metric unit of *area* is the *square meter* (m^2). A square meter is a square measuring 1 meter on each side (Fig. 6-7).

Large land areas are measured using the *square kilometer* (km^2). Smaller land areas are measured in square hectometers (hm^2). Metric measures of area, the symbols used to show them on drawings, and conversion factors are shown in Fig. 6-8.

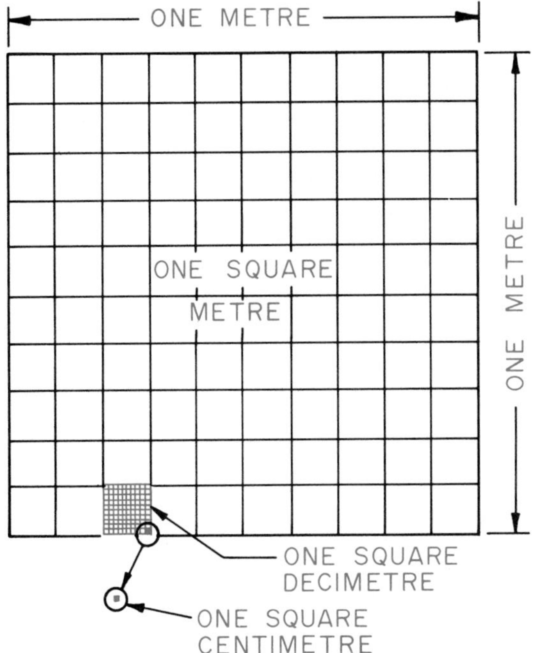

FIGURE 6-7

Metric units for measuring area.

SCALE

Construction drawings are drawn to *scale*. This means that they are drawn much smaller than actual size. A fraction of an inch is used to represent a foot of actual building. Scales used to draw residential and commercial buildings are shown in Fig. 6-9. For example, residential floor plans

Floor plan ¼″ = 1′-0″
Foundation plan ¼″ = 1′-0″
Elevations ¼″ = 1′-0″
Construction details ¾″ to 1½″ = 1′-0″
Wall sections ¾″ to 1½″ = 1′-0″
Cabinet details ⅜″ to ½″ = 1′-0″
Plot plan 1″ = 20′ or 40′

(A)

Floor plan	1:50
Foundation plan	1:50
Elevations	1:50
Construction details	1:20 and 1:10
Wall sections	1:20 and 1:10
Cabinet details	1:50 and 1:25
Plot plan	1:100 and 1:500

(B)

FIGURE 6-9

(A) Customary (English, in.) and (B) metric (mm) scales used to draw house plans.

100 square millimeters (mm^2) = 1 square centimeter (cm^2)
100 square centimeters (cm^2) = 1 square decimeter (dm^2)
100 square decimeters (dm^2) = 1 square meter (m^2)

(A)

1 square kilometer (km^2) = 1 000 000 square meters (m^2)
1 square hectometer (hm^2) = 10 000 square meters (m^2)

(B)

square inches ($in.^2$) to square millimeters (mm^2)—multiply square inches by 645.4
square inches ($in.^2$) to square centimeters (cm^2)—multiply square inches by 6.454
square centimeters (cm^2) to square inches—multiply square centimeters by 0.155
square feet (ft^2) to square meters (m^2)—multiply square feet by 0.093
square meters (m^2) to square feet—multiply square meters by 10.76
square yards (yd^2) to square meters (m^2)—multiply square yards by 0.836
square meters (m^2) to square yards—multiply square meters by 1.197
acres to square meters (m^2)—multiply acres by 4046.87
square miles (mi^2) to square meters (m^2)—multiply square miles by 2,589,998
square miles (mi^2) to square kilometers (km^2)—multiply square miles by 2.589

(C)

FIGURE 6-8

(A) Metric measures of area.
(B) Designations for large areas.
(C) Conversion factors for measurement of area.

generally are made to the scale of ¼ in. = 1 ft 0 in. of the actual house. The metric scale used for residential floor plans is usually 1:50. This means that 1 millimeter on the drawing represents 50 millimeters of the actual house.

Construction drawings are fully dimensioned. The dimensions shown are the actual size, even though the part is drawn to scale. A carpenter seldom has to measure a drawing to get a size. If the drawing is properly made, the dimension can be read off the drawing. Occasionally, a dimension is missing. The carpenter has to measure the drawing to get it. This gives only an approximate size because the drawing may not be accurate. Also, the carpenter uses a rule, which is also not as accurate as the measuring tools used by drafters. Figure 6-10 shows how to find a missing dimension using a carpenter's rule. The first two examples show the plan drawn using feet and inches. The second two examples were drawn using millimeters. Notice that the carpenter's rule is divided into centimeters and subdivided into millimeters.

Most construction projects are presently designed using feet and inches as the units of linear measure. An example is shown in Fig. 6-19. In the future metric measure will find increasing use. This will occur as manufacturers of construction materials begin to supply them in metric sizes. A metric drawing has all units in millimeters (Fig. 6-11). Construction details are also given in millimeters. The plot plan is dimensioned using meters. The meters are carried to

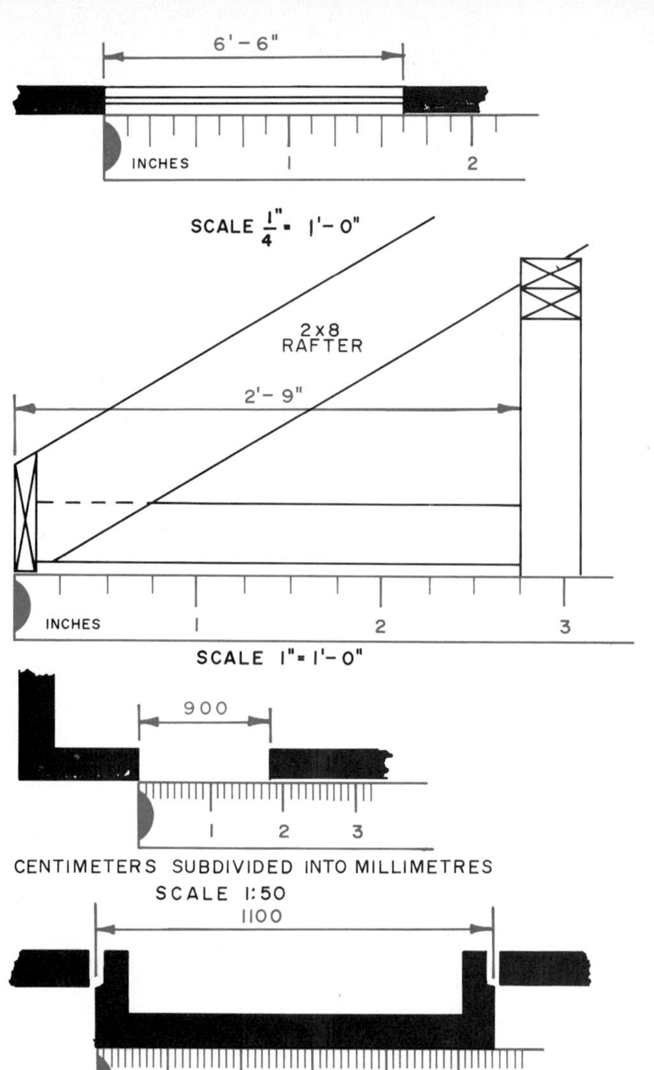

FIGURE 6-10

How to read scales used in drawing house plans.

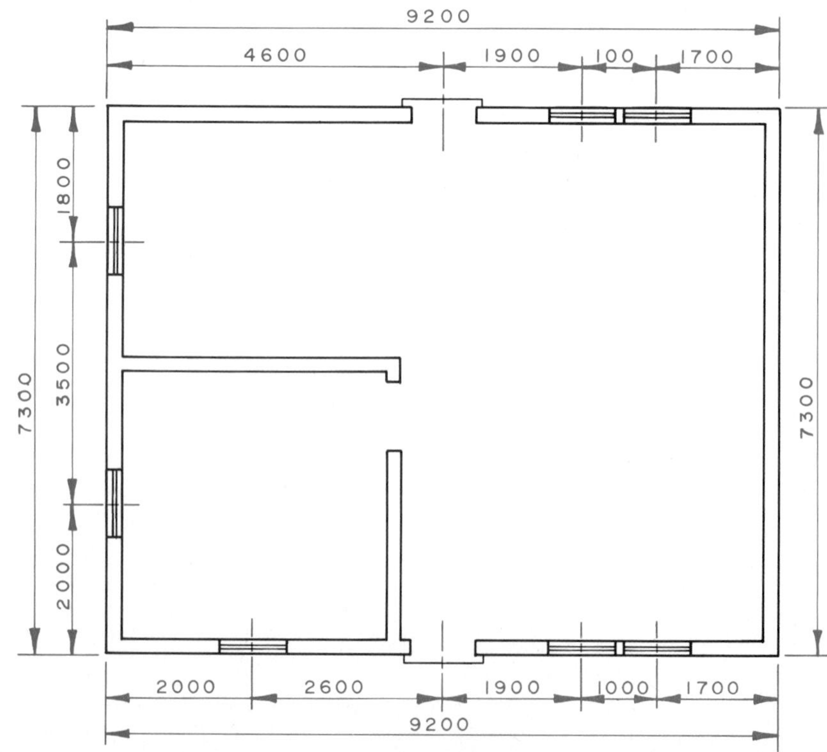

FIGURE 6-11

A floor plan with metric measurements.

three decimal places. This enables rapid conversion to millimeters if needed. For example, 25.125 meters equals 25 125 millimeters (Fig. 6-12).

ARCHITECTURAL SYMBOLS

Construction drawings are drawn to very small scales. This makes it impossible for the architect to show things as they really appear. Even if it were possible, the cost of the drafting time to do it would be excessive. *Architectural symbols* are used on drawings to represent these various elements. Selected symbols are shown in Figs. 6-13 through 6-18.

Another shortcut is to use abbreviations. This saves space and speeds up drafting time. Selected abbreviations are given inside the covers of the book.

Carpenters must know how to read these symbols and abbreviations.

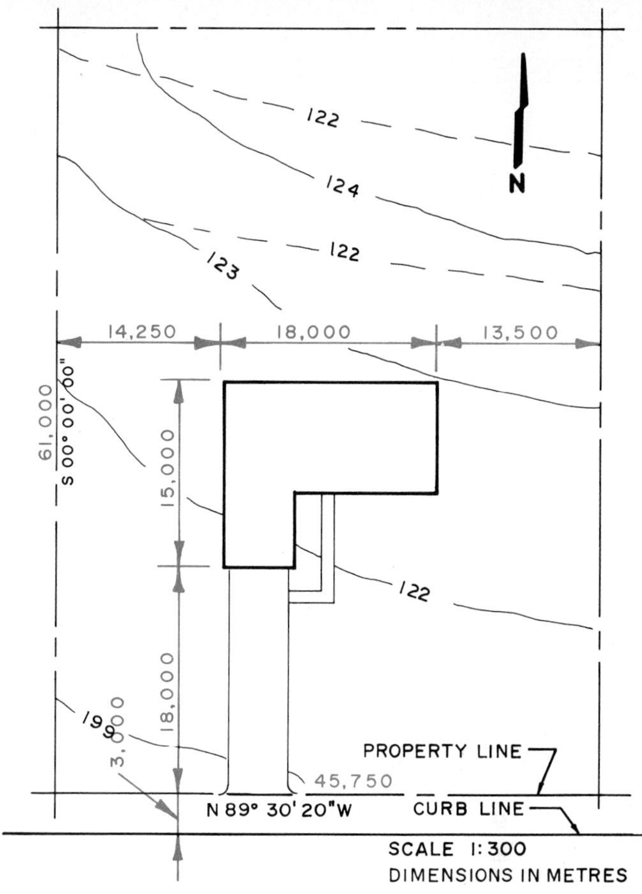

FIGURE 6-12

A plot plan in metric measurements.

FIGURE 6-13

Materials symbols used on construction drawings.

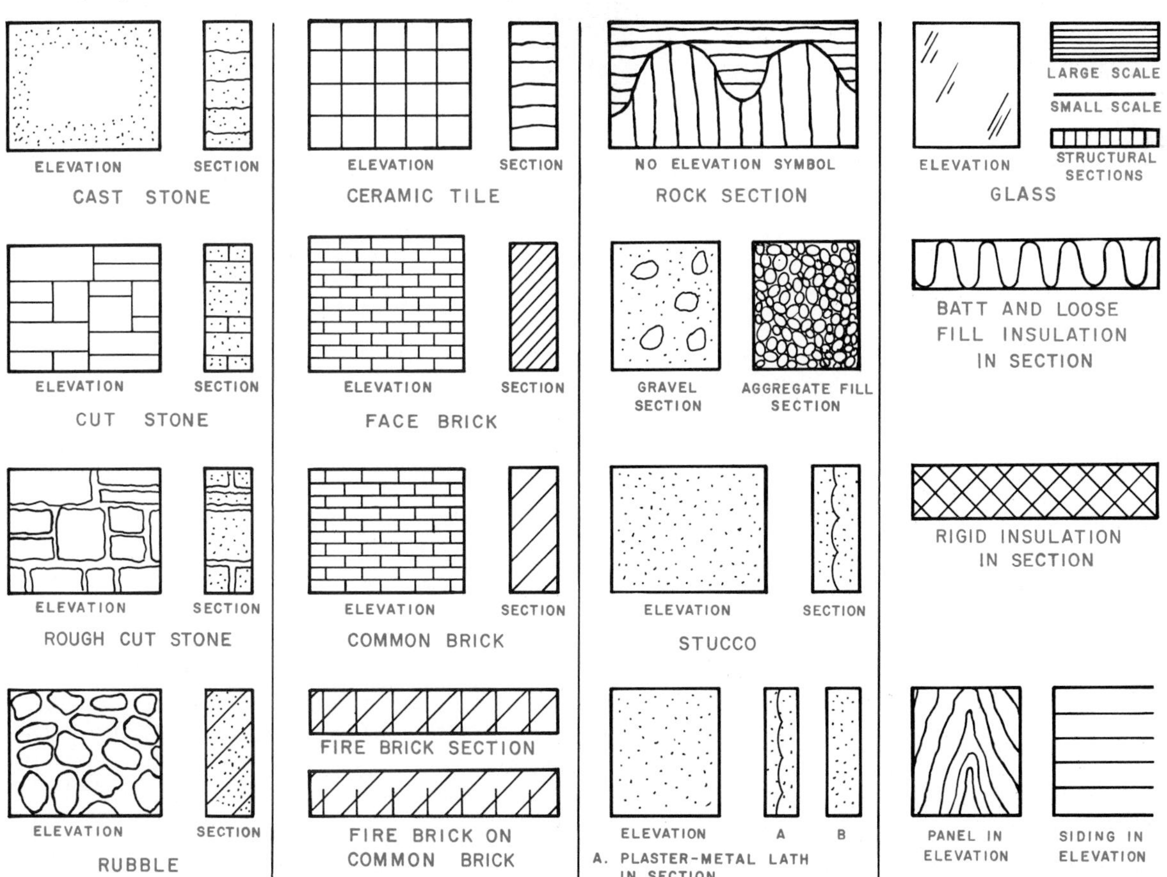

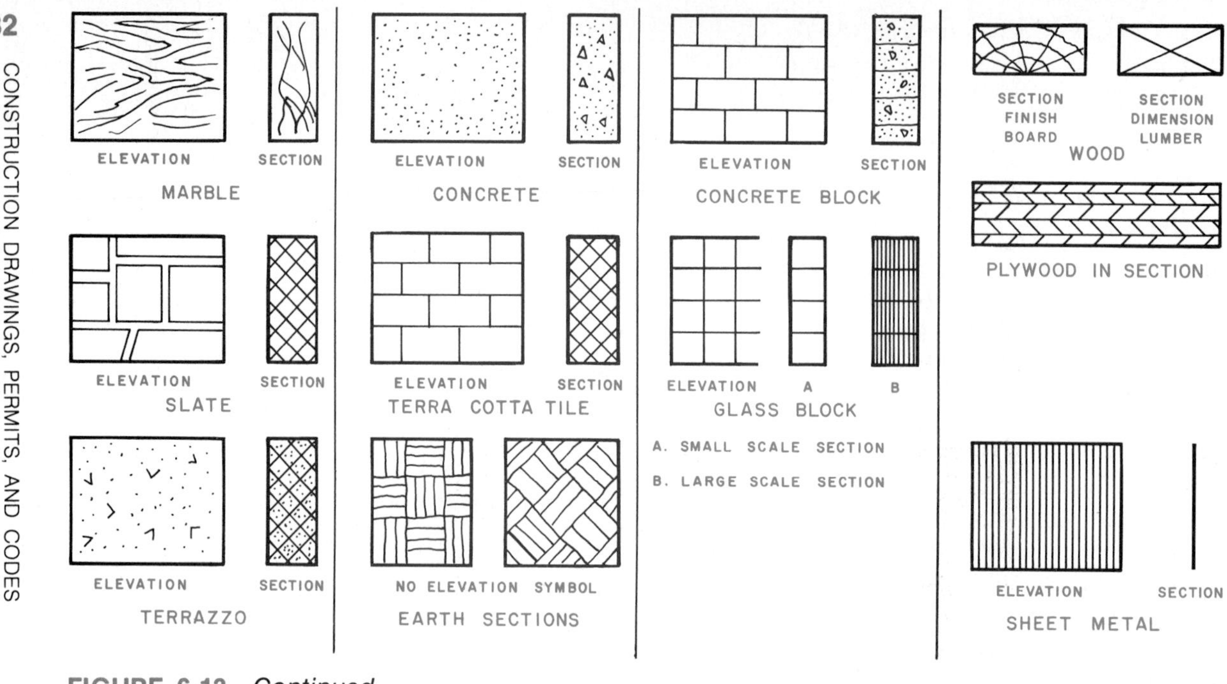

FIGURE 6-13 *Continued*

FIGURE 6-14

Window symbols used on construction drawings.

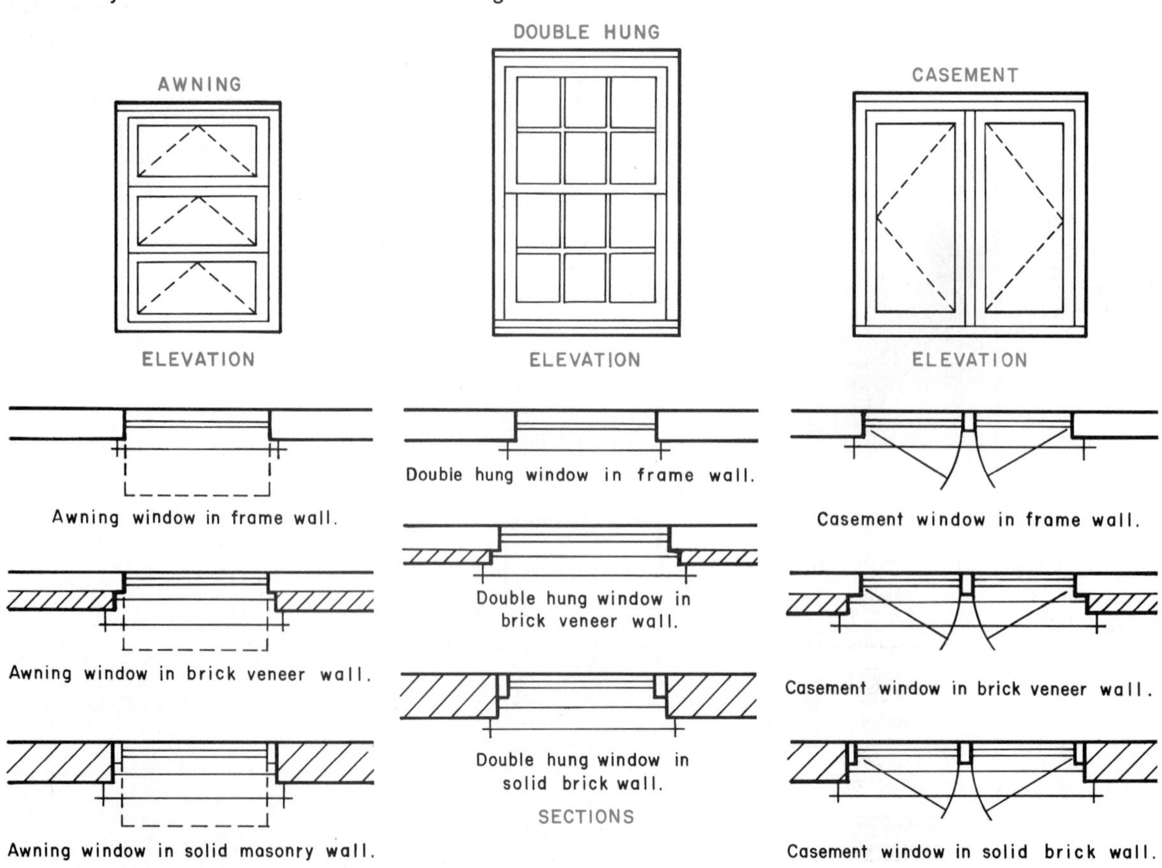

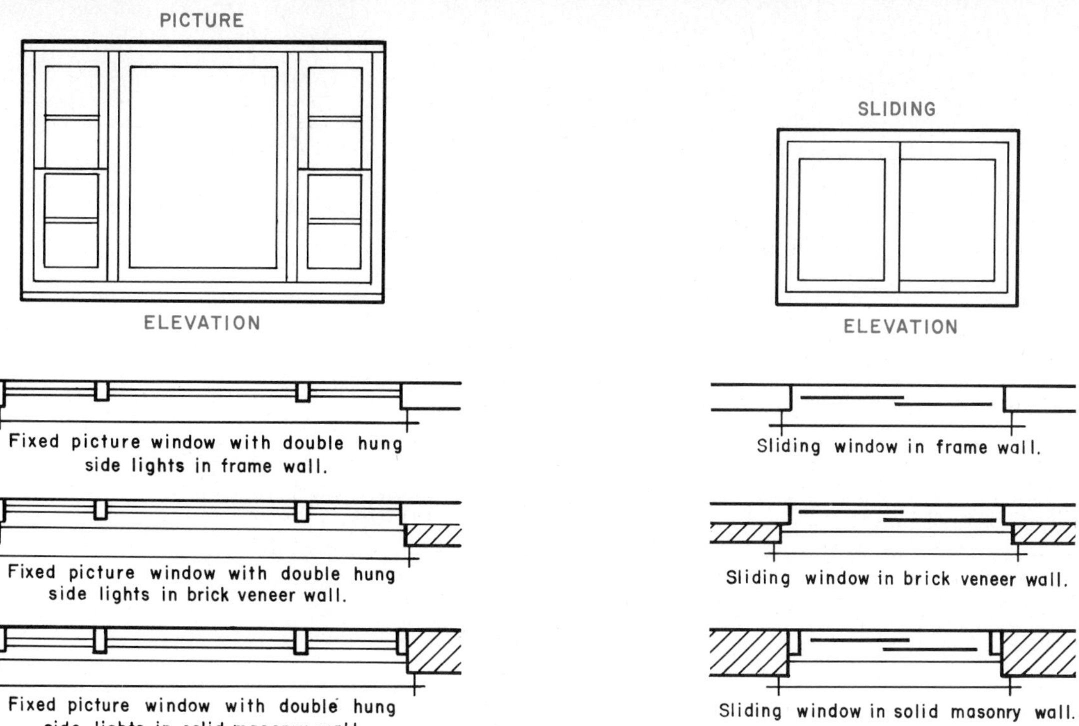

FIGURE 6-14 *Continued*

FIGURE 6-15

Door symbols in section as used on construction drawings.

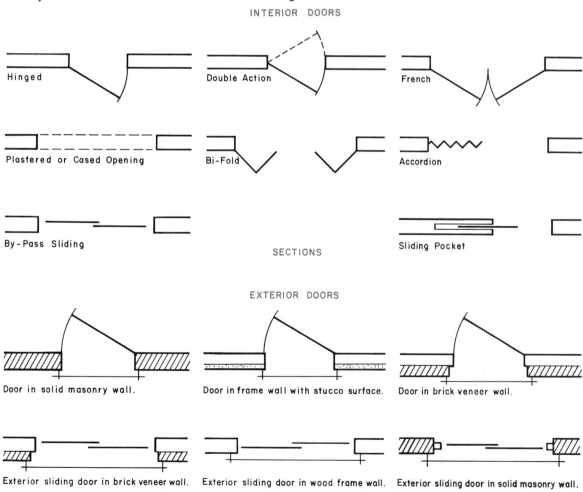

INTERIOR DOORS

Hinged

Double Action

French

Plastered or Cased Opening

Bi-Fold

Accordion

By-Pass Sliding

Sliding Pocket

SECTIONS

EXTERIOR DOORS

Door in solid masonry wall.

Door in frame wall with stucco surface.

Door in brick veneer wall.

Exterior sliding door in brick veneer wall.

Exterior sliding door in wood frame wall.

Exterior sliding door in solid masonry wall.

SECTIONS

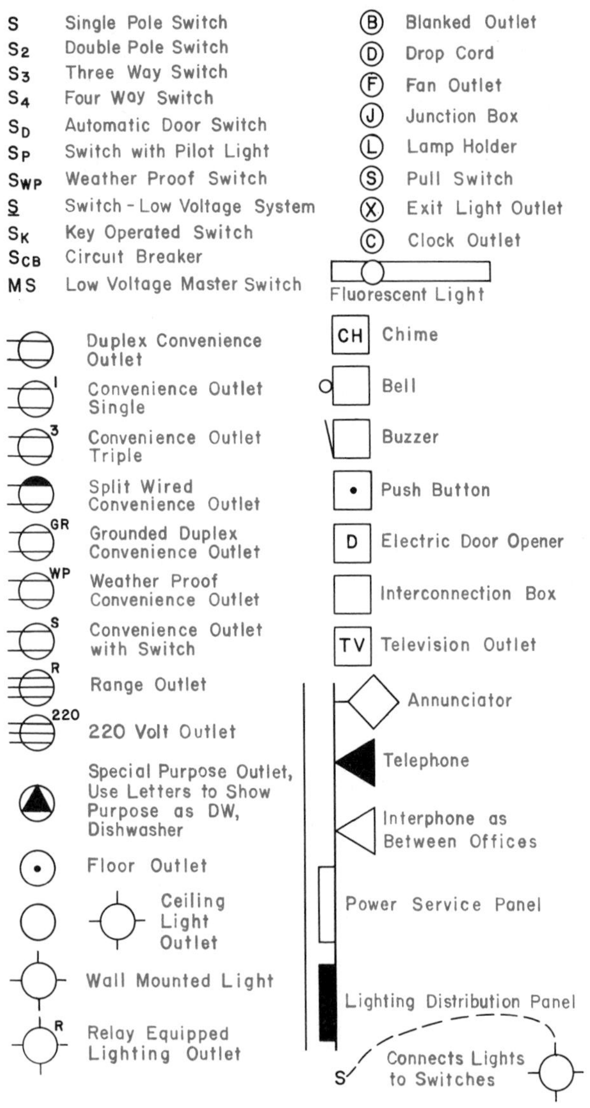

S Single Pole Switch
S_2 Double Pole Switch
S_3 Three Way Switch
S_4 Four Way Switch
S_D Automatic Door Switch
S_P Switch with Pilot Light
S_{WP} Weather Proof Switch
S Switch - Low Voltage System
S_K Key Operated Switch
S_{CB} Circuit Breaker
MS Low Voltage Master Switch
Duplex Convenience Outlet
Convenience Outlet Single
Convenience Outlet Triple
Split Wired Convenience Outlet
GR Grounded Duplex Convenience Outlet
WP Weather Proof Convenience Outlet
S Convenience Outlet with Switch
R Range Outlet
220 220 Volt Outlet
Special Purpose Outlet, Use Letters to Show Purpose as DW, Dishwasher
Floor Outlet
Ceiling Light Outlet
Wall Mounted Light
R Relay Equipped Lighting Outlet
B Blanked Outlet
D Drop Cord
F Fan Outlet
J Junction Box
L Lamp Holder
S Pull Switch
X Exit Light Outlet
C Clock Outlet
Fluorescent Light
CH Chime
Bell
Buzzer
Push Button
D Electric Door Opener
Interconnection Box
TV Television Outlet
Annunciator
Telephone
Interphone as Between Offices
Power Service Panel
Lighting Distribution Panel
S Connects Lights to Switches

FIGURE 6-16

Electrical symbols used on construction drawings.

PIPING
COLD WATER
HOT WATER
G G G G
GAS LINE
SEWAGE PIPING
SOIL, WASTE, LEADER (ABOVE GRADE)
SOIL, WASTE, LEADER (BELOW GRADE)
VENT PIPE
S-CI
SEWER - CAST IRON
S-CT
SEWER-CLAY TILE, BELL AND SPIGOT
DRAIN - CLAY TILE, BELL AND SPIGOT
DRAIN-OPEN TILE
PIPING SYMBOLS

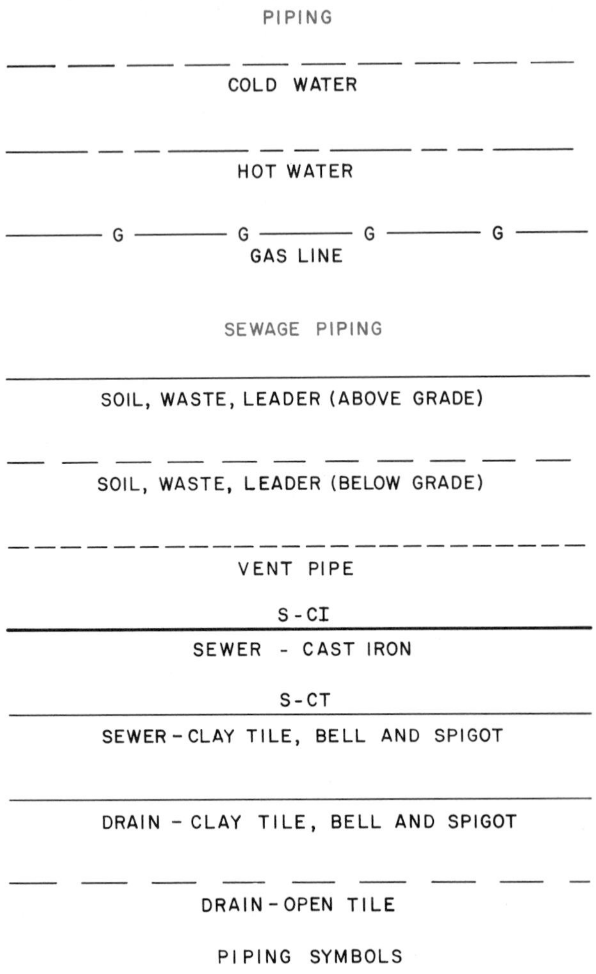

FIGURE 6-17

Piping symbols used on construction drawings.

SHOWER
WATER CLOSET
WALL HUNG LAVATORY
LAVATORIES BUILT-IN CABINETS
BUILT-IN OVEN

KITCHEN SINKS
RANGE
BATH TUB
REFRIGERATOR
BUILT-IN COOKING UNIT
WASHER
DRYER

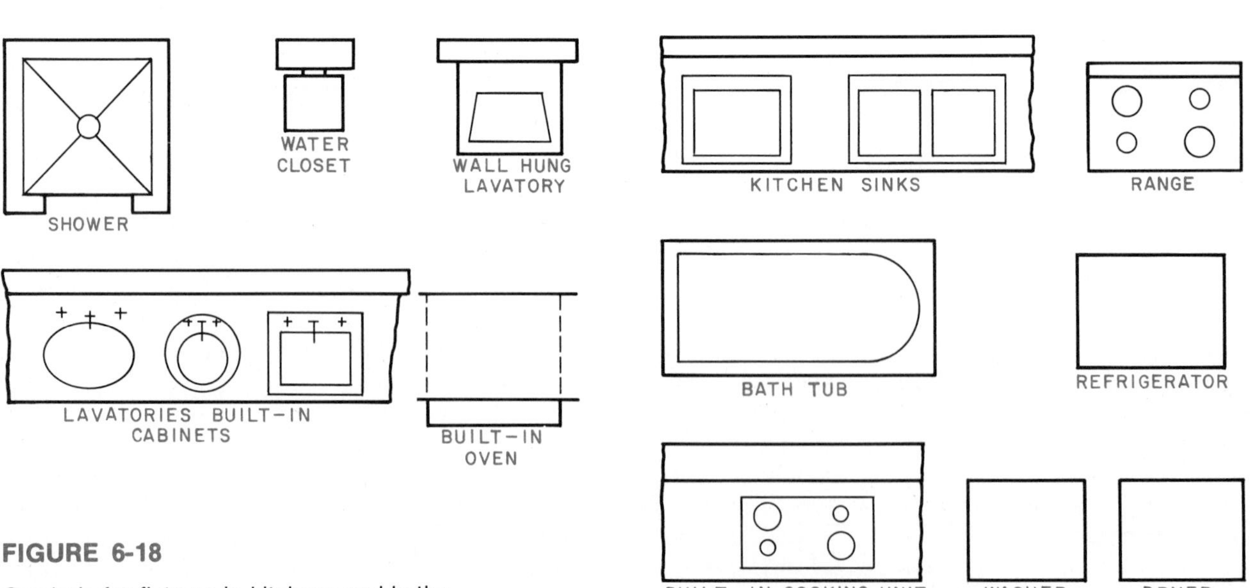

FIGURE 6-18

Symbols for fixtures in kitchens and baths.

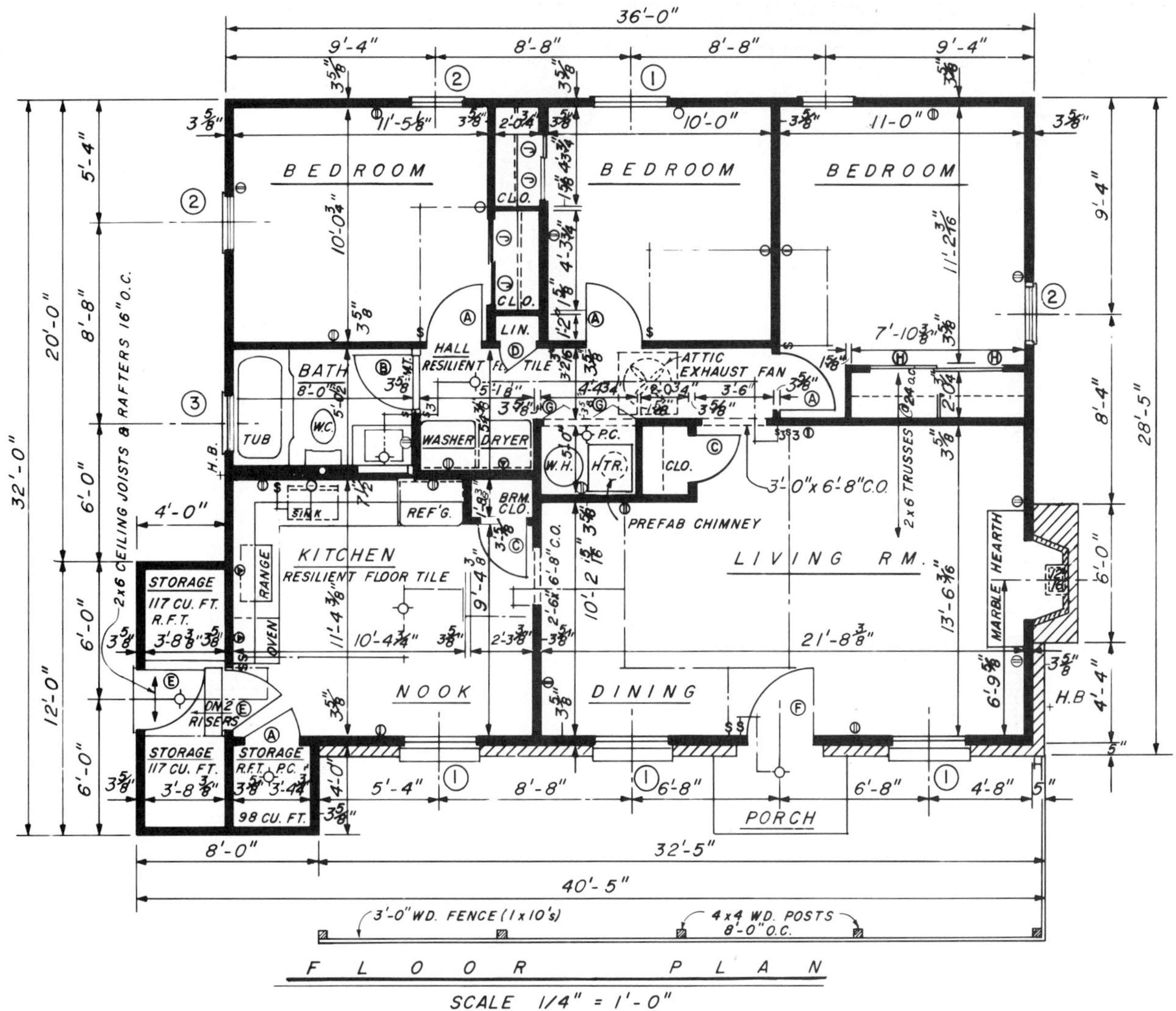

FIGURE 6-19

A typical residential floor plan. *(Copyright, Home Planners, Inc., Detroit, Mich.)*

READING A FLOOR PLAN

The *floor plan* shows how the building would look if the roof were removed. It shows the shape of the building; size and location of rooms, doors, windows, and stairs; and electrical and plumbing units. Sometimes the heating and air conditioning are on this plan. It is fully dimensioned, so everything is located or the needed sizes are given. Study the floor plan in Fig. 6-19. Look up each symbol that you do not recognize.

Notice that interior partitions are dimensioned to center on the wall. The dimensions extending to the exterior wall go to the edge of the foundation. Windows and doors are located by their centerline. The direction, size, and spacing of ceiling joists is shown. The overall dimensions are the same as the overall dimensions of the foundation.

Located on the same sheet as the floor plan are door and window schedules. These are coded to the floor plan. They show the type and size of door or window for each opening (Fig. 6-20).

READING A FOUNDATION PLAN

The *foundation plan* shows the footings and foundation wall. Figure 6-21 shows the plan for the house if it has a crawl space. Figure 6-22 shows it if it is to have a basement. Shown are any piers, columns, and beams. The sizes of these are given. If the building has a basement, the stair is shown (Fig. 6-23). Also, the electrical, plumbing, and heating and air conditioning are shown. Notice that the piers, columns, and beams are located by their centerline. The size, direction, and spacing of floor joists is given. The exterior dimensions show the size and location of the various parts of the foundation. Openings, such as vents or windows, are located by their centerlines.

MARK	SIZE	MATERIAL	NUMBER	REMARKS
A	2′-6″ × 6′-8″	Birch	3	Flush, H.C.
B	2′-4″ × 6′-8″	Birch	1	Flush, H.C.
C	2′-0″ × 6′-8″	Birch	2	Flush, H.C.
D	1′-8″ × 6′-8″	Birch	1	Flush, H.C.
E	2′-8″ × 7′-0″	Pine	2	Exterior, 6 light
F	3′-0″ × 7′-0″	Pine	1	Exterior, 6 light
G	2′-0″ × 6′-8″	Pine	2	Bifold
H	3′-0″ × 6′-8″	Birch	1	Flush, H.C., sliding
J	2′-0″ × 6′-8″	Birch	4	Flush, H.C., sliding

(A)

MARK	SIZE	MATERIAL	NUMBER	REMARKS
1	36 × 28	Wood	4	Double-hung, 6 light
2	24 × 24	Wood	4	Double-hung, 4 light
3	30 × 20	Wood	1	Sliding

(B)

FIGURE 6-20

A typical (A) door and (B) window schedule.

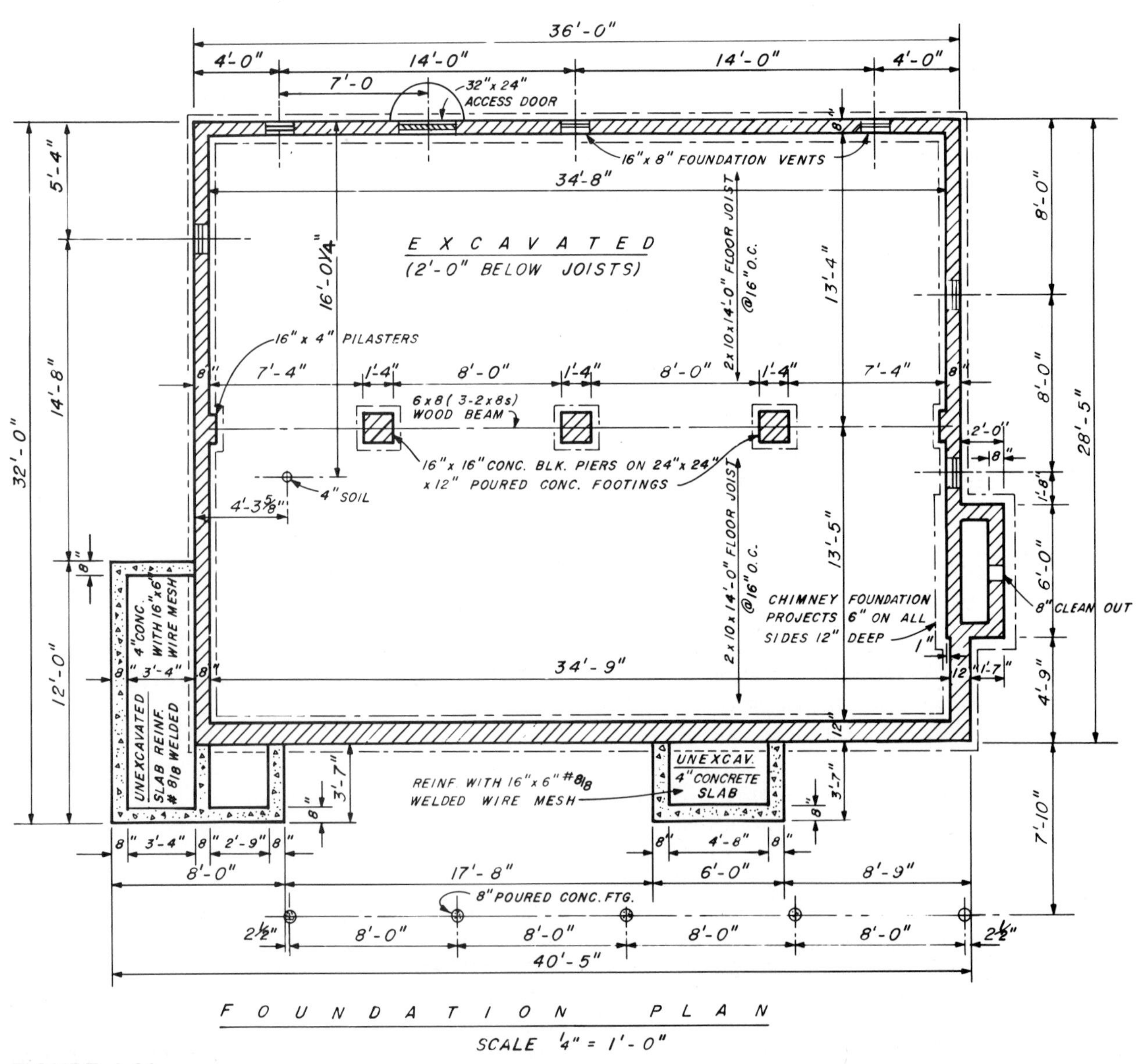

FIGURE 6-21

A basement plan. *(Copyright, Home Planners, Inc., Detroit, Mich.)*

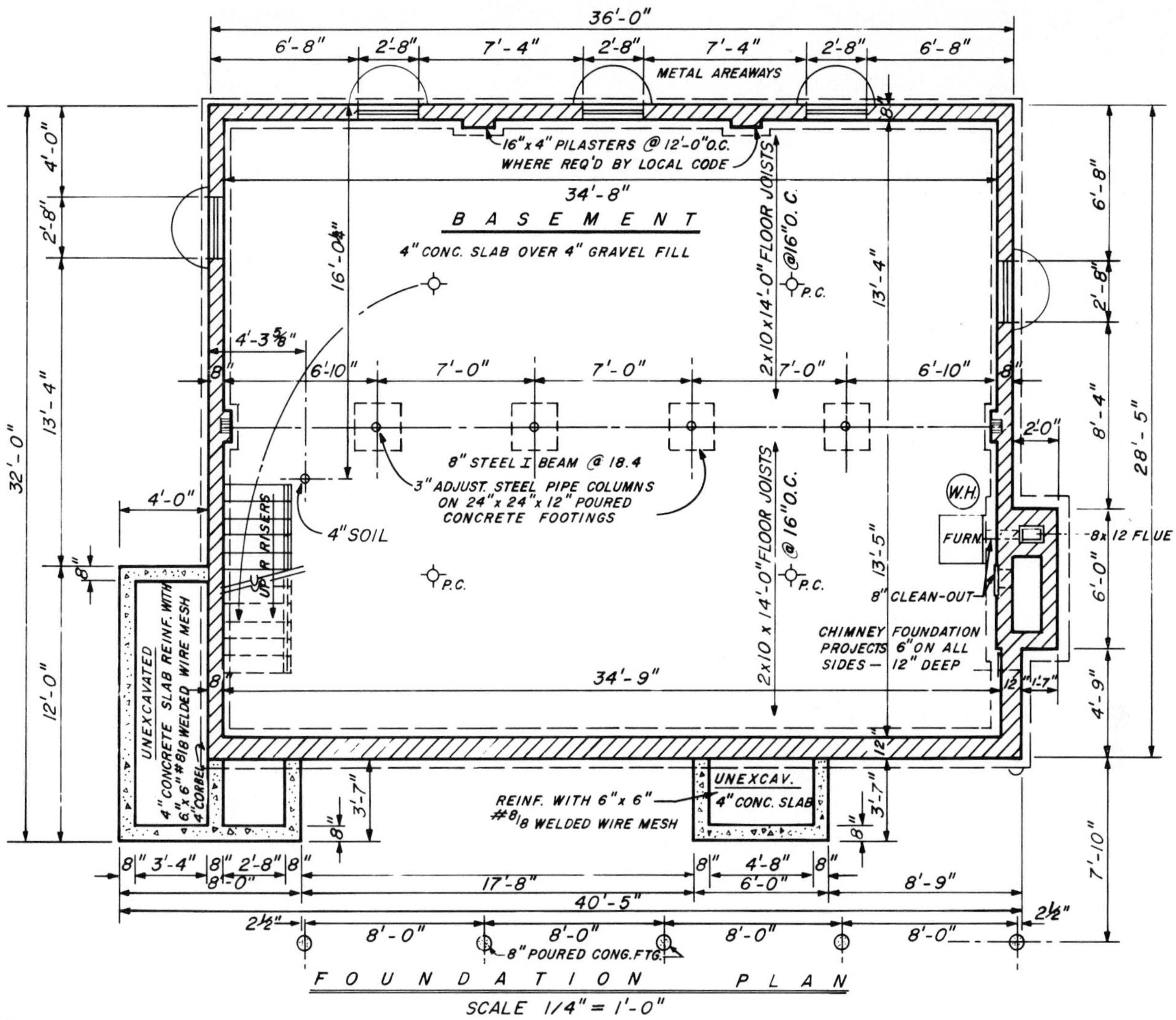

FIGURE 6-22

A foundation plan for a house with a crawl space. *(Copyright, Home Planners, Inc., Detroit, Mich.)*

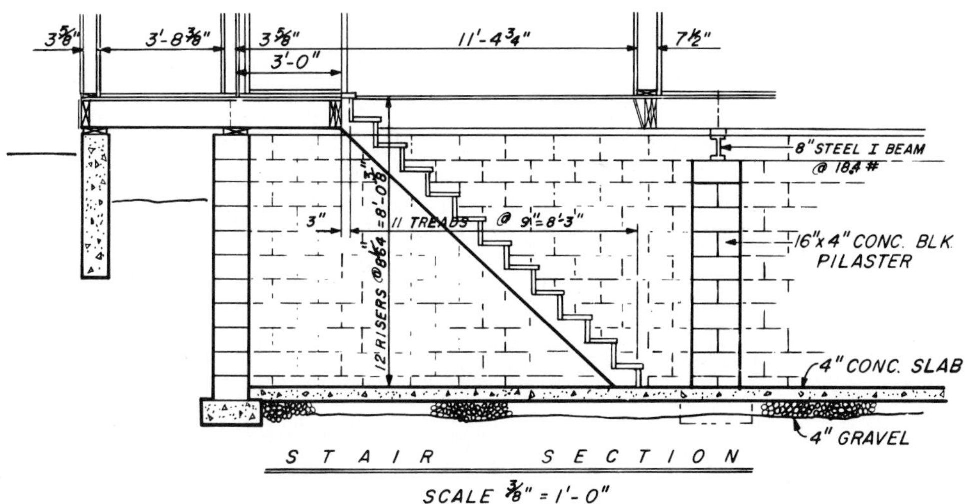

FIGURE 6-23

Basement stair detail. *(Copyright, Home Planners, Inc., Detroit, Mich.)*

READING THE ELEVATIONS

A typical building has four *elevations*, one for each side (Fig. 6-24). The elevations show how the building will look when the outside is finished. They show the roof height, windows, and doors. The finish materials, such as brick siding, are indicated. Sometimes they are shown with material symbols. Other times they are indicated by words. The building is related to the ground and the footing.

READING SECTION VIEWS

Section views are made through the exterior wall. This shows construction details from the footing to the roof. A wall section for the house if it is built with a crawl space is shown in Fig. 6-25. If it is built with a basement, the section would appear as shown in Fig. 6-26. The sections give the heights of the various members. The sizes of structural members are shown.

Other sections are drawn. These show any

FIGURE 6-24

Elevations of a residence. *(Copyright, Home Planners, Inc., Detroit, Mich.)*

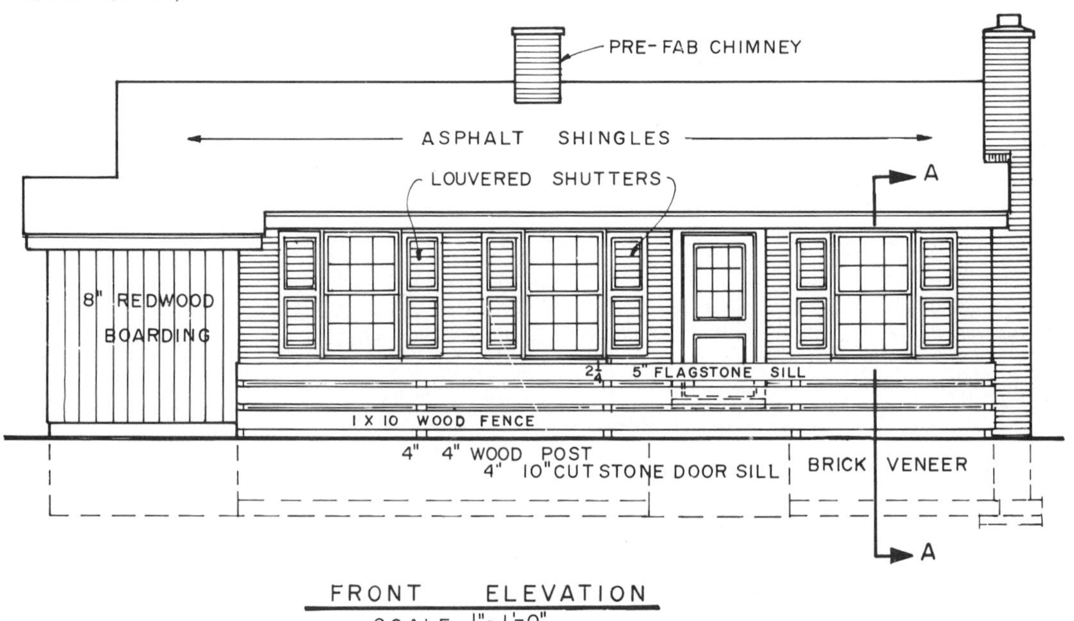

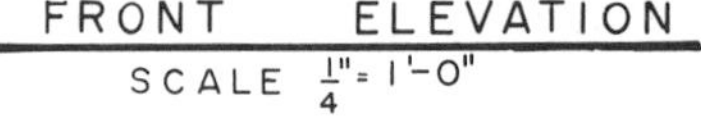

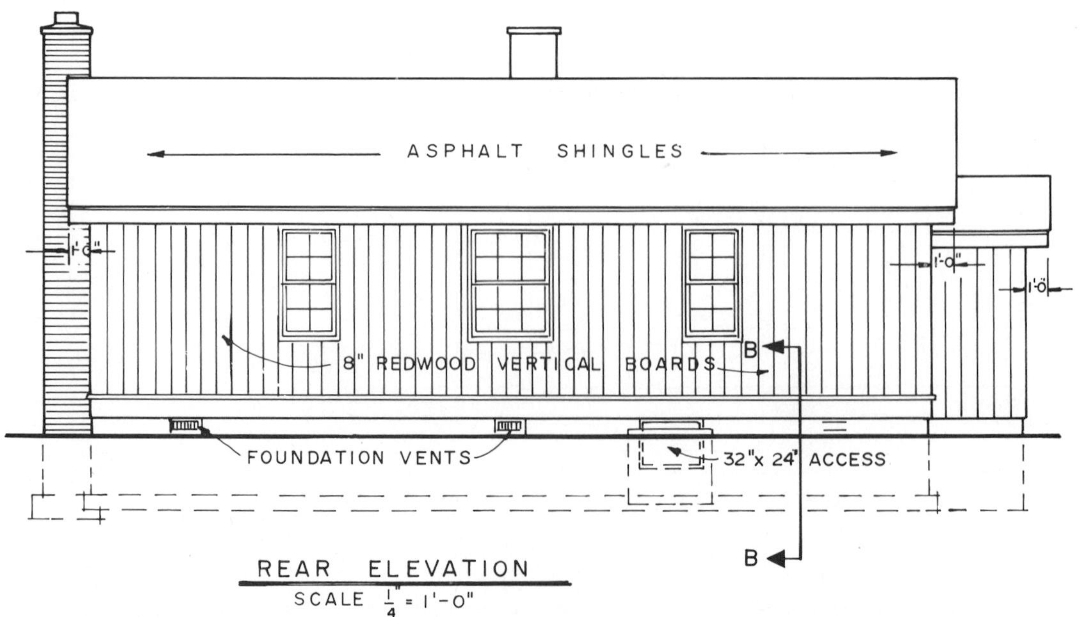

details that the architect believes are needed to construct the building (Fig. 6-27).

READING FRAMING PLANS

Some sets of construction drawings have *framing plans* (Fig. 6-28). These show the location of each wood structural member. One plan shows floor joists and another shows ceiling joists. Roof framing is also detailed. Occasionally, the framing for each wall is drawn. Structural members are indicated by a simple line. The sizes and spacing of structural members are noted.

READING A PLOT PLAN

The *plot plan* locates the building on the site (Fig. 6-29). The boundaries of the lot are located. The direction and length of each side is recorded. The overall dimensions of the building are given. Its distance for the edges of the lot are noted.

FIGURE 6-24 *Continued*

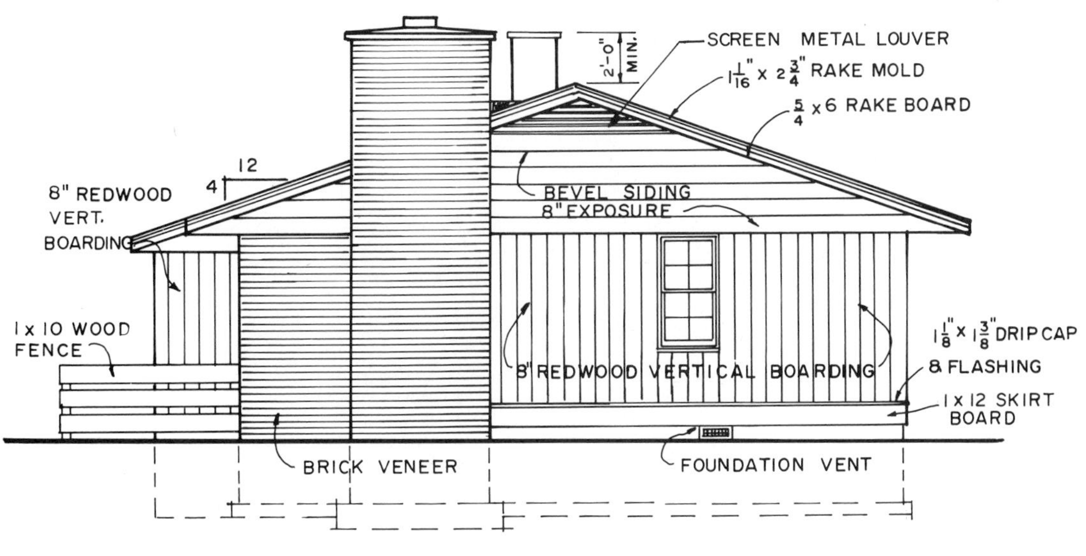

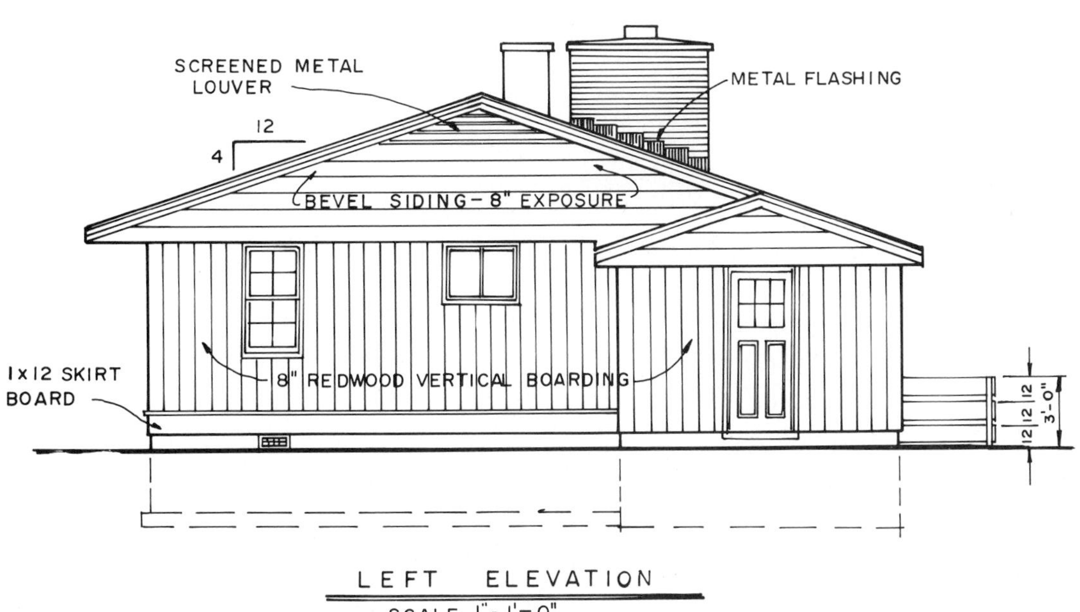

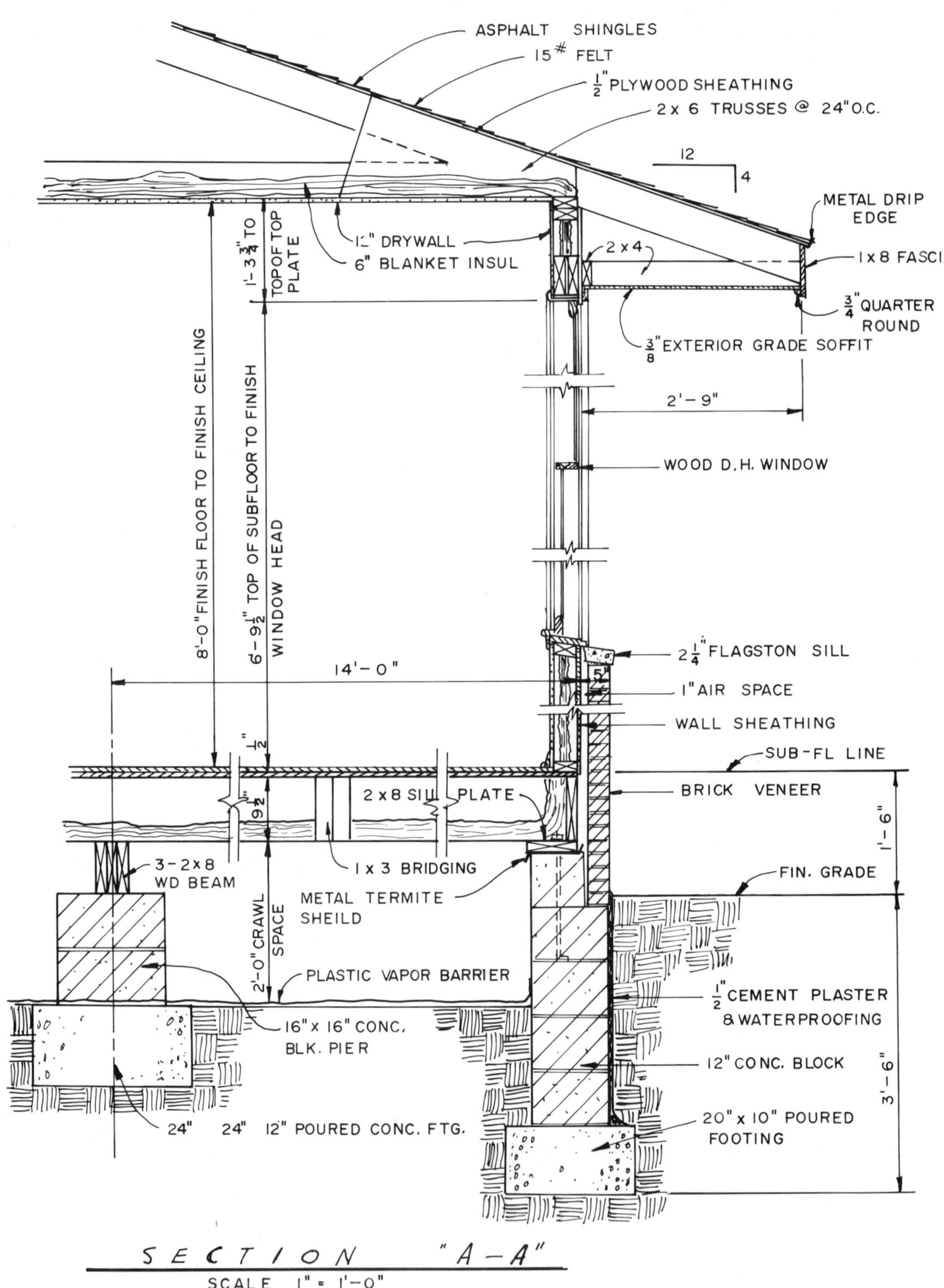

FIGURE 6-25

A typical wall section for house with crawl space.
(Copyright, Home Planners, Inc., Detroit, Mich.)

Often, a plot plan will also show location and sizes of sidewalks, driveway, detached buildings, swimming pool, and other such items. The contour of the land is recorded. Contour refers to the height of the land and how it slopes. Parts of the site that are the same height are connected with a line. The existing contours are shown with a continuous solid line. After the site is graded, new contours are established. These are shown with a dashed line.

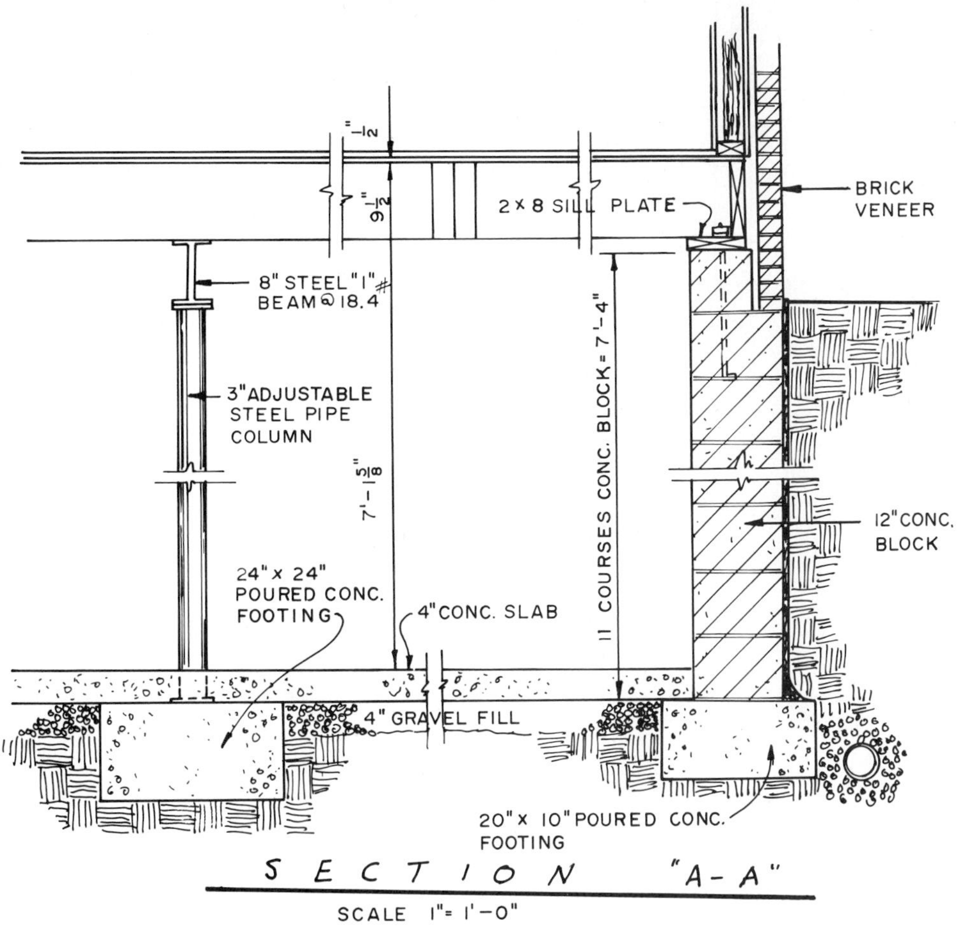

FIGURE 6-26

A section through the foundation and basement. *(Copyright, Home Planners, Inc., Detroit, Mich.)*

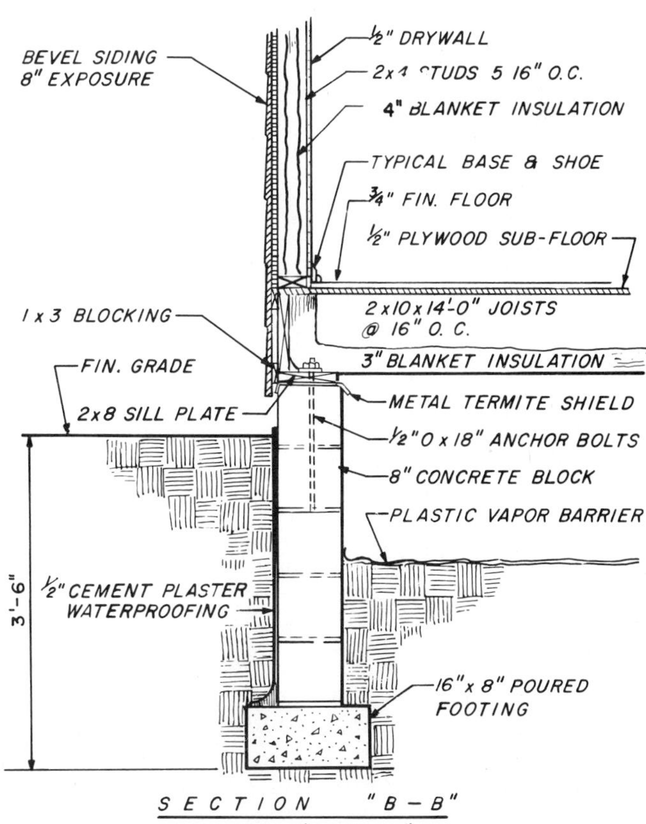

FIGURE 6-27

A section through a foundation. *(Copyright, Home Planners, Inc., Detroit, Mich.)*

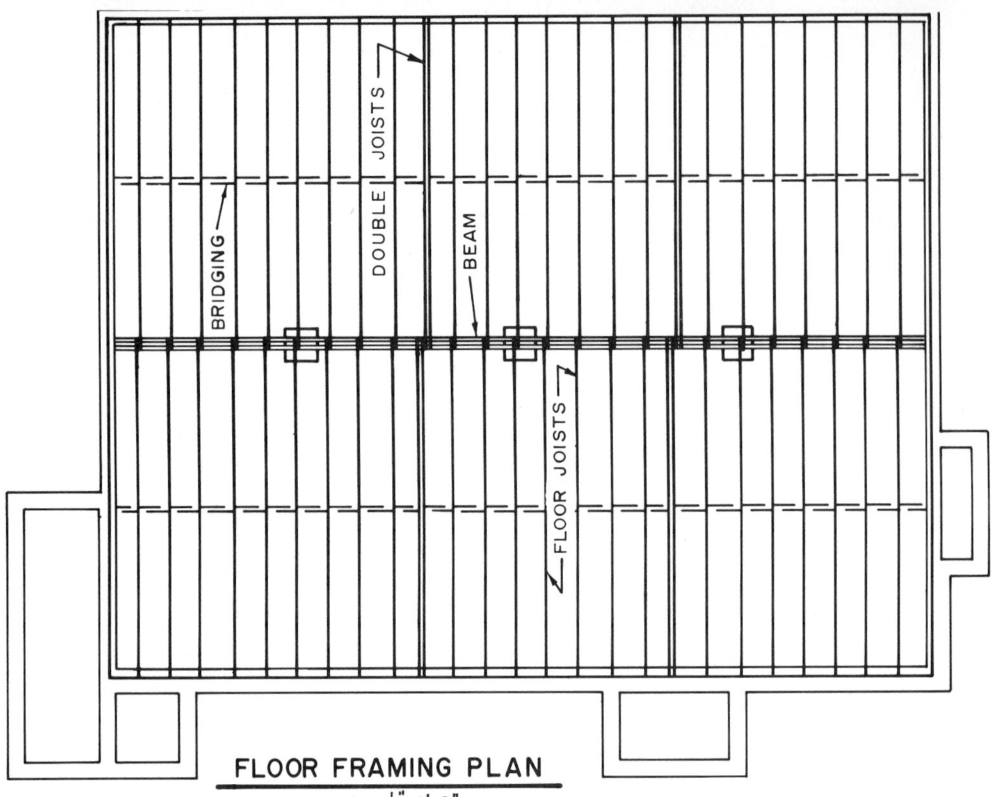

FIGURE 6-28

A floor framing plan. *(Copyright, Home Planners, Inc., Detroit, Mich.)*

OTHER DETAILS

Drawings such as fireplace elevations and sections and cabinet details are drawn (Figs. 6-30 and 6-31). These give construction details and actual sizes.

FIGURE 6-29

A plot plan dimensioned in feet.

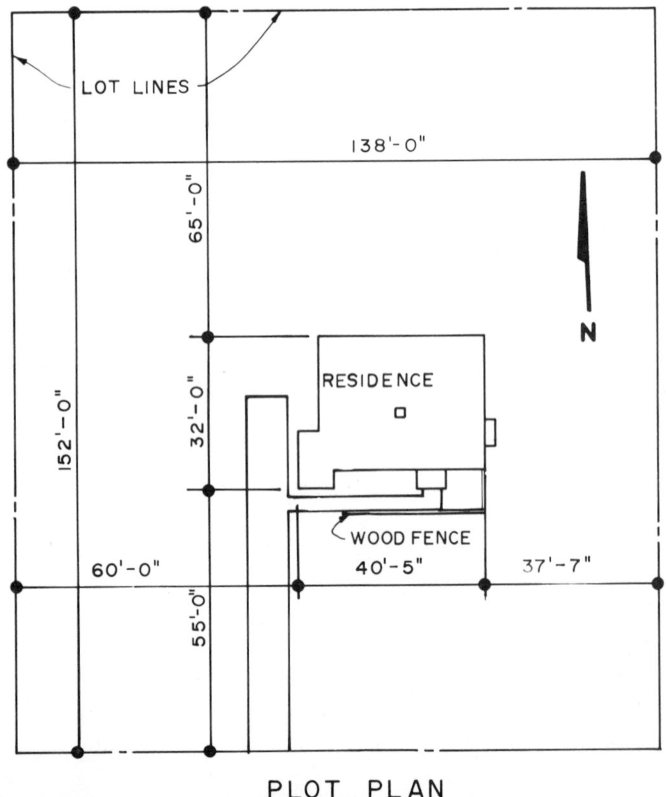

SPECIFICATIONS

The architect writes specifications to explain the many things not shown on the drawings. Before constructing a building, a carpenter should study these carefully. A typical set of specifications for residential construction would have the following sections:

1. General and special conditions
2. Excavation and grading
3. Concrete
4. Masonry
5. Carpentry and millwork
6. Interior finish
7. Sheet metal
8. Painting and finishing
9. Electrical
10. Plumbing
11. Heating and air conditioning

BUILDING CODES

A *building code* is a series of laws accepted by a governmental unit, such as a city, which indicates the minimum acceptable construction

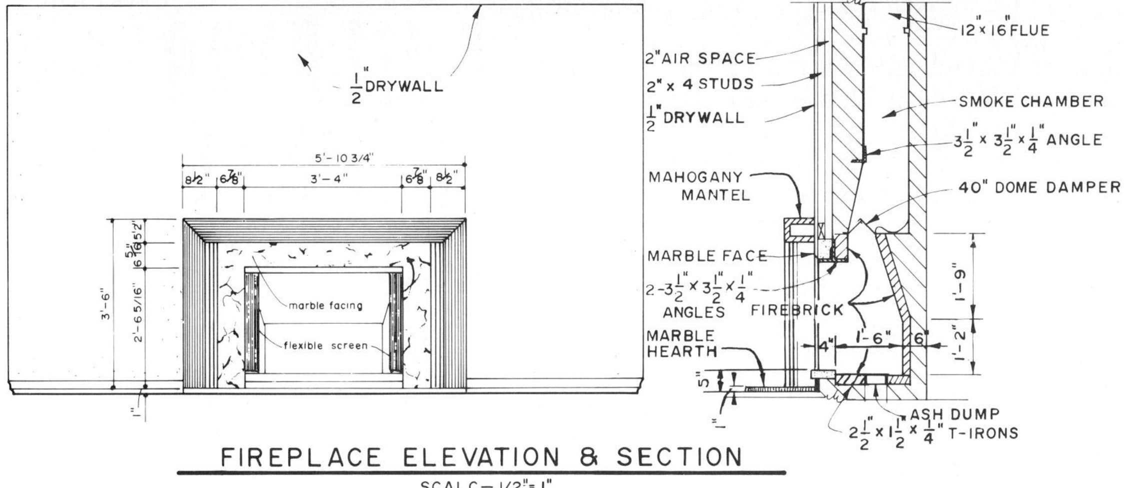

FIGURE 6-30 *(above)*

Typical fireplace construction details. *(Copyright, Home Planners, Inc., Detroit, Mich.)*

FIGURE 6-31 *(below)*

Elevation showing the cabinets and fixtures in the kitchen. *(Copyright, Home Planners, Inc., Detroit, Mich.)*

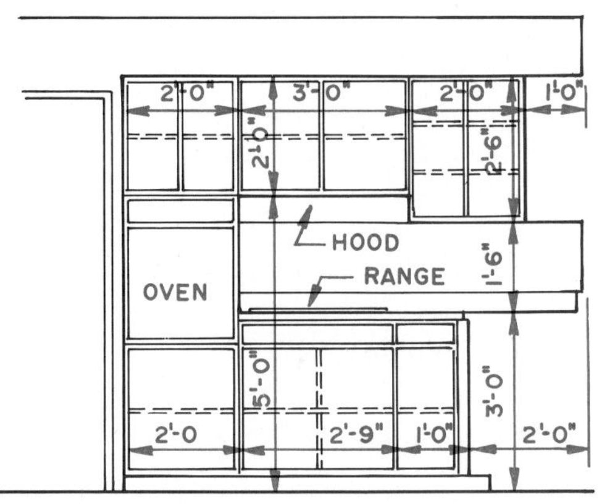

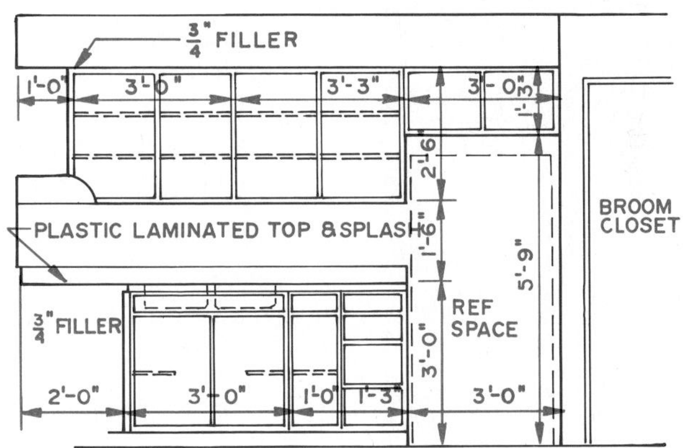

standards. A typical code will include the following sections:

1. Height and area restrictions
2. Light and ventilation
3. Means of entry
4. Requirements for various types of construction
5. Fire protection requirements
6. Design loads
7. Chimneys, flues, and vents
8. Heating, air conditioning, ventilation
9. Construction safeguards
10. Gas piping and plumbing
11. Electrical installation

There are several national organizations that prepare and update building codes. Many cities cannot keep up with the changes and new developments in construction materials and techniques. They cannot afford the cost of constantly researching and revising building codes of their own design. Therefore, many cities adopt for their use one of these national codes. Sometimes, the local situation requires special laws in addition to the national code. For example, in an area where hurricanes prevail, special bracing is needed. If earthquakes are a possibility, the locality may adopt their own special codes in addition to the national general codes.

Carpenters must be aware of local codes and perform their work accordingly.

BUILDING PERMITS

Before a contractor can begin construction, the local governmental agency, such as a city, must issue a *building permit*. The owner of the building must submit drawings and specifications to the building inspector. These are reviewed to see that they meet local building codes. The placement of the building on the lot is also studied. If the building meets local standards, a permit to build is issued (Fig. 6-32). This is posted on the site in a prominent place.

As the building is constructed, the building inspector visits the site and approves the work or requests changes. When the work is performed satisfactorily, the inspector prepares a signed record of that fact. Carpenters and others working on the building must remember that parts of the building require inspection before they are covered by other work. For example, before covering the plumbing with drywall or cabinets, the carpenter must be certain that the inspector has seen and approved the plumbing installation.

When the building is finished, the building inspector makes a final inspection. The building cannot be occupied before this final inspection.

BUILDING PERMIT

Type Construction: ALTER & ADDITION Street Address: 1323 E. QUINCY

File No.: 9547

Permit No.: 316 Date of Issue: 8-22-78

Footing Inspection: ______ Miscellaneous: ______

Plumbing: Electrical Work: Gas: ______

Rough In ______ Final ______ Sewer ______

Rough In ______ Final ______ Water ______

FIGURE 6-32

A typical building permit.

STUDY QUESTIONS

After reading the chapter, answer each of the following questions. If you do not know the answer, review the chapter.

1. What items are found on a set of construction drawings?
2. What are the specifications for a building?
3. What is the basic metric unit of length?
4. How many millimeters are in a meter?
5. How do you change 3.751 meters to millimeters?
6. What is the metric unit of mass?
7. How is volume measured in the metric system?
8. How is area measured in the metric system?
9. What are the customary and metric scales commonly used to produce residential floor plans?
10. What is shown on the foundation plan?
11. List the sections found in the specifications.
12. What is a building code?
13. Why are building permits required?

IMPORTANT TECHNICAL TERMS

Following are technical terms that you should be able to use as part of your working vocabulary. Write a brief description of the meaning of each term.

Construction drawings
Specifications
Meter
Millimeter
Cubic meter
Square meter
Square kilometer
Scale
Architectural symbols
Floor plan
Foundation plan
Elevation
Sections
Framing plans
Plot plan
Building codes
Building permit

7 WOOD—A BASIC CONSTRUCTION MATERIAL

Wood is a renewable natural resource. It is the most commonly used material for residential construction. Wood is easily cut, shaped, and assembled. It is strong yet light in weight. It has a natural beauty due to the color and grain. The insulation value of wood is good. The composition of wood is such that it can be used as a raw material for manufacturing a variety of products.

The basic use for wood in residential construction is for structural framing and finishing applications. Some species of trees produce wood that is stronger than others. Some produce wood that is heavier or more attractive. Some woods naturally resist rot and decay even when exposed to considerable moisture. The hardness of various species of wood varies considerably.

A carpenter should be familiar with the major species of wood used in construction. He needs to have a knowledge of the characteristics of each and how lumber is graded and specified.

THE STRUCTURE OF WOOD

The structure of wood is related to how a tree grows and the parts of the tree.

A tree has three main parts: the crown, trunk, and roots (Fig. 7-1). The growing parts are the leaves, root tips, and cambium layer just below the bark. The roots absorb water, which moves to the leaves through the sapwood. The leaves contain a green pigment called chlorophyll. Leaves use the energy of the sun to combine with carbon dioxide from the air and water from the roots to produce a food called carbohydrate. This is carried to the various living parts of the tree by the inner bark.

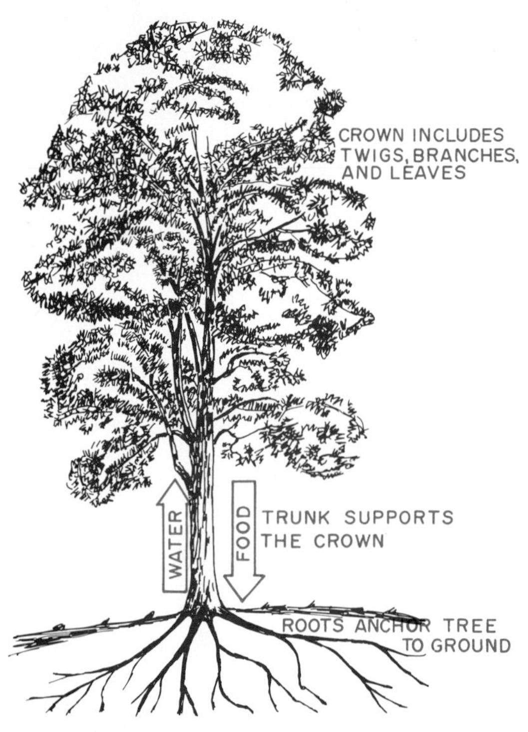

FIGURE 7-1

A tree has three main parts: the crown, trunk, and roots.

FIGURE 7-2

A sawed log showing the hard and soft growth rings. *(Courtesy of Forest Products Laboratory)*

New cells are formed on the inside of the cambium, producing wood. The cells formed on the outside of the cambium produce bark.

In the spring the cambium produces rapid growth. The growth slows in the summer. The rapid spring growth produces large, thin-walled cells. These form the soft growth rings. The summer growth produces darker, thicker-walled cells, which form the dark growth rings. The age of a tree can be ascertained rather closely by counting these rings. The grain pattern in lumber is caused when the saws cut through the growth rings (Fig. 7-2).

The parts of a tree in cross section are shown in Fig. 7-3. The *pith* is a soft center. The *heartwood* is the mature wood. It is no longer growing. It is the color characteristic of the species. The *sapwood* is the light-colored wood between the heartwood and the cambium. It carries the sap from the roots to the leaves. It is the growing area of the tree. As the older rings of sapwood mature, they become heartwood. The *cambium* layer forms the cells for sapwood and bark. It is responsible for all new growth. The *inner bark* is a thin layer just beneath the bark. It carries the food from the leaves to the growing parts of the tree. The *bark* is the outer layer of the tree. It is a layer of dead materials. Bark serves to protect the living layers directly beneath it. The *medullary rays* connect the various layers of the tree. They run from the pith to the bark. They serve as a means of moving food across the section of the tree.

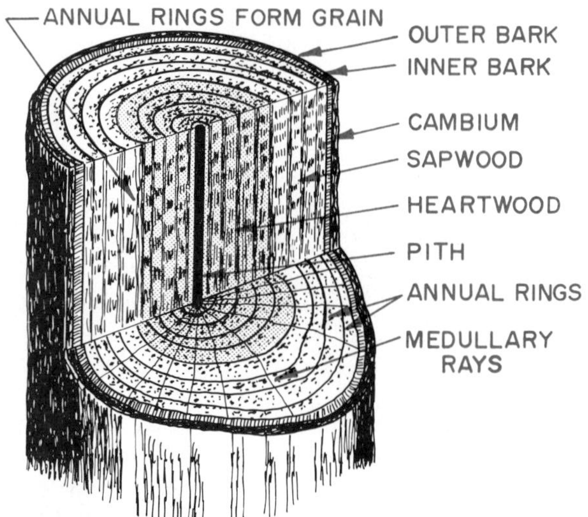

FIGURE 7-3

The parts of a tree trunk.

SPECIES

There are hundreds of species of trees in American forests. Of these, some find extensive use in the construction industry.

The various species are divided into two broad groups: hardwoods and softwoods (Fig. 7-4). The *hardwoods* have broad leaves which are dropped in the fall. Typical examples include oak, maple, walnut, gum, and poplar. The *softwoods* have cones and/or needle-like leaves. They remain green the year round. Typical examples are pine, fir, cedar, and redwood.

SOFTWOODS	HARDWOODS
Alaska cedar	Black walnut
Incense cedar	Mahogany
Port Orford cedar	Philippine mahogany
Eastern red cedar	Sugar maple
Western red cedar	Yellow birch
Northern white cedar	Black cherry
Southern white cedar	White oak
Cypress	Red oak
Balsam fir	Yellow poplar
Douglas fir	Willow
Noble fir	American elm
White fir	Sweet gum
Eastern hemlock	White ash
Mountain hemlock	American beech
West coast hemlock	Cottonwood
Western juniper	Rock elm
Western larch	Hickory
Jack pine	Basswood
Lodgepole pine	
Norway pine	
Ponderosa pine	
Sugar pine	
Idaho white pine	
Northern white pine	
Longleaf yellow pine	
Shortleaf pine	
Eastern spruce	
Englemann spruce	
Sitka spruce	
Tamarack	
Pacific yew	
Redwood	

FIGURE 7-4 *(above)*

Selected species of hardwood and softwood.

The terms "hardwood" and "softwood" do not refer to the actual hardness of the wood. Some hardwoods are actually softer than some softwoods.

Construction framing and finishing lumber is primarily from trees classified as softwoods. Some of the most used softwoods are Douglas fir, white fir, southern yellow pine, ponderosa pine, white pine, sugar pine, western larch, spruce, western red cedar, redwood, and cypress (Fig. 7-5).

Cabinets, some trim, and built-in furniture are often made from hardwoods. Common hardwoods include walnut, oak, cherry, birch, maple, beech, gum, and mahogany.

MANUFACTURING LUMBER

Logs are cut into lumber in two ways: plain sawing and quarter sawing (Fig. 7-6). To *plain-saw* lumber, the log is first squared and then the boards are cut tangent to the growth rings. Plain-sawed lumber is cheaper and the defects go through fewer boards. Plain sawing produces wider boards which are easier to kiln-dry.

FIGURE 7-5 *(below)*

Grain patterns of commonly used soft woods.

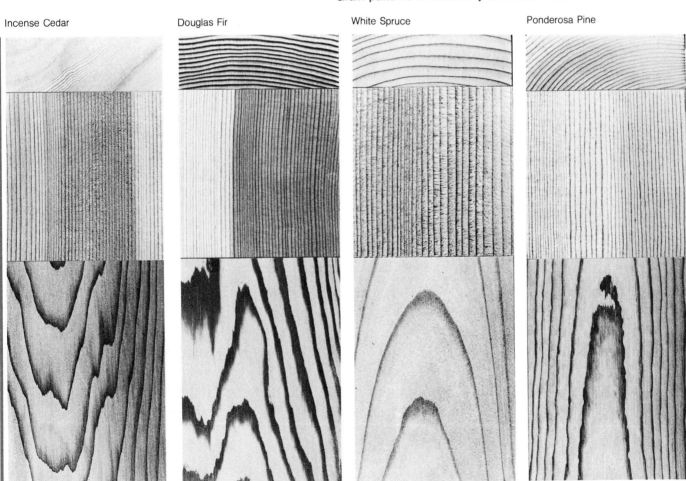

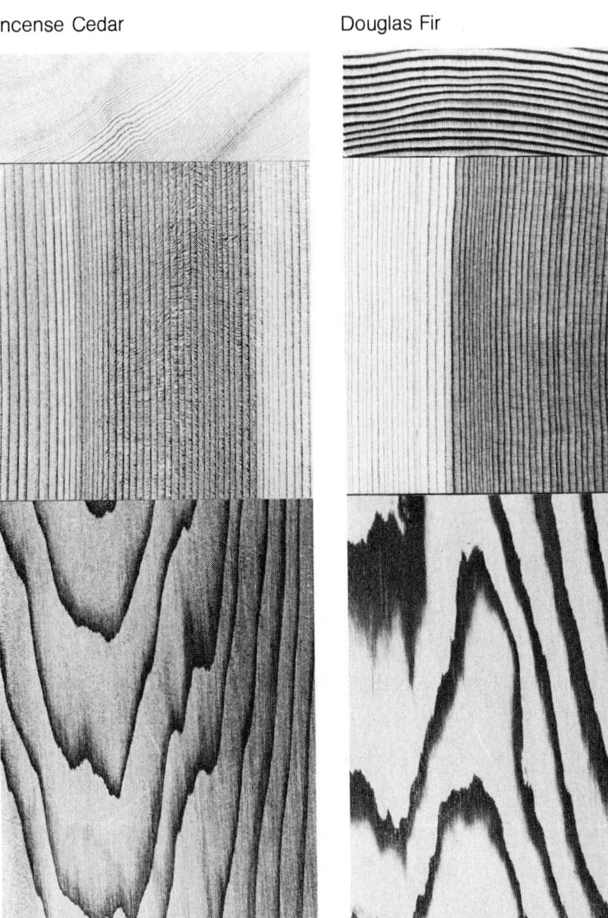

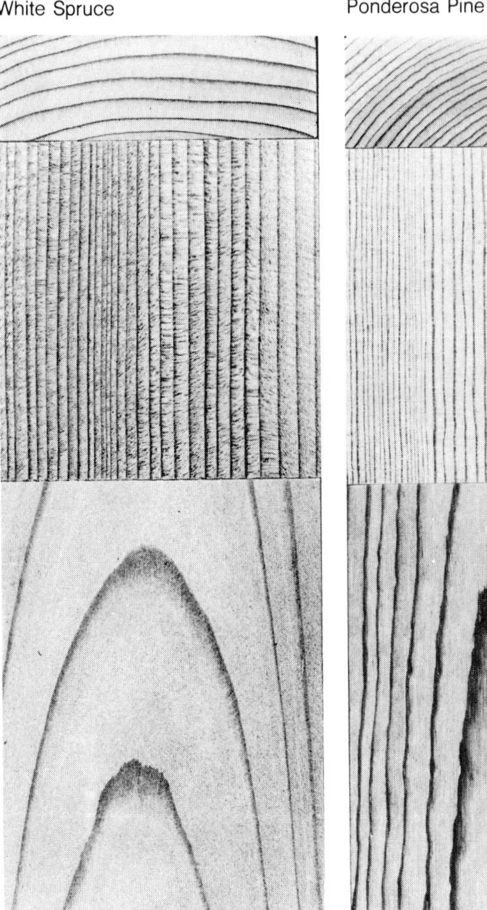

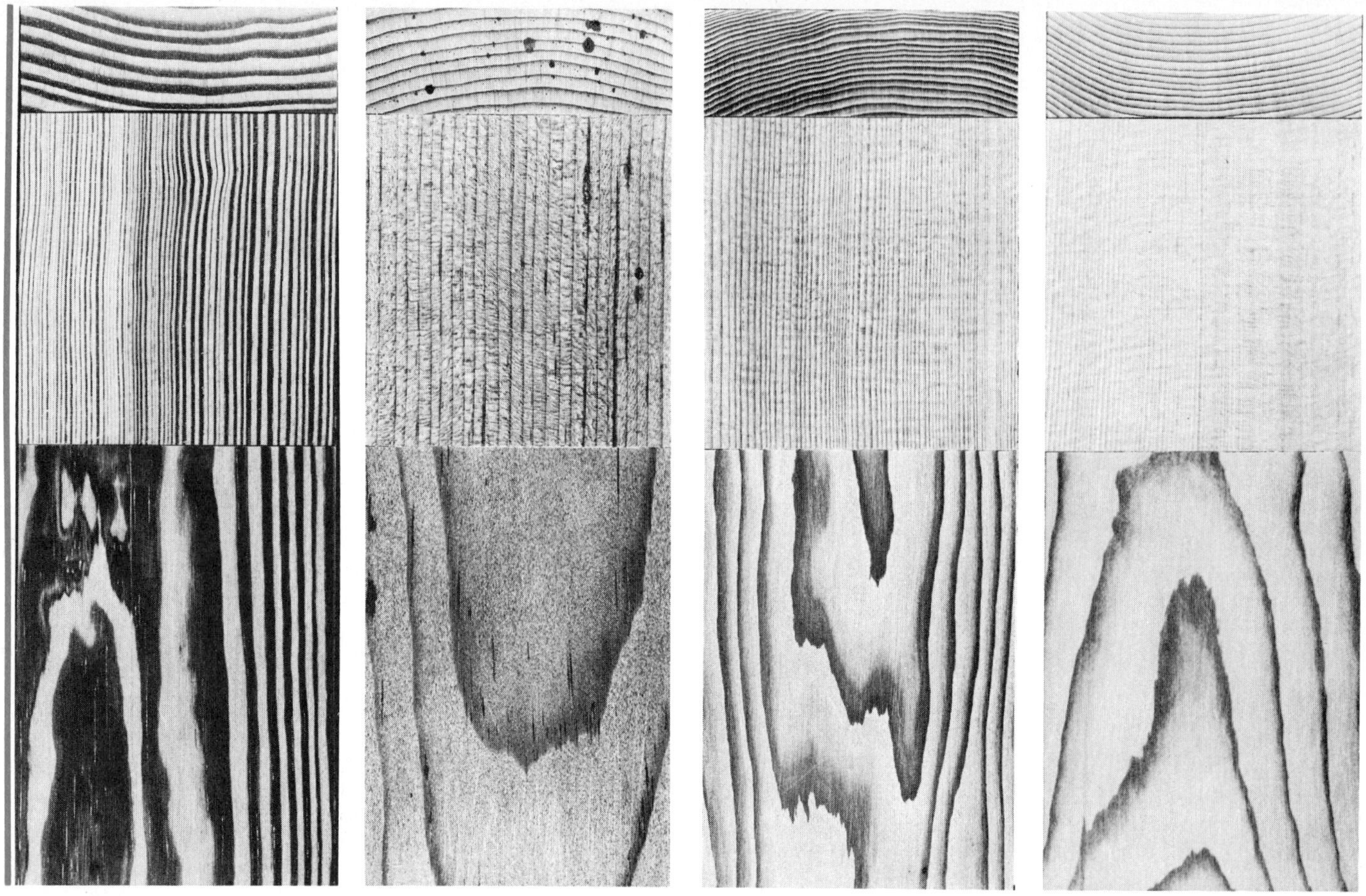

Shortleaf Pine

Sugar Pine

Western Hemlock

Engelmann Spruce

FIGURE 7-5 *Continued*

Western White Pine

Western Red Cedar

Redwood

Sitka Spruce

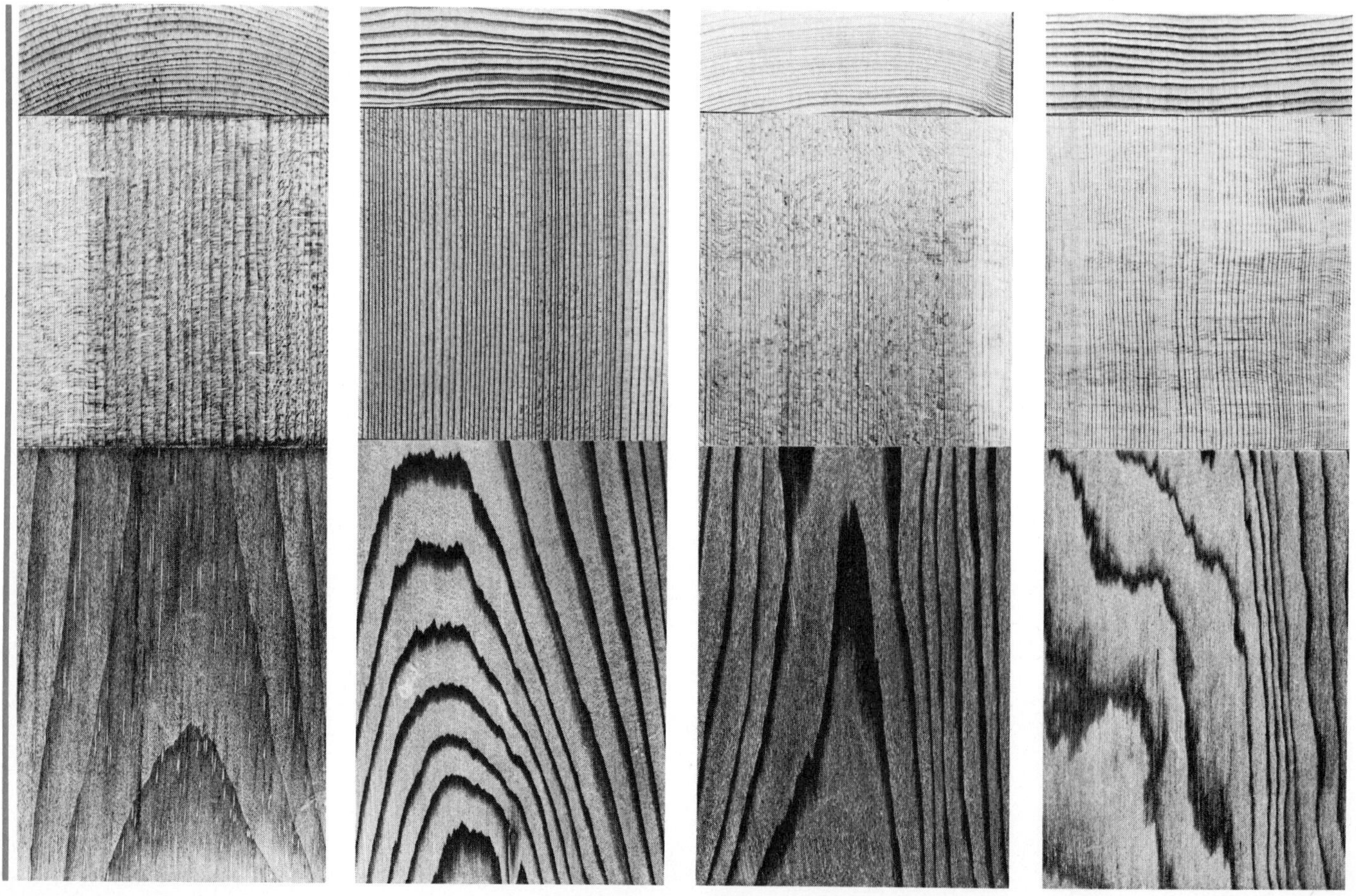

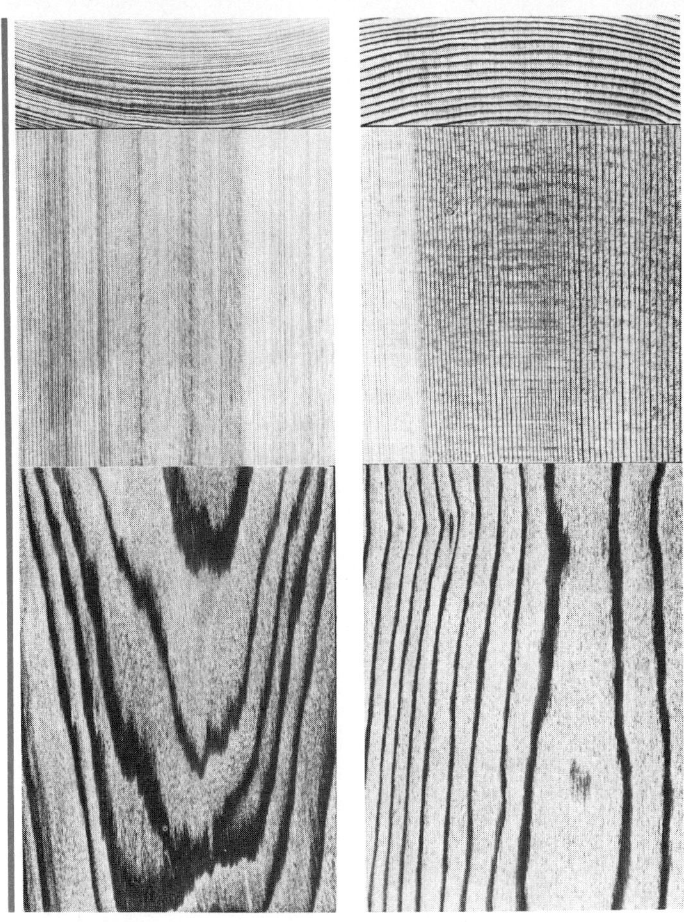

Bald Cypress

Western Larch

FIGURE 7-5 *Continued*

Plain-sawed lumber warps and shrinks more than does quarter-sawed.

Quarter-sawed lumber is produced by first cutting the log into quarters. Each of these is cut into boards. The growth rings are cut so that they form an angle of 45 to 90° with the surface. Quarter-sawed boards will resist wear because the edges of the hard-growth rings are exposed. It resists warping, twisting, and shrinking.

When a sawyer begins to cut lumber from a log, there are various ways to do it. Two typical sawing sequences are shown in Figs. 7-7 and 7-8.

In Fig. 7-7 the sawyer opened the log at side 1 and sawed off five thicknesses. Then the log was turned and five cuts were made on side 2. The log was rotated to side 3 and eight cuts were made. The log was again turned and five cuts were made on side 4. The log was then a 12-in. square. The sawyer can cut this further or send the log to other saws. There the square log could be cut as shown. The decision depends upon the sizes needed by waiting customers.

The processing of small logs is shown in Fig. 7-8. The whole log passes in one motion through a series of band or circular saws and wood chippers. These reduce it to 2- and 4-in. pieces. These pieces are turned flat and moved

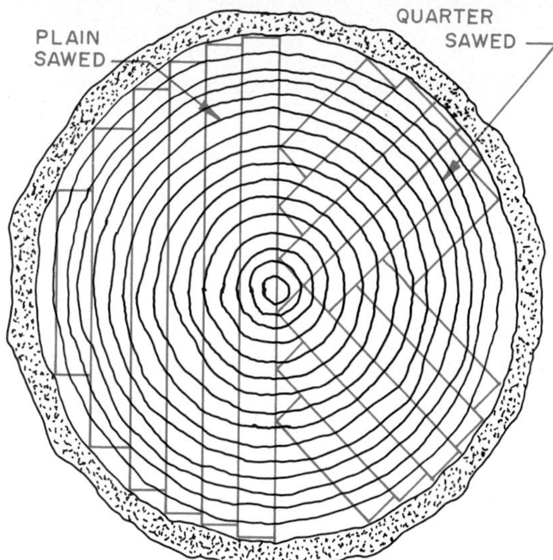

FIGURE 7-6

Lumber can be plain-sawed or quarter-sawed.

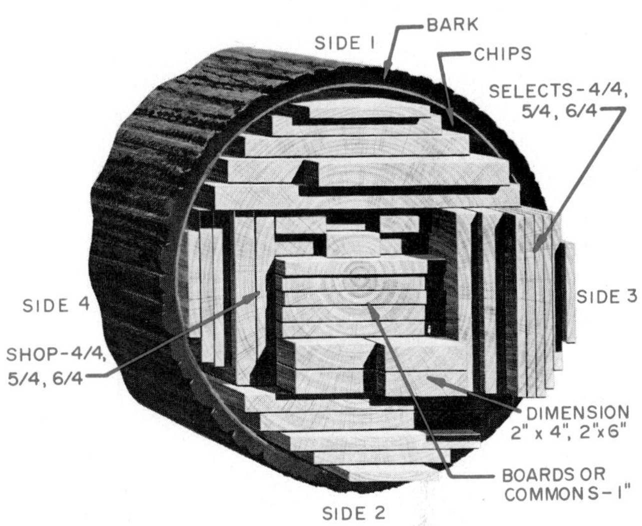

FIGURE 7-7

One sequence a sawyer might use when sawing a log into lumber. *(Courtesy of Western Wood Products Association)*

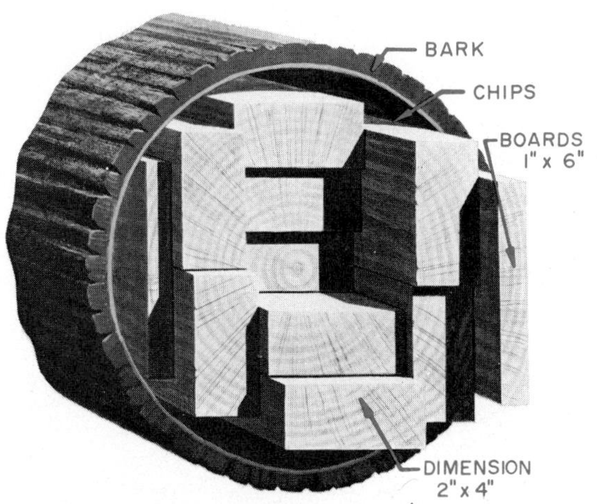

FIGURE 7-8

This is the way a sawyer would cut a small log. *(Courtesy of Western Wood Products Association)*

through another series of saws, which cuts them into 2 × 2, 2 × 3, and 2 × 4 in. sizes. Pieces 10 ft or shorter may be graded and become studs. One-inch boards may be cut from the outside cuts of the log.

SEASONING

After a tree is cut and sawed into lumber, much of the moisture in it must be removed. Often up to half the weight of a freshly sawed board is due to moisture.

For general construction purposes lumber should have a moisture content of 12 to 15%. For furniture, cabinets, and interior trim, the moisture should be 7 to 10%.

There are two ways to season lumber: air drying and kiln drying. *Air drying* is done by stacking the lumber out-of-doors with each layer separated from the next by laying narrow strips of wood between them. This permits the air to circulate between the layers of wood (Fig. 7-9). The length of time it takes to air dry lumber varies according to the species and climate. Usually, a period of 2 to 3 months is needed. Air-dried lumber can usually have the moisture content reduced to 15 to 20%.

Kiln-dried lumber is produced by stacking air-dried lumber on carts with sticks between layers. The carts are rolled into ovens called kilns (Fig. 7-10). Here the moisture, temperature, and air flow are carefully controlled. Steam is run into the kiln and saturates the wood. This replaces moisture lost during air drying. It helps prevent surface checking. Then the steam is reduced and heated air is passed over it. This removes the moisture from the wood. For general

FIGURE 7-9

Lumber stacked and air drying. *(Courtesy of Western Wood Products Association)*

FIGURE 7-10

Typical dry kilns used to dry lumber. *(Courtesy of Western Wood Products Association)*

construction use, lumber can be kiln dried in 3 to 5 days. After it is kiln dried, construction lumber has a preservative applied to the ends to prevent splitting.

As wood is seasoned, defects will appear. It might warp, crack, cup, or have surface cracks. Careful drying helps reduce this, but cannot totally prevent it.

Wood shrinks as it dries. It shrinks the most along the direction of the annual rings (width and thickness). It shrinks very little in length.

After a building is finished, the wood will continue to dry. In very dry climates, as in the Southwest, it will finally reach about 6% moisture. In most of the United States, the wood will dry to about 8%. In the wet Southeast, it will stabilize at about 11%. Since this is drier than when the house was built, some cracking is possible because of the shrinkage.

DETERMINING MOISTURE CONTENT

Moisture content is expressed as a percentage of the weight of the wood. The lower the percentage, the drier the wood.

A carpenter can check the moisture content with a moisture meter. A moisture meter will give a rapid means for getting the approximate moisture content. Meters are usually accurate to within 1% of the actual moisture content.

There are two types of moisture meters. One has two needles which are pushed into the wood. It measures the electrical resistance of the current flow through the wood between the two needles (Fig. 7-11).

The other type measures the relationship between moisture content and a fixed setting. It uses metal plates which are pressed against the surface of the wood.

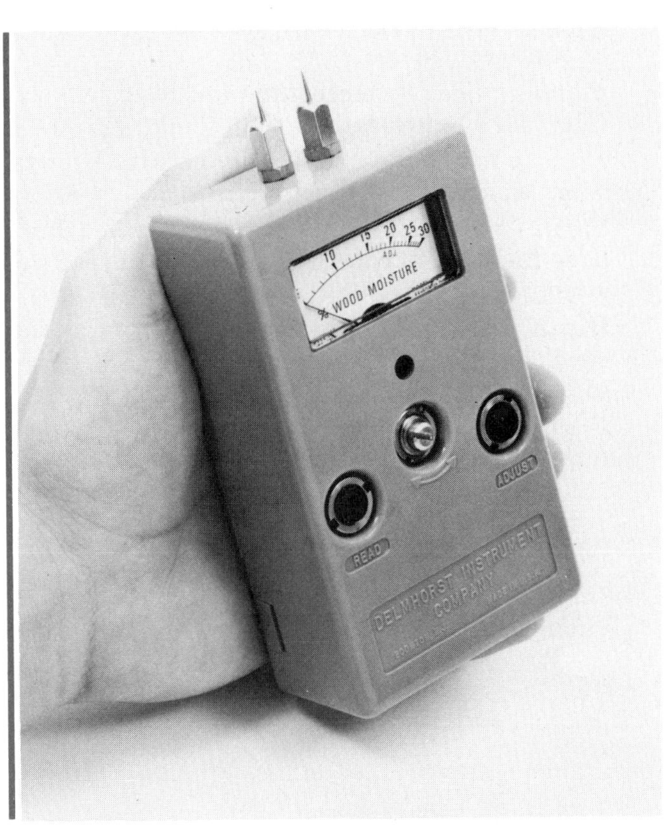

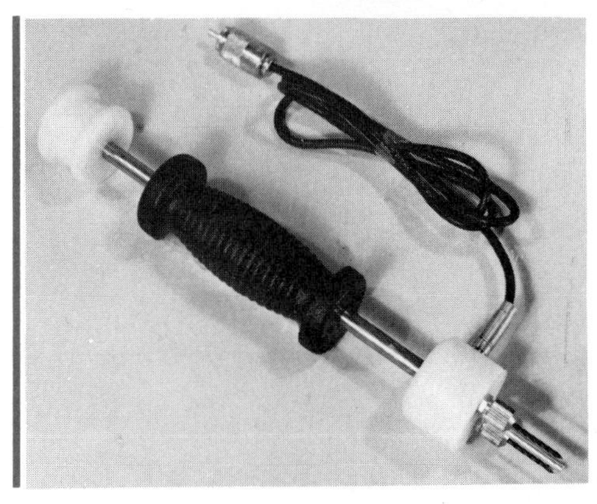

FIGURE 7-11

Moisture meters with probes. *(Courtesy of Delmhorst Instrument Co.)*

Boards need to be checked in several places along their length. The moisture could vary in the length of the board. If it is a rainy day, the meters will not register accurately.

If lumber is improperly stored, it can absorb moisture, thus increasing its moisture content.

DEFECTS IN LUMBER

Defects in lumber reduce its strength and appearance. Often, they can render a board useless. Defects are considered as lumber is graded. Grading is explained in the next section. The defects are described in rules used to grade lumber. Some frequently listed defects and descriptions follow:

Warp variations from a straight, true plane surface. It includes bow, crook, cup twist, and combinations of these (Fig. 7-12).

Bow deviations flatwise from a straight line from end to end.

Crook deviations edgewise from a straight line from end to end.

Cup deviations flatwise from a straight line across the width of the piece.

Check separations of the grain fibers running lengthwise. The cracks go into the board but not through it (Fig. 7-13).

Splits separations of the grain fibers running lengthwise which go completely through the board (Fig. 7-13).

Shakes a lengthwise grain separation between or through the annual rings.

Knot a growth defect caused by branches. The branches embedded in the trunk are cut when the boards are sawed (Fig. 7-14). When grading lumber, the size, location, and firmness of the knot are considered.

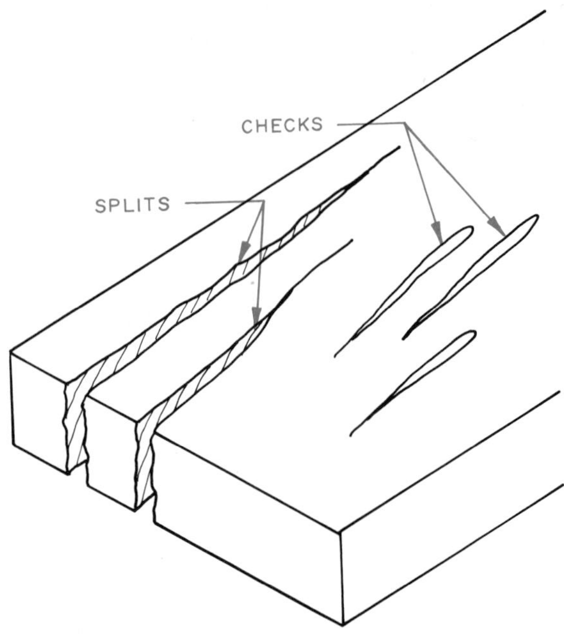

FIGURE 7-13

Splits and checks are separations in the grain.

Pitch a heavy concentration of resin in wood cells in a single location.

Pitch pocket an open cavity between growth rings which has filled with pitch.

Stain a discoloration of the board which damages the natural color. It does not influence the strength.

Wane presence of bark or a lack of wood on the edge of a manufactured board. It causes the edge to be rounded (Fig. 7-15).

Decay spongy, soft, crumbling wood due to fungi.

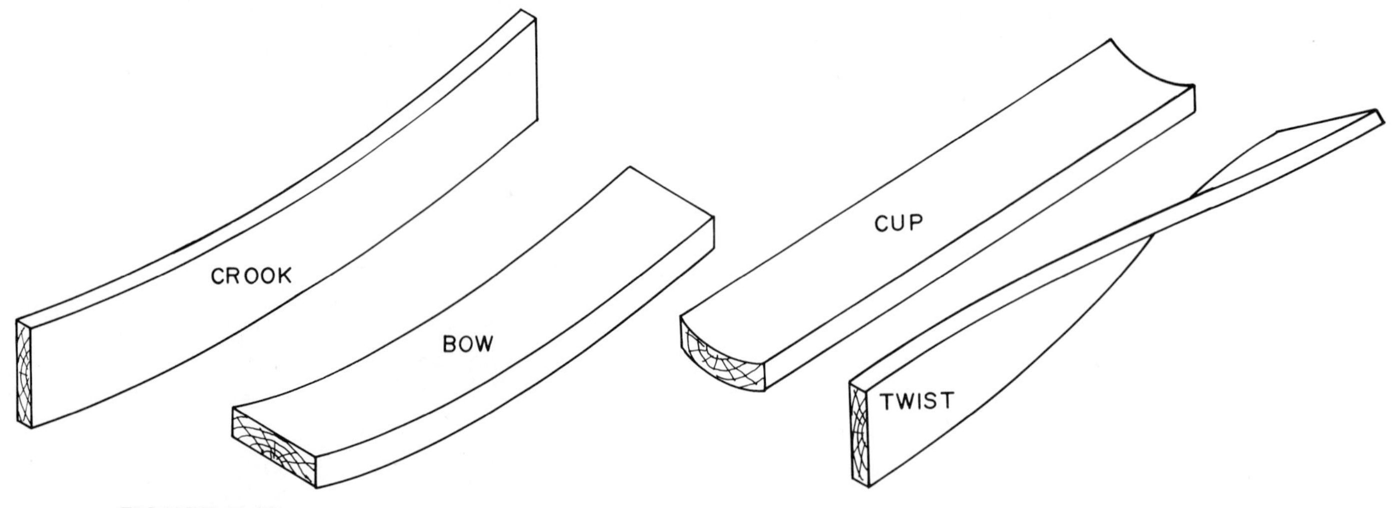

FIGURE 7-12

Ways boards can warp.

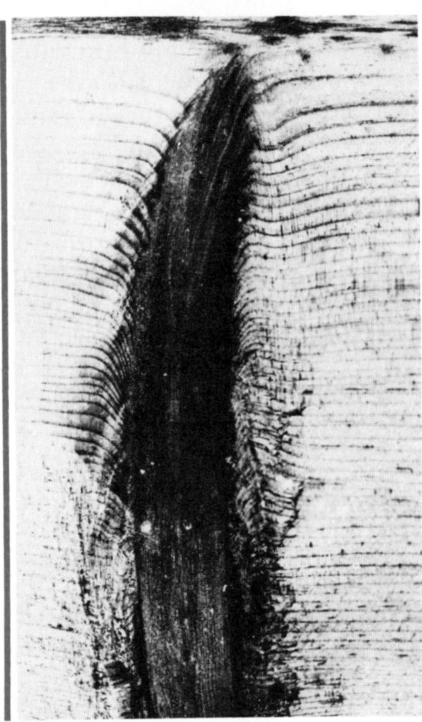

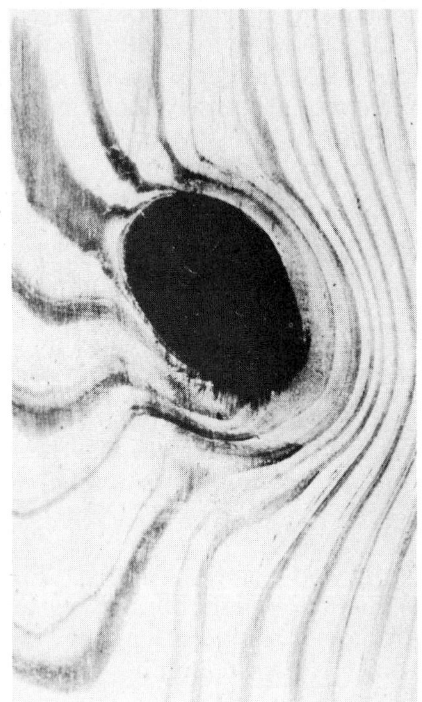

FIGURE 7-14

Most common types of knots: (1) a spike, (2) intergrown (usually solid), and (3) encased (often loose). *(Courtesy of Forest Products Laboratory)*

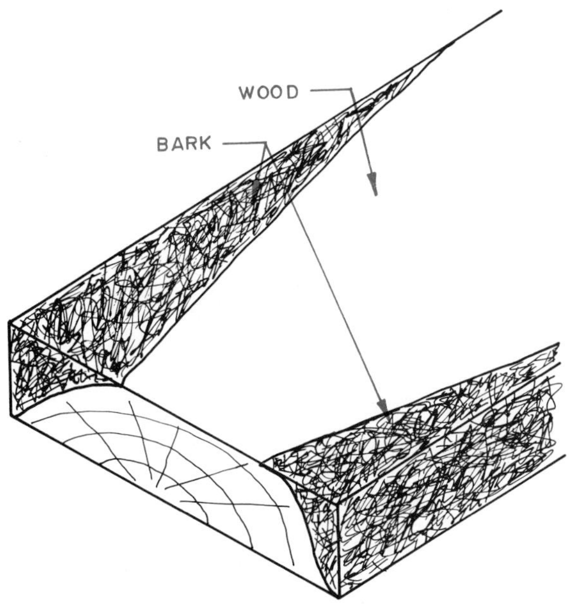

FIGURE 7-15

Wane is the presence of bark or lack of wood on the edges of a board.

SOFTWOOD LUMBER GRADING

The American Softwood Lumber Standard, PS20-70, was developed by the U.S. Department of Commerce. PS20-70 provides a National Grading Rule which specifies grade names and sizes. Grading rules are established by various associations of lumber producing companies. These use the National Grading Rule as a base. However, there are many types of lumber not covered in the National Grading Rule. The associations establishing grading rules include the Southern Pine Inspection Bureau, Western Wood Products Association, California Redwood Association, and others.

Softwood Grading Classifications Following are some of the grading classifications for softwoods commonly used in construction.

Finish lumber is used where high quality is required. It can be stained or finished natural.

Board-grade lumber is commonly used for sheathing, subflooring, shelving, boxing, and crating.

Dimension lumber is surfaced softwood lumber. It is used for joists, rafters, studs, and small timbers. Dimension lumber is divided into the following categories: structural light framing, light framing, studs, structural joists and planks, and appearance framing.

Structural light framing grades are designed to fit those engineered applications where higher-bending-strength ratios are needed.

Light framing grades are designed to provide good appearance at lower design levels for uses where both good appearance and high design levels are not needed.

Studs are suitable for all stud construction purposes.

Structural joists and planks are members that carry a load. This strength varies with the grade.

Appearance framing grade is used where it will be exposed to view. It has high strength and fine appearance.

Timbers are members 5 × 5 in. or larger used to carry loads. There are two categories: non-stress- and stress-rated.

Softwood Lumber Grades Figure 7-16 presents an explanation of the commonly used grades for softwood lumber. These are provided by the National Grading Rule.

The Southern Pine Inspection Bureau formulates and publishes standard grading rules for southern pine lumber. Grading rules are in conformance with the American Softwood Lumber Standards, PS20-70. The bureau provides training for mill lumber graders and assists mills to maintain accuracy in grading lumber. Manufacturers who observe these standards are licensed to use the SPIB grade mark on their lumber.

Examples of Southern Pine Inspection Bureau grade marks are shown in Fig. 7-17. The lumber classification and grades issued by SPIB are in Fig. 7-18.

The Western Wood Products Association provides lumber grading rules and inspection for 12 western states: Arizona, California, Colorado, Idaho, Montana, Nevada, New Mexico, Oregon, South Dakota, Utah, Washington, and Wyoming. Grading rules are in conformance with the American Softwood Lumber Standards PS20-70.

FIGURE 7-16

Southern pine softwood lumber grades. *(Courtesy of Southern Forest Products Association)*

PRODUCT	GRADE	CHARACTER OF GRADE AND TYPICAL USES
Finish	B&B	Highest recognized grade of finish. Generally clear, although a limited number of pin knots permitted. Finest quality for natural or stain finish.
	C	Excellent for painted or natural finish where requirements are less exacting. Reasonably clear but permits limited number of surface checks and small tight knots.
	C&Btr	Combination of B&B and C grades; satisfies requirements for high-quality finish.
	D	Economical, serviceable grade for natural or painted finish.
Boards S4S	No. 1	High quality with good appearance characteristics. Generally sound and tight-knotted. Largest hole permitted is 1/16″. A superior product suitable for wide range of uses, including shelving, form, and crating lumber.
	No. 2	High-quality sheathing material, characterized by tight knots. Generally free of holes.
	No. 3	Good, serviceable sheathing, usable for many applications without waste.
	No. 4	Admit pieces below No. 3 which can be used without waste or contain usable portions at least 24″ in length.
Dimension Structural light framing 2″ to 4″ thick 2″ to 4″ wide	Select Structural Dense Select Structural	High quality, relatively free of characteristics that impair strength or stiffness. Recommended for uses where high strength, stiffness, and good appearance are required.
	No. 1 No. 1 Dense	Provide high strength; recommended for general utility and construction purposes. Good appearance, especially suitable where exposed because of the knot limitations.
	No. 2 No. 2 Dense	Although less restricted than No. 1, suitable for all types of construction. Tight knots.
	No. 3 No. 3 Dense	Assigned design values meet wide range of design requirements. Recommended for general construction purposes where appearance is not a controlling factor. Many pieces included in this grade would qualify as No. 2 except for single limiting characteristic. Provides high-quality, low-cost construction.

PRODUCT	GRADE	CHARACTER OF GRADE AND TYPICAL USES
Studs 2″ to 4″ thick 2″ to 6″ wide 10′ and shorter	Stud	Stringent requirements as to straightness, strength, and stiffness adapt this grade to all stud uses, including load-bearing walls. Crook restricted in 2″ × 4″–8′ to ¼″, with wane restricted to ⅓ of thickness.
Structural joists and planks 2″ to 4″ thick 5″ and wider	Select Structural Dense Select Structural	High quality, relatively free of characteristics that impair strength or stiffness. Recommended for uses where high strength, stiffness, and good appearance are required.
	No. 1 No. 1 Dense	Provide high strength; recommended for general utility and construction purposes. Good appearance; especially suitable where exposed because of the knot limitations.
	No. 2 No. 2 Dense	Although less restricted than No. 1, suitable for all types of construction. Tight knots.
	No. 3 No. 3 Dense	Assigned stress values meet wide range of design requirements. Recommended for general construction purposes where appearance is not a controlling factor. Many pieces included in this grade would qualify as No. 2 except for single limiting characteristic. Provides high-quality, low-cost construction.
Light framing 2″ to 4″ thick 2″ to 4″ wide	Construction	Recommended for general framing puposes. Good appearance, strong, and serviceable.
	Standard	Recommended for same uses as Construction grade, but allows larger defects.
	Utility	Recommended where combination of strength and economy is desired. Excellent for blocking, plates, and bracing.
	Economy	Usable lengths suitable for bracing, blocking, bulkheading, and other utility purposes where strength and appearance not controlling factors.
Appearance framing 2″ to 4″ thick 2″ and wider	Appearance	Designed for uses such as exposed-beam roof systems. Combines strength characteristics of No. 1 with appearance of C&Btr.
Timbers 5″ × 5″ and larger	No. 1 SR No. 1 Dense SR No. 2 SR No. 2 Dense SR	No. 1 and No. 2 are similar in appearance to corresponding grades of 2″ dimension. Recommended for general construction uses. SR in grade name indicates Stress Rated.

FIGURE 7-16 *Continued*

SPIB® C&BTR KD (7)   SPIB® No.2 S-DRY (7)

SPIB® No.1KD (7)   SPIB® KD STUD (7)

FIGURE 7-17
Southern Pine Inspection Bureau grade marks. *(Courtesy of Southern Pine Inspection Bureau)*

CLASSIFICATION	THICKNESS NOMINAL (in.)	WIDTH NOMINAL (in.)	GRADES
Finish	⅜–4	2–16	B&B, C, C&Btr, D
Boards	1–1½	2–12+	No. 1, No. 2, No. 3, No. 4
Structural light framing	2–4	2–4	Select Structural, No. 1, No. 2, No. 3
Light framing	2–4	2–4	Construction, Standard, Utility
Studs	2–4	2–6	Stud
Structural joists and planks	2–4	5 and wider	Select Structural, No. 1, No. 2, No. 3
Appearance framing	2–4	2 and wider	Appearance
Timbers, nonstress	5 and larger	5 and larger	Square-edge and sound, No. 1, No. 2, No. 3
Timbers, stress-rated	5 and larger	5 and larger	No. 1 SR, No. 2 SR

FIGURE 7-18

Lumber classifications, grades, and sizes for southern pine. *(Courtesy of Southern Pine Inspection Bureau)*

CLASSIFICATION	THICKNESS NOMINAL (in.)	WIDTH NOMINAL (in.)	GRADE
Appearance grade—selects	⅜–4	2–16	B and B, C Select, D Select
Appearance grade—finish	⅜–4	2–16	Superior, Prime, E
Boards	1–1½	2–12	No. 1 Common, No. 2 Common, No. 3 Common, No. 4 Common, No. 5 Common
Structural light framing	2–4	2–4	Select Structural, No. 1, No. 2, No. 3
Light framing	2–4	2–4	Construction, Standard, Utility
Studs	2–4	2–6	Stud
Structural joists and planks	2–4	5 and wider	Select Structural, No. 1, No. 2, No. 3
Timbers	5 and thicker	2 or more	Select Structural, No. 1, No. 2, No. 3

FIGURE 7-19

Western softwood lumber classifications, grades, and sizes. *(Courtesy of Western Wood Products Association)*

The most common species involved include Douglas fir-larch, Douglas fir south, hem-fir, mountain hemlock, western hemlock, Englemann spruce, alpine fir, lodgepole pine, ponderosa pine, sugar pine, Idaho white pine, and western cedars.

All dimension lumber graded under Western Wood Products Association rules can have working stresses applied. These are available from the association.

The grades established for boards, dimension stock, and timbers are in Fig. 7-19. The grades are stamped on the lumber using standardized abbreviations. These are shown in Fig. 7-20.

Lumber meeting WWPA standards is stamped with a grade mark. Examples are shown in Fig. 7-21. The work includes the WWPA symbol, the grade, the species, dryness, and specie.

The "dry" symbol can specify three levels. The symbol S-DRY indicates that the moisture content does not exceed 19%. MC-15 indicates a maximum of 15% moisture. S-GRN indicates that the moisture content exceeds 19%.

Machine-stress-rated lumber is marked with a stamp. It gives the modulus of elasticity and includes the word "machine rated" (Fig. 7-22).

GRADE	ABBREVIATION
B and Better	B&B
C and Better	C&Btr
Select Structural	Sel Str
Construction	Const
Standard	Stand
Utility	Util
Appearance	A
Air Seasoned	S-Dry
Kiln Dried	KD
Stress Rated	SR

FIGURE 7-20

Abbreviations used on lumber grading stamps.

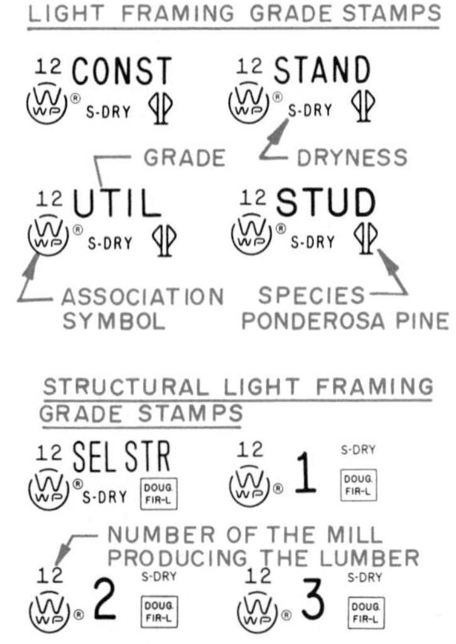

FIGURE 7-21

Western softwood lumber grade marks. *(Courtesy of Western Wood Products Association)*

Grade Stamp Example
Machine Stress-Rated Lumber

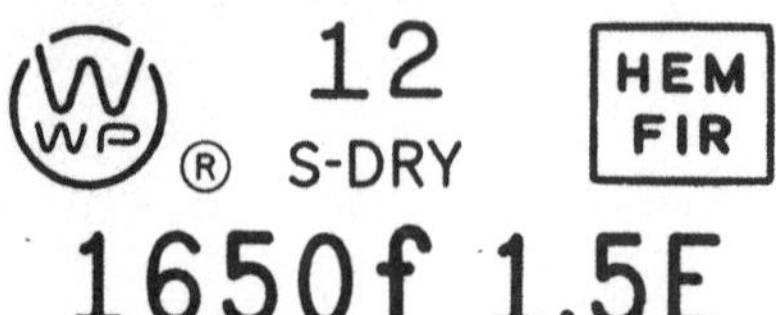

FIGURE 7-22

A typical machine stress-rated lumber mark. *(Courtesy of Western Wood Products Association)*

For complete details on grading, the various associations should be contacted and the National Grading Rule should be secured from the U.S. Department of Commerce.

Seasoning Requirements The National Grading Rule also relates lumber size to moisture content. Separate size schedules have been established for dry and green lumber. This is necessary so that both products will be about the same size when in use. The green lumber is slightly larger, so that when it dries it will be about the same size as dried lumber. Softwood grading rules specify the moisture content limit. Figure 7-23 shows the moisture requirement for the manufacture of southern yellow pine.

Stress Grade Lumber Wood structural members used to carry heavy loads must have their load-carrying capabilities known. This is called their stress rating. Structural members are subject to live and dead loads. *Live loads* are due to the weight of materials placed on the member. This includes things such as flooring, roofing, and furniture. The dead load is the weight of the wood member itself. These forces act on the member, thus putting stress on it. Stress tends to force the member to bend.

Wood structural members can be tested on a machine that measures the modulus of elasticity. The *modulus of elasticity* is a measure of the stiffness of the piece. The amount a structural member may deflect (bend) when under a known load depends upon the size of the member, the load applied, the distance it spans, and the modulus of elasticity. Research has established the modulus of elasticity of each grade in each species of wood used for structural members. The stress rating is marked on each piece of stress-rated lumber (Fig. 7-22).

	MOISTURE CONTENT LIMIT (%)	
ITEMS (Nominal)	Maximum (Dry)	Kiln-Dried (KD or MC15)
D&Btr Grades		
1 and 1¼ in.	15	12 on 90% of pieces, 15 on remainder
1½, 1¾, and 2 in.	18	15
Over 2 in., not over 4 in.	19	15
Over 4 in.	20	18
Paneling, 1 in.		12
Boards 2 in. and less Dimension, 2–4 in.	19	15
Decking		
2 in. thick	19	15
3 and 4 in. thick		15 on 90% of pieces, 18 on remainder
Heavy Dimension, over 2 in., not over 4 in.	19	15
Timbers, 5 in. and thicker	23	20

FIGURE 7-23

Seasoning requirements for southern yellow pine.

Grade Utilization It must be decided which grade of lumber to use for each application in a building. Following are a few recommendations for using southern pine.

Framing

USE	GRADE
Sills, joists, rafters	Utility
Headers	No. 3
Plates	Utility
Studs	Stud grade
Boards	No. 2
Subflooring	No. 3
Wall sheathing	No. 3
Roof, decking, flat	No. 2 KD[a]
Roof sheathing, sloped	No. 3
Exposed decking (3 × 4 in.) (appearance important)	Dense standard DT & G Deck[b]
Industrial roof decking (3 × 4 in.)	Commercial DT & G
Roof trusses (2 to 4 in.), upper and lower chords	No. 2
Roof trusses—other parts 2 × 4 in. stock	No. 3
Roof trusses—5 in. and thicker parts	No. 2 SR[c]
Heavy timber beams, built-up (2 × 4 in.)	No. 2
Heavy timber beams solid (over 5 in.)	No. 2 SR
Posts and columns (2 to 4 in.)	No. 2
Posts and columns (over 5 in.)	No. 2 SR

[a]*KD, kiln-dried.*
[b]*Double tongue and groove.*
[c]*SR, stress-rated.*

HARDWOOD LUMBER GRADES

Grading rules for hardwood lumber are set by the National Hardwood Lumber Association. The grading standards differ slightly for each species of hardwood. Generally, four grades are used: firsts and seconds (FAS), select, No. 1 common, and No. 2 common. Firsts and seconds is the best grade. Pieces should be not less than 6 in. wide and 8 ft long. They should give at least 83⅓% clear cuttings. Selects is a lower grade and permits pieces 4 in. wide and 6 ft long. One face can have more defects than the other.

No. 1 common and No. 2 common should be at least 3 in. wide and 4 ft long. No. 1 common should produce 66⅔% clear cuttings and No. 2 common 50% clear cuttings.

THICKNESS[a] (in.)			WIDTH (in.)		
	Standard ALS Minimum Dressed			Standard ALS Minimum Dressed	
Nominal	Dry	Green	Nominal	Dry	Green
2	1½		2	1½	
2½	2[b]	2⅟₁₆	3	2½	2⁹⁄₁₆[c]
3	2½[b]	2⁹⁄₁₆	4	3½	3⁹⁄₁₆[c]
3½	3[b]	3⅟₁₆	5	4½	4⅝[c]
4	3½[b]	3⁹⁄₁₆	6	5½	5⅝[c]
			8	7¼	7½
			10	9¼	9½
			12	11¼	11½
			14	13¼	13½
			16	15¼	15½
			18	17¼	17½
			20	19¼	19½
5 and thicker	½ off nominal	½ off nominal	5 and wider	½ off nominal	½ off nominal

[a]2″ dressed green thickness of 1⁹⁄₁₆ applies to widths of 14 in. and over.

[b]Not required to be dry unless specified.

[c]These green widths apply to thicknesses of 3 and 4 in. only, except as provided in footnote a.

FIGURE 7-24

Softwood lumber sizes for dressed dimensional and structural lumber.

LUMBER SIZES

Lumber sizes are given in tables in inches. In actual practice carpenters refer to nominal sizes by "quarters." For example, a 1-in.-thick board is referred to as 4/4 (four quarters). A 1½-in. board is 6/4 (six quarters).

Softwood lumber sizes have been standardized. The widths and thicknesses are shown in Figs. 7-24 through 7-27. The lengths are 6, 8, 10, 12, 14, 16, 18, and 20 ft. Although these are recognized standard sizes, they are not all easily available. The demand for some sizes is greater than others; therefore, they are most likely to be available. For example, 1-in. boards are stocked by every lumber dealer, but 1¼-in. boards are difficult to find in many localities.

Hardwoods have no standard widths or lengths. Common thicknesses of rough boards include ¼, ½, 1, 1¼, 1½, 2, 2½, 3, and 4 in. Hardwoods are specified as random widths and lengths with the thickness specified.

Lumber sizes are given in two ways: nominal and actual. The *nominal size* is the rough, unfinished size after the lumber is sawed but before it is dried and dressed. The dressed size is after the lumber is planed smooth. The dressed size is smaller than the rough size. For example, a typical member that is 2 × 4 in. nominal is actually 1½ × 3½ in. dried and dressed. Figure 7-24 shows sizes for structural and dimension lumber. Notice that both dry and green sizes are given. The green size is slightly larger than the dry size. After green lumber is used in a building, it naturally air dries and shrinks. The difference in size makes it close to the same size as dry lumber.

The nominal and dressed sizes of finish lumber and boards are given in Fig. 7-25. These are sizes for dry lumber.

FIGURE 7-25
Softwood lumber sizes for finish lumber and boards.

	THICKNESS (in.)		WIDTH (in.)	
	Nominal	Dressed	Nominal	Dressed
Finish	⅜	⁵⁄₁₆	2	1½
	½	⁷⁄₁₆	3	2½
	⅝	⁹⁄₁₆	4	3½
	¾	⅝	5	4½
	1	¾	6	5½
	1¼	1	7	6½
	1½	1¼	8	7¼
	1¾	1⅜	9	8¼
	2	1½	10	9¼
	2½	2	11	10¼
	3	2½	12	11¼
	3½	3	14	13¼
	4	3½	16	15¼
Boards	1	¾	2	1½
	1¼	1	3	2½
	1½	1¼	4	3½
			5	4½
			6	5½
			7	6½
			8	7¼
			9	8¼
			10	9¼
			11	10¼
			12	11¼
			Over 12	Off ¾

TERMS USED TO DESCRIBE LUMBER

Lumber is the product produced by a sawmill and planing mill. It is not manufactured further after it is sawed and planed. *Rough lumber* is that produced by the sawmill. It is rough-sawed to nominal sizes. *Yard lumber* refers to lumber of all sizes and patterns which is used for general building purposes. *Dimension lumber* is at least

2 in. but less than 5 in. thick and 2 in. or more wide. This includes joists, rafters, studding, planks, and small timbers.

A *board* is lumber less than 2 in. thick and 1 in. wide. A *plank* is a broad board, usually more than 1 in. thick. It is laid with its wide dimension in a horizontal position. It carries a load such as planks used on scaffolding. A *timber* has a cross section greater than 5 in. in its smallest dimension. It is used for framing buildings. *Structural lumber* is 2 in. or more thick and 4 in. or more wide. It is used where structural strength is needed.

Matched lumber is shaped on the edge to form a tongue-and-groove joint. *Shiplapped lumber* is shaped on the edge to form a lap joint.

SIZES OF MANUFACTURED STRUCTURAL PRODUCTS

The manufacturers of wood products in various regions have standardized the sizes of the items produced. In Fig. 7-26 are shown the sizes for selected products recommended by the Southern Forest Products Association.

	THICKNESS (in.)		WIDTH (in.)		
	Nominal	Worked	Nominal	Face	Overall
Bevel siding	½	3/16 × 7/16	4	3½	3½
	⅝	3/16 × 9/16	5	4½	4½
	¾	3/16 × 11/16	6	5½	5½
	1	3/16 × ¾	8	7¼	7¼
Drop siding	⅝	9/16	4	3⅛	3⅜
Rustic and drop siding (dressed and matched)	1	23/32	5	4⅛	4⅜
			6	5⅛	5⅜
			8	6⅞	7⅛
			10	8⅞	9⅛
Rustic and drop siding (shiplapped)	⅝	9/16	4	3	3⅜
	1	23/32	5	4	4⅜
			6	5	5⅜
			8	6⅝	7⅛
			10	8⅝	9⅛
			12	10⅝	11⅛
Flooring	⅜	5/16	2	1⅛	1⅜
	½	7/16	3	2⅛	2⅜
	⅝	9/16	4	3⅛	3⅜
	1	¾	5	4⅛	4⅜
	1¼	1	6	5⅛	5⅜
	1½	1¼			
Ceiling	⅜	5/16	3	2⅛	2⅜
	½	7/16	4	3⅛	3⅜
	⅝	9/16	5	4⅛	4⅜
	¾	11/16	6	5⅛	5⅜
Partition	1	23/32	3	2⅛	2⅜
			4	3⅛	3⅜
			5	4⅛	4⅜
			6	5⅛	5⅜
Paneling	1	23/32	3	2⅛	2⅜
			4	3⅛	3⅜
			5	4⅛	4⅜
			6	5⅛	5⅜
			8	6⅞	7⅛
			10	8⅞	9⅛
			12	10⅞	11⅛
Shiplap	1	¾	4	3⅛	3½
			6	5⅛	5½
			8	6⅞	7¼
			10	8⅞	9¼
			12	10⅞	11¼
Dressed and matched	1	¾	4	3⅛	3⅜
	1¼	1	5	4⅛	4⅜
	1½	1¼	6	5⅛	5⅜
			8	6⅞	7⅛
			10	8⅞	9⅛
			12	10⅞	11⅛

FIGURE 7-26
Sizes for other softwood products. *(Courtesy of Southern Forest Products Association)*

THICKNESS (mm)	WIDTH (mm) 75	100	125	150	175	200	225	250	300
16	X	X	X	X					
19	X	X	X	X					
22	X	X	X	X					
25	X	X	X	X	X	X	X	X	X
32	X	X	X	X	X	X	X	X	X
36	X	X	X	X					
38	X	X	X	X	X	X	X		
40	X	X	X	X	X	X	X		
44	X	X	X	X	X	X	X	X	X
50	X	X	X	X	X	X	X	X	X
63		X	X	X	X	X	X		
75		X	X	X	X	X	X	X	X
100		X		X		X		X	X
150				X		X			X
200						X			
250								X	
300									X

Note: Lengths range from 1.8 m to 6.3 m in steps of 0.3 m. Reduce these sizes by 4 mm for planing to get dressed lumber.

FIGURE 7-27

Metric sizes for softwood lumber.

Metric Lumber Sizes Most of the world uses metric sizes. Lumber produced in the United States for shipment overseas must be in metric measure. All sizes are stamped on them in millimeters. The metric sizes are almost the same physical size as those made in inches. The basic U.S. size is 1 in. The basic metric size is 25 millimeters (mm). One inch equals 25.4 mm; therefore, they are almost the same. The metric widths increase from one size to another by 25 mm. This is basically 1 in.

The nominal sizes for metric lumber are listed in Fig. 7-27.

Metric lumber lengths start at 1.8 meters (m); (about 6 ft) and increase in steps of 0.3 m (about 1 ft) to 6.3 m (about 20 ft).

When metric lumber is dressed, it is reduced in size 4 mm. For example, a 100-mm board (3.94 in.) is actually 96 mm (3.78 in.) when dressed. A 4-in. board when dressed is actually 3.5 in.

When specifying lumber sizes and figuring the amount of material, the nominal size is used.

Board Feet Lumber is sold by the *board foot.* A board foot is equal to 144 cubic inches of wood. For example, a piece 1 in. thick and 12 in. square is a board foot (Fig. 7-28).

Board feet are figured by multiplying the thickness in *inches* by the width in *inches* by the length in *feet* by the number of boards this size and dividing this sum by 12. See the following example for six boards 2 in. thick × 10 in. wide × 12 ft long.

$$\text{board feet} = \frac{\text{no. boards} \times T\,(\text{in.}) \times W\,(\text{in.}) \times L\,(\text{ft})}{12}$$

$$= \frac{6 \times 2\text{ in.} \times 10\text{ in.} \times 12\text{ ft}}{12} = \frac{1440}{12} = 120\text{ board feet}$$

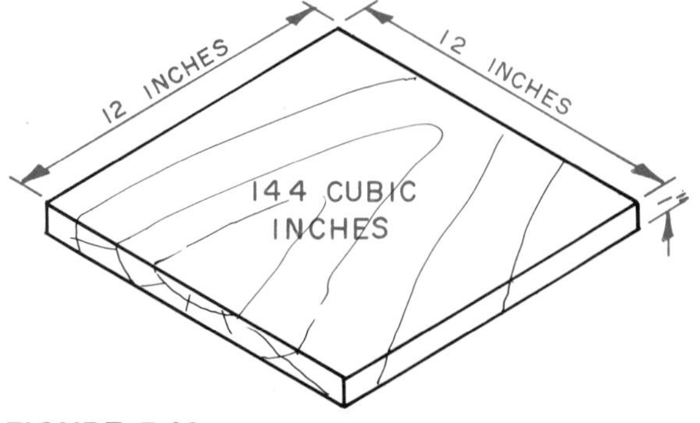

FIGURE 7-28

A board foot.

Boards less than 1 in. in thickness are figured as 1 in. thick.

There are tables available showing the board feet in various size boards (Fig. 7-29).

Sometimes a carpenter needs to estimate board feet quickly. The proportions in Fig. 7-30 are helpful.

Another help in figuring board feet is found on the blade of many framing squares (Fig. 7-31). This device is based on stock being 1 in. thick. If figuring thicker stock, find the board feet for 1 in. and multiply the answer by the number of inches of thickness.

Study Fig. 7-31. The inch markings on the outer edge of the blade are the width sizes. The length is given in the vertical column under the figure "12" on the outer edge of the blade. The other figures are board feet.

Following is how to use this device. Find the width on the outer scale. Find the length on the vertical scale under the figure "12." Run down the width line until it crosses the horizontal line from the length. The figure where they cross is the board feet. The number to the left of the line is board feet. The number to the right is

FIGURE 7-29

A ready reference chart of board-feet measure.

NOMINAL SIZE (in.)	LENGTH (ft) 8	10	12	14	16	18	20	22	24
1 × 2	1⅓	1⅔	2	2⅓	2⅔	3	3⅓	3⅔	4
1 × 3	2	2½	3	3½	4	4½	5	5½	6
1 × 4	2⅔	3⅓	4	4⅔	5⅓	6	6⅔	7⅓	8
1 × 5	3⅓	4⅙	5	5⅚	6⅔	7½	8⅓	9⅙	10
1 × 6	4	5	6	7	8	9	10	11	12
1 × 7	4⅔	5⅚	7	8⅙	9⅓	10½	11⅔	12⅚	14
1 × 8	5⅓	6⅔	8	9⅓	10⅔	12	13⅓	14⅔	16
1 × 10	6⅔	8⅓	10	11⅔	13⅓	15	16⅔	18⅓	20
1 × 12	8	10	12	14	16	18	20	22	24
1¼ × 4	3⅓	4⅙	5	5⅚	6⅔	7½	8⅓	9⅙	10
1¼ × 6	5	6¼	7½	8¾	10	11¼	12½	13¾	15
1¼ × 8	6⅔	8⅓	10	11⅔	13⅓	15	16⅔	18⅓	20
1¼ × 10	8⅓	10 5/12	12½	14 7/12	16⅔	18¾	20⅚	22 11/12	25
1¼ × 12	10	12½	15	17½	20	22½	25	27½	30
1½ × 4	4	5	6	7	8	9	10	11	12
1½ × 6	6	7½	9	10½	12	13½	15	16½	18
1½ × 8	8	10	12	14	16	18	20	22	24
1½ × 10	10	12½	15	17½	20	22½	25	27½	30
1½ × 12	12	15	18	21	24	27	30	33	36
2 × 2	2⅔	3⅓	4	4⅔	5⅓	6	6⅔	7⅓	8
2 × 4	5⅓	6⅔	8	9⅓	10⅓	12	13⅓	14⅔	16
2 × 6	8	10	12	14	16	18	20	22	24
2 × 8	10⅔	13⅓	16	18⅔	21⅓	24	26⅔	29⅓	32
2 × 10	13⅓	16⅔	20	23⅓	26⅔	30	33⅓	36⅔	40
2 × 12	16	20	24	28	32	36	40	44	48
3 × 3	6	7½	9	10½	12	13½	15	16½	18
3 × 6	12	15	18	21	24	27	30	33	36
3 × 8	16	20	24	28	32	36	40	44	48
3 × 10	20	25	30	35	40	45	50	55	60
3 × 12	24	30	36	42	48	54	60	66	72
4 × 4	10⅔	13⅓	16	18⅔	21⅓	24	26⅔	29⅓	32
4 × 6	16	20	24	28	32	36	40	44	48
4 × 8	21⅓	26⅔	32	37⅓	42⅔	48	53⅓	58⅔	64
4 × 10	26⅔	33⅓	40	46⅔	53⅓	60	66⅔	73⅓	80
4 × 12	32	40	48	56	64	72	80	88	96

WIDTH (in.)	THICKNESS[a] (in.)	BOARD FEET EQUALS:
3	1 or less	¼ of the length
4	1 or less	⅓ of the length
6	1 or less	½ of the length
9	1 or less	¾ of the length
12	1 or less	Same as the length
15	1 or less	1¼ of the length

[a]For thicker stock, find board feet for 1 in. of thickness and multiply by the thickness.

FIGURE 7-30

A method for estimating board feet.

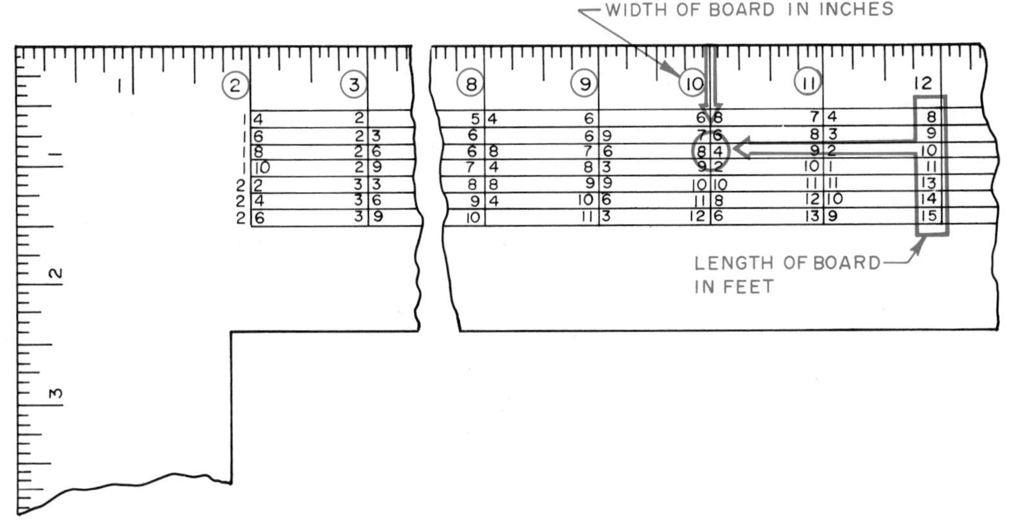

FIGURE 7-31

How to figure board feet using a carpenter's square.

the number of $^1/_{12}$s of a board foot. How many board feet in a board 10 in. wide and 10 ft long? The answer is $8^4/_{12}$ or $8^1/_3$ *board feet.*

PLYWOOD

Plywood is a manufactured wood product. It is produced in large sheets. The most commonly produced size sheet is 4 ft 0 in. × 8 ft 0 in.

Plywood is heavily used in building construction. Following are some typical uses:

1. Wall sheathing
2. Subfloors and underlayment for tile and carpets
3. Roof decking
4. Interior wall paneling
5. Cabinets and other built-in furniture units
6. Concrete forms
7. Fences and patio screens
8. Finished exterior siding
9. Soffits

Plywood is made by gluing together layers of veneer (Fig. 7-32). Veneer is a thin layer of wood that has been sawed or peeled from a log. It is cut in thicknesses ranging from $^1/_{10}$ to $^5/_{16}$ in.

Plywood most commonly used has three, five, or seven layers. Certain special types are made with four and six layers.

Plywood with an odd number of layers has the grain of each layer at right angles to the one next to it. Plywood with an even number of layers has the center layers glued together with the grain running in the same direction.

The outer layers of veneer are called the face and back. The layer of highest quality is the *face*. The other layer is the *back*. The inner layers whose grain is perpendicular to the face are called *cross-bands*. The center layer may be a veneer or solid lumber.

FIGURE 7-32

Plywood is made of layers of wood.

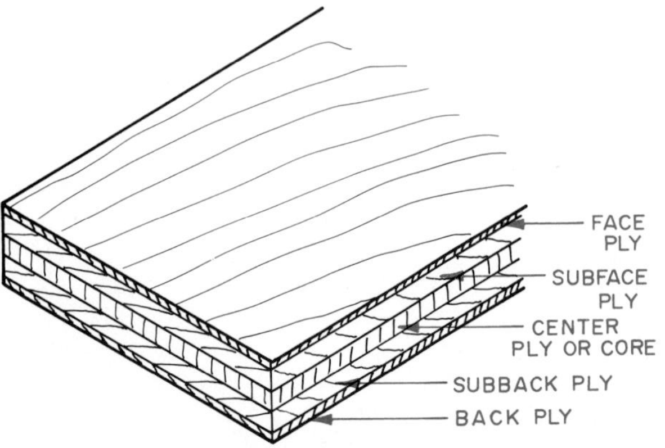

Hardwood Plywood Plywood having a face veneer of hardwood is classified as hardwood plywood. The inner plies are also often all hardwood.

Hardwood Plywood Grades Hardwood and decorative plywood are graded under Voluntary Product Standard PS 51-71. It was developed by the Office of Engineering Standards Services of the National Bureau of Standards of the U.S. Department of Commerce. It provides marketing classifications, quality criteria, test methods, definitions, grade marking, and certification practices.

The grades of veneer and the symbol used to identify them follow:

GRADE	SYMBOL	
Premium grade	A	Carefully matched veneers
Good grade	1	Face veneer good but not carefully matched
Sound grade	2	Face veneer free from open defects; to be painted
Utility grade	3	Face veneer can have cracks, knotholes
Backing grade	4	Face veneer has knots and splits unselected for grain
Specialty grade	SP	Matched grain panels and special veneers on order

Types of Hardwood Plywood The types of plywood in descending order of water-resistance capability follow:

Technical—exterior
Type I—exterior
Type II—interior
Type III—interior

Constructions The constructions of hardwood plywood based on the kind of core follow:

Hardwood veneer core—odd number of plies such as 3, 5, 7, etc.
Softwood veneer core—odd number of plies such as 3, 5, 7, etc.
Hardwood lumber core—3, 5, and 7 ply
Softwood lumber core—3, 5, and 7 ply
Particleboard core—3 and 5 ply
Hardboard—3 ply
Special core—3 ply

Sizes and Thicknesses The common panel sizes are 48 × 84 in., 48 × 96 in., and 48 × 120 in. Thicknesses range from ⅛ to ¾ in.

Softwood Plywood Plywood having a face ply of a softwood is classified as softwood plywood. The inner plies are made from a variety of softwoods.

Softwood Plywood Definitions

Back the outer layer of a panel that has a lower-quality veneer than the other side (Fig. 7-32).

Centers inner layers whose grain direction runs parallel to the outer plies (Fig. 7-32).

Cross-band inner layers whose grain direction runs perpendicular to that of the outer plies. The grain may be parallel to laminated plies. Sometimes it is called the *core* (Fig. 7-30).

Face the outer layer of a panel that has a higher-quality veneer than the other side (Fig. 7-32).

Group a term used to classify species of wood for softwood plywood. For example, birch and southern pine are both in Group 1.

Identification index a set of numbers used in the marking of sheathing grades of plywood.

Inner plies other than the face and back. This includes subface, subback, cross-bank, and center plies.

Patches inserts of sound wood or synthetic material in veneers for replacing defects. "Boat" patches are oval shaped. "Router" patches have parallel sides and round ends. "Sled" patches are rectangular with feathered ends.

Softwood Plywood Grades The standard for softwood plywood grades is U.S. Product Standard PSI-74. It was developed by the Product Standards Section, National Bureau of Standards. The purpose of the standard is to establish nationally recognized requirements for the principal types and grades of construction and industrial plywood. The standard covers wood species, veneer grading, glue bonds, panel construction, dimensions and tolerances, marking, moisture content, and packing of plywood for construction and industrial uses.

Classification of Plywood Plywood is classified into *interior* and *exterior*. This involves veneer grade and adhesive durability.

Interior plywood has three durability levels of adhesive.

1. Bonded with interior adhesive. This type is intended only for interior use.
2. Bonded with intermediate adhesive. This type is intended for use in protected construction and industrial use where moderate delays in providing protection may be expected or conditions of high humidity or water leakage may exist.
3. Bonded with waterproof exterior adhesive. This type is intended for use where protection against moisture exposure due to long construction delays is required.

Exterior plywood is adhered with an adhesive that will retain its bond when repeatedly wetted and dried. It is intended for permanent exterior exposure.

Species The classification of species by groups is listed in Fig. 7-33. The species of face

FIGURE 7-33

Wood species are classified into five groups. *(Courtesy of American Plywood Association)*

GROUP 1	GROUP 2		GROUP 3	GROUP 4	GROUP 5
Apitong	Cedar, Port Orford	Maple, black	Alder, red	Aspen	Basswood
Beech, American	Cypress	Mengkulang	Birch, paper	Bigtooth	Fir, balsam
Birch	Douglas fir 2	Meranti, red	Cedar, Alaska	Quaking	Poplar, balsam
Sweet	Fir	Mersawa	Fir, subalpine	Cativo	
Yellow	California red	Pine	Hemlock, eastern	Cedar	
Douglas fir 1	Grand	Pond	Maple, bigleaf	Incense	
Kapur	Noble	Red	Pine	Western red	
Keruing	Pacific silver	Virginia	Jack	Cottonwood	
Larch, western	White	Western white	Lodgepole	Eastern	
Maple, sugar	Hemlock, western	Spruce	Ponderosa	Black (western poplar)	
Pine	Lauan	Red	Spruce		
Caribbean	Almon	Sitka	Redwood	Pine	
Ocote	Bagtikan	Sweetgum	Spruce	Eastern white	
Pine, southern	Mayapis	Tamarack	Black	Sugar	
Loblolly	Red Lauan	Yellow poplar	Engelmann		
Longleaf	Tangile		White		
Shortleaf	White Lauan				
Slash					
Tanoak					

N	Smooth surface "natural finish" veneer. Select, all heartwood or all sapwood. Free of open defects. Allows not more than 6 repairs, wood only, per 4 × 8 panel, made parallel to grain and well matched for grain and color.	**C Plugged**	Improved C veneer with splits limited to ⅛ in. width and knotholes and borer holes limited to ¼ × ½ in. Admits some broken grain. Synthetic repairs permitted.
A	Smooth, paintable. Not more than 18 neatly made repairs, boat, sled, or router type, and parallel to grain, permitted. May be used for natural finish in less demanding applications.	**C**	Tight knots to 1½ in. Knotholes to 1 in. across grain and some to 1½ in. if total width of knots and knotholes is within specified limits. Synthetic or wood repairs. Discoloration and sanding defects that do not impair strength permitted. Limited splits allowed.
B	Solid surface. Shims, circular repair plugs, and tight knots to 1 in. across grain permitted. Some minor splits permitted.	**D**	Knots and knotholes to 2½ in. width across grain and ½ in. larger within specified limits. Limited splits are permitted. Limited to Interior grades of plywood.

FIGURE 7-34

Symbols used to identify veneer grades on plywood. *(Courtesy of American Plywood Association)*

and back plies can be from any of the five groups. The entire face or back ply should be of the same species.

Inner plies of panel may be of any species listed in groups 1, 2, 3, or 4. Group 5 panels may use any of the species in group 5.

Veneer Grades The veneer grades are identified as N, A, B, C, and D. Grade N is the highest. The other grades run from A through D, with D being the lowest. Grade specifications are given in Fig. 7-34.

Panel Grades Panel grades are made of a combination of veneer grades. The panel grades for interior panels are shown in Fig. 7-35. The panel grades for exterior panels are listed in Fig. 7-36.

Panel Constructions Plywood panels are made of three to seven plies. A ply is a single thickness of veneer. Panels are always made of an odd number of layers, such as three, five, or seven. A layer may be one or more veneers glued together with the grain in the same direction. For example, a panel may have four plies and three layers. Two inner plies are glued together, with the grain in the same direction forming one layer.

Recommended panel constructions are shown in Fig. 7-37.

FIGURE 7-35

Panel-grade designations for interior types of softwood plywood. *(Courtesy of American Plywood Association)*

	MINIMUM VENEER QUALITY			
PANEL GRADE DESIGNATIONS	Face	Back	Inner Plies	SURFACE
N-N	N	N	C	Sanded 2 sides
N-A	N	A	C	Sanded 2 sides
N-B	N	B	C	Sanded 2 sides
N-D	N	D	D	Sanded 2 sides
A-A	A	A	D	Sanded 2 sides
A-B	A	B	D	Sanded 2 sides
A-D	A	D	D	Sanded 2 sides
B-B	B	B	D	Sanded 2 sides
B-D	B	D	D	Sanded 2 sides
Underlayment	C Plugged	D	C & D	Touch-sanded
C-D Plugged	C Plugged	D	D	Touch-sanded
Structural I C-D[a]	Must meet special structural requirements			Unsanded[b]
Structural I C-D Plugged, Underlayment[a]				Touch-sanded
Structural II C-D[a]				Unsanded[b]
Structural II C-D Plugged, Underlayment[a]				Touch-sanded
C-D	C	D	D	Unsanded[b]
C-D with exterior glue	C	D	D	Unsanded[b]

[a]Structural panels—these panels are especially designed for engineered applications such as structural components where design properties, including tension, compression, shear, cross-panel flexural properties, and nail bearing, may be of significant importance.

[b]Except for decorative grades, panels must not be sanded, touch-sanded, surface-textured, or thickness-sized by any mechanical means.

PANEL GRADE DESIGNATIONS[a]	MINIMUM VENEER QUALITY Face	Back	Inner Plies	Surface
Marine, A-A, A-B, B-B, HDO, MDO[b]				Sanded 2 sides
Special Exterior, A-A, A-B, B-B, HDO, MDO[c]				Sanded 2 sides
A-A	A	A	C	Sanded 2 sides
A-B	A	B	C	Sanded 2 sides
A-C	A	C	C	Sanded 2 sides
B-B (concrete form)[d]				
B-B	B	B	C	Sanded 2 sides
B-C	B	C	C	Sanded 2 sides
C-C Plugged	C Plugged	C	C	Touch-sanded
C-C	C	C	C	Unsanded[e]
A-A High Density Overlay	A	A	C Plugged	—
B-B High Density Overlay	B	B	C Plugged	—
B-B High Density Concrete Form Overlay	B	B	C Plugged	—
B-B Medium Density Overlay	B	B	C	—
Special Overlays	C	C	C	—

[a]Also available in Structural I and Structural II classifications.

[b]Marine grades must meet the requirements of Exterior type and be of one of the following grades: A-A, A-B, B-B, High Density Overlay, or Medium Density Overlay.

[c]Special Exterior—an Exterior-type panel that may be produced of any species covered by this Standard. Except in regard to species, it must meet all of the requirements for Marine panels (see footnote b) and be produced in one of the following grades: A-A, A-B, B-B, High Density Overlay, or Medium Density Overlay.

[d]B-B concrete form panels—face veneers must be not less than B grade and always be from the same species group. Inner plies must be not less than "C" grade. This grade of plywood is produced in two classes and panels of each class must be identified accordingly. Panels will be sanded two sides and mill-oiled unless otherwise agreed upon.

[e]Except for decorative grades, panels will not be sanded, touch-sanded, surface-textured, or thickness-sized by any mechanical means.

FIGURE 7-36 *(above)*

Panel-grade designations for exterior-type softwood plywood. *(Courtesy of American Plywood Association)*

FIGURE 7-37 *(below)*

Panel constructions for softwood plywood. *(Courtesy of American Plywood Association)*

PANEL GRADES	FINISHED PANEL NOMINAL THICKNESS RANGE (in.)	MINIMUM NUMBER OF PLIES	MINIMUM NUMBER OF LAYERS
Exterior			
Marine	Through ⅜	3	3
Special Exterior	Over ⅜, through ¾	5	5
B-B (concrete form)	Over ¾	7	7
High Density Overlay			
High Density concrete form overlay			
Interior			
N-N, N-A, N-B, N-D, A-A, A-B, A-D			
B-B, B-D			
Structural I (C-D, C-D Plugged, and			
Underlayment)	Through ⅜	3	3
Structural II (C-D, C-D Plugged, and	Over ⅜, through ½	4	3
Underlayment)	Over ½, through ⅞	5	5
Exterior	Over ⅞	6	5
A-A, A-B, A-C, B-B, B-C			
Structural I and Structural II			
Medium Density and special overlays			
Interior			
(including grades with exterior glue)			
Underlayment	Through ½	3	3
Exterior	Over ½, through ¾	4	3
C-C Plugged	Over ¾	5	5
Interior			
(including grades with exterior glue)			
C-D	Through ⅝	3	3
C-D Plugged	Over ⅝, through ¾	4	3
Exterior	Over ¾	5	5
C-C			

Note: The proportion of wood based on nominal finished panel thickness and dry veneer before layup, as used, with grain running perpendicular to the panel face grain must fall within the range 33 to 70%. The combined thickness of all inner layers must be not less than ½ of panel thickness based on nominal finished panel thickness and dry veneer thickness before layup, as used, for panels with 4 or more plies.

Panel Thickness and Size Softwood plywood panels are available in widths of 36, 48, and 60 in. and in lengths from 60 to 144 in. in 12-in. segments. The most commonly used size is 48 × 96 in. Metric sizes have not been standardized, but a typical panel will probably be 1200 × 2400 mm. This is almost 48 × 96 in.

The thickness of panels runs from 1/4 to 1 1/4 in. and greater in 1/8-in. steps. Unsanded panels run from 5/16 to 1 1/4 in. in steps of 1/8 in. for thicknesses over 3/8 in.

Grade Markings Grade trademarks are stamped on the face and edge of construction and industrial softwood plywood that meets the PS-1-74 requirements. This stamp shows that the product has been subject to the rigid inspection and testing program of the American Plywood Association. Typical grade markings are shown in Fig. 7-38.

A guide to engineered grades of plywood is shown in Fig. 7-39. A guide to appearance grades of plywood is given in Fig. 7-40.

Some of grades most commonly used in general carpentry include the following:

- C-D Interior—wall and roof sheathing, subflooring
- Underlayment Interior—underlayment or combination subfloor-underlayment
- C-C Exterior—waterproof for flooring and roof decking
- B-B Plyform—concrete forms

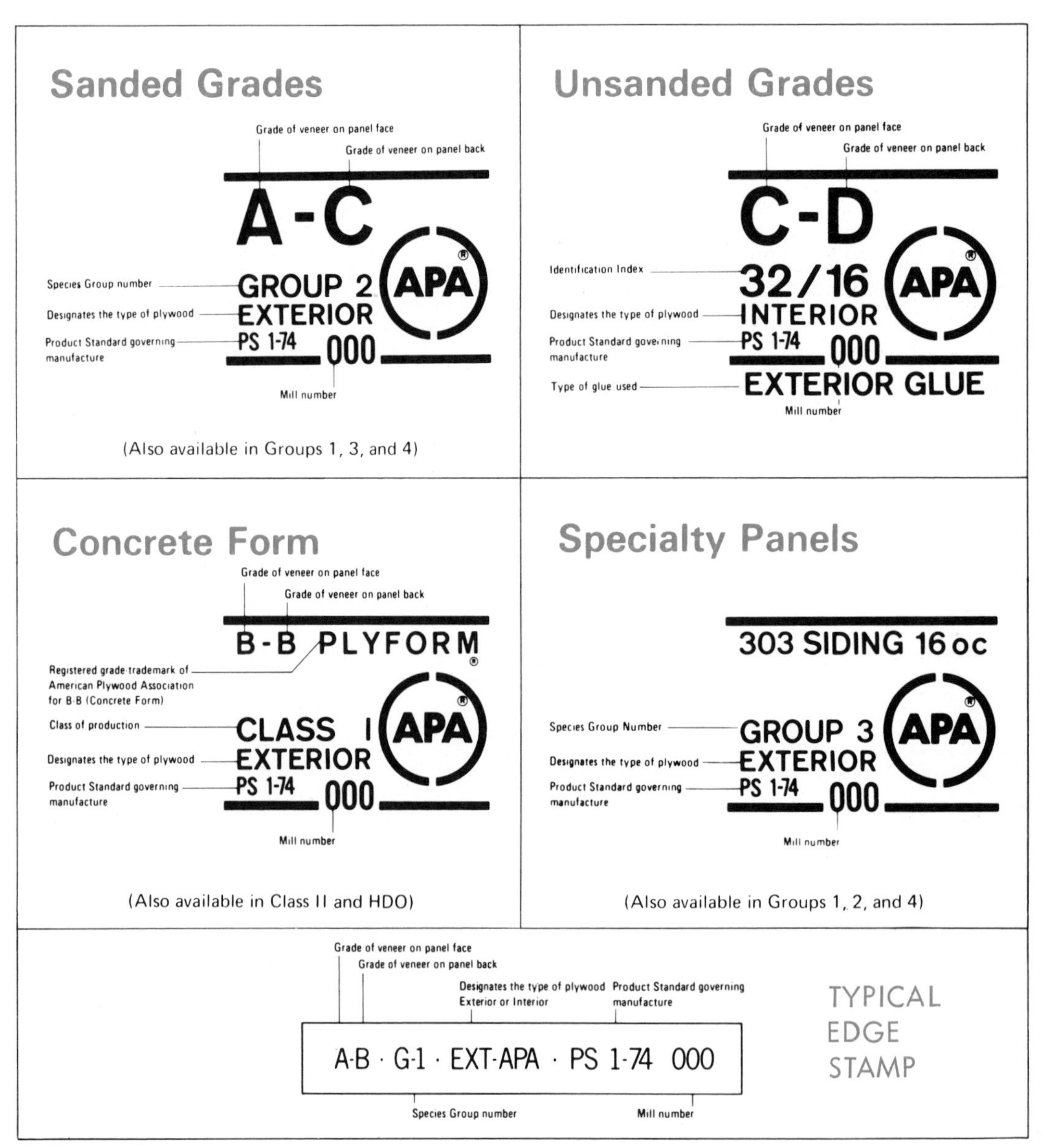

FIGURE 7-38

Typical grade marks on plywood. *(Courtesy of American Plywood Association)*

Guide to engineered grades of plywood

Type	Grade Designation	Description and Most Common Use	Typical Grade-trademarks	Veneer Grade: Face	Veneer Grade: Back	Veneer Grade: Inner Plies	Most Common Thicknesses (inch) (1)					
Interior Type	C-D INT-APA	For wall and roof sheathing, subflooring, industrial uses such as pallets. Also available with intermediate glue or exterior glue. Specify intermediate glue for moderate construction delays; exterior glue for better durability in somewhat longer construction delays, and for treated wood foundations. (2) (10)	C-D 32/16 APA INTERIOR 000	C	D	D		5/16	3/8	1/2	5/8	3/4
Interior Type	STRUCTURAL I C-D INT-APA and STRUCTURAL II C-D INT-APA	Unsanded structural grades where plywood strength properties are of maximum importance: structural diaphragms, box beams, gusset plates, stressed-skin panels, containers, pallet bins. Made only with exterior glue.	STRUCTURAL C-D 24/0 APA INTERIOR 000 EXTERIOR GLUE	C[6]	D[7]	D[7]		5/16	3/8	1/2	5/8	3/4
Interior Type	UNDERLAYMENT INT-APA	For underlayment or combination subfloor-underlayment under resilient floor coverings, carpeting in homes, apartments, mobile homes. Specify exterior glue where moisture may be present, such as bathrooms, utility rooms. Touch-sanded. Also available in tongue and groove. (2) (3) (9)	UNDERLAYMENT GROUP 1 APA INTERIOR 000	C Plugged	D	C[8] & D	1/4		3/8	1/2	5/8	3/4
Interior Type	C-D PLUGGED INT-APA	For built-ins, wall and ceiling tile backing, cable reels, walkways, separator boards. Not a substitute for UNDERLAYMENT as it lacks UNDERLAYMENT's indentation resistance. Touch-sanded. (2) (3) (9)	C-D PLUGGED GROUP 2 APA INTERIOR 000	C Plugged	D	D		5/16	3/8	1/2	5/8	3/4
Interior Type	2·4·1 INT-APA	Combination subfloor-underlayment. Quality base for resilient floor coverings, carpeting, wood strip flooring. Use 2·4·1 with exterior glue in areas subject to moisture. Unsanded or touch-sanded as specified. (2) (5) (11)	2·4·1 GROUP 1 APA INTERIOR 000	C Plugged	D	C & D	1-1/8"					
Exterior Type	C-C EXT-APA	Unsanded grade with waterproof bond for subflooring and roof decking, siding on service and farm buildings, crating, pallets, pallet bins, cable reels. (10)	C-C 42/20 APA EXTERIOR 000	C	C	C		5/16	3/8	1/2	5/8	3/4
Exterior Type	STRUCTURAL I C-C EXT-APA and STRUCTURAL II C-C EXT-APA	For engineered applications in construction and industry where full Exterior type panels are required. Unsanded. See (9) for species group requirements.	STRUCTURAL C-C 32/16 APA EXTERIOR 000	C	C	C		5/16	3/8	1/2	5/8	3/4
Exterior Type	UNDERLAYMENT C-C Plugged EXT-APA C-C PLUGGED EXT-APA	For underlayment or combination subfloor-underlayment under resilient floor coverings where severe moisture conditions may be present, as in balcony decks. Use for tile backing where severe moisture conditions exist. For refrigerated or controlled atmosphere rooms, pallets, fruit pallet bins, reusable cargo containers, tanks and boxcar and truck floors and linings. Touch-sanded. Also available in tongue and groove. (3) (9)	UNDERLAYMENT C-C PLUGGED GROUP 2 APA EXTERIOR 000 C-C PLUGGED GROUP 3 APA EXTERIOR 000	C Plugged	C	C[8]	1/4		3/8	1/2	5/8	3/4
Exterior Type	B-B PLYFORM CLASS I & CLASS II EXT-APA	Concrete form grades with high re-use factor. Sanded both sides. Mill-oiled unless otherwise specified. Special restrictions on species. Also available in HDO. (4)	B-B PLYFORM CLASS I APA EXTERIOR 000	B	B	C					5/8	3/4

(1) Panels are standard 4x8-foot size. Other sizes available.
(2) Also made with exterior or intermediate glue.
(3) Available in Group 1, 2, 3, 4, or 5.
(4) Also available in STRUCTURAL I.
(5) Made only in woods of certain species to conform to APA specifications.
(6) Special improved C grade for structural panels.
(7) Special improved D grade for structural panels.
(8) Special construction to resist indentation from concentrated loads.
(9) Also available in STRUCTURAL I (all plies limited to Group 1 species) and STRUCTURAL II (all plies limited to Group 1, 2, or 3 species).
(10) Made in many different species combinations. Specify by Identification Index.
(11) Can be special ordered in Exterior type for porches and patio decks, roof overhangs, and exterior balconies.

FIGURE 7-39

Description and use of engineered grades of plywood.

A-C Exterior—soffits, fences

MDO Exterior—medium-density overlay for exterior siding, signs

303 Siding Exterior—exterior siding

Identification Index for Sheathing The identification index is part of the grade trademark stamped on softwood plywood panels. See the unsanded grades listed in Fig. 7-41. These are used on sheathing grades C-C, structural C-C, C-D, and structural C-D.

The index in Fig. 7-42 shows the acceptable spans for roof and floor structural members for panels of various species and thicknesses. For example, a panel made of a Group 1 species that is ½ in. thick will span roof rafters at 32 in. on center and floor joists at 16 in. on center.

Structural Panels Structural panels are designed for engineered applications where design properties are important. A typical example would be in a load-carrying structural member. This includes data on species, grade, and glue bond requirements. The standards governing structural panels also specify other requirements that must be met. Grade, glue bond, and species specifications are shown in Fig. 7-43.

Guide to appearance grades of plywood

	Grade Designation (2)	Description and Most Common Uses	Typical Grade-trademarks	Veneer Grade			Most Common Thicknesses (inch) (3)					
				Face	Back	Inner Plies						
Interior Type	N-N, N-A, N-B INT-APA	Cabinet quality. For natural finish furniture, cabinet doors, built-ins, etc. Special order items.	NN G1 INT APA PS 1 74 / NA G2 INT APA PS 1 74	N	N,A, or B	C						3/4
	N-D-INT-APA	For natural finish paneling. Special order item.	ND G3 INT APA PS 1 74	N	D	D	1/4					
	A-A INT-APA	For applications with both sides on view. Built-ins, cabinets, furniture and partitions. Smooth face; suitable for painting.	AA G4 INT APA PS 1 74	A	A	D	1/4		3/8	1/2	5/8	3/4
	A-B INT-APA	Use where appearance of one side is less important but two smooth solid surfaces are necessary.	AB G4 INT APA PS 1 74	A	B	D	1/4		3/8	1/2	5/8	3/4
	A-D INT-APA	Use where appearance of only one side is important. Paneling, built-ins, shelving, partitions, and flow racks.	A-D GROUP 1 INTERIOR PS 1-74 000 APA	A	D	D	1/4		3/8	1/2	5/8	3/4
	B-B INT-APA	Utility panel with two smooth sides. Permits circular plugs.	BB G3 INT APA PS 1 74	B	B	D	1/4		3/8	1/2	5/8	3/4
	B-D INT-APA	Utility panel with one smooth side. Good for backing, sides of built-ins. Industry: shelving, slip sheets, separator boards and bins.	B-D GROUP 3 INTERIOR PS 1-74 000 APA	B	D	D	1/4		3/8	1/2	5/8	3/4
	DECORATIVE PANELS—APA	Rough-sawn, brushed, grooved, or striated faces. For paneling, interior accent walls, built-ins, counter facing, displays, and exhibits.	DECORATIVE BD G1 INT APA PS 1 74	C or btr.	D	D		5/16	3/8	1/2	5/8	
	PLYRON INT-APA	Hardboard face on both sides. For counter tops, shelving, cabinet doors, flooring. Faces tempered, untempered, smooth, or screened.	PLYRON INT APA PS 1 74			C & D				1/2	5/8	3/4

	Grade Designation (2)	Description and Most Common Uses	Typical Grade-trademarks	Face	Back	Inner Plies						
Exterior Type (7)	A-A EXT-APA	Use where appearance of both sides is important. Fences, built-ins, signs, boats, cabinets, commercial refrigerators, shipping containers, tote boxes, tanks, and ducts. (4)	AA G3 EXT APA PS 1 74	A	A	C	1/4		3/8	1/2	5/8	3/4
	A-B EXT-APA	Use where the appearance of one side is less important. (4)	AB G1 EXT APA PS 1 74	A	B	C	1/4		3/8	1/2	5/8	3/4
	A-C EXT-APA	Use where the appearance of only one side is important. Soffits, fences, structural uses, boxcar and truck lining, farm buildings. Tanks, trays, commercial refrigerators. (4)	A-C GROUP 1 EXTERIOR PS 1-74 000 APA	A	C	C	1/4		3/8	1/2	5/8	3/4
	B-B EXT-APA	Utility panel with solid faces. (4)	BB G1 EXT APA PS 1 74	B	B	C	1/4		3/8	1/2	5/8	3/4
	B-C EXT-APA	Utility panel for farm service and work buildings, boxcar and truck lining, containers, tanks, agricultural equipment. Also as base for exterior coatings for walls, roofs. (4)	B-C GROUP 2 EXTERIOR PS 1-74 000 APA	B	C	C	1/4		3/8	1/2	5/8	3/4
	HDO EXT-APA	High Density Overlay plywood. Has a hard, semi-opaque resin-fiber overlay both faces. Abrasion resistant. For concrete forms, cabinets, counter tops, signs and tanks. (4)	HDO 60/60 BB PLYFORM I EXT APA PS 1 74	A or B	A or B	C or C plgd			3/8	1/2	5/8	3/4
	MDO EXT-APA	Medium Density Overlay with smooth, opaque, resin-fiber overlay one or both panel faces. Highly recommended for siding and other outdoor applications, built-ins, signs, and displays. Ideal base for paint. (4)	MDO BB G4 EXT APA PS 1 74	B	B or C	C			3/8	1/2	5/8	3/4
	303 SIDING EXT-APA	Proprietary plywood products for exterior siding, fencing, etc. Special surface treatment such as V-groove, channel groove, striated, brushed, rough-sawn. (6)	303 SIDING 16 oc GROUP 1 EXTERIOR PS 1-74 000 APA	(5)	C	C			3/8	1/2	5/8	
	T 1-11 EXT-APA	Special 303 panel having grooves 1/4" deep, 3/8" wide, spaced 4" or 8" o.c. Other spacing optional. Edges shiplapped. Available unsanded, textured, and MDO. (6)	303 SIDING 16 oc T1-11 GROUP 1 EXTERIOR PS 1-74 000 APA	C or btr.	C	C					5/8	
	PLYRON EXT-APA	Hardboard faces both sides, tempered, smooth or screened.	PLYRON EXT APA PS 1 74			C				1/2	5/8	3/4
	MARINE EXT-APA	Ideal for boat hulls. Made only with Douglas fir or western larch. Special solid jointed core construction. Subject to special limitations on core gaps and number of face repairs. Also available with HDO or MDO faces.	MARINE AA EXT APA PS 1 74	A or B	A or B	B	1/4		3/8	1/2	5/8	3/4

(1) Sanded both sides except where decorative or other surfaces specified.
(2) Available in Group 1, 2, 3, 4, or 5 unless otherwise noted.
(3) Standard 4x8 panel sizes, other sizes available.
(4) Also available in Structural I (all plies limited to Group 1 species) and Structural II (all plies limited to Group 1, 2, or 3 species).
(5) C or better for 5 plies; C Plugged or better for 3-ply panels.
(6) Stud spacing is shown on grade stamp.
(7) For finishing recommendations, see form V307.
(8) For strength properties of appearance grades, refer to "Plywood Design Specification," form Y510.

FIGURE 7-40

Description and uses of appearance grade of plywood.

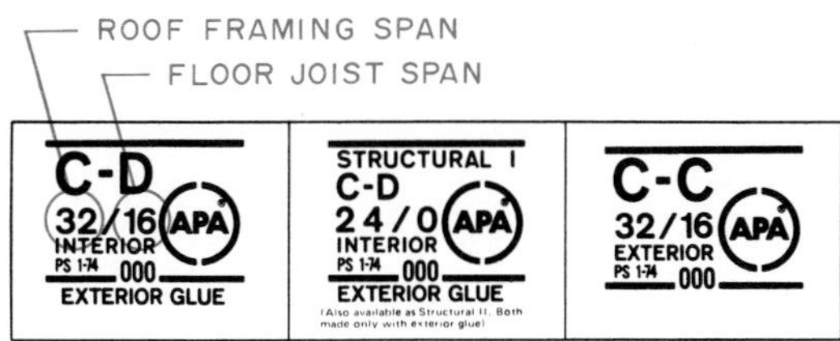

FIGURE 7-41

Typical grade markings for unsanded grades used as sheathing panels. *(Courtesy of American Plywood Association)*

Identification Index[(a)] table for sheathing panels

Species of face and back	Grade		
Group 1	C-C Str. I C-C, C-D Str. II C-C, C-D[(c)] C-D — (b)		
Group 2 (d)	C-C Str. II C-C, C-D C-D	C-C Str. II C-C, C-D C-D — (b)	
Group 3		C-C Str. II C-C, C-D C-D — (b)	
Group 4		C-C C-D — (d)	C-C C-D — (b)
Nominal Thickness 5/16	20/0	16/0	12/0
3/8	24/0	20/0	16/0
1/2	32/16	24/0	24/0
5/8	42/20	32/16	30/12
3/4	48/24	42/20	36/16
7/8		48/24	42/20
(e)			

(a) Identification Index refers to the numbers in the lower portion of the table which are used in the marking of sheathing grades of plywood. The numbers are related to the species of panel face and back veneers and panel thickness in a manner to describe the bending properties of a panel. They are particularly applicable where panels are used for subflooring and roof sheathing to describe recommended maximum spans in inches under normal use conditions and to correspond with commonly accepted criteria. The left hand number refers to spacing of roof framing with the right hand number relating to spacing of floor framing. Actual maximum spans are established by local building codes. See reference source given in section 2 for complete description and product use information.

(b) Panels of standard nominal thickness and construction.

(c) Panels manufactured with Group 1 faces but classified as Structural II by reason of Group 2 or Group 3 inner plys.

(d) Panels conforming to the special thickness and panel construction provisions of 3.8.6.

(e) Panels thicker than 7/8 inch shall be identified by group number.

FIGURE 7-42

Identification index of sheathing panels. *(Courtesy of American Plywood Association)*

Grade	Glue Bond	Species
Structural I C-D[(a)] C-D Plugged[(a)] Underlayment[(a)]	Shall meet the requirements of 3.7.3	Face, back and all inner plys limited to Group 1 species
Structural II C-D[(a)] C-D Plugged[(a)] Underlayment[(a)]	Shall meet the requirements of 3.7.3	Face, back and all inner plys may be of any Group 1, 2, or 3 species
Structural I All Exterior grades (see table 3)	Exterior	Face, back and all inner plys limited to Group 1 species
Structural II All Exterior grades (see table 3)	Exterior	Face, back and all inner plys may be of any Group 1, 2, or 3 species

(a) Special limitations applying to Structural (C-D, C-D Plugged, Underlayment) grade panels are:

- In D grade veneers white pocket in any area larger than the size of the largest knothole, pitchpocket or split specifically permitted in D grade shall not be permitted in any ply.
- Sound tight knots in D grade shall not exceed 2-1/2 inches measured across the grain, except as provided in table 5.
- Plugs, including multiple repairs, shall not exceed 4 inches in width.
- Panel construction shall be as specified in 3.8.

FIGURE 7-43

Structural panel standards. *(Courtesy of American Plywood Association)*

Ordering Softwood Plywood When ordering sanded grades of plywood, specify the species group, number of pieces, width, length, number of plies, type, grade, and finished thickness. Width is across the grain and is always given before length. An example for sanded plywood would be Group 2 plywood, 100 pieces, 48 × 96 in., three ply, Interior type, A-D grade, sanded two sides to ¼-in. thickness.

When ordering unsanded plywood sheathing, specify the grade, identification index, number of pieces, width, length, number of plies, and the thickness. An example for unsanded plywood would be C-D, 24/0, 100 pieces, 48 × 96 in., five ply, ⅜ in. thickness. This would assume interior glue. If exterior glue is wanted, it must be added to the specifications.

When ordering concrete form plywood, specify the class, number of pieces, width, length, thickness, and grade. An example for concrete form plywood would be Concrete Form, Class 1, 100 pieces, 48 × 96 in. × ⅝ in. thickness, B-B Exterior type. Panels are mill-oiled unless specified otherwise.

STUDY QUESTIONS

After reading the chapter, answer each of the following questions. If you do not know the answer, review the chapter.

1. What are the main parts of a tree?
2. What is the difference between a hardwood and a softwood?
3. What wood species are used for construction framing?
4. Explain the difference between plain-sawed and quarter-sawed lumber.
5. How is lumber seasoned?
6. What is the proper moisture content for construction lumber and interior trim?
7. In which direction does a board shrink the most?
8. How can moisture content of lumber be found?
9. List and describe the defects found in lumber.
10. What are the grading classifications for softwoods?
11. What standard is used to grade softwoods?
12. How can you tell the grade and moisture content of lumber at the lumberyard?
13. What is stress-rated lumber?
14. What organization sets up grading rules for hardwoods?
15. List the inch thicknesses and widths of softwood lumber.
16. What is the difference between the nominal and actual lumber size?
17. List the metric thickness of lumber.
18. How many board feet are there in a board 1 in. × 10 in. × 16 ft 0 in.?
19. What is plywood?
20. What are the parts of a sheet of plywood?
21. What are the two classifications of softwood plywood?
22. How can you tell the grade of plywood sheets?

IMPORTANT TECHNICAL TERMS

Following are technical terms that you should be able to use as part of your working vocabulary. Write a brief description of the meaning of each term.

Heartwood
Sapwood
Cambium
Plain-sawed
Quarter-sawed
Kiln-dried
Warp
Bow
Crook
Cup
Check
Split
Shake
Knot
Pitch
Pitch pocket
Stain
Wane
Decay
Finish lumber
Dimension lumber
Timbers
Machine-stress-rated lumber
Live loads
Modulus of elasticity
Nominal size
Rough lumber
Board
Plank
Matched lumber
Shiplapped lumber
Board feet
Plywood
Cross-bands
Identification index

8 LOCATING AND LEVELING A BUILDING

OPTICAL LEVELING INSTRUMENTS

The carpenter is most likely to use some type of *level.* A level consists of an accurate spirit level and a telescope mounted on a circular base. The base has *vernier scales* used for measuring angles. The dumpy or *builder's level* does not have a tilting mechanism. It can be used only for measuring horizontal angles (Fig. 8-1). The *transit level* can be moved vertically and horizontally. This enables it to be used to measure vertical and horizontal angles (Fig. 8-2).

Leveling instruments are easily damaged. When not in use, always store in their carrying case. Keep free of dust, dirt, and moisture. When on a tripod, cover the instrument if it is to be left out but not used for awhile. When fastening the head to the tripod, grip it by the base. Do not lift by the scope. Be certain that the tripod is firmly set in the ground and the legs locked so that the instrument will not fall over.

LINEAR MEASURING TOOLS

Steel measuring tapes are used to measure linear distances. They are available on winding reels that store 50 to 300 ft of tape (Fig. 8-3). Tapes are available with a variety of graduations. Some are graduated in feet and inches, others feet and decimal parts of a foot, while others are in meters with subdivisions in centimeters (Fig. 8-4).

SETTING UP THE LEVEL

The level is set on a three-leg *tripod.* The legs of the tripod are adjustable. Set the tripod so that the legs spread about 3½ ft apart and are

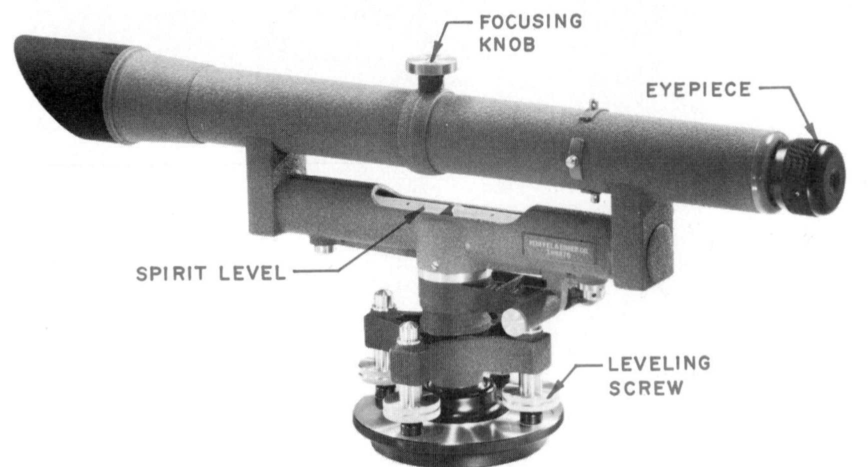

FIGURE 8-1

A dumpy or builder's level is used to check for levelness. *(Courtesy of Keuffel and Esser Co.)*

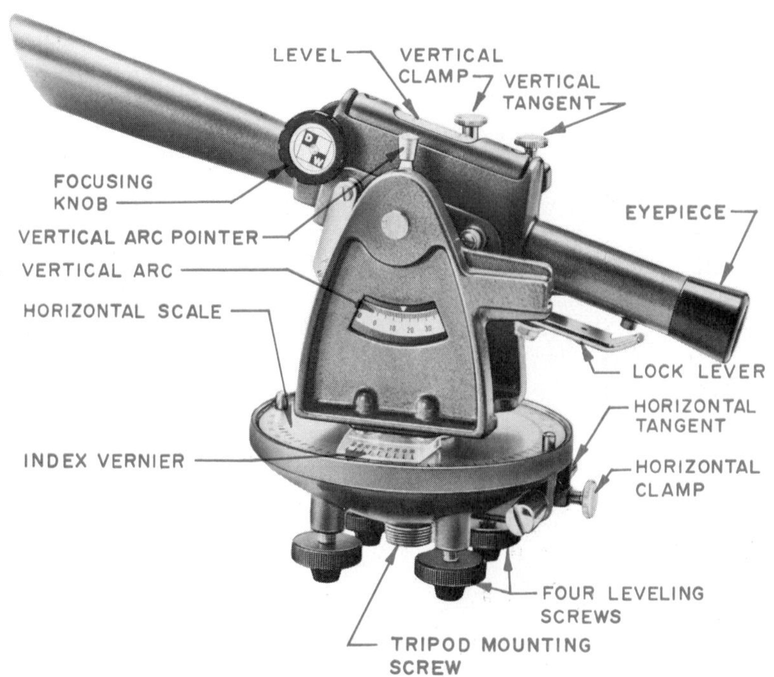

FIGURE 8-2

A level transit measures horizontal and vertical angles. *(Courtesy of David White Instruments)*

FIGURE 8-3

Steel tapes are used to measure linear distances. *(Courtesy of Keuffel and Esser Co.)*

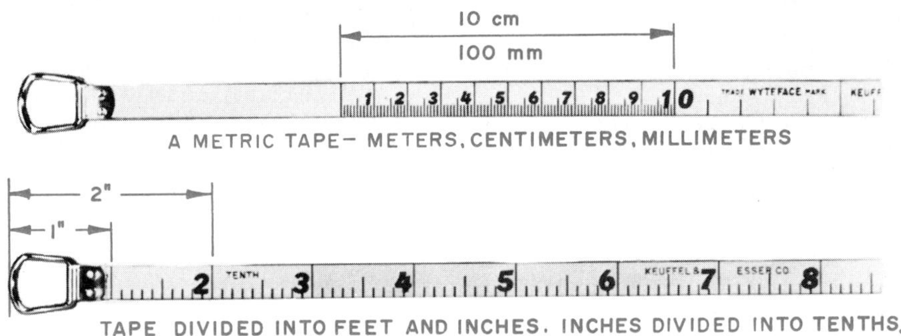

FIGURE 8-4

Steel tapes are available in a variety of graduations. *(Courtesy of Keuffel and Esser Co.)*

firmly embedded in the ground. Adjust the lengths of the legs so that the head is about level (Fig. 8-5).

Now place the level on the tripod. Lift the level by the base. Screw it on to the tripod. Tighten until it is firm. Do not overtighten.

Now set the instrument over the *station mark*. A plumb bob hangs below the level. The instrument is set so that the plumb bob points directly to the station mark. The station mark is a known location from which all measurements are taken. It is established by the surveyor. Often, it is one of the corner stakes of the lot (Fig. 8-6). Once the instrument is in place, the head must be adjusted so that it is level. The instrument will have a screw that holds the head in position. Loosen it and move the head until the spirit level is in line with two opposite *leveling screws*. Level the head by turning these leveling screws. They are turned simultaneously. As you tighten one, you must loosen the other (Fig. 8-7). Adjust until the spirit level shows that the head is level in that direction (Fig. 8-8). Now rotate the head so that the spirit level is in line with the other two leveling screws. Repeat the leveling process in this direction. Turn the head

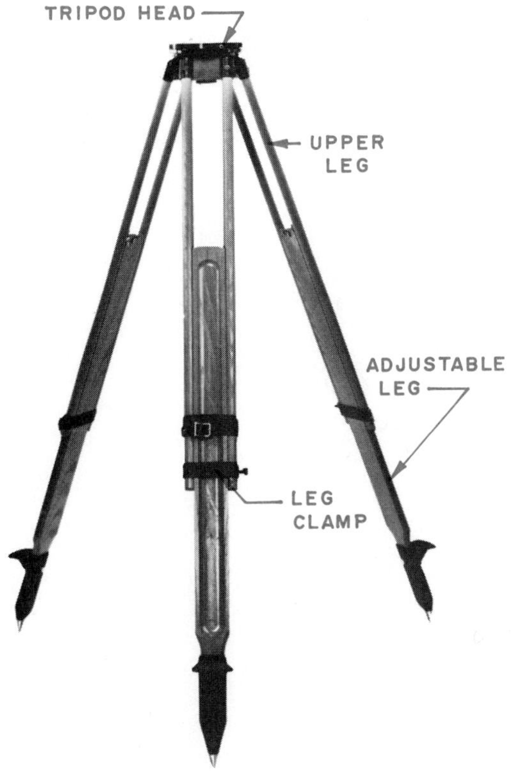

FIGURE 8-5

A tripod used to hold a level or level transit. *(Courtesy of Keuffel and Esser Co.)*

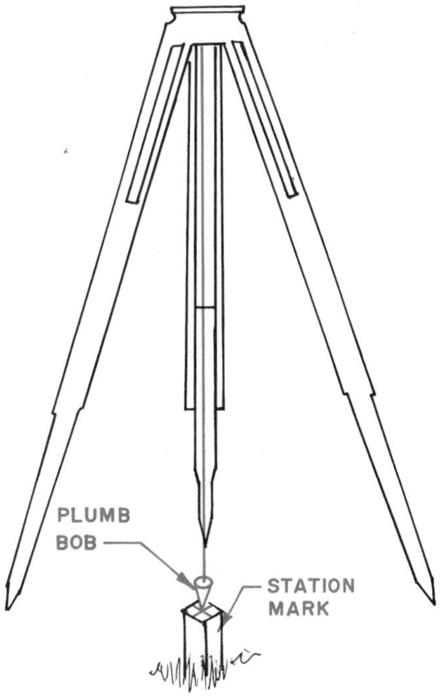

FIGURE 8-6

Set the instrument plumb over the plumb bob.

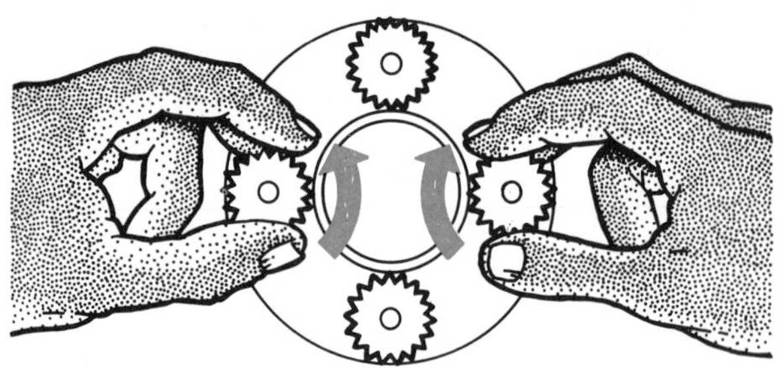

FIGURE 8-7

Set the instrument head so that it is level. To do this, turn the two leveling screws simultaneously in opposite directions.

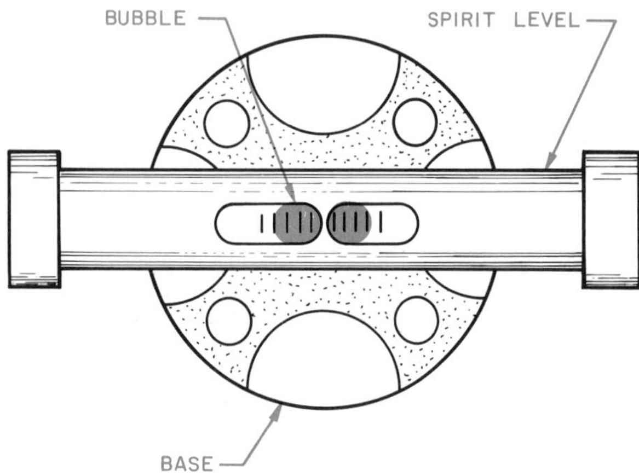

FIGURE 8-8

The spirit level shows when the instrument is level.

FIGURE 8-9

A leveling rod is used when sighting longer distances. *(Courtesy of David White Instruments)*

back to the first leveling screws and recheck. Readjust for levelness if necessary. Then return the head to the second position and recheck. Continue until the spirit level is within one graduation on either side of the center mark.

Now that the instrument head is level, be very careful that it is not bumped or jarred.

THE LEVELING ROD

When it is necessary to sight over long distances, a *leveling rod* is used as the target (Fig. 8-9). The rod is in two sections and can be raised to 12 ft. This makes it possible to sight over a small hill or obstruction. The rod has large graduations in feet and inches which can be easily read. When sighting distances less than 100 ft, a folding rule is often used. It can be taped to a strip of wood to make it rigid.

SIGHTING A LEVEL LINE

Sight the target along the outside of the barrel. This gets the target in range. Sight the target through the telescope (Fig. 8-10). Adjust the

FIGURE 8-10

Sight the target through the scope. *(Courtesy of David White Instruments)*

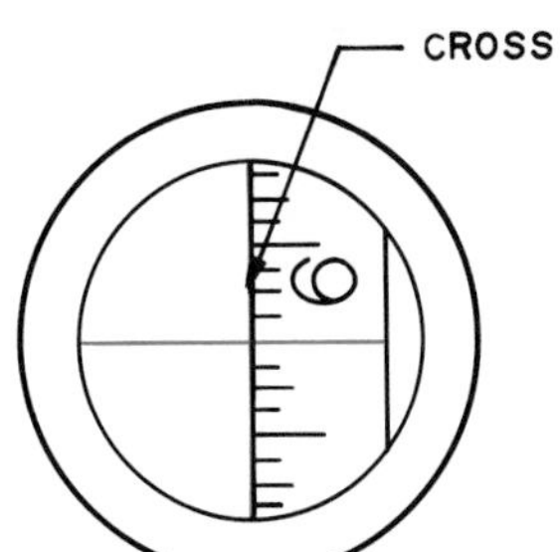
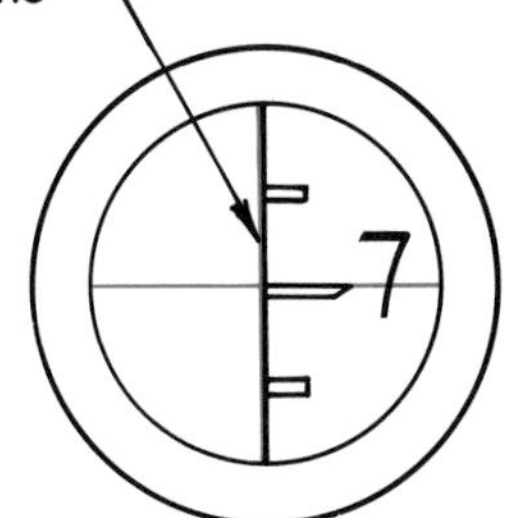
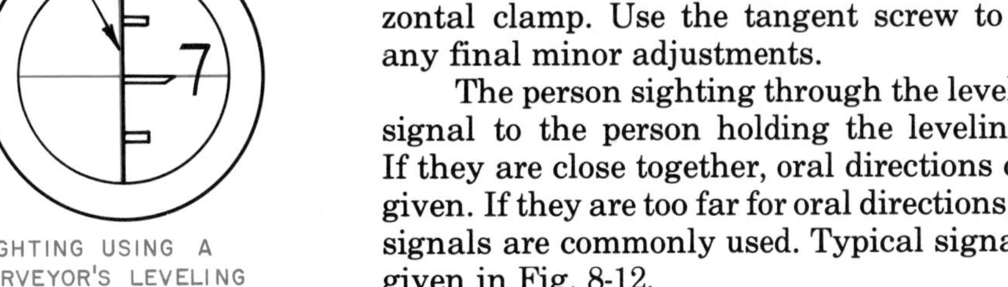
CROSS HAIRS
SIGHTING USING A CARPENTER'S RULER AS A ROD. MEASUREMENT READ IN INCHES.
SIGHTING USING A SURVEYOR'S LEVELING ROD. MEASUREMENT READ IN HUNDREDTHS OF A FOOT.

FIGURE 8-11

The cross hairs in the scope are lined up with the markings on the rod.

focus so that the target is clear. Move the telescope horizontally until the cross hairs align with the target (Fig. 8-11). Tighten the horizontal clamp. Use the tangent screw to make any final minor adjustments.

The person sighting through the level must signal to the person holding the leveling rod. If they are close together, oral directions can be given. If they are too far for oral directions, hand signals are commonly used. Typical signals are given in Fig. 8-12.

FIGURE 8-12

The surveyor uses hand signals to tell the person holding the rod which way to move it.

Move the rod to the left of the person holding it.

Move the rod to the right of the person holding it.

Raise the target.

Lower the target.

The target is plumb.

The target is on grade.

Often, it is necessary to find the difference in levels between several points. To do this, set the instrument level. Place a leveling rod at the first location and record the reading. Place it at the other points and record those readings. The difference between readings is the difference in the levels of these areas (Fig. 8-13).

A common application is setting the *grade stakes* at the corners of a building. The grade stake locates the desired elevation of the soil at each corner. To set the grade stakes, place the level near the center of the building (Fig. 8-14). Establish the desired elevation of the grade at one corner. Sight it and note the reading. Drive grade stakes at each corner and mark the location of the elevation on them.

Sometimes, the land slopes or rises so that the elevation requires some soil be removed or a low area filled. In these areas the grade stake is driven and a line (the grade stake mark) is drawn across it. The grade stake mark is the point from which the cut-and-fill measurements are taken.

If the area requires a fill to get the grade the proper height, a stake is driven and marked F (fill). The amount of fill required is written above a line on the stake. If the area requires some soil be removed to get to grade, a cut is needed. The stake is marked C and the amount of cut is written on it (Fig. 8-15).

If the building site slopes a great deal, it may not be possible to establish grade level with one sighting. This is done by establishing a height at the first point and an intermediate point up the slope. Move the level up the slope and take a second sighting (Fig. 8-16).

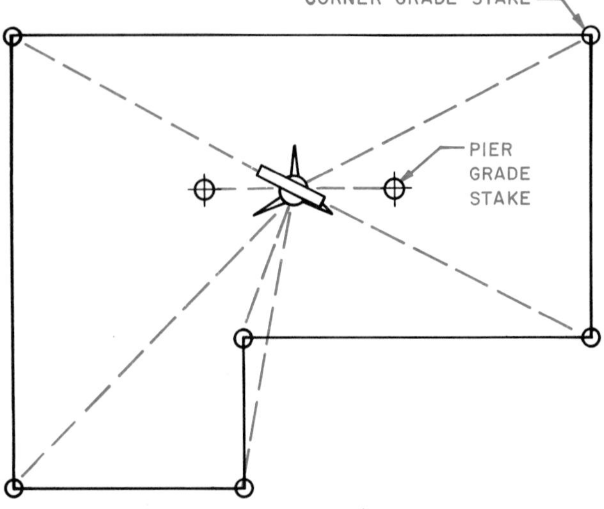

FIGURE 8-14

How to set the grade stakes at the corners of the house and for each pier.

LAYING OUT HORIZONTAL ANGLES

Set the level over the station mark. Line up the plumb bob with the mark. Level the instrument. Sight the leveling rod at one corner of the house (Fig. 8-17). Set the horizontal circle scale to zero to align with 0 on the vernier scale. Swing the level the required number of degrees to locate

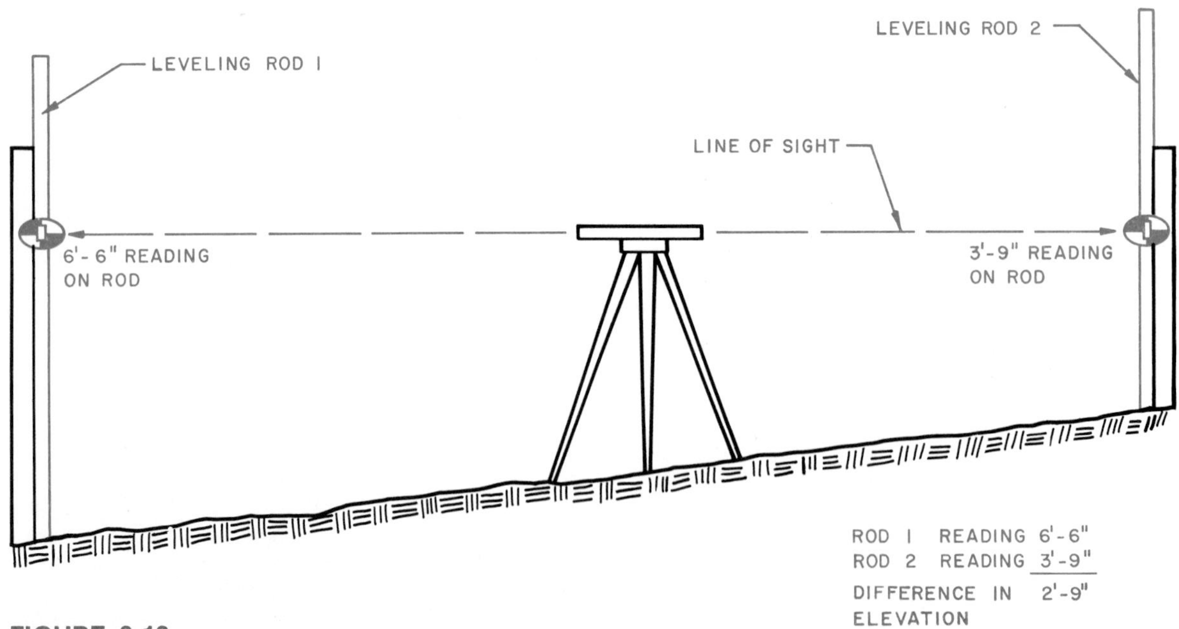

FIGURE 8-13

Using a level to find the difference between the elevations of two points.

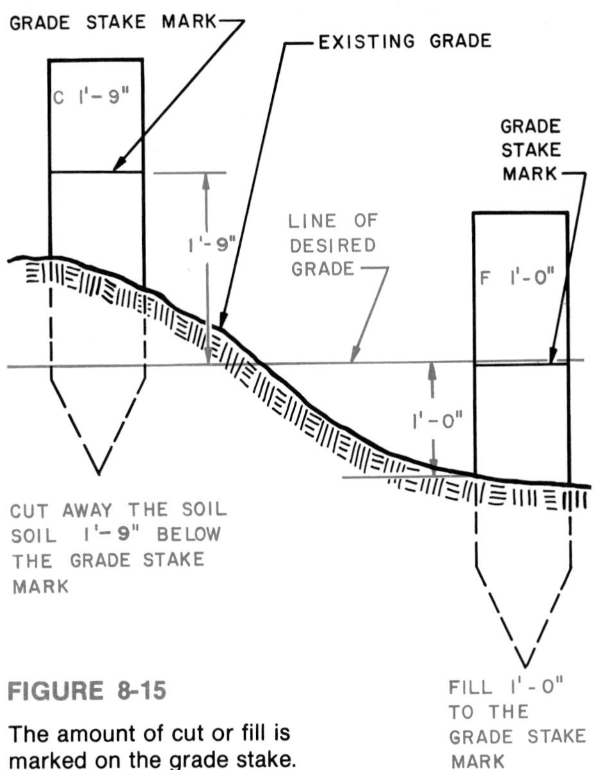

FIGURE 8-15

The amount of cut or fill is marked on the grade stake.

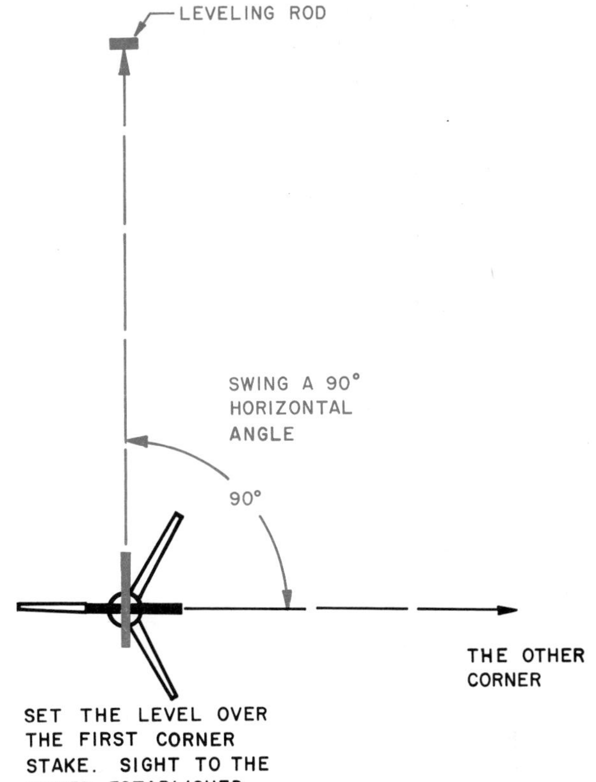

FIGURE 8-17

To lay out the angles of a corner first sight one corner and swing the level transit through the degrees forming the corner.

the other corner. In Fig. 8-18 it was 90°. Read the degrees on the horizontal scale. Position the leveling rod so that the cross hairs align with it. This produces an accurate horizontal angle.

Full degrees are read on the horizontal scale. The vernier is divided in 5-minute intervals. It is used to lay out an angle that is not in full degrees, such as 65°10′.

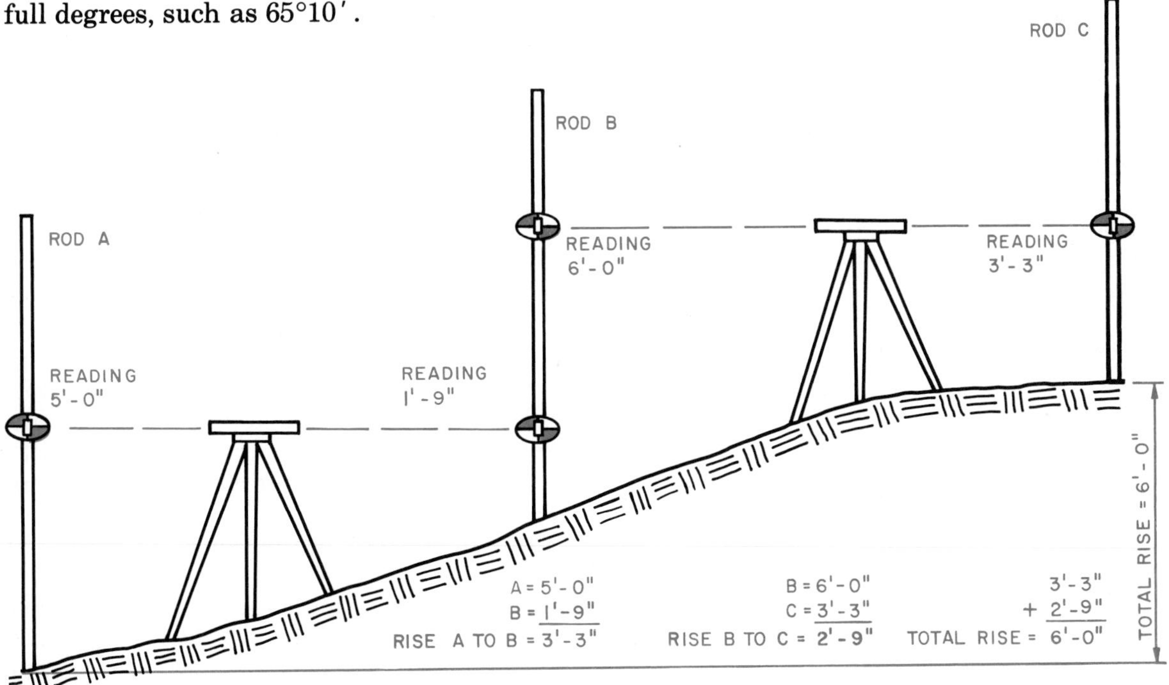

FIGURE 8-16

It may take several sightings to establish a grade level on a steep slope.

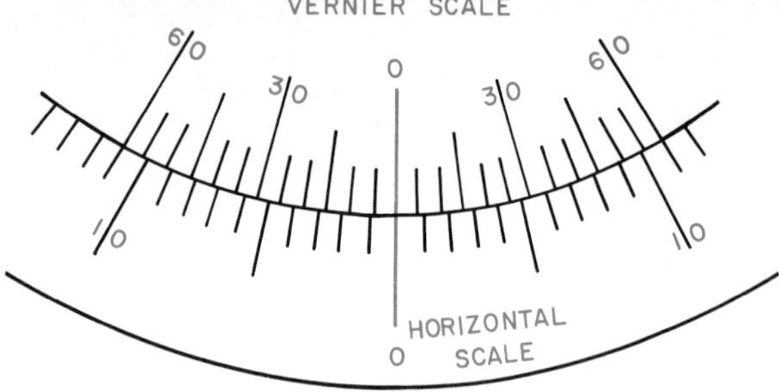

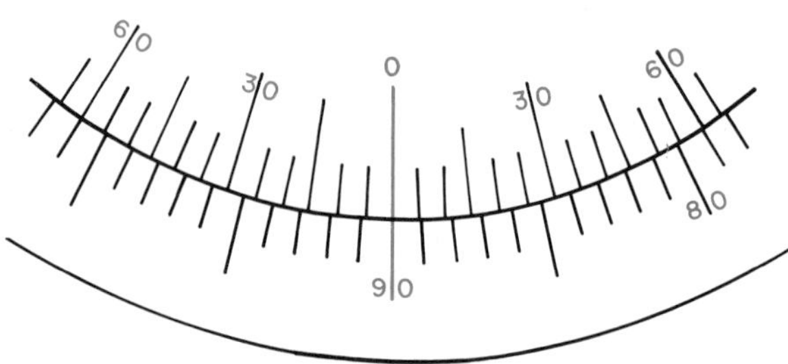

FIGURE 8-18

The angle of the corner is found by measuring the degrees on the vernier scale.

MEASURING VERTICAL ANGLES

First level the instrument as previously described. Swing the level horizontally until it is in line with the vertical surface involved. Lock it in position. Rotate the level vertically to the angles desired. The angles are shown on a vertical scale (Fig. 8-19). The points located will be in a vertical line.

A vertical member can be checked to see if it is plumb by sighting it as just described. Then move the level 90° from the first location and sight it again (Fig. 8-20).

FIGURE 8-19

Vertical angles are measured from the horizontal.

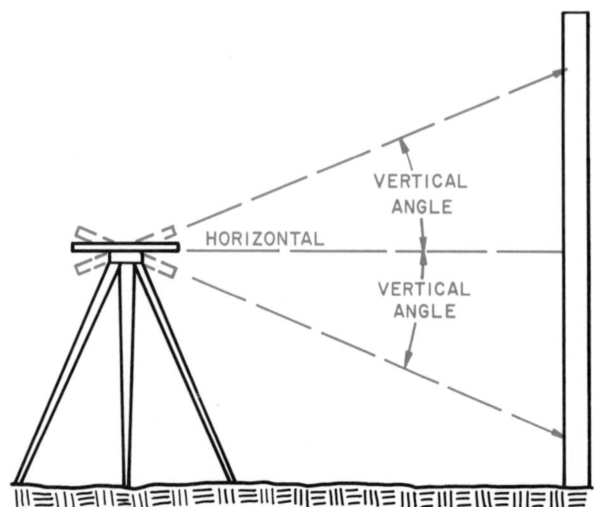

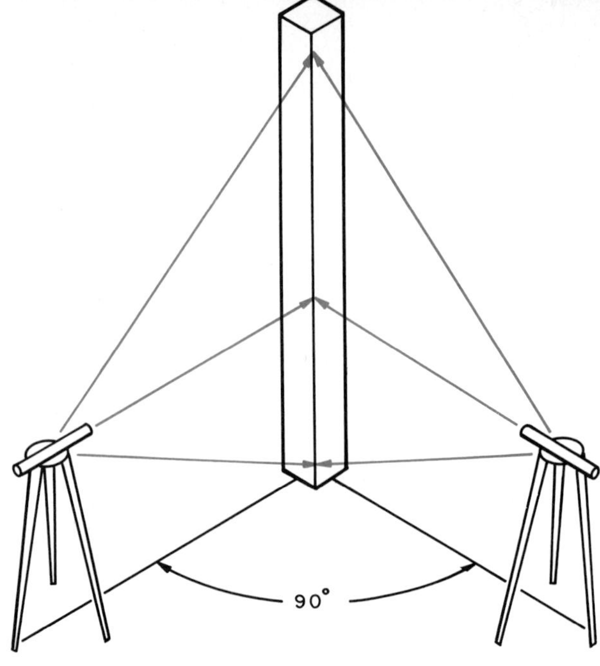

FIGURE 8-20

Vertical columns can be checked to see if they are plumb by sighting points on them from two sides.

LOCATING A BUILDING ON THE SITE

The plot plan shows the location of the sides of the lot. It also locates the house on the lot. Usually, a surveyor is employed to find and drive stakes at the corners of the lot. The surveyor may also locate the corners of the building on the lot. Sometimes, the carpenter or general supervisor must do this.

The building can be located in one of two ways:

1. Measurements can be made from an established *reference line* with a steel tape.
2. The corners can be located using a builder's level or a transit level.

LAY OUT A BUILDING WITH A LEVEL

First locate the front of the house by measuring from the front property line with a steel tape. Then locate the two front corners by measuring in from the sides of the property (Fig. 8-21). Drive a stake at each corner with a nail locating the exact location.

Now set the level at the first corner, A. Sight corner B. Set the horizontal and the vernier scales on 0. Swing the level 90° to find corner C. The distance from A to C is measured with a steel tape. Drive a stake at the corner (Fig. 8-22).

Move the level to corner B. Sight corner A and set the scales on 0. Swing the level 90° to find corner D. The distance from B to D is measured with a steel tape.

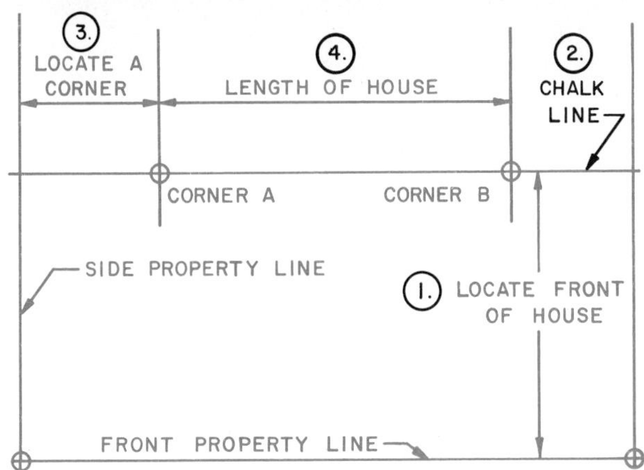

1. MEASURE IN FROM THE FRONT PROPERTY LINE WITH A STEEL MEASURING TAPE TO LOCATE THE FRONT OF THE HOUSE.
2. RUN A CHALK LINE ACROSS THE PROPERTY MARKING THE FRONT OF THE HOUSE.
3. LOCATE THE FRONT CORNER FROM THE SIDE OF THE PROPERTY WITH A MEASURING TAPE.
4. MEASURE THE LENGTH OF THE HOUSE WITH A MEASURING TAPE.

FIGURE 8-21

The steps to locate the front corners of a building on a lot.

FIGURE 8-22

The steps to locate the back corners of a building.

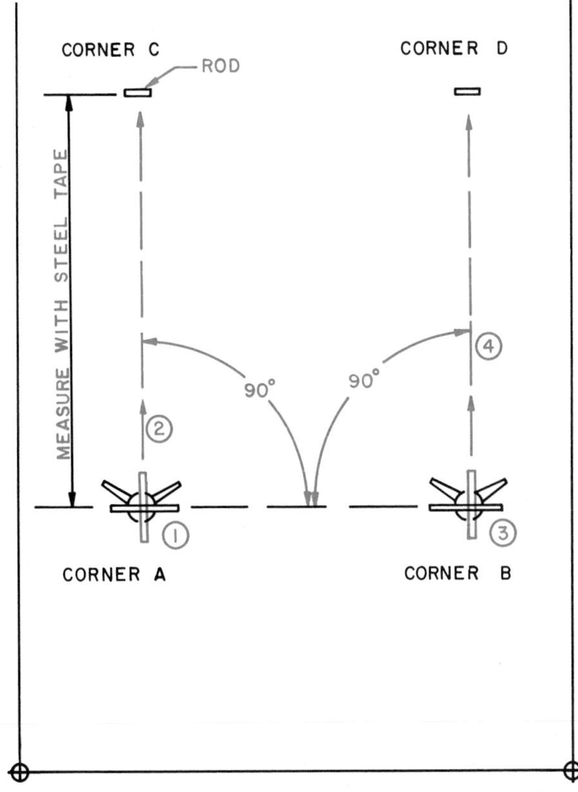

1. SET THE LEVEL ON THE FIRST CORNER. SIGHT CORNER B. SET THE HORIZONTAL SCALE AND THE VERNIER ON 0.
2. ROTATE THE LEVEL 90° TO FIND CORNER C. MEASURE THE DISTANCE TO THE CORNER WITH A STEEL TAPE.
3. MOVE LEVEL TO CORNER B. SIGHT CORNER A. SET SCALE ON 0.
4. ROTATE LEVEL 90° TO FIND CORNER D. MEASURE WITH A TAPE.

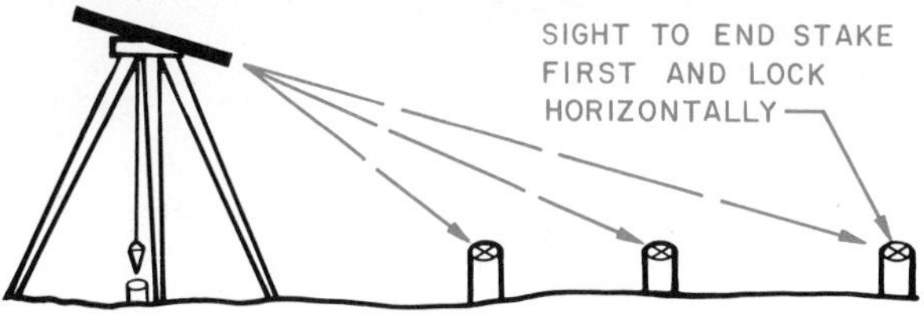

FIGURE 8-23

Several stakes can be set in line with a level transit.

Sometimes, it is necessary to locate several stakes along the line between two corners. After the corner is located with the level, rotate it in a vertical plane. By changing the angle, any number of stakes can be located (Fig. 8-23). This is useful when locating fence posts or staking the sides of a driveway.

MEASURING FROM A REFERENCE LINE

The following steps are shown in Fig. 8-24.

1. Locate a side of the building from a known reference line. In Fig. 8-24, the front of the house was located by measuring from the edge of the lot next to the street. Measure on both sides of the lot.
2. Run a chalk line to locate the front of the house.
3. Locate one side by measuring the distance it is from the side of the lot. Drive a stake

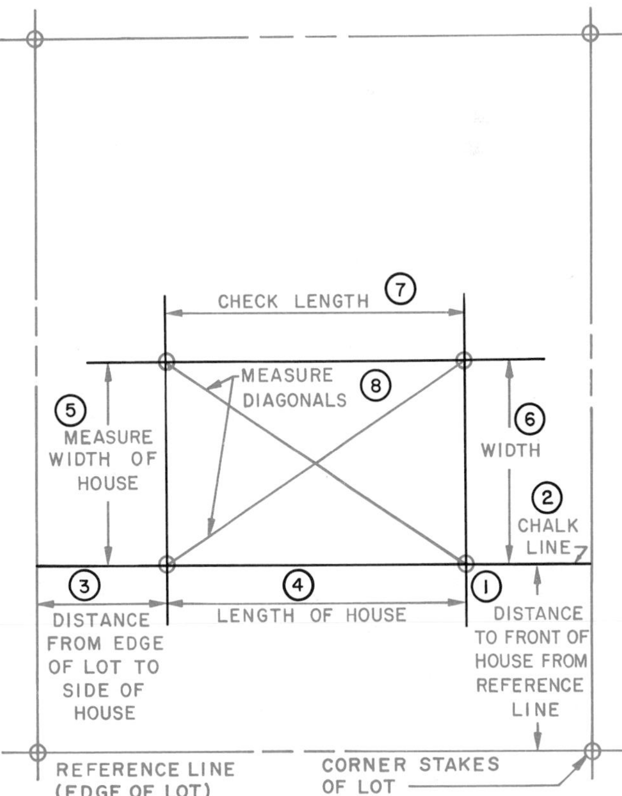

FIGURE 8-24

How to lay out a house measuring from a reference line.

at this point. Drive a nail in the stake at the exact location.

4. From this stake, measure the length of the house. Drive another stake with a nail giving the exact location.
5. From one front corner stake, run a chalk line perpendicular to the line locating the front of the house. Measure the width of the house along this line. Drive a stake at this corner with a nail in the exact location. Check for squareness using the 6, 8, 10 method (Fig. 8-25).
6. Repeat step 5 on the other side of the house.
7. Check by measuring the length of the house between the rear corner stakes. It should equal the front length.
8. Check the layout for squareness by measuring the diagonals. If the layout is square, the diagonals will measure exactly the same.

If a building is not rectangular, use the same procedures for locating the other corners. It could be broken down into several rectangular areas and each of these can be located.

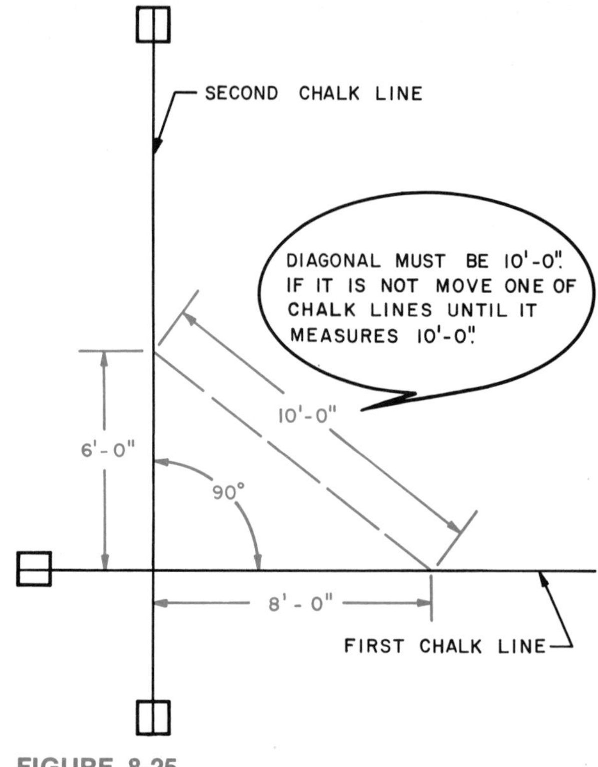

FIGURE 8-25
How to check to see if a corner is square.

STUDY QUESTIONS

After reading the chapter, answer each of the following questions. If you do not know the answer, review the chapter.

1. What is the difference between a builder's level and a transit level?
2. What are the steps to set up a level?
3. Explain how to find the difference in height of two places using a level.
4. What steps are followed to lay out a horizontal angle?
5. How can a vertical angle be measured with a transit level?
6. What is a reference line?
7. What is a station mark?

IMPORTANT TECHNICAL TERMS

Following are technical terms that you should be able to use as part of your working vocabulary. Write a brief description of the meaning of each term.

Vernier scales
Builder's level
Transit level
Tripod
Station mark
Leveling screws
Leveling rod
Grade stake
Reference line

9 FOOTINGS AND FOUNDATIONS

Building *concrete formwork* is one of the jobs a carpenter is expected to do. Forms are the units built to hold the concrete in place as it hardens. Some carpenters specialize in form construction. When small buildings are constructed, the carpenters generally do the formwork and all other carpentry on the building. When large multistory buildings are built, the carpenters are more likely to specialize in one area, such as formwork. Carpenters are needed anywhere concrete is to be formed and poured. This includes large projects such as bridges, dams, and sewage disposal projects (Fig. 9-1).

In addition to building forms that can be reused, carpenters build single-use forms. Figure 9-2 shows a draft tube form for a hydroelectric powerhouse. It is used only once but represents the ultimate in carpentry craftsmanship in heavy construction.

Properly constructed formwork is essential to the satisfactory construction of the building. The footings and foundation must be of the proper size and in the proper location. They must be level and of the proper height because they form the base upon which the building is built. The forms must be strong enough to withstand the pressures from the wet concrete and not bow or buckle.

A carpenter must carefully read the blueprints so that the forms are the proper size. Also, the footings must be below the frost line of the ground. The *frost line* is the depth to which ground freezes in the winter. This varies from as little as 1 ft in warm climates to 4 or 5 ft in cold climates. The architect will indicate the depth of the footing on the drawings.

FIGURE 9-1

Site-built forms are constructed and erected by carpenters. *(Courtesy of American Plywood Association)*

FIGURE 9-2

Single-use forms are often complex. *(Courtesy of Morrison-Knudsen Co., Inc.)*

CONCRETE

Concrete is a mixture of portland cement, sand, gravel, and water. When the cement is mixed with water, it sets and then hardens. This binds the sand and gravel into a firm substance.

The proportion of each ingredient in concrete varies depending upon its use. Most concrete used in construction is mixed in a plant and delivered to the site in specially designed trucks.

Concrete for foundations is made from regular type 1 portland cement. Ingredients can be added which *retard* the amount of heat produced when concrete dries. This is helpful when large quantities are poured because it greatly reduces the heat generated. *Air-entrained* portland cement concrete is resistant to frost. It has millions of air bubbles trapped inside. *High early strength* concrete sets quickly. This lessens the time it has to be protected from freezing, thus reducing construction costs. It also speeds up construction time.

General all-purpose concrete is completely cured in about 28 days. It is stronger if it is kept moist as it cures.

MATERIALS IN FORMWORK

The carpenter generally uses plywood, wood structural members, and metal ties to build formwork.

A special plywood called *plyform* is made especially for the form facing material. The plywood grades are shown in Fig. 9-3.

As forms are designed, the plywood panels must be of the proper thickness to support the pressure. In addition, the structural members, joists, studs, and walls must be of the proper size. For most residential work ⅝- or ¾-in. plyform is used. When a form is properly designed, the architect considers the rate at which concrete is poured, the job temperature, concrete slump, cement type, concrete density, the method of vibration, and the height of the form. The faster concrete is poured, the greater the pressure on the form. A pour of 1 to 2 ft in depth per hour is slow. A pour of 3 to 7 ft per hour is fast. Design tables for computing the structural design of concrete forms are available from the American Plywood Association, P.O. Box 2277, Tacoma, Wash. 98401.

There are various metal ties used to hold the forms together. They use some form of wedge, screw, or cam to tighten them (Fig. 9-4).

The wood structural members are the same as those used in building construction. The standard sizes are discussed in Chapter 7.

Steel reinforcing bars are placed inside the forms before the concrete is poured. They increase the ability of the concrete to withstand bending and tension stresses (Fig. 9-5). Wire mesh is used in concrete floor slabs. Reinforcing bars are specified by the architect by number. Each number represents ⅛ in. For example, a No. 4 bar is 4/8 or ½ in. in diameter (Fig. 9-6). Wire mesh is commonly available in 6 × 6 in. and 10 × 10 in. squares.

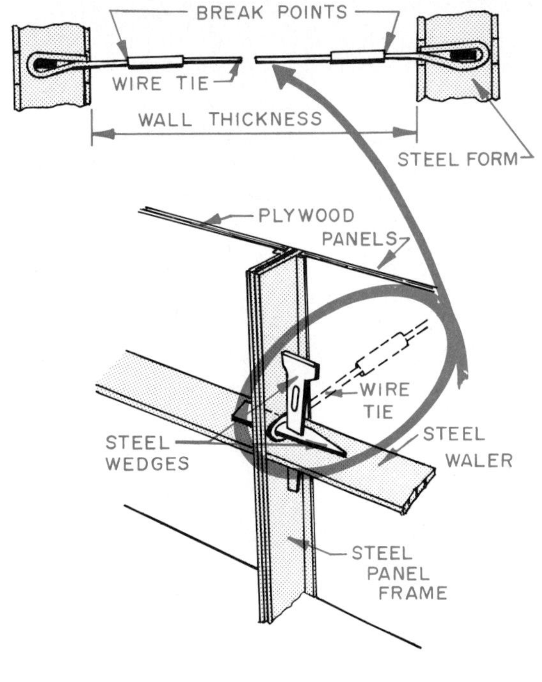

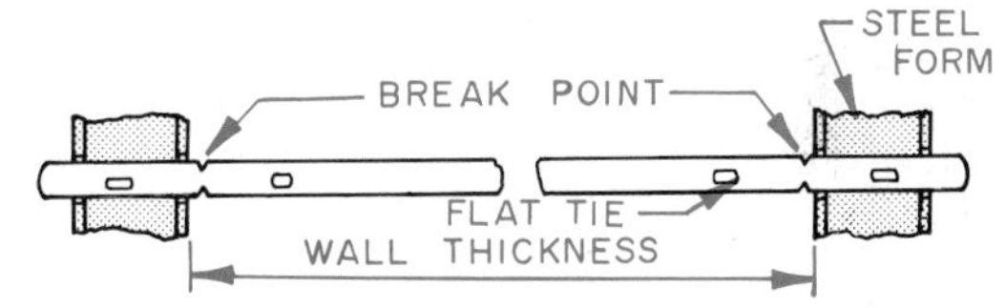

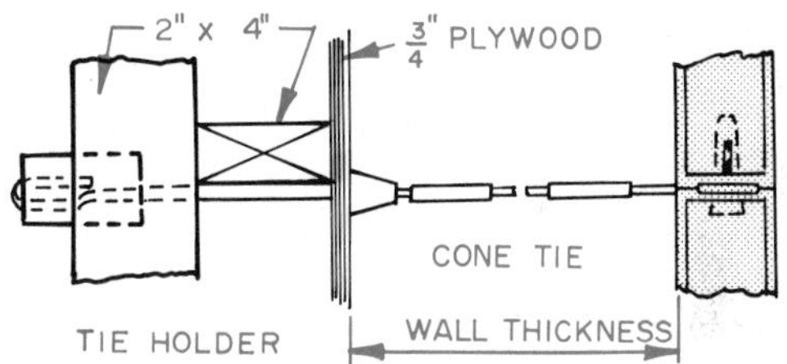

FIGURE 9-4

Typical metal ties for spacing forms. *(Courtesy of Symons Corporation)*

Grade-Use Guide for Concrete Forms (a)

Use These Terms When You Specify Plywood	Description	Typical Grade-Trademarks	Veneer Grade: Faces	Veneer Grade: Inner plies
B-B PLYFORM Class I & II (b) APA	Specifically manufactured for concrete forms. Many reuses. Smooth, solid surfaces. Mill-oiled unless otherwise specified.	B-B PLYFORM CLASS I APA EXTERIOR PS 1-74 000	B	C
High Density Overlaid PLYFORM Class I & II APA	Hard, semi-opaque resin-fiber overlay heat-fused to panel faces. Smooth surface resists abrasion. Up to 200 reuses. Light oiling recommended between pours.	HDO · PLYFORM-I · EXT-APA · PS 1-74	B	C Plugged
STRUCTURAL I PLYFORM APA	Especially designed for engineered applications. All Group 1 species. Stronger and stiffer than PLYFORM Class I and II. Recommended for high pressures where face grain is parallel to supports. Also available with High Density Overlay faces.	STRUCTURAL I B-B PLYFORM CLASS I APA EXTERIOR PS 1-74 000	B	C or C Plugged

(a) Commonly available in ⅝" and ¾" panel thickness (4'x8' size).

FIGURE 9-3

A guide for the selection of plywood for concrete-form construction. *(Courtesy of American Plywood Association)*

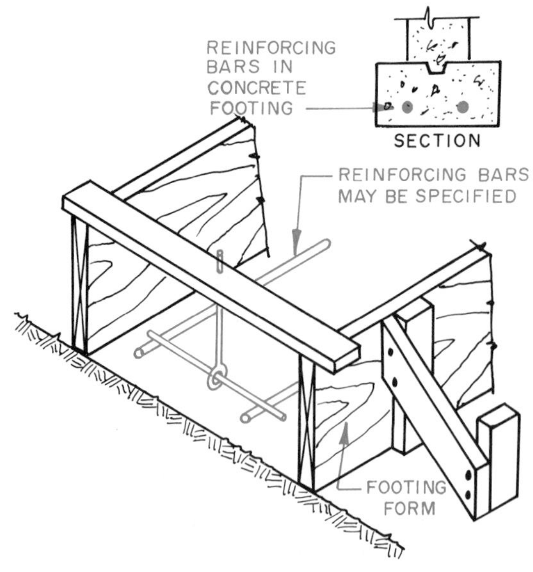

FIGURE 9-5

Steel reinforcing rods are used to strengthen footings.

FIGURE 9-6

A footing with reinforcing bars set to go in the foundation wall. *(Courtesy of Precise Forms, Inc.)*

SOIL TYPE	SAFE LOAD (tons/ft²)[a]
Soft clay; sandy loam	1
Firm clay; sand and clay mix; fine sand, loose	2
Hard dry clay; fine sand, compact; sand and gravel mixtures; coarse sand, loose	3
Coarse sand, compact; stiff clay; gravel, loose	4
Gravel; sand and gravel mixtures, compact	6
Soft rock	8
Exceptionally compacted gravels and sands	10
Hardpan or hard shale; sillstones; sandstones	15
Medium hard rock	25
Hard, sound rock	40
Bedrocks, such as granite, gneiss, traprock	100

[a]Safe load in tons per square foot of bearing surface of the footing.

FIGURE 9-7

Recommended safe loads for selected types of soils.

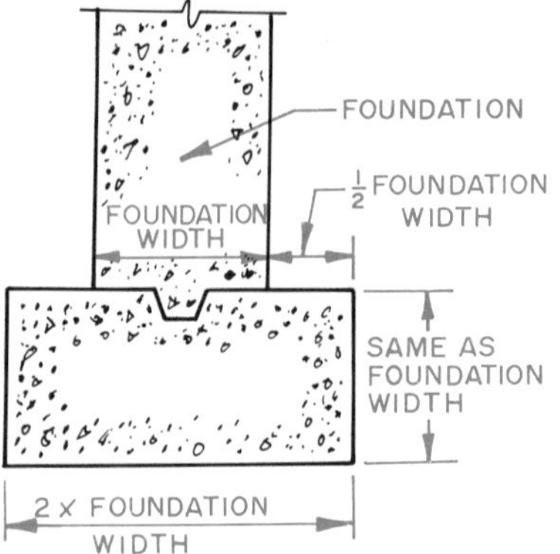

FIGURE 9-8

A typical rectangular footing.

TYPES OF FOOTINGS

The architectural engineer shows the design sizes of footings on the construction drawings. The carpenter builds them as drawn.

The *footing* is a concrete pad placed on the soil. The *foundation wall* is built on the footing. The footing holds the weight of the entire building and resists its sinking into the soil. The size of the footings depends upon the type of soil and the weight of the building. Soft clay and sandy loam will support only 1 ton per square foot. Other safe loads for soils are given in Fig. 9-7. The architectural engineer designs the width of the footing to support the calculated weight. The thickness of the footing depends upon the weight to be carried and the width of the footing. For most single-story buildings a rectangular footing is used. The width is twice the thickness of the foundation and the thickness is equal to the foundation thickness (Fig. 9-8).

Simple rectangular footings are sometimes formed by carefully digging a trench the size of the footing. The dirt walls serve as the form. Often this is not possible and wood forms are built. The following steps are typical of those used to build the wood forms.

1. Drop a line with a plumb bob from the *batterboard* lines that locate the building. This locates the corner of the building (Fig. 9-9).

2. Drive a wood 2 × 2 × 16 in. stake below the plumb bob. Drive a nail in the stake at the point where the plumb bob touches it. This is the corner of the building.

3. Drive the stake into the ground until its top is the exact level of the top of the footing to be poured. Check this height with the point of beginning on the batterboards. The leveling instrument is used to make this check. This is an important step because the elevation of the entire building depends upon it.

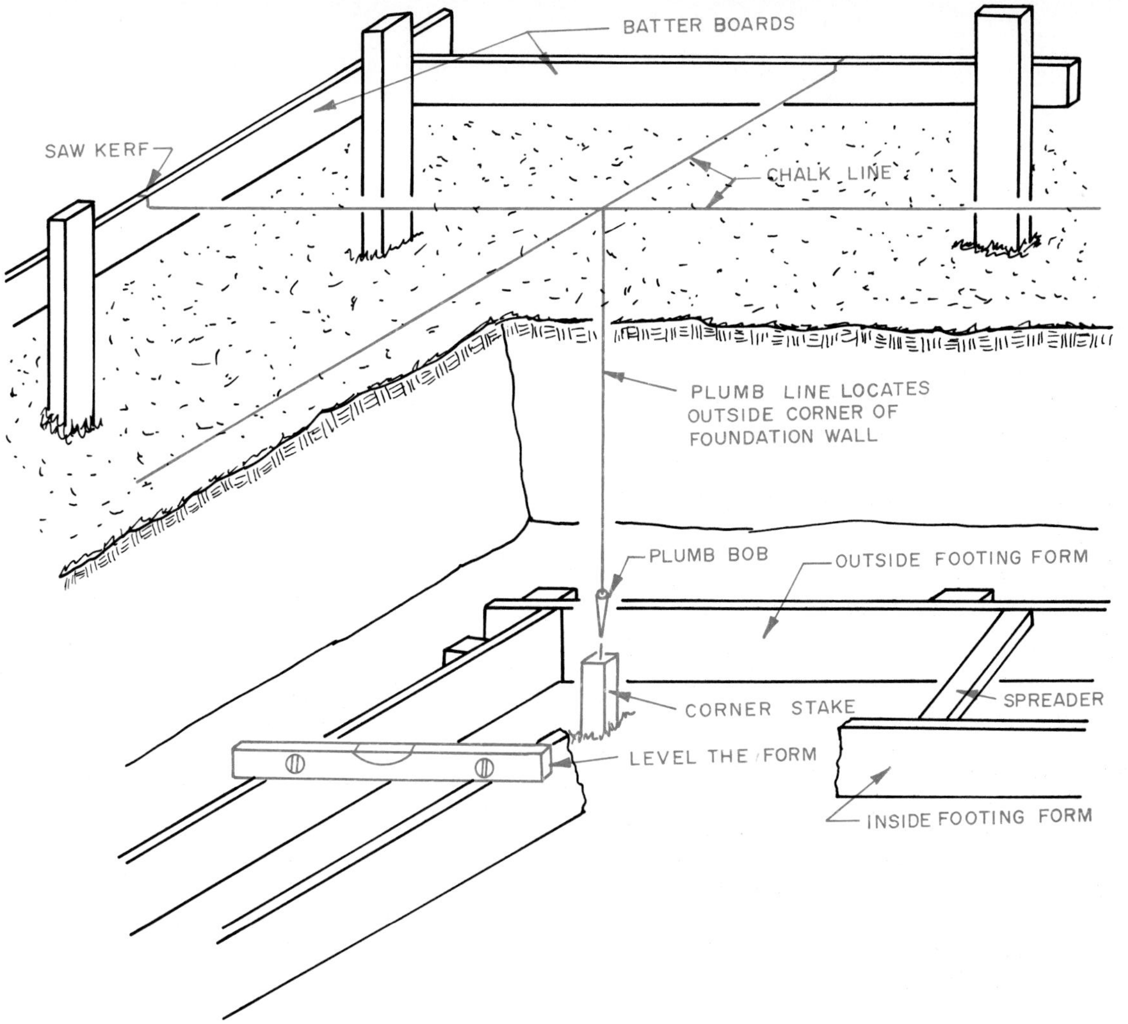

FIGURE 9-9

Batterboards, chalk line, and plumb bob are used to locate the corner stake.

4. Repeat these steps at the other corners of the building. Each corner will then have a stake locating it, and the top of the stake will be the exact height of the top of the footing.

5. Connect the nails in the corner stakes with a line. This forms the outer line of the building (Fig. 9-10).

6. Now place the footing form boards. Their location is measured from the building line. The distance of the outside form board from the building line is the width of the footing projection. The inside form board is located by measuring the width of the footing from the outside form board. The inside form board is located using spreaders. A spreader is a wood member whose length equals the width of the footing (Fig. 9-11). The spreaders are removed as the concrete is poured.

The top of the form boards must be on the same level as the top of the corner stake. These can be lined up with a level.

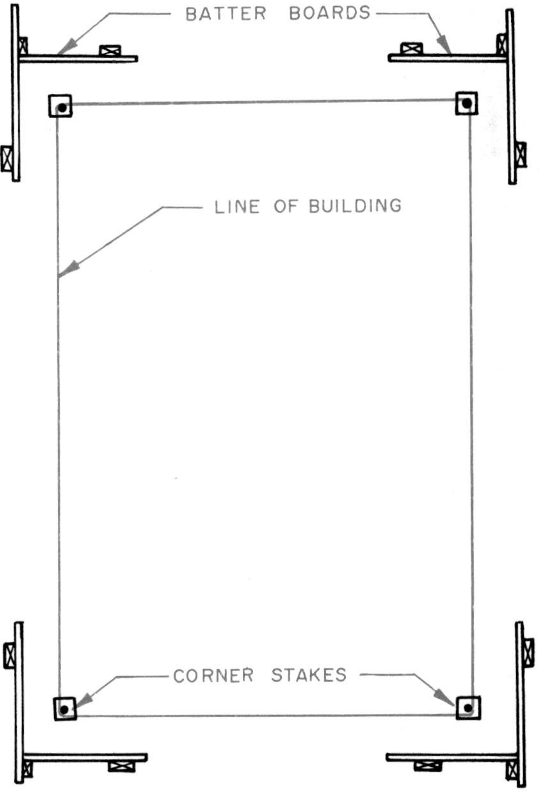

FIGURE 9-10

Corner stakes locate the line of the building.

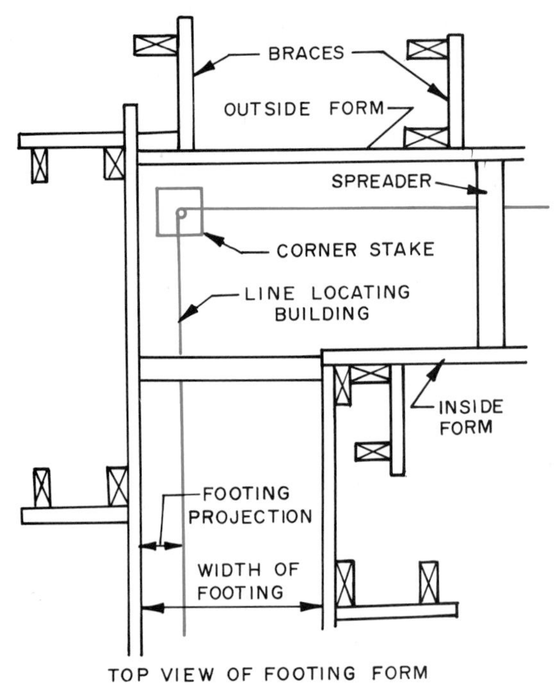

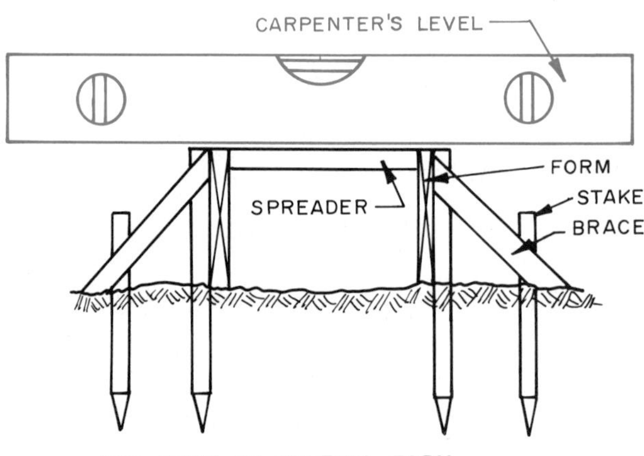

FIGURE 9-11

Form boards are staked and leveled.

The form boards are supported by stakes driven into the ground every 3 or 4 ft. They are braced (Fig. 9-11).

Often, *metal ties* are used to help hold the footing forms to the proper width. They are placed about every 4 ft.

7. Before pouring the concrete, recheck the level of the form boards.

8. Now the footings can be poured. Care should be taken when placing the concrete so that the forms are not moved or damaged.

9. After the footings are poured and the concrete has begun to set, a *keyway* can be formed. A keyway is used to tie the foundation wall to the footing (Fig. 9-12). The keyway is usually made by pressing a 2 × 4 in. wood member into the concrete before it sets. The member is slanted on the sides to make removal easier.

10. After the concrete is hard, the form boards are removed.

If the width of the footing exceeds certain design limits it must be made thicker. A wide but thin footing will tend to crack. The thickness is added by forming a step in the footing (Fig. 9-13). Typical wood forms for forming this spread footing are shown in Fig. 9-14.

If a building is built into the side of a hill, the footing might be stepped (Fig. 9-15). This reduces the amount of excavation needed and saves money because the foundation wall requires less forming and concrete. Stepped footings have horizontal and vertical steps. The vertical step is generally not higher than three-fourths of the horizontal distance between steps. The horizontal distance between steps should be not less than 2 ft 0 in. (Fig. 9-16).

Piers and columns rest on footings. A *pier* is a masonry unit that rests on a footing. It is used to support a section of a building, such as

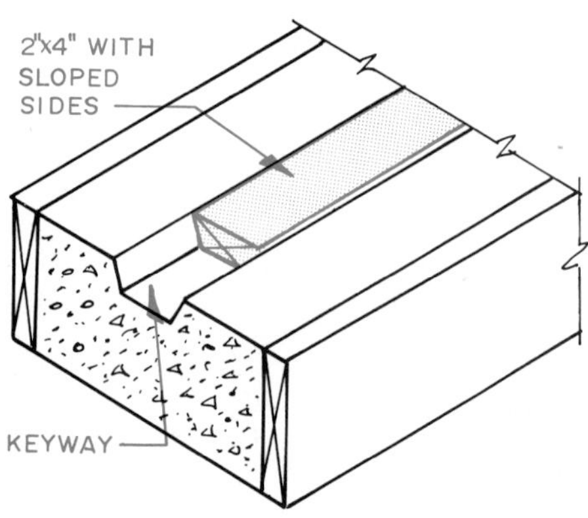

FIGURE 9-12

The keyway in the footing is formed with a tapered 2 × 4 in. member.

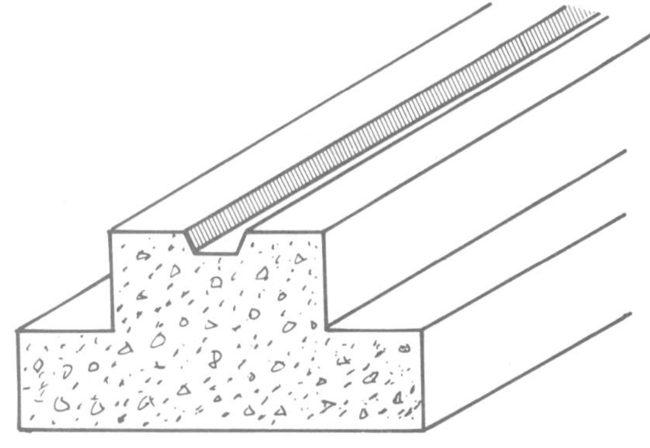

FIGURE 9-13

A footing thickness must be stepped if the projection beyond the foundation wall exceeds the thickness of the footing.

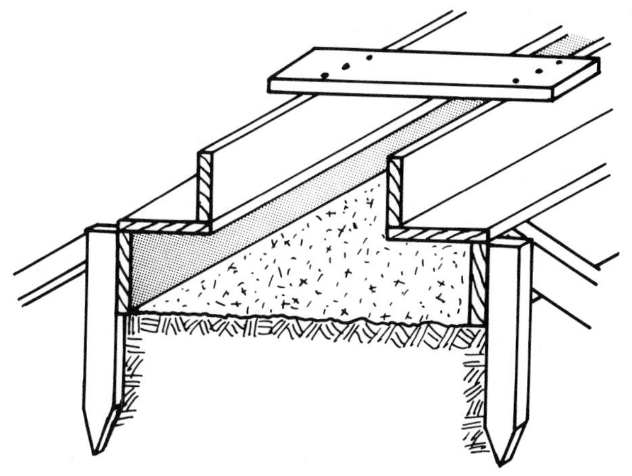

FIGURE 9-14

Typical site-built forms for a stepped footing.

part of the floor (Fig. 9-17). The footing is designed in the same manner as just described. A *column* is a wood, steel, or concrete post used to support a beam (Fig. 9-18). A typical wood footing form is shown in Fig. 9-17.

Sometimes, a pier or column footing must be stepped. This is needed to get sufficient strength to carry the design load. A typical form for a stepped pier or column footing is shown in Fig. 9-19.

Tapered column footings are formed as shown in Fig. 9-20. The end forms are made the exact size of the footing. The side forms are larger so that the cleats can be attached.

Some structures are built using piers and a grade beam. This is usually done where the soil is not stable and it is necessary to go deep to

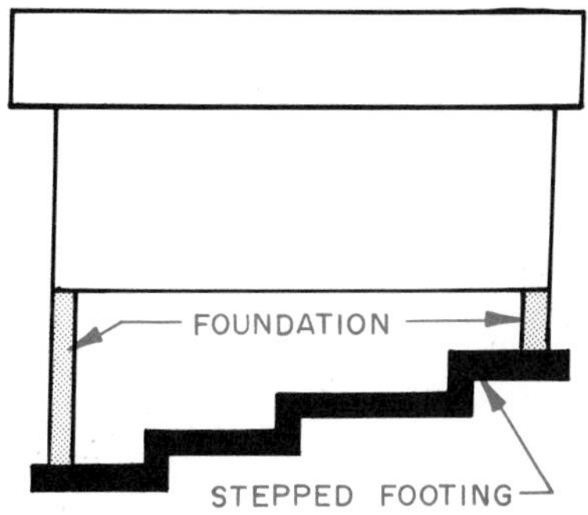

FIGURE 9-15

Footings may be stepped up a slope to save labor and materials.

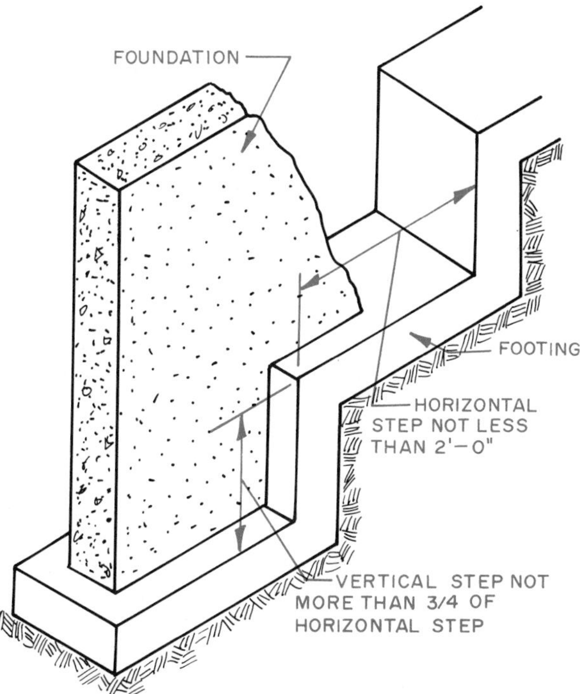

FIGURE 9-16

Typical design details for a footing stepped up a slope.

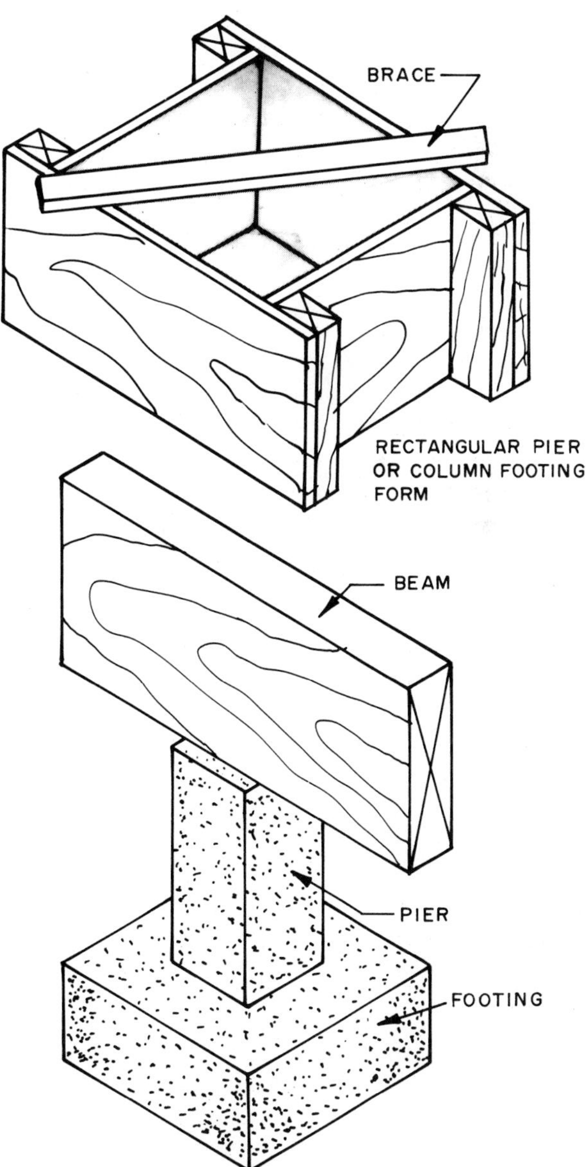

FIGURE 9-17

Typical pier footing and its forms.

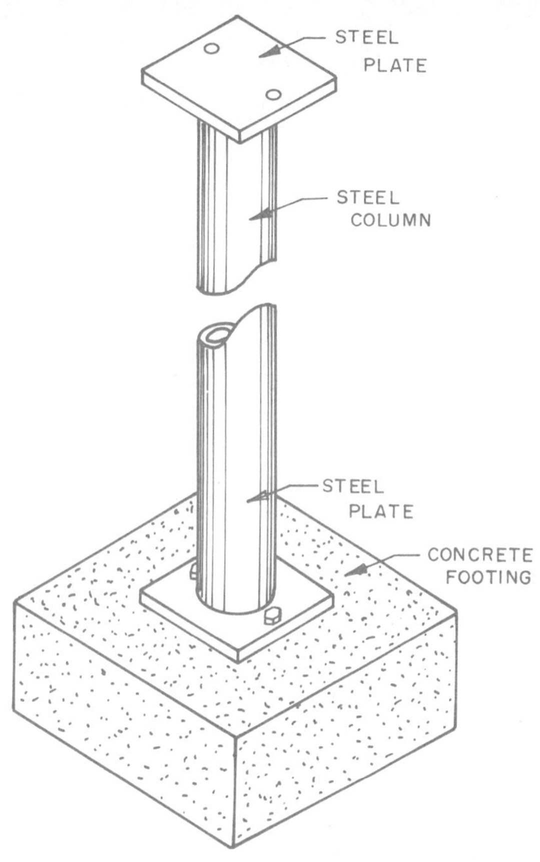

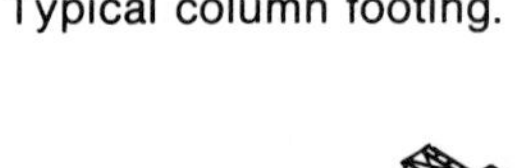

FIGURE 9-18

Typical column footing.

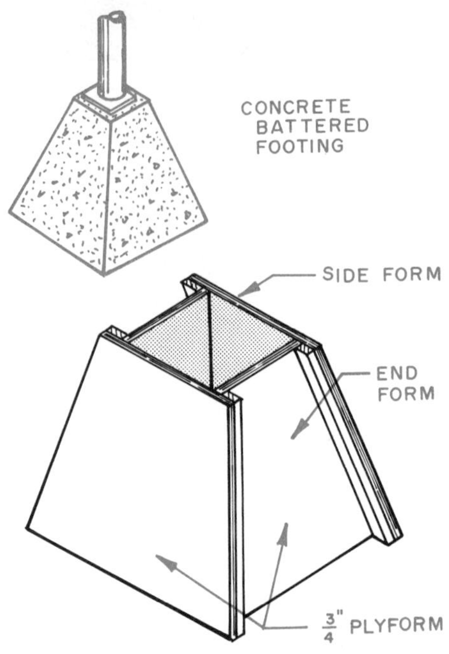

FIGURE 9-20

Forms for a tapered footing.

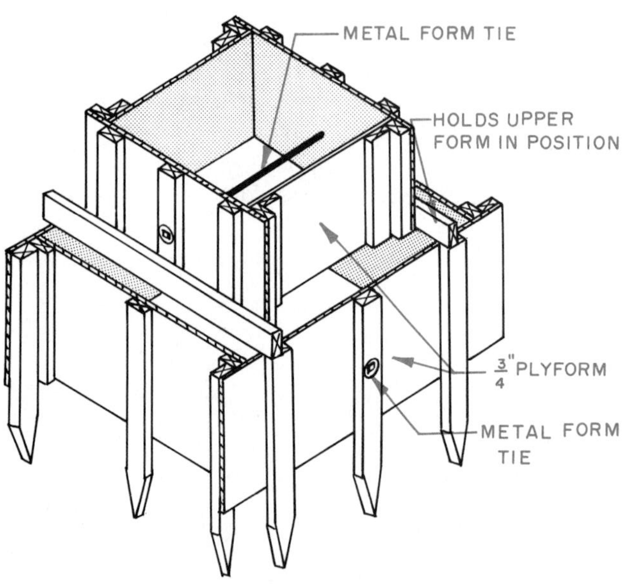

FIGURE 9-19

Forms for a stepped column footing.

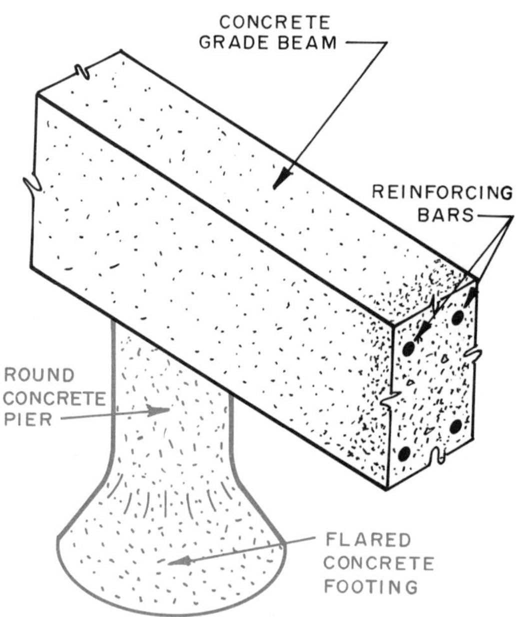

FIGURE 9-21

A typical grade beam.

reach a solid soil layer. Piers are set into the ground. They are often formed by using a large auger to drill holes in the soil. A 12-in.-diameter hole is common. The grade beam is a concrete member formed and cast on top of the pier. Together they form the foundation upon which the building is built (Fig. 9-21).

FORMING WALLS

There are several types of wall-forming systems in use. For most work prefabricated panels are used. These may be wood or metal (Fig. 9-22). Some special foundation designs require carpenters to custom build forms.

FIGURE 9-22

Metal forms are used to form concrete walls. *(Courtesy of Precise Forms, Inc.)*

FIGURE 9-23

Metal forms with patterned surfaces are used to produce textured concrete surfaces. *(Courtesy of Precise Forms, Inc.)*

Forms must be strong enough to support loads without bending or breaking. They should not leak.

The finished surface of the concrete depends upon the surface of the inside of the form. Textures are produced using special plywood or metal forms with textured surfaces (Fig. 9-23). Textured form surfaces tend to increase the force needed to strip the forms. This limits the number of times the form can be used. Sometimes film coatings, such as lacquer, polyurethane, or epoxy, are used with a release agent to make stripping of plywood forms easier.

Most complex, special-built forms use plywood as the surface material. See the materials section earlier in the chapter for details. Since plywood has good insulating qualities, it helps level out temperatures during curing. This provides more consistent curing temperatures. It can be easily cut and formed to build special forms. No special equipment is needed.

After plywood forms are removed, they should immediately be cleaned and repaired. Holes should be patched. A fiber brush is used to clean the surface. The surface could be lightly coated with mill oil. A frequently used mill oil is a 100 or higher viscosity pale-colored oil. A parting agent may be used instead of oil. Some parting agents, such as wax or silicones, affect the surface, so that painting the concrete is a problem.

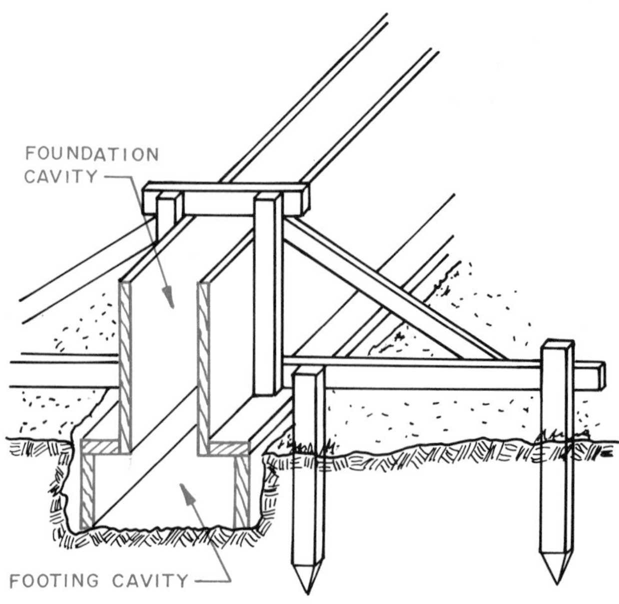

FIGURE 9-25

A wood form for monolithically casting a footing and a low wall.

Low Wall Forms Low wall forms are those under 4 ft in height. They commonly are made from ¾-in. plywood and 2 × 4 in. wood framing material. The studs are generally placed about 2 ft apart. It is important to brace the form at the bottom since the pressures from the liquid concrete are greatest there. Inclined braces are needed to support the top of the form.

A wood form tie is nailed to the studs at the top of the form. This holds the form at the proper width and prevents it from opening as the concrete is poured.

A typical wood form is shown in Fig. 9-24. The footing is cast first and the wall is formed after it has hardened.

Forms for casting the footing and low wall *monolithically* are shown in Fig. 9-25. "Monolithically" refers to casting them at the same time.

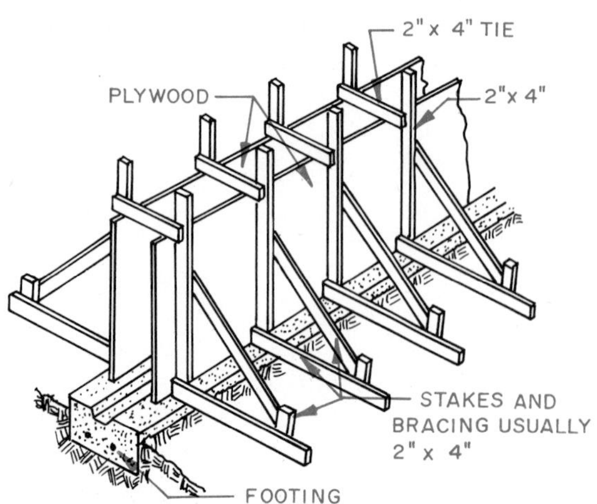

FIGURE 9-24

A typical wood form for a low wall.

Wales can be used to keep low wall forms in line (Fig. 9.26). A finished low foundation wall with the forms removed is shown in Fig. 9-27.

Full Wall Forms Forms for a conventional basement wall are usually made in 2 × 8 and 4 × 8 ft panels. Designs for wood panels are shown in Fig. 9-28. These are often held together with carriage bolts or duplex nails. This makes it easy to

FIGURE 9-26

Wales are used to keep the forms in line.

FIGURE 9-27

A low foundation with the forms removed. *(Courtesy of Precise Forms, Inc.)*

remove the forms (Fig. 9-29). Wales are used to stiffen wood forms. A *wale* is a horizontal structural member placed outside the form. A wall form using these panels is shown in Fig. 9-30.

Another form design uses vertical studs (Fig. 9-31). This system also uses wales for support and stiffening. Wales may be single or double studs, depending upon the strength needed. The wales are reinforced at the corner by corner ties (Fig. 9-32). The 2 × 4 in. members are nailed to the wales to hold the corner tight. Usually, *duplex nails* are used.

A typical double wall form of this type is shown in Fig. 9-33.

Metal form ties are used to help keep the forms the proper distance apart and strengthen them when concrete is poured. There are many different types available. One type is a rod with snap points (Fig. 9-34). The form tie goes through holes in the sheathing and studs. Some go through the wales or between double wales. This type is held tight by wedges pressing against the wales. The forms usually have the holes for the ties predrilled before they are set up on the footing.

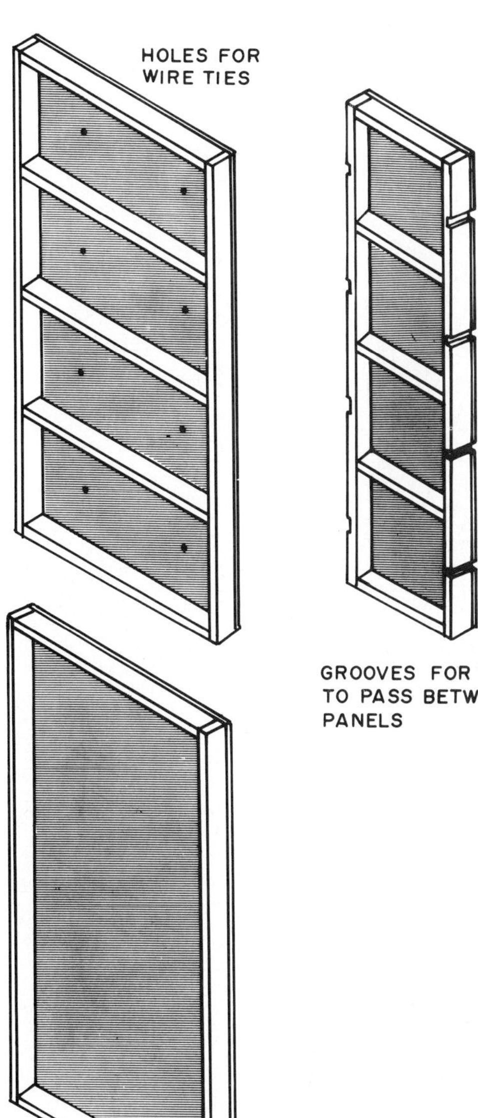

FIGURE 9-28

Types of wood forms.

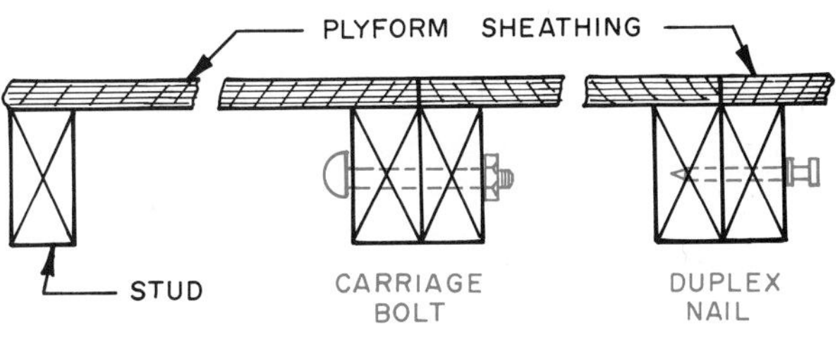

FIGURE 9-29

Typical ways of joining wood forms.

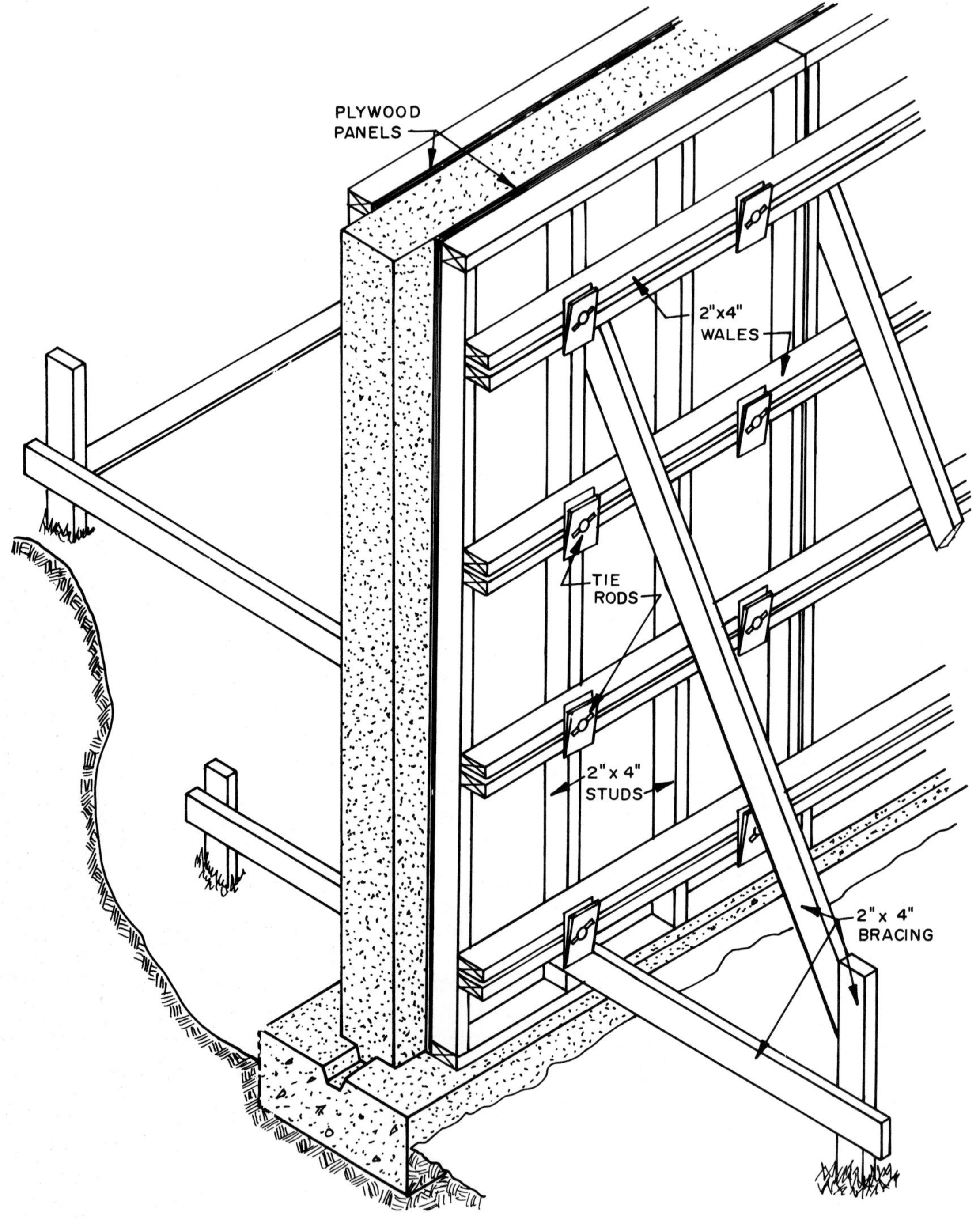

FIGURE 9-30

Full wall forms require several wales.

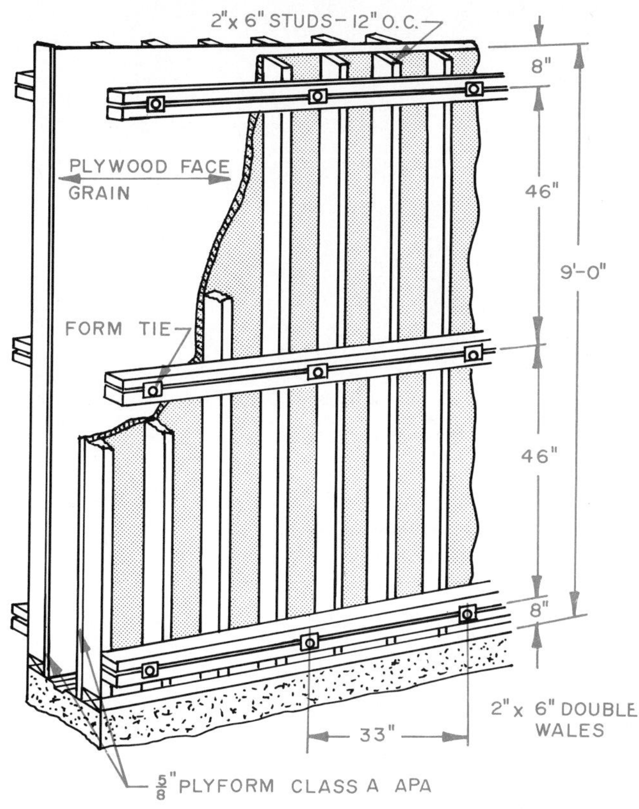

FIGURE 9-31

Another type of wood full wall form. *(Courtesy of American Plywood Association)*

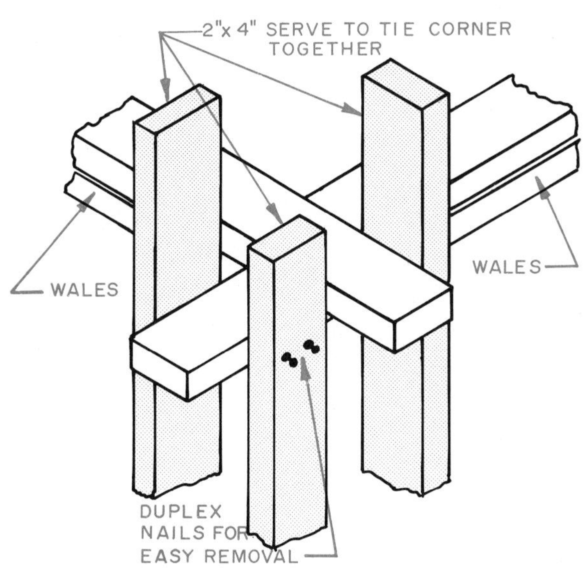

FIGURE 9-32

Wales are reinforced at the corners with corner ties.

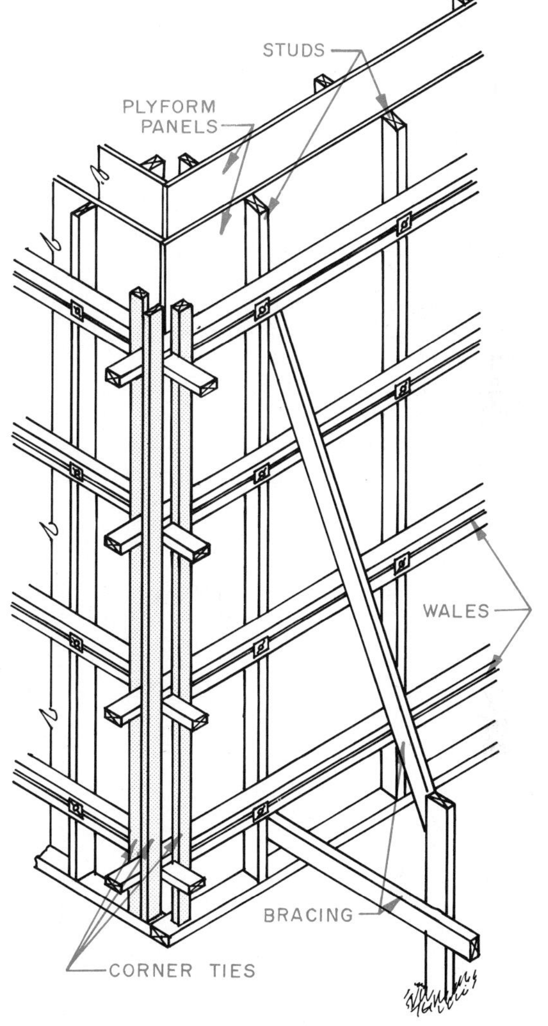

FIGURE 9-33

A full wall form with wales and corner ties.

Face form with ties in place.

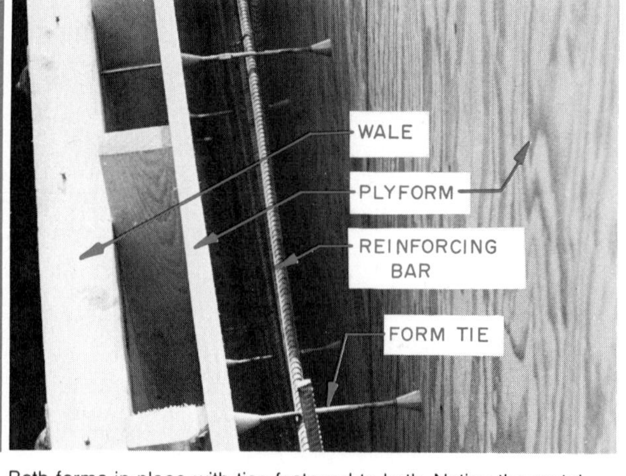

Both forms in place with ties fastened to both. Notice the metal reinforcing rods.

FIGURE 9-34

Metal ties are used to space and strengthen forms.

FIGURE 9-35

Metal forms with horizontal reinforcement built into the form. The steel wales are placed vertically to align the form. *(Courtesy of Symons Corporation)*

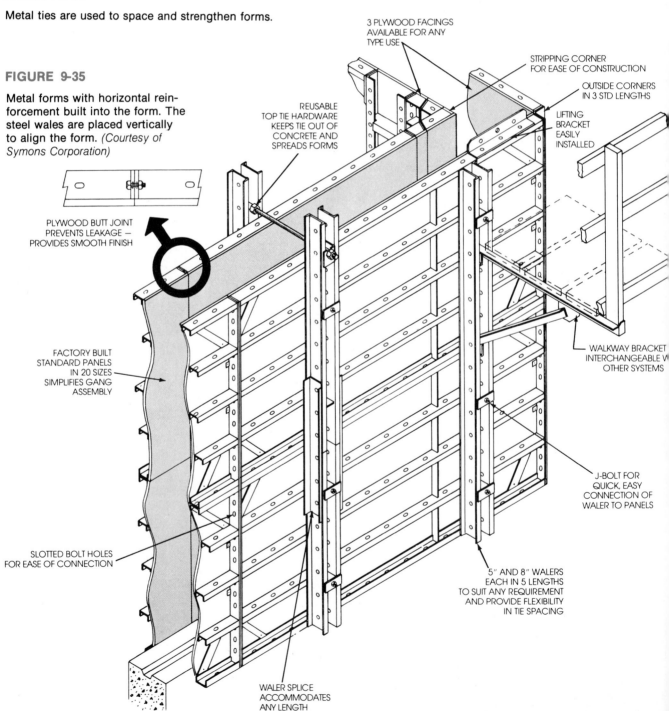

3 PLYWOOD FACINGS AVAILABLE FOR ANY TYPE USE

STRIPPING CORNER FOR EASE OF CONSTRUCTION

OUTSIDE CORNERS IN 3 STD LENGTHS

REUSABLE TOP TIE HARDWARE KEEPS TIE OUT OF CONCRETE AND SPREADS FORMS

LIFTING BRACKET EASILY INSTALLED

PLYWOOD BUTT JOINT PREVENTS LEAKAGE — PROVIDES SMOOTH FINISH

WALKWAY BRACKET INTERCHANGEABLE W OTHER SYSTEMS

FACTORY BUILT STANDARD PANELS IN 20 SIZES SIMPLIFIES GANG ASSEMBLY

J-BOLT FOR QUICK, EASY CONNECTION OF WALER TO PANELS

SLOTTED BOLT HOLES FOR EASE OF CONNECTION

5″ AND 8″ WALERS EACH IN 5 LENGTHS TO SUIT ANY REQUIREMENT AND PROVIDE FLEXIBILITY IN TIE SPACING

WALER SPLICE ACCOMMODATES ANY LENGTH REQUIREMENTS

Some types of forms have strong horizontal members built into the form. These systems use vertical wales (Fig. 9-35).

When the forms are stripped from the concrete wall, the rods are broken at the snap point. This leaves a small hole usually about 1 in. deep. The hole is filled with mortar.

A job that is being stripped is shown in Fig. 9-36.

Setting Up the Forms Using a plumb bob, drop a line from the intersection of the building lines on the batterboards to the footing. Mark the location of each corner on the footing.

Run a chalk line from one corner mark to the next on the footing. Chalk the line and snap a chalk line on the footing. Repeat this around all sides of the building. The chalk line mark on the footing locates the outside face of the foundation wall (Fig. 9-37).

Next, stand a preassembled plyform panel on the footing. Start by setting up a corner. Line up the face of the panel with the chalk mark. Nail the sole into the green concrete footing. Add a temporary brace. Then set the next panel. Join it to the first with bolts or duplex nails. Nail the sole to the footing and brace it.

When one wall is set, insert the form ties in the holes. Some forms have the ties pass through holes in the form. In this case the inner and outer panels must be placed so that the holes line up. Other forms have the ties pass in grooves between the panels. The panels must be placed so that their edges line up, to permit the ties to pass through them. Usually, the wall will require small panels to fill out the space evenly.

FIGURE 9-36

Forms being stripped from a foundation. *(Courtesy of Precise Forms, Inc.)*

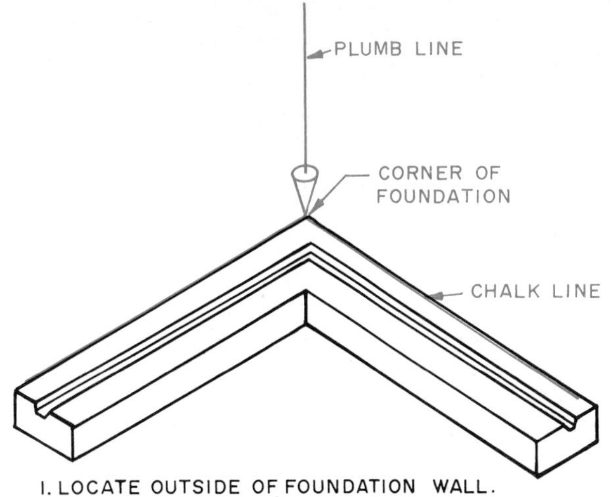

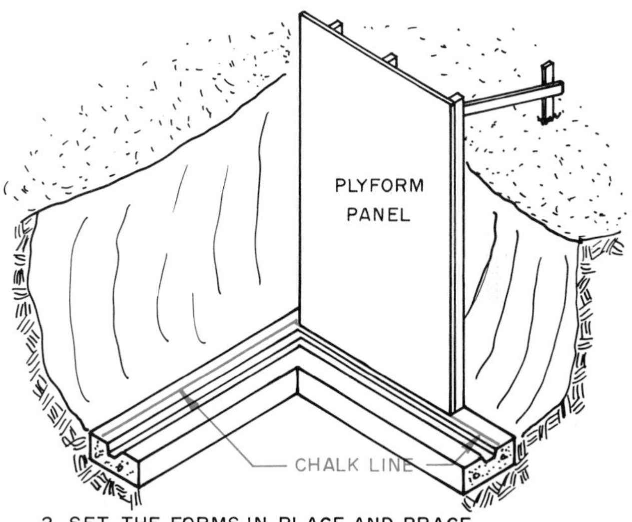

FIGURE 9-37

A chalk line is used to locate the outer line of the foundation on the footing.

Now set the wales in the wale brackets (Fig. 9-38). Place the form wedge lightly in place.

Set the permanent bracing as the wall is lined up. The verticalness can be checked with a large level. The straightness can be checked by running a chalk line the length of the wall (Fig. 9-39). Adjust it so that it is plumb and true and nail the permanent bracing tight.

Next, reinforce the corners by nailing 2 × 4 in. corner ties to the wales.

Set the forms for the inside face of the wall in the same manner as those for the outside. Insert the form ties through the holes in the form. Install the wales and wedges (Fig. 9-40).

Bracing the Forms Typical bracing systems are shown in Figs. 9-30 and 9-33. As the forms are put in place, temporary braces are often used. These are removed when the permanent bracing

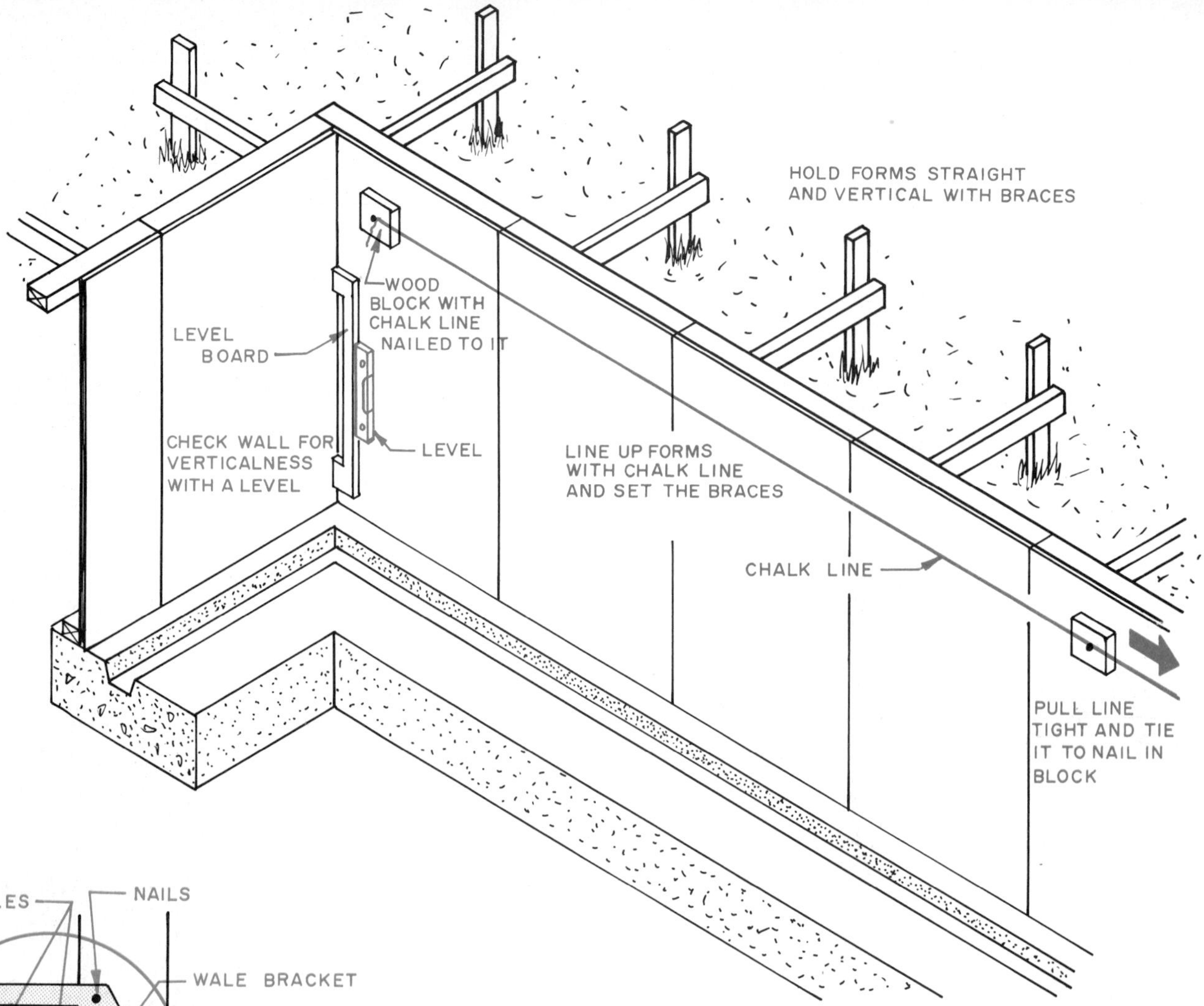

FIGURE 9-39

A chalk line can be used to line up a wall as the permanent braces are set.

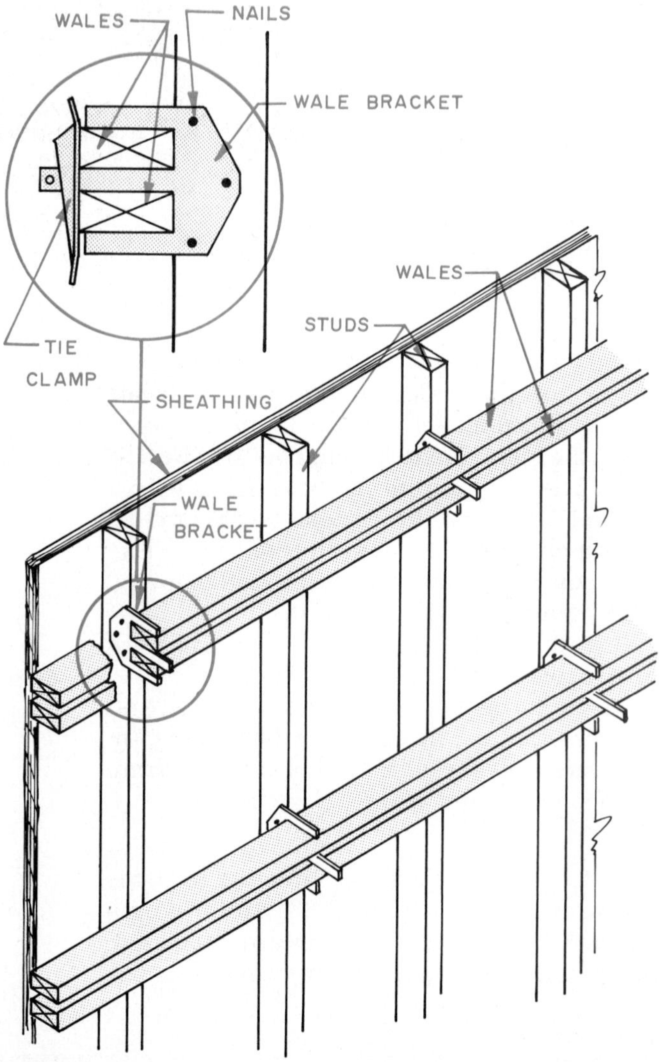

FIGURE 9-38

Walls can be held with wale brackets.

is installed. The bottom of the form has the greatest pressure and must be strongly braced to keep the form from slipping. The top of the form tends to spread and must also be braced. The top can be held with horizontal or diagonal braces. Often both are used (Fig. 9-41).

An adjustable brace is available (Fig. 9-42). It helps to hold the form in place as it is assembled. It makes it easy to adjust the alignment of the form since its length can be easily changed with the screw.

The stakes in the ground should be 2 × 4 in. material and driven several feet deep. If the ground is soft, they must be even deeper. Watch for cracked or split material. If so weakened, they may break and permit a portion of the form to buckle.

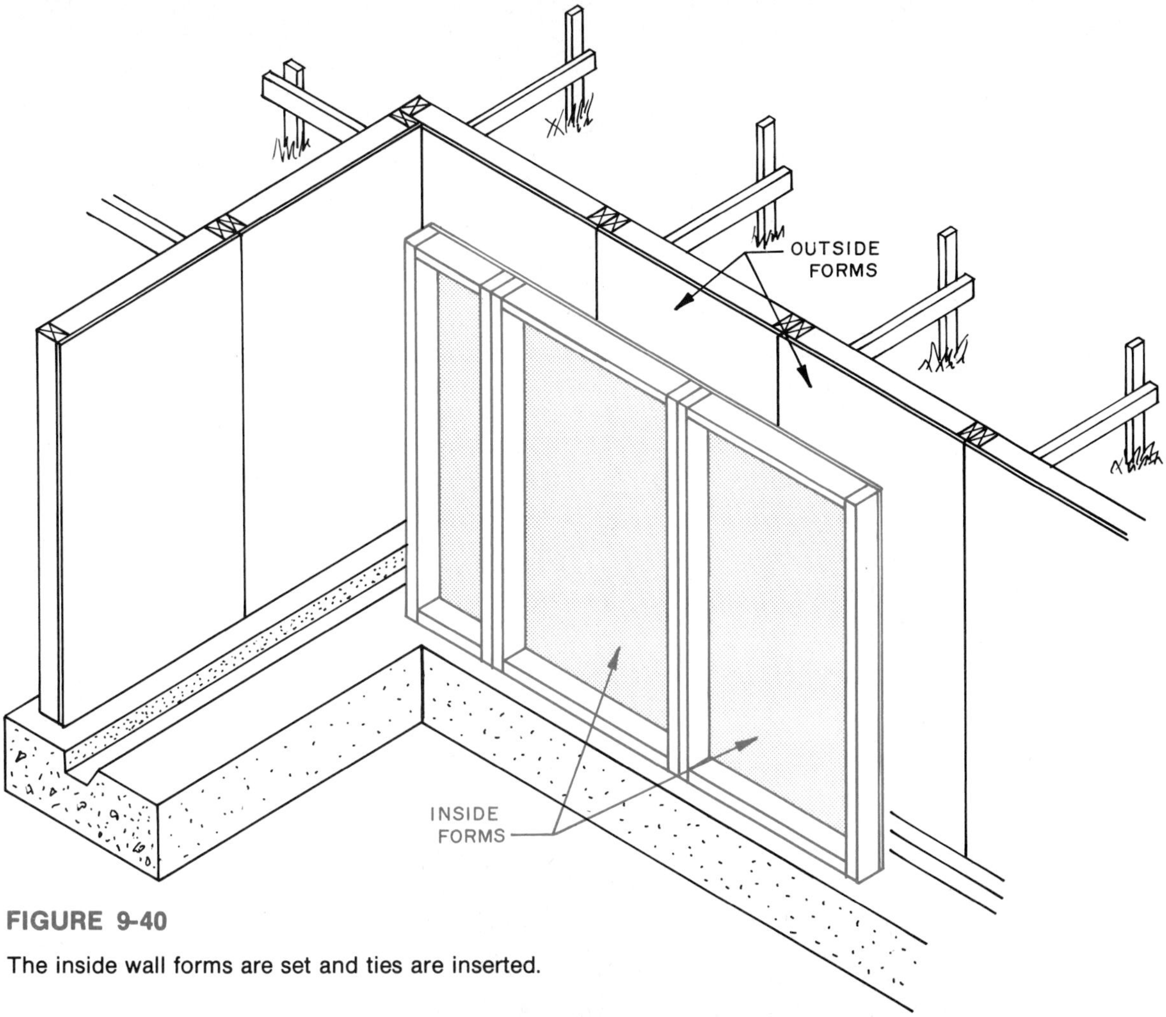

FIGURE 9-40

The inside wall forms are set and ties are inserted.

FIGURE 9-41

Filling the form with concrete. *(Courtesy of Portland Cement Association)*

FIGURE 9-42

Adjustable braces can be used to secure forms. *(Courtesy of Symons Corporation)*

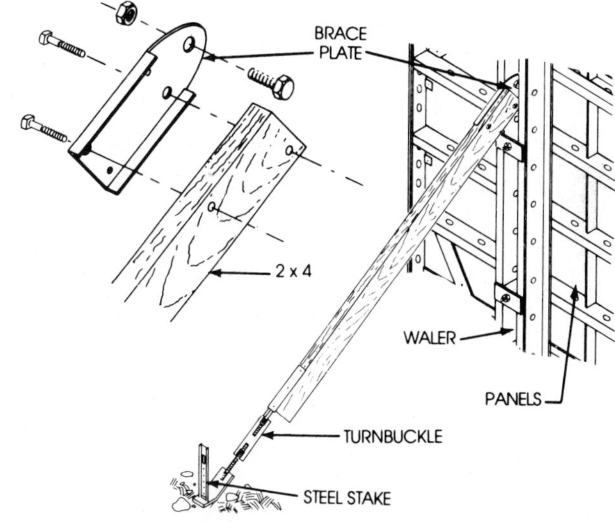

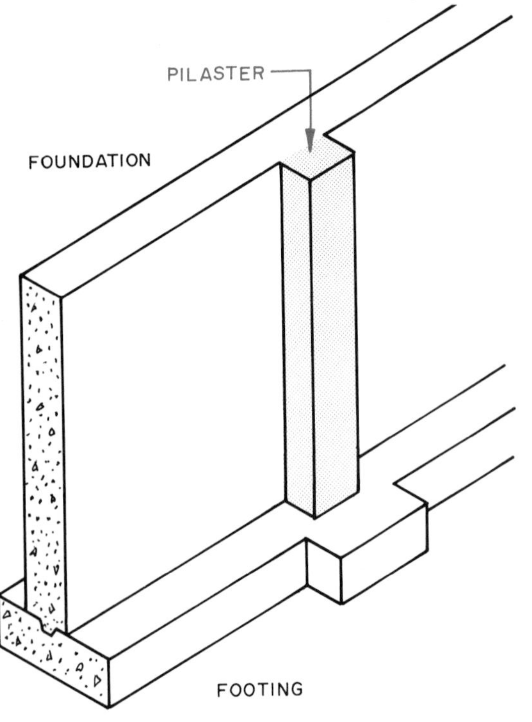

FIGURE 9-43

A pilaster strengthens a long wall.

Sometimes, a foundation wall will have pilasters. A *pilaster* is used to support the ends of beams and stiffen a long foundation wall (Fig. 9-43). They are usually formed when the wall is formed. In this way they become a part of the wall. A typical way to form pilasters is shown in Fig. 9-44.

WALL OPENINGS

Concrete walls often have openings for doors, windows, pipes, or pockets to carry beams. The work to form these must be done before the inner foundation wall form is erected.

Small openings, as for a pipe, can be made by placing fiber, plastic, or metal tubes between the forms. They are held in place by wood blocks or plastic rosettes nailed to the form (Fig. 9-45). Rectangular openings, such as for a ventilation pipe, can be made using a wood box-like form. The form is tapered to help remove it from the wall when the forms are stripped. It is nailed to the form with duplex nails (Fig. 9-46).

Large openings in concrete foundations, such as that for a window, are made in several ways. If an opening the size of the unit, as a window, is wanted a box is built. Its outside dimensions equal the size of the opening. After the concrete has hardened, the form is removed (Fig. 9-47). To make it easy to remove the form, a saw kerf or notch can be cut in the sides (Fig. 9-48). A cleat is used to hold the sides in place. When the form is removed, the sides can be pried in until they collapse.

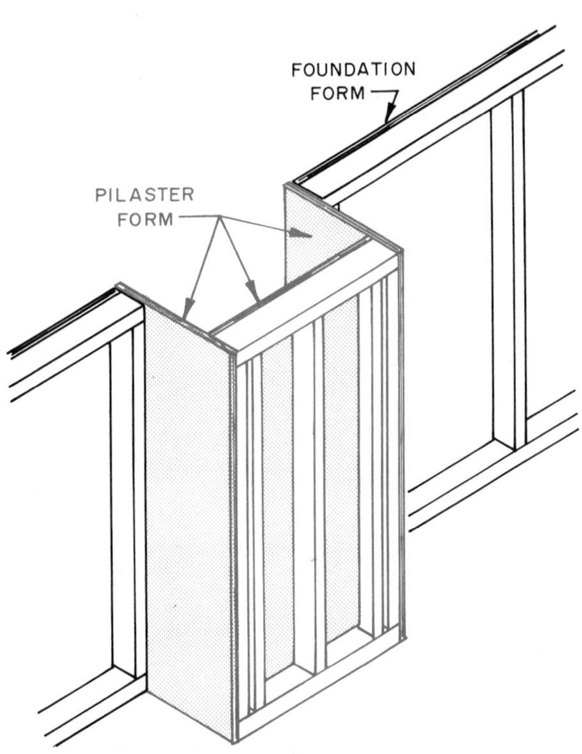

FIGURE 9-44

One way to form a pilaster.

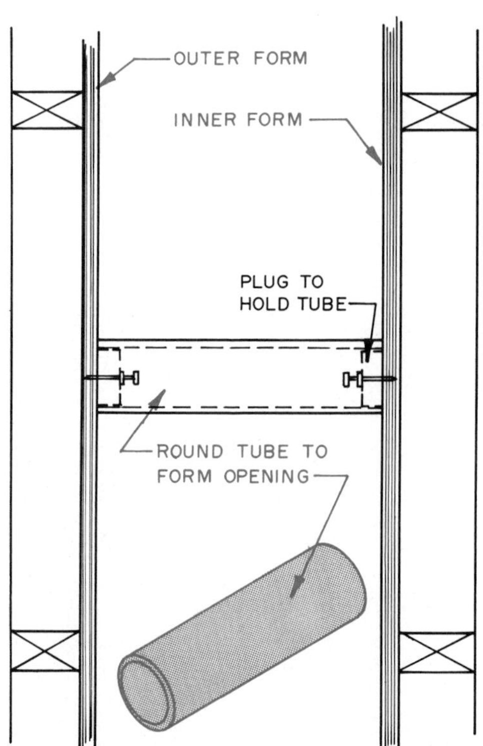

FIGURE 9-45

Tubes are inserted between forms to form a round hole in the wall.

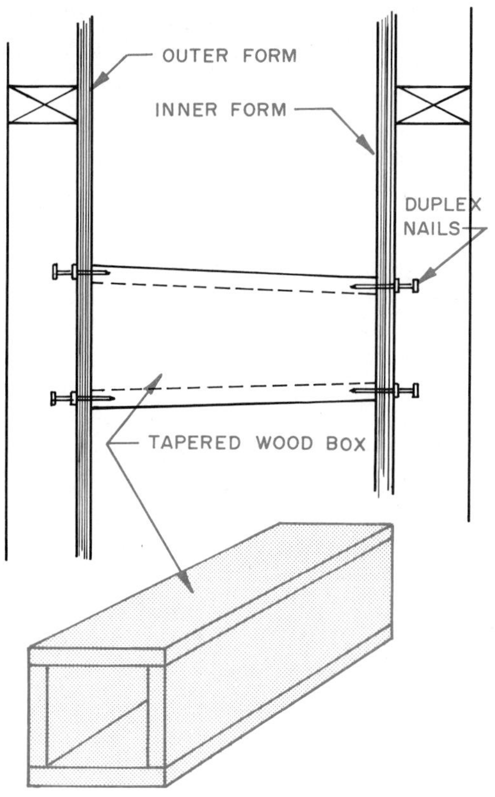

FIGURE 9-46

A tapered wood form is inserted between forms to form a square or rectangular hole.

FIGURE 9-47

A worker removing a metal form used to make a window opening. *(Courtesy of Precise Forms, Inc.)*

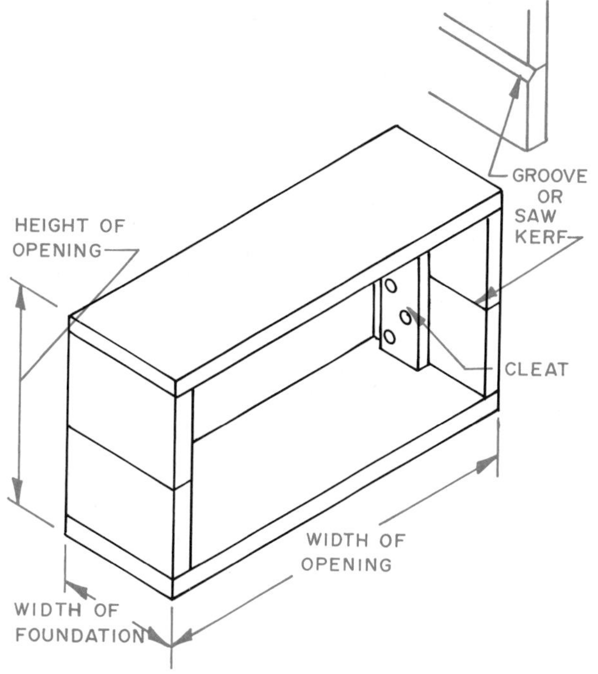

FIGURE 9-48

A removable wood form is built the size needed for a window opening in a wall.

The form must have internal bracing. These are usually 2 × 4 or 2 × 6 in. members. The larger the opening, the more bracing is needed. If the form is poorly braced, the weight of the concrete will cause it to sag (Fig. 9-49).

Sometimes it is desired to leave a nailing strip in the concrete wall. This can be done by lightly nailing a wedge-shaped member to the box frame (Fig. 9-50). When the box is removed, the wedge remains in the concrete.

Sometimes, the door frame is installed in the foundation wall before the concrete is poured.

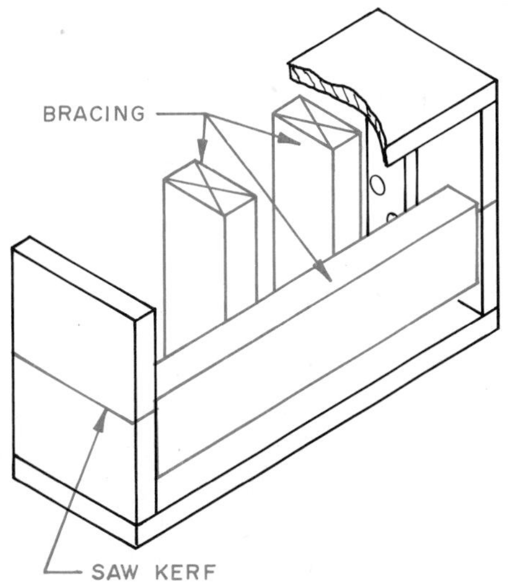

FIGURE 9-49

Internal bracing strengthens the wood form.

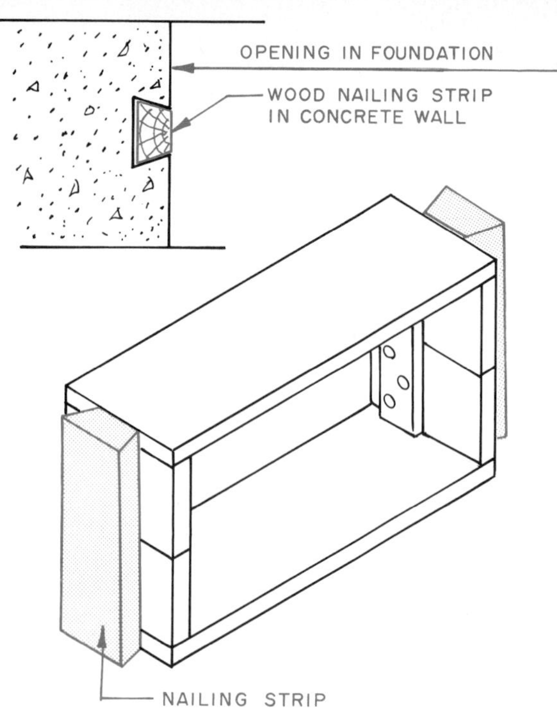

FIGURE 9-50

Wedge-shaped wood blocks are used for nailing strips in concrete walls.

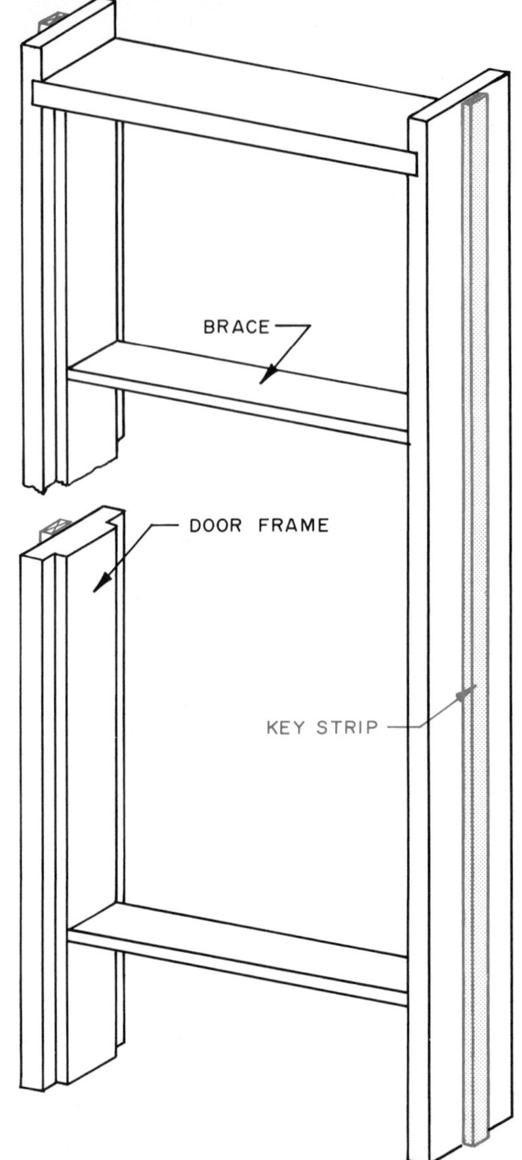

FIGURE 9-51

A door frame with a key strip designed to remain in the wall after the concrete hardens.

A wood key strip is used to hold the frame in place (Fig. 9-51). If the width of the foundation is greater than that of the door frame, a filler strip must be used (Fig. 9-52). A door opening is being installed in Fig. 9-53.

Another way is to use metal-framed window or door units. The metal window frame is placed in the form. The concrete is poured around it, thus locking it in place. When the forms are stripped the metal frame remains. The glass units are placed inside the frame (Fig. 9-54).

The concrete above an opening must be reinforced. This is often done by adding steel reinforcing rods above the opening (Fig. 9-55).

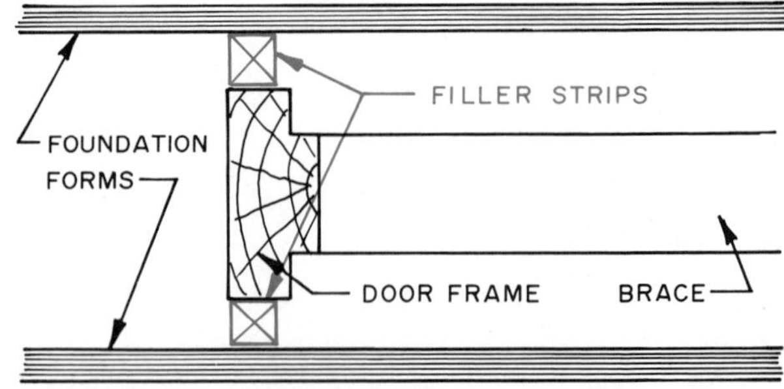

FIGURE 9-52

Filler strips are used between the door frame and the form.

FIGURE 9-53 (*left*)

A wood door form being placed in the foundation form. *(Courtesy of Precise Forms, Inc.)*

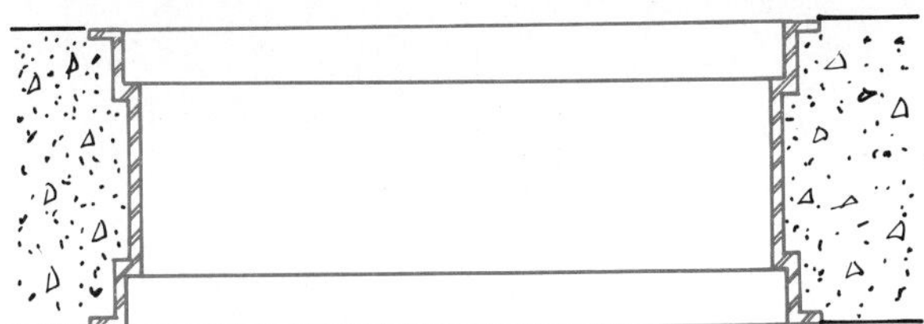

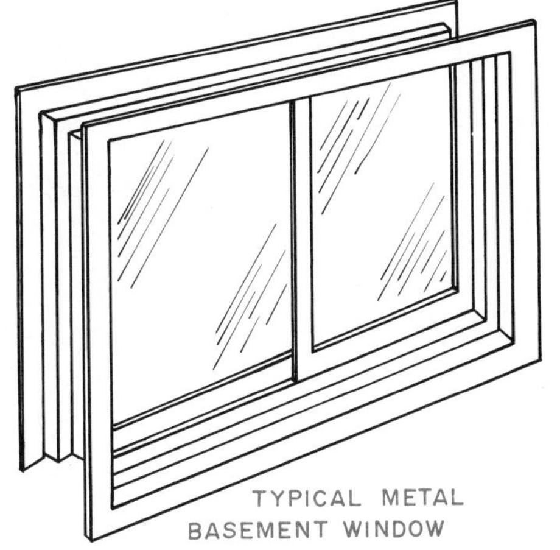

FIGURE 9-54

Steel-framed windows can be inserted into the form and cast in place.

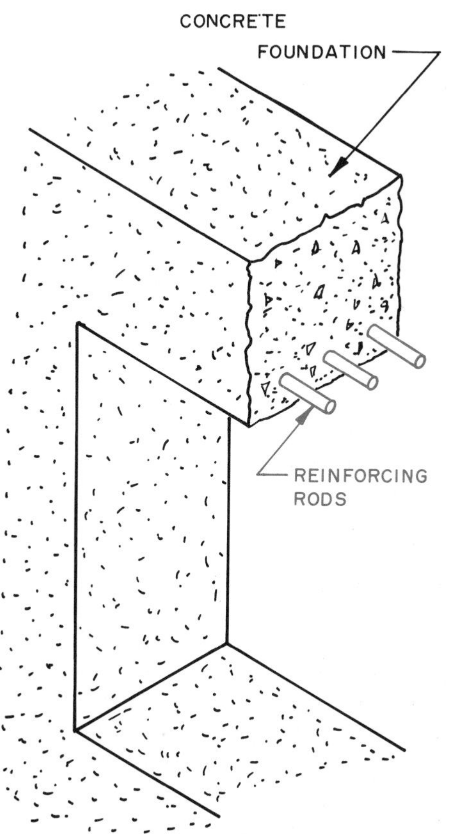

FIGURE 9-55

The concrete above an opening must be reinforced with reinforcing rods.

When beams are required to be set into the foundation, they are set in pockets (Fig. 9-56). A wood box the size of the desired opening is made. It is nailed to the inside of the plyform wall panel. Usually, beams must rest on at least 4 in. of concrete. This is the usual pocket depth.

Anchor bolts must be placed in the foundation wall. Anchor bolts are used to bolt the frame wall of the first floor of the house to the foundation. Anchor bolts are usually ½ in. in diameter and 18 in. long. They are spaced 4 ft apart. They usually have a large washer, metal plate, or bend on the end to be in the concrete. As soon as the concrete wall is poured and leveled, the anchor bolts are pushed into the concrete. They are held in place by running them through a board (Fig. 9-57). Enough bolt must be above the concrete to go through a 2-in.-thick sill and leave room for a washer and the nut. Another type of anchor is shown in Fig. 9-58.

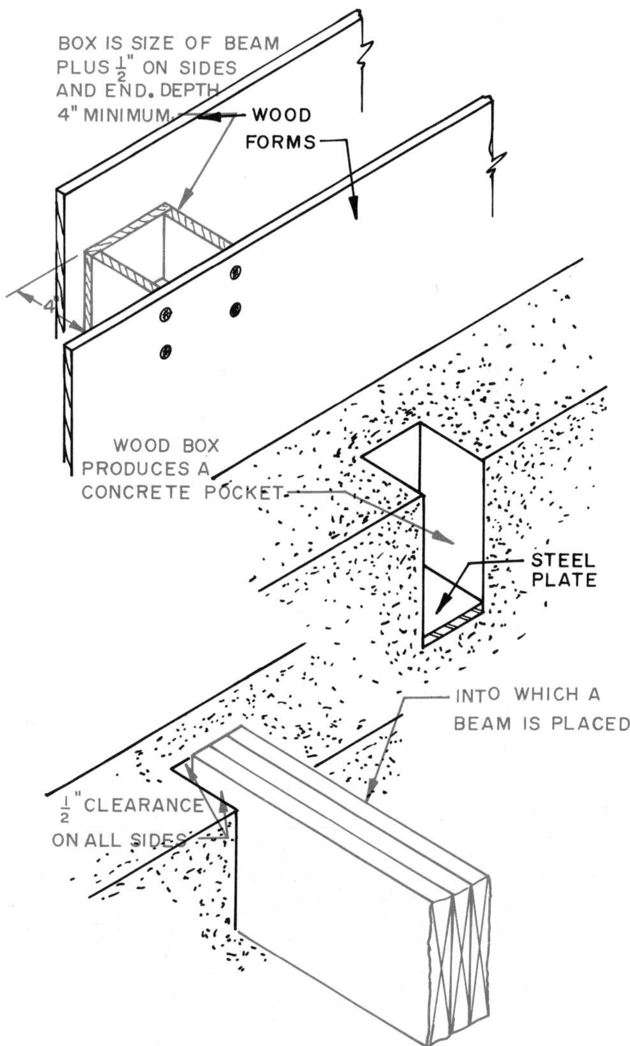

FIGURE 9-56

A wood form can be used to form a pocket for a beam.

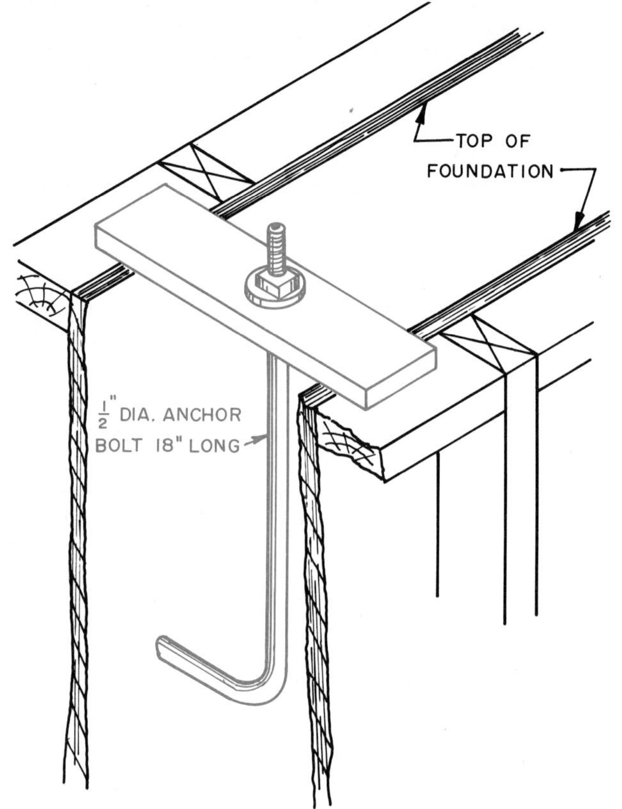

FIGURE 9-57

Anchor bolts are set in place after the concrete is poured.

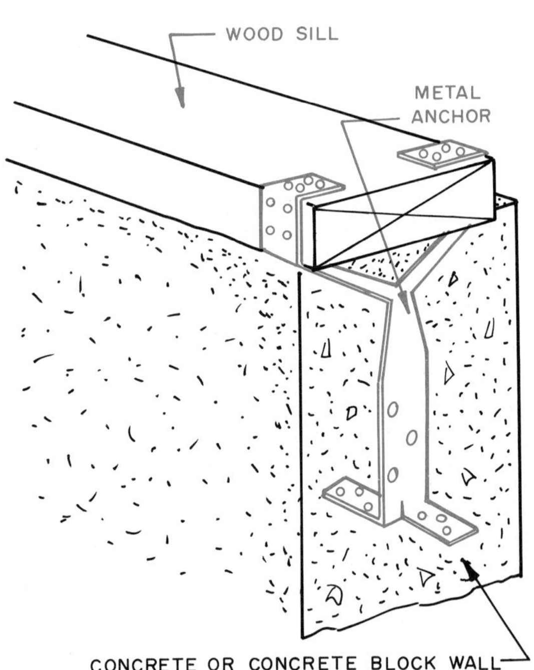

FIGURE 9-58

Another type of sill anchor.

FORMING CONCRETE STAIRS

The forming for stairs that are open on the sides is shown in Fig. 9-59. The side forms can be dimension lumber, such as 2 × 6 in. members or

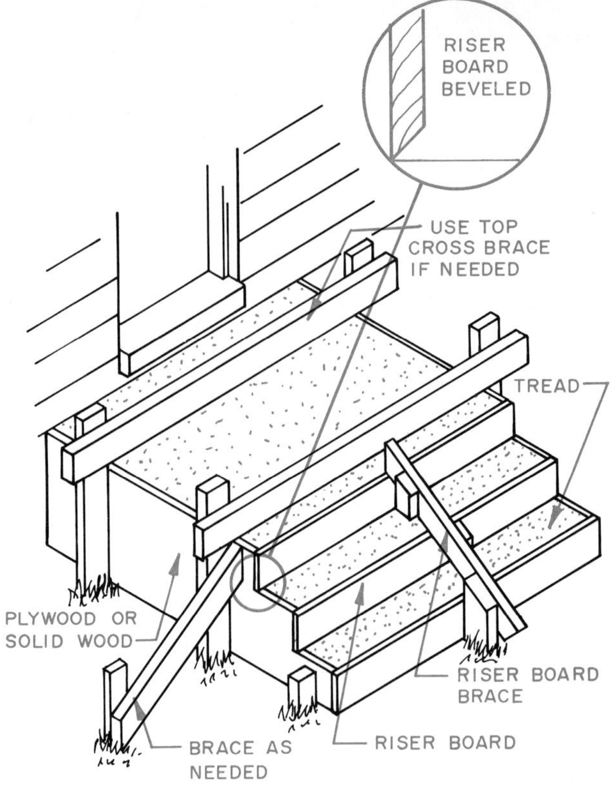

FIGURE 9-59

One way to form a landing and steps.

¾-in. plyform. The side forms have the sizes of the landing, treads, and risers marked on them. They are cut to size. The riser boards are cut to length. This is the width of the steps plus material to nail them to the side forms.

Notice in Fig. 9-59 how the lower edge of the riser board is beveled. This provides room for the mason to trowel the step surface smooth.

Adequate bracing is necessary. One way to do this is shown in Fig. 9-59.

Sometimes concrete steps are built where the sides are not open. This might be on an outside entrance to a basement. They are formed by using side braces to retain the riser boards (Fig. 9-60). Braces can be nailed to the top edges of the riser form boards if needed.

FORMING PORCHES

Many houses will have a small concrete porch at each exterior door. This will consist of a landing and concrete steps. The porch floor needs support to keep it from settling and pulling away from the foundation. Often, concrete braces are poured which are attached to the foundation wall (Fig. 9-61).

Larger porches are formed using a regular footing and foundation wall. The open area is filled with soil or gravel and serves as a base for the concrete. A wood form is built to form the

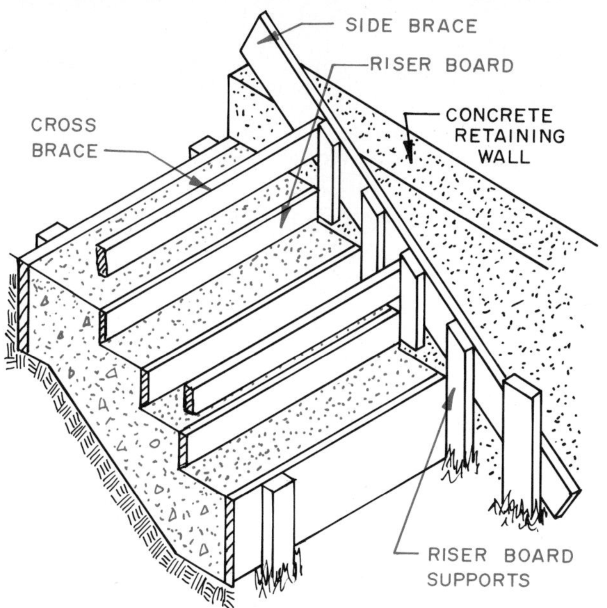

FIGURE 9-60

Some steps are formed between concrete walls.

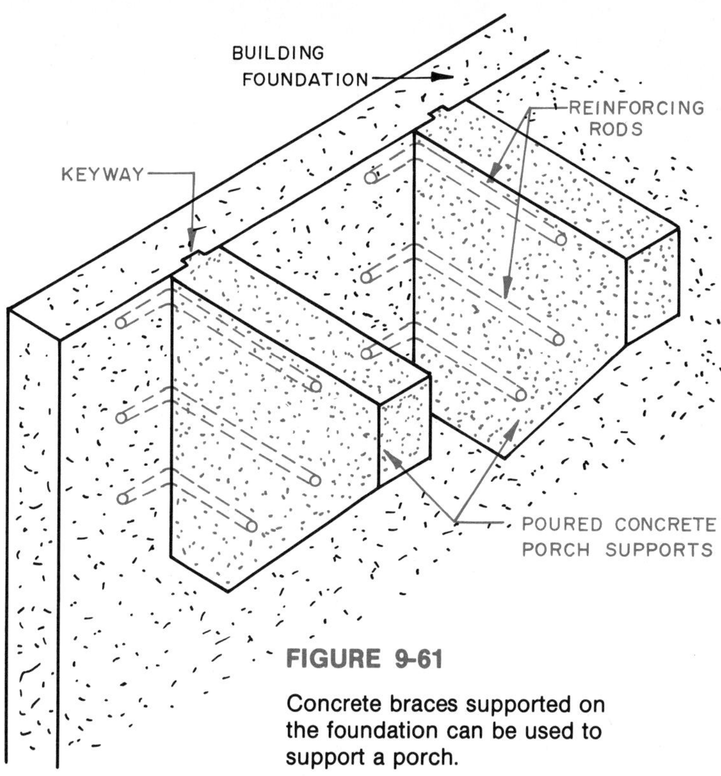

FIGURE 9-61

Concrete braces supported on the foundation can be used to support a porch.

floor (Fig. 9-62). *Reinforcing wire* is placed in the floor. If the span is large, steel reinforcing bars are used. Piers could be used to support the center of the slab if the porch is very large.

FORMING SIDEWALKS AND DRIVES

Carpenters often are expected to build the forms for sidewalks and drives. Usually, they are built after the building is almost finished. This keeps them from being damaged.

Under normal conditions the concrete is placed on the soil. It should be firmly packed. If fill is needed to level the walk, it should be thoroughly packed. In areas where moisture may be a problem, 4 in. of rock or sand is used. This helps drainage, provides a firm unpinning, and helps in areas where cold may cause the walk to crack.

Stakes are driven at the corners of the walk. They should be carefully located so that the size is correct. A chalk line is run between stakes on one side of the walk. A leveling instrument is used to level the string. Mark the level on the stakes.

Stakes are driven into the ground along the chalk line. The side form boards are nailed to the stakes. These are usually 2 × 4 in. members. The top of the form board is lined up with the chalkline (Fig. 9-63).

Spreaders are cut the width of the sidewalk. They are used to locate the form boards on the other side. The walk normally slopes ⅛ in. per foot of width to one side. This helps drain the surface of the walk. This is usually done with a carpenter's level.

FIGURE 9-62

Forming a porch using a conventional footing and foundation.

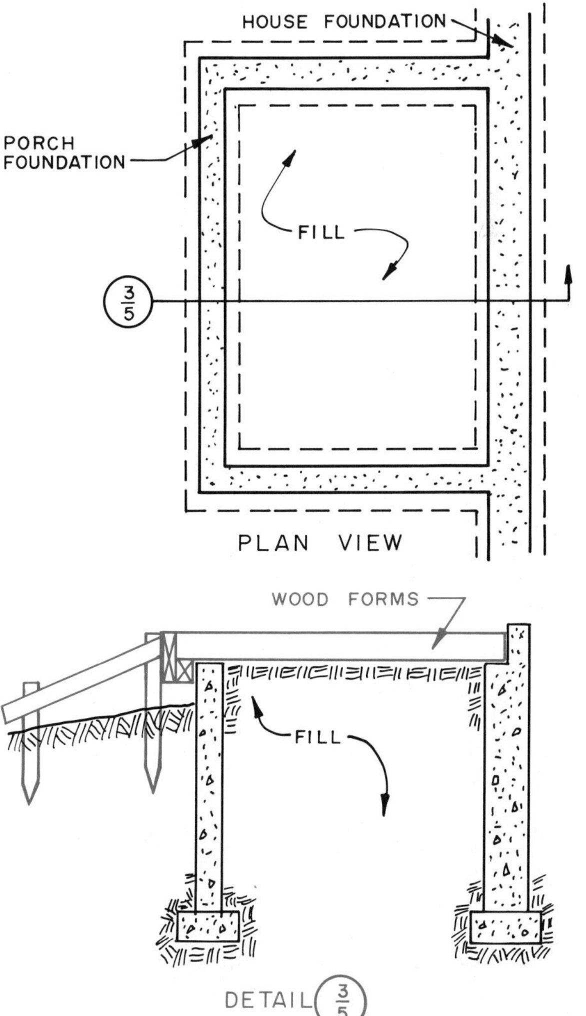

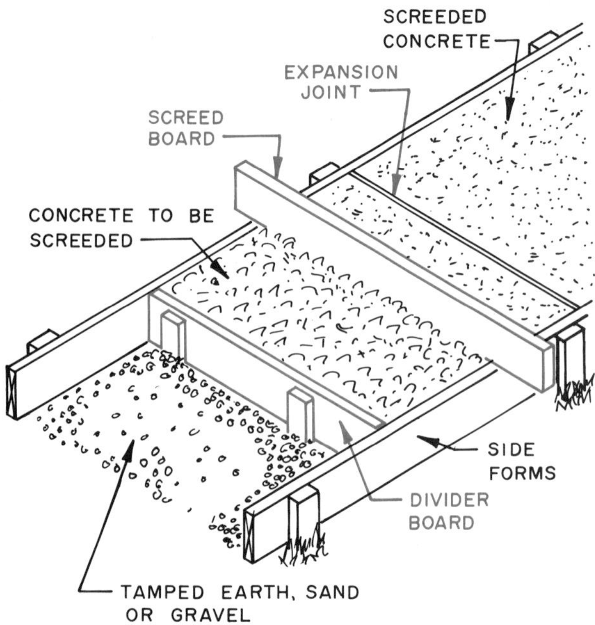

FIGURE 9-63

Side forms are used to form sidewalks and driveways.

FIGURE 9-65

The cement worker uses hand tools to finish the concrete surface.

Wherever a walk or drive meets a building, an *isolation joint* is needed. This is made of an asphalt-impregnated composition material. It is about ½ in. thick. This permits the walk to expand and contract. It also permits the walk to settle without cracking. If it was firmly joined to the foundation and settled, it would break.

Walks generally have reinforcing wire mesh. This strengthens them and helps reduce cracking.

When the concrete is poured, a screed is used to level it. The forms are overfilled. The screed is placed on top of the side-form boards. It is moved from side to side as it is pulled along the form boards. This levels the concrete to the correct height (Fig. 9-64).

After the concrete is level, the surface is troweled. Various degrees of smoothness can be had, depending upon the troweling done. Finally, control joints are made. These are cuts that go into the concrete about ½ in. They provide weakened places for the concrete to crack if needed (Fig. 9-65). They are placed every 4 or 5 ft on sidewalks and every 10 ft on driveways.

The edges of the walk are rounded with an edging tool. This is done before the concrete hardens. It makes a rounded edge and helps prevent the edge from breaking off.

FIGURE 9-64

After the concrete is poured, it is screeded.

FORMING CONCRETE SLAB HOUSE FLOORS

Concrete slab floors are placed on a bed of gravel that is directly on the ground. *Polyethylene-film moisture barriers* are placed over the gravel and beneath the concrete. This keeps the moisture in the ground from penetrating the slab.

Such buildings are heated in several ways. Pipes can be placed in the slab. Hot water is pumped through the pipes. Hot-air ducts can be laid on the gravel and the concrete is poured over them. Electric heat is widely used. This eliminates the need for ducts or pipes.

The forming of concrete floor slabs uses the procedures discussed in this chapter. The footings and foundations are dug and placed as those just discussed. The floor is laid on the ground or over a gravel bed much the same as a sidewalk.

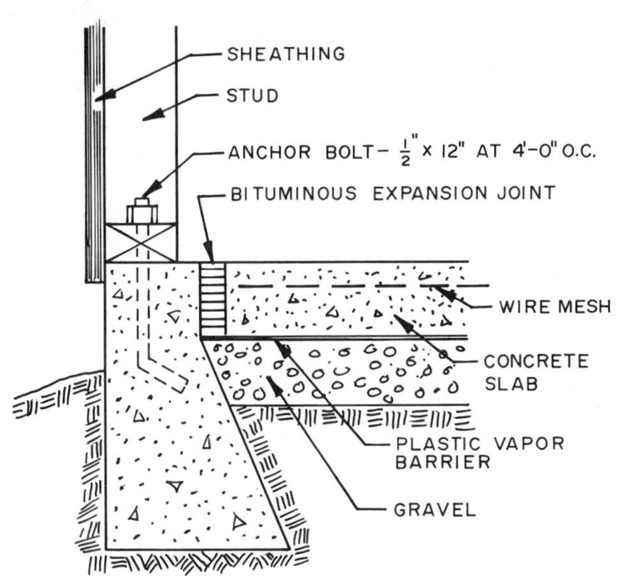

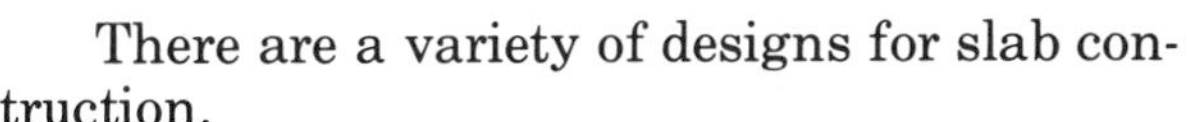

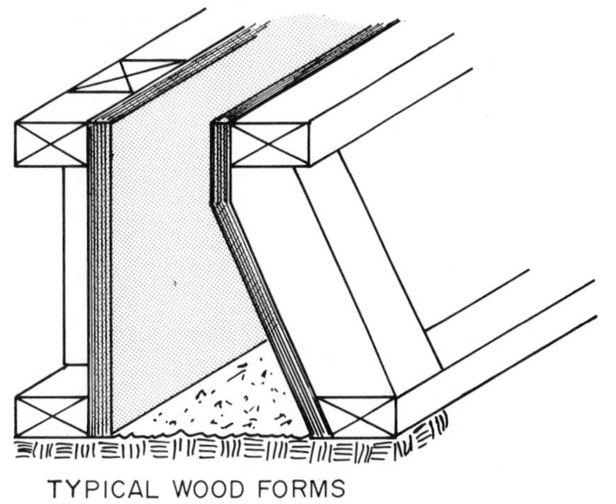

FIGURE 9-66

Design details for a monolithically cast flared footing and foundation for slab construction.

There are a variety of designs for slab construction.

In warm climates the footing does not have to go deep into the soil. A flared footing with floating slab can be used (Fig. 9-66). The slab and wall are separate. The wall supports the weight of the house. The slab can rise and fall independently from the wall. This avoids cracks at the edges.

The footing, wall support, and slab can be cast as a monolithic unit (Fig. 9-67). Reinforcing wire mesh is placed in the slab. The footing is also reinforced. Notice the footing under interior supporting walls. The exterior of the wall can be insulated if desired. A rigid waterproof insulation board protected with asbestos cement board can be used.

In colder climates the footing must go below the frost line. Insulation around the edge of the slab is required. One such design is in Fig. 9-68. The ground is excavated so that the footing is below the frost line. A normal foundation is

FIGURE 9-67

In warm climates the footing and slab can be poured monolithically.

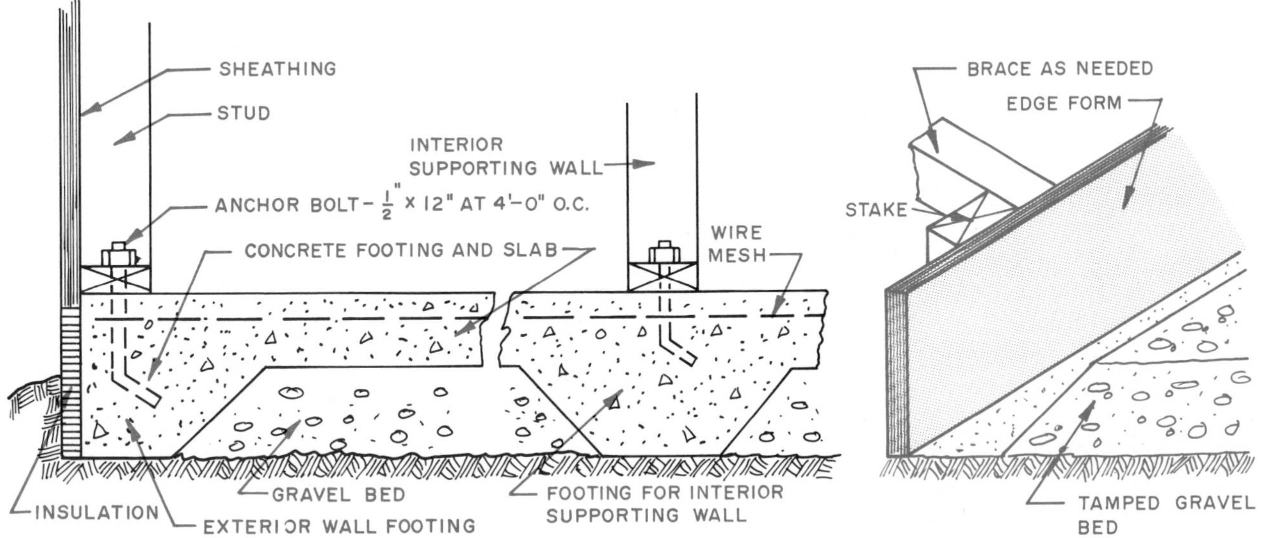

NOTE:
FOUNDATIONS FORMED AS EXPLAINED FOR LOW WALL FORMS.

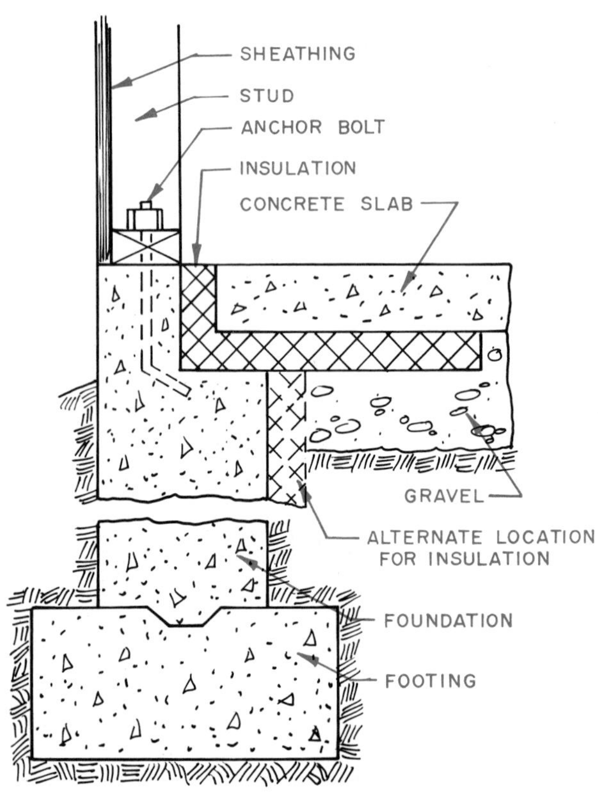

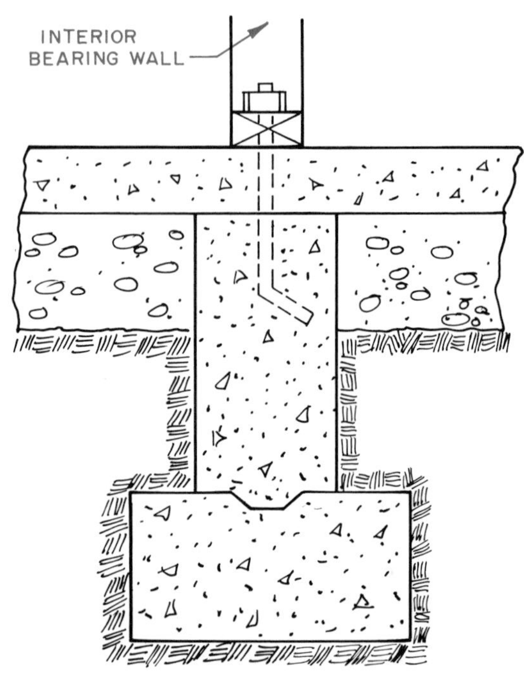

FIGURE 9-68

Design details for a foundation for a slab house for a cold climate. Notice the support under the interior supporting wall.

formed on top of it. The insulation is placed around the edge of the slab. The foundation supports the edge of the slab. Special footings are cast beneath interior supporting walls. The foundation wall forms the outer wall against which the slab is poured. The forms for the footing and wall are the same as those discussed earlier in the chapter.

FORMING GRADE BEAMS

Grade beams are structural members resting on piers or pilings that serve as the foundation wall (Fig. 9-69). They are used for concrete slab (Figs. 9-70 and 9-71), and conventional construction (Fig. 9-72).

The piers are formed by boring holes into the soil. The diameter is usually 12 to 16 in. The holes are bored until solid ground is reached. The hole is filled with concrete. A ⅝-in. metal dowel rod is placed into the pier before the concrete hardens. It extends up into the grade beam. Piers may be 8 to 12 ft apart.

The grade beam is formed the same as a low wall foundation. It rests on top of the piers. Steel reinforcing rods are placed at the bottom and top of the grade beam. The top of the form is the finished height of the foundation. It must be level. This is leveled in the same way as explained for foundation walls.

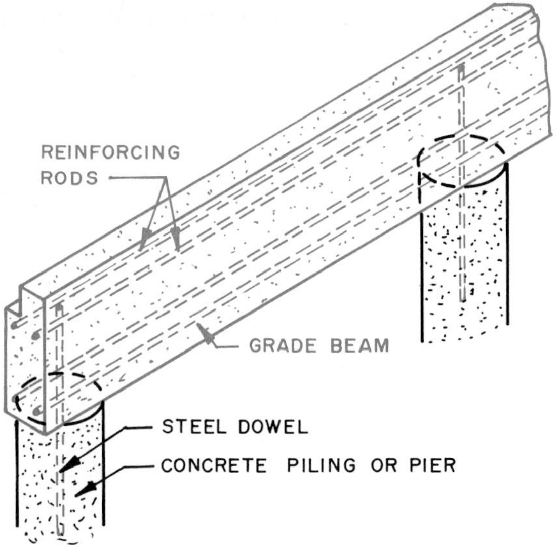

FIGURE 9-69

Grade beams are concrete structural members supported by piers or pilings.

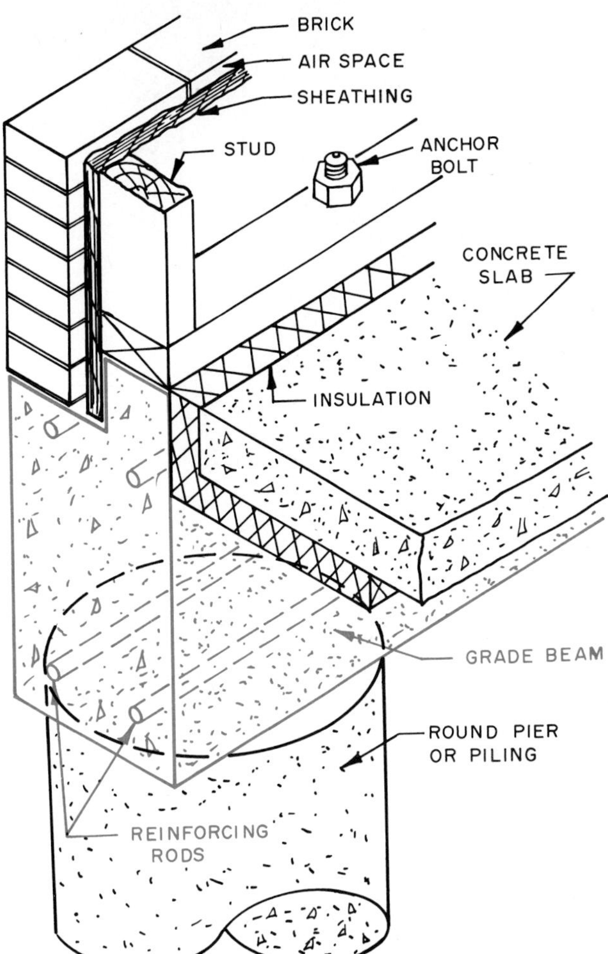

FIGURE 9-70

Design details for a grade beam with concrete slab construction.

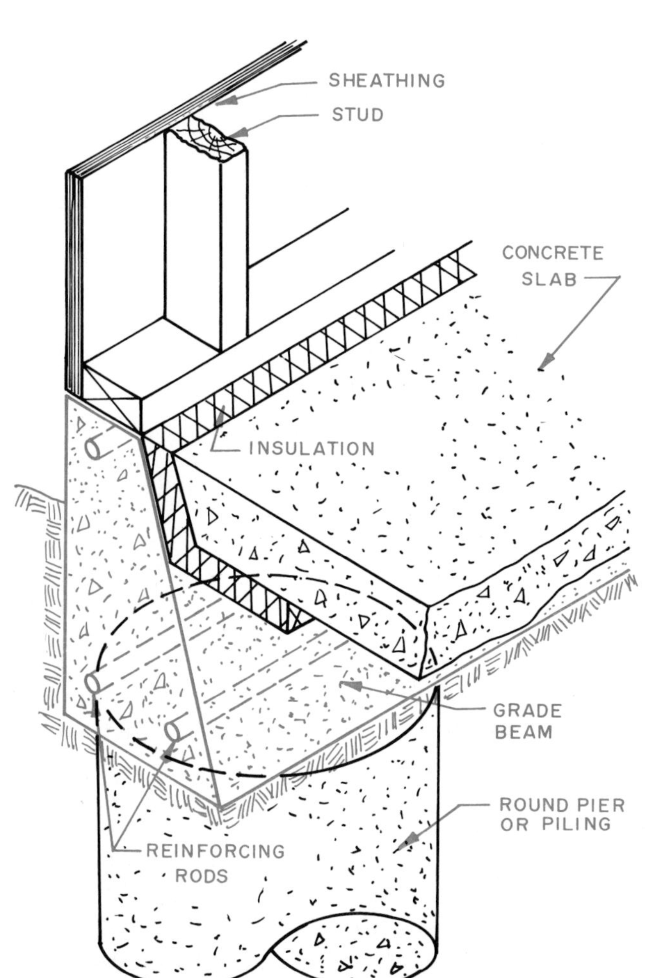

FIGURE 9-71

Another grade-beam design with slab construction.

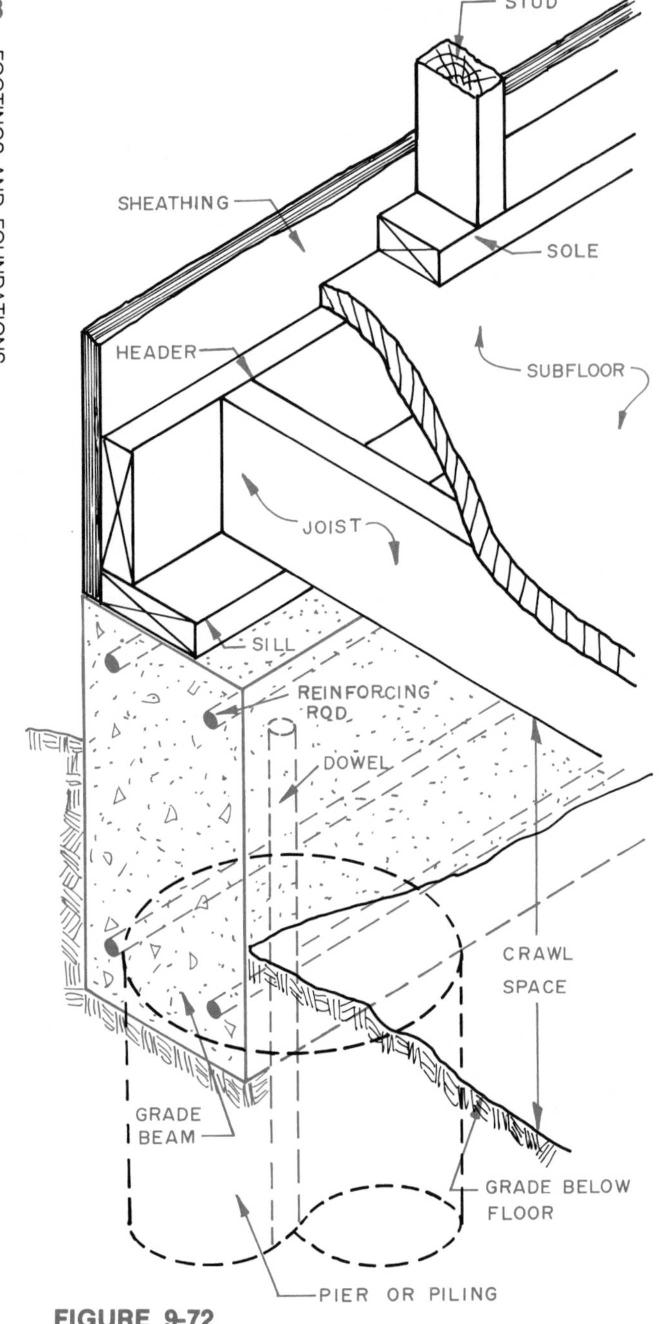

FIGURE 9-72

A typical detail using a grade beam and a wood-framed floor.

ALL-WEATHER WOOD FOUNDATION

The *all-weather wood foundation* is a load-bearing wood-frame wall system designed for below-grade use as a foundation for light-frame construction. It is basically the same as above-ground frame construction except for several factors. The lumber and plywood used in framing are stress-graded to withstand the lateral soil pressures. It also carries the usual live and dead loads. Vertical loads on the foundation are distributed to the supporting soil by a footing made of a wood footing plate and a structural gravel layer.

Following are some design specifications.

1. All lumber and plywood in contact with or close to the soil are pressure-treated with wood preservatives. This protects against decay and insects. The preservative treatment must meet the requirements set by the American Wood Preservers Bureau. If lumber is cut after treatment, the cut surface should be brush-coated with preservative.
2. Extensive moisture control measures are used. Moisture reaching the upper part of the wall is deflected by polyethylene sheeting or the plywood sheathing. This moisture flows into a porous gravel layer built around the lower part of the basement. The moisture flows through the gravel to a positively drained sump which removes the water.
3. Lumber should be of a species and grade for which allowable unit stresses are set forth in The National Design Specification for Stress Grade Lumber and Its Fastenings, by the National Forest Products Association.
4. Plywood must be bonded with exterior glue and be grade-marked as meeting U.S. Department of Commerce Product Standard PSI, Construction and Industrial Plywood.
5. The fasteners used should be silicon, bronze, copper, or stainless steel types 304 or 316 as defined by the American Iron and Steel Institute Classification. They may be hot-dipped zinc-coated steel meeting requirements of the American Society for Testing and Materials Standard A153.
6. Framing anchors are zinc-coated sheet steel which conform to Grade A, set forth by the American Society for Testing and Materials Standard A446.
7. The gravel used should be washed and graded. The largest size stone acceptable is ¾ in. It must be free from organic-, clay-, or silt-type soils.
8. Sand should be coarse and have grains not smaller than $^{1}/_{16}$ in. It should be free from organic-, clay-, or silt-type soils.
9. Crushed rock should have a maximum size of ½ in.

The all-weather wood foundation does not require the customary concrete footing. This reduces the overall weight and cost of construction. Since it can be assembled in a shop and erected on the site it permits builders to install foundations in any kind of weather. Installation is fast and does not require the usual crew of concrete workers.

The basement tends to be warmer. The wood wall provides good insulation. In addition, it can be insulated on the inside in the same manner as above-grade frame construction. This tends to reduce heating costs.

Since considerable attention is given to waterproofing and handling below-surface water, the basement produced is dry. This reduces the mildew often associated with basements.

The basement area is easier to finish into a living area. Normal procedures are used to apply wall finish over the insulation. It is easy to install electrical outlets around the basement walls.

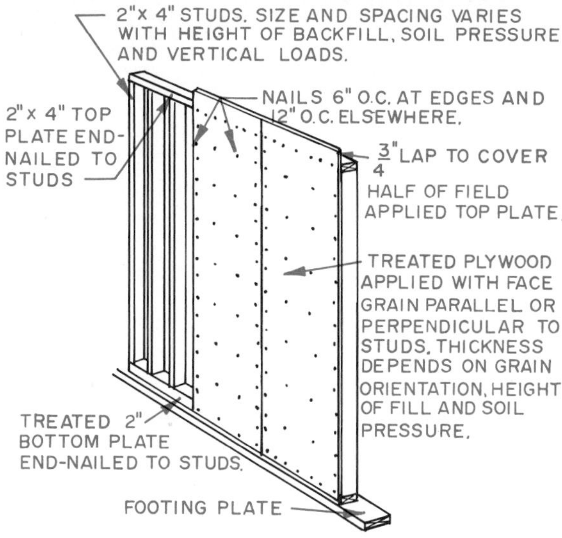

FIGURE 9-73

A typical all-wood foundation wall panel. *(Courtesy of National Forest Products Association)*

Installation Procedure

1. Excavate for the footings, pipes, conduit, and sump system. The footings should go below the frost line.
2. Fill the footings with the required gravel or coarse sand and level them. Fill the floor area to the proper depth needed for pouring the concrete floor.
3. Set the preassembled foundation wall panels on the footings. Plumb and square them and fasten together. Install the upper top plates to tie the units together.
4. Now install the floor joists and subfloor for the first floor of the house.
5. Caulk the joints between the panels and the nailing strip and the plywood panels where they are above the ground.
6. Glue a 6-mil polyethelene film to the exterior of the panels where they will be below grade.
7. Pour the concrete floor before backfilling. This braces the bottom of the panels against the lateral thrust of the earth.
8. Backfill the foundation and the job is finished. First fill in at least 1 ft of gravel above the footings and then backfill with earth.

Typical Construction Details

A typical foundation panel is shown in Fig. 9-73. The thickness of the plywood depends upon the grain direction, height of soil, and soil pressure. Recommended design data are given in Fig. 9-74.

If the wall has openings for doors or windows, support members around them should be doubled.

In Fig. 9-75 the details for forming a corner are shown. Notice that after the panels are joined, a top plate is nailed in place. The method for joining panels in the wall is shown in Fig. 9-76.

Minimum structural requirements for exterior foundation wall framing are shown in Fig. 9-77. This figure gives data about lumber grade, structural member size, spacing, and footing sizes. Detailed structural design procedures to cover things such as soil types, transverse lateral loads against the wood foundation, racking loads (loads parallel to the wall), footing design, and design of load-bearing walls are available from the National Forest Products Association, 1619 Massachusetts Ave., N.W., Washington, D.C. 20036.

Recommended minimum footing plate sizes are given in Fig. 9-78.

Complete construction details for a foundation wall for frame and a masonry veneer wall for houses with basements are shown in Figs. 9-79 and 9-80. Details for houses with crawl spaces are shown in Figs. 9-81 and 9-82.

The wall for frame construction rests on a footing plate that sets on a gravel bed. The masonry veneer wall is much the same except that it adds a knee wall to hold the masonry veneer. The footing plate is considerably larger because it has to support the foundation wall and the knee wall. The knee wall should also be used to support a concrete porch (Fig. 9-83).

HEIGHT OF FILL (in.)	STUD SPACING (in.)	FACE GRAIN ACROSS STUDS[a]			FACE GRAIN PARALLEL TO STUDS		
		Grade[b]	Minimum Thickness (in.)[c]	Identification Index	Grade[b]	Minimum Thickness (in.)	Identification Index
24	12	B	½	32/16	B	½	32/16
	16	B	½	32/16	B	½ (4, 5 ply)	32/16
48	12	B	½	32/16	B	½[d] (5 ply)	32/16
					A	½	32/16
	16	B	½	32/16	A	⅝	42/20
					B	¾	48/24
72	12	B	½	32/16	B	⅝[d] (5 ply)	42/20
	16	A[e]	½[d]	32/16	A	¾[d]	48/24
86	12	B	½[d]	32/16	A	⅝[d]	42/20
					B	¾[d]	48/24

[a]Blocking between studs required at all horizontal panel joints more than 4 ft below adjacent ground level.

[b]Minimum grade: A. Structural I C-D: B. C-D (exterior glue). If a major portion of the wall is exposed above ground, a better appearance may be desired. In this case, the following Exterior grades would be suitable: A. Structural I A-C, Structural I B-C, or Structural I C-C (Plugged); B. A-C Exterior Group 1, B-C Exterior Group 1, C-C (Plugged) Exterior Group 1, or MDO Exterior Group 1. All plywood to carry the grade trademarks of the American Plywood Association.

[c]For crawl-space construction, use ⅜ in. minimum thickness.

[d]For this combination of fill height and minimum panel thickness, panels that are continuous over less than three spans (across less than three stud spacings) require blocking 2 ft above bottom plate. Offset adjacent blocks and fasten through studs with two 16d corrosion-resistant nails at each end.

[e]Only Structural I A-C may be substituted.

FIGURE 9-74

Minimum APA plywood grade and thickness for exterior foundation wall sheathing (30 lb/ft³ equivalent fluid density). *(Courtesy of American Plywood Association)*

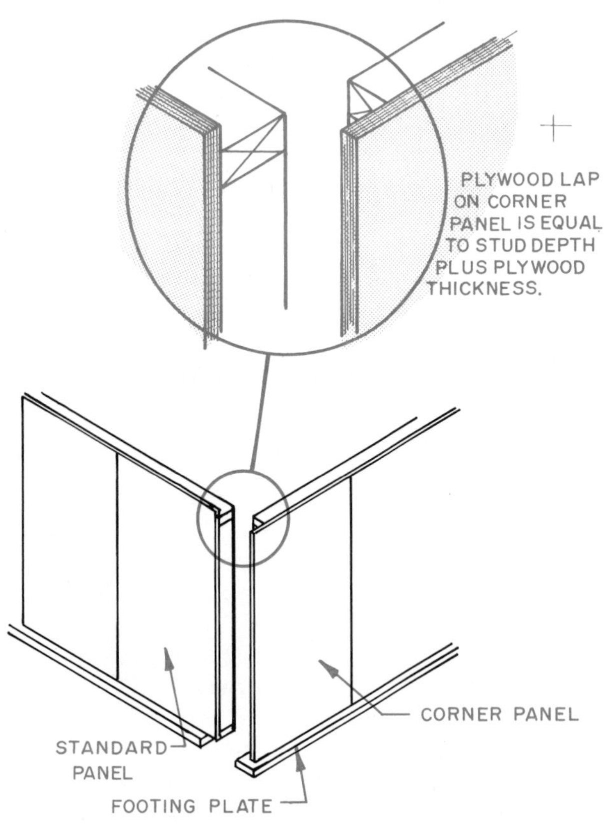

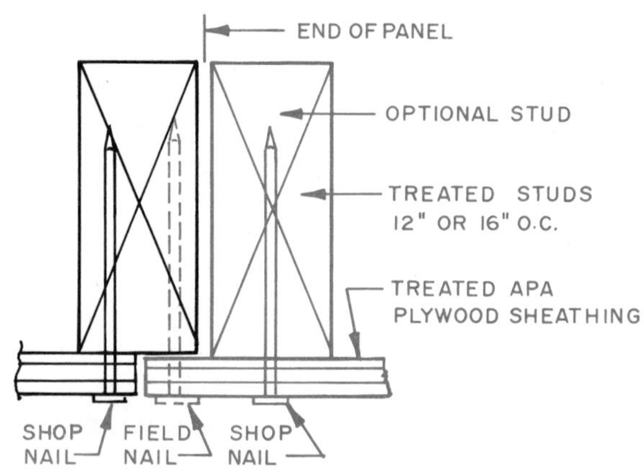

FIGURE 9-76

Method for joining wood foundation panels. *(Courtesy of National Forest Products Association)*

FIGURE 9-75 *(left)*

Details for forming the wood foundation corner. *(Courtesy of National Forest Products Association)*

UNIFORM LOAD CONDITIONS										
			ROOF, 40 PSF LIVE; CEILING, 10 PSF LIVE; 1ST FLOOR, 50 PSF LIVE AND DEAD; 2ND FLOOR, 50 PSF LIVE AND DEAD				ROOF, 30 PSF LIVE: CEILING, 10 PSF LIVE; 1ST FLOOR, 50 PSF LIVE AND DEAD; 2ND FLOOR, 50 PSF LIVE AND DEAD			
HOUSE WIDTH (ft)	NUMBER OF STORIES	HEIGHT OF FILL (in.)	Lumber Species and Grade[a]	Stud and Plate Size (Nominal)	Stud Spacing (in.)	Size of Footing[b] (Nominal)	Lumber Species and Grade[a]	Stud and Plate Size (Nominal)	Stud Spacing (in.)	Size of Footing[b] (Nominal)
Basement Construction										
24-28	1	24	C	2 × 4	12	2 × 8	C	2 × 4	16	2 × 6
			B	2 × 4	16	2 × 8				
		48	C	2 × 6	16	2 × 8	B	2 × 4	12	2 × 6
							C	2 × 6	16	2 × 8
		72	B	2 × 6	12	2 × 8	B	2 × 6	12	2 × 8
			A	2 × 6	16	2 × 8	A	2 × 6	16	2 × 8
		86	A	2 × 6	12	2 × 8	A	2 × 6	12	2 × 8
29-32	1	24	C	2 × 4	12	2 × 8	C	2 × 4	12	2 × 8
			B	2 × 4	16	2 × 8	B	2 × 4	16	2 × 8
		48	C	2 × 6	16	2 × 8	C	2 × 6	16	2 × 8
		72	B	2 × 6	12	2 × 8	B	2 × 6	12	2 × 8
			A	2 × 6	16	2 × 8	A	2 × 6	16	2 × 8
		86	A	2 × 6	12	2 × 8	A	2 × 6	12	2 × 8
24-32	2	24	C	2 × 6	16	2 × 10	C	2 × 6	16	2 × 10
		48	C	2 × 6	16	2 × 10	C	2 × 6	16	2 × 10
		72	B	2 × 6	12	2 × 10	B	2 × 6	12	2 × 10
			A	2 × 6	16	2 × 10	A	2 × 6	16	2 × 10
		86	A	2 × 6	12	2 × 10	A	2 × 6	12	2 × 10
Crawl-Space Construction										
24-28	1		B	2 × 4	16	2 × 8	B	2 × 4	16	2 × 6
		Max. 2-ft	C	2 × 6	16	2 × 8	C	2 × 6	16	2 × 6
29-32	1	differ-	B	2 × 4	12	2 × 8	B	2 × 4	16	2 × 8
		ence in	C	2 × 6	12	2 × 8	C	2 × 6	16	2 × 8
		outside	B	2 × 6	16	2 × 8				
24	2	and	B	2 × 6	16	2 × 8	B	2 × 6	16	2 × 8
		inside fill	C	2 × 6	12	2 × 8	C	2 × 6	12	2 × 8
25-32	2	height	B	2 × 6	16	2 × 10	B	2 × 6	16	2 × 10
			C	2 × 6	12	2 × 10	C	2 × 6	12	2 × 10

Notes: (1) 2000 lb/ft^2 (psf) allowable soil bearing pressure, 30 lb/ft^3 soil equivalent fluid density. (2) Loading and framing conditions are based on the assumption that roof framing spans between exterior walls and floor framing spans from interior bearing walls located midway between exterior walls. (3) Interior bearing walls support floor loads only and must be at least grade C studs on grade C foundation plates 2 in. wider than studs. Studs must be 2 by 4 in. at 16 in. on center where supporting one floor and 2 by 6 in. at 16 in. on center where supporting two floors.

[a]Species groups and grades having the following minimum properties (surfaced dry or surfaced green) [lb/in.^{2} (psi)]:

	GRADE		
	A	B	C
F_b (repetitive member)			
2 × 6	1750	1450	1150
2 × 4	—	1650	1300
F_c			
2 × 6	1250	1050	850
2 × 4	—	1000	800
$F_{c\perp}$	385	385	245
F_v	90†	90	75
E	1,800,000	1,600,000	1,400,000

†Length of end splits or checks at lower end of studs not to exceed width of piece.

Examples of wood grades are: A, No. 1 grade Douglas fir-larch or southern pine; B, No. 2 grade Douglas fir-larch or No. 2 medium grain southern pine; and C, No. 2 grade western hemlock, No. 2 grade hem-fir, No. 1 grade Douglas fir south, or No. 2 grade southern pine.

[b]Where width of footing plate is 4 in. or more wider than that of stud plate, use ¾-in.-thick continuous treated plywood strips under footing; minimum grade C-D with exterior glue. Use plywood of same width as footing with face grain perpendicular to wall run. Fasten to footing with two 16d nails spaced 16 in. on center.

FIGURE 9-77

Minimum sizes of structural members for wood foundation. *(Courtesy of National Forest Products Association)*

HOUSE WIDTH (ft)	ROOF, 40 PSF LIVE; 10 PSF DEAD. CEILING, 10 PSF; 1ST FLOOR, 50 PSF LIVE AND DEAD; 2ND FLOOR, 50 PSF LIVE AND DEAD		ROOF, 30 PSF LIVE; 10 PSF DEAD. CEILING, 10 PSF; 1ST FLOOR, 50 PSF LIVE AND DEAD; 2ND FLOOR, 50 PSF LIVE AND DEAD	
	2 Stories	1 Story	2 Stories	1 Story
32	2 × 10	2 × 8	2 × 10[a]	2 × 8
28	2 × 10	2 × 8	2 × 8	2 × 6
24	2 × 8	2 × 6	2 × 8	2 × 6

Notes: Footing plate must be not less than species and grade combination "D" as recommended by the National Forest Products Association. Where width of footing plate is 4 in. or more wider than that of stud and bottom plate, use ¾-in.-thick continuous treated plywood strips with face grain perpendicular to footing; minimum grade C-D (exterior glue). Use plywood of same width as footing and fasten to footing with two 6d nails spaced 16 in.

[a]This combination of house width and height may have 2 × 8 in. footing plate when second floor design load is 40 psf live and dead total load.

FIGURE 9-78

Minimum footing plate sizes for wood foundation (in., nominal). *(Courtesy of National Forest Products Association)*

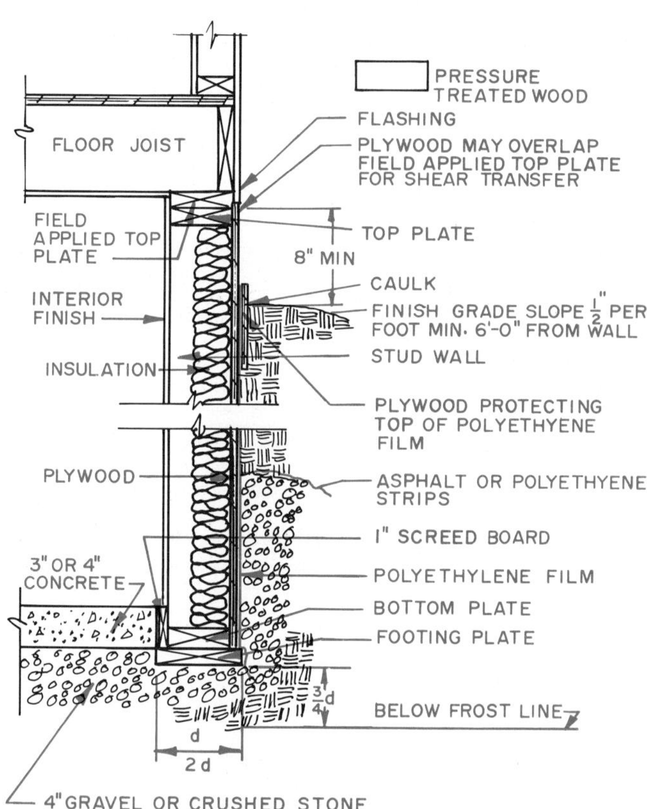

FIGURE 9-79

Construction details for wood foundation used with frame construction and full basement. *(Courtesy of National Forest Products Association)*

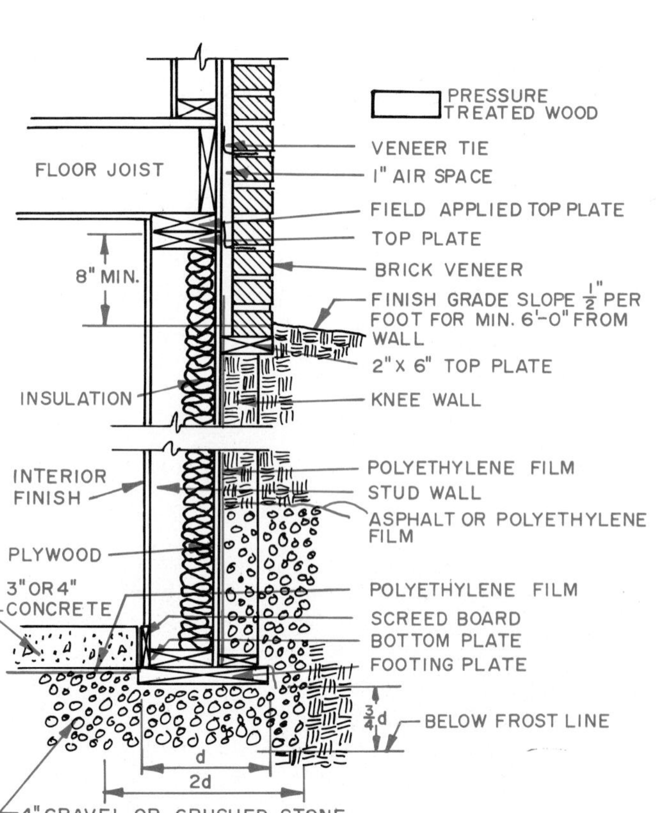

FIGURE 9-80

Construction details for wood foundation with brick-veneer construction and full basement. *(Courtesy of National Forest Products Association)*

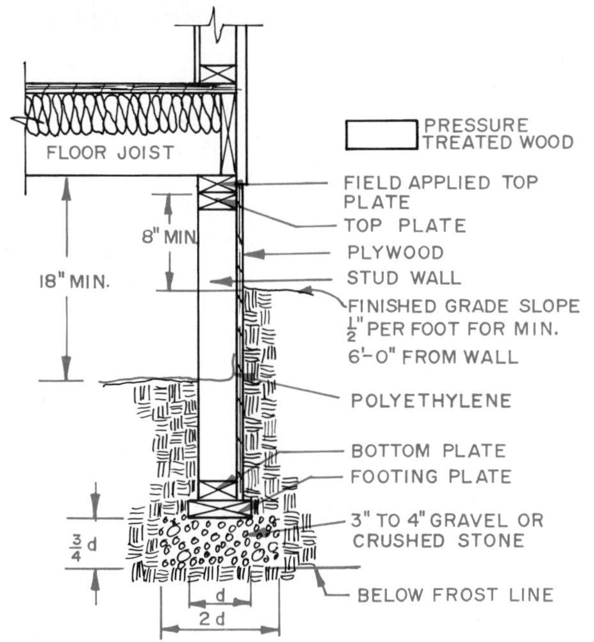

FIGURE 9-81

Construction details for wood foundation with frame construction over a crawl space. *(Courtesy of National Forest Products Association)*

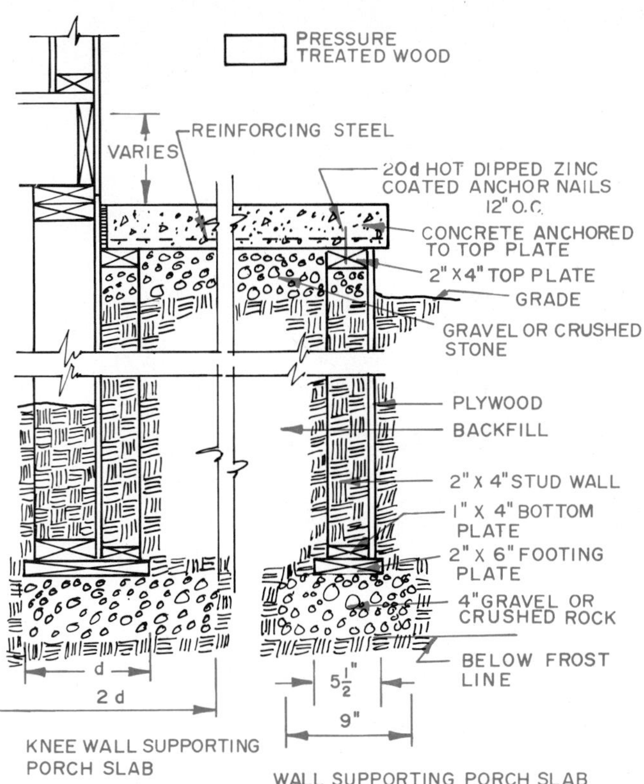

FIGURE 9-83

A wood foundation can support a concrete porch. *(Courtesy of National Forest Products Association)*

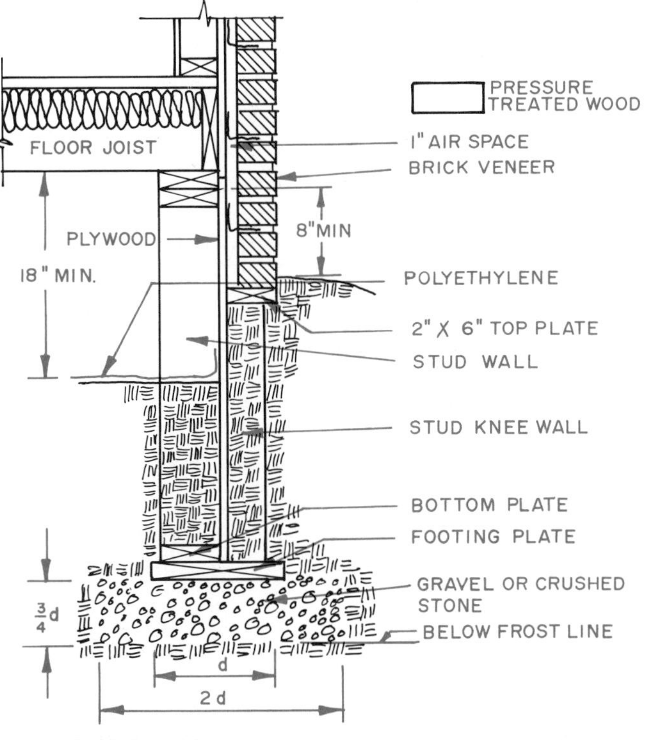

FIGURE 9-82

Construction details for wood foundation with brick-veneer construction over a crawl space. *(Courtesy of National Forest Products Association)*

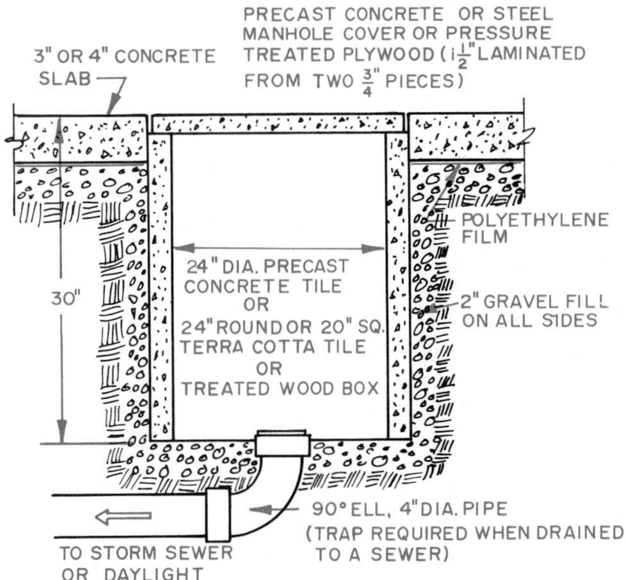

FIGURE 9-84

A recommended pit for collecting water to be removed with a sump pump. *(Courtesy of National Forest Products Association)*

A section through a typical sump is shown in Fig. 9-84. This concrete-lined unit could require an electric sump pump to keep the water level down. This is required in poorly drained soil. In areas where the soil permits easy drainage of moisture, a perforated pipe in a pit of gravel is adequate. The water collects in the pit and flows out the drain by gravity (Fig. 9-85).

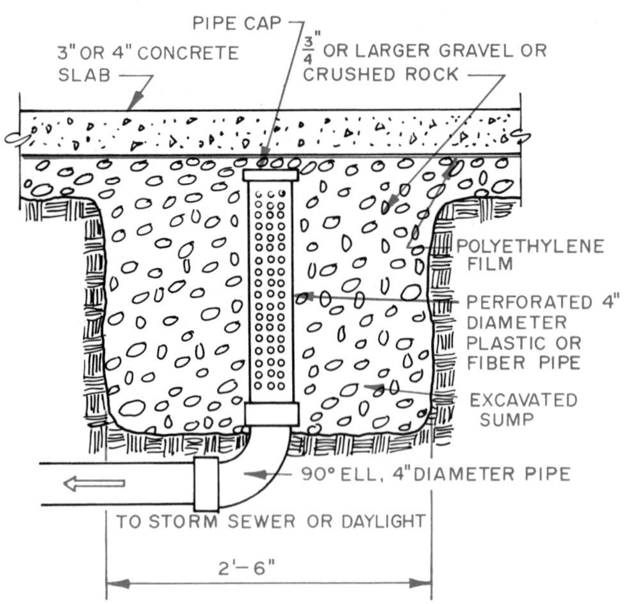

FIGURE 9-85

A gravel-filled pit permits water to drain into the soil and into the tile by gravity flow. *(Courtesy of National Forest Products Association)*

Essential to the construction of the all-weather wood foundation is the use of proper nails and recommended nail spacing. These are available in table form from the National Forest Products Association.

ESTIMATING FORM MATERIALS

Forms built on the site are of wood. The structural members are usually 2 × 4 in. stock. On heavy work 2 × 6 in. stock is needed. The sheathing can be boards but is almost always plywood.

To estimate the plywood needed:

1. Calculate the lineal feet of the outside of the building. Multiply by 2 because there are inside and outside forms.
2. Note the height of the foundation.
3. Calculate the number of sheets of plywood to sheath the forms. If sheets 4 ft 0 in. wide are used and the wall is 8 ft high, divide the lineal feet of wall (found in step 1) by 4. This gives the number of sheets.

   If the wall is low, as 4 ft or lower, figure the sheets as being placed on their 8-ft edge. To find the number of sheets, divide the lineal feet of wall by 8.

To estimate the structural members needed:

1. Decide how many 2 × 4 studs will be used per panel and multiply by the number of panels. This gives the number of studs. If walls are less than 8 ft, figure 8-ft studs anyway. Often, they are not cut in length but are permitted to stick beyond the upper edge of the form. This saves them for reuse as 8-ft pieces.
2. Wales are usually 2 × 4 in. × 16 ft 0 in. To figure the number of pieces needed, divide the lineal feet of wall by 16. Multiply this by the number of rows planned for the foundation.
3. To figure stock for bracing, decide how many braces will be used for each 4 ft 0 in. × 8 ft 0 in. panel. One brace is commonly used. The horizontal brace member is often 8 ft 0 in. and the inclined member is 10 ft 0 in. Multiply each size by the number of braces.

ESTIMATING CONCRETE

Concrete is sold by the cubic yard. A *cubic yard* contains 27 cubic feet. It is easiest when figuring the cubic yards of concrete needed to keep all sizes in feet.

The formula for figuring cubic yards is:

$$\text{cubic yards} = \frac{\text{width (ft)} \times \text{length (ft)} \times \text{thickness (ft)}}{27}$$

As an example, figure the cubic yards of concrete needed for a driveway 40 ft long, 8 ft wide, and 6 in. thick.

$$\text{cubic yards} = \frac{8 \times 40 \times \frac{1}{2}}{27}$$

$$= \frac{8 \times \overset{20}{\cancel{40}} \times 1}{27 \quad \underset{1}{\cancel{2}}} = \frac{160}{27}$$

$$= 5.93 \text{ cubic yards}$$

This does not allow for waste, unexpected variances in the form, or irregularities in the earth. Usually, the order is increased 5 to 10% to cover the unexpected.

There are tables available that can be used to determine the amount of concrete needed.

Most contractors use tables because it is faster and there is less chance to make a mistake.

STUDY QUESTIONS

After reading the chapter, answer each of the following questions. If you do not know the answer, review the chapter.

1. What is a frost line?
2. What materials make up concrete?
3. What type of plywood is used to build concrete forms?
4. What thickness of plywood is used for most concrete forms?
5. How are the sides of a concrete form held together?
6. Why is reinforcing steel placed inside a concrete foundation wall?
7. What is a concrete footing?
8. What purpose does a pier serve?
9. How are wood concrete forms treated after they are removed from the hardened wall?
10. Explain how forms are set up on a footing.
11. Why are pilasters used in foundation walls?
12. Why is sand or gravel placed below concrete sidewalks?
13. What is an all-weather wood foundation?
14. How many cubic yards of concrete are there in a wall 8 ft 0 in high × 8 in. thick × 20 ft 0 in. long?

IMPORTANT TECHNICAL TERMS

Following are technical terms that you should be able to use as part of your working vocabulary. Write a brief description of the meaning of each term.

Concrete forms
Frost line
Concrete
Footing
Foundation
Batterboards
Ties
Keyway
Pier
Column
Monolithically
Wale
Duplex nails
Pilaster
Anchor bolts
Reinforcing wire
Isolation joint
Polyethylene moisture barrier
Grade beam
All-weather wood foundation
Cubic yard

10 FLOOR CONSTRUCTION

When the foundation is completed, the carpenter begins to construct the floor framing. This includes beams and posts, sills, joists and subfloor. The same system is used for houses with basements or crawl spaces (Fig. 10-1).

SILL PLATE CONSTRUCTION

Sill plates are usually made from 2 × 6 in. lumber. They are bolted to the foundation. Anchor bolts ½ in. in diameter by 10 in. long are used for poured concrete foundations (Fig. 10-2). In concrete block foundations 18-in.-long anchor bolts are used. The cavity to hold the bolt is filled with concrete. A screen is placed under the cavities to be filled with concrete (Fig. 10-3). Anchor bolts are used to join frame walls to concrete slabs in the same manner described for sills (Fig. 10-4).

The sill plate must be level and true. A thin layer of mortar can be placed under it to correct irregularities. A *sill sealer* is often placed below the sill plate. It is a fibrous pad. It fills small irregularities between the top of the foundation and the wood sill plate. A termite shield is placed below the sill sealer (Fig. 10-5).

Following are the steps to frame the sill plate.

1. Place the sill plate on top of the foundation wall. Use pieces 14 to 16 ft long. Locate them so that no joints occur over openings in the foundation. The sill plate members meet with a square butt joint.
2. Lay the sill plate next to the anchor bolts. Draw square lines on the plate from each anchor bolt. Then measure in from the out-

FIGURE 10-1

The floor is framed using wood joists. This house has the joists lapped and has plywood subflooring. The subfloor is being glued and nailed to the joists. *(Courtesy of American Plywood Association)*

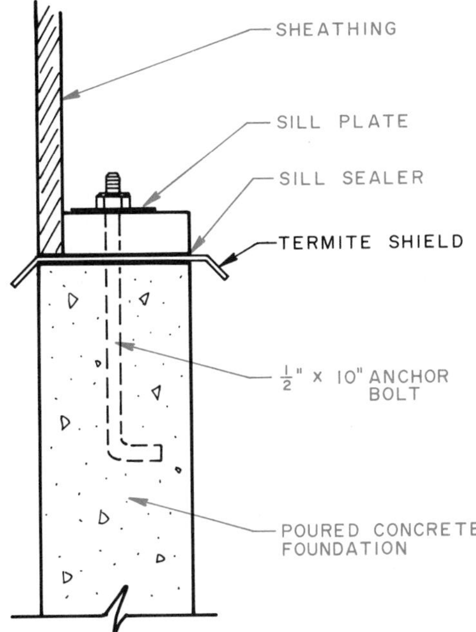

FIGURE 10-2

The sill plate is bolted to the foundation.

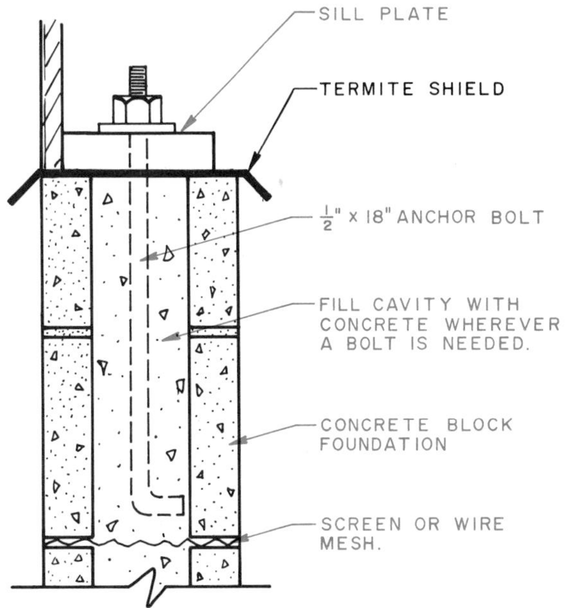

FIGURE 10-3

Concrete block foundations require the cavity holding the anchor bolt to be filled with concrete.

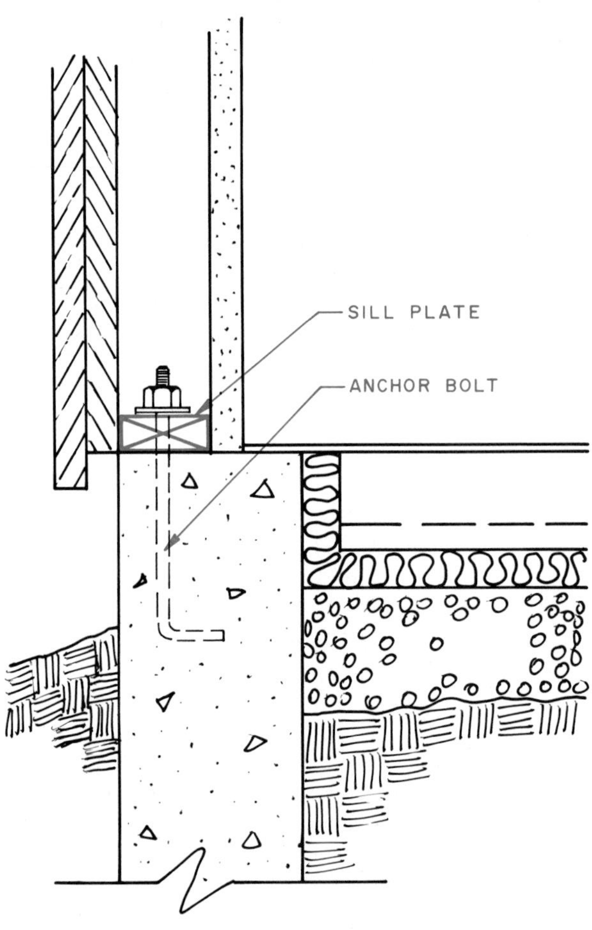

FIGURE 10-4

Frame walls are bolted to the concrete foundation in slab construction.

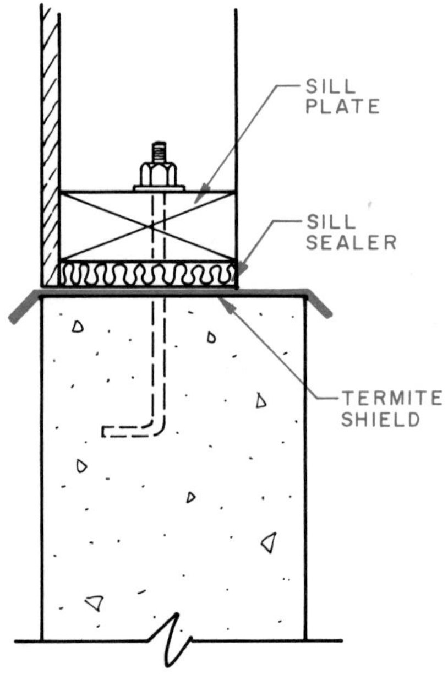

FIGURE 10-5

A termite shield and sill sealer are placed under the sill plate.

side edge to locate the center of the anchor bolt hole (Fig. 10-6). Remember that the sill plate is set in from the edge of the foundation the thickness of the sheathing. The sheathing goes over the sill plate and should be flush with the face of the foundation.

3. Bore the anchor bolt holes with a bit larger than the ½-in. diameter of the anchor bolt. This permits the sill plate to be adjusted so that it is straight. It also makes it easier to set the sill plate over the bolts.
4. Place the sill plate over the bolts. Cut them at the corners so that they form square butt joints. Check the sill plates at the corners to make certain the corners are square. (Fig. 10-7). This can be done using a framing square or by laying out a 3-4-5 triangle. This is the preferred method. Lay out 3 feet on one sill and 4 feet on the other. The corner is square when the diagonal measures 5 ft. (See Fig. 10-7).
5. Place washers and nuts on the anchor bolts and screw down so that they are firm.
6. Check the top of the sill to see that it is straight and level. If it is not, loosen the nuts and shim up the sill plate. Place wood shims every 3 or 4 ft between the sill plate and the foundation. If the top of the foundation is very untrue, place a thin mortar bed on it and level the sill plate in the mortar. Tighten the nuts just enough to hold it straight. Do not put a load on the sill plate until the mortar is completely dry.

Under no conditions should you overtighten the anchor bolt nuts. Tighten just enough so that they are firm.

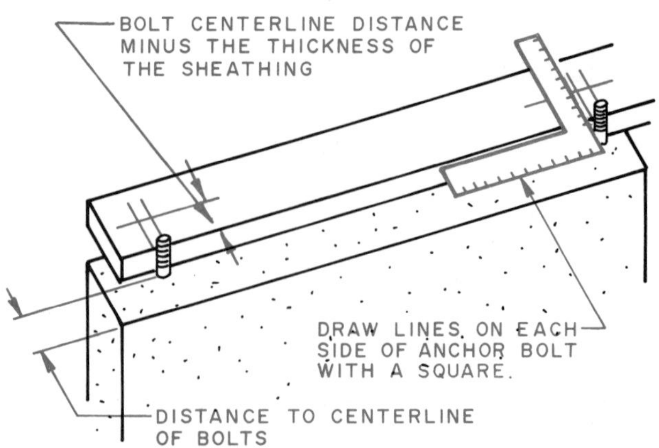

FIGURE 10-6

Use a square to locate anchor bolt holes in the sill plate.

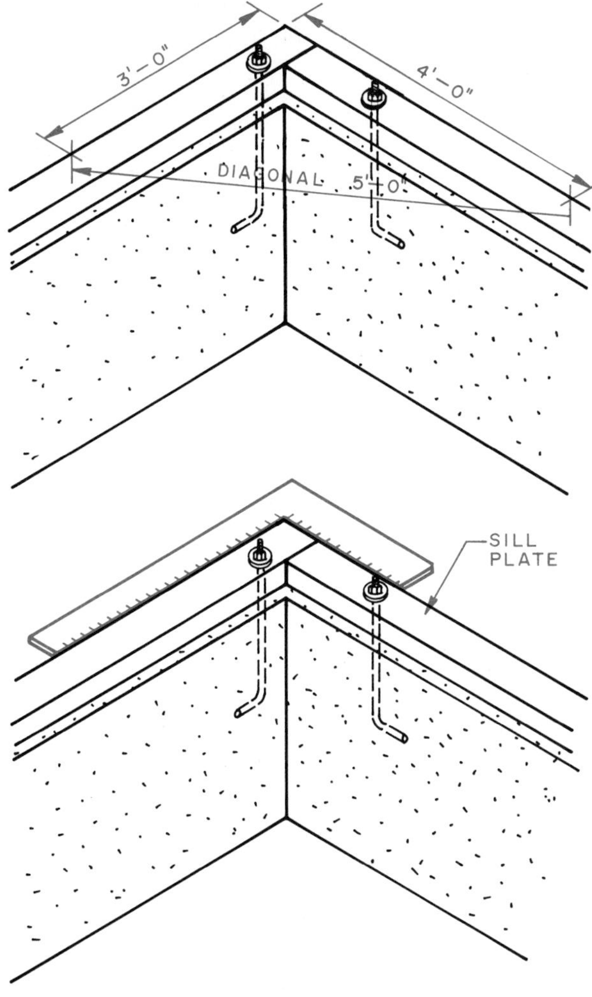

FIGURE 10-7

Check the sill plates at the corners to make certain they are square.

BEAMS AND GIRDERS

A *beam* or *girder* is a structural member used to support loads occurring between supporting walls or columns. Generally, the terms are used interchangeably. Sometimes wood members are called girders and steel members are called beams. In other cases the girder is considered the largest structural member. Beams are considered smaller structural members and are often supported by a girder. For purposes of this text, terms will be used interchangeably to represent a large structural member of any material that spans a void and carries a load.

Steel beams used in construction are usually the S-beam and the W-beam (Fig. 10-8). The *W-beam* is preferred because it has wide flanges. This gives it broader support on the masonry wall. An *S-beam* often has a wider metal plate welded to the bottom where it rests on the foundation. The size of beam to use depends upon the span and load to be carried.

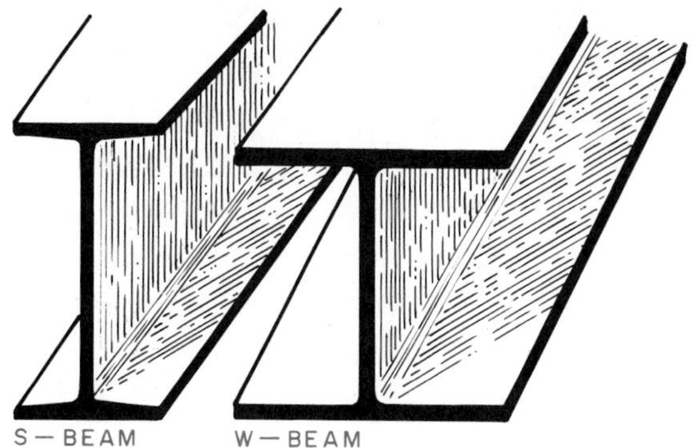

FIGURE 10-8

Steel beams used in floor construction.

This information is given on the plans for the building.

A sample of the type of loads S-beams and W-beams carry is shown in Fig. 10-9. Detailed design data is available from the American Institute of Steel Construction, 1221 Avenue of the Americas, New York, NY 10020.

Wood beams can be solid, built-up, laminated or box type (Fig. 10-10).

Solid or *laminated wood beams* are used where they are to be visible because they are more attractive. Built-up beams are most often used because they can be made from kiln-dried dimension lumber and are more dimensionally stable than solid wood beams. Laminated beams are of high quality but more expensive.

Built-up girders are made by joining together two or more pieces of 2-in. stock. The joints in the girder must be staggered and occur

SPAN (ft)	S4 BEAM (in.)		S5 BEAM (in.)	
	4 × 7.7	4 × 9.5	5 × 10	5 × 14.75
4	12.2	13.6	19.7	24.4
6	8.1	9.0	13.1	16.2
8	6.1	6.8	9.8	12.2
10	—	—	7.9	9.7

(A)

SPAN (ft)	W8 BEAM (in.)		
	8 × 10	8 × 13	8 × 15
4	30.5	39.6	47.2
6	20.3	26.4	31.5
8	15.3	19.8	23.6
10	12.2	15.8	18.9
12	10.2	13.2	15.7
14	8.7	11.3	13.5
16	7.6	9.9	11.8

(B)

Note: A kip is 1000 lb.

FIGURE 10-9

Typical load-carrying capabilities (kips) of selected (A) S-beams and (B) W-beams.

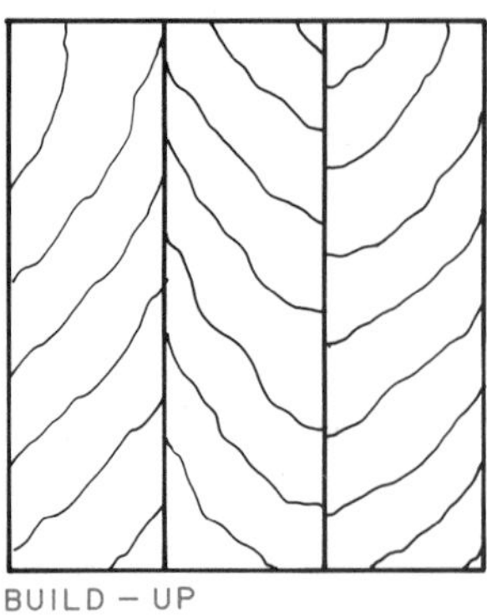

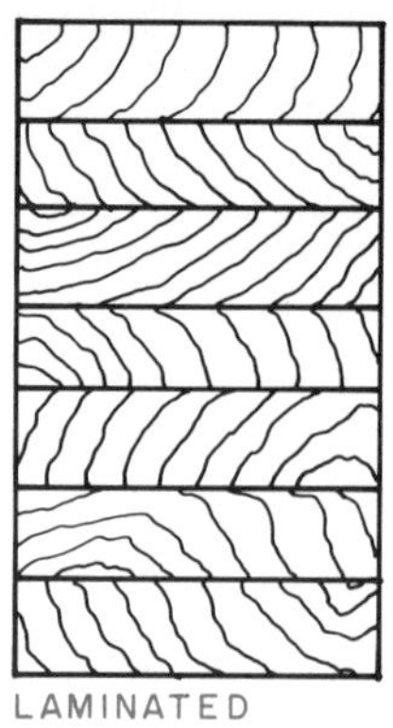

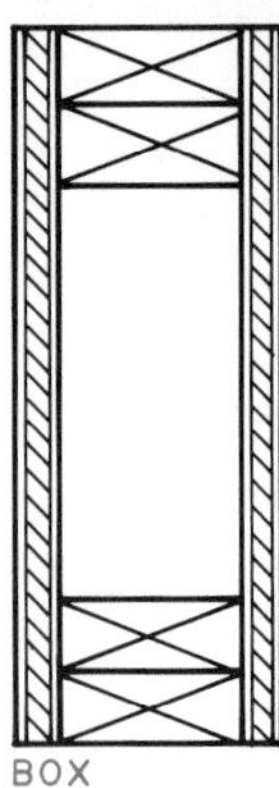

FIGURE 10-10

Types of wood beams.

over posts. Some built-up girders have the layers glued together. This produces a high-quality product. They may also be made by nailing or bolting. If bolted use at least ⅝-in.-diameter bolts. They should be staggered and placed 20 in. apart.

Two-piece nailed girders are joined with 10d nails all driven from one side. Two nails are driven at the end of each piece. The others are driven near the top and bottom of the girder, staggered and spaced 16 in. apart (Fig. 10-11).

Three-piece girders are nailed from both sides using 20d nails. Two nails are driven near the end of each piece. They are also driven near the top and bottom of the girder, staggered and spaced 32 inches apart (Fig. 10-12).

The ends of the wood girder must rest at least 4 in. on the foundation wall. If the lumber is not preservative treated leave a ½-in. air space at the end and on all sides (Fig. 10-13). This requires that a 4½-in.-deep pocket be prepared in the foundation wall.

The concrete pocket should be lined with sheet metal to protect the beam from termites. It is best if the girder is set on a ⅜-in.-thick metal bearing plate. If the girder needs to be raised a little, it should be shimmed with metal pieces the size of the bottom of the concrete pocket.

To select the proper size of solid or built-up wood beam the span, total load on the beam, the species and grade of lumber to be used and the strength or allowable working stresses of the lumber must be known. Tables showing the maximum allowable loads on various sizes and spans of beams are available in timber construction manuals. An example of design data is shown in Fig. 10-14. This is for solid wood beams. Built-up beams of the same size will carry a slightly smaller load.

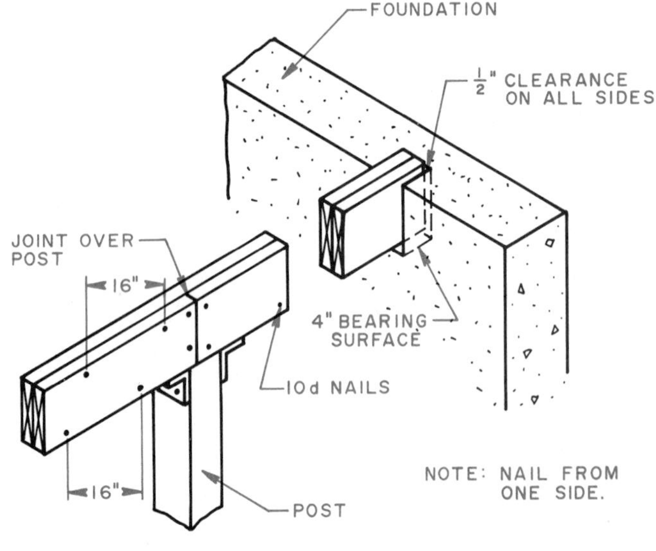

FIGURE 10-11

Nailing pattern of a two-member built-up wood girder.

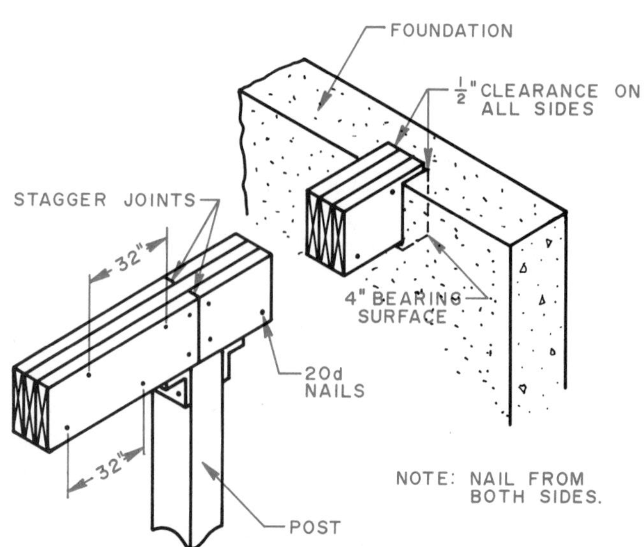

FIGURE 10-12

Nailing pattern of a three-member built-up wood girder.

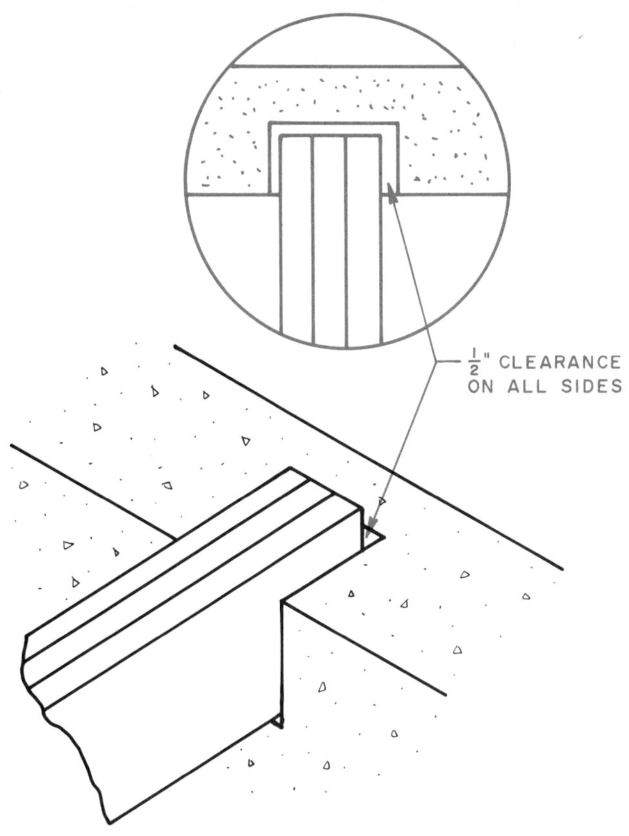

FIGURE 10-13

Wood girders fit in pockets cast in the foundation wall.

Glued, laminated structural members are manufactured in standard sizes and lengths. They are designed to carry specified loads for various spans. They are about one-third stronger than solid timber members made from equal quality seasoned material.

Laminated beams to be used for interior purposes are made with a water resistant glue. Those to be exposed to the weather use a waterproof glue.

A few of the many sizes available are shown in Fig. 10-15. These are designed to carry specified pounds per square foot when spaced at the distances shown. For example, a 16-ft beam spaced 6 ft apart that is 3¼ × 8 in. in cross section will carry 30 lb per square foot.

Box beams are made by combining lumber and plywood. They are made following carefully prescribed designs. They are hollow structural units that develop high strength-to-weight ratios. There are two types, glued and nailed beams.

Glued beams are made by combining one or more vertical plywood webs laminated to seasoned lumber flanges (Fig. 10-16). The flanges are spaced at intervals along the beam's length by vertical spacers which distribute concentrated loads and prevent web buckling. The lumber flanges of more than one lamination must be glued under positive mechanical pressure. Plywood may be glued to other plywood or lumber under positive mechanical pressure or with pressure applied with nails or staples. They are manufactured under controlled conditions to rigid specifications. Design details are available from the American Plywood Association, 7011 South 19th St., Box 11700, Tacoma, Wash. 98411.

Nailed box beams can often be used in construction. They can be constructed on the site. One example of a common use for nailed box beams is a garage door header (Fig. 10-17). It is made of 2 × 4 in. lumber and C-D INT-APA plywood with exterior glue (Fig. 10-18). Typical allowable loads for several sizes are in Fig. 10-19. They are based on using No. 1 Douglas Fir or southern pine kiln-dried lumber. The lumber members must be full length. They cannot be spliced.

To build these box beams follow the following steps.

1. Figure the load on the beam. Assume a roof load of 355 lb per lineal foot on an 18 ft beam.

NOMINAL SIZE (solid) (in.)	ACTUAL SIZE (in.)	EASTERN HEMLOCK SELECT STR NO. 1 (lb)	CALIF. REDWOOD SELECT STR NO. 1 (lb)	DOUGLAS FIR SELECT STR NO. 1 (lb)	EASTERN SPRUCE SELECT STR NO. 1 (lb)	SOUTHERN PINE SELECT STR NO. 1 (lb)
4 × 8	3½ × 7¼	1,975	2,140	2,630	1,940	2,960
4 × 10	3½ × 9¼	3,826	3,990	4,490	3,160	5,480
4 × 12	3½ × 11¼	5,660	5,900	6,640	4,670	8,120
6 × 10	5½ × 9½	6,340	6,610	7,440	5,230	9,090
6 × 12	5½ × 11½	9,294	9,690	10,910	7,670	13,330
6 × 14	5½ × 13½	12,800	13,350	15,020	10,570	18,360
8 × 12	7½ × 11½	12,670	13,220	14,870	10,460	18,170
8 × 14	7½ × 13½	17,460	18,210	20,490	14,420	25,050

Note: Figures are the number of pounds a beam will carry when spanning 10 ft.

FIGURE 10-14

Design data for selected wood beams.

SPAN OF BEAM (ft)	SPACING BETWEEN BEAMS (ft)	BEAM SIZE (in.)		
		30*	40*	50*
16	6	3¼ × 8	3¼ × 9⅝	3¼ × 9⅝
	8	3¼ × 9⅝	3¼ × 11¼	3¼ × 11¼
	10	3¼ × 9⅝	3¼ × 11¼	3¼ × 12⅞
	12	3¼ × 11¼	3¼ × 12⅞	3¼ × 14½
	14	3¼ × 11¼	3¼ × 12⅞	3¼ × 14½
	16	3¼ × 12⅞	3¼ × 14½	5¼ × 12⅞
20	8	3¼ × 11¼	3¼ × 12⅞	3¼ × 14½
	10	3¼ × 12⅞	3¼ × 14½	5¼ × 12⅞
	12	3¼ × 12⅞	5¼ × 12⅞	5¼ × 14½
	14	3¼ × 14½	5¼ × 12⅞	5¼ × 14½
	16	3¼ × 14½	5¼ × 14½	5¼ × 16⅛
	18	5¼ × 12⅞	5¼ × 14½	5¼ × 16⅛
24	8	3¼ × 14½	5¼ × 12⅞	5¼ × 14½
	10	3¼ × 14½	5¼ × 14½	5¼ × 14½
	12	5¼ × 14½	5¼ × 14½	5¼ × 16⅛
	14	5¼ × 14½	5¼ × 16⅛	5¼ × 17¾
	16	5¼ × 14½	5¼ × 16⅛	5¼ × 17¾
	18	5¼ × 16⅛	5¼ × 17¾	5¼ × 19⅜

*Pounds per square feet.

FIGURE 10-15

Design data (lb/ft²) for selected glued laminated wood beams. *(Courtesy of Timber Structures, Inc.)*

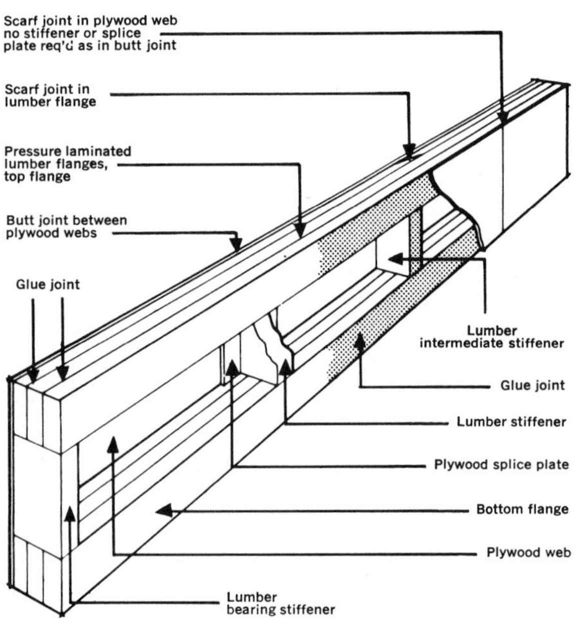

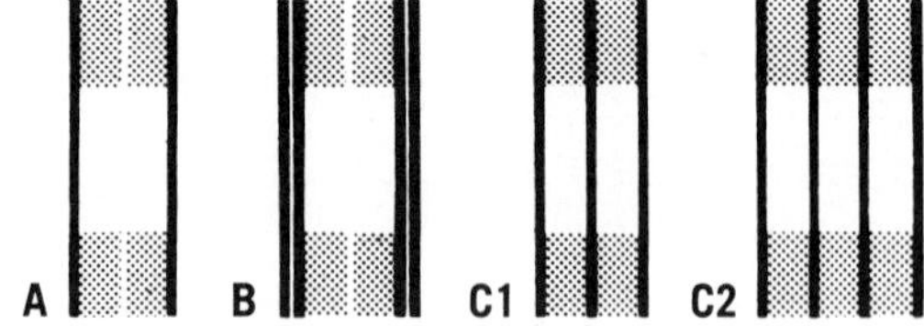

FIGURE 10-16

A carefully manufactured and engineered box beam. *(Courtesy of American Plywood Association)*

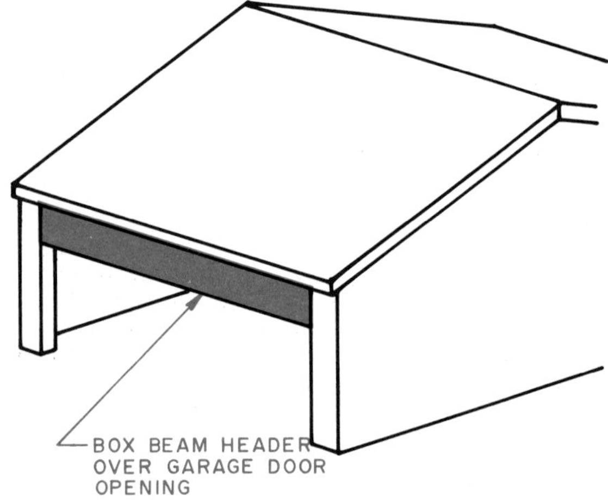

FIGURE 10-17

A box beam used as a header over a garage door opening.

2. Select the proper beam design. Look at the load design in Fig. 10-19. The 18-ft column shows that a beam 20 in. deep with two 2 × 4 in. flange members top and bottom and ¾-in. plywood webs will carry 362 lb per lineal foot. This is the closest size equaling 355 or more. The proper design is therefore cross section B as shown in Fig. 10-19.

To construct this box beam follow three steps.

1. Determine the layout of stiffeners and the plywood butt joints. These are shown in Fig. 10-20. The plywood joint locations require a minimum of a 2-ft stagger between the panel butt joints on opposite sides of the beam. All butt joints are located in the middle half of the beam. Vertical stiffeners should be added to these layouts so that they are no farther apart than 4 ft.

The 6 in. added to the clear span of the beams in Fig. 10-20 represent the double 2 × 4 in.

FIGURE 10-18

An exploded view of a carpenter built box beam. *(Courtesy of American Plywood Association)*

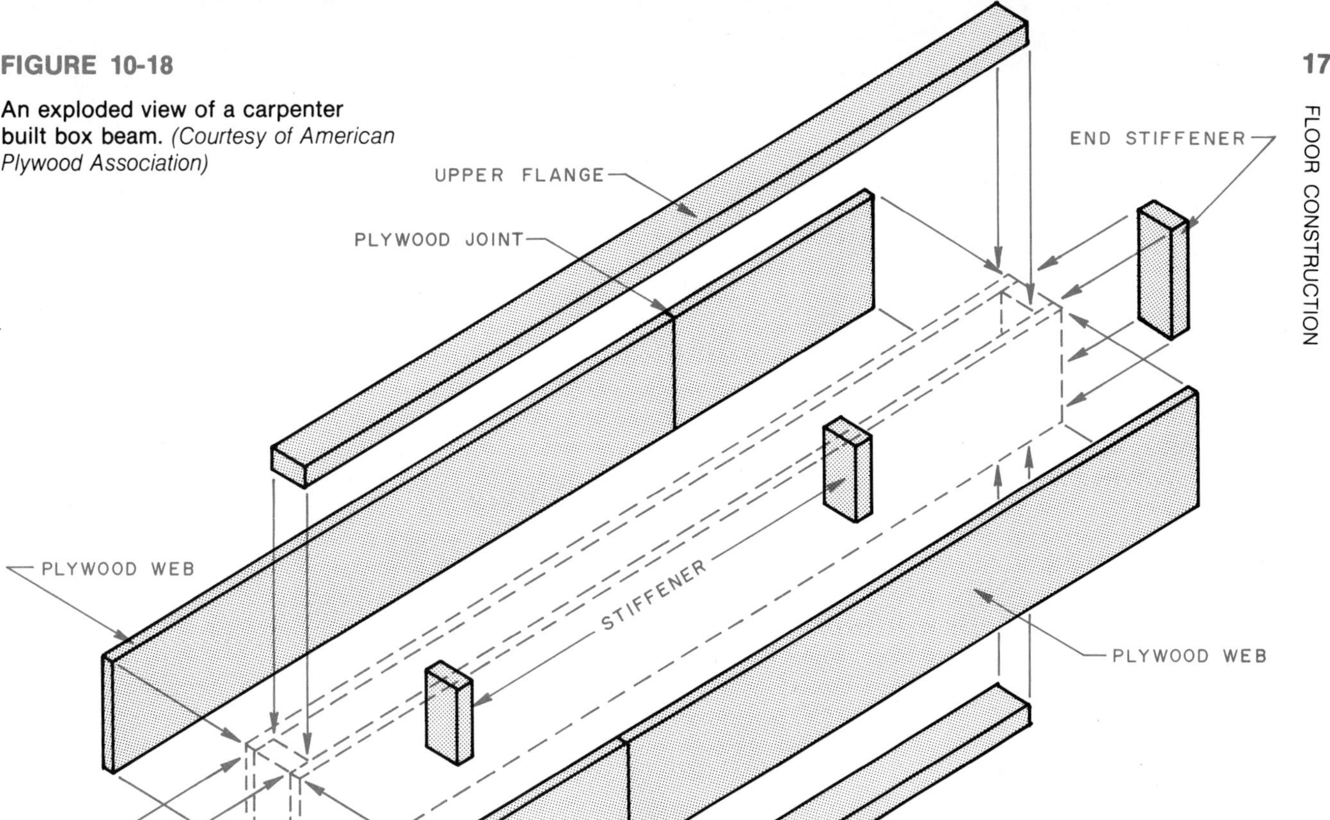

FIGURE 10-19

Allowable loads on box beams for headers or roof beams. *(Courtesy of American Plywood Association)*

PLYWOOD	CROSS SECTION	APPROX. WT. PER FOOT (lb)	SPAN (ft) 10	12	14	16	18	20	22	24
	Allowable Load[a] for 12-in.-Deep Roof Beam or Header (lb/lin ft)									
½-in. 32/16	A	6	304	253	189	145	114	92	76	64
½-in. 32/16	B	8	332	276	237	207	173	140	116	97
¾-in. 48/24	B	10	536	391	287	220	173	140	116	97
¾-in. 48/24	C	12	—	410	323	247	195	158	130	110
	Allowable Load[a] for 16-in.-Deep Roof Beam or Header (lb/lin ft)									
½-in. 32/16	A	7	436	364	270	207	163	132	109	92
½-in. 32/16	B	9	465	387	332	290	258	216	178	150
¾-in. 48/24	B	11	749	600	441	338	267	216	178	150
¾-in. 48/24	C	14	—	—	499	409	323	262	216	182
	Allowable Load[a] for 20-in.-Deep Roof Beam or Header (lb/lin ft)									
½-in. 32/16	A	8	579	479	352	269	212	172	142	119
½-in. 32/16	B	10	597	498	426	373	332	293	242	204
¾-in. 48/24	B	13	957	798	599	459	362	293	242	204
¾-in. 48/24	C	15	—	—	650	569	460	372	308	258
	Allowable Load[a] for 24-in.-Deep Roof Beam or Header (lb/lin ft)									
½-in. 32/16	A	9	732	590	433	332	262	212	175	147
½-in. 32/16	B	11	—	606	519	454	404	363	307	258
¾-in. 48/24	B	14	1164	970	759	581	459	372	307	258
¾-in. 48/24	C	17	—	—	802	701	600	486	402	337

CROSS SECTIONS

A

B

C

[a]Includes 15% snow loading increase.

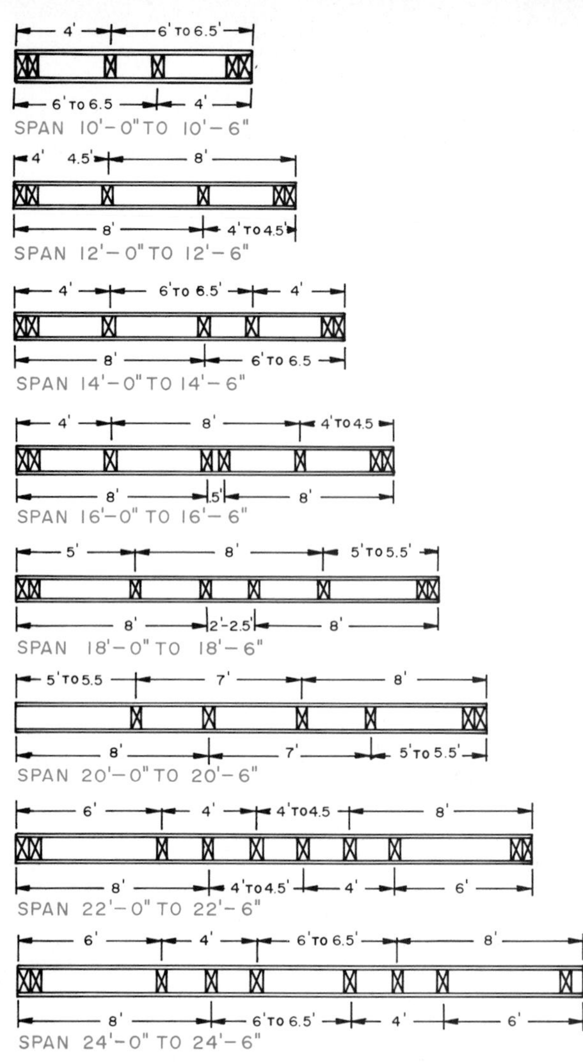

FIGURE 10-20

Joint and stiffener layout for box beams used as headers and roof beams. *(Courtesy of American Plywood Association)*

end stiffeners. These are the bearing ends which are used to support the beam.

2. Build the framework of lumber flanges and stiffeners.

Use dry lumber having a moisture content of 19% or less for Douglas fir and 15% or less for southern pine. Lumber should be free of warp or characteristics that would produce gaps greater than ⅛ in. between the lumber and plywood.

Nail the flanges to the stiffeners with 8d common nails. When two or more flanges or stiffeners are to be together they are nailed one at a time with 10d common nails.

3. Fasten the plywood webs to the framework. Plywood should be installed with its face grain in the same direction as the flanges. Butt joints must occur over stiffeners.

The plywood is nailed to the frame with 8d common nails. The prescribed nailing pattern is shown in Fig. 10-21. If the beam is to be exposed to the weather corrosion resistant nails must be used.

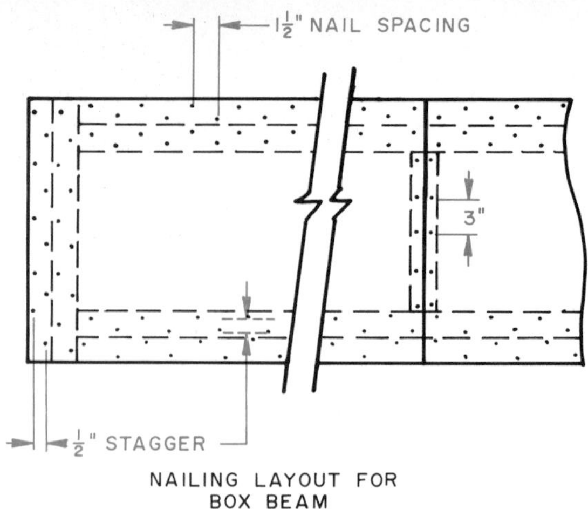

FIGURE 10-21

Nailing pattern for wood box beams used as headers and roof beams. *(Courtesy of American Plywood Association)*

COLUMNS

Columns are made from wood or steel. They are used to support beams spanning long distances. The column shortens the distance the beam must span. This makes it possible to use a smaller beam.

Steel columns are used to support both steel and wood beams. They have a steel base that rests on a concrete footing. The top of the column has a steel plate to hold the beam (Fig. 10-22).

FIGURE 10-22

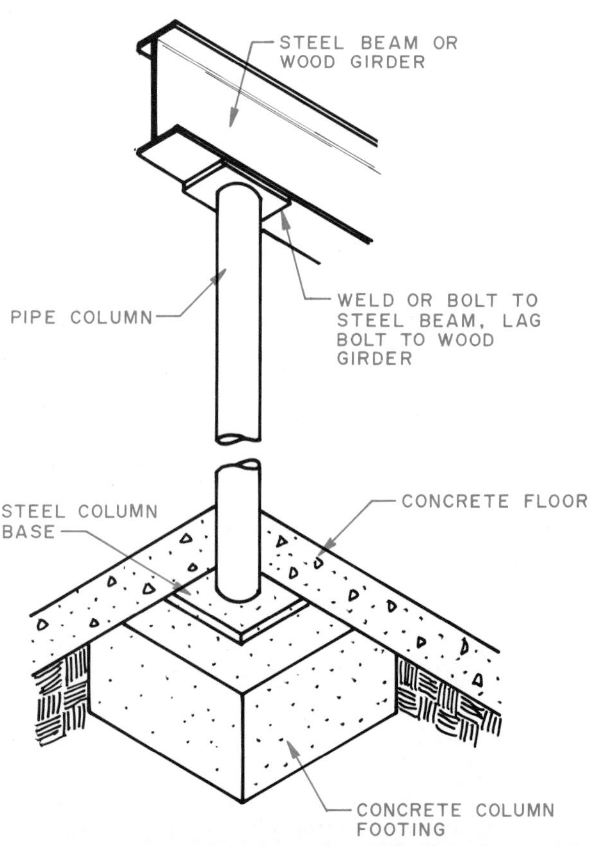

Wood columns are held to their footing and the wood beam with a variety of metal devices (Fig. 10-23). They should be solid lumber at least 6 × 6 in. in size. The bottom of the post should be raised 2 to 3 in. above the floor. This helps prevent decay. It is best if the bottom of the post is treated with a preservative.

WOOD FRAMING SYSTEMS

Light wood framing used in typical residential construction generally follows one of three commonly used systems: balloon framing, western or platform framing, or post-and-beam framing. While there may be variations of these

FIGURE 10-23

Selected ways to connect wood posts to beams and footings.

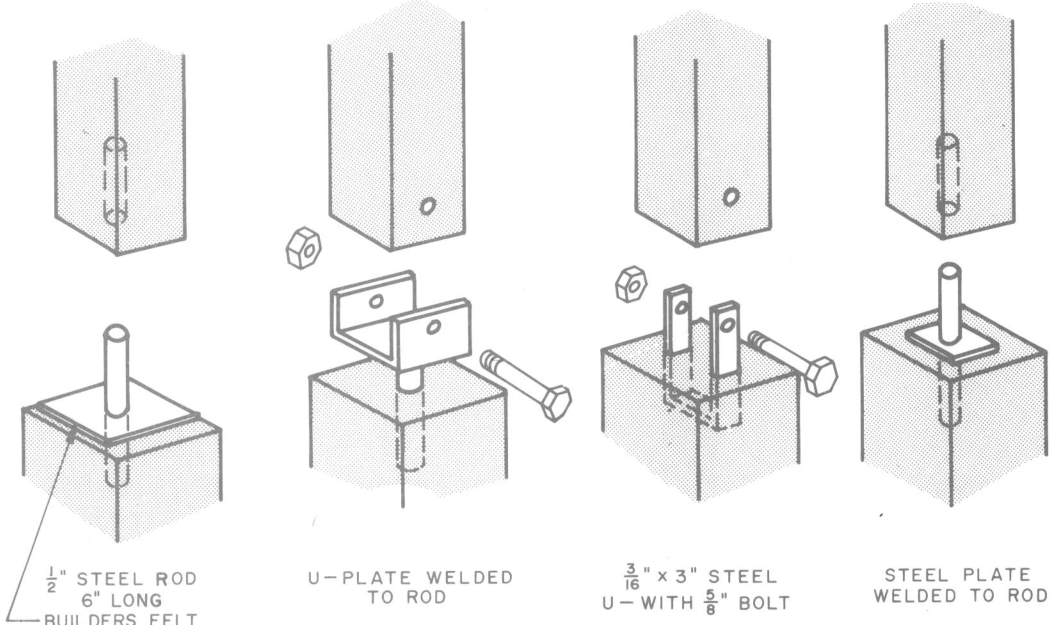

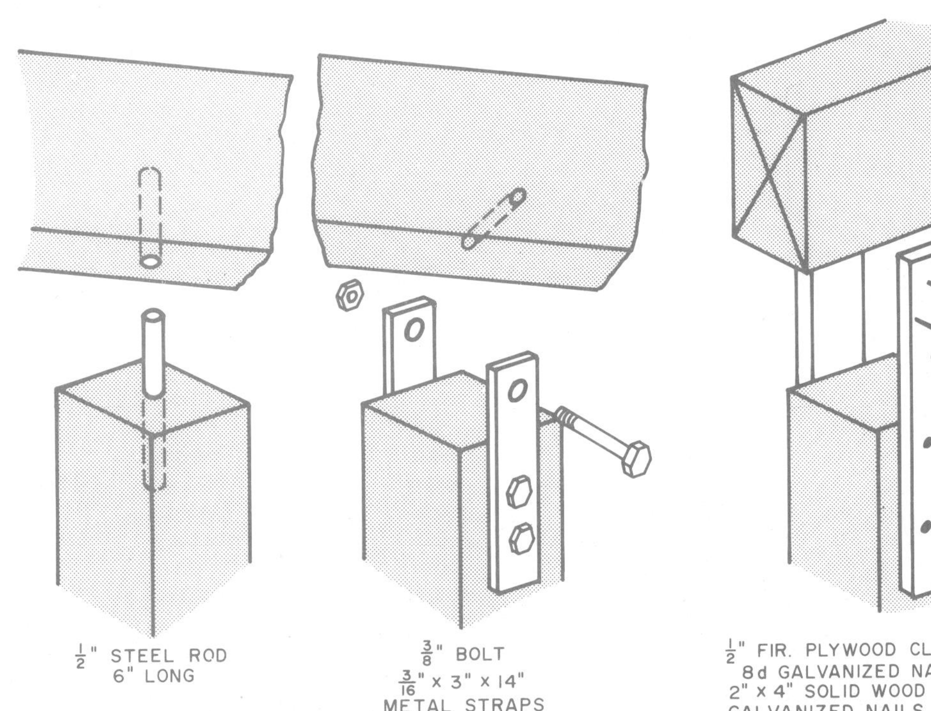

used in some localities, these are the basic systems from which the variations are taken.

Balloon Framing *Balloon framing* uses a continuous stud from the sill to the roof plate (Fig. 10-24). The weight of the exterior wall and roof are transmitted directly to the sill and foundation. The studs rest on the sill. The sill should be 2 in. or thicker. The joists are toenailed to the sill with three 8d or two 10d nails. The joist is also face-nailed to a stud with two 10d nails. Each stud is toenailed to the sill with two 10d nails. Box nails are often used because they are thinner and are less likely to split the wood. Common nails are also frequently used.

Since the exterior wall in balloon framing is open from the sill to the roof, *fire stops* must be nailed in place. Fire stops are wood members nailed between joists or studs which will break the natural draft or chimney like effect. This retards the spread of a fire inside the wall to the roof. Fire stops are shown in Figs. 10-25 and 10-26.

Two types of sill construction are common. The standard sill has the floor joists face nailed to the studs with two 10d nails. They are toenailed to the sill with two 10d nails (Fig. 10-25). Fire stops the same size as the joists are nailed next to the studs. They must touch the sill below and the subflooring above.

The T-sill has the joists nailed to a *header* (Fig. 10-25). The header is end nailed to the joist with three 16d nails. The joist is toenailed to the sill with two 10d nails. The header is toenailed to the sill with 8d nails spaced about 16 in. apart.

Construction details at the second floor exterior wall are shown in Fig. 10-26. A *ribbon* is let in (notched into) the studs in the exterior

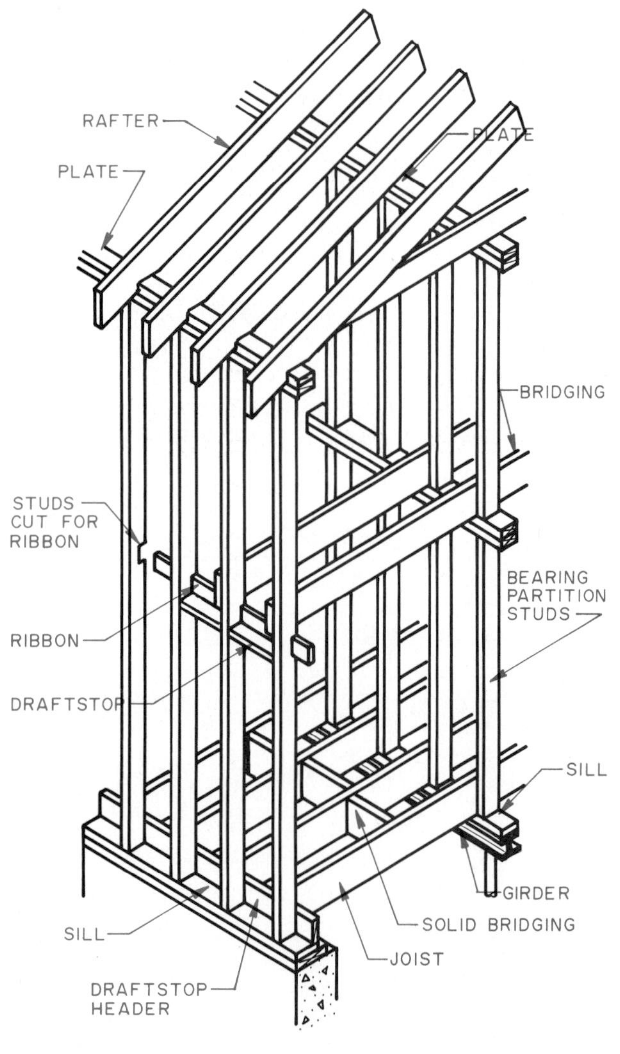

FIGURE 10-24

Typical balloon framing system.

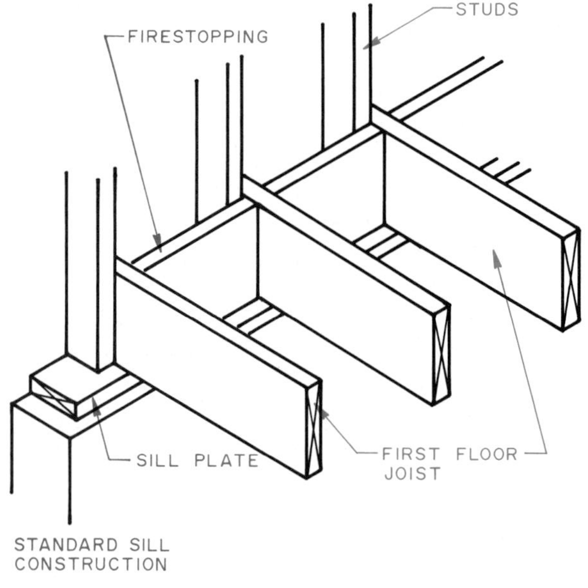

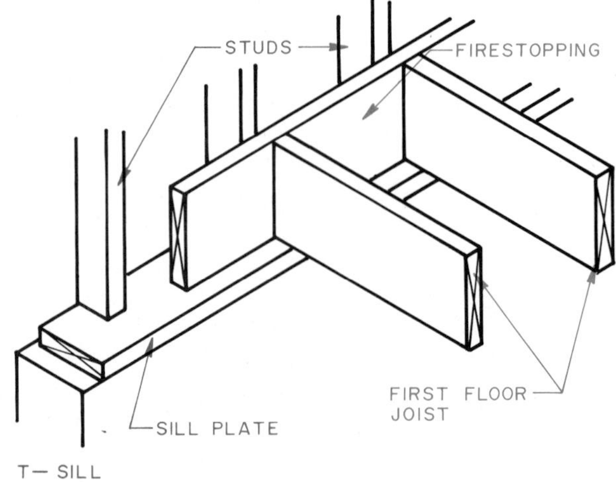

FIGURE 10-25

Sill construction details for balloon framing.

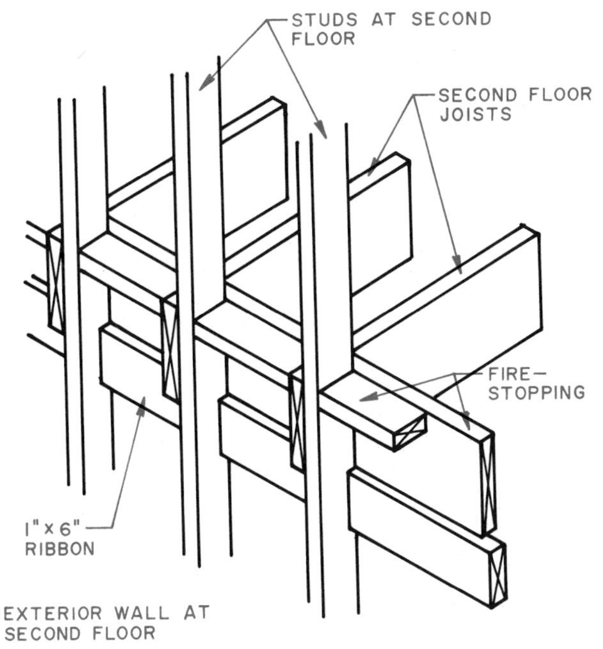

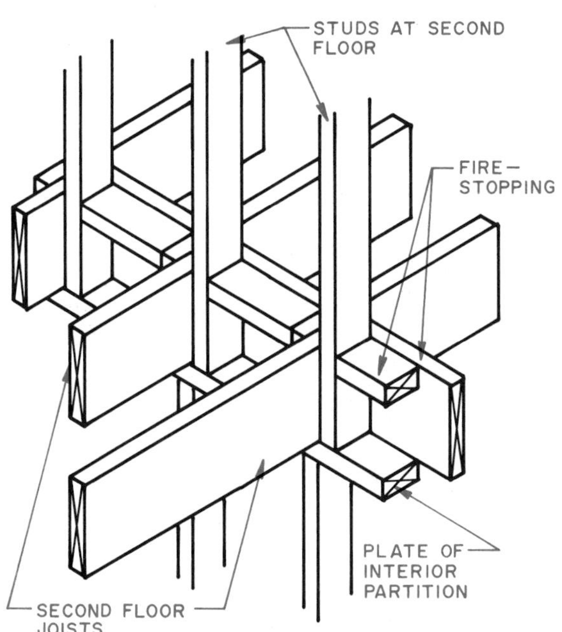

FIGURE 10-26

Construction details for balloon framing at the second floor.

wall. The floor joists rest on the ribbon. They are face nailed to the studs with three 10d nails. The ribbon is face nailed to the studs with two 8d nails. A 2 × 4 in. fire stop is nailed between studs. Another fire stop is nailed between joists in the same manner as was done at the sill.

Generally the second floor joists will rest on an *interior bearing partition* (Fig. 10-26). Each joist will be placed directly above a stud in the partition. A *single* 2 × 4 in. *plate* can be used, although some persons prefer a *double plate.* The joists overlap the thickness of the stud. They are nailed to each other with three 16d nails. Each joist is toenailed to the plate with two 10d nails. Notice the fire stopping used at this interior supporting wall.

Platform Framing

Platform framing is most commonly used in residential construction. Each floor or wall is one story high. It resembles building a one-story house on top of another one-story house (Fig. 10-27). It has the advantage of using shorter pieces of lumber and at each floor level it provides a solid platform upon which the carpenter can work.

The first floor is built as a complete platform. It uses box sill construction (Fig. 10-28). The header is end nailed to the joists with two 16d nails. A header is a member that runs perpendicular to the joists. Each joist is toenailed to the sill with two 10d nails. The header is toenailed to the sill with 8d nails placed every 16 in. The placement of the *stringer* is shown in Fig. 10-28. A stringer is the outer member of the platform frame which runs parallel with the floor joists. The stringer is set flush with the outer edge of the sill. It is toenailed to the sill with 8d nails spaced every 16 in.

After the floor joists are in place, the subfloor is nailed in place. The first floor wall rests on top of this (Fig. 10-28). If the house has a second floor another box sill platform is built on top of the first floor wall (Fig. 10-29). The exterior wall has a double 2 × 4 in. plate. The floor joists are placed over the studs in the wall. Headers and stringers are used. The second floor joists are usually supported inside the structure by bearing partitions (Fig. 10-29).

Post, Plank, and Beam Framing

This system uses vertical wood posts to support the floor and floor load. The floor is decked with 2-in. wood planks. There are two basic framing systems. One system has the floor and roof beams running the width of the building. The floor and roof planks run the length of the building. The other system has the floor and roof beams running the length of the building. The floor and roof planking runs the width of the building. In both systems posts are needed in the center of the structure to support the ridge beam.

The exterior walls are built between the posts supporting the roof. These are non-load-bearing walls. They can be of any suitable material.

See Chapter 23 for additional information.

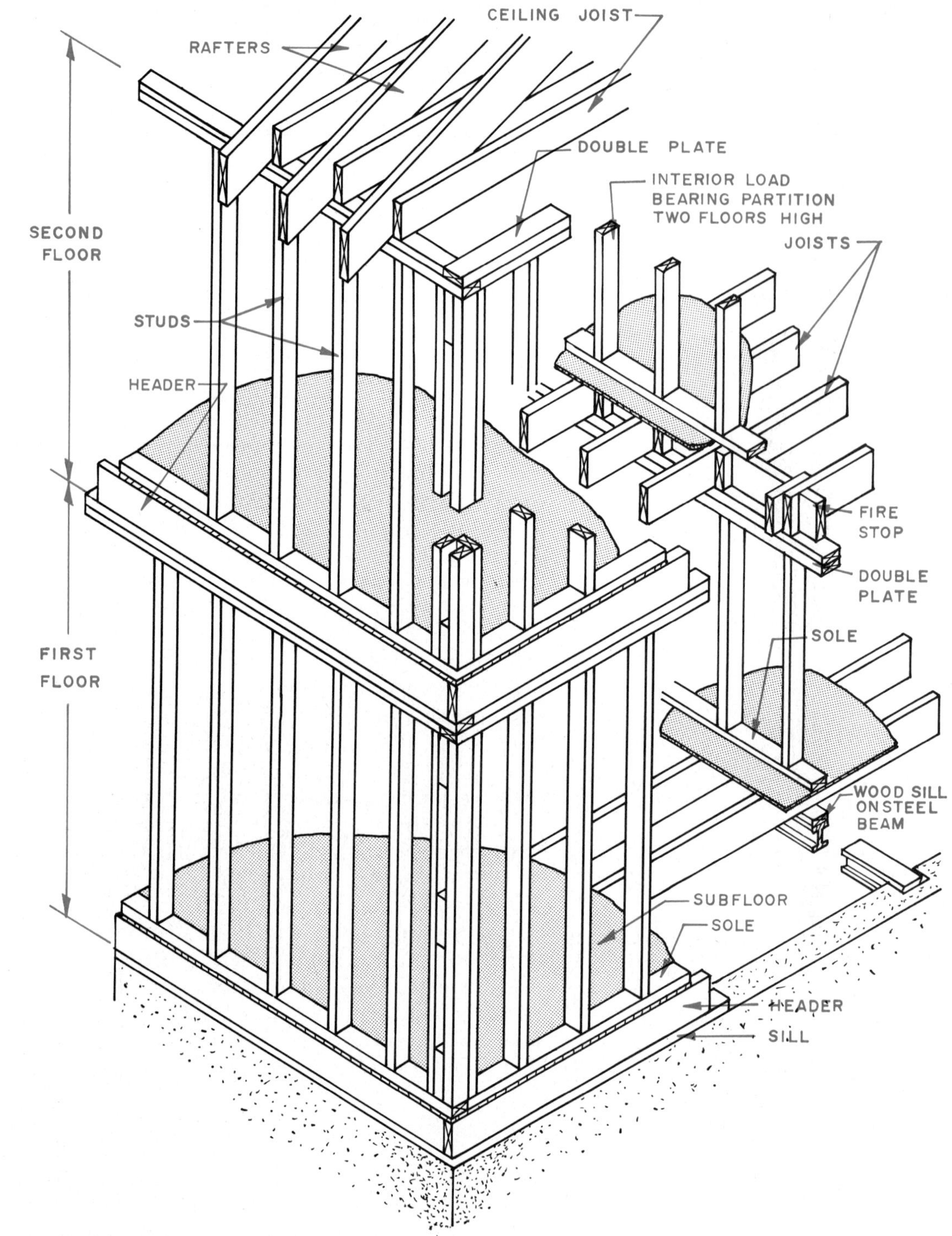

FIGURE 10-27

Typical platform framing system.

JOIST TO BEAM OR GIRDER INSTALLATION

Joists are generally supported with steel or wood beams. A steel beam is set flush with the top of the foundation. It has a wood member fastened to it the same thickness as the sill (Fig. 10-30). Metal fasteners are used to join the wood plate to the steel beam. Often it is bolted by drilling holes through the flange of the beam every 24 in. The floor joists rest on this and overlap at least 4 in. but not more than 12 in. They are face-nailed to each other using three 16d nails. They are toenailed to the wood member on the beam with two 10d nails.

Sometimes it is desired to have the joists

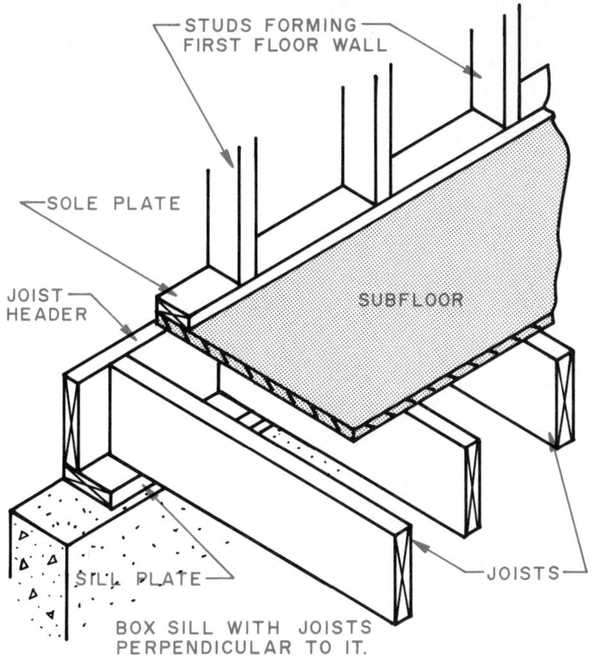

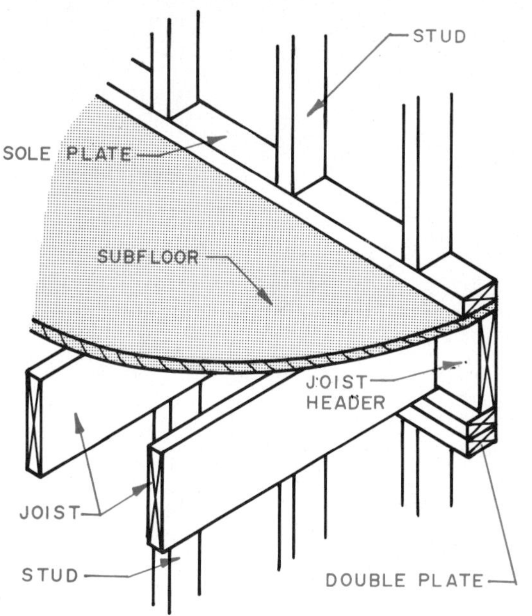

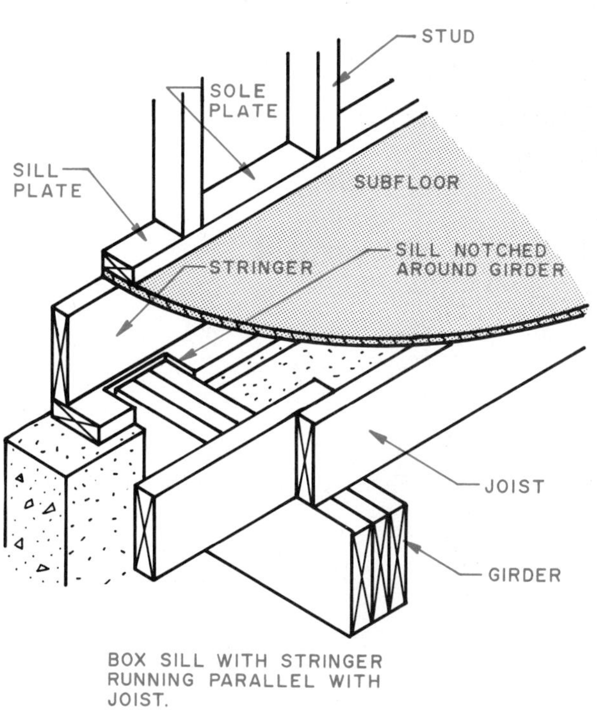

FIGURE 10-28

Sill construction details for platform framing.

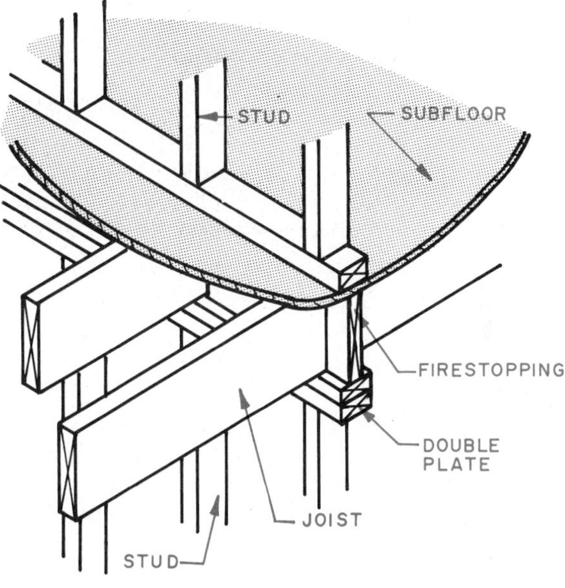

FIGURE 10-29

Construction details for platform framing at the second floor.

in-line. They meet end to end over the beam rather than lapping. (Fig. 10-31). A 2-in. thick scab is nailed across the joint. At least three 16d nails are used on each side of the joint.

There are designs available from the American Plywood Associations for in-line joist systems. This utilizes joists of uneven lengths. One joist cantilevers (overhangs) the beam (Fig. 10-32). The two parts are connected with a metal connector or plywood splice plate. The cantilevered joint is alternated on every other joist.

Wood beams are usually set in the foundation so the top is flush with the top of the sill. Joists can rest on this and be nailed as just explained (Fig. 10-33).

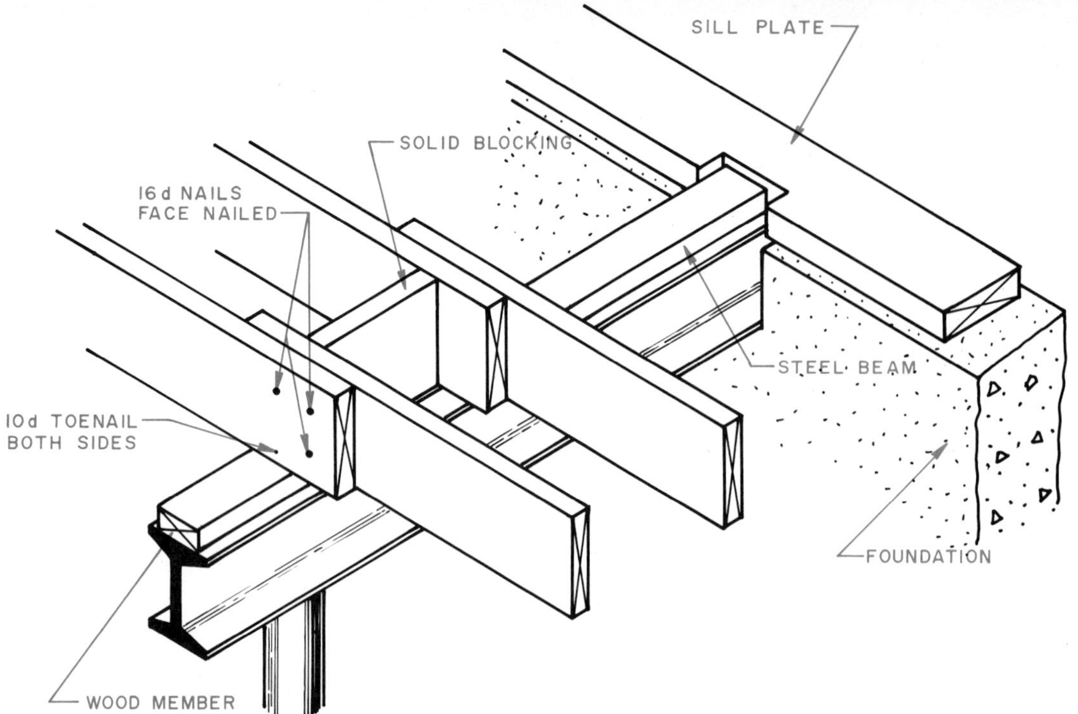

FIGURE 10-30

Joists are toenailed to a wood member fastened on top of a steel beam.

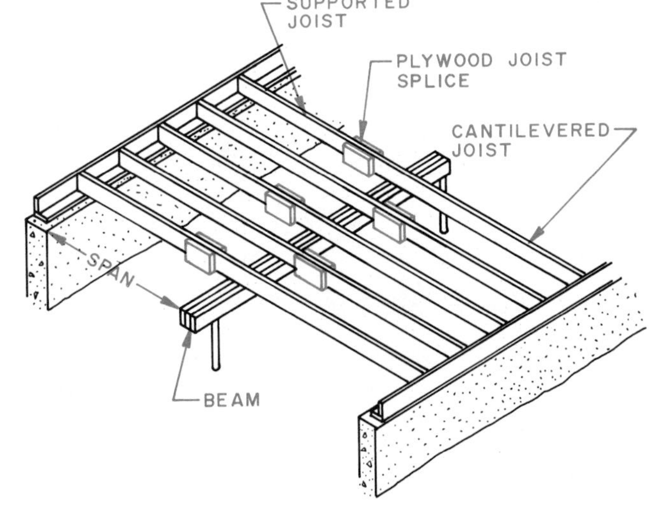

FIGURE 10-32

An in-line joist system using cantilevered joists.

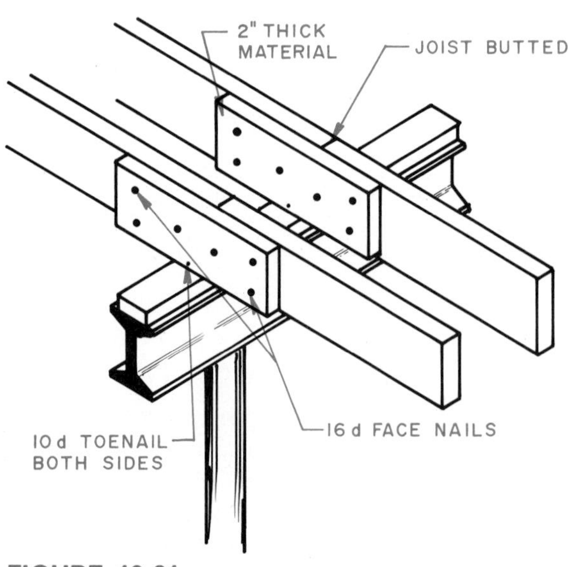

FIGURE 10-31

Joists can be butted over the beam and joined with a wood scab.

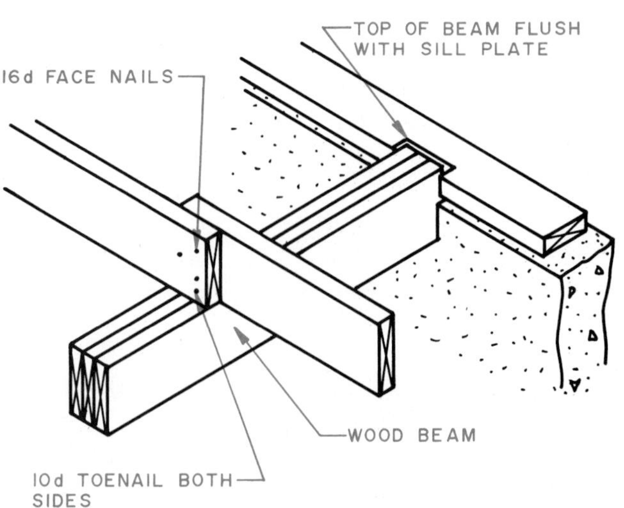

FIGURE 10-33

Wood beams are set so their top is flush with the sill plate.

If it is desired for the joists to be flush with beams metal *joist hangers* are used (Fig. 10-34). This permits the surface below to be flat as needed for a ceiling. If a flush ceiling is desired and a metal beam is used the joist is framed around the beam (Fig. 10-35).

Ceiling height can be adjusted by using ledgers. A ledger is a 2 × 2 in. or larger wood member nailed to a wood beam (Fig. 10-36). It is face-nailed with three 16d nails at each point where a joist will rest. It helps support the joist which is notched to fit over the beam. The notched portion overlaps at least the thickness of the beam. The joists are toenailed to the wood beam with at least two 10d nails and nailed to each other with two 10d nails.

The joists can also be butted and supported with a ledger. A wood cross tie is face nailed to each joist with 10d nails (Fig. 10-36). It is impor-

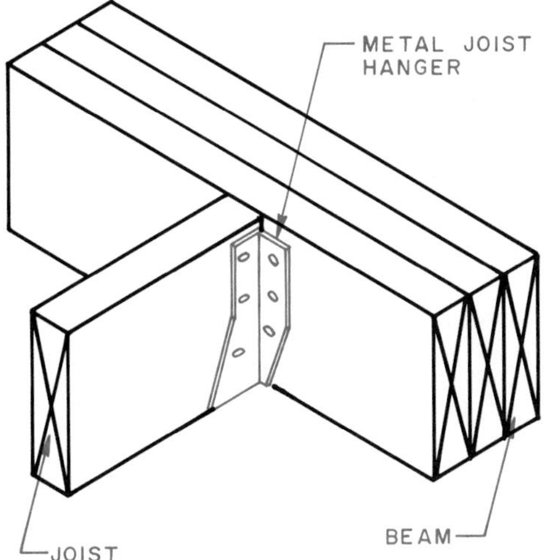

FIGURE 10-34

Metal joist hangers are used to join joists to beams.

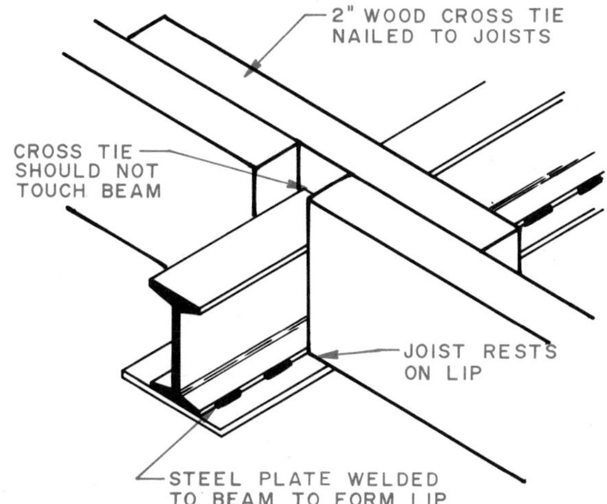

FIGURE 10-35

A joist can be framed around a steel beam to give a flush ceiling in the room below.

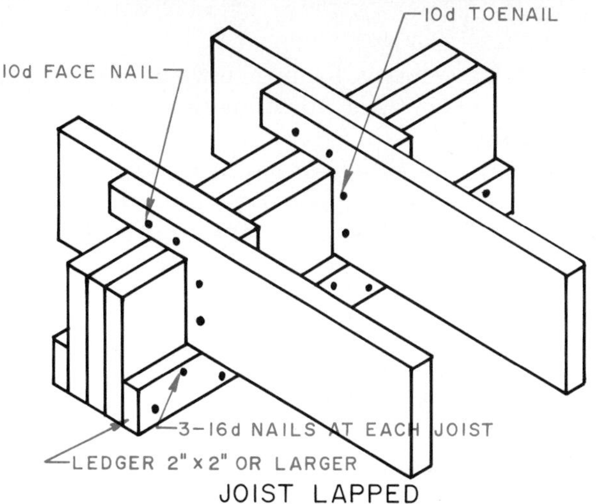

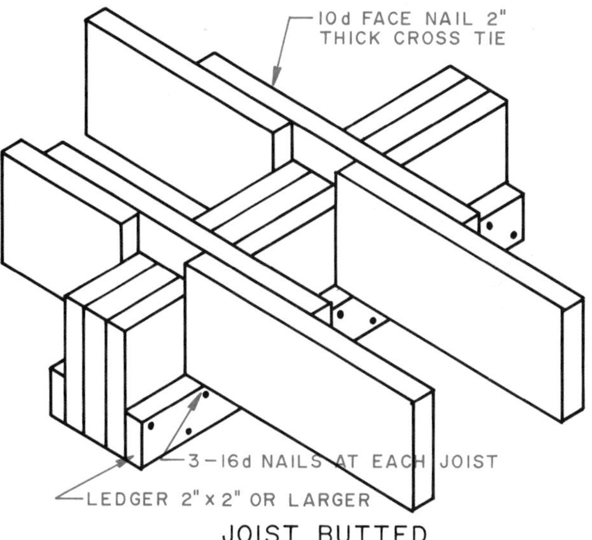

FIGURE 10-36

Ledges can be used to support floor joists at the beam.

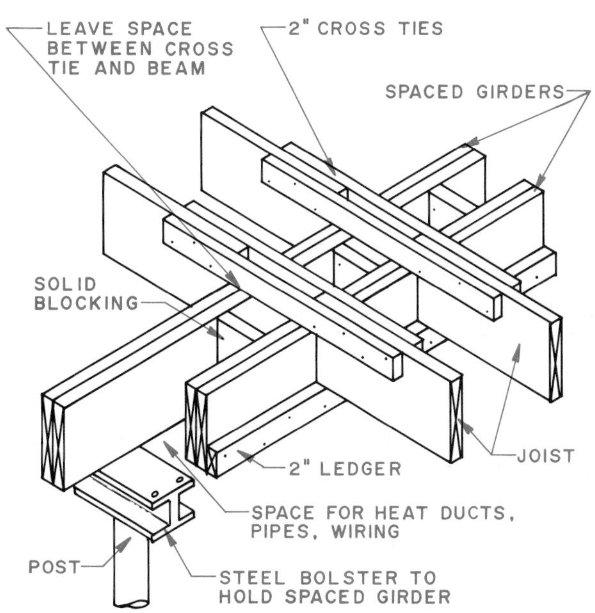

FIGURE 10-37

A spaced girder can be used when it is necessary to put heat ducts through it.

tant to leave a small space between the crosstie and the beam. This will allow for shrinkage in the joists.

If it is necessary to put heat ducts in a partition, the girder supporting it may have to be spaced (Fig. 10-37). Solid blocking is used between the two girders. If the girder has a single post support, it will require a wide bolster. A *bolster* is a support.

LAYING OUT THE JOISTS

In balloon framing the layout for the joists is made on the sill. This is because the joist is nailed to the sill (Fig. 10-38). For platform framing the layout for the joists is made on the header. The joist is located with a vertical line and an X. The X shows which side of the line the joist should be placed (Fig. 10-39). Remember, when joists are lapped the spacing between the stringer and the first joist differs from the regular spacing on one side (Fig. 10-40).

If a joist has a crown (is curved), it should be installed with the crown up. When the joist is under load, it will tend to straighten (Fig. 10-41).

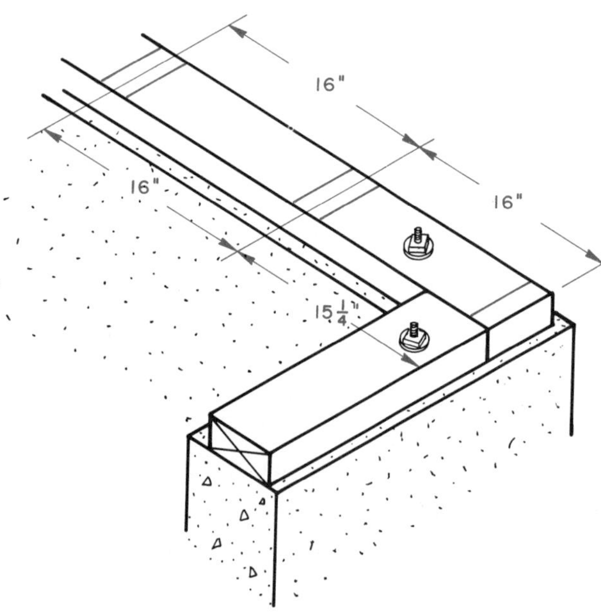

FIGURE 10-38

Joists are laid out on the sill when using balloon framing.

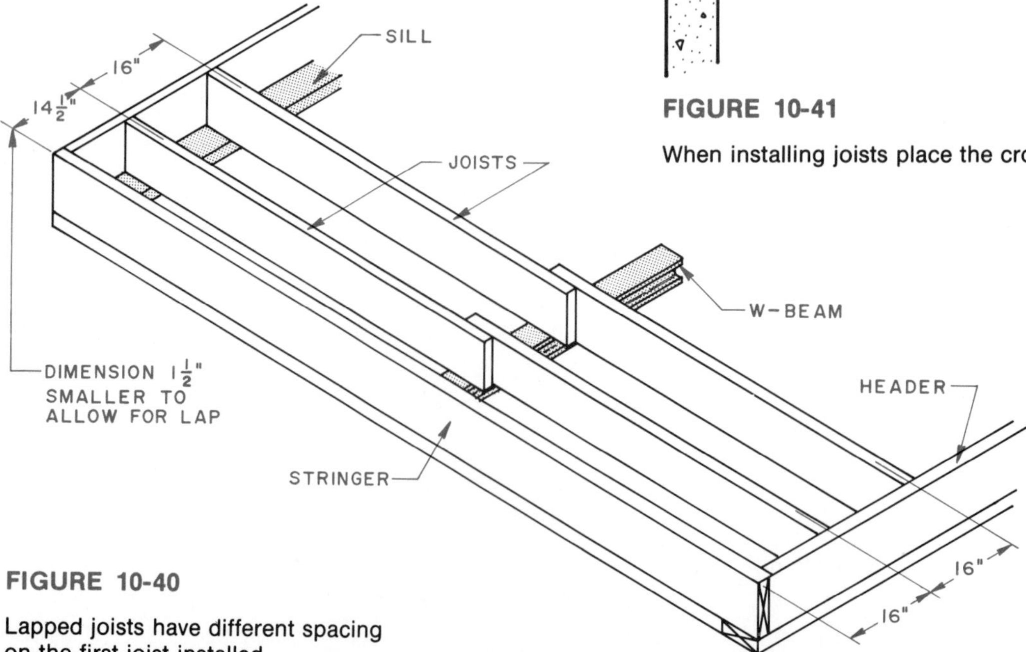

FIGURE 10-40

Lapped joists have different spacing on the first joist installed.

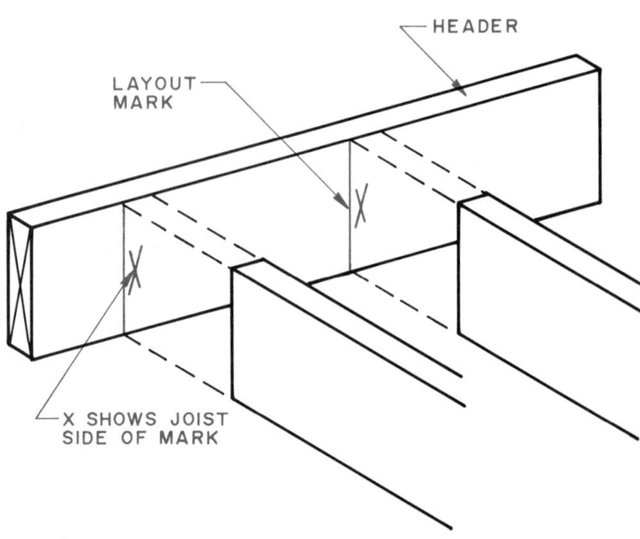

FIGURE 10-39

Joists are laid out on the header when using platform framing.

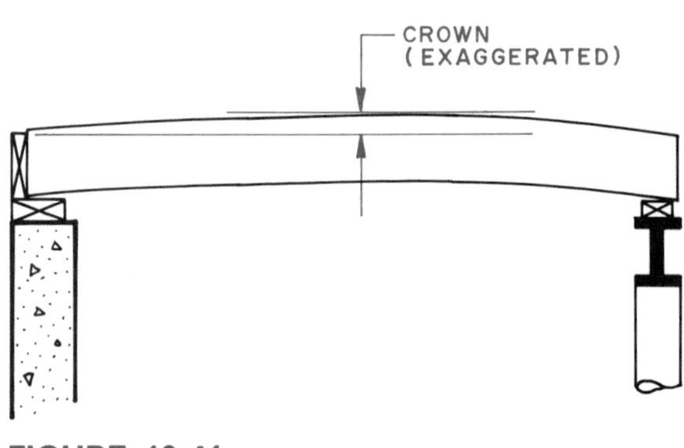

FIGURE 10-41

When installing joists place the crown up.

FLOOR JOIST SPANS

The size of floor joist to use depends upon the size of the span, the spacing between joists, the species and grade of lumber used, and the load to be carried on the floor (called the *live load*). Live load includes items placed on the floor, such as furniture. In residential use common loads used are 30 and 40 pounds per square foot. The span is the clear distance between supporting members.

Joist-load-carrying data are available in tables for various species and grades of lumber. Figure 10-42 shows data for several grades of southern pine. The grades of lumber are shown across the top of the table. The figures in the columns below are the maximum clear spans for the size lumber shown in the left column. For example a 2 × 10 in. floor joist of Dense Select Structural Kiln Dried with a 30-lb per square foot live load will span 19 ft 1 in. if the joists are spaced 16 in. on center. If they are spaced 12 in. on center, it will span 21 ft 0 in.

PARTITION SUPPORT

Joists are doubled under interior partitions that run parallel with the joists (Fig. 10-43). Solid 2 × 4 in. spacers are used between the joists. This provides structural support and space for heating ducts and plumbing which may go in the wall (Fig. 10-44).

FRAMING OPENINGS

It is sometimes necessary to provide an opening in a floor for a stair or chimney. The joists are doubled on each side of the opening. These are called *trimmer joists*. Joists that are shortened by the opening, called tail joists, are nailed to

FIGURE 10-42

Floor joists. *(Courtesy of Southern Forest Products Association)*

		30-PSF LIVE LOAD[a]			40-PSF LIVE LOAD[b]				
SIZE (in.)	SPACING (in. on center)	Dense Sel Str KD and No. 1 Dense KD	Dense Sel Str, Sel Str KD, No. 1 Dense and No. 1 KD	Sel Str, No. 1 and No. 2 Dense KD	No. 2 Dense, No. 2 KD and No. 2	Dense Sel Str KD and No. 1 Dense KD	Dense Sel Str, Sel Str KD, No. 1 Dense and No. 1 KD	Sel Str, No. 1 and No. 2 Dense KD	No. 2 Dense, No. 2 KD and No. 2
2 × 5	12.0	10′-3″	10′-0″	9′-10″	9′-8″	9′-3″	9′-1 ″	8′-11″	8′-9″
	13.7	9′-9″	9′-7″	9′-5″	9′-3″	8′-11″	8′-9″	8′-7″	8′-5″
	16.0	9′-3″	9′-1″	8′-11″	8′-9″	8′-5″	8′-3″	8′-2″	8′-0″
	19.2	8′-9″	8′-7″	8′-5″	8′-3″	7′-11″	7′-10″	7′-8″	7′-6″
	24.0	8′-1″	8′-0″	7′-10″	7′-8″	7′-4″	7′-3″	7′-1″	7′-0″[c]
2 × 6	12.0	12′-6″	12′-3″	12′-0″	11′-10″	11′-4″	11′-2″	10′-11″	10′-9″
	13.7	11′-11″	11′-9″	11′-6″	11′-3″	10′-10″	10′-8″	10′-6″	10′-3″
	16.0	11′-4″	11′-2″	10′-11″	10′-9″	10′-4″	10′-2″	9′-11″	9′-9″
	19.2	10′-8″	10′-6″	10′-4″	10′-1″	9′-8″	9′-6″	9′-4″	9′-2″
	24.0	9′-11″	9′-9″	9′-7″	9′-4″	9′-0″	8′-10″	8′-8″	8′-6″[c]
2 × 8	12.0	16′-6″	16′-2″	15′-10″	15′-7″	15′-0″	14′-8″	14′-5″	14′-2″
	13.7	15′-9″	15′-6″	15′-2″	14′-11″	14′-4″	14′-1″	13′-10″	13′-6″
	16.0	15′-0″	14′-8″	14′-5″	14′-2″	13′-7″	13′-4″	13′-1″	12′-10″
	19.2	14′-1″	13′-10″	13′-7″	13′-4″	12′-10″	12′-7″	12′-4″	12′-1″
	24.0	13′-1″	12′-10″	12′-7″	12′-4″	11′-11″	11′-8″	11′-5″	11′-3″[c]
2 × 10	12.0	21′-0″	20′-8″	20′-3″	19′-10″	19′-1″	18′-9″	18′-5″	18′-0″
	13.7	20′-1″	19′-9″	19′-4″	19′-0″	18′-3″	17′-11″	17′-7″	17′-3″
	16.0	19′-1″	18′-9″	18′-5″	18′-0″	17′-4″	17′-0″	16′-9″	16′-5″
	19.2	18′-0″	17′-8″	17′-4″	17′-0″	16′-4″	16′-0″	15′-9″	15′-5″
	24.0	16′-8″	16′-5″	16′-1″	15′-9″[c]	15′-2″	14′-11″	14′-7″	14′-4″[c]
2 × 12	12.0	25′-7″	25′-1″	24′-8″	24′-2″	23′-3″	22′-10″	22′-5″	21′-11″
	13.7	24′-5″	24′-0″	23′-7″	23′-1″	22′-3″	21′-10″	21′-5″	21′-0″
	16.0	23′-3″	22′-10″	22′-5″	21′-11″	21′-1″	20′-9″	20′-4″	19′-11″
	19.2	21′-10″	21′-6″	21′-1″	20′-8″	19′-10″	19′-6″	19′-2″	18′-9″
	24.0	20′-3″	19′-11″	19′-7″	19′-2″[c]	18′-5″	18′-1″	17′-9″	17′-5″[c]

Note: Data for southern pine structural members.

[a]Sleeping rooms and attic floors.

[b]All rooms except sleeping rooms and attic floors.

[c]Deflection limitation of *l*/360.

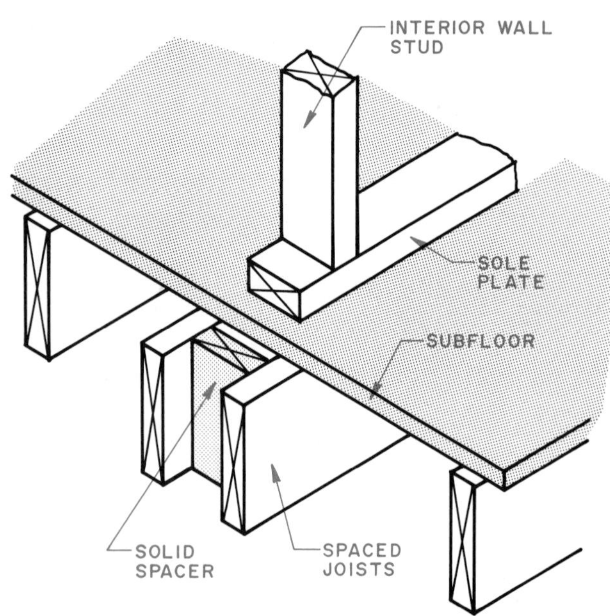

FIGURE 10-43

Joists are doubled under interior partitions.

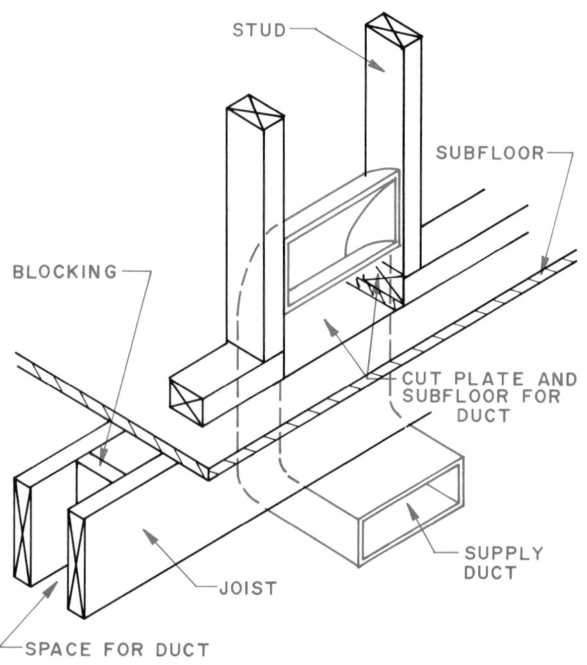

FIGURE 10-44

Spaced joists can be used to provide room for heat ducts.

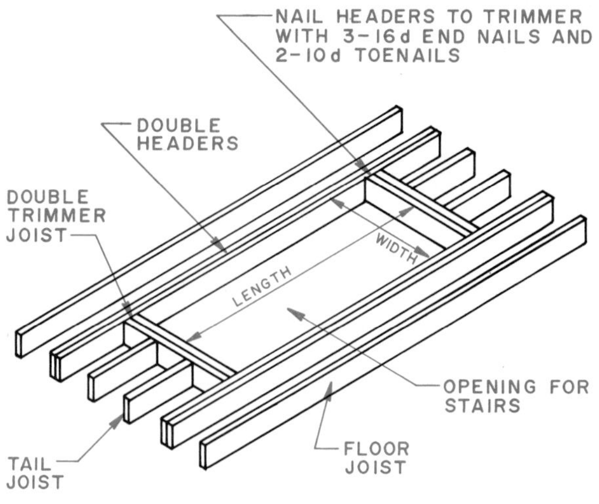

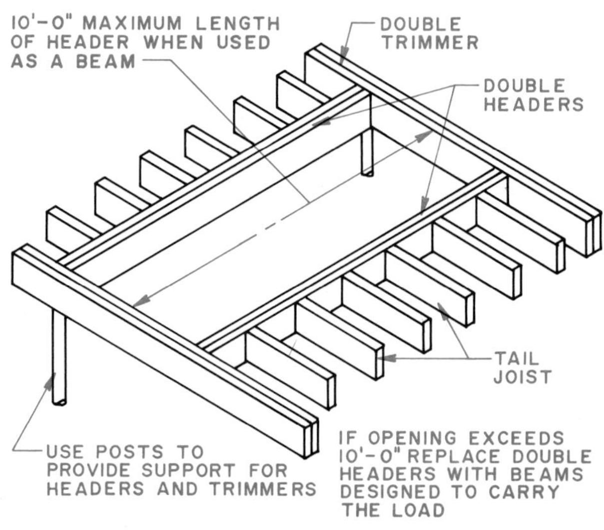

FIGURE 10-45

Openings in floors are framed with double joists and double headers.

headers. Usually, the headers are doubled. They are nailed to the trimmer joists. (Fig. 10-45).

A single header can be used if the opening along the header is less than 4 ft. A single trimmer can be used if the length along the trimmer is less than 4 ft.

Tail joists shorter than 6 ft are connected to the header with three 16d end nails and two 10d nails toenailed. Tail joists longer than 6 ft should be hung with metal joist anchors or ledger strips.

Headers are joined to trimmers with three 16d end nails and two 10d toenails.

A suggested nailing sequence for openings with double headers is shown in Fig. 10-46. Notice that the opening is first framed using single headers and trimmers. Then the second headers and trimmers are face nailed to those in place.

Stair openings are usually the largest opening in a floor. They can run parallel or perpendicular with the joists (Fig. 10-45). When the opening runs perpendicular to the joists and the opening is over ten feet the header should be designed as a beam. It also requires a post or supporting wall below it.

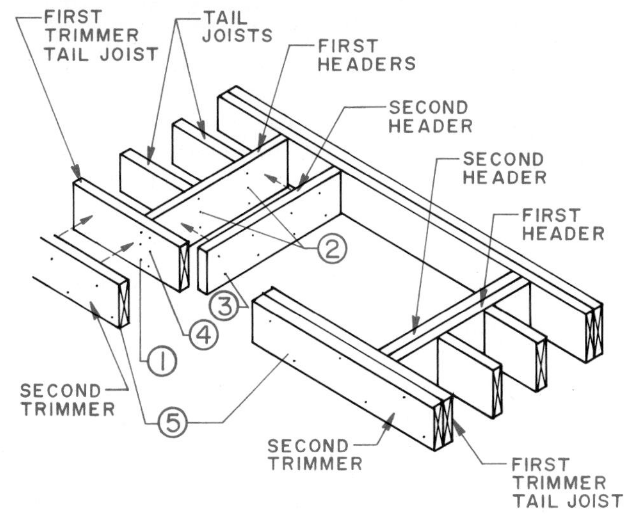

FIGURE 10-46

Nailing sequence for framing openings in floors.

FLOOR PROJECTIONS

Floor projections are used for balconies, bay windows or cantilevered second floors. When the projection runs perpendicular to the joists, it is formed by extending the joists (Fig. 10-47). Under normal conditions a 2-ft projection is maximum. Anything larger needs to be specially designed. The joists on the outside edges are doubled. A single header is adequate. They are nailed using the same pattern as openings in the floor. The subfloor should be cut so that it is flush with the outside edge of the joists.

When a projection runs parallel with the joists, it is framed as in Fig. 10-48. Parallel projections are limited to small areas and extend not more than 2 ft. The joists forming the floor of the projection extend past the stringer to the second joist. This joist is doubled. The joist cut and the projection joists are joined using metal joist hangers or a ledger strip. The outer joists are doubled.

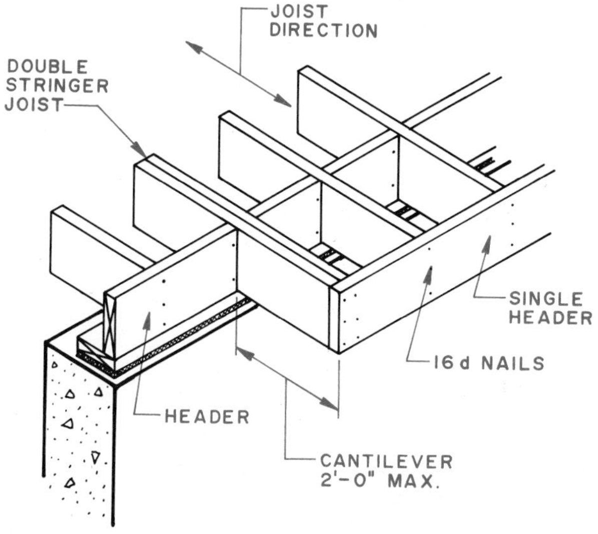

FIGURE 10-47

A floor projection formed by extending the joists.

FRAMING FOR BATH

The bathroom floor often supports heavier than normal loads. These include a bathtub full of water, ceramic fixtures and sometimes a ceramic tile floor set in concrete. This requires extra framing material. The joists under each side of the bathtub are doubled (Fig. 10-49). If the wall is to hold plumbing the joists are spaced apart with 2 × 4 in. blocking. Sometimes the joists are spaced closer together for additional strength.

If a tile floor is to be laid on a concrete base extra support is needed. The joists will have to be smaller than the others to make room for the floor. This will mean that all joists will have to be doubled and perhaps spaced closer together (Fig. 10-50). Heavier joists can be used and the floor dropped down between them (Fig. 10-51). The concrete is poured on plywood subfloor held

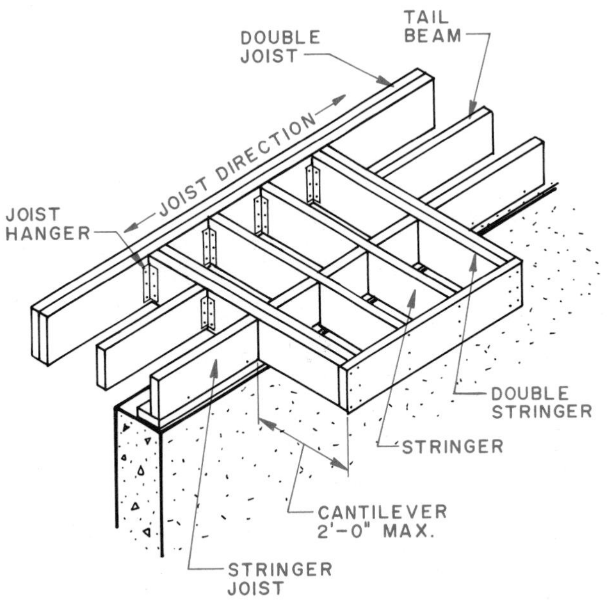

FIGURE 10-48

A floor projection formed with stringers laid perpendicular to the joists.

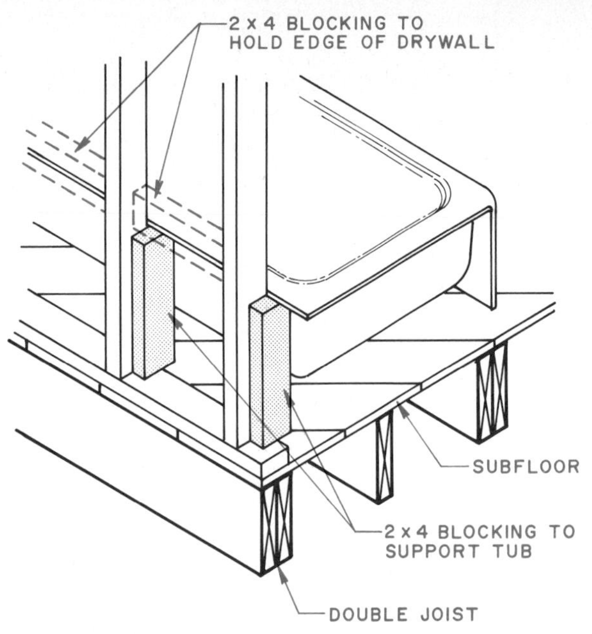

FIGURE 10-49

How to frame to support a bathtub.

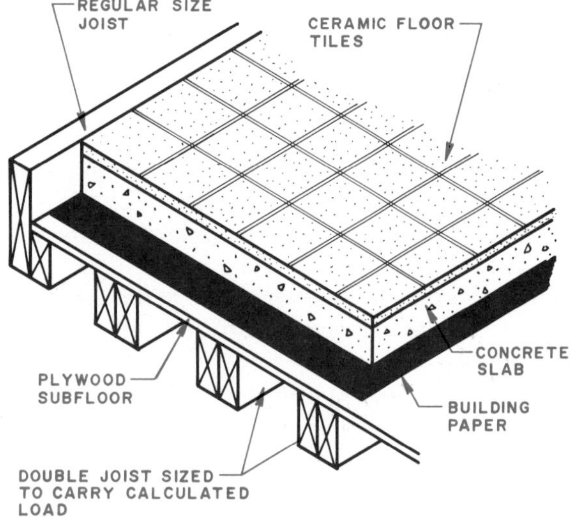

FIGURE 10-50

FIGURE 10-51

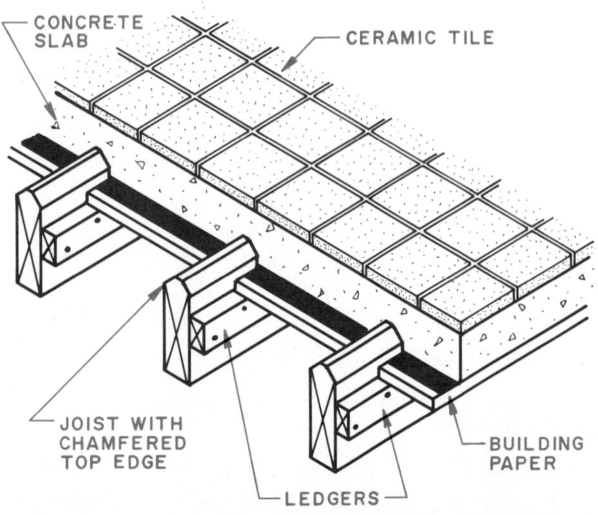

by ledger strips. The joists are chamfered to keep from cracking the concrete layer. The size and spacing of the joists depends upon the load.

The same techniques are used for entrances covered with slate or brick. The fireplace hearth can be formed in the same manner.

NOTCHING JOISTS

It is necessary to notch or drill a hole in joists to permit the passage of pipes, plumbing and electric wires. This should be kept to a minimum. A joist should be notched from the top or bottom only in the end one-third of the span. The depth of the notch should not be more than one-sixth of the joist depth. Holes bored in joists should not be over 2 in. in diameter and not less than 2½ in. from either edge (Fig. 10-52).

If necessary to exceed these limits, strengthen the joist by adding scabs on both sides or doubling the joist. If a large opening is needed, such as for a large duct, treat the problem as a regular opening and install headers.

FIGURE 10-52

Maximum allowances for notching or boring holes in joists.

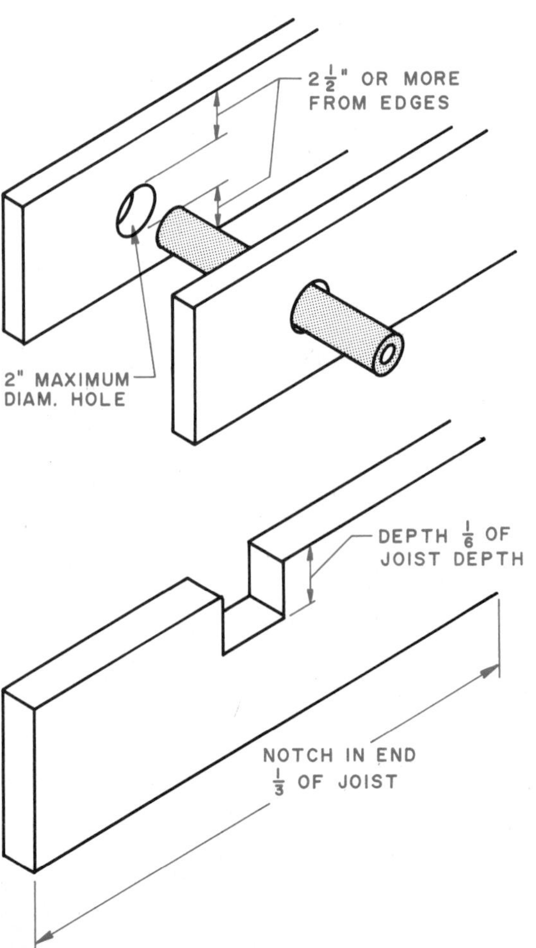

INSTALLING BRIDGING

Once the floor joists are nailed in place bridging can be installed. *Bridging* is wood or metal bracing placed between floor joists. The value of bridging is often questioned. It does not improve the floor's ability to transfer vertical loads after the subfloor and finish floor are installed. It may increase resistance to lateral (horizontal) loads imposed upon it by the foundation wall. Many building codes still require bridging installed in the center of the joist span or at least every 8 ft. The U.S. Department of Housing and Urban Development standards require bridging between joists with a maximum spacing of 10 ft when the nominal depth of the joist is more than six times the nominal thickness. For example, a joist 2 × 14 in. would require bridging.

Two types are in common use. These are solid blocking and cross-bridging. Solid blocking is made of 2-in. material the same size as the floor joists. The pieces are staggered to help nail them in place (Fig. 10-53). Nail with at least two 16d common nails in each end. Solid blocking often shrinks as the wood dries. This causes it to pull away from the joists, leaving a crack.

To install solid blocking first measure the location of the row of blocking on each end of the house. Then snap a chalk line across the tops of the joists. Nail each piece on alternate sides of the chalk line (Figs. 10-53 and 10-54).

FIGURE 10-53

Solid bridging is staggered to help when nailing.

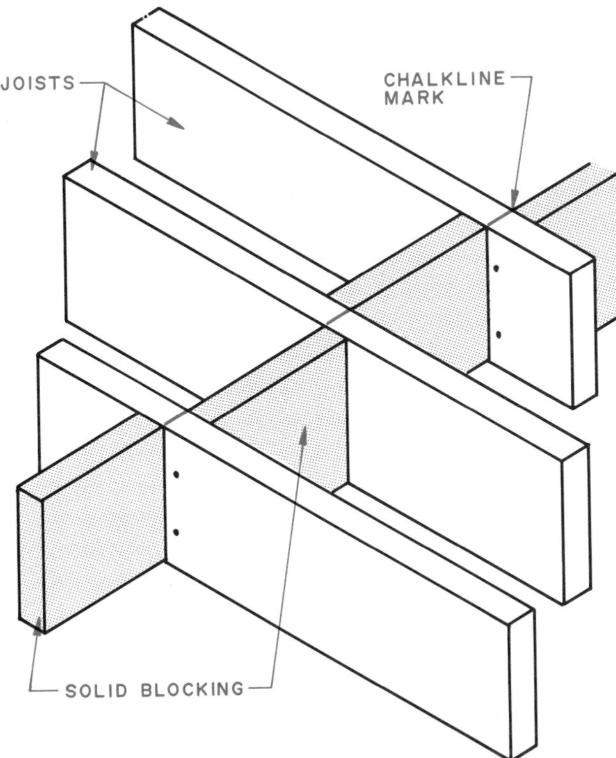

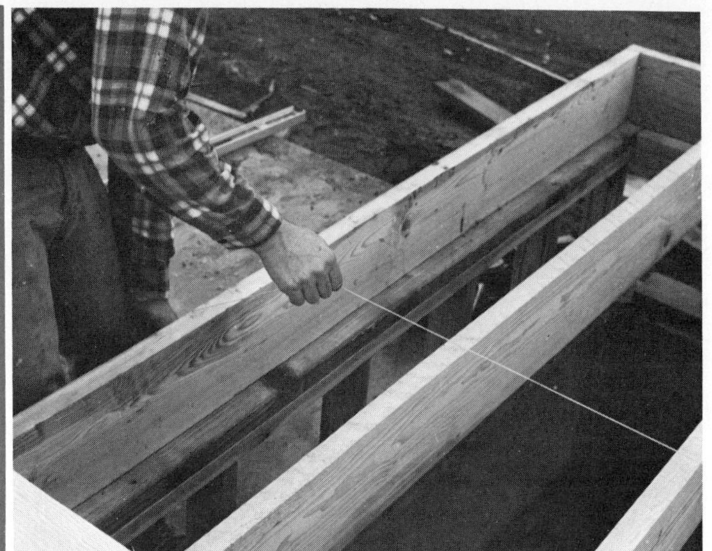

FIGURE 10-54

Snapping a chalkline to mark the location of bridging on the joists. *(Courtesy of American Plywood Association)*

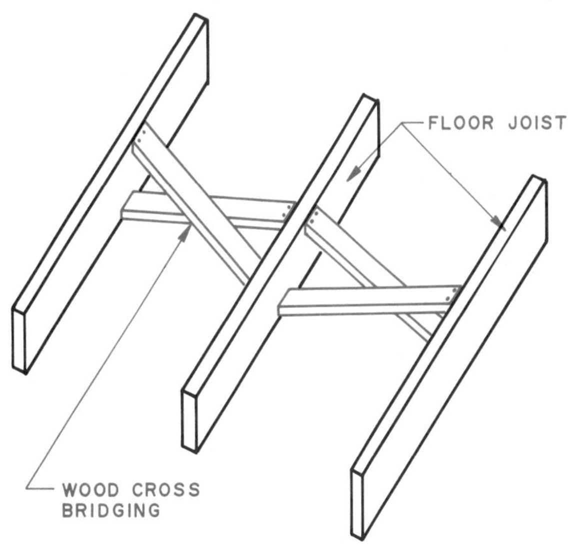

FIGURE 10-55

Wood cross-bridging is nailed between joists.

Wood cross-bridging is made from 1 × 3 or 1 × 4 in. lumber (Fig. 10-55). They are cut on the job. Lay out one piece by placing it across two scraps from the joists spaced the same distance apart as the joists (Fig. 10-56). This gives the length and angles. Set up a radial arm saw to cut the rest of them. Cut two pieces for each space between joists.

Another method uses the framing square. Place one blade of the square on the face of a joist. Read the size on that blade. On the other blade read the distance between joists. Lay a 1 × 3 in. member across the square and mark the angles and lengths. Figure 10-57 is a layout for 2 × 12 in. joists spaced 16 in. on center.

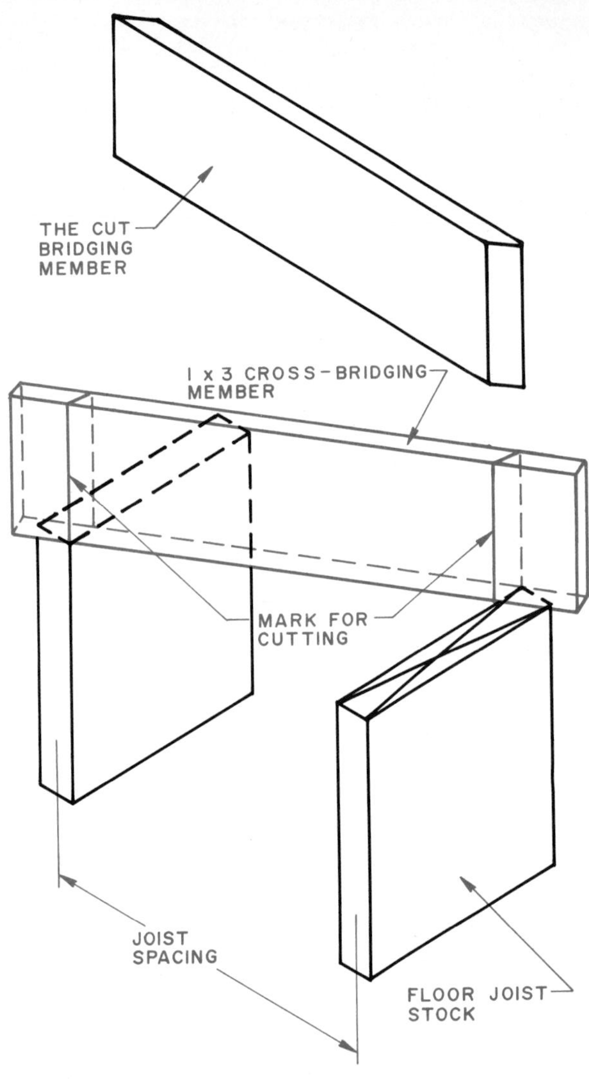

FIGURE 10-56

One way to mark the length and angle of wood cross-bridging.

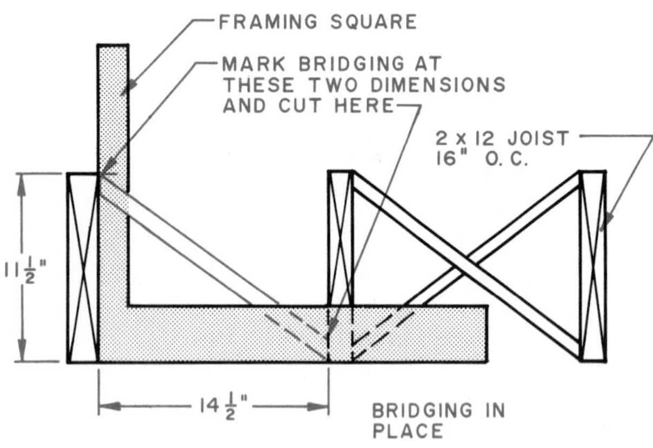

FIGURE 10-57

Wood cross-bridging can be laid out with a framing square.

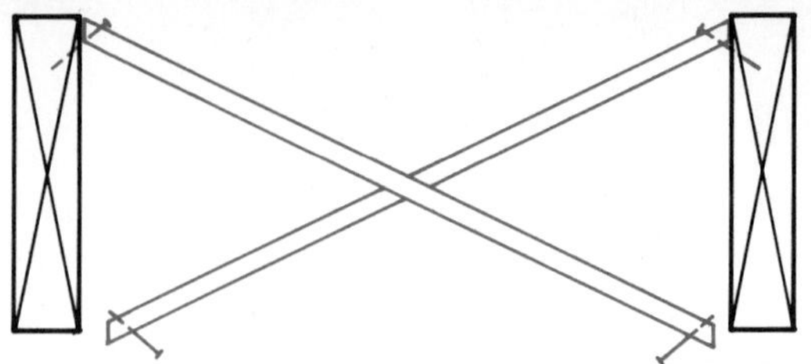

1. CROSS-BRIDGING NAILED AT THE TOP
BEFORE SUBFLOOR IS LAID.

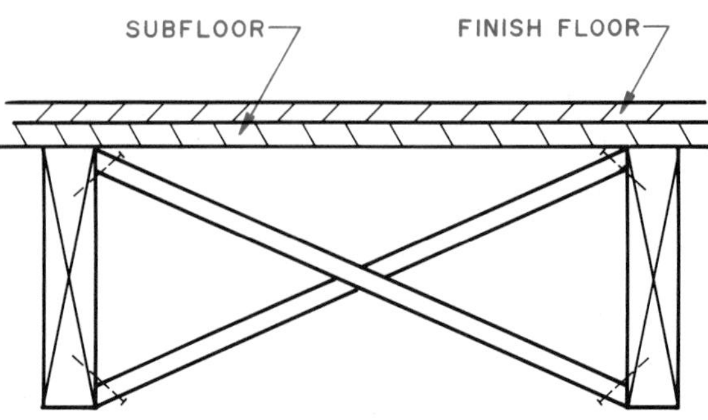

FIGURE 10-58

First nail the top end of the cross-bridging in place. Install the subfloor. Then nail the bottom end.

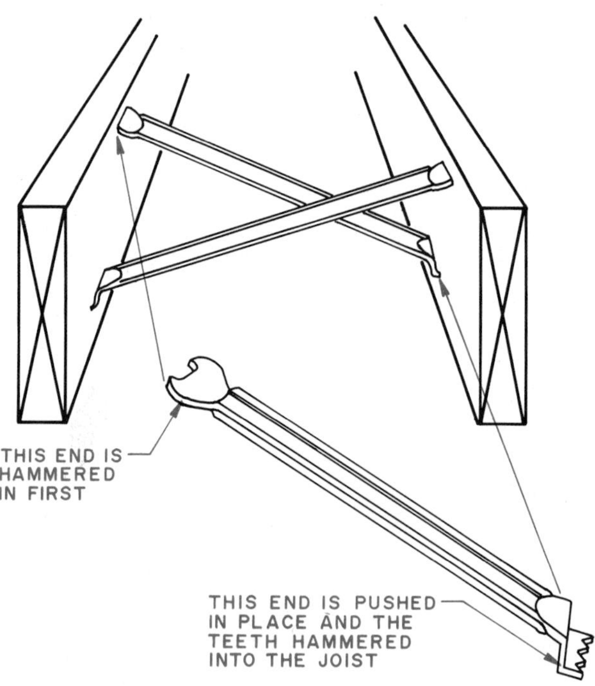

FIGURE 10-59

How to install metal cross-bridging.

To install, snap a chalk line across the top of the joists as previously mentioned. Nail the top ends of the bridging with two 8d nails. Make certain that the end of the bridging is flush with the top of the joist. It must never stick above the joist. Nail one piece on each side of the chalk line. Do not nail the bottom end (Fig. 10-58). This is nailed later after the floor is installed and the lumber has dried some.

Metal cross-bridging is also available. It has finger-like prongs cut in each end. The top end is driven into the top edge of the joist. The bottom edge is driven into place and the prongs set into the bottom of the joist (Fig. 10-59).

INSTALLING THE SUBFLOOR

The *subfloor* is the layer of sheathing placed directly over the floor joists. It is usually nominal 1-in. thick wood boards 4 to 8 in. wide or plywood ½ to ¾ in. thick.

Solid wood boards can be applied at right angles to the joists or on a 45° degree diagonal. When applied at right angles to the joists, finish flooring of wood strips or planks must be applied at right angles to the subfloor (Fig. 10-60). Diagonal installation is preferred because it allows the finish wood flooring to be laid at right angles or parallel to the joists (Fig. 10-61).

The ends of the boards should always rest on a joist. If the ends are tongue-and-grooved, they do not have to end on a joist.

Subfloor boards should not be over 8 in. wide. They are nailed to each joist they cross with two 8d common nails if they are under 8 in. wide. Boards 8 in. wide require three 8d common nails at each joist. If desired, 7d annularly threaded nails can be used.

If the subfloor is to be covered with 25/32 in. or thinner finish wood flooring laid parallel to the joists, the joists should not be wider than 16 in. on center. If the finish wood flooring is laid perpendicular to the joists, they may be spaced as wide as 24 in. on center.

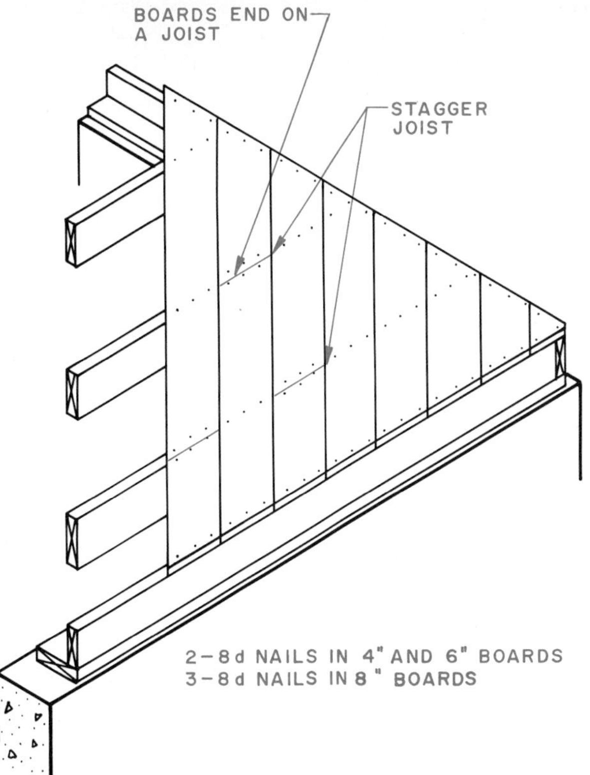

FIGURE 10-61

It is preferred to apply solid wood subflooring on a diagonal.

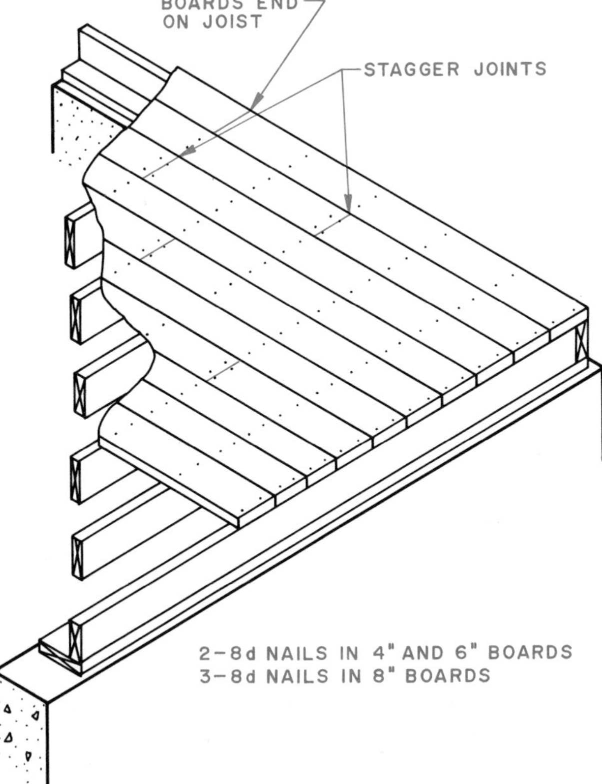

FIGURE 10-60

Solid wood subflooring can be applied at right angles to the joists.

Plywood subflooring has its grade and spacing marked on each sheet. (See Chapter 7 for types and span data.)

Plywood subflooring is applied with the grain of the outer ply at right angles to the joists. The panels should be staggered so that the end joints occur at different joists (Fig. 10-62). The edge joints must have 2-in. blocking below them unless they are tongue-and-grooved.

Conventional plywood subflooring is fastened with common nails. The sizes and spacing are shown in Fig. 10-63. The end joints should have a 1/16-in. space. This type of subflooring panel is usually covered with a separate *underlayment* panel. Underlayment is not needed if the subfloor is to be covered with lightweight concrete or tongue and groove strip finish wood flooring.

Plywood underlayment should be laid with the face grain across the joists and the end joints of the panels occurring over a joist. Apply the underlayment just before laying the finished floor. Protect it from moisture or physical damage. Space panel ends about 1/32 in. apart. No blocking is required for underlayment joints if they do not fall over the subfloor joints. The nails to use and their spacing are in Fig. 10-63. It is best to countersink the heads of nails 1/16 in. below the surface of the underlayment just be-

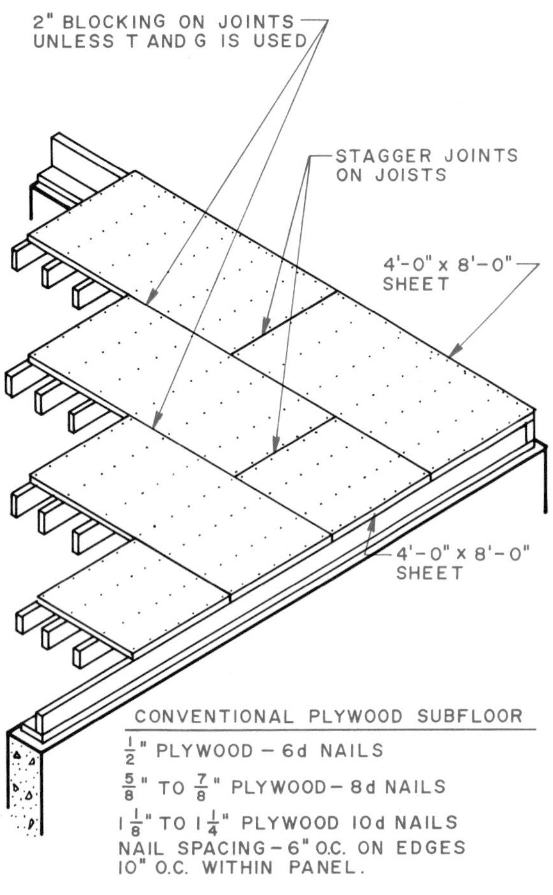

FIGURE 10-62

The joints in plywood subfloor sheets must be staggered.

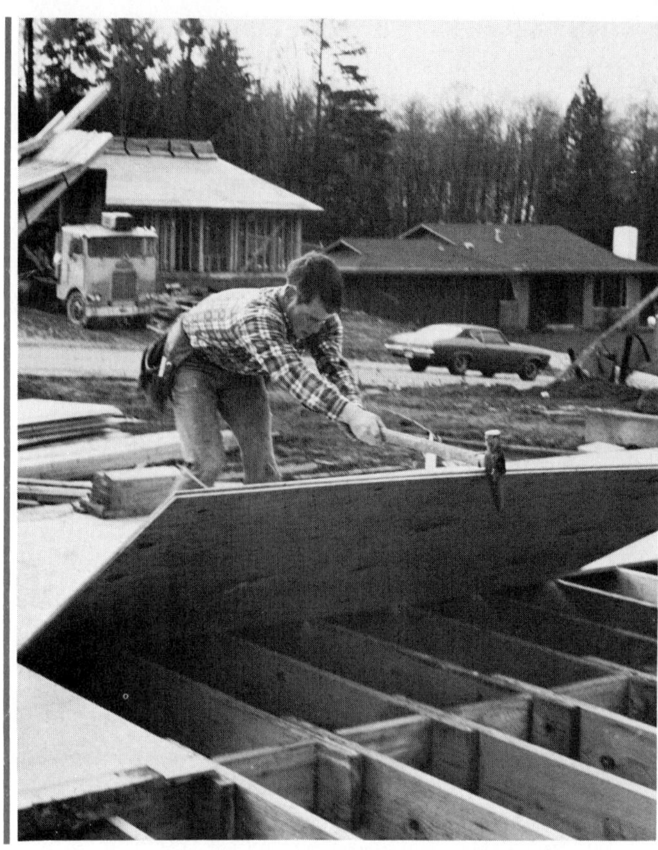

FIGURE 10-64

Installing a sheet of subfloor-underlayment. *(Courtesy of American Plywood Association)*

FIGURE 10-63

Nails and nailing patterns for plywood subfloor and underlayment panels. *(Courtesy of American Plywood Association)*

	NAILS	SPACING (in. on center)
Conventional plywood subfloor	½-in.-thick 6d common. ⅝- to ⅞-in.-thick 8d common. 1⅛- to 1¼-in.-thick 10d common.	6 on panel edge 10 on panel interior
Plywood underlayment	Up to ½-in.-thick 3d ring-shank. ⅝ in. and thicker 4d ring-shank. 16-gauge staples with ⅜-in. crown 1⅝ in. long.	3 on panel edges 6 on panel interior
Subfloor-underlayment	¾-in.-thick 6d deformed shank. ⅞ in. and thicker 8d deformed shank.	6 on panel edges 10 on panel interior

fore laying the finish floors. This avoids nail popping. Staples should be countersunk 1/32 in.

The combined subfloor-underlayment panel can be used in place of the two just mentioned. It is a single sheet providing strength and a smooth surface (Fig. 10-64). The size of nails and nailing pattern are shown in Fig. 10-63. Countersink the nails 1/16 in. just before laying the finish floor. If resilient floor, such as vinyl tile, is to be applied, fill and sand the joints.

If the panels used have a tongue and groove edge leave 3/32-in. space between the upper plies of the panel. If butt joint panels are used, leave 1/16-in. between the panels.

A special adhesive is available for gluing plywood subfloors to joists. This system increases the stiffness of the floor since it forms a T-beam unit with the joist. It also reduces the squeaks that result when the wood shrinks as it dries. The adhesive is applied to the joists.

The framing and blocking of the subfloor for gluing is the same as for nailed floors. Leave a 1/16-in. space between end joints and a 3/32-in. space between tongue-and-groove edges.

After the adhesive is applied to the joist

FIGURE 10-65

Installing plywood subflooring with an adhesive. *(Courtesy of American Plywood Association)*

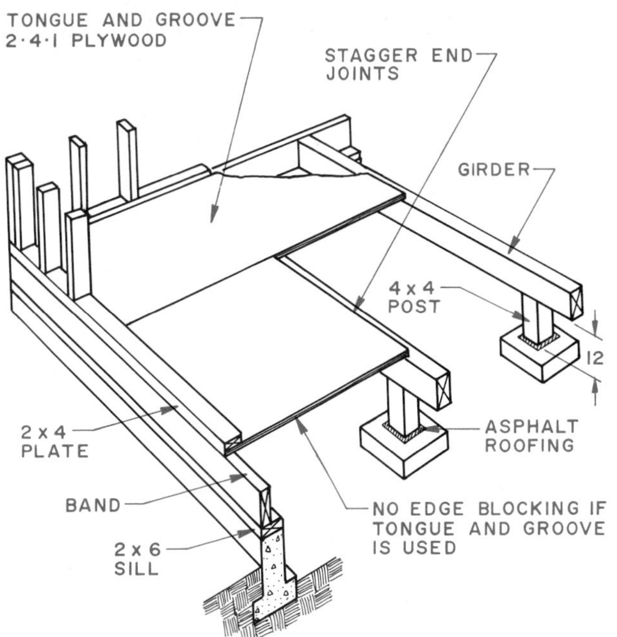

FIGURE 10-66

The 2-4-1 flooring system using beams and tongue-and-groove 2-4-1 plywood. *(Courtesy of American Plywood Association)*

the plywood sheet is set in place. It is nailed with 6d annularly threaded nails spaced 12 in. on center at all joists (Fig. 10-65).

2-4-1 Subfloor-Underlayment This floor system uses girders spaced 48 inches on center which are supported with piers or a central beam. The floor is made of a special 1⅛-in. thick plywood, called 2-4-1, which will span 48 in. between supports (Fig. 10-66). A typical box sill can be used at the foundation.

The panels are installed with the grain on the face ply perpendicular to the girders. The end joints are staggered. All joints must have a 1/16-in. space. The panels are nailed with 8d annularly threaded nails spaced 6 in. on center on all edges (Fig. 10-67).

Since the 2-4-1 plywood panel edges are

FIGURE 10-67

Installing a 2-4-1 flooring system. *(Courtesy of American Plywood Association)*

tongue-and-grooved, blocking of panel edges is not necessary.

Plank Subfloors Planks are 2-in.-thick wood members which are laid flat. The 2-in. dimension takes the bending stresses. Usually they are tongue and grooved on the 2-in. dimension. In lumber grading this tongue and grooved material is called decking. Decking is available from 2 to 4 in. thick.

The floor framing system for plank subfloors is much like that for 2-4-1 plywood floors. The span between floor girders is 6 ft or more. The plank decking provides the structural floor. The underside of the plank provides a suitable finished ceiling for the area below. Generally the plank subfloor is covered with an underlayment before resilient tile or adhesively set ceramic tile is applied. If it is to be carpeted, an underlayment is not necessary.

Typical construction details are shown in Figs. 10-68 through 10-70. Planks are toenailed to the girder with one 10d common nail. If they are 4 to 6 in. wide they are face-nailed with one 16d nail at each support. Eight-in. planks are face-nailed with two 16d nails, whereas 10- and 12-in. planks require three 16d nails.

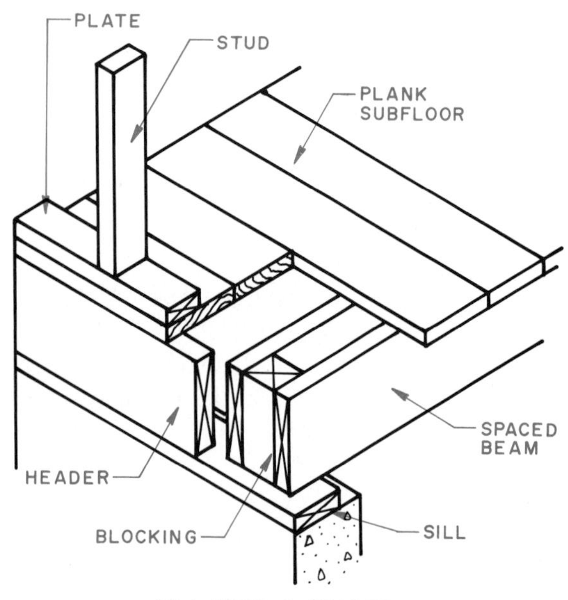

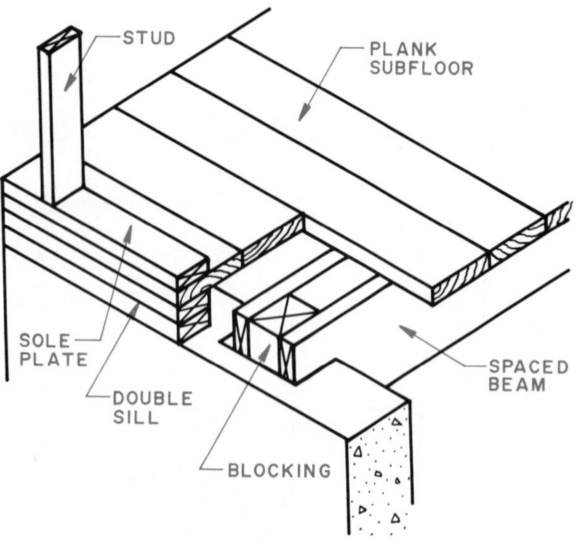

FIGURE 10-68

Plank subfloor can be held with a spaced beam or girder.

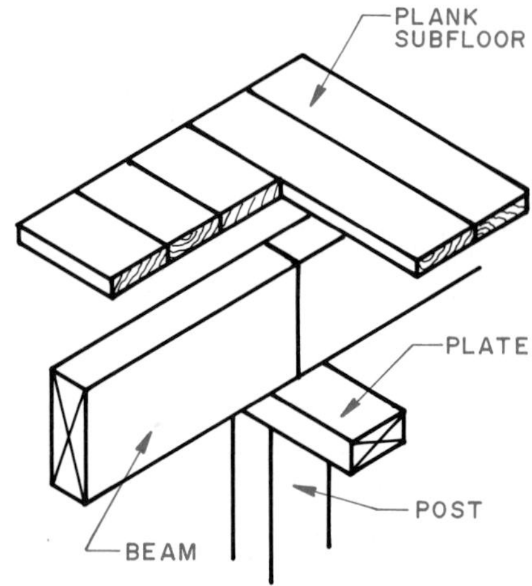

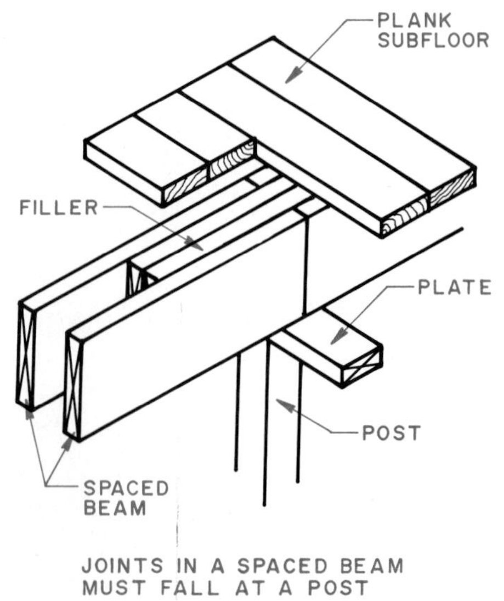

FIGURE 10-69

Intermediate support for plank subfloor can be given with built-up or spaced beams supported by posts.

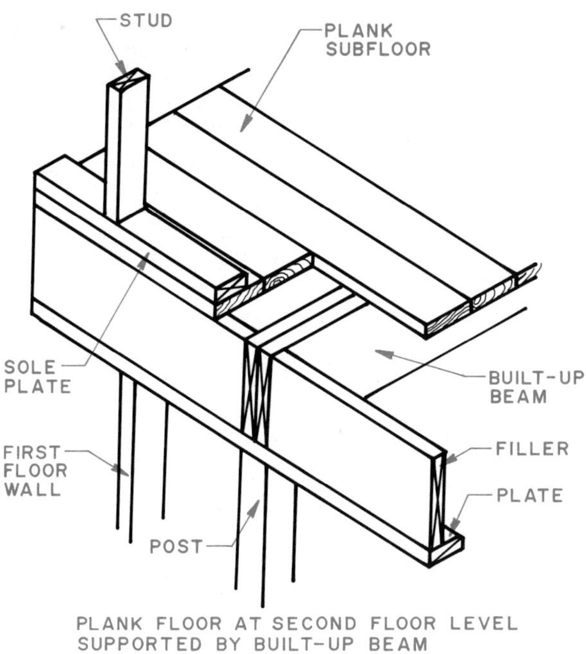

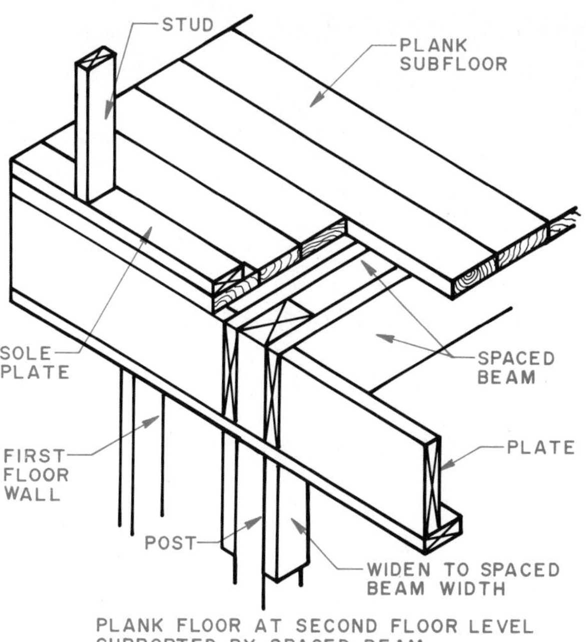

FIGURE 10-70

A second-floor level subfloor can be supported by built-up beams or spaced beams over posts.

STUDY QUESTIONS

After reading the chapter, answer each of the following questions. If you do not know the answer, review the chapter.

1. What is a beam?
2. What types of steel beams are used in residential construction?
3. How are built-up girders made?
4. How much of a beam or girder should rest on the foundation?
5. How is the foundation prepared to receive a beam that is to be flush with the top of the foundation?
6. What types of columns are used in residential construction?
7. How are cracks between a sill and the top of the foundation wall filled?
8. What size bolts are used to hold the sill to the foundation?
9. How does balloon framing differ from platform framing?
10. What are the two basic systems for post-and-beam framing?
11. What is a live load?
12. What change is made in the frame when an opening for a stair is necessary?
13. What is the maximum length of floor projections beyond the foundation without special support?
14. Under what conditions can floor joists be notched?
15. What types of bridging can be used?
16. In what direction should plywood subfloor be laid?
17. What is the value of using adhesive to join a subfloor to the joists?

STUDY QUESTIONS

IMPORTANT TECHNICAL TERMS

Following are technical terms that you should be able to use as part of your working vocabulary. Write a brief description of the meaning of each term.

Beam
Girder
W-beam
S-beam
Laminated girders
Built-up girders
Box beam
Column
Sill sealer
Balloon framing
Stud
Fire stops
Header
Ribbon
Interior bearing partition
Sole plate
Double plate
Platform framing
Stringer
Post, plank, and beam framing
Scab
In-line joists
Cantilever
Joist hangers
Ledger
Bolster
Live load
Trimmer joists
Bridging
Subfloor
Underlayment

11 WALL FRAMING

After the floor joists are set and the subflooring is in place, the carpenters begin on the wall framing. Review the section on balloon and platform framing in Chapter 10. Most residential work uses platform framing.

The exterior wall frame serves as the base for sheathing and siding. It also carries the weight of the ceiling and roof. Interior walls, called *partitions*, support the interior wall finish such as dry wall or paneling. Some partitions support the ceiling. These are called load-bearing partitions.

THE BASIC WALL STRUCTURE

A typical frame wall is made up of a *sole plate*, *top plate*, *studs*, and *headers*. Some type of bracing is needed. Most conventionally framed walls use 2 × 4 in. material. The studs are generally spaced 16 in. on center; however, 24 in. on center is also used. Some energy-efficient designs are using 2 × 6 in. studs and plates spaced 24 in. on center. This enables more insulation to be placed in the wall (Fig. 11-1).

The spacing of studs on 16- or 24-in. centers must be done accurately. Sheet material such as sheathing, plywood, and drywall are made in 4 ft 0 in. × 8 ft 0 in. sizes. Some are 4 ft 0 in. × 12 ft 0 in. These are designed to be nailed to the studs without cutting (Fig. 11-2).

When a wall is framed there should always be a stud every 16 in. (or 24 in.) even though other studs, as for a window, may break the pattern (Fig. 11-3).

Study Chapter 7 for information on the grades and species of lumber for wall framing. Stud and No. 3 dimension grades are generally used. These have the required strength and straightness to carry the loads.

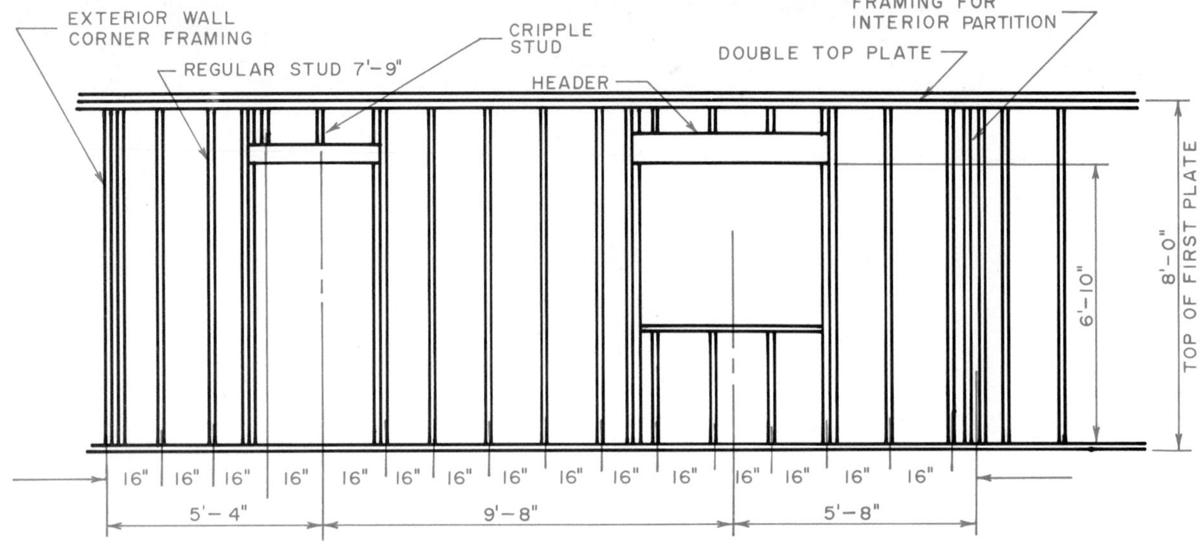

FIGURE 11-1

The parts of a wood-framed wall.

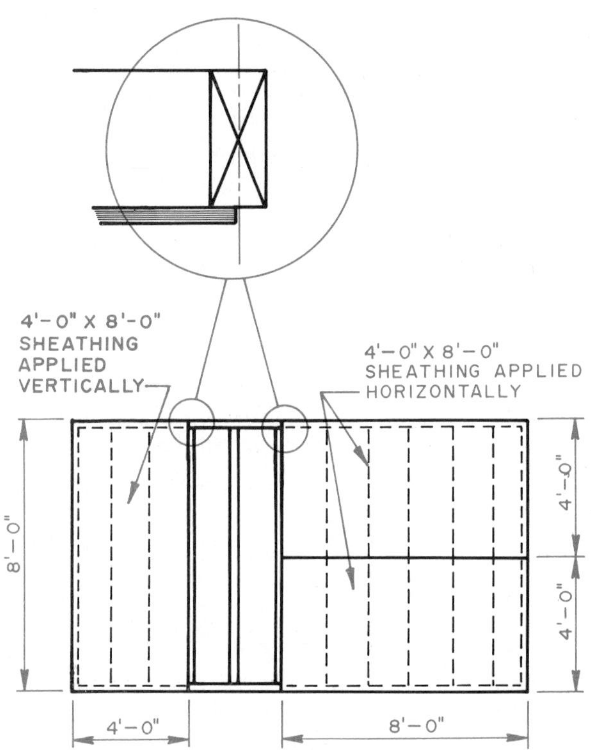

FIGURE 11-2

Studs must be accurately located so that covering material can be applied without cutting.

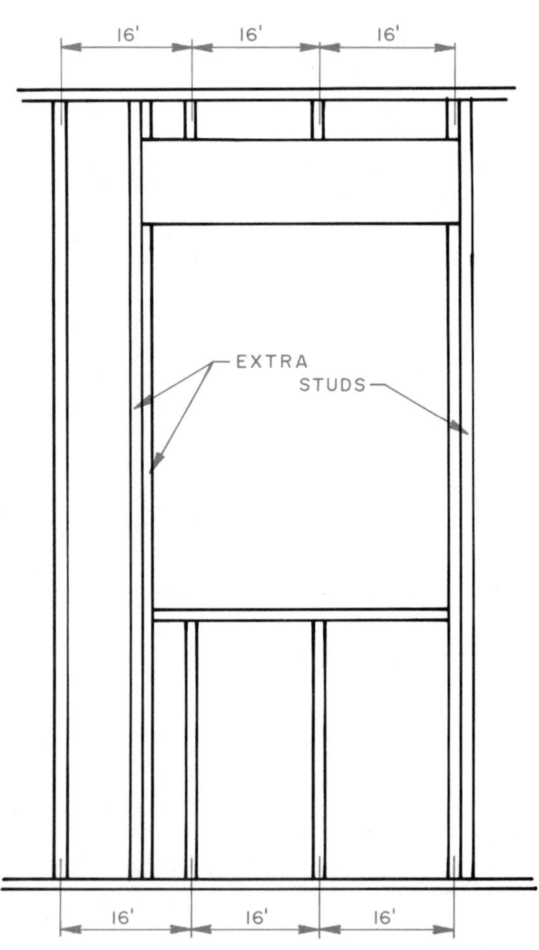

FIGURE 11-3

Standard stud spacing is maintained even when extra studs are needed in the wall.

FRAMING DOOR AND WINDOW OPENINGS

Openings in walls have to be spanned by headers. Headers are structural members that will carry the load over the opening. Openings for doors and windows are examples. The door or window unit should not carry any load.

Framing for a typical window opening is shown in Fig. 11-4. The studs on each side of the opening are doubled. The header rests on top of the trimmer stud. Short studs called *cripple studs* are placed 16 in. on center above

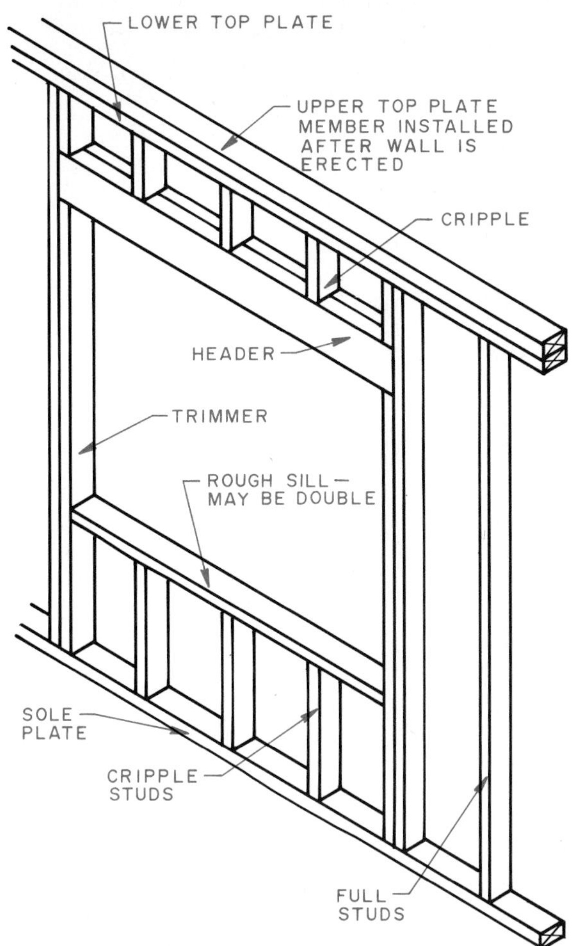

FIGURE 11-4

A typical framing plan for a window opening.

and below the window opening. A single rough sill member is used. Some prefer to use a double rough sill. The size of the opening is determined by the rough opening size specified for the window to be set in place. If the opening is large, the header will be large and leave little space for cripples above it. Generally, the carpenter will fill this void with solid wood or increase the size of the header (Fig. 11-5).

Framing for a typical door opening is shown in Fig. 11-6. It is framed just like a window opening. The size of the rough opening is specified by the size of the door frame to be placed in the opening. Notice that the sole plate continues across the door opening. This is needed to stiffen the wall frame until it is braced by other walls and is sheathed. This part of the sole plate is then sawed out.

Framed openings in balloon framing are shown in Fig. 11-7. The header must carry the weight of the second floor as well as the ceiling and roof. The actual framing can be done several ways. The studs, which extend from the sole plate to the roof, could be erected first. Then the openings could be cut and the headers, rough sills, and cripple studs installed. Another way is to erect only the full-length studs. Then install the headers, rough sills, and cripple studs in the open areas.

Notice in Fig. 11-7 that the studs are all

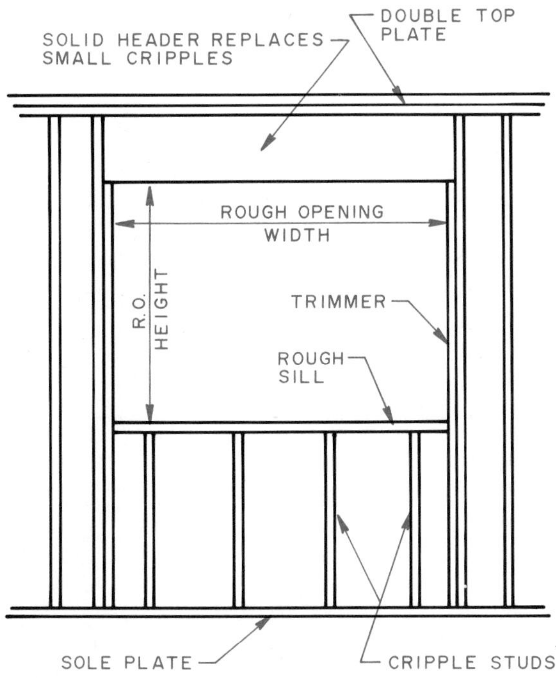

FIGURE 11-5

A framed window opening with a solid header rather than cripples.

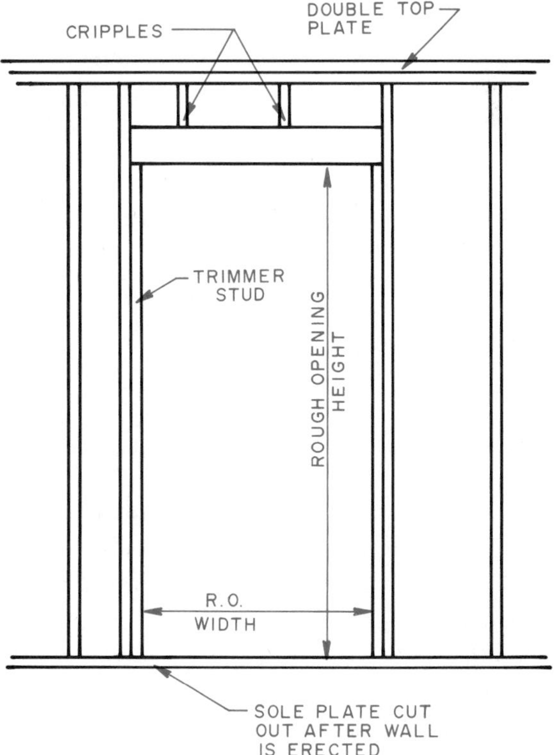

FIGURE 11-6

Framing for a typical door opening in an exterior wall.

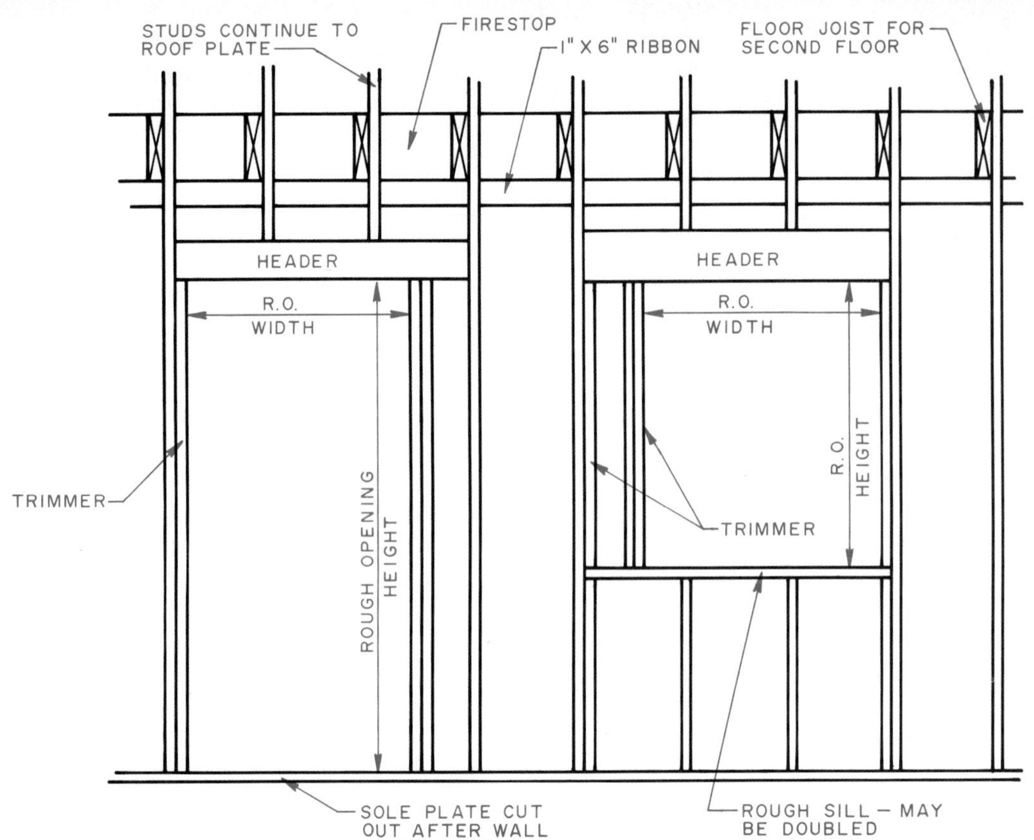

FIGURE 11-7

Typical framing for a balloon-framed exterior wall.

spaced evenly. The headers continue past the side of the rough opening to the next full stud. The side of the rough opening is framed with a double stud.

Headers Door and window openings in walls have a header above them to carry the overhead load. The size of the door or window to be in the opening is given on the architectural drawing. It may be lettered by each unit on the floor plan or be in a door and window schedule (Fig. 11-8). This shows the finished door size without the door frame. The window size shown is usually the sash size. The carpenter must know the rough opening sizes. If this is not given on the door and window schedule, the supplier of the doors and windows can give the required rough opening sizes. They will vary with the type of unit and the manufacturer.

Additional information on doors and windows is in Chapter 15.

The header must span the rough opening plus 3 in. needed to rest on the *trimmer studs.* The size of lumber used to make a header will depend upon the width of the opening. Typical header sizes are shown in Fig. 11-9. Some building codes specify header sizes. The carpenter should know the local building codes.

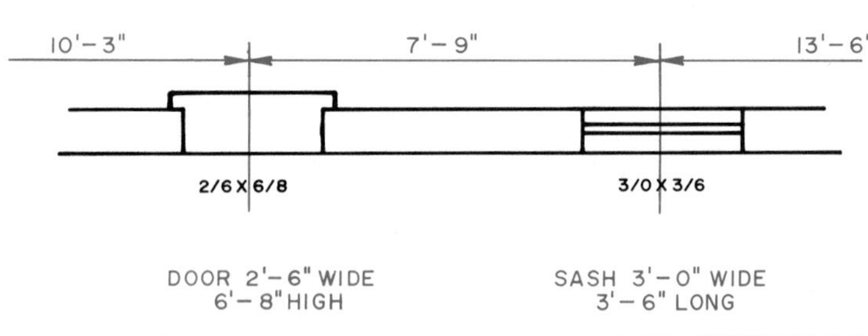

WINDOW SCHEDULE				
SYM	NO.	TYPE	ROUGH OPENING	SASH SIZE
A	5	DOUBLE HUNG	1'-10⅛" x 3'-1⅛"	1'-8" x 2'-10"
B	4	DOUBLE HUNG	2'-10⅛" x 4'-1⅛"	2'-8" x 3'-10"
C	8	DOUBLE HUNG	3'-2⅛" x 4'-9⅛"	3'-0" x 4'-6"
D	2	GLIDING	4'-7¼" x 4'-2⅛	4'-4⅛ x 3'-11"

FIGURE 11-8

Door and window openings are located by their centerline.

MATERIAL FORMING HEADER (in.)[a]	SUPPORTING ONE FLOOR, CEILING, AND ROOF	SUPPORTING ONLY CEILING AND ROOF
2 × 4	3′-0″	3′-6″
2 × 6	5′-0″	6′-0″
2 × 8	7′-0″	8′-0″
2 × 10	8′-0″	10′-0″
2 × 12	9′-0″	12′-0″

[a]Two pieces this size nailed together to form the header.

FIGURE 11-9

Typical header spans for normal residential construction.

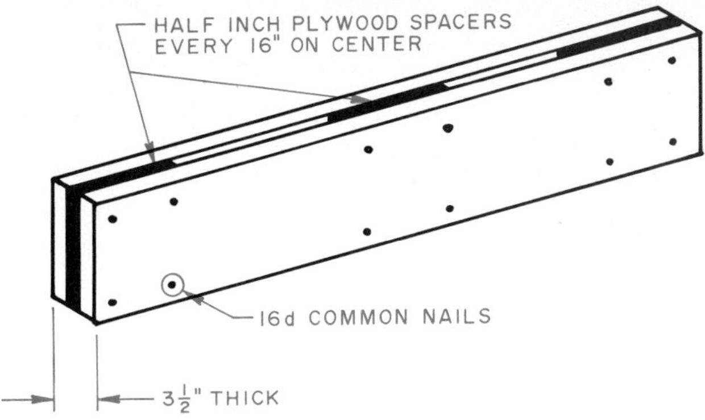

FIGURE 11-10

The nailing pattern for a header.

To make a header, nail the two 2-in.-thick pieces together. They should have ½-in. plywood spacers between them every 16 in. on center. This makes the thickness 3½ in., which is the same as a stud. Use 16d nails around the outside (Fig. 11-10).

FRAMING AT EXTERIOR CORNERS

There are several ways the exterior wall corners can be formed. When using platform framing the corners are made as a part of the wall, which is nailed together lying on the subfloor. The walls are raised to a standing position and braced. The corner is formed by nailing the ends of the walls together. In Fig. 11-11 are two ways to frame the ends of walls to form the corners. In each case the corner provides a surface to nail the interior wall finish material. One corner uses two studs separated by 2 × 4 in. blocking. In the other an extra stud is turned sideways to provide a nailing surface in the corner.

In balloon framing the corner is usually formed with an extra stud and some blocking (Fig. 11-12).

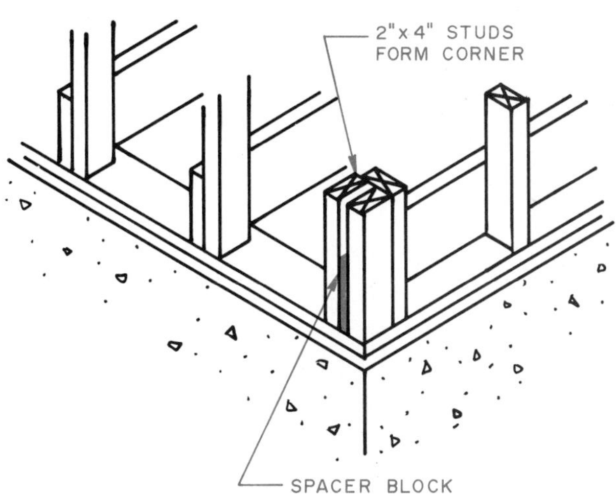

FIGURE 11-12

How to form the exterior corner when using balloon framing.

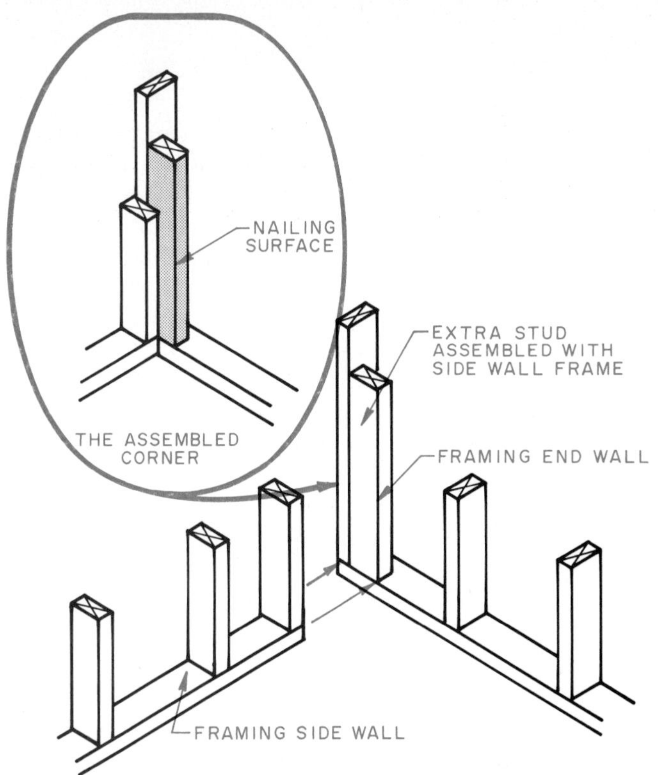

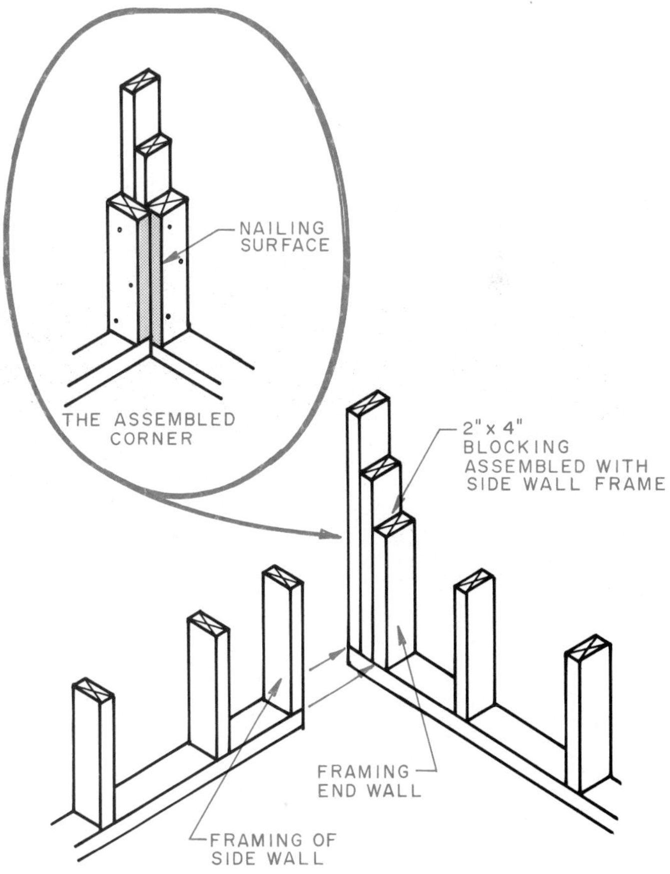

FIGURE 11-11

Two ways to form the exterior corner.

Interior partitions are framed in the same way as are exterior partitions. If it is a bearing partition, a double top plate is needed. Headers over openings are needed in interior bearing partitions. If it is a nonbearing partition, a single 2 × 4 in. over the opening is used. Cripples are not needed except to provide nailing for drywall or paneling. It can be built with a single top plate. If this is done, the studs must be cut 1½ in. longer to bring the top plate flush with the sides with two top plates (Fig. 11-13).

It is possible to frame non-load-bearing walls from 2 × 2 or 2 × 3 in. material. This is often done between closets. When soundproofing is needed, a 2 × 6 or 2 × 8 in. wall is built. Most commonly, the studs are staggered (Fig. 11-14). This reduces the amount of sound that hits the wall from being transferred to the other wall.

Interior partitions in baths and kitchens must be made wider for plumbing. The size depends upon the plumbing. Usually, a 2 × 6 in. stud will be adequate. Some cast-iron sewer pipes require a 2 × 8 in. wall (Fig. 11-15).

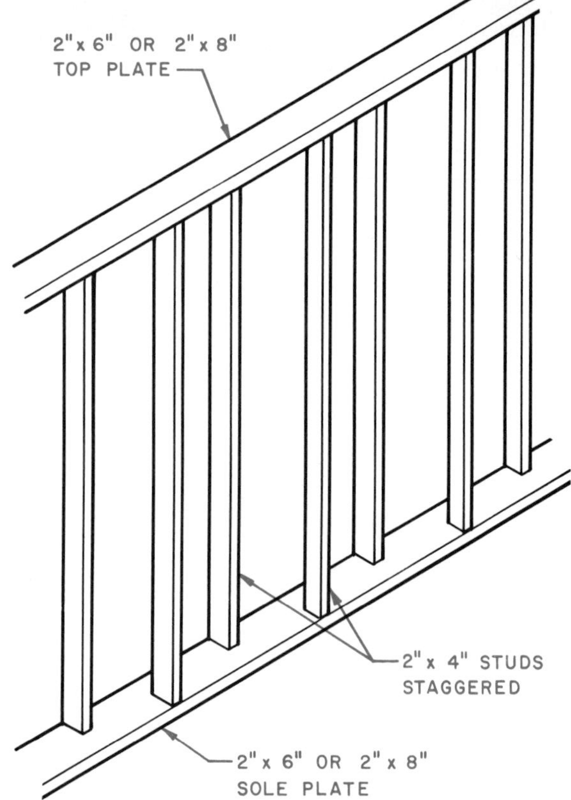

FIGURE 11-14

Framing a wall with staggered studs.

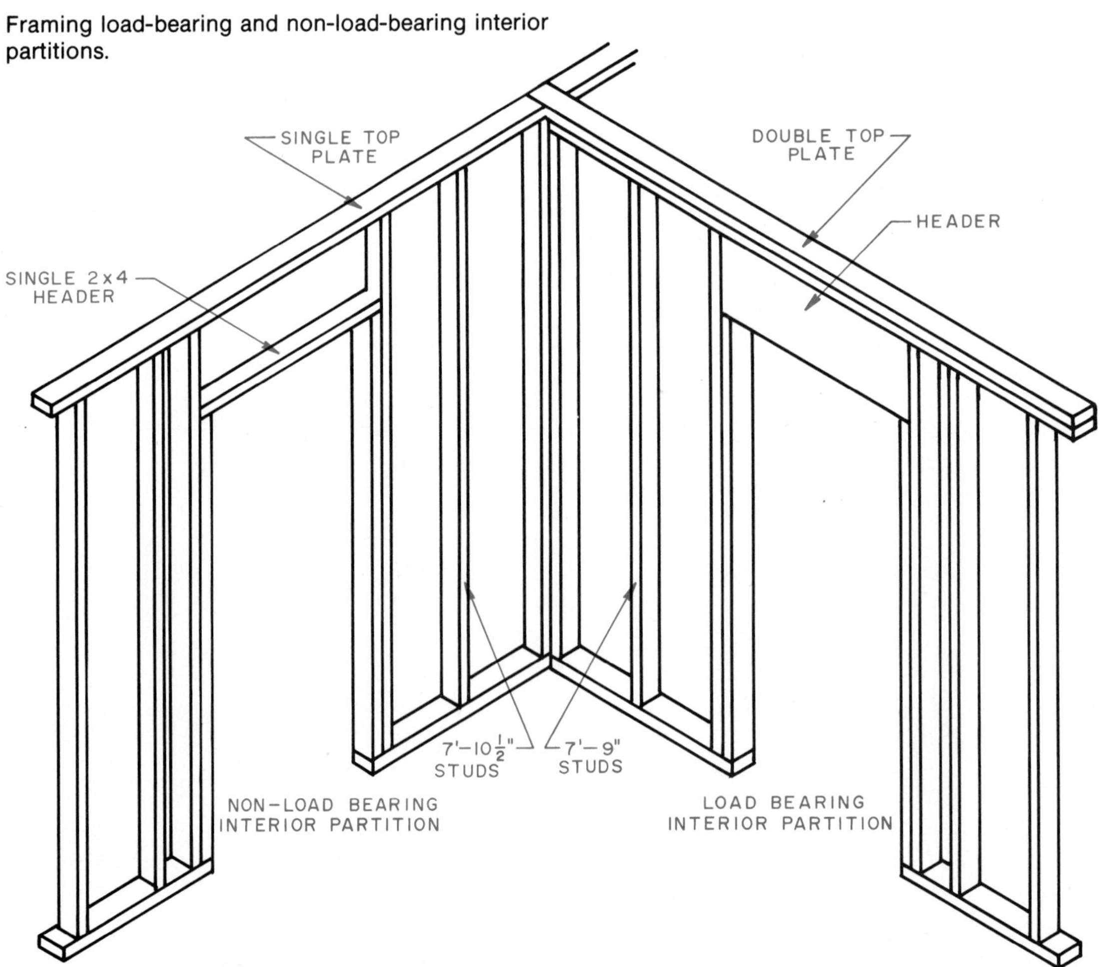

FIGURE 11-13

Framing load-bearing and non-load-bearing interior partitions.

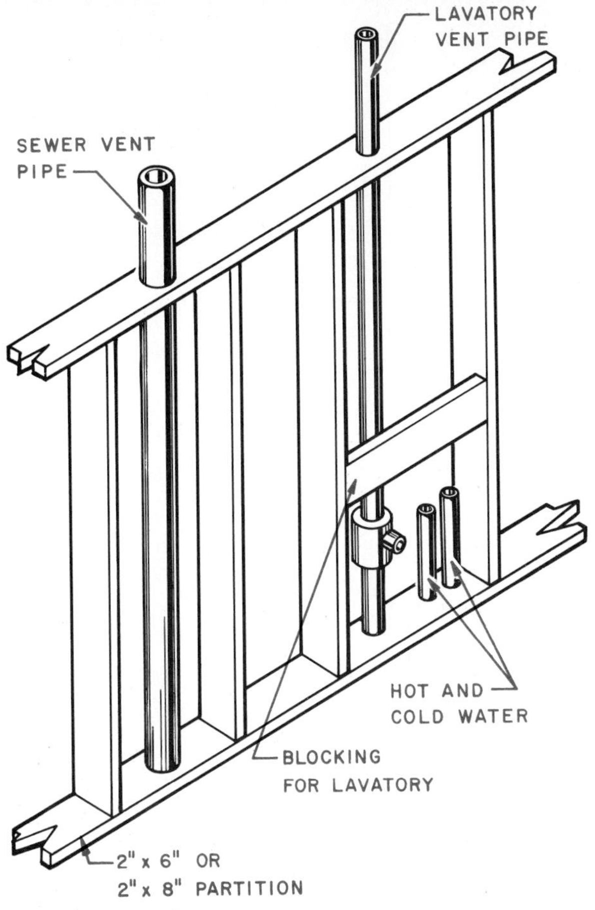

FIGURE 11-15

A wider wall is required to handle plumbing.

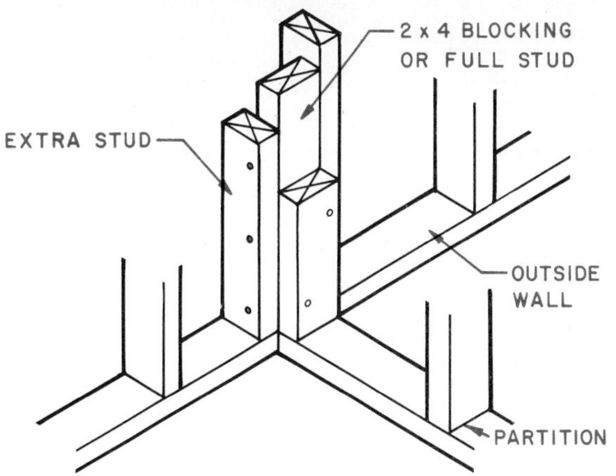

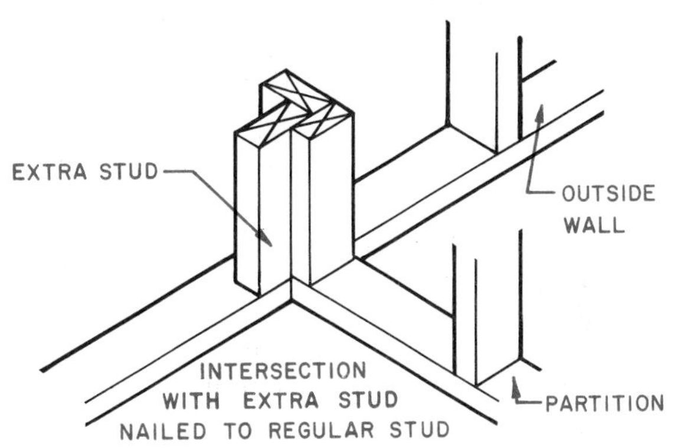

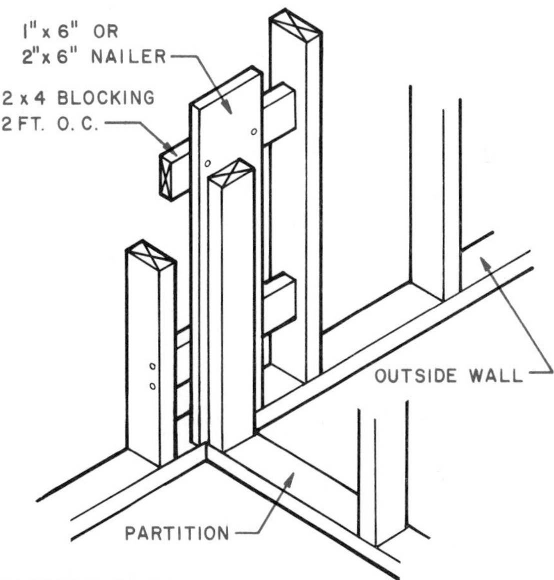

FIGURE 11-16

Three ways to join a partition to the exterior wall.

PARTITION INTERSECTIONS

Interior partitions must be firmly nailed to the exterior frame. The method used depends upon where the walls meet. If they meet at a regular stud, a second stud can be erected (Fig. 11-16). The interior partition is nailed to the spacing blocks. The extra stud provides a nailing surface for the interior finish wall.

If the wall does not meet a stud, 2 × 4 in. blocking is nailed between two regular studs. A 1 × 6 in. board is nailed to the blocking. The partition is nailed to the board (Fig. 11-16).

LAYING OUT THE PLATE

The first step in constructing the wall is to mark the sole and top plates. These marks show the location of the studs, the rough openings, trimmers, and cripple studs. Select straight 2 × 4 in. members and place them along the edge of the subfloor. These are the sole plate. Usually, several 2 × 4 in. pieces are needed to run the length of a building. Be certain that they join in the center of a stud. Place a second piece beside this for the top plate. The sole and top plates are marked at the same time. Nail them lightly to the subfloor so that they will not move while being marked. First lay out the wall to have the corner post framing as part of it. Starting at a corner, mark 1½-in. wide spaces for the corner studs. The first stud from the end will be 15¼ in. from the outside corner. All other studs will be

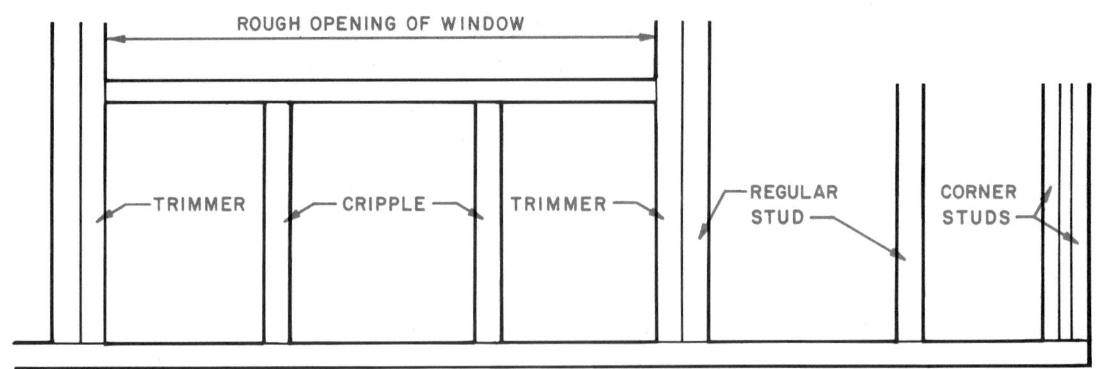

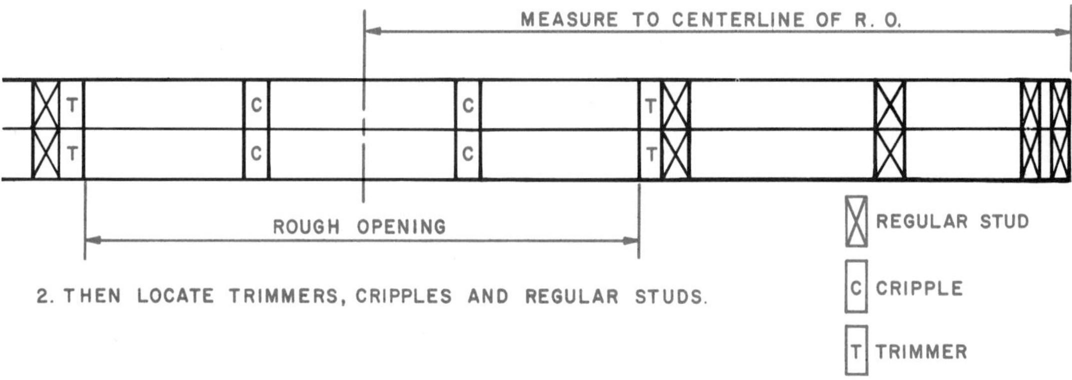

FIGURE 11-17

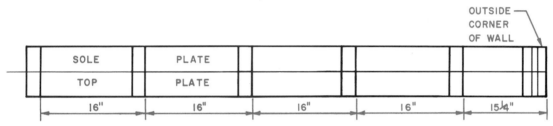

Procedure for laying out the sole and top plates for the wall to contain the corner post framing.

16 in. from edge to edge. Now lay out the studs on the 16-in. spacing (Fig. 11-17).

Refer to the architectural drawings to locate the sizes and locations of rough openings. Then draw the trimmer and cripple studs. Notice how the regular, trimmer, and cripple studs are marked on the plates. An X marks the regular studs, a T a trimmer, and a C a cripple. Place a full-length stud by each trimmer stud. If the wall has a partition intersecting it, locate the necessary studs.

The exterior wall which meets the wall with the corner framing is laid out as in Fig. 11-18. The same procedure is used except for locating the second stud. It will be 11¾ in. from the end of the wall. This allows for the 3½-in. thickness of the corner framing.

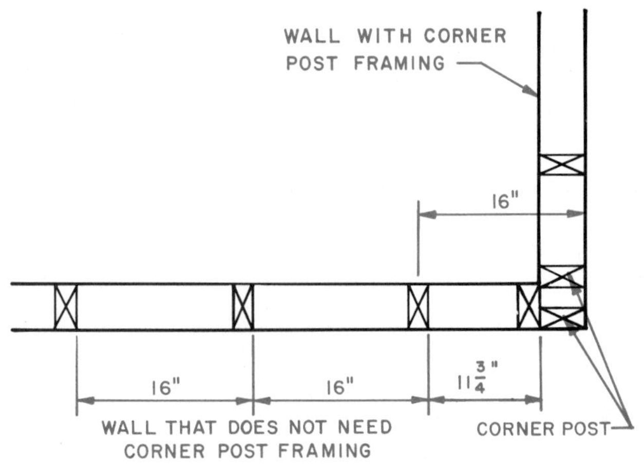

FIGURE 11-18

How to lay out the sole and top plates for the exterior wall without the corner post framing.

THE STORY POLE

The *story pole* is used to locate the horizontal members in a building. It is made by the carpenter from a straight 1 × 2 or 1 × 4 in. board. It is marked to show the floor level, door and window rough openings, ceiling height, and top and sole plates (Fig. 11-19). It reduces the chance for error because the carpenter does not have to measure each thing with a rule every time that one is to be located.

The story pole is laid on the wall fame and horizontal members are located on the studs.

DOUBLE PLATE
CRIPPLE
HEADER
ROUGH SILL
ROUGH SILL
ROUGH SILL
SOLE PLATE
SUBFLOOR TO TOP OF PLATE
SUBFLOOR TO TOP OF ROUGH OPENING
R.O. LIVING ROOM WINDOWS
R.O. BEDROOM, DEN WINDOWS
R.O. BATH, KITCHEN WINDOWS

FIGURE 11-19

A story pole is used to locate members vertically on a wall frame.

A balloon-framed building will require a story pole about 20 ft long. This is needed to locate the second-floor level and the ceiling on the second floor. Split-level houses generally require a story pole for each level.

ASSEMBLING WALL SECTIONS

The sole and top plates are marked and placed on edge 8 ft apart. The marked side is placed inward. Full-length regular studs are laid between the plates. Usually, these are accurately cut to length at the lumber mill and need no trimming on the building site. Nail a sole plate and top plate to each full-length stud. Drive two 16d common nails through the plates into the ends of the studs (Fig. 11-20). Some carpenters prefer to use box nails. They are slightly thinner and are less likely to split the wood.

Next set the trimmer studs in place and nail to full-length studs. Use 10d nails spaced 16 in. on center. Place them on alternate edges of the stud (Fig. 11-21).

Set the header on the trimmer studs. Nail through the full-length studs into the ends of each board forming the header. Use two 16d nails in each piece (Fig. 11-21).

Set the upper cripples in place and nail. Use two 16d nails through the top plate into each cripple. Toenail the cripple to the header with 8d nails (Fig. 11-21).

If there is a window opening, mark the location of the lower cripples on the rough sill.

FIGURE 11-20

Wall sections are assembled by laying them on the floor and nailing them in this position. *(Courtesy of Morrison-Knudsen Co., Inc.)*

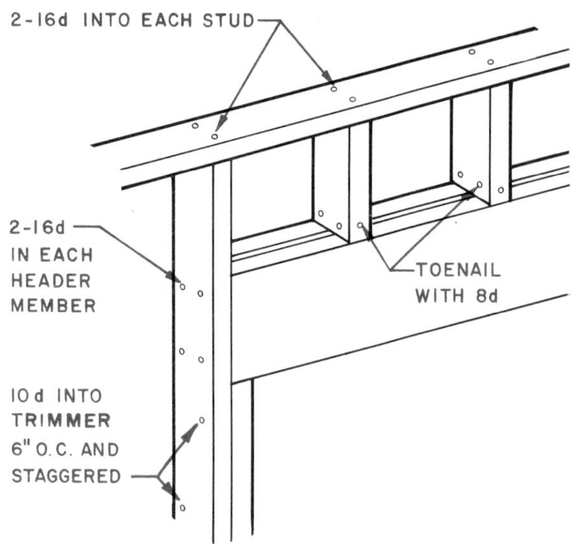

FIGURE 11-21

Recommended frame wall nailing patterns.

Nail the sill in position with two 16d nails through the full-length studs and two more into the cripple through the sole plate.

If the wall is to be met by a partition, nail the needed studs and blocking.

If diagonal *bracing* is required mark the studs, cut the recesses and nail the bracing in place. Be certain that the wall is square. This is checked by measuring the diagonals. If it is square, the diagonals will be the same length (Fig. 11-22).

Finally, apply the *sheathing*. Again, be certain the wall is square. To help hold it square, nail it lightly to the floor. The side of the wall next to the floor is the interior side. The sheathing is nailed to the side that is up (Fig. 11-23).

Some carpenters prefer to install the sheathing after the wall is raised to its vertical position. It is usually necessary to nail some temporary diagonal braces across the wall to strengthen it as it is raised.

ERECTING A WALL SECTION

After the wall section is assembled on the subfloor, it is raised into position. Small sections can be lifted by several carpenters. Large sections require a small crane or forklift (Fig. 11-24). After it is raised, temporary bracing is nailed from the wall to the subfloor. The braces should be nailed near the top of the wall (Fig. 11-25). After making certain that the sole plate is in the proper location and is straight, nail it through the subfloor into the joists. Use 20d nails at each joist. The sole can be checked for straightness with a chalk line.

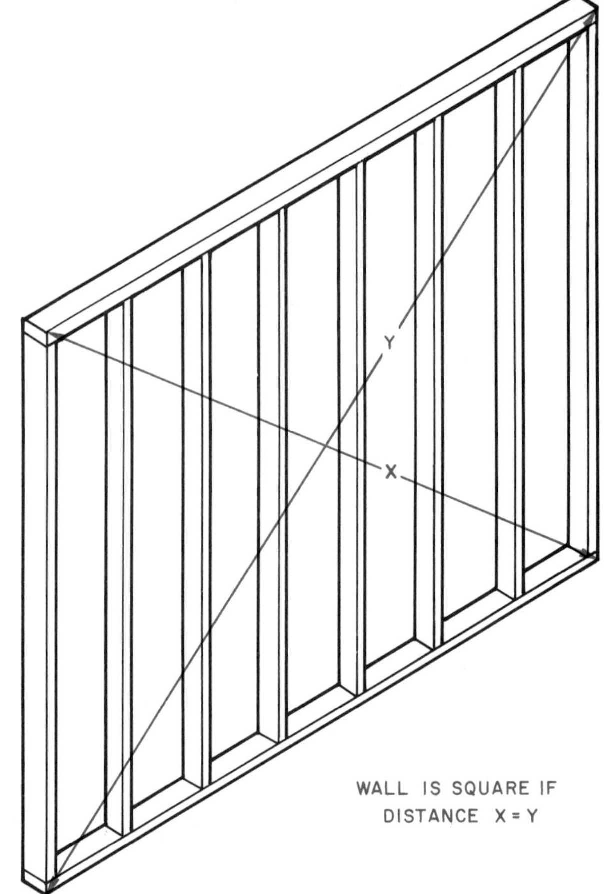

FIGURE 11-22

A wall can be checked for squareness by measuring its diagonals.

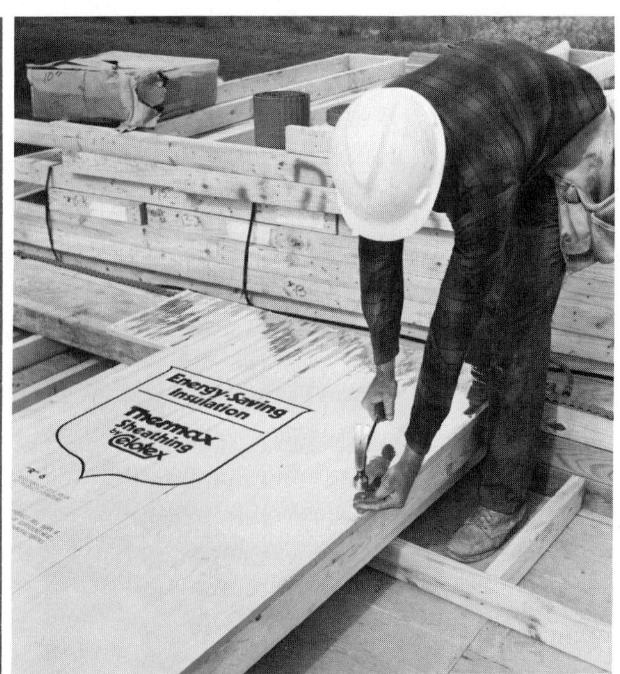

FIGURE 11-23

Sheathing may be applied to the wall before it is raised into position. *(Courtesy of The Celotex Corporation)*

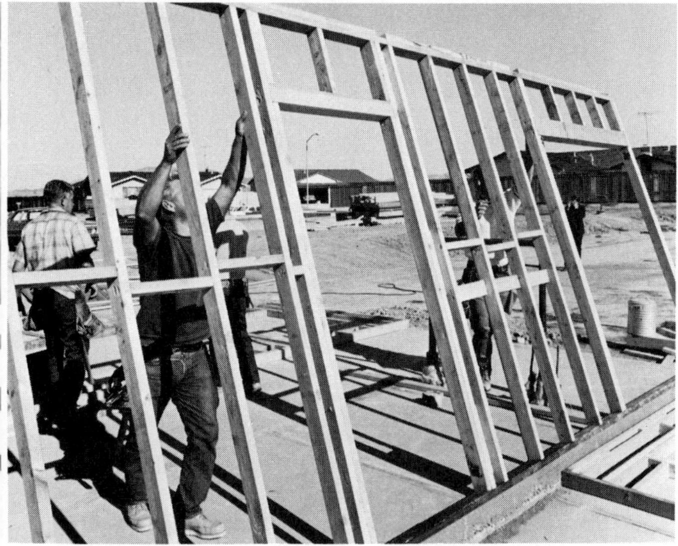

FIGURE 11-24

The assembled wall section is lifted into position.

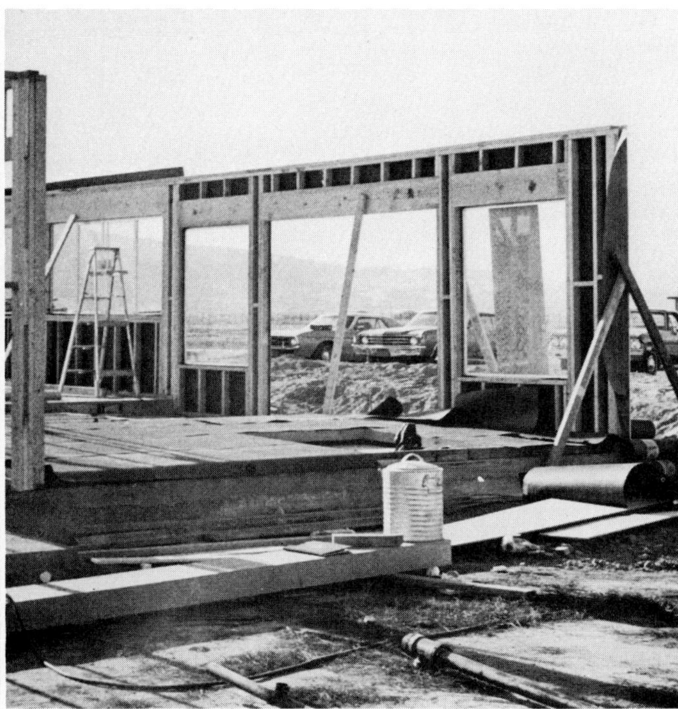

FIGURE 11-25

Bracing is used to hold the wall section erect. *(Courtesy of Western Wood Products Association)*

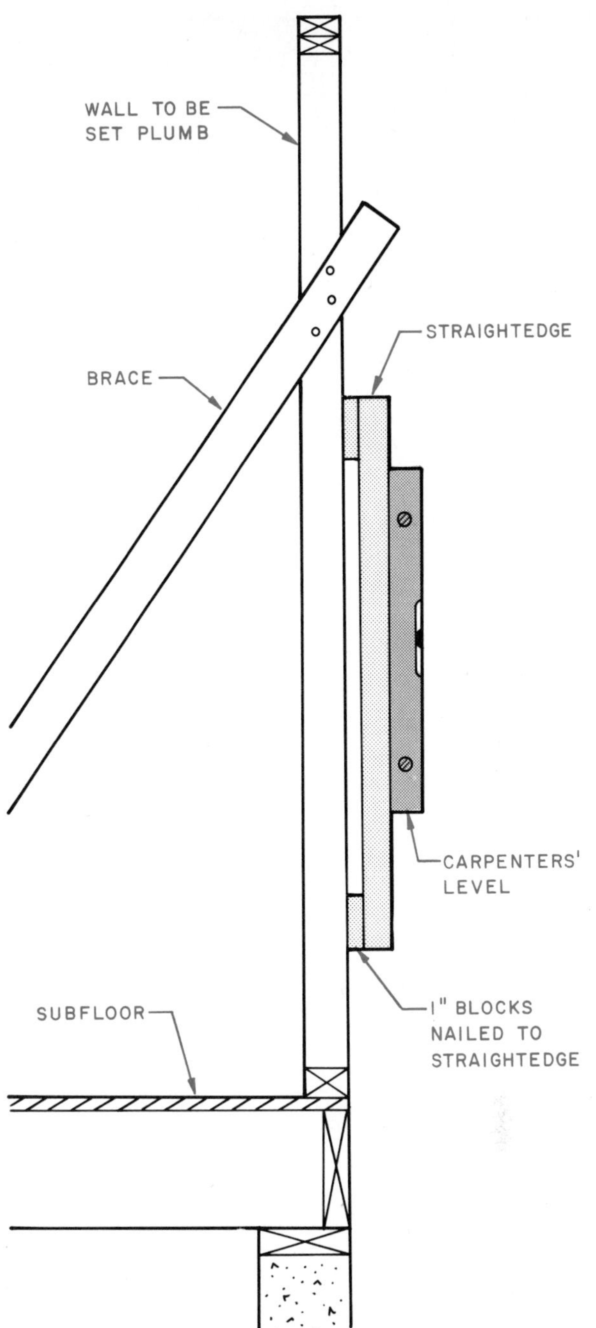

FIGURE 11-26

A carpenter's level and straightedge are used to plumb a wall.

Now the wall must be checked to see if it is plumb vertically. A carpenter's level and a straightedge can be used (Fig. 11-26). To adjust the wall, loosen the braces and move it until it is plumb. Then nail the braces back to the subfloor.

This procedure is repeated for each wall section. Generally, the larger sections are built and erected first. The smaller sections and projections then fill in the wall.

ERECTING THE PARTITIONS

After the exterior walls are erected straight and plumb, the partitions are built. The load-bearing partitions are built first. The others can be delayed until after the building is weather-tight, if so desired. If the building has roof trusses, no interior portions are load-bearing. The partitions are usually not built until after the building is weather-tight (Fig. 11-27).

The carpenter locates the partitions on the

FIGURE 11-27

Buildings with roof trusses do not need partitions installed until after the roof is in place. *(Courtesy of Western Wood Products Association)*

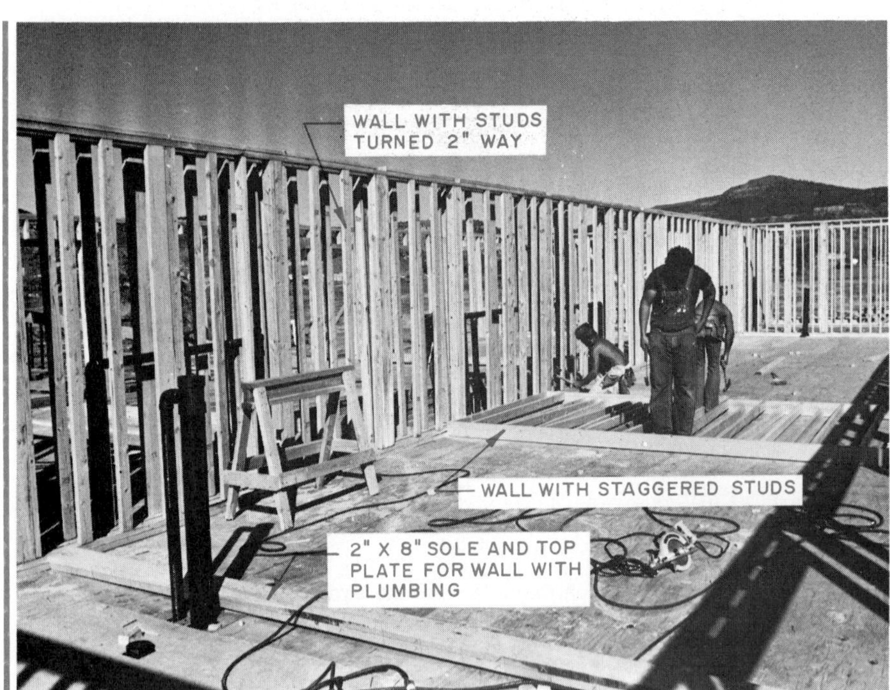

FIGURE 11-28

This job has a 2-in. interior partition. The carpenters are building a staggered stud wall. Notice the plates in place for a 2 × 8 in. wall for the plumbing. *(Courtesy of Morrison-Knudsen Co., Inc.)*

subfloor. The distances are read from the architectural drawings. Chalk lines are run, chalked, and snapped to the subfloor. This leaves a line that locates the sole plate. The partitions are built as discussed for exterior walls. The longer partitions are built first while there is room. Short partitions and closets are built last. When partitions meet the corners, they must have extra studs and blocking to provide for nailing the interior wall finish. This is the same as discussed with exterior wall-partition joining.

Remember to construct wider walls to handle plumbing. Stagger studs to provide sound barriers, as around a bath. Turn studs sideways to save floor space (Fig. 11-28). Also remember to cut openings for heat ducts and frame the opening.

Various bathroom fixtures require block-

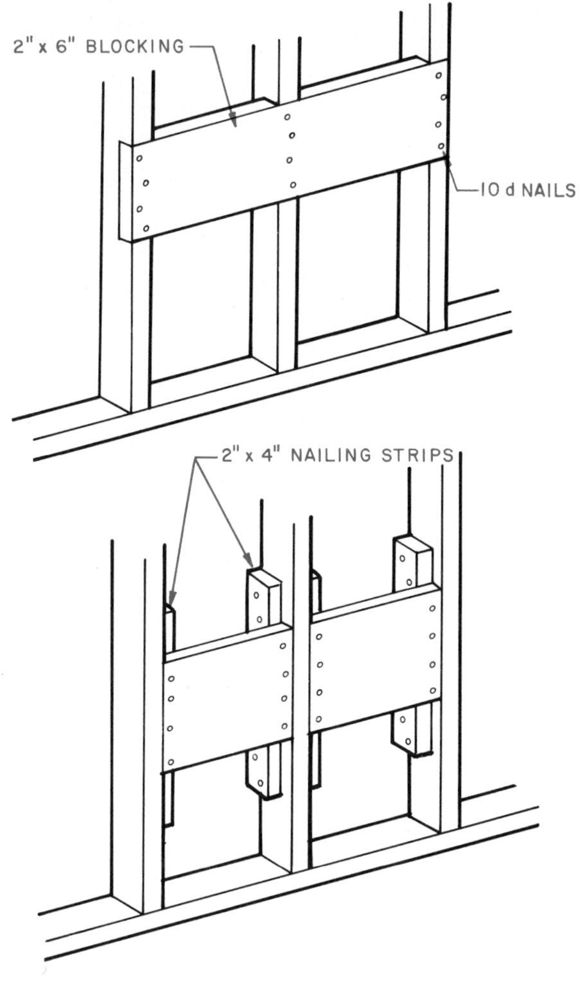

FIGURE 11-29

Typical framing for blocking to support fixtures in a kitchen or bath.

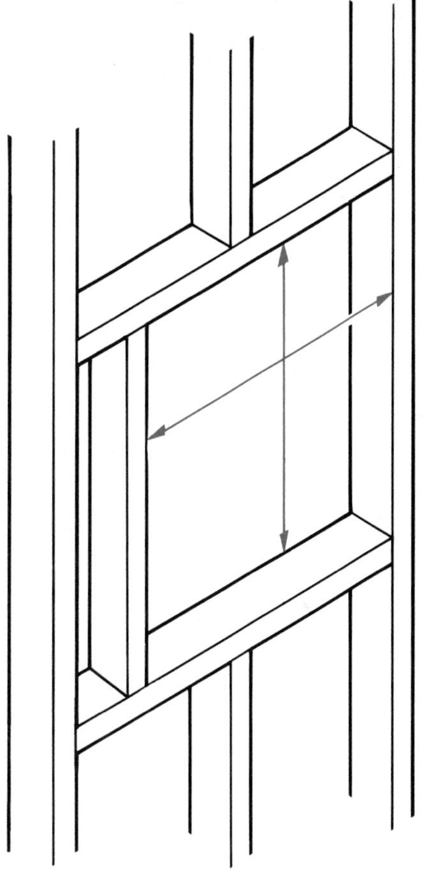

FIGURE 11-30

Framing for a medicine cabinet.

ing. These include towel racks, toilet tanks, and shelving (Fig. 11-29). A wall may require a recess for a medicine cabinet (Fig. 11-30). Bathtubs require extra blocking. These can be installed at this time.

BRACING EXTERIOR WALLS

Exterior wall construction must provide resistance to racking. This can be done by using diagonal solid wood sheathing, 1 × 4 in. diagonal braces let in the studs running from the sole plate to the top plate or plywood sheathing. Local building codes sometimes specify the needed bracing.

Diagonal bracing is laid across the studs from sole plate to top plate and lightly tacked in place. The location is marked on the studs and the brace is removed. The plates and studs have saw cuts made into them the thickness of the brace. The wood between the cuts is removed with a wood chisel (Fig. 11-31). The brace is nailed in the grooves with two 8d nails at each stud (Fig. 11-32). Trim the ends of the brace flush with the plates. Another way to brace the walls is with a metal strap. It is nailed to the studs on a 45° angle (Fig. 11-33).

Once the exterior walls are up, the second top plate can be added. It needs this added strength because the ceiling and roof joists rest on it. It also ties the sections of the wall together.

Joints in the second top plate should be at least 4 ft from joints in the first top plate. The second plate is face nailed to the plate below it with 10d nails spaced about 16 in. apart. Stagger them from side to side. Put two 10d nails in each end of each piece of the plate.

When top plates meet at a corner, they are overlapped. The partition top plate also overlaps the exterior-wall top plate (Fig. 11-34).

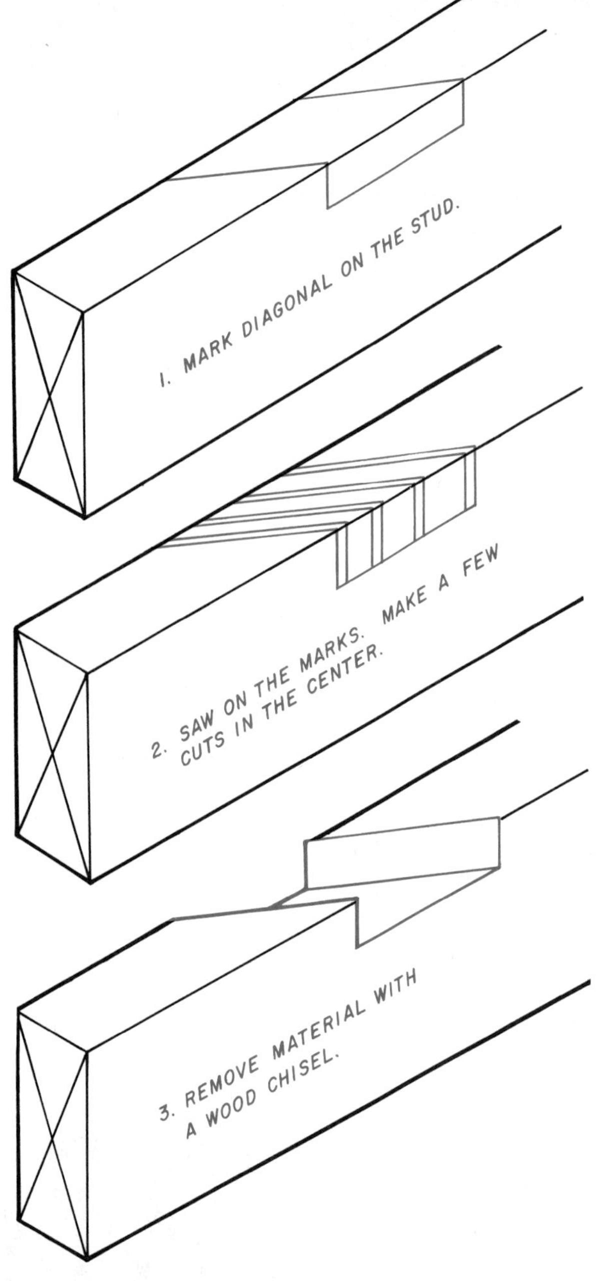

FIGURE 11-31

Steps to cut slots in studs and plates for diagonal bracing.

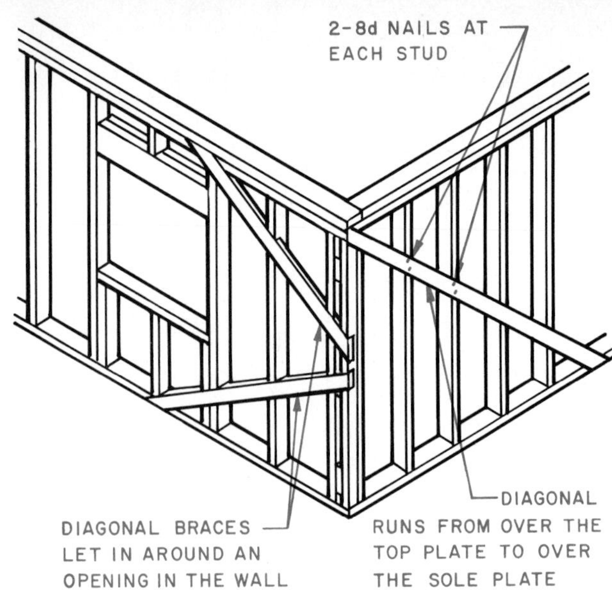

FIGURE 11-32

Diagonal bracing for walls with and without openings.

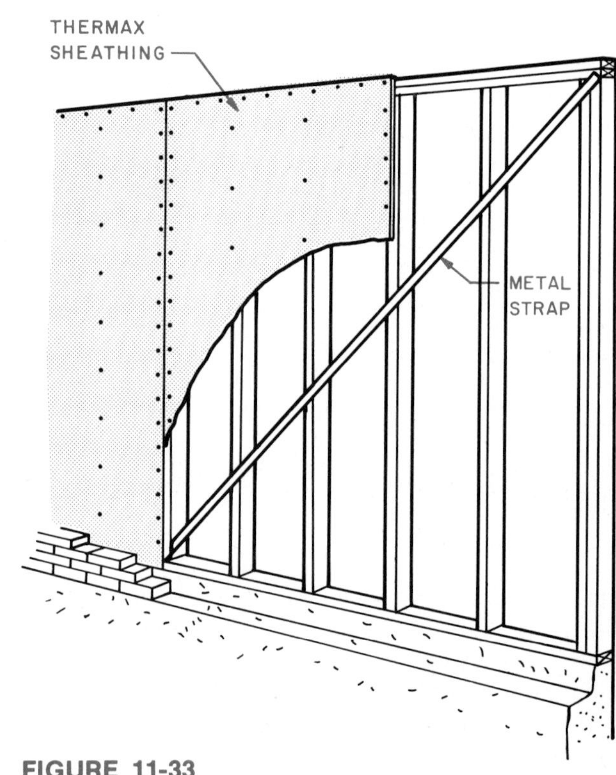

FIGURE 11-33

Metal straps are used as diagonal bracing. *(Courtesy of The Celotex Corporation)*

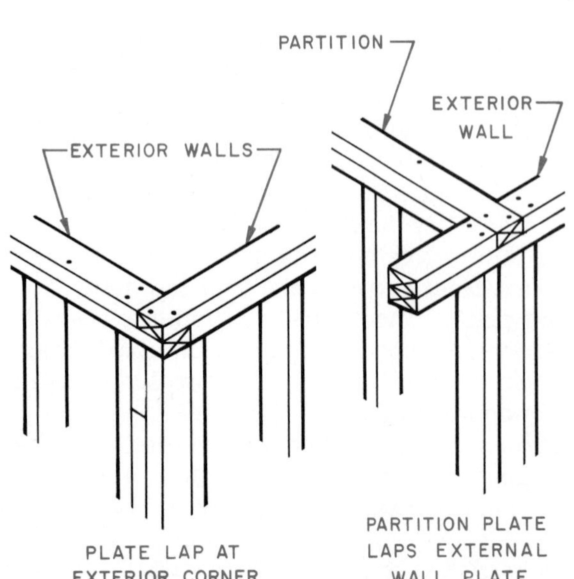

FIGURE 11-34

Top plates must lap wherever walls join.

SHEATHING THE WALL

The exterior walls should have sheathing applied before the ceiling and roof is started. Sheathing adds to the stiffness and strength of the wall.

Fiberboard sheathing is available in ½- and ²⁵⁄₃₂-in.-thick sheets. They are 4 ft 0 in. × 8 ft 0 in. in size. The ½-in. sheets are applied with 1½-in. galvanized roofing nails. The ²⁵⁄₃₂-in. sheets require 1¾-in. galvanized roofing nails. The nails are placed at least ⅜ in. from the edge of the sheet. They are spaced every 3 in. on the outside edges of the sheet. They are spaced every 6 in. on the inside studs.

Walls sheathed with ½-in.-thick fiberboard sheets require diagonal bracing. Those with ²⁵⁄₃₂-in. sheets do not require bracing. When ½-in. sheets are used, 4 ft 0 in. × 8 ft 0 in. plywood sheathing is applied at each corner. This provides the necessary bracing (Fig. 11-35).

Gypsum sheathing is available in ½-in.-thick sheets. It requires that the wall have some form of corner bracing. Usually, 1¾-in. nails are used. They are spaced 4 in. on the outer edges and 6 to 8 in. on the interior studs.

Plastic foam is available in a wide range of thicknesses and sizes. Typically, these are from ½ to 2 in. in thickness. It has a high insulation value but no structural strength. Both sides are covered with aluminum foil. The wall requires corner bracing. The use of metal diagonal bracing is recommended. Refer to Fig. 11-33.

Fiberboard, gypsum, and plastic foam insulation will not hold nails. Siding applied over these materials must be nailed into the studs.

Plywood sheathing is available in ⁵⁄₁₆-in. thickness for studs spaced 16 in. on center and ⅜ in. for studs spaced 24 in. on center. Often, ½-in.-thick sheathing is used because the finish siding can be nailed directly into the plywood. With the thinner sheets the exterior finish must be nailed into the studs. See Chapter 7 for more information.

Plywood sheathing can be applied with the long direction of the side vertically or horizontally (Fig. 11-36). It should be nailed with 6d box nails spaced about 6 in. apart on the edges and 12 in. apart on the center studs.

A plywood-sheathed wall is strong and requires no corner bracing.

When solid wood boards are used for wall sheathing, they should be dressed ¾-in.-thick lumber from 6 to 12 in. wide. It may be applied horizontally or diagonally (Fig. 11-37). If applied diagonally, no corner bracing is needed. The sheets should be slanted in opposite direc-

FIGURE 11-35

Plywood sheets at each corner provide adequate bracing.

FIGURE 11-36

This plywood sheathing was applied horizontally. *(Courtesy of American Plywood Association)*

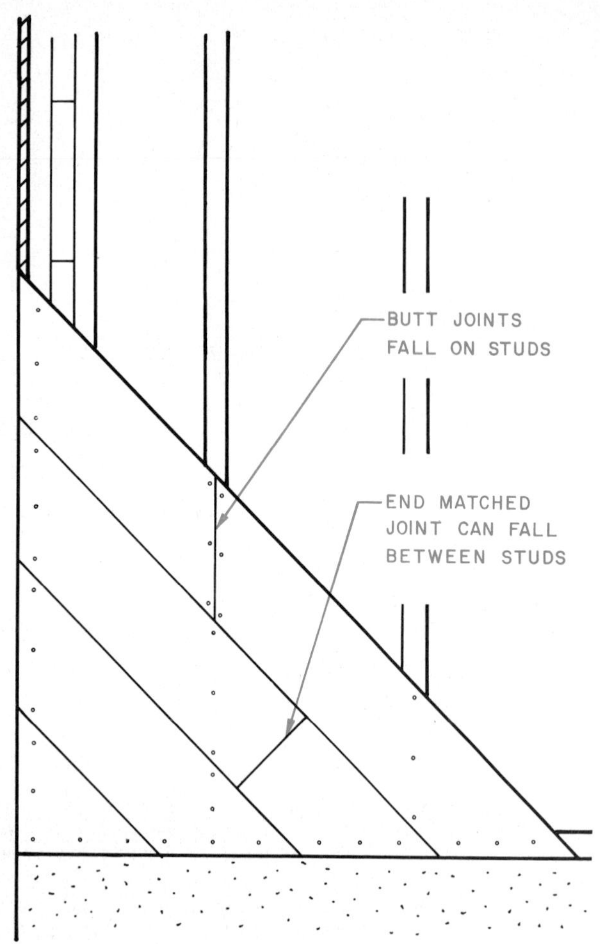

FIGURE 11-37

How to apply diagonal wood sheathing.

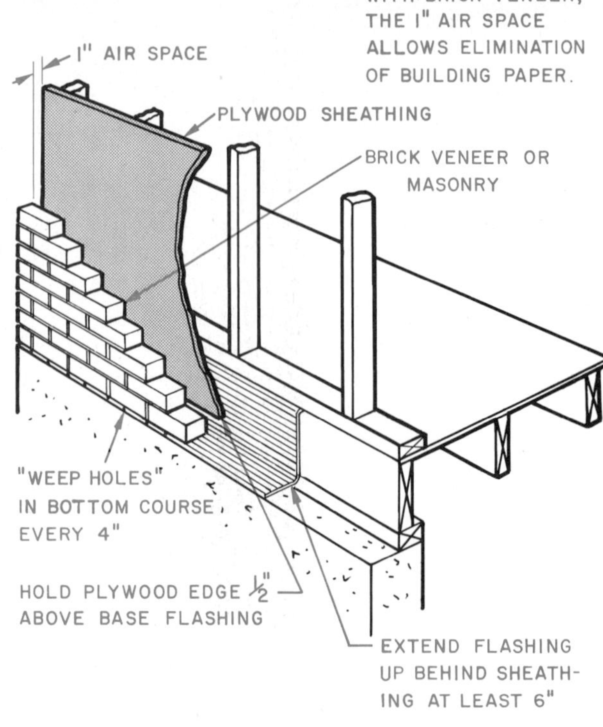

FIGURE 11-38

Sheathing a wall to be covered with brick veneer.

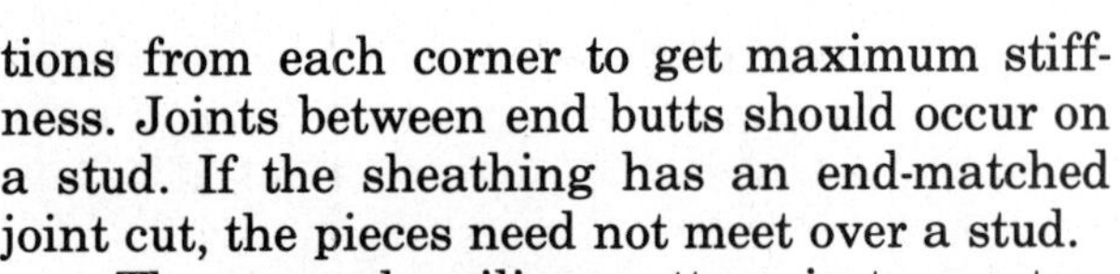

tions from each corner to get maximum stiffness. Joints between end butts should occur on a stud. If the sheathing has an end-matched joint cut, the pieces need not meet over a stud.

The normal nailing pattern is to use two 8d box nails for boards 6 to 8 in. wide. Use three nails for boards over 8 in. wide (Fig. 11-37).

If the solid wood sheathing is applied horizontally, corner bracing is required.

Construction details for a brick veneer over plywood sheathing are in Fig. 11-38. A 1-in. air space is left between the sheathing and bricks. Building paper is not needed. The sheathing extends to within ½ in. of the concrete foundation. It covers the header. Metal flashing is extended at least 6 in. up under the sheathing.

When building a stucco wall, the plywood sheathing is covered with building paper (Fig. 11-39). This is nailed or stapled to the plywood sheathing. The self-furring metal lath or other wire base is nailed or stapled to the plywood sheathing (Fig. 11-40).

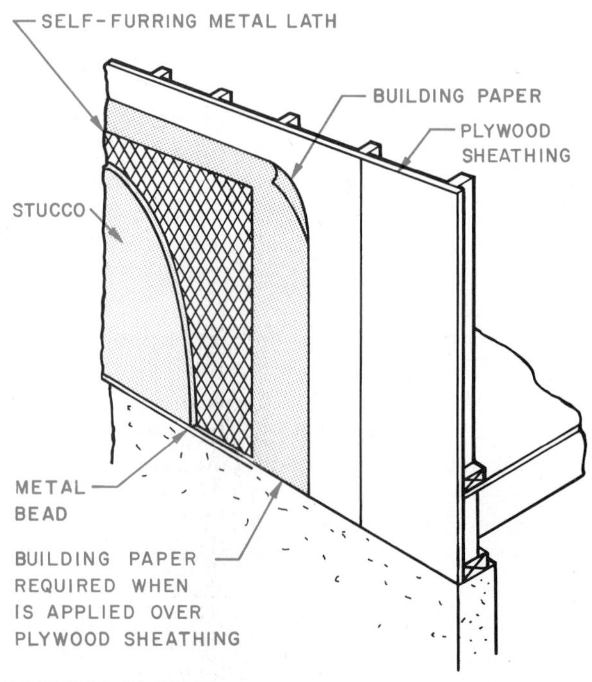

FIGURE 11-39

Sheathing a wall to be covered with stucco.

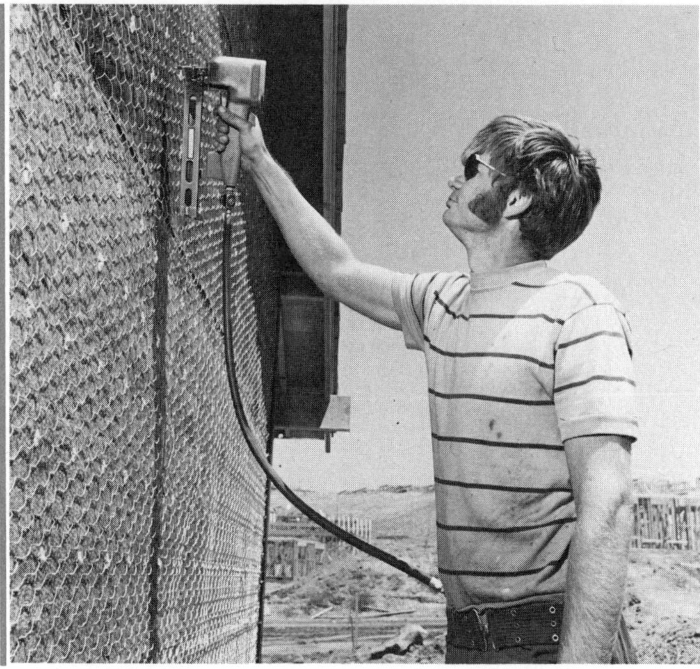

FIGURE 11-40

Using a power stapler to apply wire mesh to sheathing forming the base for a stucco wall. *(Courtesy of Bostitch Textron)*

STUDY QUESTIONS

After reading the chapter, answer each of the following questions. If you do not know the answer, review the chapter.

1. What are the parts of a typical frame wall?
2. What is the normal spacing of studs?
3. Why is it important to space studs accurately?
4. Why do openings in walls need headers?
5. How is the length of a header determined?
6. How is an interior partition joined to the exterior wall?
7. Why are the sole and top plates marked at the same time?
8. How do carpenters use a story pole?
9. How are walls checked to see if they are erected plumb?
10. What ways are exterior walls braced?
11. What types of sheathing are being used?
12. Which type of sheathing eliminates the need for wall bracing?

IMPORTANT TECHNICAL TERMS

Following are technical terms that you should be able to use as part of your working vocabulary. Write a brief description of the meaning of each term.

Partitions
Sole plate
Top plate
Studs
Headers
Cripples
Trimmers
Story pole
Bracing
Sheathing

12 CEILING FRAMING

The ceiling framing supports the interior finish material that forms the ceiling. Under normal conditions the ceiling framing is supported by the exterior wall frame and interior load-bearing partitions. See Chapter 11 for framing details for these walls.

The ceiling frame is very much like that used for floors. (See Chapter 10). The ceiling members are usually smaller because they do not generally carry the loads expected of a floor. The first-floor joists serve as the ceiling framing for a basement ceiling. In a two-story house the floor joists forming the floor for the second story also serve as the framing for the first-floor ceiling.

SELECTION OF CEILING JOIST SIZES

Ceiling joists are selected in the same manner as floor joists. The spacing between the joists and the distance they span must be known. The closer together the joists are placed and the shorter the span, the smaller the wood member that can be used. Another factor to be considered is the live load to be placed on the ceiling joists. If there is a possibility that the area above the ceiling joists might be made into rooms at some future time, the ceiling joists are selected using floor joist data. If an unusual load, such as mechanical equipment, is to be placed in the attic on the ceiling joists, the joists must be strong enough to support the extra weight.

The carpenter can find the size on the house plan. These should conform to local building codes. Recommended sizes are shown in Fig. 12-1.

SIZE (in.)	SPACING (in. on center)	10-PSF LIVE LOAD[a]				20-PSF LIVE LOAD[b]			
		Dense Sel Str KD and No. 1 Dense KD	Dense Sel Str, Sel Str KD, No. 1 Dense and No. 1 KD	Sel Str, No. 1 and No. 2 Dense KD	No. 2 Dense, No. 2 KD and No. 2	Dense Sel Str KD and No. 1 Dense KD	Dense Sel Str, Sel Str KD, No. 1 Dense and No. 1 KD	Sel Str, No. 1 and No. 2 Dense KD	No. 2 Dense
2 × 4	12.0	13'-2"	12'-11"	12'-8"	12'-5"	10'-5"	10'-3"	10'-0"	9'-10"
	13.7	12'-7"	12'-4"	12'-1"	11'-10"	10'-0"	9'-9"	9'-7"	9'-5"
	16.0	11'-11"	11'-9"	11'-6"	11'-3"	9'-6"	9'-4"	9'-1"	8'-11"
	19.2	11'-3"	11'-0"	10'-10"	10'-7"	8'-11"	8'-9"	8'-7"	8'-5"
	24.0	10'-5"	10'-3"	10'-0"	9'-10"	8'-3"	8'-1"	8'-0"	7'-10"
2 × 5	12.0	16'-11"	16'-7"	16'-3"	15'-11"	13'-5"	13'-2"	12'-11"	12'-8"
	13.7	16'-2"	15'-10"	15'-7"	15'-3"	12'-10"	12'-7"	12'-4"	12'-1"
	16.0	15'-4"	15'-1"	14'-9"	14'-6"	12'-2"	11'-11"	11'-9"	11'-6"
	19.2	14'-5"	14'-2"	13'-11"	13'-8"	11'-5"	11'-3"	11'-0"	
	24.0	13'-5"	13'-2"	12'-11"	12'-8"[c]	10'-8"	10'-5"[c]	10'-3"	
2 × 6	12.0	20'-8"	20'-3"	19'-11"	19'-6"	16'-4"	16'-1"	15'-9"	15'-6"
	13.7	19'-9"	19'-5"	19'-0"	18'-8"	15'-8"	15'-5"	15'-1"	14'-9"
	16.0	18'-9"	18'-5"	18'-1"	17'-8"	14'-11"	14'-7"	14'-4"	14'-1"
	19.2	17'-8"	17'-4"	17'-0"	16'-8"	14'-0"	13'-9"	13'-6"	
	24.0	16'-4"	16'-1"	15'-9"	15'-6"[c]	13'-0"	12'-9"[c]	12'-6"	
2 × 8	12.0	27'-2"	26'-9"	26'-2"	25'-8"	21'-7"	21'-2"	20'-10"	20'-5"
	13.7	26'-0"	25'-7"	25'-1"	24'-7"	20'-8"	20'-3"	19'-11"	19'-6"
	16.0	24'-8"	24'-3"	23'-10"	23'-4"	19'-7"	19'-3"	18'-11"	18'-6"
	19.2	23'-3"	22'-10"	22'-5"	21'-11"	18'-5"	18'-2"	17'-9"	
	24.0	21'-7"	21'-2"	20'-10"	20'-5"[c]	17'-2"	16'-10"[c]	16'-6"	
2 × 10	12.0	34'-8"	34'-1"	33'-5"	32'-9"	27'-6"	27'-1"	26'-6"	26'-0"
	13.7	33'-2"	32'-7"	32'-0"	31'-4"	26'-4"	25'-10"	25'-5"	24'-11"
	16.0	31'-6"	31'-0"	30'-5"	29'-9"	25'-0"	24'-7"	24'-1"	23'-8"
	19.2	29'-8"	29'-2"	28'-7"	28'-0"	23'-7"	23'-2"	22'-8"	
	24.0	27'-6"	27'-1"	26'-6"	26'-0"[c]	21'-10"	21'-6"[c]	21'-1"	

Note: Data for southern pine structural members.
[a]With no future sleeping rooms and no attic storage.
[b]With no future sleeping rooms and limited attic storage.
[c]Deflection limitation of $l/240$.

FIGURE 12-1

Span tables for ceiling joists from southern pine.
(Courtesy of Southern Forest Products Association)

JOIST LAYOUT

Ceiling joists usually are run across the width of a building (Fig. 12-2). The ends that rest on the exterior wall must have the end sloped so that it does not extend above the rafter (Fig. 12-3). This cut is laid out as shown in Fig. 12-4. The height of the rafter is marked on the end of the ceiling joist. This is the height from the birdsmouth to the top of the rafter. A framing square is used to set the slope by locating the rise and run of the rafter. The joist end is marked and cut. This can form the pattern for marking the other ceiling joists. Cut the joist so that its sloped surface will be slightly below the top of the rafter.

A large room may have a span beyond that which a ceiling joist can reach. To frame this ceiling a beam is placed in the center. The joists are framed to the beam (Fig. 12-5). They can be fastened with a ledger or any of several metal joist hangers. (See Chapter 10 for these details as used on floor joists.)

When ceiling joists run parallel to the edge of the roof, they will extend above the top of the rafters. This is especially true with low pitched roofs. The ceiling framing in this area is made of short members run perpendicular to the ceiling joists (Fig. 12-6).

INSTALLING THE CEILING JOISTS

The ceiling joists are located on the plate in the same manner as floor joists. (See Chapter 10.) Since the supporting wall has a double plate, it is not necessary to line up the ceiling joists with the wall studs. It is important to locate them so

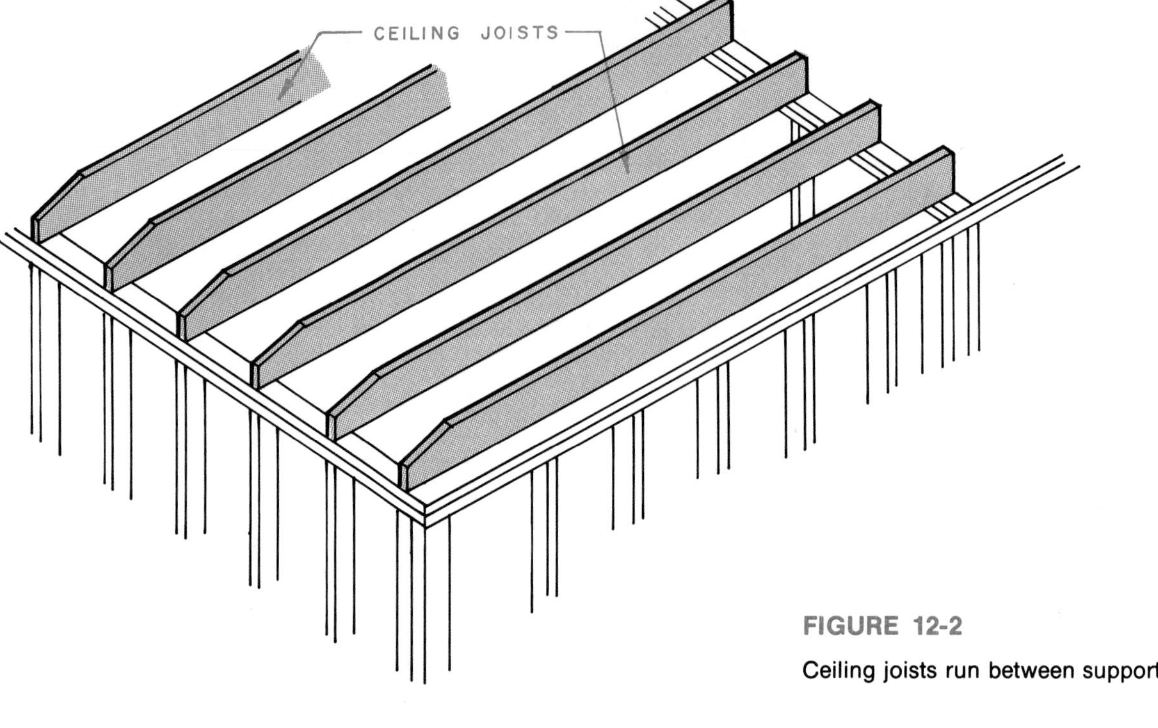

FIGURE 12-2

Ceiling joists run between supporting walls.

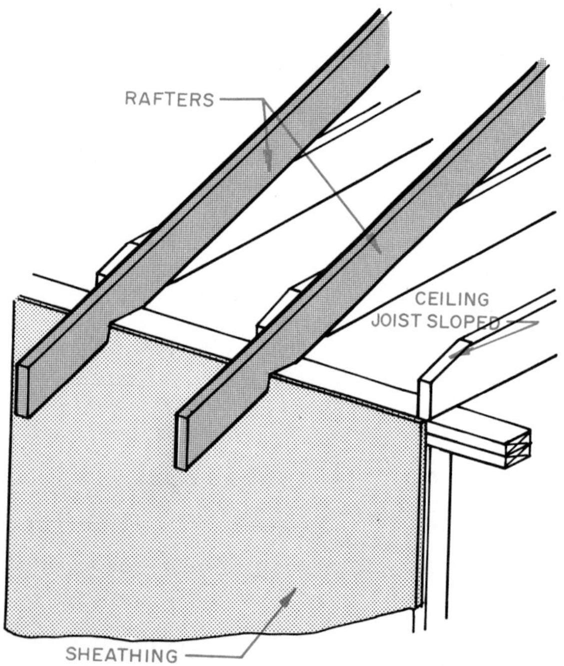

FIGURE 12-3

The ends of ceiling joists are sloped to prevent them from sticking above the rafter.

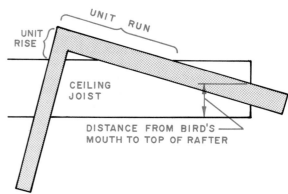

UNIT RUN
UNIT RISE
CEILING JOIST
DISTANCE FROM BIRD'S MOUTH TO TOP OF RAFTER

FIGURE 12-4

How to lay out the slope on the end of a ceiling joist.

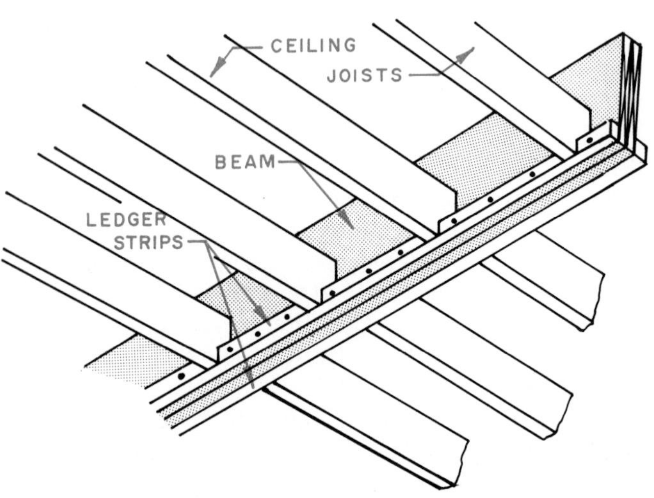

FIGURE 12-5

Ceiling joists can be framed to a beam with a ledger strip.

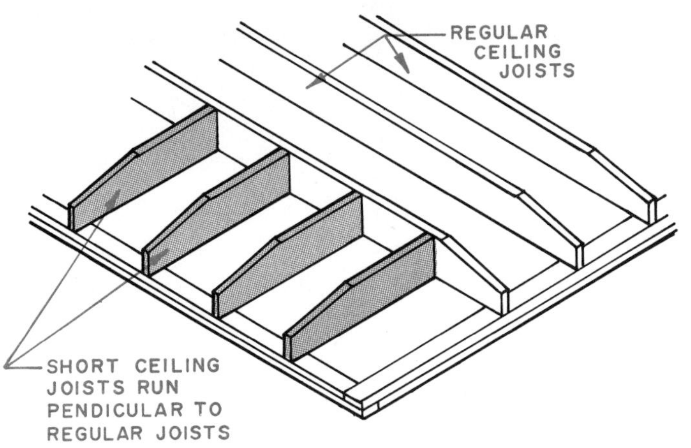

FIGURE 12-6

Low-pitched roofs require short ceiling joists run perpendicular to the regular joists.

they are next to a rafter. The joist and rafter can then be nailed together. The ceiling joists are installed before the rafters are set in place, so rafter location must be decided before installing the ceiling joists. Usually, the end joist is set in 3½ in. from the edge of the plate (Fig. 12-7).

Ceiling joists can be butted or lapped on the interior bearing wall. If they are to be butted, they need a wood splice nailed on each side. The splice can be ⅜-in. plywood or ¾-in. solid wood and should be 24 in. long (Fig. 12-8).

When joists are lapped, they should be face-nailed with three 16d nails (Fig. 12-9). Ceiling joists should be toenailed to the plates with two 10d nails on each side. As each joist is set in place, sight down its length. Place it with the crowned side up.

Partitions that run parallel with the joists must be fastened to them. This requires a nailing strip and blocking. The nailing strip is

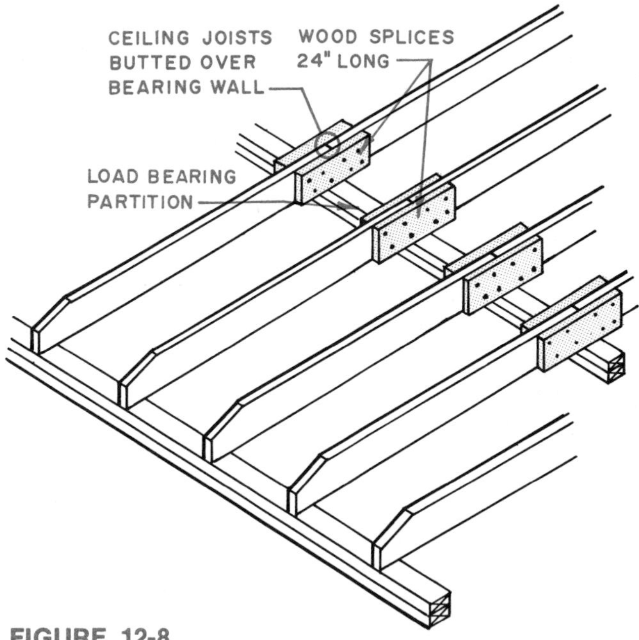

FIGURE 12-8

Ceiling joists can be butted and spliced over load-bearing partitions.

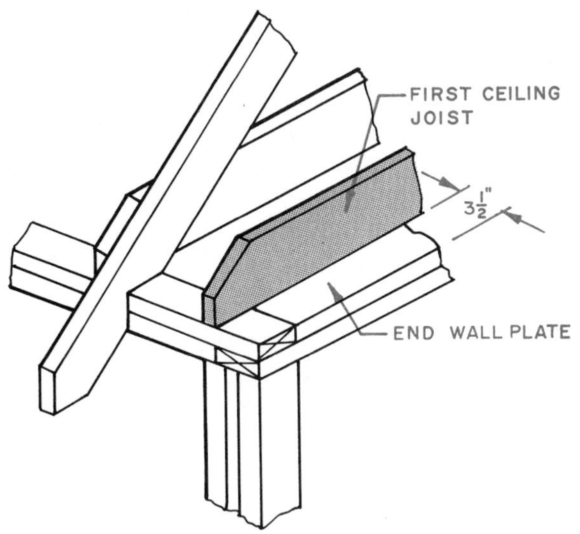

FIGURE 12-7

The first ceiling joist is set 3½ in. from the face of the exterior wall plate.

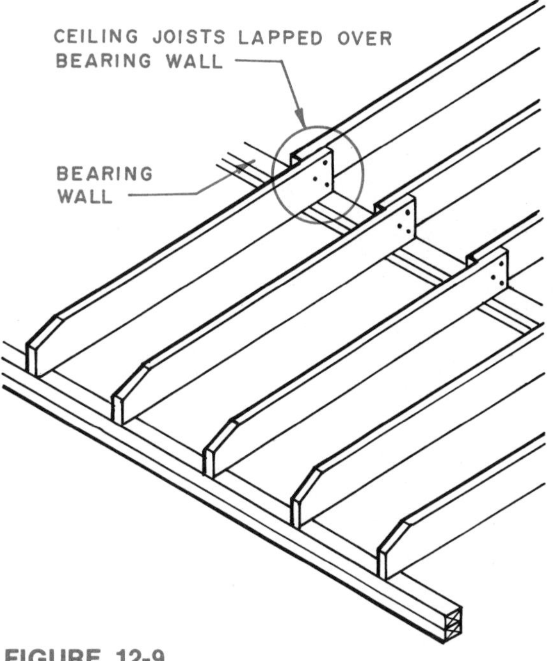

FIGURE 12-9

Ceiling joists can be lapped over load-bearing partitions.

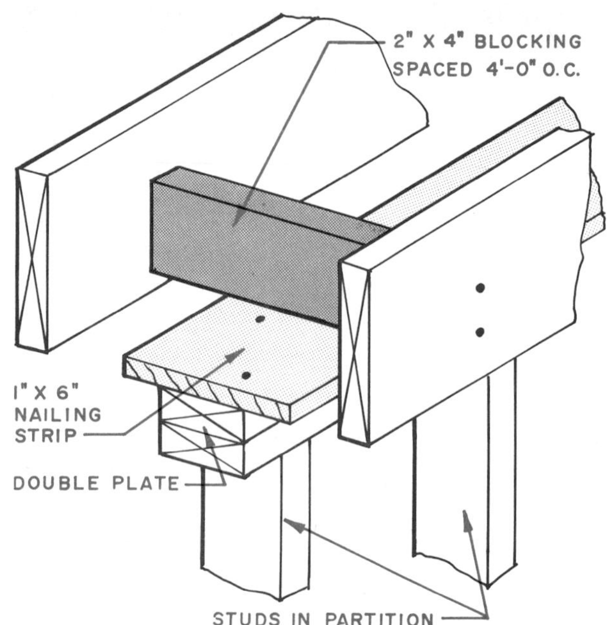

FIGURE 12-10

Partitions running parallel to ceiling joists are anchored to them with blocking.

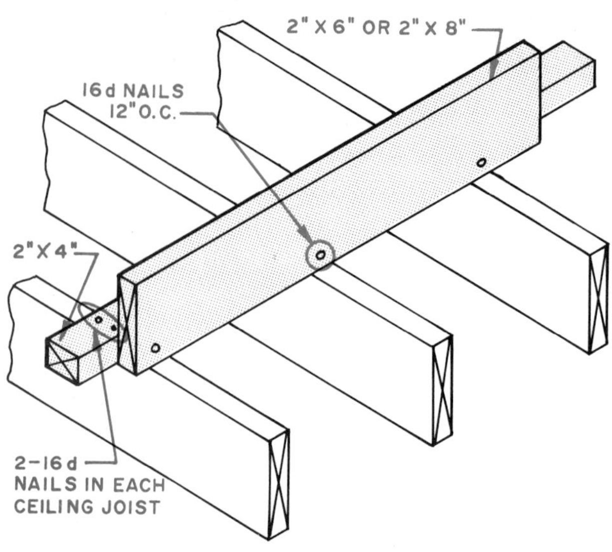

FIGURE 12-11

A strongback is used to stiffen ceiling joists.

needed for holding the interior finish material (Fig. 12-10).

Openings in the ceiling, such as an access door, require the use of headers. These are handled the same as in floor framing. (See Chapter 10.)

If joists span a long area and it is desired to stiffen the ceiling, a strong back can be formed (Fig. 12-11). A 2 × 4 in. member is nailed to the top of the joists. A 2 × 6 or 2 × 8 in. member is set vertically and nailed to the 2 × 4. This forms a beam, thus stiffening the ceiling framing.

TRUSSES

When a roof is framed with trusses, the lower chord forms the ceiling joist. After the exterior walls are in place, the trusses are raised. This forms the ceiling framing. Trusses are discussed in Chapter 13.

STUDY QUESTIONS

After reading the chapter, answer the following questions. If you do not know the answer, review the chapter.

1. How can ceiling joists be adjusted to span wide openings which are beyond the distance a single joist can reach?
2. What are the two ways in which ceiling joists can meet on an interior supporting partition?
3. How are ceiling joists joined to the top plates?
4. How are ceiling joists joined to partitions that run parallel with the joists?
5. What part of a truss forms the ceiling joist?

13 FRAMING THE ROOF

The roof framing provides the structure to support the roofing materials and snow, and to withstand rain and wind. It must be securely fastened to the wall framing. On most residential buildings the roof is a dominant design feature and should be carefully and accurately built.

ROOF TYPES

The typical roof types used in home construction and some small commercial buildings include (Fig. 13-1):

Flat roof: flat or has a slight slope (Fig. 13-2).
Shed roof: a single sloping surface providing drainage.
Butterfly roof: sheds water to the center of the house. It is two shed roofs meeting at their lower edges. The valley requires special waterproofing. It is not recommended for areas having heavy snow.
Gable roof: two sloping surfaces that meet at a ridge above the house. This forms a gable at each end of the house. It is the most commonly used type of roof (Fig. 13-3).
Hip roof: four sloping sides. It is much like a gable roof, but the end sloping surfaces replace the gable end (Fig. 13-4).
Gambrel roof: variation of the gable roof. It has each slope broken into two different pitches. This provides more headroom in the space under the roof. Dormers are generally used to provide light and ventilation (Fig. 13-85).
Mansard roof: much like the gambrel roof except that it also slopes on the ends of the house. This replaces the siding needed on the end of the house with a gambrel roof. It also replaces the siding found on

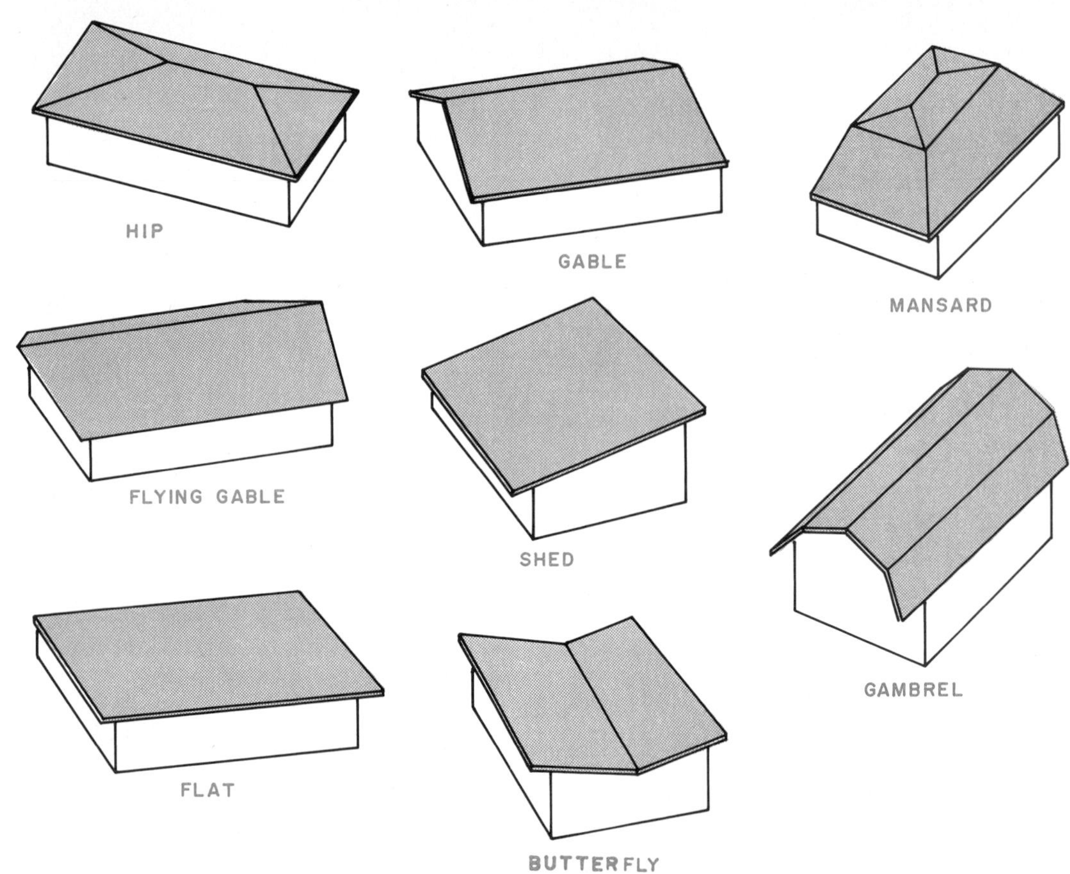

FIGURE 13-1

Basic roof types.

FIGURE 13-2

A residence with a flat roof. *(Courtesy of Home Planners, Inc.)*

FIGURE 13-3

This house has two gable roofs. *(Courtesy of The Garlinghouse Co., Inc.)*

FIGURE 13-4

A hipped roof is used on the garage and house.

the second floor of a two-story house with a gable roof. Dormers are usually used (Fig. 13-90).

TECHNICAL TERMS

Span: total distance between the outside face of the double plates (Fig. 13-5).

Run: distance from the outside face of a double plate to the centerline of the ridge. It is half the span (Fig. 13-5).

Measuring line: imaginary line running from the outside corner of the double plate parallel with the edge of the rafter to the centerline of the ridge.

Rise: vertical distance from the top of the double plate to the measuring line at the ridge (Fig. 13-5).

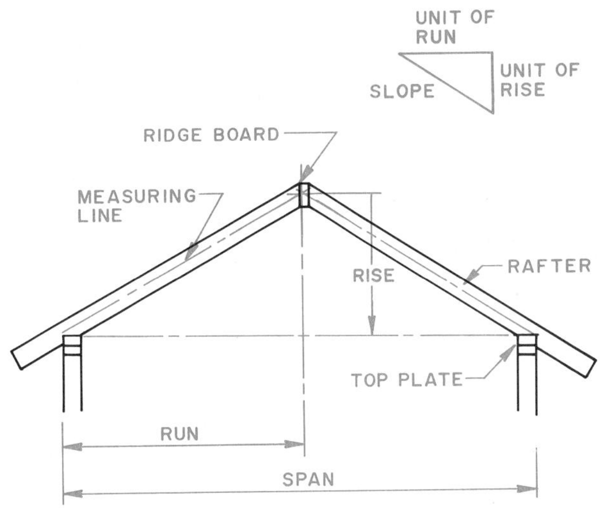

FIGURE 13-5

The terms used to describe the slope and pitch of a roof.

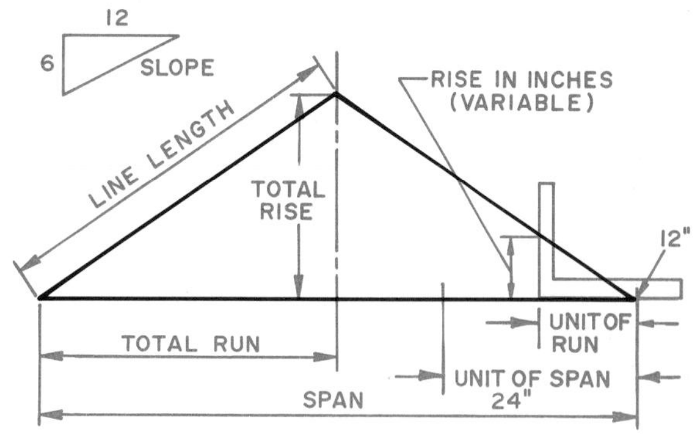

UNIT OF RUN IS ALWAYS 12"

UNIT OF RISE VARIES

UNIT OF SPAN IS ALWAYS 24"

$$\text{SLOPE} = \frac{\text{UNIT OF RISE}}{\text{UNIT OF RUN}}$$

$$\text{PITCH} = \frac{\text{UNIT OF RISE}}{\text{UNIT OF SPAN}}$$

FIGURE 13-6

How to determine roof slope and pitch.

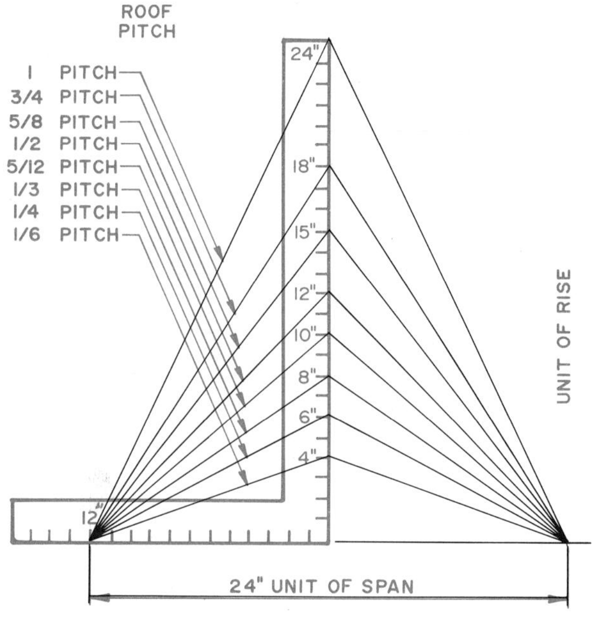

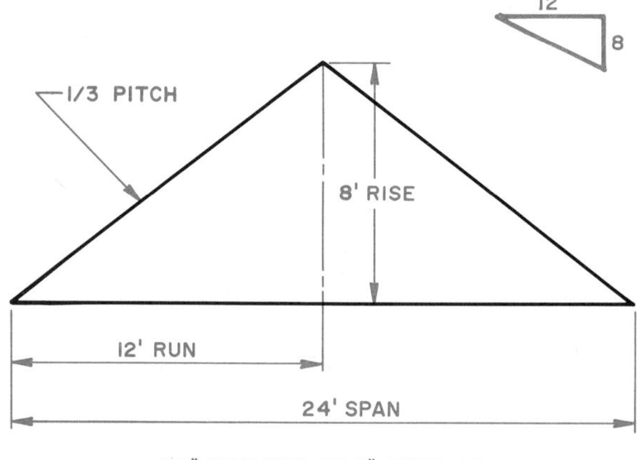

96" RISE FOR 288" SPAN OR
8" TO THE FOOT PITCH

FIGURE 13-7

How to use a framing square to determine roof pitch.

Unit of rise: number of inches a roof rises per foot of run (Fig. 13-6).

Unit of run: 12-in. horizontal distance used to describe slope (Fig. 13-6).

Slope: incline of a roof expressed in *units of rise to units of run.* For example, a roof with a unit of rise of 3 in. has a slope of 3 in 12 (Fig. 13-6). The slope is often called the "cut of the roof."

Pitch: angle the roof surface makes with the horizontal. It is the ratio of the *unit of rise to the unit of span.* For example, a roof with a rise of 8 in. to a unit span of 24 in. has a ratio of 8:24 and is called a 1/3 pitch (Fig. 13-7). The unit of span used to figure pitch is 24 in. (twice the unit of run).

KINDS OF RAFTERS

A *rafter* is a structural member used to carry the weight of the roof plus other loads, such as wind and snow. Typical examples are listed in Fig. 13-8.

Common rafters are those which run from the exterior wall plate to the ridge at a right angle to both.

Hip rafters run from the plate to the ridge but are at a 45° angle. They form the line of intersection between each side of the hip-type roof.

Valley rafters run from the plate to the ridge on an angle. They form the intersection between two parts of a roof.

Hip jack rafters run between the plate and a hip rafter.

Valley jack rafters run between the plate and a valley rafter.

Cripple jack rafters run between hip jack and valley jack rafters.

PARTS OF A RAFTER

The rafter meets the ridge with a *plumb cut.* A *bird's-mouth* (a notch) is cut so that it can fit on the plate. The *tail* provides an overhang. The *tail cut* shapes the end of the rafter. If there is no overhang, the end of the rafter is formed with a seat cut (Fig. 13-9).

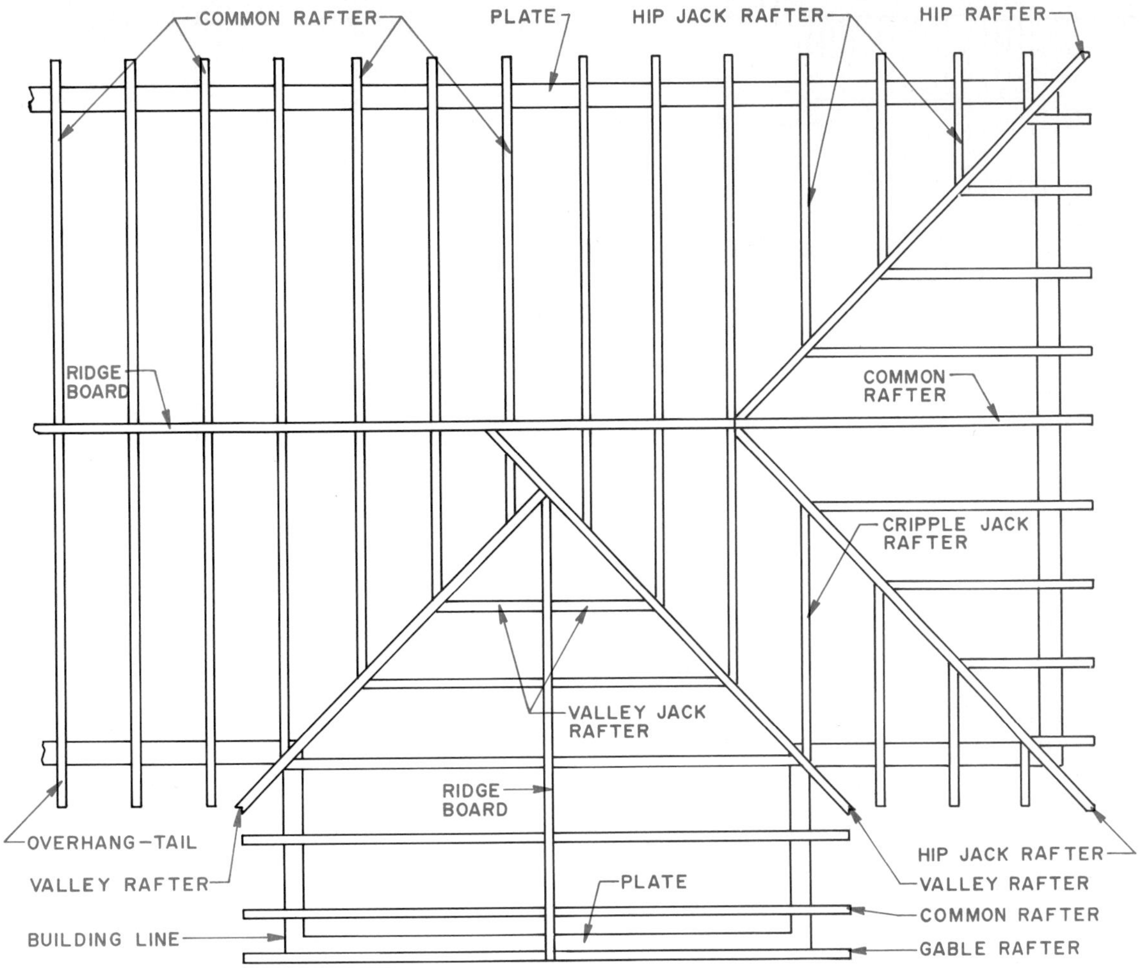

COMMON RAFTER
PLATE
HIP JACK RAFTER
HIP RAFTER
RIDGE BOARD
COMMON RAFTER
CRIPPLE JACK RAFTER
VALLEY JACK RAFTER
RIDGE BOARD
OVERHANG—TAIL
VALLEY RAFTER
BUILDING LINE
PLATE
HIP JACK RAFTER
VALLEY RAFTER
COMMON RAFTER
GABLE RAFTER

FIGURE 13-8

The kinds of rafters used in residential roof construction.

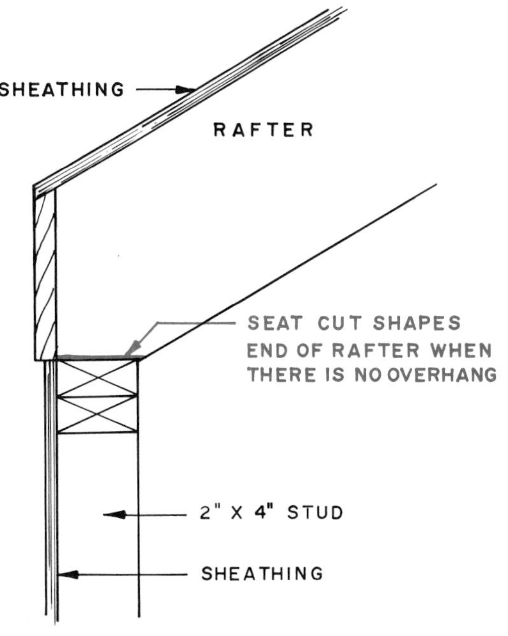

FIGURE 13-9

A rafter with no overhang has the end formed with the seat cut.

TYPICAL FRAMING PLANS

Most house plans will have a roof framing plan. If not, the carpenter will have to draw one. This can be done by placing tracing paper over the floor plan and locating each member. Rafters are shown with a single line. Following are typical plans for gable and hip roofs and some with valleys (Fig. 13-10).

SELECTING RAFTERS

The size of the wood member to form a rafter depends upon the distance it must span, the slope, the distance between rafters, and the weight of the roofing material plus wind and snow loads common for the area. These are live loads. Local building codes often specify these roof loads.

Typical sizes for low-sloped roofs, 3 in 12 or less, are in Fig. 13-11. Sizes for high-sloped roofs, over 3 in 12, are shown in Fig. 13-12. These do not include an interior finished ceiling.

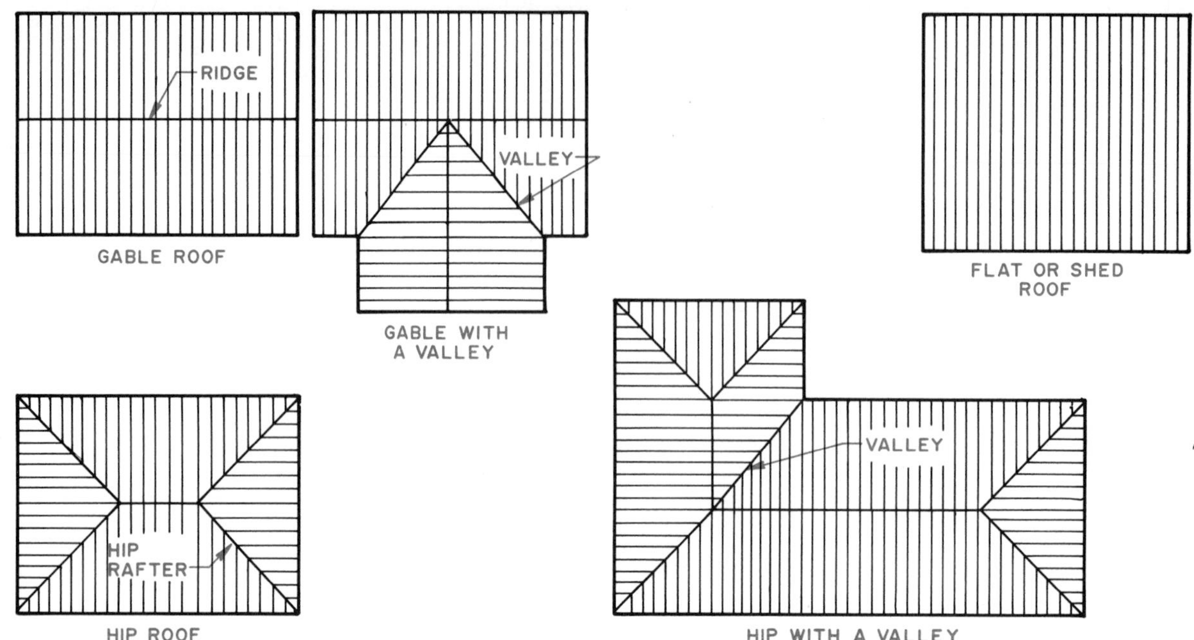

FIGURE 13-10

Typical roof framing plans.

		20-PSF LIVE LOAD			30-PSF LIVE LOAD				40-PSF LIVE LOAD			
SIZE (in.)	SPACING (in. on center)	Dense Sel Str KD and No. 1 Dense KD	Dense Sel Str, Sel Str KD, No. 1 Dense and No. 1 KD	Sel Str	Dense Sel Str KD and No. 1 Dense KD	Dense Sel Str and Sel Str KD	No. 1 Dense	Sel Str	Dense Sel Str KD and No. 1 Dense KD	Dense Sel Str and Sel Str KD	No. 1 Dense	No. 1 KD
2 × 5	12.0	13′-5″	13′-2″	12′-11″	11′-8″	11′-6″	11′-6″	11′-13″	10′-8″	10′-5″	10′-5″	10′-5″
	13.7	12′-10″	12′-7″	12′-4″	11′-2″	11′-0″	11′-0″	10′-9″	10′-2″	10′-0″	10′-0″	10′-0″
	16.0	12′-2″	11′-11″	11′-9″	10′-8″	10′-5″	10′-5″	10′-3″	9′-8″	9′-6″	9′-6″	9′-6″
	19.2	11′-5″	11′-3″	11′-0″	10′-0″	9′-10″	9′-10″	9′-8″	9′-1″	8′-11″	8′-11″	
	24.0	10′-8″	10′-5″[a]	10′-3″	9′-3″	9′-1″	9′-1″	8′-11″	8′-5″	8′-3″		
2 × 6	12.0	16′-4″	16′-1″	15′-9″	14′-4″	14′-1″	14′-1″	13′-9″	13′-0″	12′-9″	12′-9″	12′-9″
	13.7	15′-8″	15′-5″	15′-1″	13′-8″	13′-5″	13′-5″	13′-2″	12′-5″	12′-3″	12′-3″	12′-3″
	16.0	14′-11″	14′-7″	14′-4″	13′-0″	12′-9″	12′-9″	12′-6″	11′-10″	11′-7″	11′-7″	11′-7″
	19.2	14′-0″	13′-9″	13′-6″	12′-3″	12′-0″	12′-0″	11′-9″	11′-1″	10′-11″	10′-11″	
	24.0	13′-0″	12′-9″[a]	12′-6″	11′-4″	11′-2″	11′-1″	10′-11″	10′-4″	10′-2″		
2 × 8	12.0	21′-7″	21′-2″	20′-10″	18′-10″	18′-6″	18′-6″	18′-2″	17′-2″	16′-10″	16′-10″	16′-10″
	13.7	20′-8″	20′-3″	19′-11″	18′-0″	17′-9″	17′-9″	17′-5″	16′-5″	16′-1″	16′-1″	16′-1″
	16.0	19′-7″	19′-3″	18′-11″	17′-2″	16′-10″	16′-10″	16′-6″	15′-7″	15′-3″	15′-3″	15′-3″
	19.2	18′-5″	18′-2″	17′-9″	16′-1″	15′-10″	15′-10″	15′-6″	14′-8″	14′-5″	14′-5″	
	24.0	17′-2″	16′-10″[a]	16′-6″	15′-0″	14′-8″	14′-8″	14′-5″	13′-7″	13′-4″		
2 × 10	12.0	27′-6″	27′-1″	26′-6″	24′-1″	23′-8″	23′-8″	23′-2″	21′-10″	21′-6″	21′-6″	21′-6″
	13.7	26′-4″	25′-10″	25′-5″	23′-0″	22′-7″	22′-7″	22′-2″	20′-11″	20′-6″	20′-6″	20′-6″
	16.0	25′-0″	24′-7″	24′-1″	21′-10″	21′-6″	21′-6″	21′-1″	19′-10″	19′-6″	19′-6″	19′-6″
	19.2	23′-7″	23′-2″	22′-8″	20′-7″	20′-2″	20′-2″	19′-10″	18′-8″	18′-4″	18′-4″	
	24.0	21′-10″	21′-6″[a]	21′-1″	19′-1″	18′-9″	18′-8″	18′-5″	17′-4″	17′-0″		
2 × 12	12.0	33′-6″	32′-11″	32′-3″	29′-3″	28′-9″	28′-9″	28′-2″	26′-7″	26′-1″	26′-1″	26′-1″
	13.7	32′-0″	31′-6″	30′-10″	28′-0″	27′-6″	27′-6″	27′-0″	25′-5″	25′-0″	25′-0″	25′-0″
	16.0	30′-5″	29′-11″	29′-4″	26′-7″	26′-1″	26′-1″	25′-7″	24′-2″	23′-9″	23′-9″	23′-9″
	19.2	28′-8″	28′-2″	27′-7″	25′-0″	24′-7″	24′-7″	24′-1″	22′-9″	22′-4″	22′-4″	
	24.0	26′-7″	26′-1″[a]	25′-7″	23′-3″	22′-10″	22′-8″	22′-5″	21′-1″	20′-9″		

Note: Data for southern pine structural members. 3 in 12 or less; no finished ceiling.

[a]Deflection limitation of *l*/240.

FIGURE 13-11

Rafter size and spans for low-slope roofs. *(Courtesy of Southern Forest Products Association)*

SIZE (in.)	SPACING (in. on center)	20-PSF LIVE LOAD[a]				30-PSF LIVE LOAD[a]				40-PSF LIVE LOAD[a]			
		Dense Sel Str KD	Sel Str KD and Dense Sel Str	No. 1 Dense KD	Sel Str	Dense Sel Str KD	Sel Str KD and Dense Sel Str	No. 1 Dense KD	Sel Str	Dense Sel Str KD	Sel Str KD and Dense Sel Str	No. 1 Dense KD	Sel Str
2 × 4	12.0	11'-6"	11'-3"	11'-6"	11'-1"	10'-0"	9'-10"	10'-0"	9'-8"	9'-1"	8'-11"	9'-1"	8'-9"
	13.7	11'-0"	10'-9"	11'-0"	10'-7"	9'-7"	9'-5"	9'-7"	9'-3"	8'-8"	8'-7"	8'-8"	8'-5"
	16.0	10'-5"	10'-3"	10'-5"	10'-0"	9'-1"	8'-11"	9'-1"	8'-9"	8'-3"	8'-1"	8'-3"	8'-0"
	19.2	9'-10"	9'-8"	9'-10"	9'-5"	8'-7"	8'-5"	8'-7"	8'-3"	7'-9"	7'-8"	7'-9"	7'-6"
	24.0	9'-1"	8'-11"	9'-1"	8'-9"	7'-11"	7'-10"	7'-11"	7'-8"	7'-3"	7'-1"	7'-3"	7'-0"
2 × 5	12.0	14'-9"	14'-6"	14'-9"	14'-3"	12'-11"	12'-8"	12'-11"	12'-5"	11'-8"	11'-6"	11'-8"	11'-3"
	13.7	14'-1"	13'-10"	14'-1"	13'-7"	12'-4"	12'-1"	12'-4"	11'-10"	11'-2"	11'-0"	11'-2"	10'-9"
	16.0	13'-5"	13'-2"	13'-5"	12'-11"	11'-8"	11'-6"	11'-8"	11'-3"	10'-8"	10'-5"	10'-8"	10'-3"
	19.2	12'-7"	12'-5"	12'-7"	12'-2"	11'-0"	10'-10"		10'-7"	10'-0"	9'-10"[b]		
	24.0	11'-8"	11'-6"			10'-3"	10'-1"[b]			9'-3"	9'-1"		
2 × 6	12.0	18'-0"	17'-8"	18'-0"	17'-4"	15'-9"	15'-6"	15'-9"	15'-2"	14'-4"	14'-1"	14'-4"	13'-9"
	13.7	17'-3"	16'-11"	17'-3"	16'-7"	15'-1"	14'-9"	15'-1"	14'-6"	13'-8"	13'-5"	13'-8"	13'-2"
	16.0	16'-4"	16'-1"	16'-4"	15'-9"	14'-4"	14'-1"	14'-4"	13'-9"	13'-0"	12'-9"	13'-0"	12'-6"
	19.2	15'-5"	15'-2"	15'-5"	14'-10"	13'-6"	13'-3"		13'-0"	12'-3"	12'-0"[b]		
	24.0	14'-4"	14'-1"			12'-6"	12'-3"[b]			11'-4"	11'-2"[b]		
2 × 8	12.0	23'-9"	23'-4"	23'-9"	22'-11"	20'-9"	20'-5"	20'-9"	20'-0"	18'-10"	18'-6"	18'-10"	18'-2"
	13.7	22'-9"	22'-4"	22'-9"	21'-11"	19'-10"	19'-6"	19'-10"	19'-2"	18'-0"	17'-9"	18'-0"	17'-5"
	16.0	21'-7"	21'-2"	21'-7"	20'-10"	18'-10"	18'-6"	18'-10"	18'-2"	17'-2"	16'-10"	17'-2"	16'-6"
	19.2	20'-4"	19'-11"	20'-4"	19'-7"	17'-9"	17'-5"		17'-1"	16'-1"	15'-10"[b]		
	24.0	18'-10"	18'-6"			16'-6"	16'-2"[b]			15'-0"	14'-8"[b]		
2 × 10	12.0	30'-4"	29'-9"	30'-4"	29'-2"	26'-6"	26'-0"	26'-6"	25'-6"	24'-1"	23'-8"	24'-1"	23'-2"
	13.7	29'-0"	28'-6"	29'-0"	27'-11"	25'-4"	24'-11"	25'-4"	24'-5"	23'-0"	22'-7"	23'-0"	22'-2"
	16.0	27'-6"	27'-1"	27'-6"	26'-6"	24'-1"	23'-8"	24'-1"	23'-2"	21'-10"	21'-6"	21'-10"	21'-1"
	19.2	25'-11"	25'-5"	25'-11"	25'-0"	22'-8"	22'-3"		21'-10"	20'-7"	20'-2"[b]		
	24.0	24'-1"	23'-8"			21'-0"	20'-8"[b]			19'-1"	18'-9"[b]		

Note: Data for southern pine structural members. Over 3 in 12; no finished ceiling.

[a]Plus 7 psf dead load.

[b]Deflection limitation of *l*/180.

FIGURE 13-12

Rafter size and spans for high-slope roofs. *(Courtesy of Southern Forest Products Association)*

LAYOUT COMMON RAFTERS

There are three commonly used methods for figuring the length of the rafter. These include step-off, rafter tables, and graphic layout. The carpenter usually finds the step-off method the most useful. The rafter tables can be used to check the results of the step-off method.

The terms used to describe the parts of a common rafter for layout purposes are given in Fig. 13-13. The line length of the rafter is from

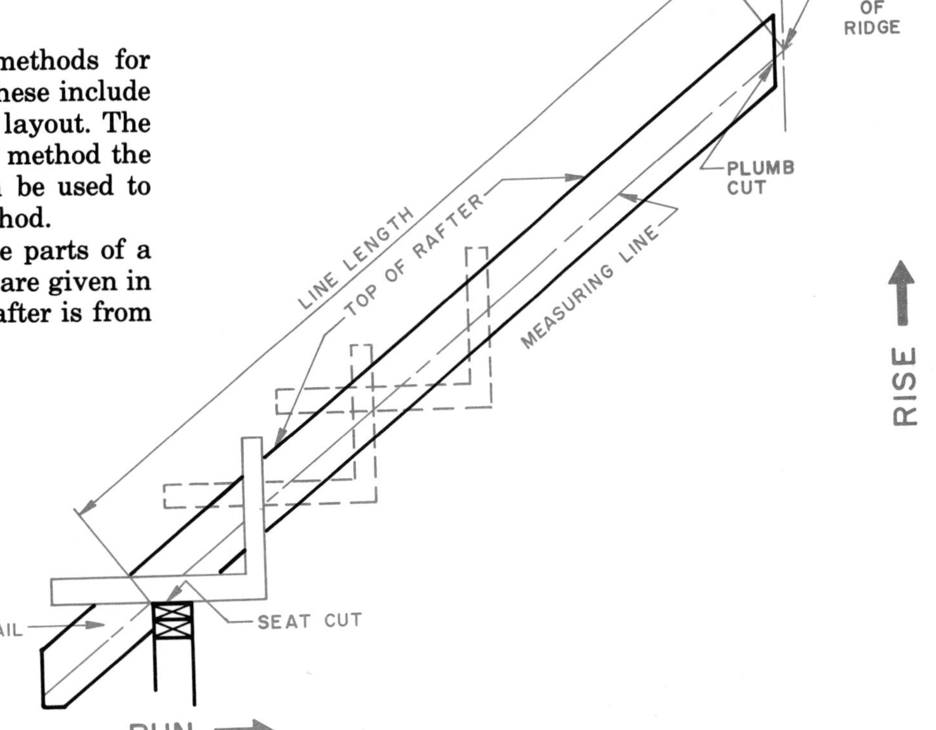

FIGURE 13-13

Terms used when laying out a rafter.

where the measuring line crosses the center of the ridge to the outer corner of the double plate. When the framing square is used, the body represents the run. The tongue of the square is the rise.

Finding the Rough Length of a Common Rafter The approximate length of a common rafter is found using a framing square (Fig. 13-14). Let each inch on the tongue represent 1 ft of rafter rise. Run is measured on the body. Then measure the diagonal from the ends of the rise and run. Each inch represents a foot of rafter length. For example, a roof with a rise of 6 ft 0 in. and a run of 12 ft 0 in. will have an approximate rafter length of 13 ft 6 in. This does not include overhang. This is the distance from the center of the ridge to the outside of the double plate.

This measurement is accurate enough to select the stock from which the rafter is cut. If the rafter above, which is 13 ft 6 in. long, has a 1 ft 6 in. overhang, the total length will be 15 ft 0 in. The carpenter knows that 16 ft 0 in. stock must be purchased.

Locate the Rafter Measuring Line The measuring line is an imaginary line running parallel to the top edge of the rafter and through the corner of the bird's-mouth.

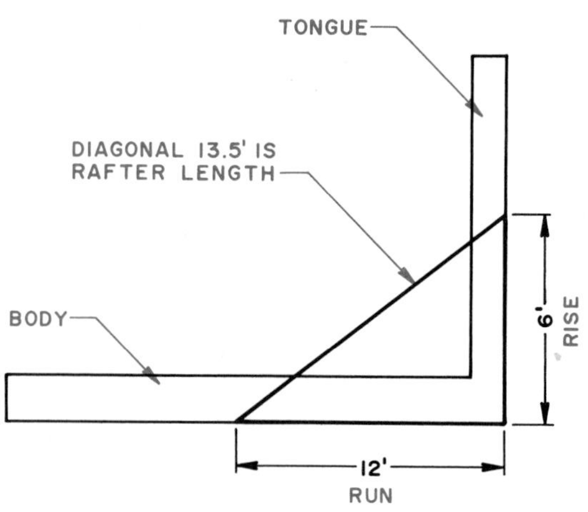

FIGURE 13-14

A framing square can be used to find the approximate length of a rafter.

To locate the measuring line (Fig. 13-15):

1. Mark the top edge of the rafter and face it toward you.
2. Place the framing square on the right end of the rafter with the blade to the right. Set it with the run (always 12 in.) on the blade and the rise on the tongue touching the top of the rafter. Be certain to leave enough material for the rafter tail.
3. Measure 3½ in. (the width of the wall stud)

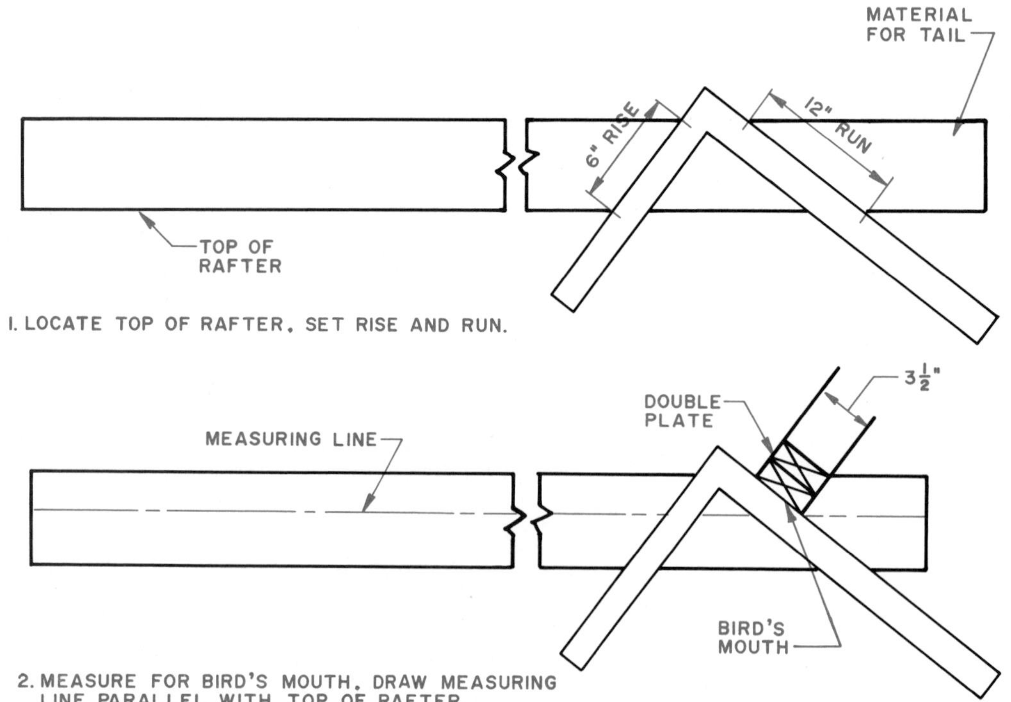

FIGURE 13-15

To begin a rafter layout, locate the top of the rafter, set the rise and run, and mark the bird's-mouth and measuring line.

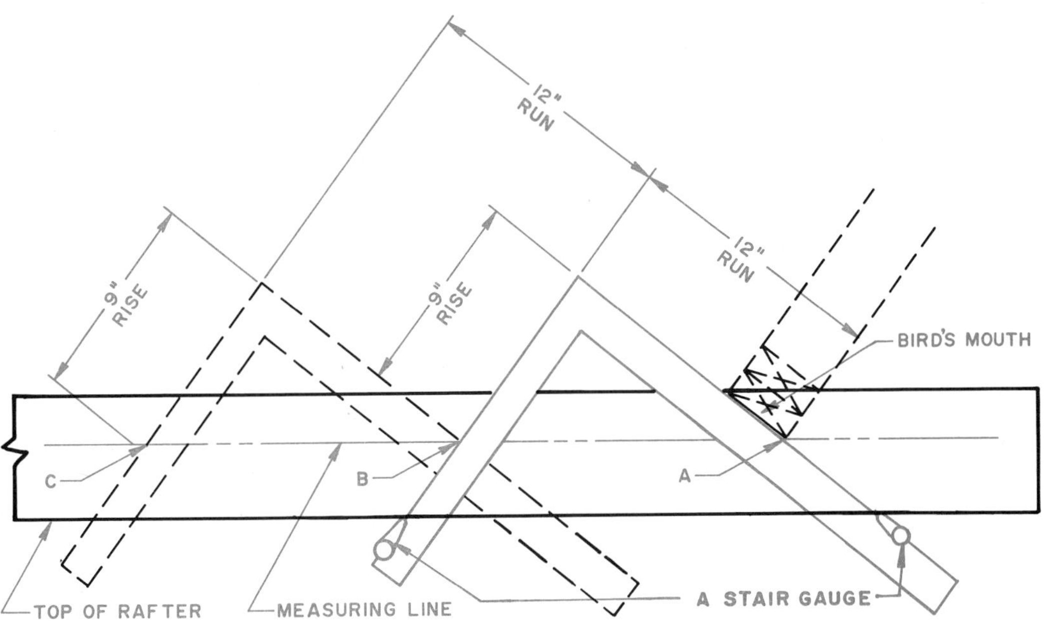

FIGURE 13-16

The step-off method for laying out a rafter.

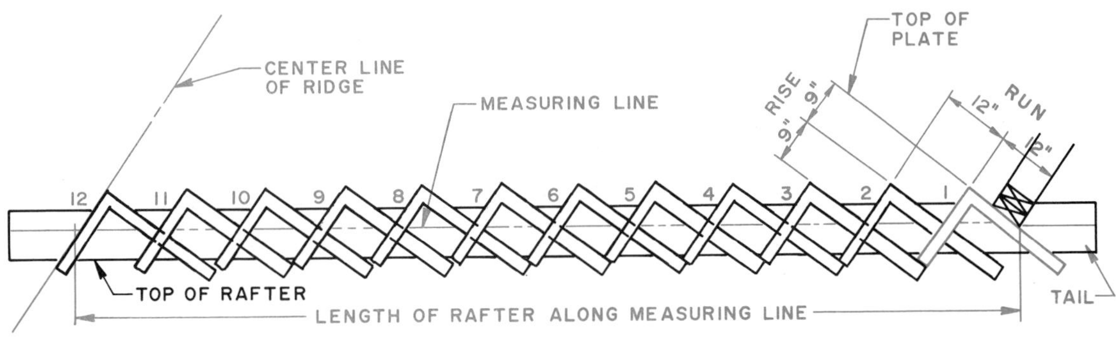

FIGURE 13-17

Step off one step for each foot of run.

from the bottom of the rafter. This locates the bird's-mouth. The inside corner is the location of the measuring line.

4. Draw it through this point parallel with the top edge of the rafter.

Step-off Method for Rafter Layout The following description is for rafters for a building whose run is an even number of feet. Assume that a building has a span of 24 ft with a rise of 6 ft. The run is therefore 12 ft. The slope of the rafter is 6 in. rise to 12 in. run.

To lay out the rafter (Fig. 13-16):

1. Locate the measuring line and the bird's-mouth, and leave stock for the rafter tail.

2. Place the framing square as shown in Fig. 13-16. The run and rise in inches are placed on the measuring line. Mark the measuring line at points A and B.

3. Slide the square to the left until the 12-in. run mark is over point B. Mark point C on the measuring line. Stair gauges are clamped to the framing square. They slide along the edge of the rafter and keep the measurements accurate.

4. Mark one step for each foot of run for the rafter. In this example the run is 12 ft (Fig. 13-17). At the twelfth position draw a line across the rafter. This is the centerline of the ridge.

5. Mark the allowance for the *ridge*. The ridge is usually made from 2-in.-thick material (1½ in. actually). Measure to the right of the centerline of the ridge one-half the thickness of the ridge (Fig. 13-18). Measure this perpendicular to the centerline of the ridge. This new line is the plumb cut forming the top of the rafter.

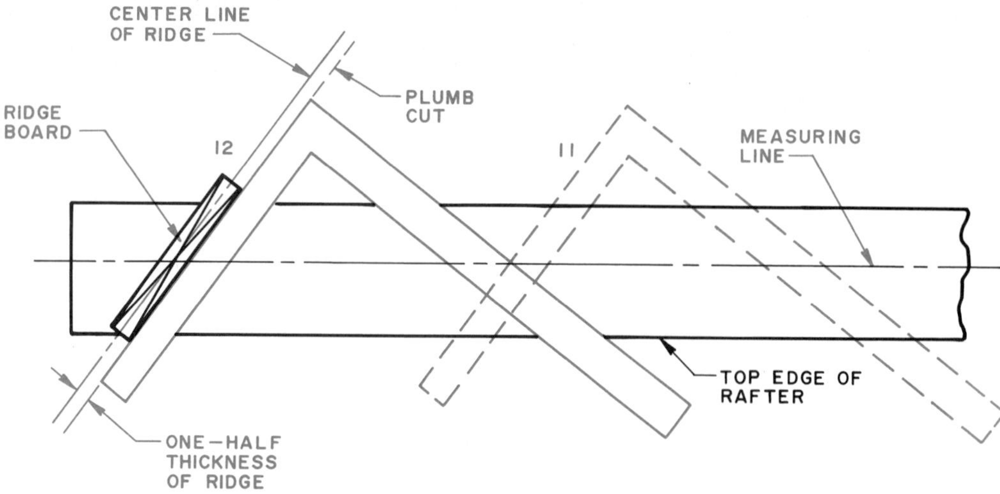

FIGURE 13-18

After locating the centerline of the ridge, mark the plumb cut.

6. Mark the bird's-mouth. The seat cut of the bird's-mouth is located by the first step. The plumb cut is located by drawing a line perpendicular to the top line, starting where it meets the measuring line. It got the name "plumb cut" because it is plumb (vertical) (Fig. 13-19).

Check the typical wall section in the drawings to see if the bird's-mouth is to be nailed directly to the plate or over the sheathing (Fig. 13-20). If it is over the sheathing, allow for this extra thickness.

7. Mark the tail of the rafter. Although there are a variety of rafter tails, the following is typical. Measure the amount of overhang perpendicular to the bird's-mouth. This distance is shown on the working drawings. Allow for the thickness of the sheathing. Draw a plumb cut, called the tail cut, along the tongue of the square (Fig. 13-21). Along this edge measure the size desired for the facia board. Mark the bottom cut of the tail.

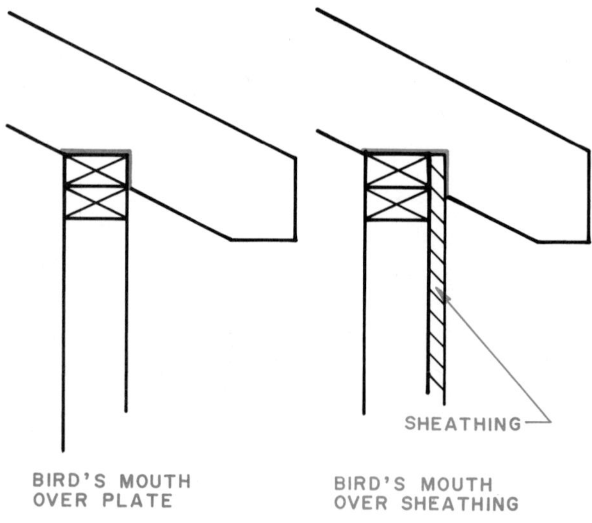

FIGURE 13-20

The bird's-mouth may be placed over the sheathing or over the double plate.

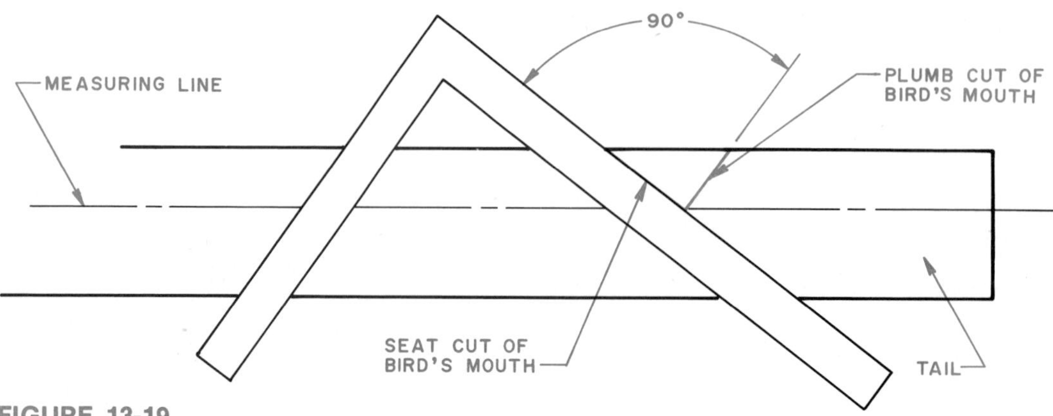

FIGURE 13-19

Mark the bird's-mouth at the first layout step.

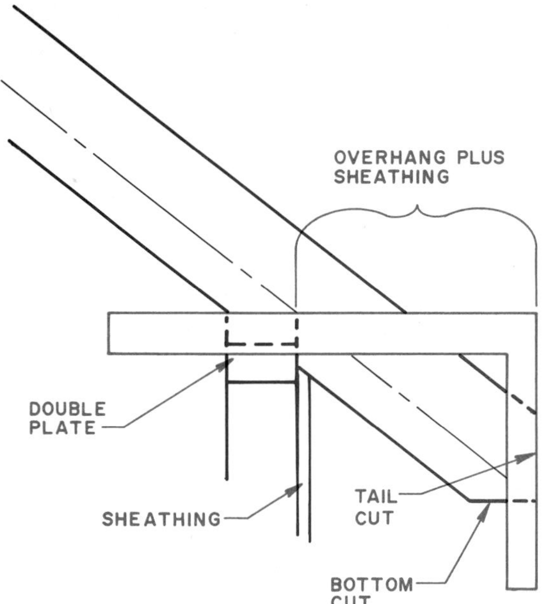

FIGURE 13-21

Mark the rafter tall cut measuring from the bird's-mouth plumb cut.

Following is how to step-off a common rafter when the span is an odd number of feet. Assume that the span is 25 ft. The procedure is the same as just described for the even-foot span except that an additional one-half step, 6 in. of run, is added (Fig. 13-22). The size of this will vary depending upon the proportion the extra step is to the run.

Using Rafter Tables Steel framing squares have rafter tables stamped into them. These can be used to find rafter length. Although these vary slightly from one company to another, the following example is typical. Refer to Fig. 13-23. Assume a rise of 6 ft.

The first row of numbers is for finding the length of common rafters. Under the 6-in. mark

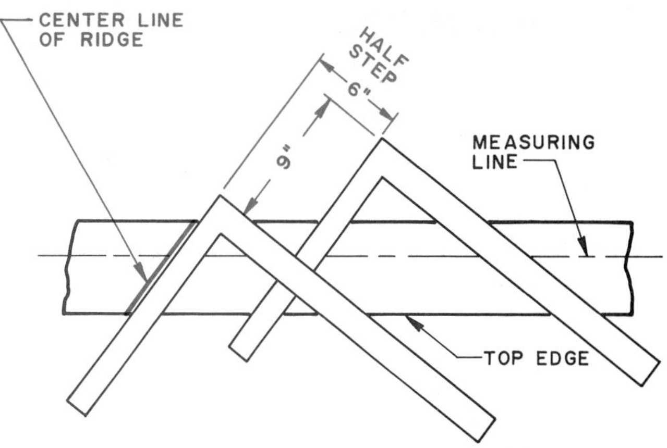

FIGURE 13-22

How to step off a common rafter when the span is an odd number of feet.

(which represents the 6-ft rise) the figure "13.42" is given. This is the line length of the rafter in inches for each foot of run. If the run were 12 ft, the rafter length would be 12 × 13.42 in. or 161.04 in. or 13.42 ft or 13 ft 5 in.

Remember, this length is from the center of the ridge to the outside of the wall plate. Overhang must be added and half the thickness of the ridge must be subtracted.

DUPLICATING AND CUTTING RAFTERS

Before cutting the rafters, check the rafter to be used as a pattern for fit. A rafter can be checked by setting it up on the floor of the building. The height of the rise can be measured and the fit to the ridge and plate can be checked. If all is in order, the other rafters can be cut.

You will need one more pair of rafters than spaces because a set is used on each end wall. For example, a 48-ft house with rafters spaced 24 in. apart needs 25 pairs. In addition, two additional pairs are needed as fascia rafters on

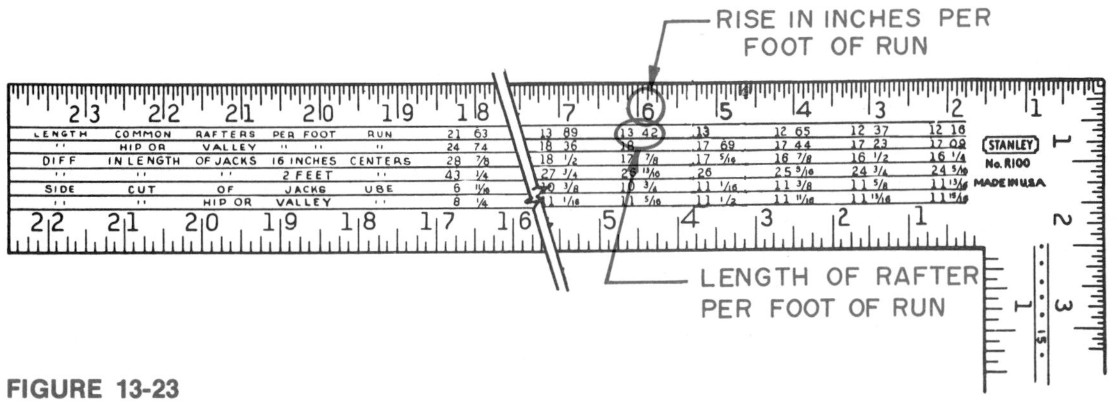

FIGURE 13-23

The rafter scales on a framing square are used to find the length of rafters.

the overhang. These must be made half the width of the ridge longer because they cover the ridge. They do not have a seat cut.

The first rafter laid out becomes the pattern. Cut it to final shape. Place it on the stock for each rafter and mark the cuts. Always place the crowned side of the stock so that it becomes the top of the rafter. A jig can be built to help hold the stock while it is being marked. The pattern can have two small wood pieces nailed to it to serve as guides (Fig. 13-24). A wood jig can be built to help line up the members as they are marked (Fig. 13-25).

The cuts forming the rafter can be made with a handsaw, portable electric circular saw, or a radial arm saw. The radial arm saw is the most accurate. (Review Chapter 4.)

ERECTING COMMON RAFTERS

The erection of a roof is best accomplished if a carpenter has at least two helpers. If one or two persons attempt it, special temporary bracing is necessary.

1. Mark the rafter locations on the plate. These will normally be next to a ceiling joist. The rafter is nailed to the joist. The first rafter is flush with the outside of the end wall plate. The first interior rafter is 16 or 24 in. on center (as specified on the drawing) from the outside of the plate. Space all other rafters the same specified spacing. Let the space between the end rafters at the other end of the house come out as it will.

Framing for an overhang at the gable end is discussed later.

2. Lay the boards to form the ridge next to the plate. Using a square, mark the rafter locations on the ridge.

Usually, 2 × 8-in. stock is used for 2 × 6-in. rafters. The stock should be of good quality and straight.

The pieces forming the ridge should either butt at the end of a rafter or be spliced. A plywood splice can be used (Fig. 13-26). Do not nail the splice on until the *ridge boards* are erected. It is easier to erect it in pieces than try to lift the entire thing at one time.

If there is to be an overhang at the gable end, allow the ridge board to extend beyond the end wall an amount equal to the overhang.

3. Move the ridge board to the plate on the other side of the house. Using a square, mark the rafter locations on this plate.

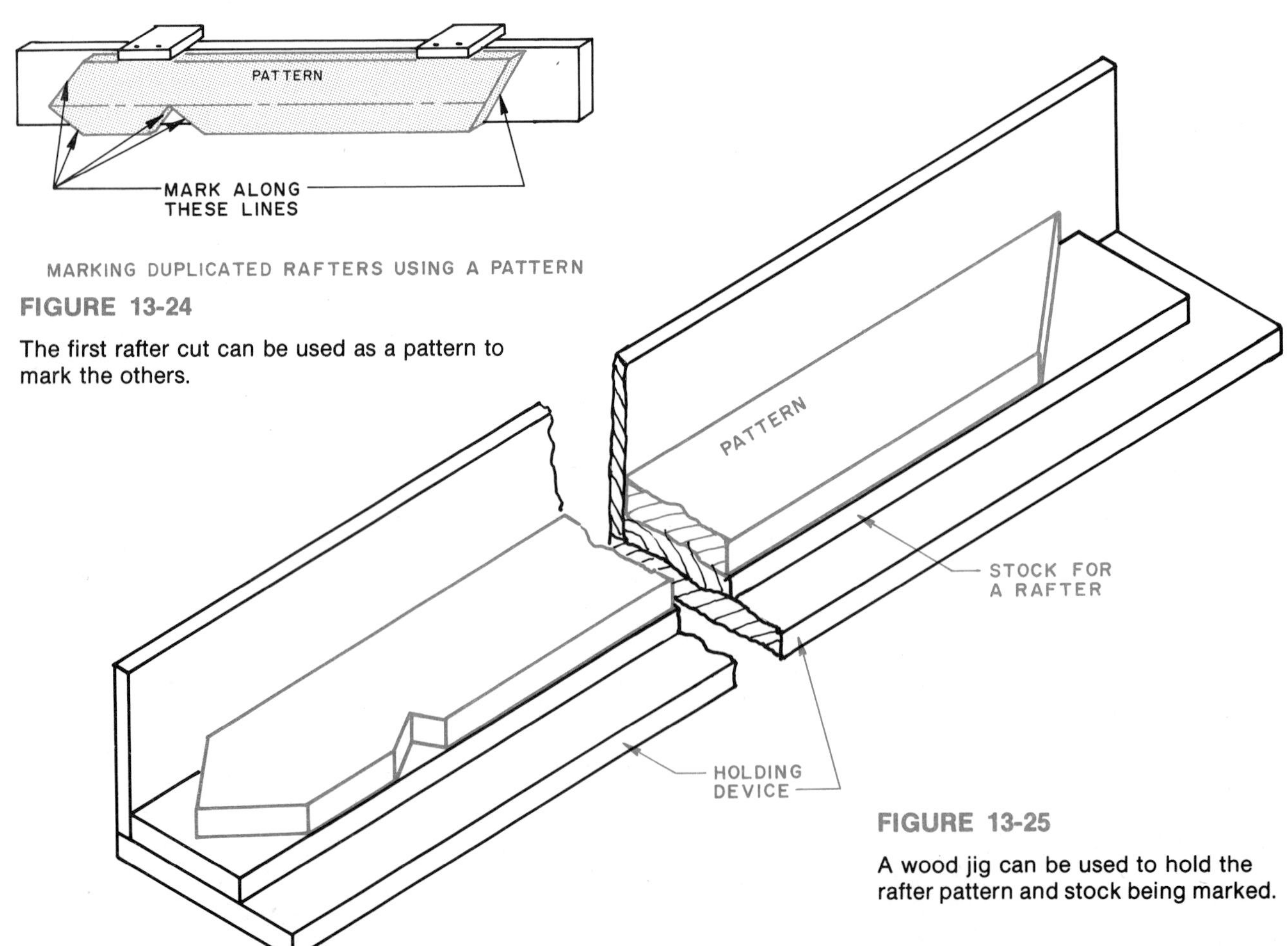

FIGURE 13-24

The first rafter cut can be used as a pattern to mark the others.

FIGURE 13-25

A wood jig can be used to hold the rafter pattern and stock being marked.

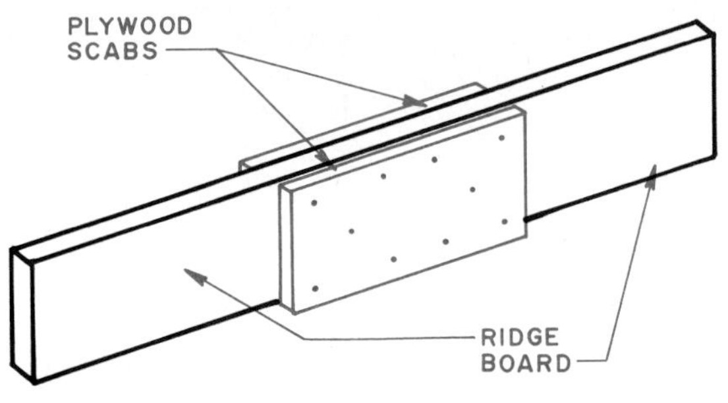

FIGURE 13-26

The ridge board can be spliced with plywood scabs.

4. Now erect a section of the ridge. Use 2 × 4-in. stock to hold it in place. The prop can be supported with a plywood gusset and an angle brace (Fig. 13-27). Check the prop with a level to be certain that it is vertical. Check the ridge to be certain that it is level and in the center of the house.

Plywood sheets can be laid on the ceiling joists to form a temporary floor. Nail them to the joists to prevent them from moving.

Erect the ridge board starting at one end of the building.

5. Nail the end rafter to the ridge board. Face-nail it through the ridge with two 16d nails. Then toenail it to the plate. Use two 8d nails on each side.

Next erect a rafter opposite this on the other side of the house (Fig. 13-28). Toenail it to the ridge with two 8d nails on each side and one 10d on top (Fig. 13-29). Toenail it to the plate.

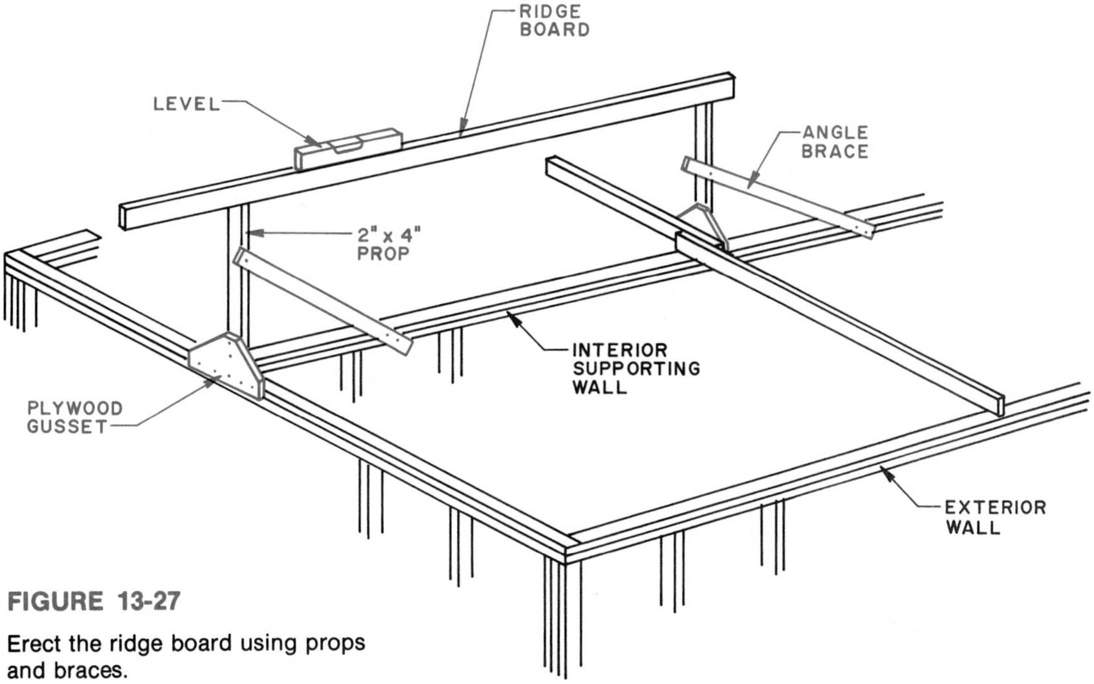

FIGURE 13-27

Erect the ridge board using props and braces.

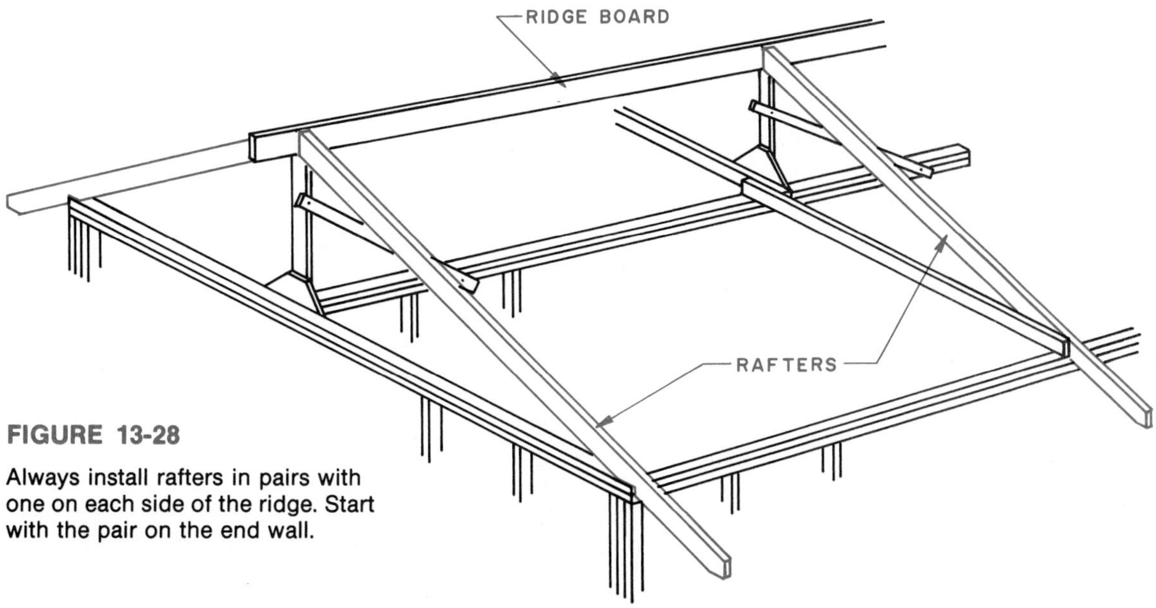

FIGURE 13-28

Always install rafters in pairs with one on each side of the ridge. Start with the pair on the end wall.

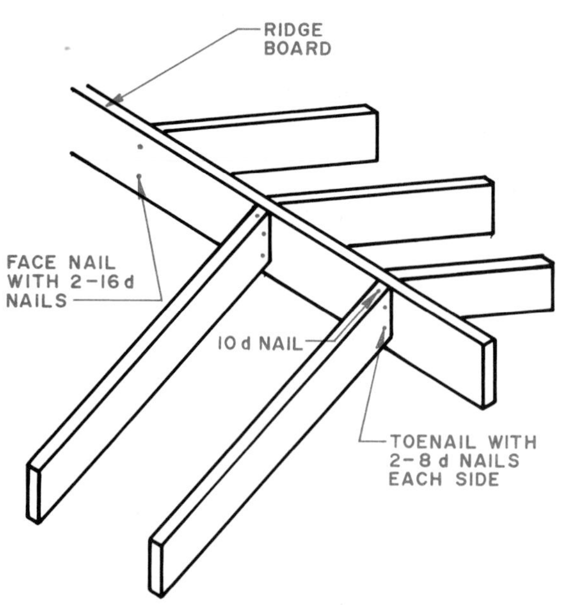

FIGURE 13-29

How to nail rafters to the ridge board.

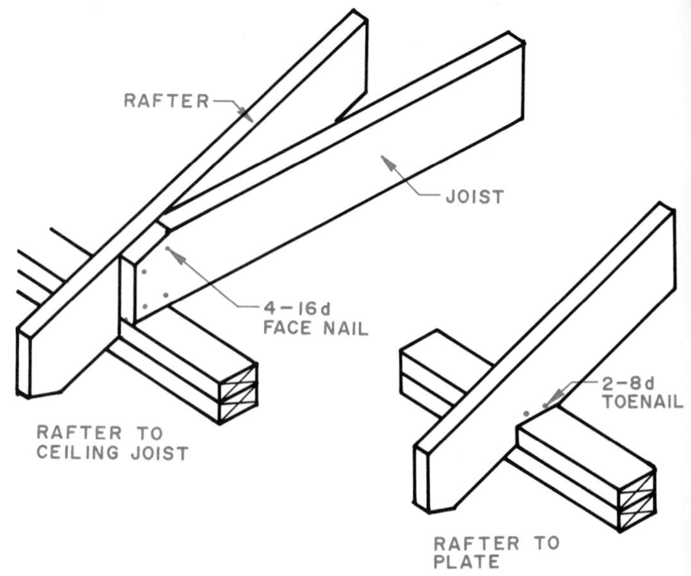

FIGURE 13-31

Face-nail the rafters to the ceiling joist or to the plate if they do not join a joist.

Metal anchors can also be used to join rafters to the plate (Fig. 13-30).

Erect a pair of rafters near the center of the section of the ridge and a pair at the end. Check to be certain that the ridge is still straight. Then fill in the other rafters. Do not put too many on one side without balancing them with some from the other side. This helps keep the ridge from being pushed out of line.

Face-nail the rafters to the ceiling joists. Use four 16d nails at each rafter (Fig. 13-31). If the area experiences high winds, metal anchors are used to provide resistance against uplifting of the roof.

If rafters are spaced 24 in. on center and ceiling joists are 16 in. on center, space as shown in Fig. 13-32. Notice that a ceiling joist is nailed to every other rafter.

It is a big help to have the rafters near where they are needed. They are usually leaned against the exterior wall (Fig. 13-33).

6. Install *collar beams* on every other pair of rafters. These are usually 1 × 6-in. stock. They are nailed about one-third of the rise below the ridge. If the rise is 6 ft, they would be about 2 ft below the ridge. The temporary bracing can be removed.

7. In areas subject to high winds and snow loads, rafters are supported with diagonal braces. These are usually 2 × 4-in. stock. They are di-

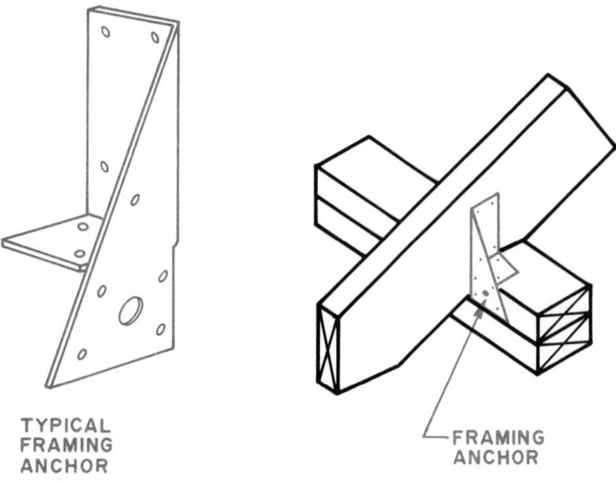

FIGURE 13-30

Metal anchors can be used to join rafters to the plate.

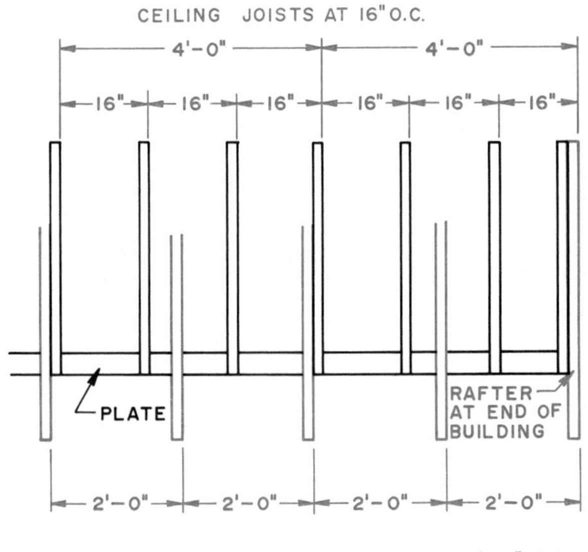

FIGURE 13-32

How to place rafters and ceiling joists when rafters are 24 in. on center and joists are 16 in. on center.

rected so that the load is applied to a load-bearing partition (Fig. 13-34). These diagonals act as trusses by reducing the span of the rafters.

8. Now frame the gable end. If a *louver* is to be installed, cut and nail horizontal members to frame it. The vertical framing is usually 2 × 4-in. stock notched on the top to fit around the end rafter (Fig. 13-35). Usual spacing is 16 in. on center. A recommended nailing pattern is shown in Fig. 13-36.

9. Cut the fascia board to size and length. The top edge is beveled to match the roof slope. It is 2-in. stock. Nail it to the rafter ends. Sections should butt on the end of a joist. Typical *open* and *boxed soffits* are shown in Fig. 13-37.

The roof is now ready for sheathing.

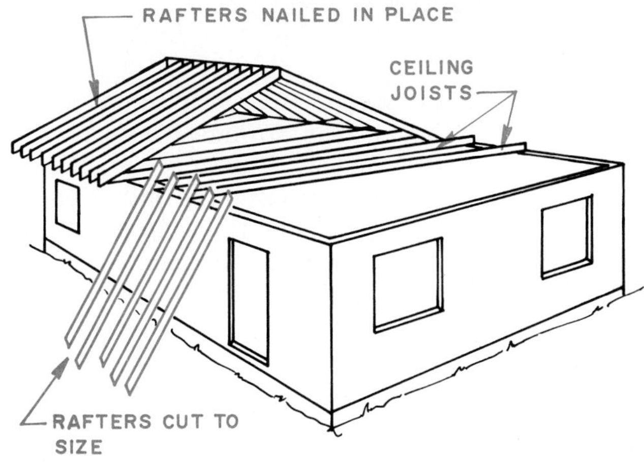

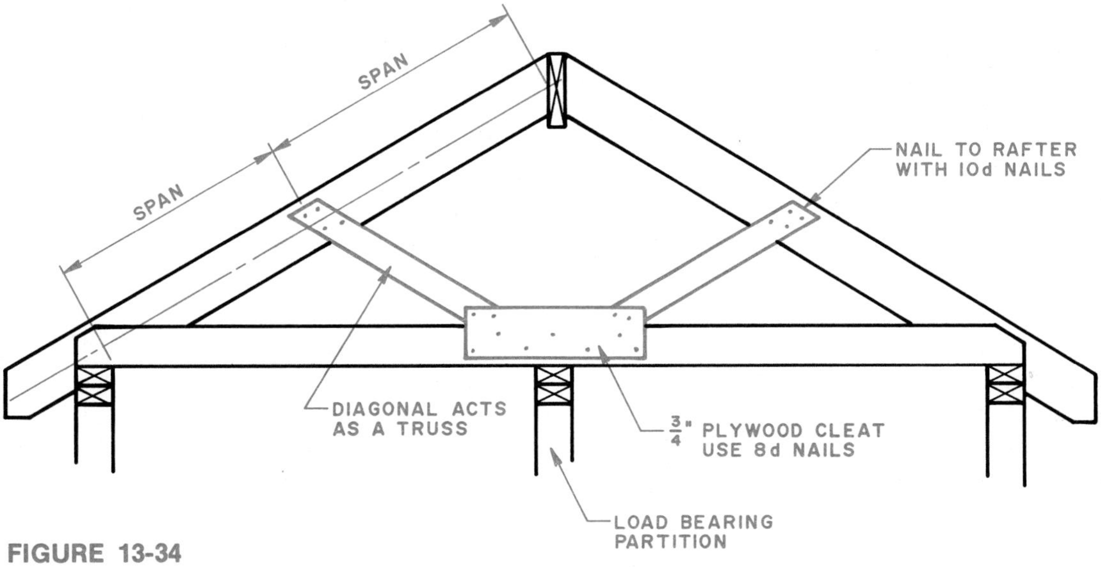

FIGURE 13-33

As the rafters are cut, lean them against the wall so they are available when needed.

FIGURE 13-34

Two × 4 in. braces are used to provide extra support.

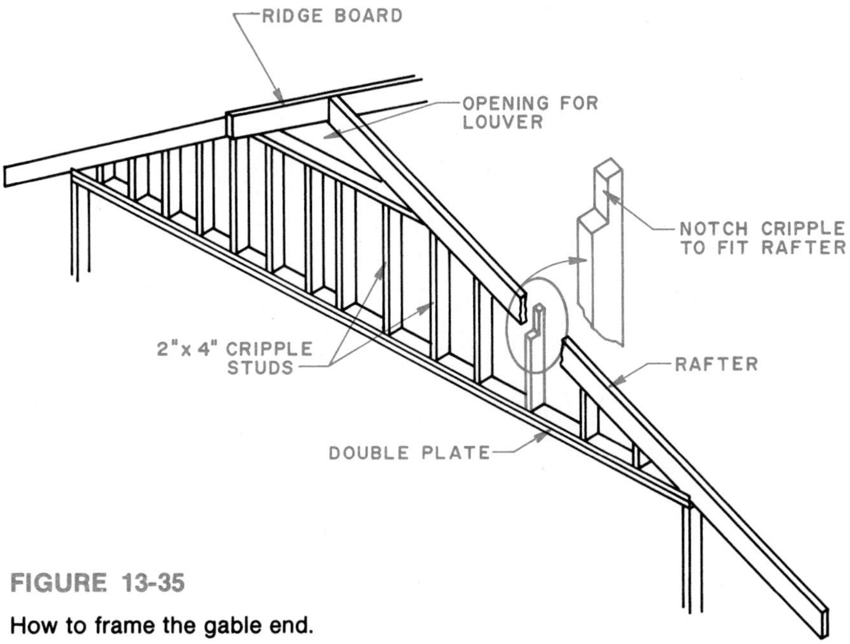

FIGURE 13-35

How to frame the gable end.

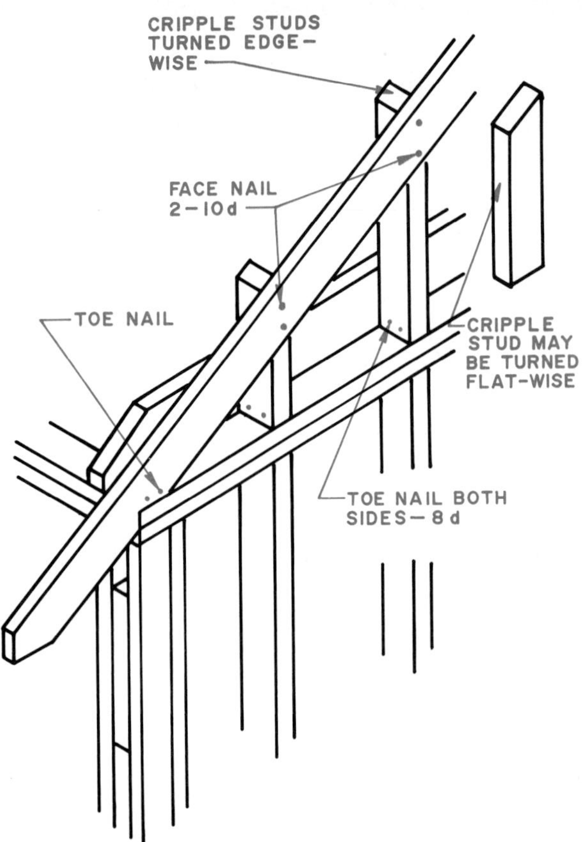

FIGURE 13-36 *(above)*

Proper nailing pattern for framing the gable end.

FRAMING AN OVERHANG AT THE GABLE END

There are several ways an overhang at the *gable end* is framed. The strongest way is to omit the end rafter. Instead, use a 2 × 4-in. plate nailed to the framing forming the gable end (Fig. 13-38). This is set below the ridge so that the *lookouts* can be installed. The lookouts are usually the same size as the rafters. Another way is to notch the end rafter. Usually, 2 × 4-in. members form the lookouts (Fig. 13-39). Small overhangs, such as 12 in. or less, are often framed as shown in Fig. 13-40. The roof sheathing adds greatly to the strength of the overhang.

HIP AND VALLEY RAFTERS

A hip roof uses common, hip, and hip jack rafters. If a valley exists, a valley and valley jack rafters are needed. A gable roof uses common rafters and valley and valley jacks if gable roofs intersect.

FIGURE 13-37 *(below)*

A soffit may be open or boxed.

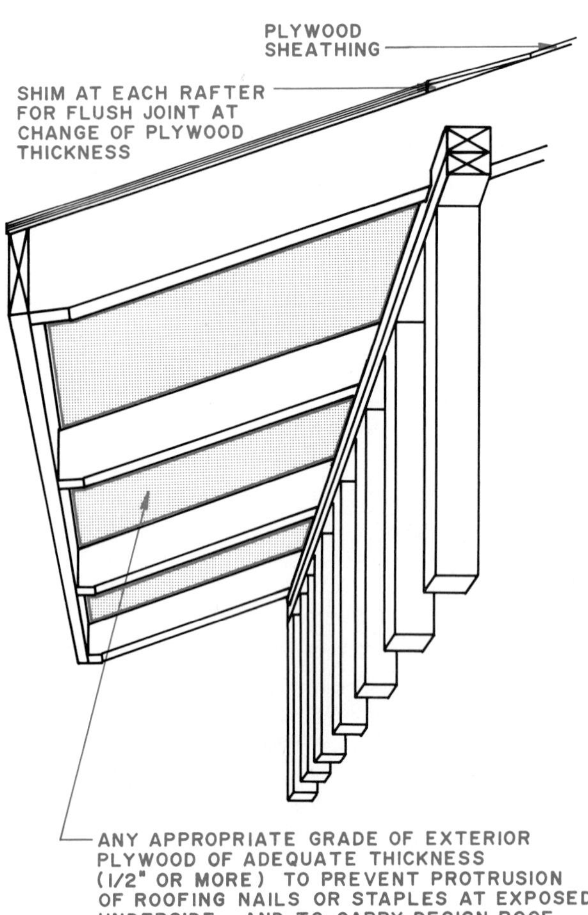

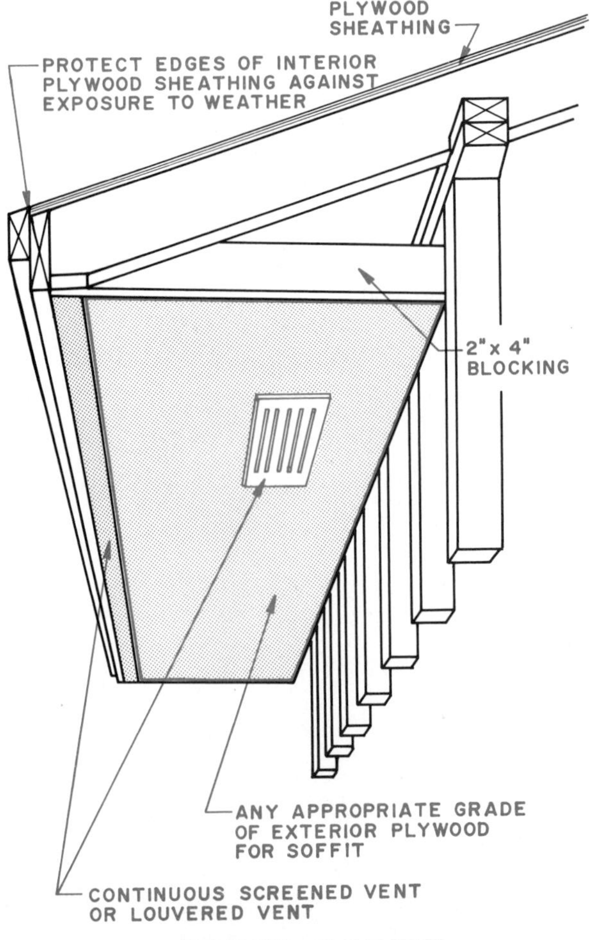

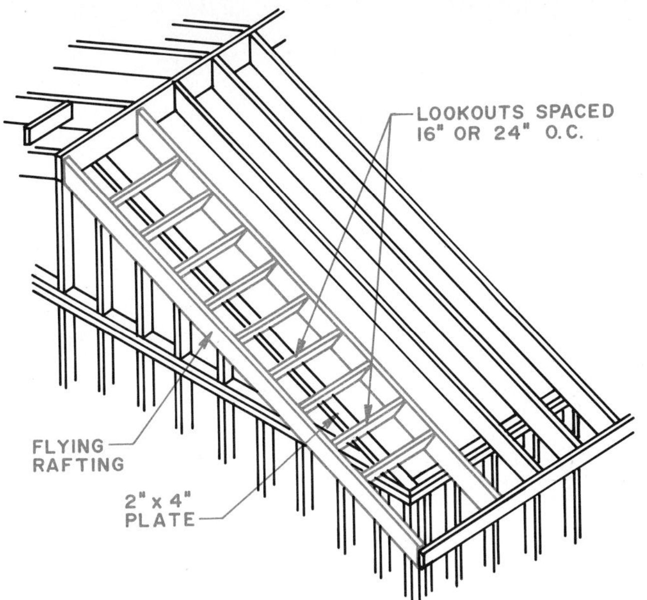

FIGURE 13-38

Lookouts supported by a 2 × 4-in. plate are used to frame an overhang at the gable end.

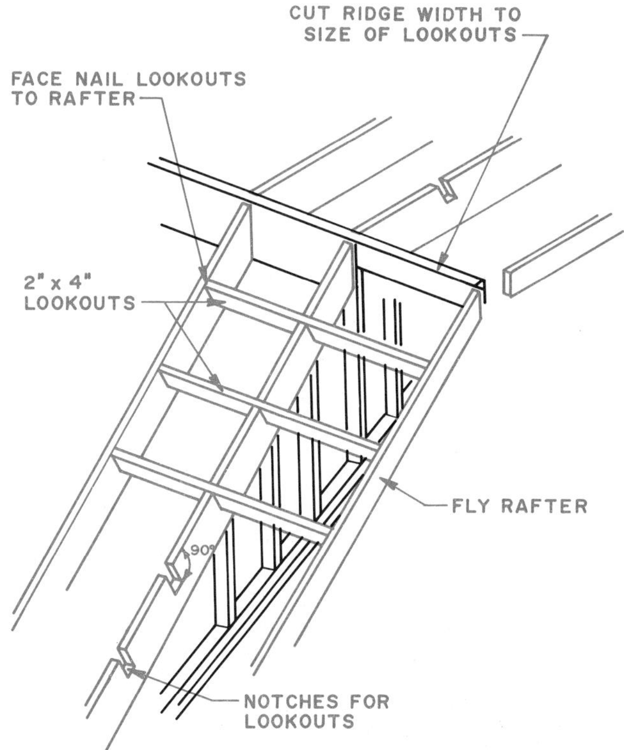

FIGURE 13-39

The end rafter can be notched to support lookouts used for an overhang on a gable end.

Hip Rafters

Most commonly, all surfaces of hip roofs have the same pitch. The following discussion relates to this type of roof.

To build a hip roof, first lay out, cut, and install the ridge and all common rafters. The common rafters are cut and installed as discussed earlier in this chapter (Fig. 13-41).

The center of the first common rafter is located from the end of the building a distance equal to the run of the common rafter. The length of the ridge is therefore equal to the length of the building minus twice the run of the common rafters plus the thickness of the rafter material (Fig. 13-42).

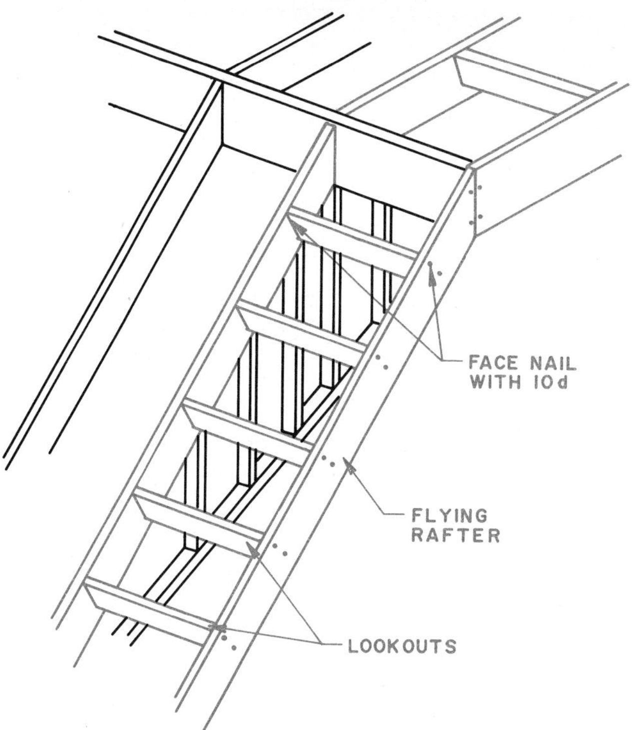

FIGURE 13-40

A small overhang can be nailed directly to the end rafter.

The *hip rafter* runs from the corner of the building to the end of the ridge. It forms a 45° angle with the adjoining common rafter. It is the diagonal of a square. The unit of run for common rafters is 12 in. The hip rafter unit of run is the diagonal of this square. This is 16.97 in. Therefore, the unit of run used for a hip rafter is rounded to 17 in. (Fig. 13-43).

The hip rafter is laid out in the same manner as described for the common rafter except that the unit of run used is 17 in.

The rafter length is found on the framing square. The second line is marked "length of hip or valley per foot run." The figures under the inch marks are the length of the hip rafter per foot run of the common rafter. The inch marks are the rise of the roof.

To find the length of the hip rafter, multiply the figure given in the table by the number of feet of run in the common rafter (Fig. 13-44). Assume that a building has a run of 10 ft and a unit rise of 6 in. Under the 6-in. mark on the scale is the figure "18." Multiply this by 10, the run of the building, to get the hip rafter length. This will be 18 × 10 = 180 in. Divide this by 12 to get feet. The length of the hip rafter would be 15 ft 0 in.

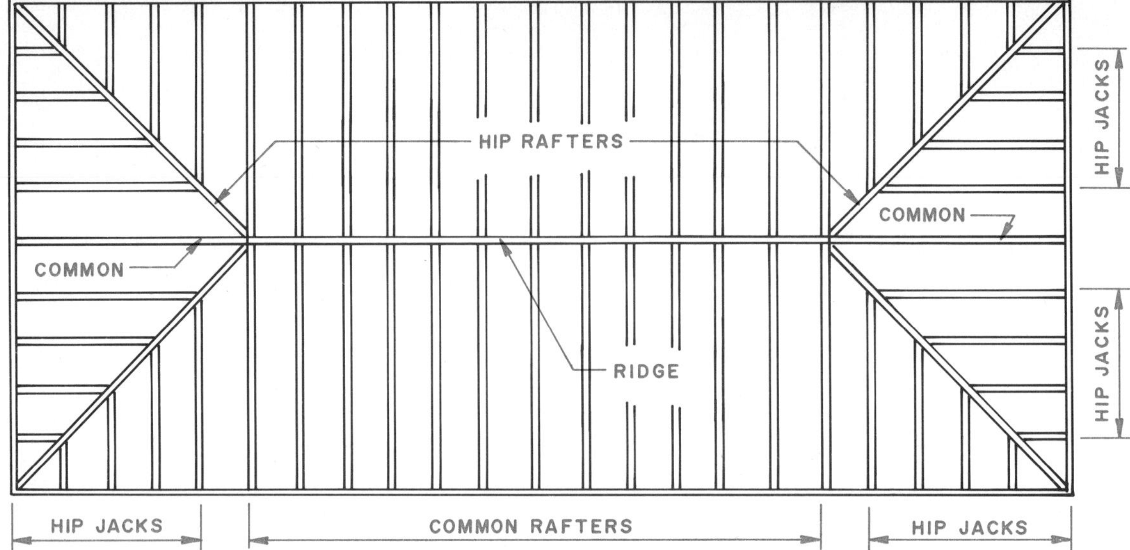

FIGURE 13-41

Rafters used to frame a hip roof.

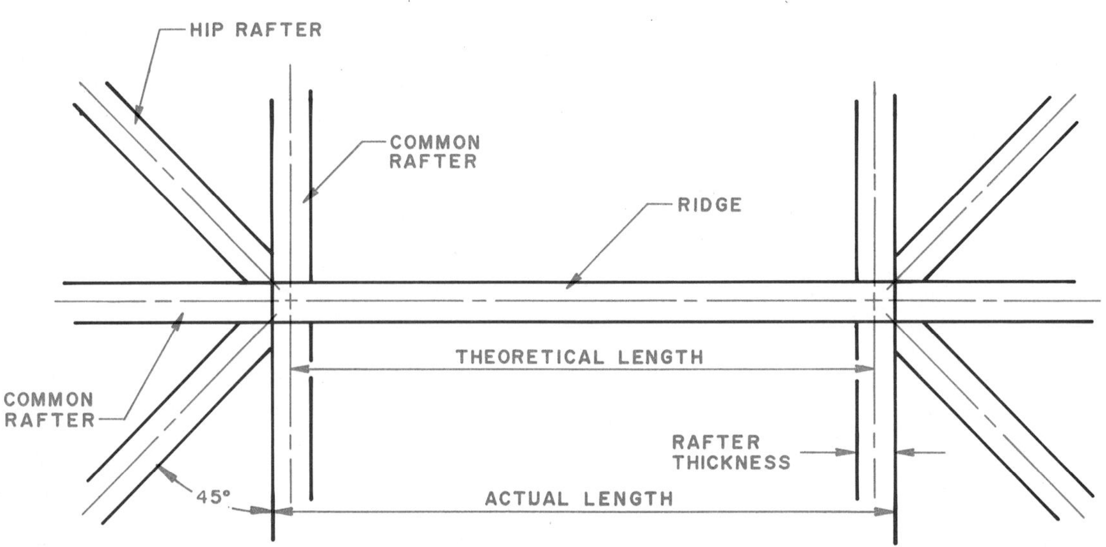

FIGURE 13-42 *(above)*

The relationship between the ridge, hip rafters, and adjoining common rafters.

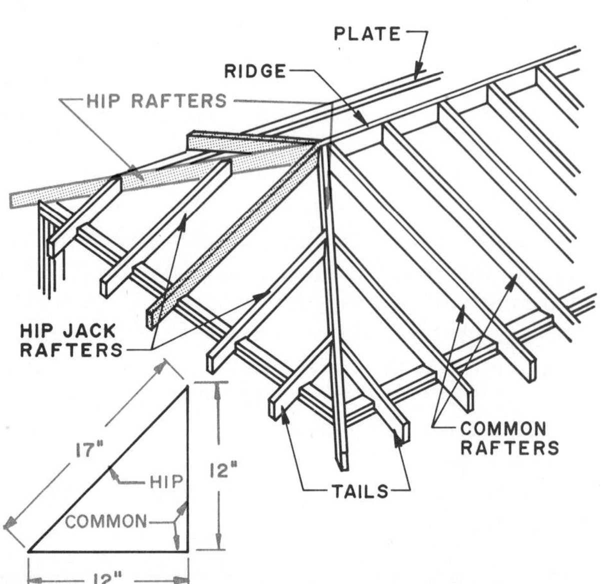

FIGURE 13-43 *(below)*

How to locate the rise and run of a hip rafter.

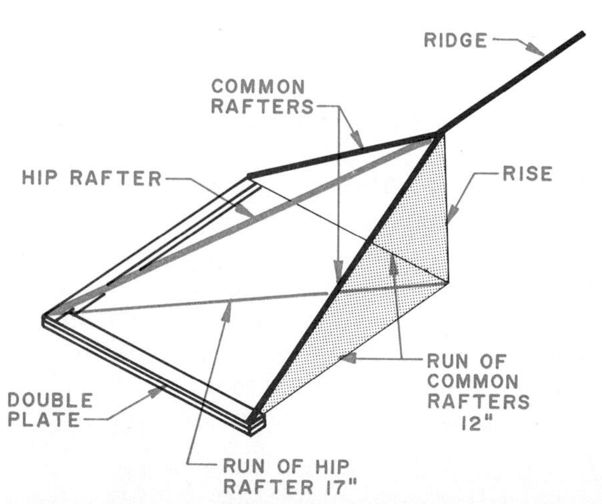

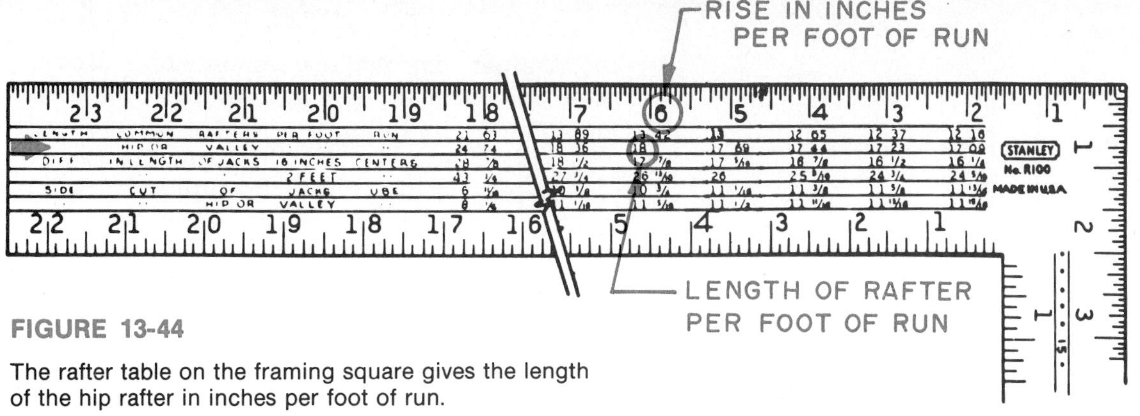

FIGURE 13-44

The rafter table on the framing square gives the length of the hip rafter in inches per foot of run.

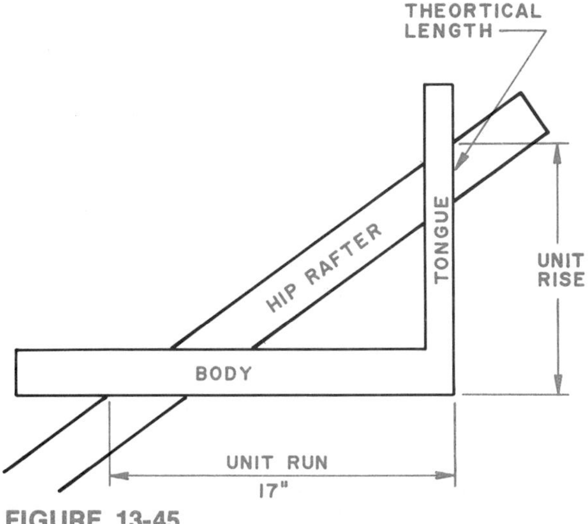

FIGURE 13-45

The hip rafter is laid out with the framing square in the same manner as a common rafter.

The length of the hip rafter can also be laid out with the framing square. This is the same as for common rafters (Fig. 13-45). The span of buildings is such that there are usually a few inches left over after laying out the units of run (17 in. for every 12 in. of run). These left-over inches are often called "odd units." To lay out the odd units, locate the distance left over on the body and tongue of the square. Measure the diagonal. This is the unit of run for the odd units. For example, if a span were 10 ft 6 in., then 6 in. in run would be left over. This is located on the square. The diagonal of these measurements is 8½ in. This is the odd unit of run for the hip rafter (Fig. 13-46).

The hip rafter must now be shortened a distance equal to one-half the 45° thickness of the ridge (Fig. 13-47). This is necessary because the measurements were from the true ridge

FIGURE 13-46

How to lay out a hip rafter having an odd unit of run.

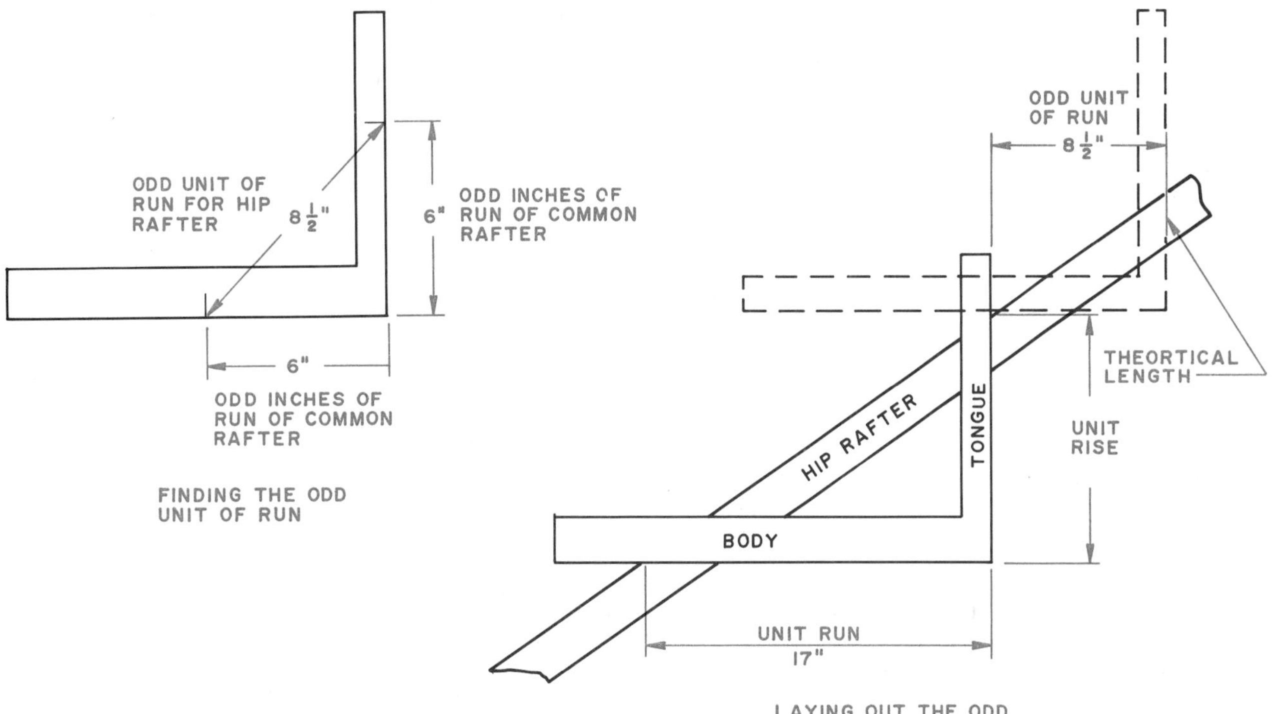

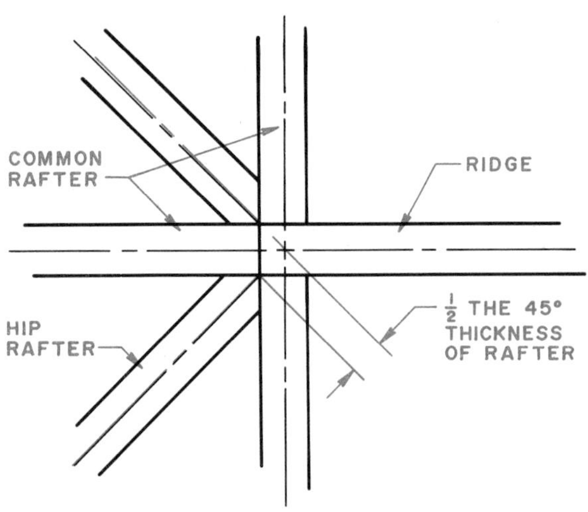

FIGURE 13-47

How to figure the amount to shorten the hip rafter at the ridge.

length rather than the actual ridge length. This shortening line is drawn parallel to the plumb cut. A carpenter can lay out the ridge and rafter on a piece of cardboard and measure this distance.

Next, mark the top and seat cuts. Mark each on the face of the rafter as shown in Fig. 13-48. They both are located using the run, 17 in., and the rise per foot run of the common rafter. These are located on the framing square, which is placed on the face of the rafter. Measurements are taken on the top edge of the rafter. The seat cut is marked with the body of the square. The top cut is marked with the tongue of the square.

Next, the top cut must be marked on an angle where it meets the ridge. This is called a cheek or side cut. Look at the last rafter table on the square marked "side cut hip or valley use." Read the figure given under the column having the rise per foot of the common rafter used in the roof. This is located on the body of the framing square. The figure used on the tongue is always 12. (Study Fig. 13-49). These are located on the measuring line of the rafter. Mark the right- and left-side cuts on the edge of the rafter. For example, a roof with a rise of 8 in. per foot run would have a reading of 10⅞ in. on the framing square table. This is located on the body of the square as shown in Fig. 13-49.

If the hip roof has an overhang, the hip rafter must have a tail. The normal overhang used is 6, 12, or 24 in. The extra length of the hip rafter to form the projection is found by drawing a square the size of the overhang. The length of the diagonal is the projected length of the rafter tail. The end of the rafter is cut using the same angle as used on the top cuts (Fig. 13-50).

A hip rafter sits above the hip jack rafters which meet it (Fig. 13-51). The hip rafter can be beveled or dropped so that it is flush with the jack rafters. This is called "*backing off*" the hip rafter (Fig. 13-52). In this way the roof sheathing will be flat over the hip rafter.

To find the amount of bevel, place the framing square on the top edge (Fig. 13-53). The unit of rise and unit of run are located on the edge. Measure one-half the thickness of the hip rafter from the unit run. This gives the depth of the bevel. Draw a line parallel to the edge through this point.

To use the drop technique, place the square as shown in Fig. 13-54. Measure in one-half the thickness of the hip rafter. Draw a line from this point. The distance between it and the square is the amount of drop. This amount is cut out of the seat of the bird's-mouth.

To find the overhang, draw a square with the sides equal to the desired overhang. The length of the diagonal is the length of the hip rafter overhang (Fig. 13-55).

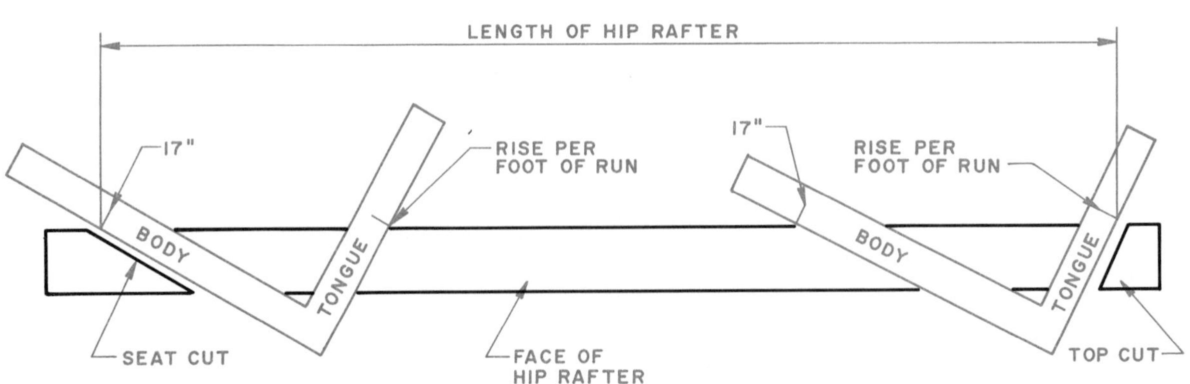

FIGURE 13-48

Mark the top and seat cuts of the hip rafter.

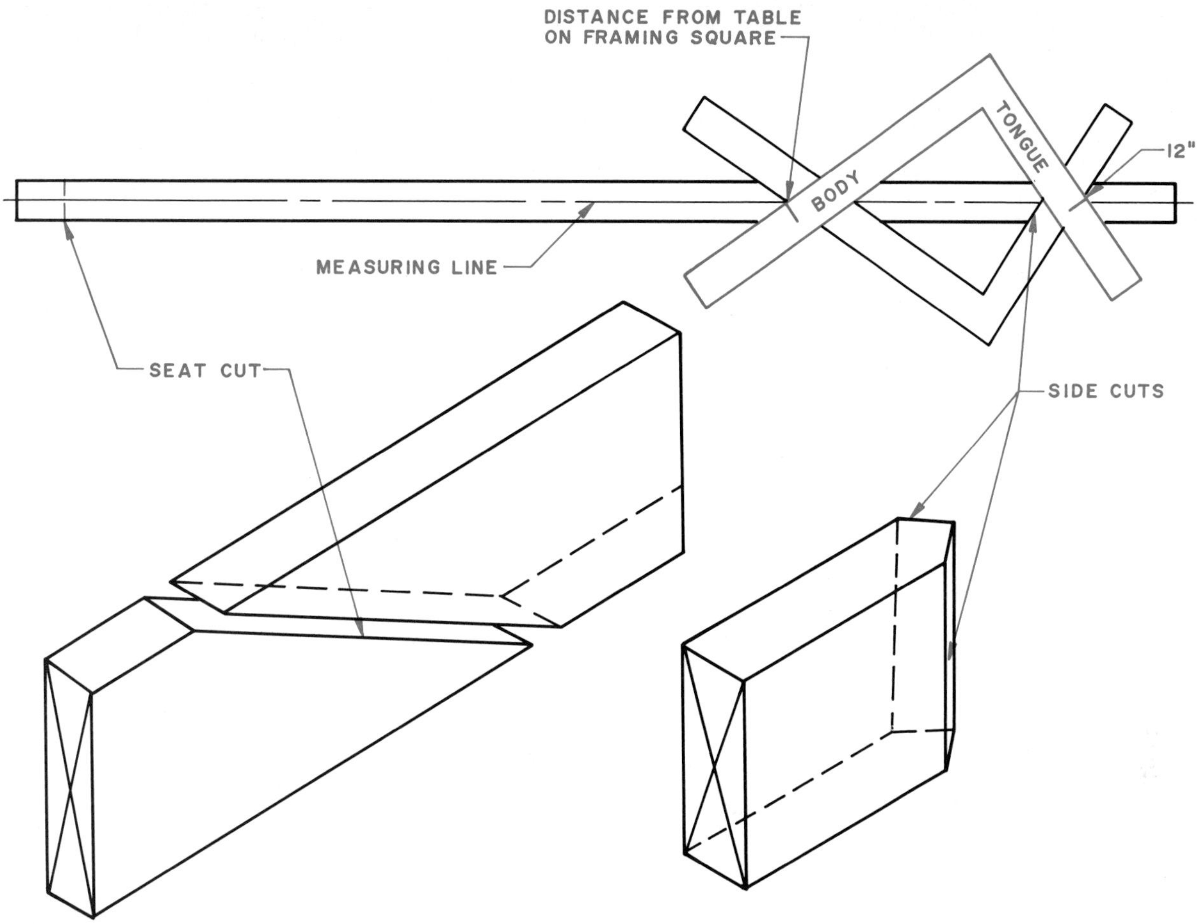

DISTANCE FROM TABLE ON FRAMING SQUARE
TONGUE
BODY
12"
MEASURING LINE
SEAT CUT
SIDE CUTS

FIGURE 13-49

Mark the angles of the top cut of a hip rafter so that it fits against the common rafters.

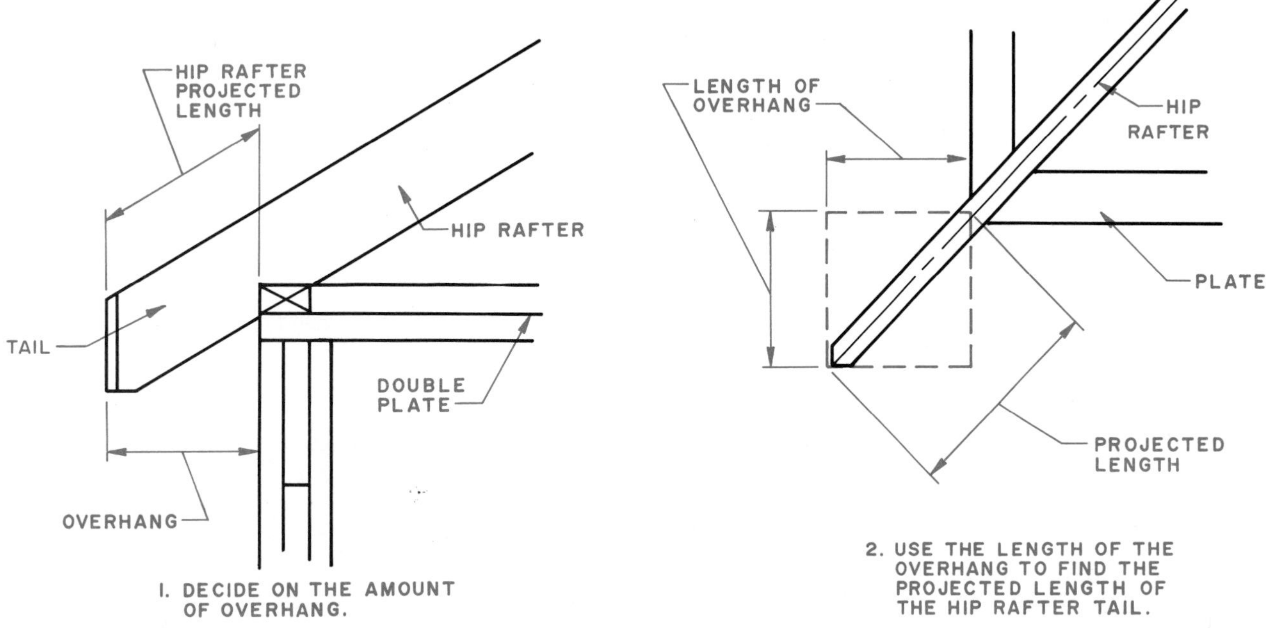

FIGURE 13-50

Mark and cut the angled plumb cuts on the hip rafter tail.

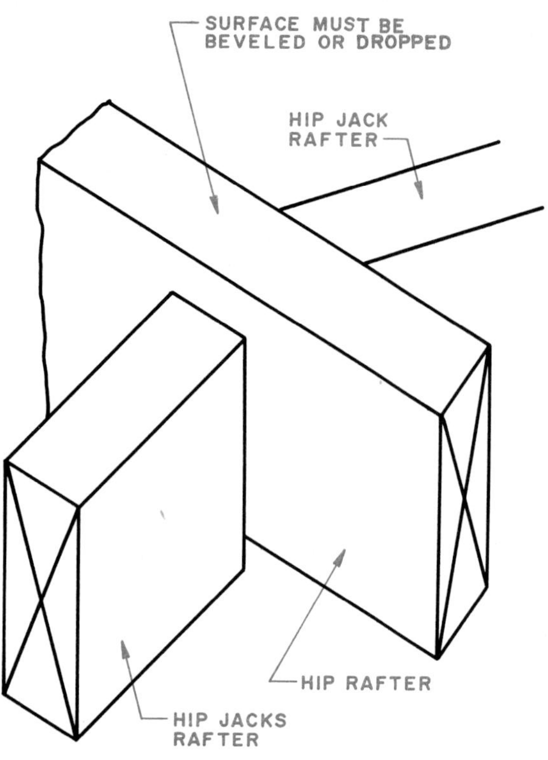

FIGURE 13-51

The hip rafter sits above hip jack rafters.

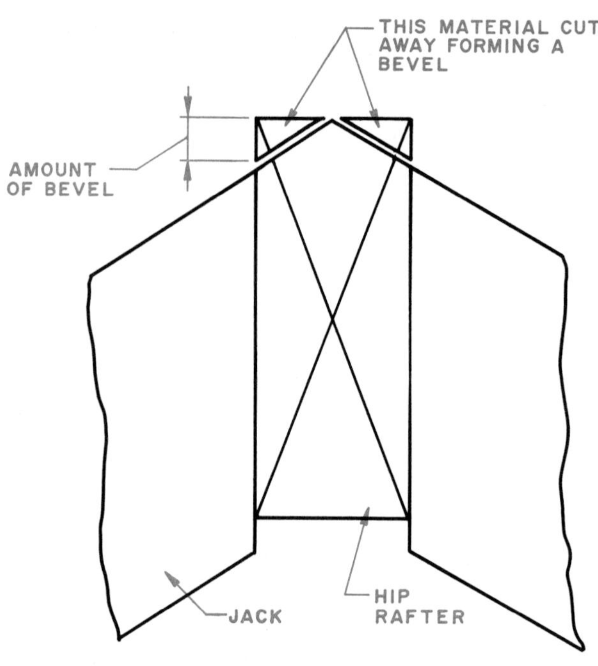

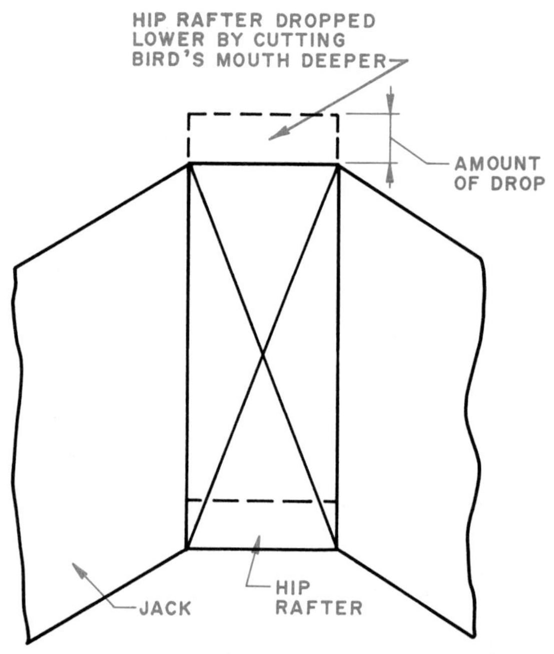

FIGURE 13-52

The hip rafter must be beveled or dropped by lowering the bird's-mouth.

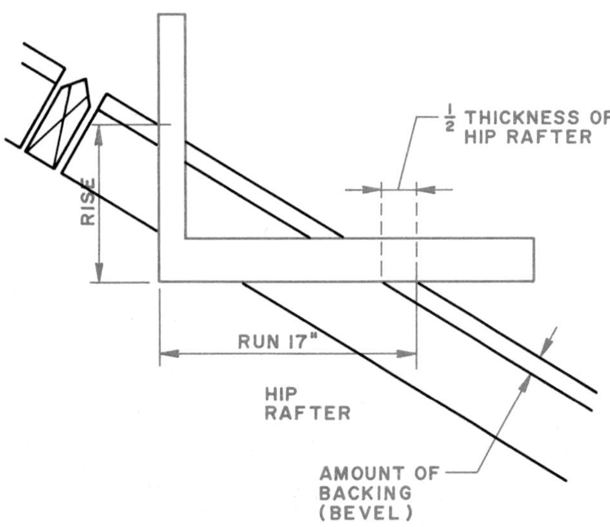

FIGURE 13-53

Using the framing square to lay out the bevel on a hip rafter.

Valley Rafters *Valley rafters* are much like hip rafters. They form the intersection between two gable roofs (Fig. 13-56). These roofs may have equal or unequal spans (Fig. 13-57). If they have the same span and pitch, the valley rafters butt against both ridges (Fig. 13-58). If the span of the intersecting roofs is different or the pitch is different, one valley rafter extends to the main ridge. It is the supporting valley. The other, the shortened valley rafters, meets the supporting valley rafter (Fig. 13-59). The ridge of this second roof begins at the intersection of the two valley rafters.

The supporting valley rafter and valleys for roofs with equal spans are the same as a hip rafter. They are laid out exactly the same using a 17-in. unit run.

The length of shortened valley rafters on

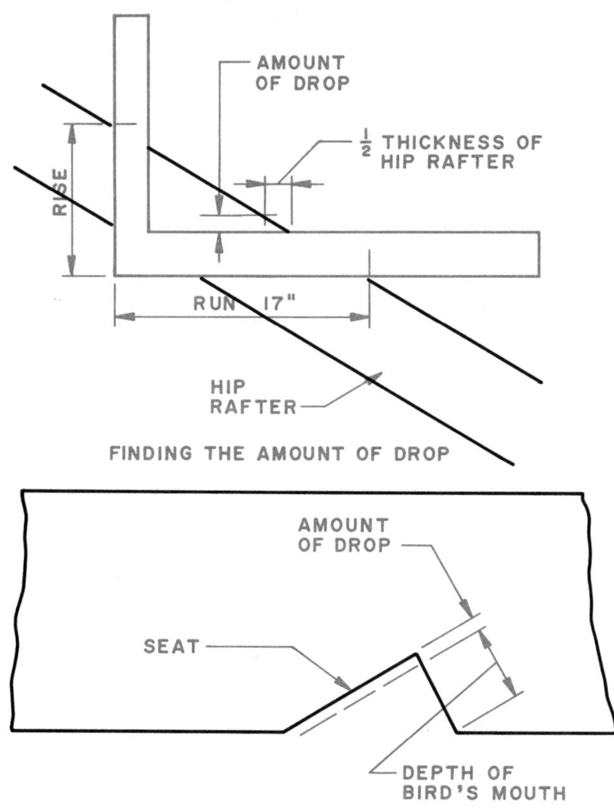

FIGURE 13-54

How to figure the amount of drop in the bird's-mouth of a hip rafter.

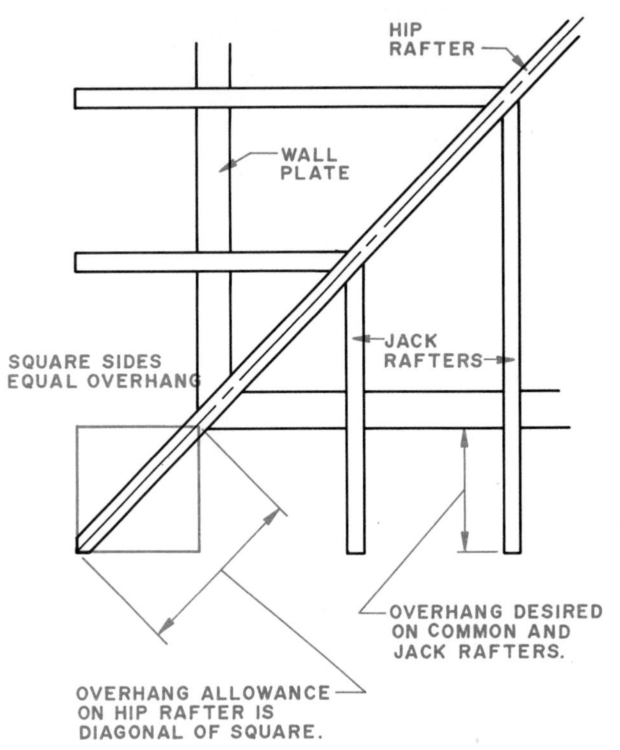

FIGURE 13-55

To find the length of the overhang of a hip rafter, make it the diagonal of a square formed by the overhang of the common rafters.

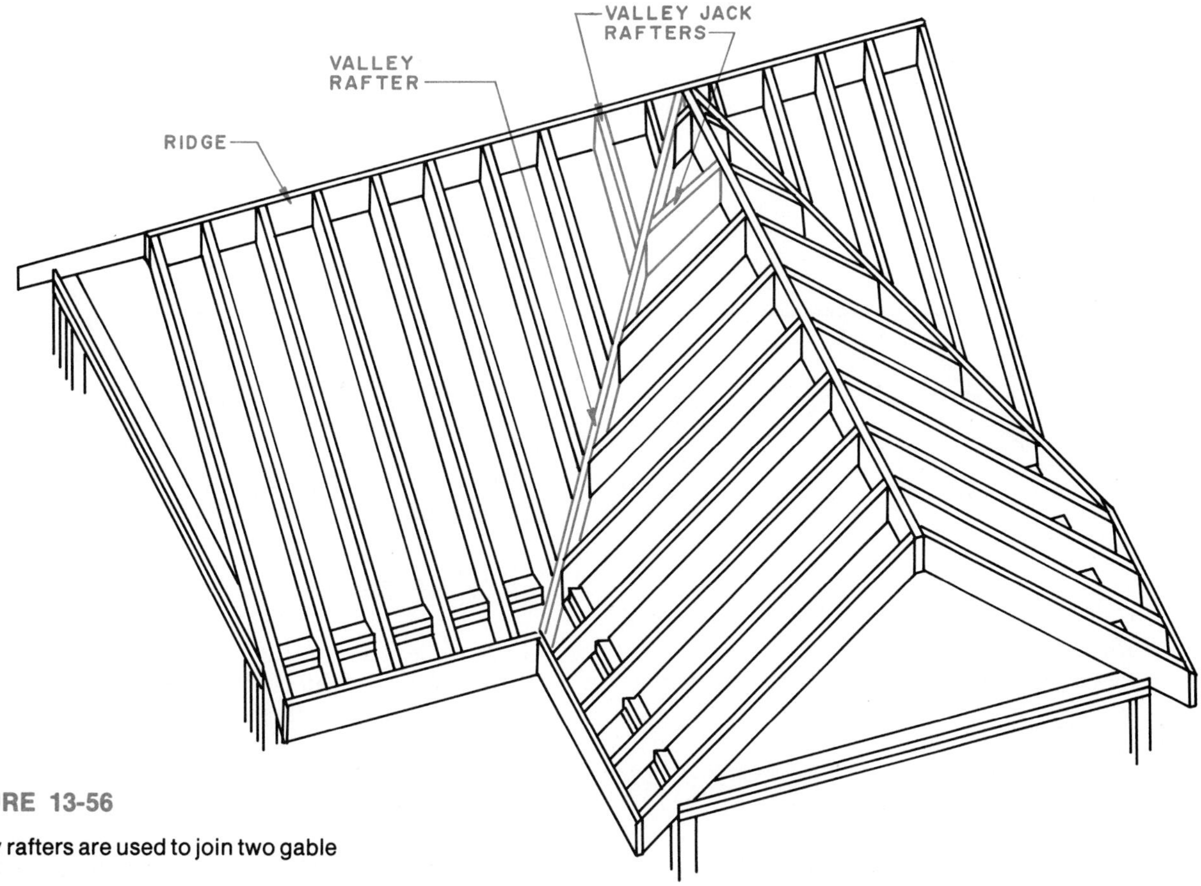

FIGURE 13-56

Valley rafters are used to join two gable roofs.

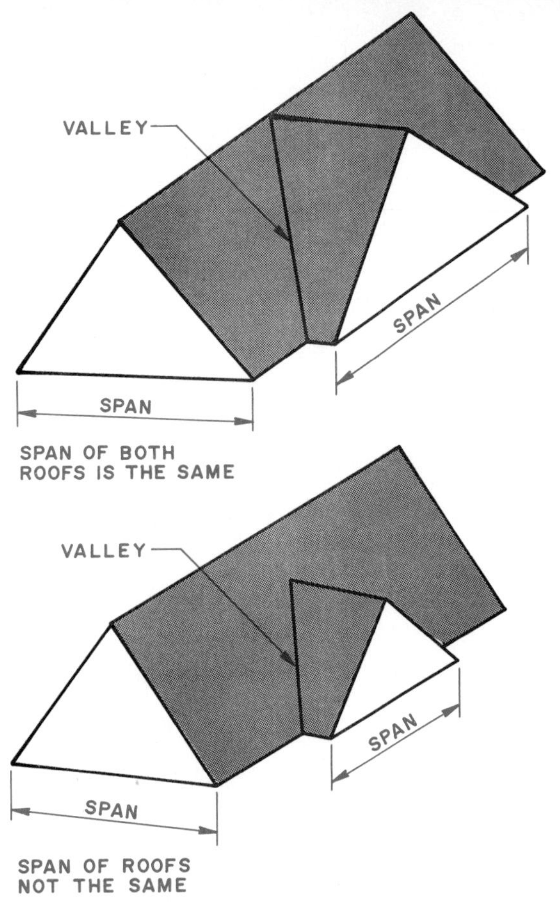

FIGURE 13-57

Intersecting gable roofs may or may not have the same size span.

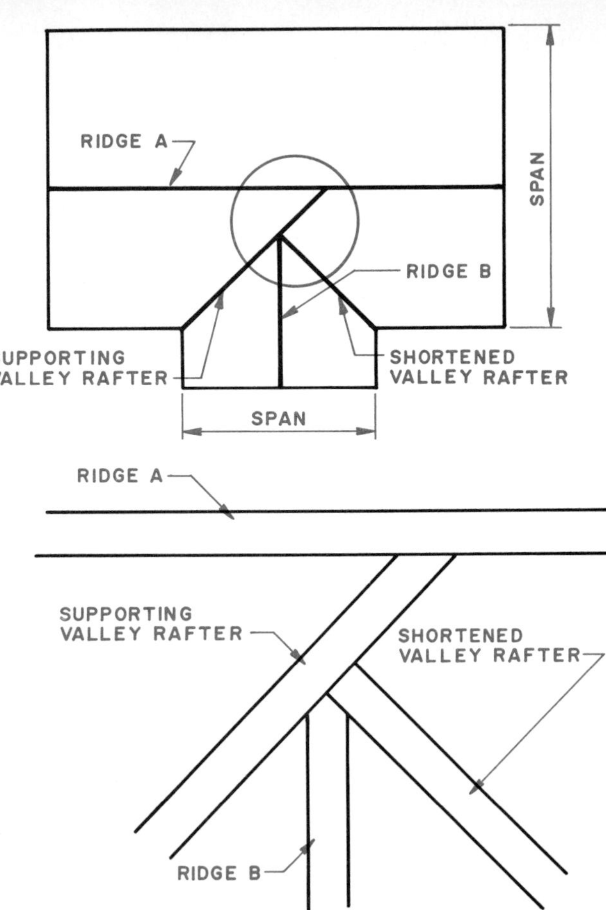

FIGURE 13-59

If the gables have different spans, a supporting valley rafter and a shortened valley rafter are used.

FIGURE 13-58

The valley rafter joins the ridges of both gables if their spans are the same.

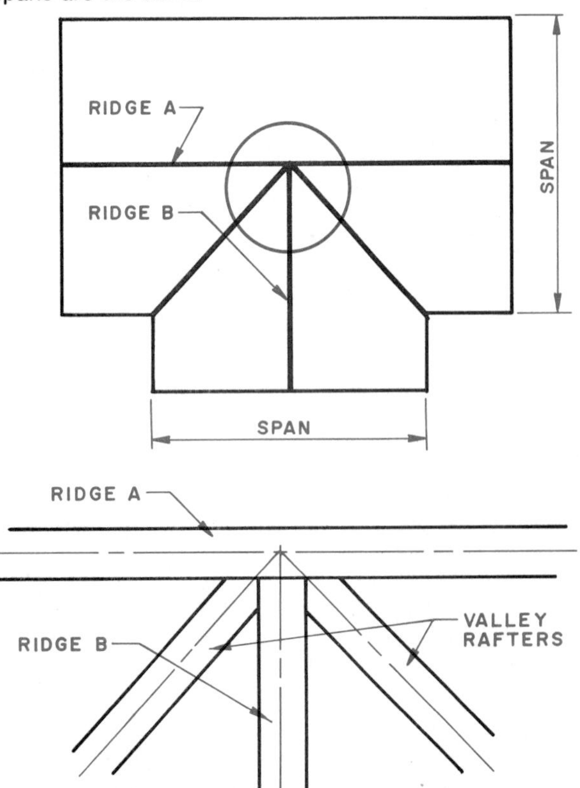

an unequal span roof is found using the feet of run of the common rafter in the shorter roof span. The unit of run for the valley rafter is 17 in.

Both supporting valley and shortened valley rafters must be shortened at the ridge after their theoretical length is found. The supporting valley rafter is shortened one-half of the 45° thickness of the ridge. The shortened valley rafter is reduced one-half of the supporting valley rafter thickness (Fig. 13-60).

The valley rafter tail has a double cut. It is cut to form an inside corner (Fig. 13-61). The overhang and bird's-mouth are figured in the same manner as described for hip rafters.

JACK RAFTERS

Jack rafters serve the same purpose as common rafters. Those which run from the plate to a hip rafter are called *hip jacks.* If they run from the ridge to a valley rafter, they are called *valley jacks.* Those running from a hip rafter to a valley rafter are called *cripple jacks* (Fig. 13-62). Hip jacks have a bird's-mouth on one end and a side cut on the other. Valley jacks have a plumb

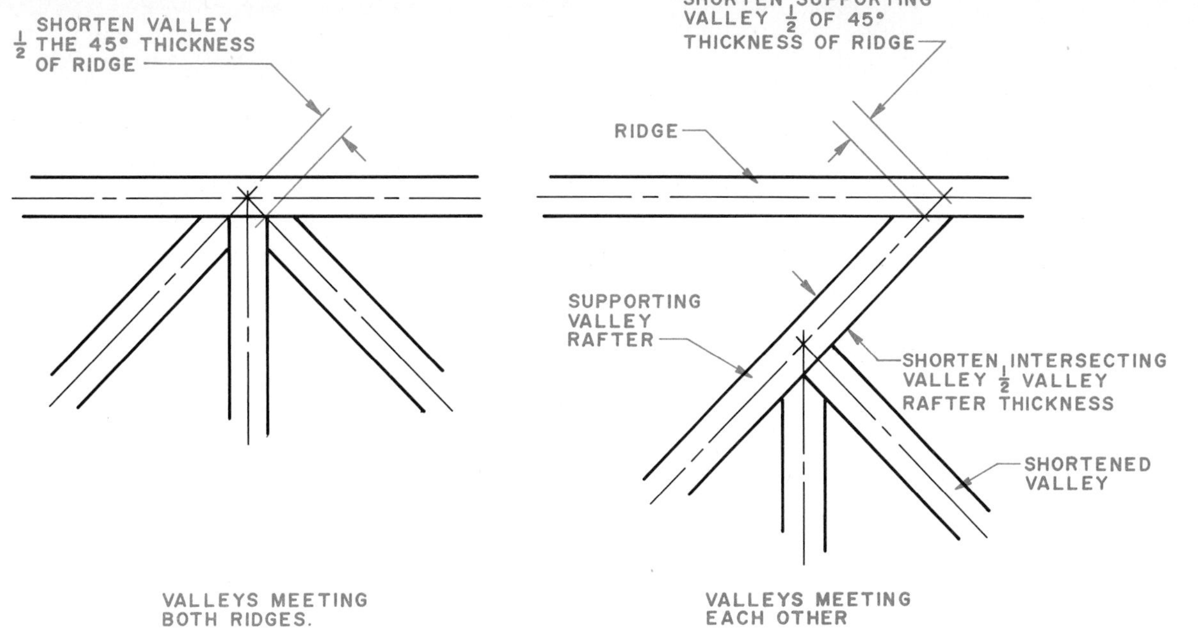

FIGURE 13-60

How to figure the amount to shorten the valley rafter to accommodate the thickness of the ridge board.

FIGURE 13-61

The tail of the valley rafter has a double cut to hold the fascia.

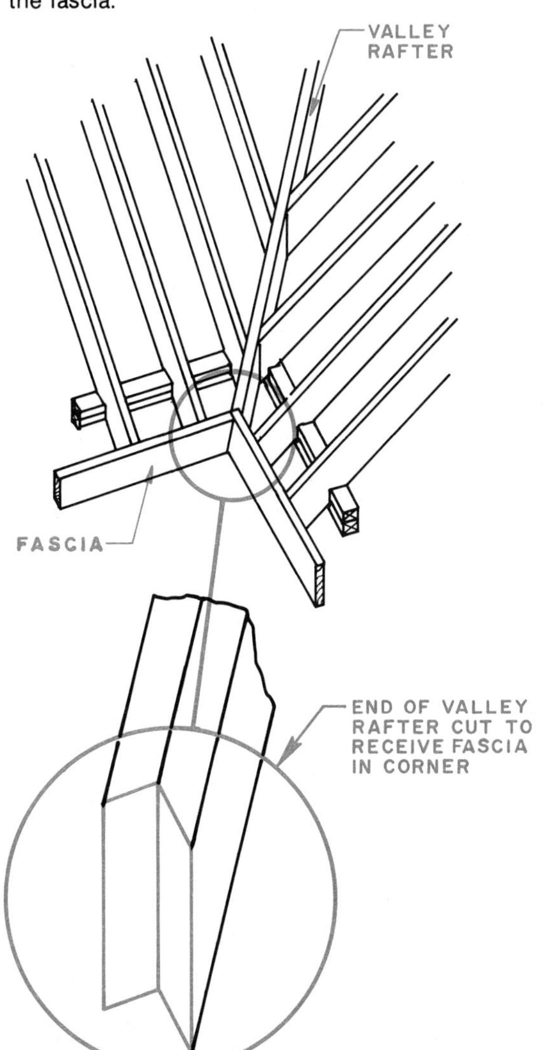

FIGURE 13-62

Uses for jack rafters.

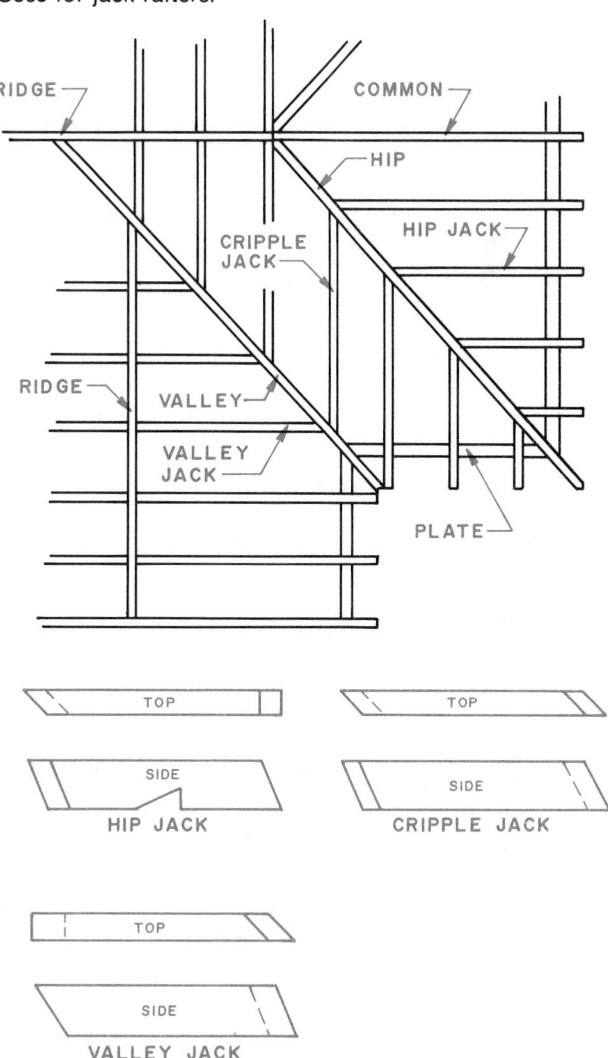

cut for the ridge and a side cut on the other end. Cripple jacks have side cuts on both ends.

Hip Jack Rafters *Hip jack rafters* have the same overhang and tail as a common rafter. They have a bird's-mouth. Their length is measured from the bird's-mouth to their intersection with the hip rafter. This length is different for each jack rafter. The amount of difference between each is the same. This is called a *common difference*. The second jack rafter is twice the length of the first. The third rafter is three times the length of the first. This common difference exists for all hip jack rafters in that roof (Fig. 13-63).

The common difference is found on the rafter table on the framing square. There are scales marked "difference in length of jacks." These are for rafters spaced 16 in. and 24 in. on center. For a roof with a 6 in 12 slope spaced 16 in. on center, the common difference is 17⅞ in. (Fig. 13-64).

The common difference can also be found by the layout method (Fig. 13-65). Place the framing square on the edge of a rafter. Locate the rise and run on the edge. Draw a line along the blade. Then slide the square along this line until the rafter spacing dimension meets the

FIGURE 13-63

How to lay out a hip rafter using the common difference.

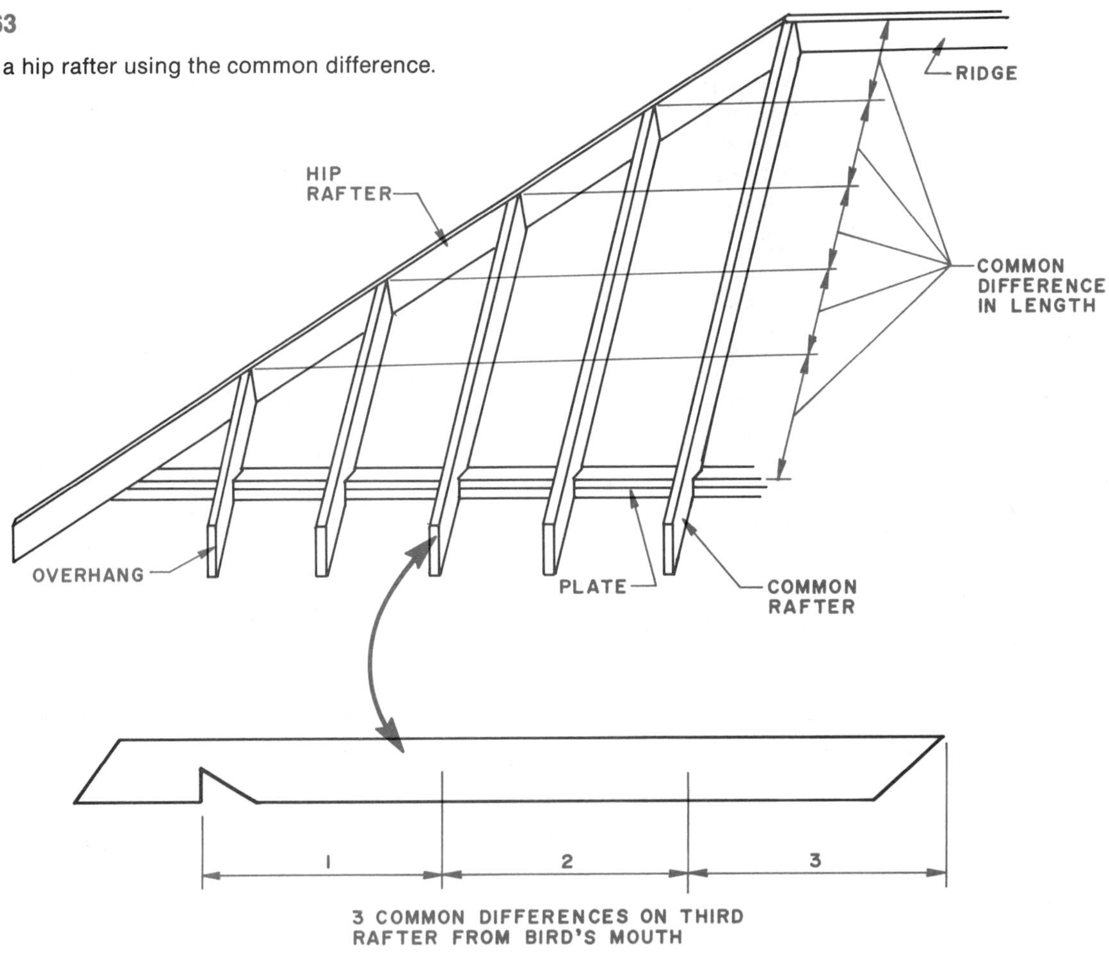

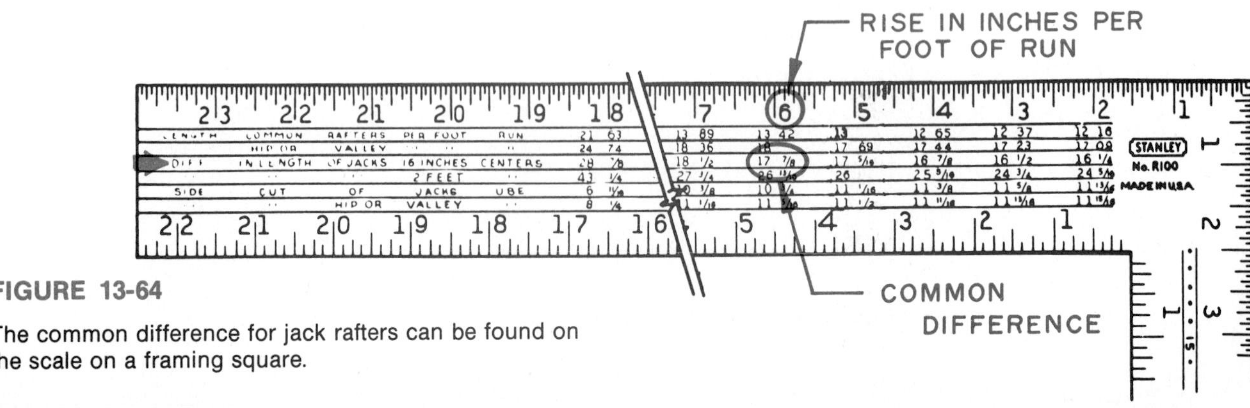

FIGURE 13-64

The common difference for jack rafters can be found on the scale on a framing square.

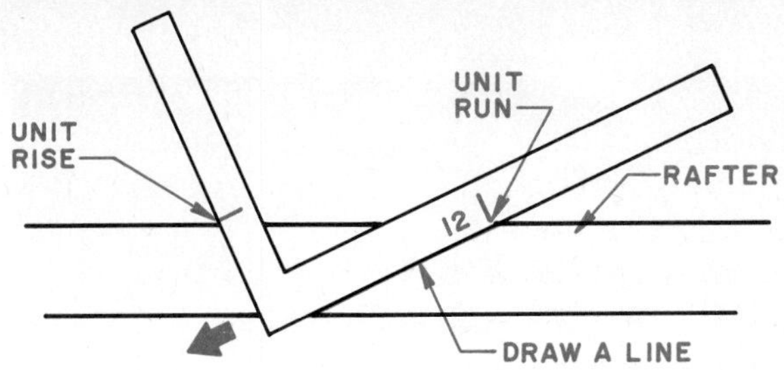

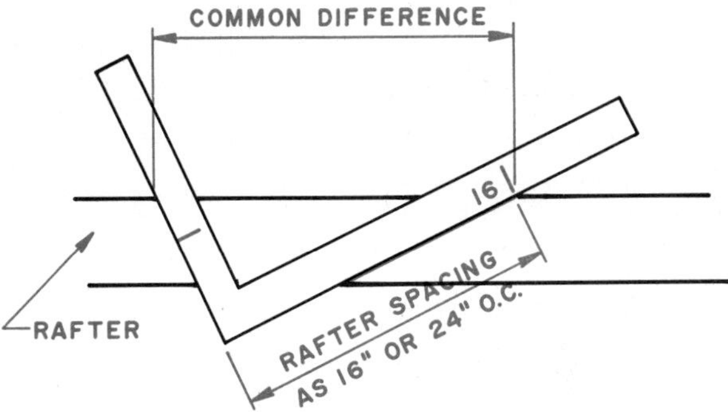

FIGURE 13-65

How to find the common difference for a jack rafter by the layout method.

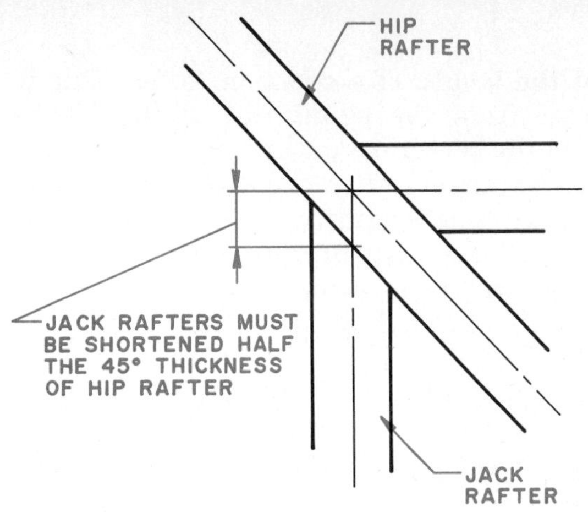

FIGURE 13-66

Shorten the jack rafter by half the 45° thickness of the hip rafter.

edge of the rafter. The horizontal distance between the unit rise location and the rafter spacing is the common difference in inches.

To find the length of the hip jack, find the common difference. This is the length of the shortest rafter. The second rafter is twice as long. The third is three times as long, and so on. Since the hip jack meets the hip rafter, it is necessary to subtract one-half of the 45° thickness of the hip rafter (Fig. 13-66).

It is best if a hip jack pattern is made. Use the common rafter pattern and lay out the bird's-mouth and overhang (Fig. 13-67). Then

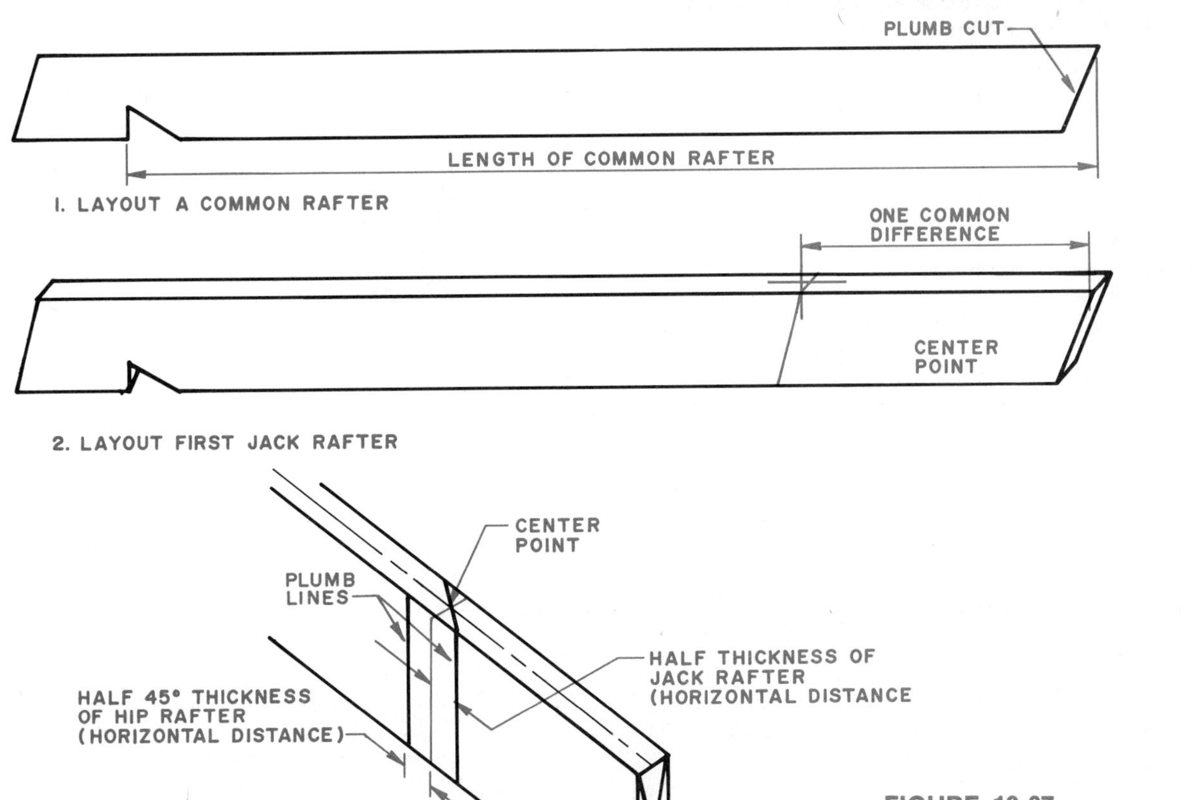

FIGURE 13-67

Use the common rafter pattern to lay out the jack rafters.

lay out the length of a common rafter. This is measured from the plumb cut of the bird's-mouth to the centerline of the ridge. Now measure one common difference from the plumb cut. Shorten this length by one-half of the 45° thickness of the hip rafter (Fig. 13-66). Draw a line across the rafter at this point and mark the center of the rafter. Now lay out the side cut as shown in Fig. 13-67.

To lay out the other hip jacks, set a T-bevel on the side cut angles. Shorten the length of each one common difference from the previous rafter (Fig. 13-68). Lay out the rafters in pairs. The side cuts are made opposite each other (Fig. 13-69). The side cuts on hip, valley, and cripple jack rafters are compound angles. The best way to cut them is with a radial arm saw (Fig. 13-70).

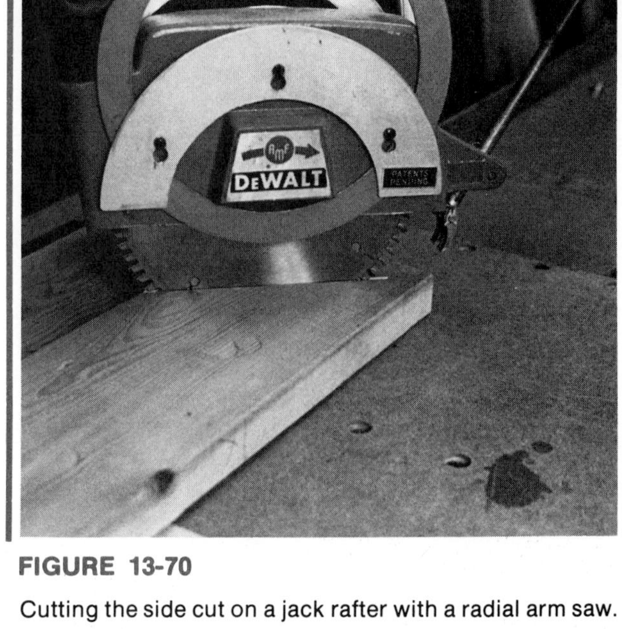

FIGURE 13-70

Cutting the side cut on a jack rafter with a radial arm saw.

Valley Jacks The *valley jack* is laid out in the same manner as the hip jack. It uses the same tables for figuring the common difference (Fig. 13-71). The top end of the rafter has a plumb cut to meet the ridge. This is the same cut used for the common rafter. The lower end has a side cut. This meets the valley rafter.

When laying out a valley jack, use the same allowances for shortening and side cuts as are used for hip jacks. Valley jacks are also cut in pairs.

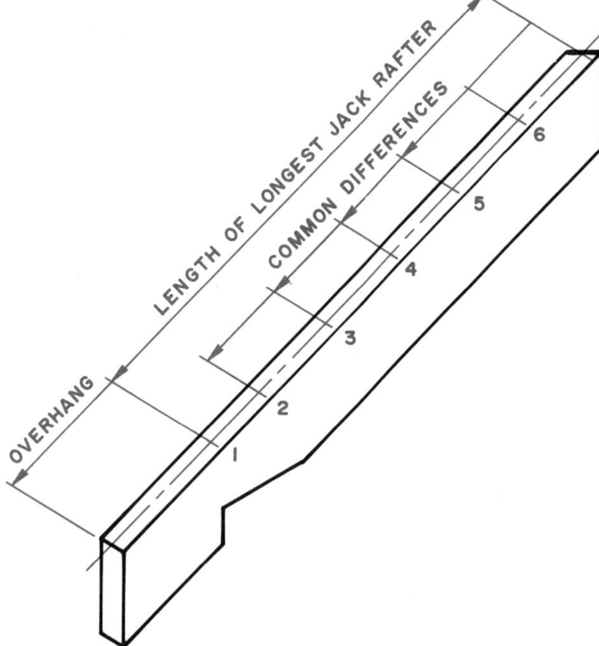

FIGURE 13-68

The layout of a series of jack rafters.

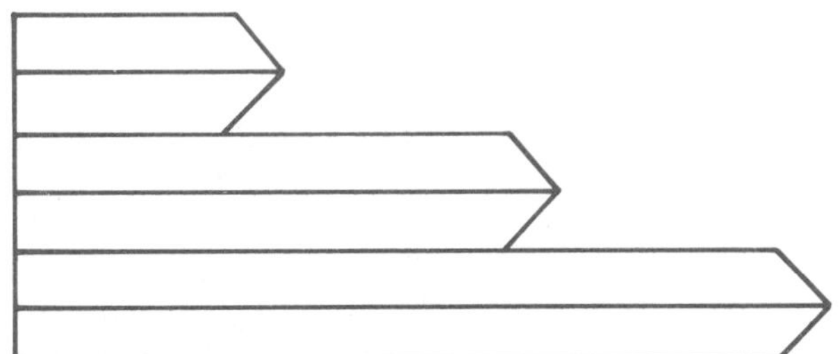

FIGURE 13-69

Jack rafters are cut in pairs.

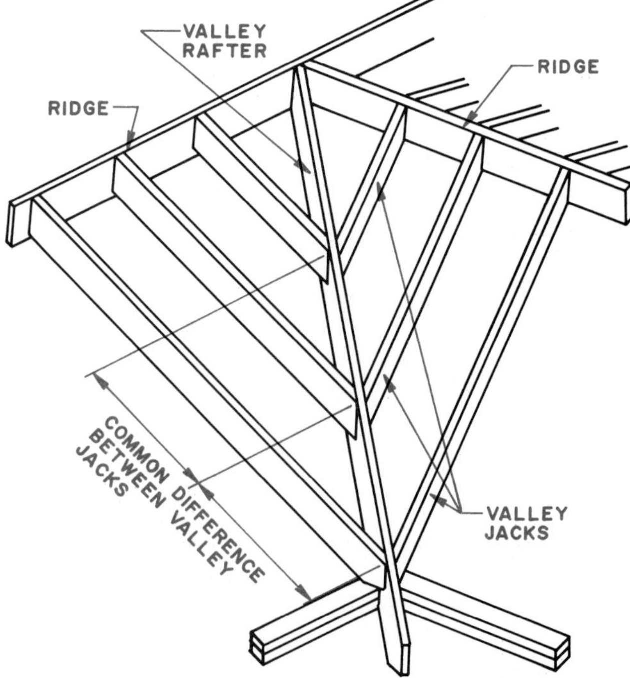

FIGURE 13-71

Valley jack rafters are laid out the same as jack rafters.

Hip Valley Cripple Jacks *Hip valley cripple jacks* run between valley and hip rafters. Since these are parallel, all cripple jacks are the same length (Fig. 13-72). Each end of the cripple jack has a side cut. These are parallel.

The run of the cripple jack is found by measuring from the center of the hip rafter to the center of the valley rafter along the plate (Fig. 13-73).

The cripple jack is laid out the same as a common rafter using the run just mentioned. Then shorten *each* end by one-half the 45° thickness of the hip and valley rafters. Locate the side cuts as described for hip jacks. Use one-half the thickness of the cripple jack to locate the side cut.

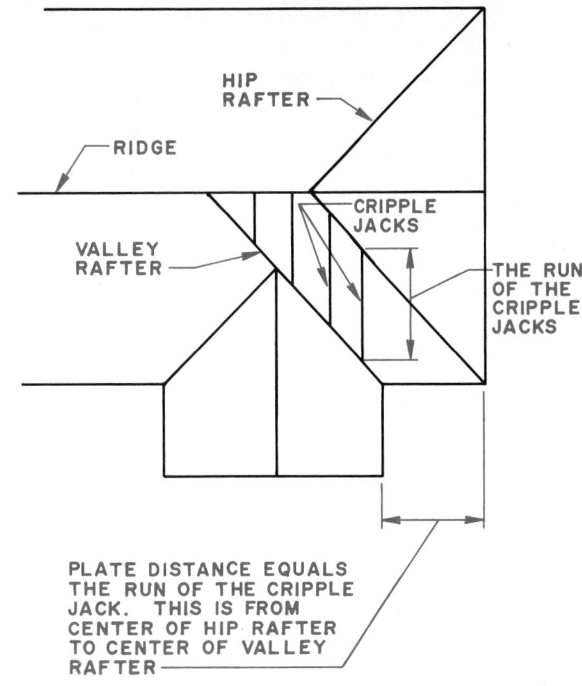

FIGURE 13-73

The run of a cripple jack rafter is found by measuring from the center of the hip rafter to the center of the valley rafter along the plate.

INSTALLING JACK RAFTERS

The spacing between rafters is laid out on the hip and valley rafters and the plate and ridge. Nail them in place with 10d nails. Start near the center of the hip and valley rafter. If they need support, add braces. Make certain that the hip and valley rafters remain straight. Do not seat the nails until all jacks are in place and the hip

FIGURE 13-72

Hip valley cripple jacks run between the hip and valley rafters.

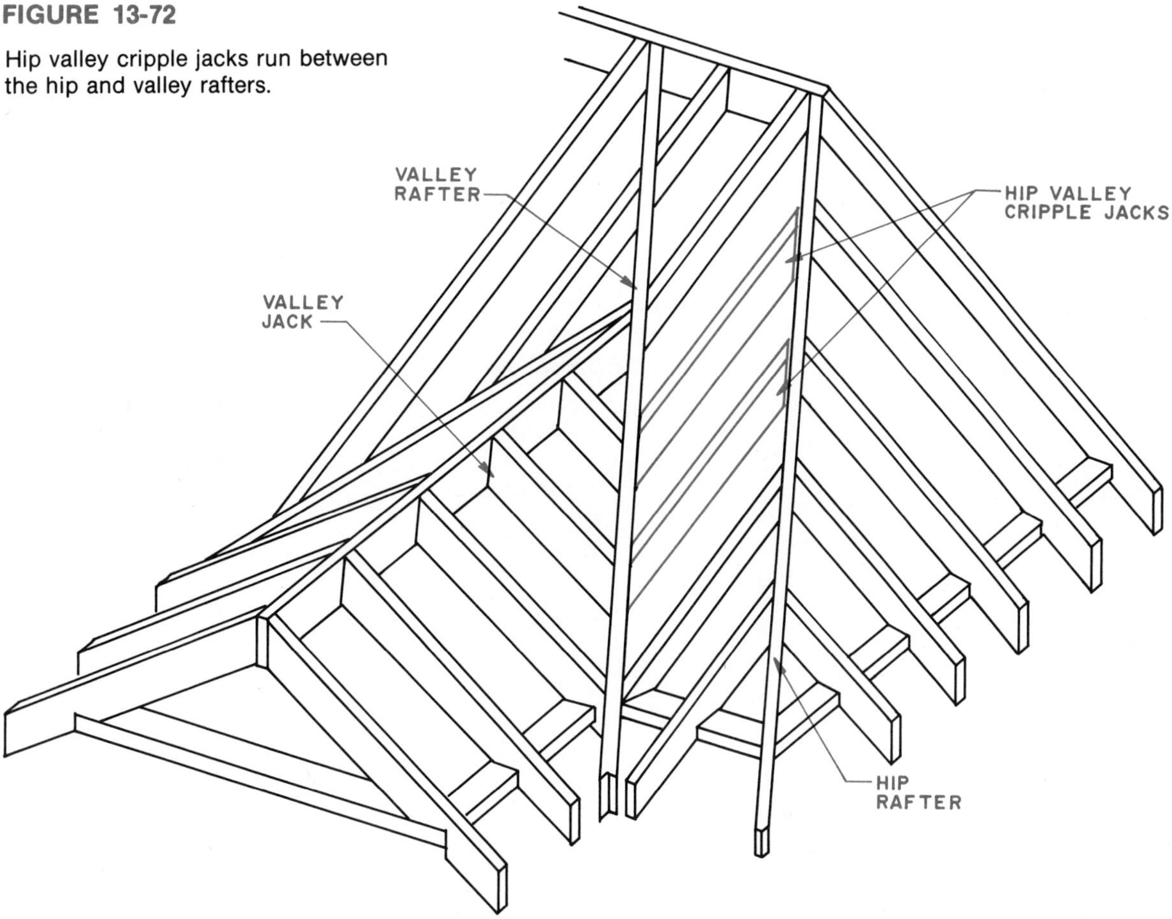

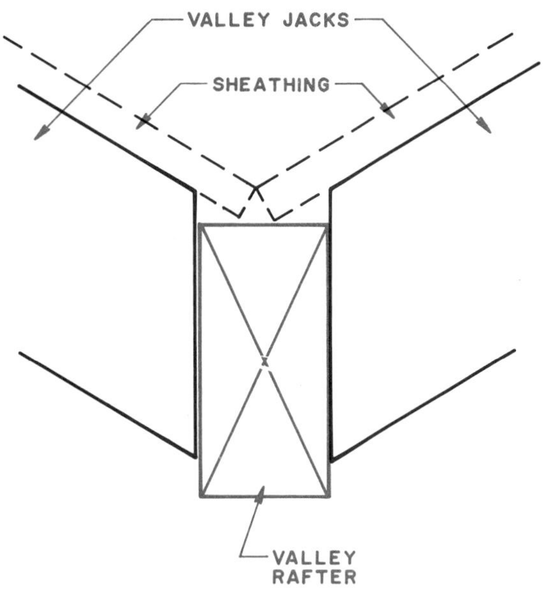

FIGURE 13-74

The valley jack rafters are installed a little above the valley rafter.

or valley rafter is straight. Then set them flush with the surface of the jack.

When installing valley jacks raise them slightly above the valley rafter. This will permit the sheathing to meet forming a smooth, tight corner (Fig. 13-74).

FRAMING DORMERS

Dormers are framed openings in a roof used to admit light, ventilation, and to increase head room. They make the area under the roof usable for living space.

The common types are shed and gable dormers. *Shed dormers* are larger and can run the entire length of a house (Fig. 13-75). They use a shed roof which often starts at the ridge.

Typical framing is shown in Fig. 13-76. The rafters are cut as described for common rafters. The side walls are framed like a gable end. Notice that the main roof rafters under the side walls are doubled. Ceiling joists are face-nailed to the dormer roof rafters. The end wall has a double plate. Window openings are framed in the normal manner.

Framing for a shed dormer roof which does not go to the ridge is shown in Fig. 13-77.

The *gable roof dormer* normally contains one window (Fig. 13-78). It has a ridge and short rafters. They are like common rafters but do not usually have a bird's-mouth. The ridge meets the roof at a double header. Valley rafters meet the ridge at the header. The shortening required for these rafters is shown in Fig. 13-79.

Framing for gable and hip roof dormers is shown in Fig. 13-80.

FIGURE 13-75

A shed dormer increases the living space provided by a gable roof. *(Courtesy of Home Planners, Inc.)*

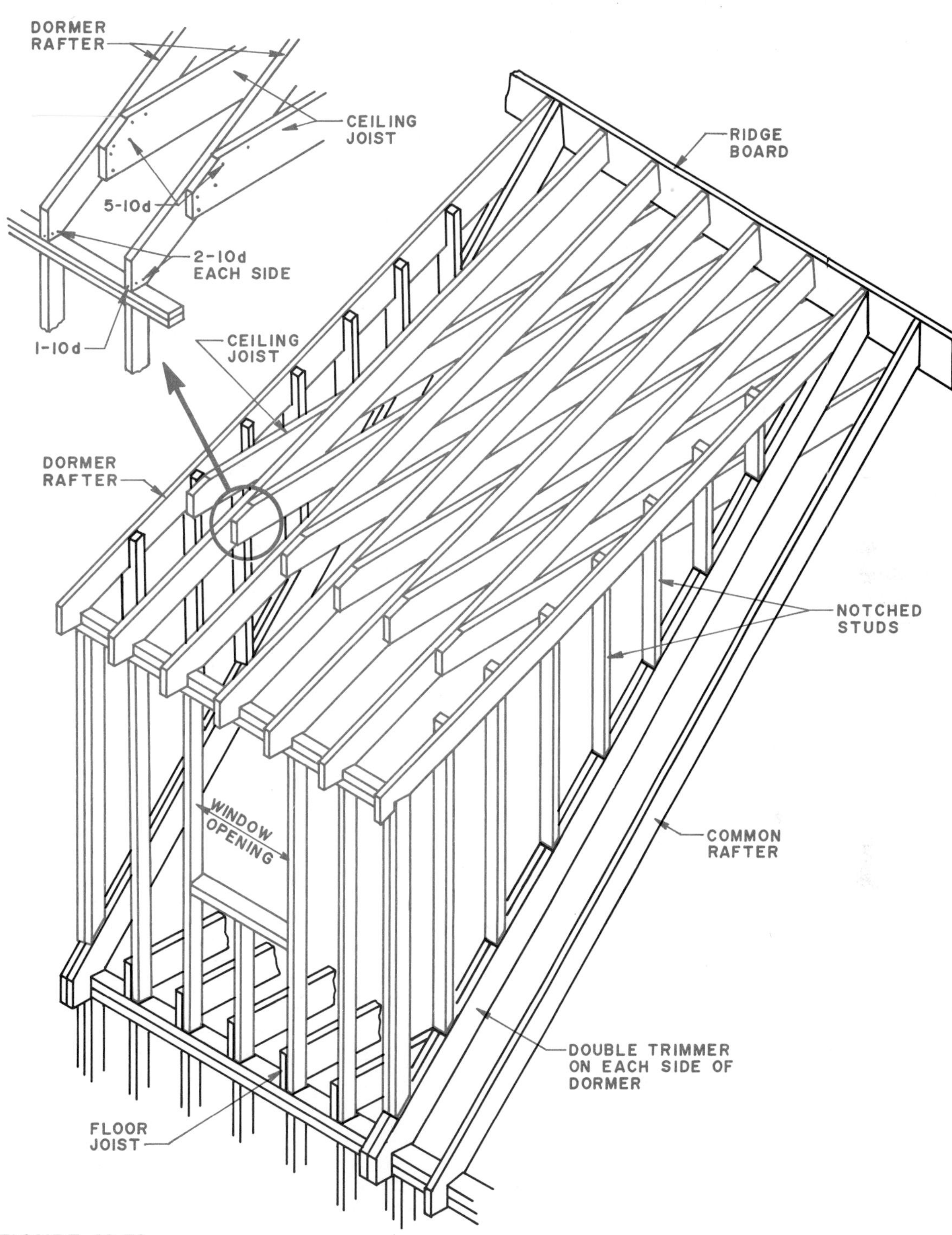

FIGURE 13-76

Typical framing for a shed dormer that starts at the ridge.

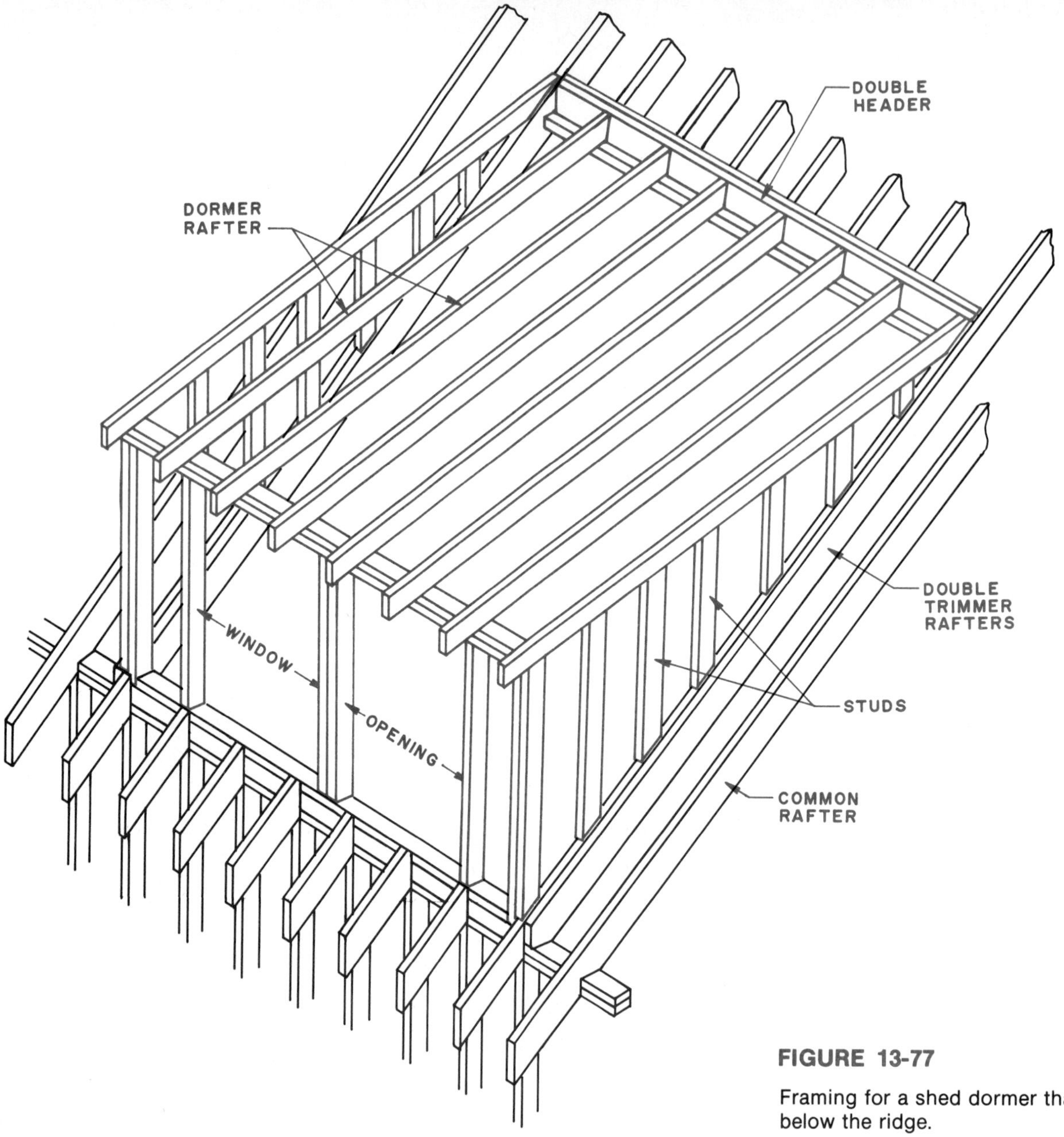

FIGURE 13-77

Framing for a shed dormer that starts below the ridge.

FIGURE 13-78

Single dormers off a gable roof. *(Courtesy of Home Planners, Inc.)*

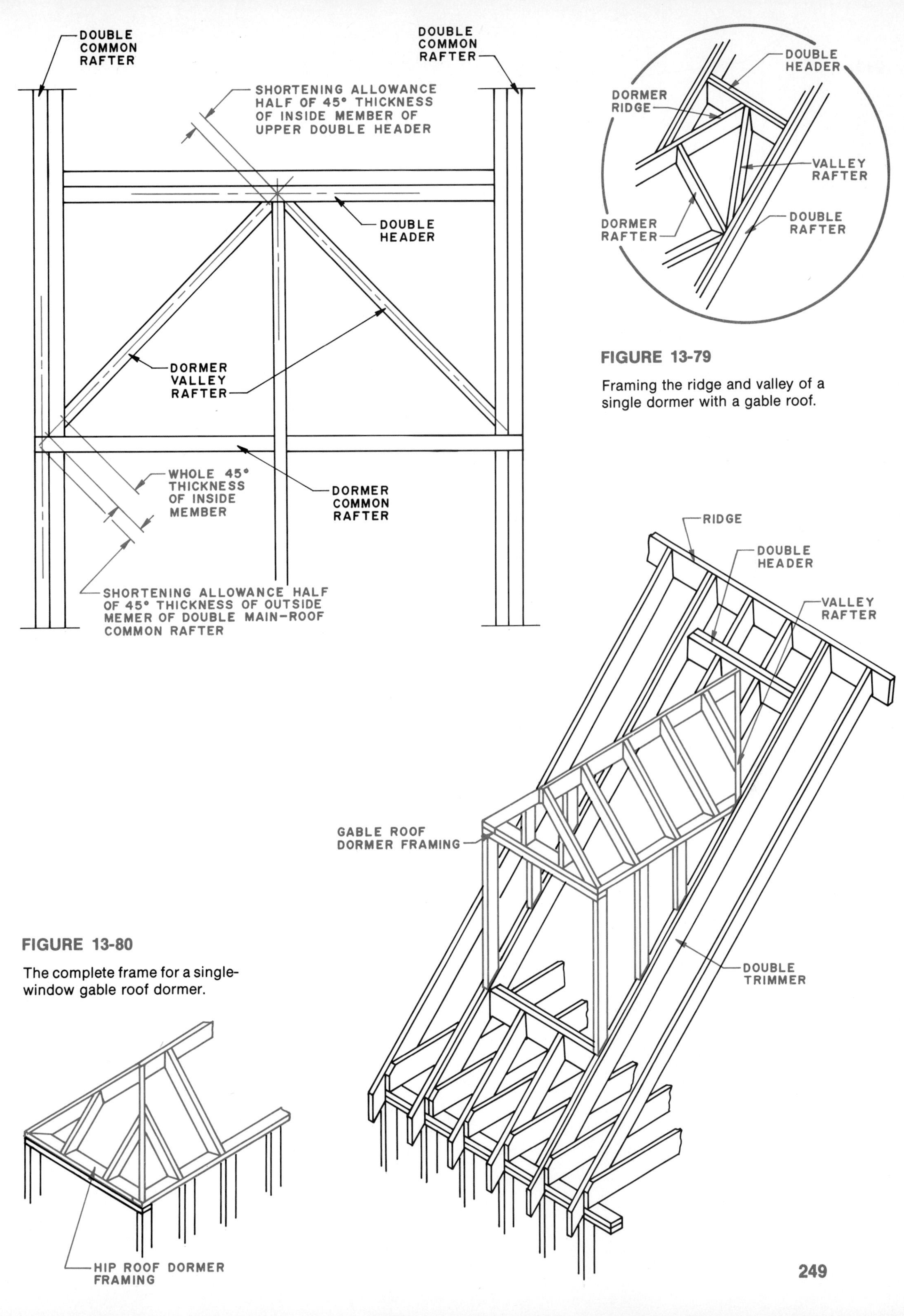

FIGURE 13-79

Framing the ridge and valley of a single dormer with a gable roof.

FIGURE 13-80

The complete frame for a single-window gable roof dormer.

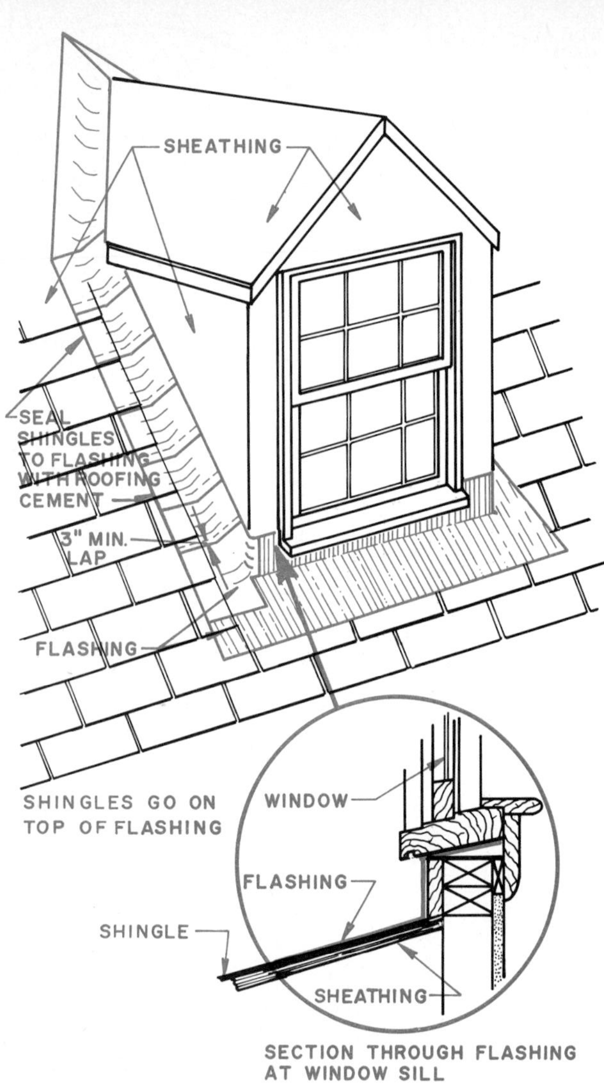

FIGURE 13-81 *(above)*

Dormer flashing runs up the sheathing and under the shingles.

A 2 × 4-in. member is nailed to the doubled rafters to provide a place to nail the sheathing. The walls are sheathed and covered with siding the same as other exterior walls. It is necessary to install flashing to repel water from the roof (Fig. 13-81).

SHED ROOFS

A *shed roof* is half a gable roof. The rafters are like common rafters except that they have a bird's-mouth and tail on each end (Fig. 13-82). The run of the rafters is the distance between the bird's-mouth plumb cuts. The difference in height of the two plates is the rise.

FLAT ROOFS

The framing of a *flat roof* is like framing a floor. The roof joists serve to support the actual roofing material and the interior ceiling finish material.

If an overhang is desired, lookout rafters are needed on two sides. The roof joists are extended to form the overhang on the other two sides (Fig. 13-83).

If the width of the roof joist is greater than the desired fascia, the joist can be sloped on the end (Fig. 13-84).

FIGURE 13-82 *(below)*

A rafter for a shed roof has a bird's-mouth and tail on each end.

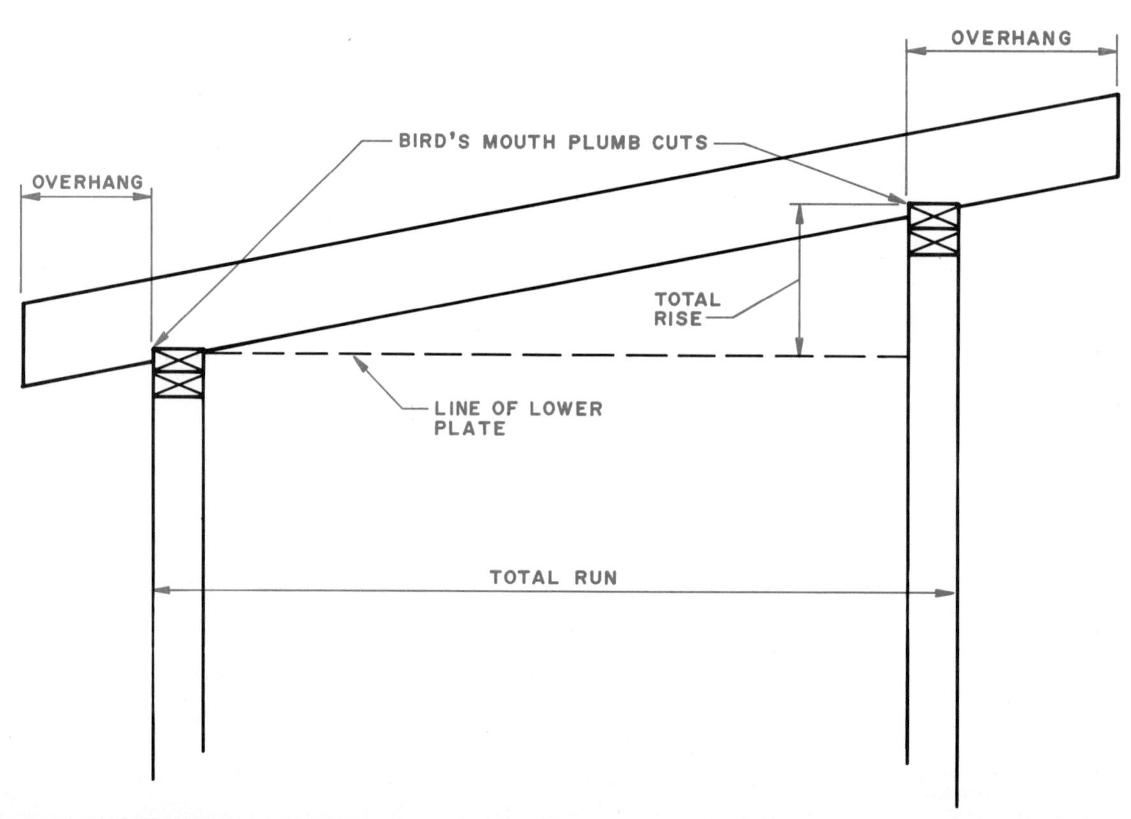

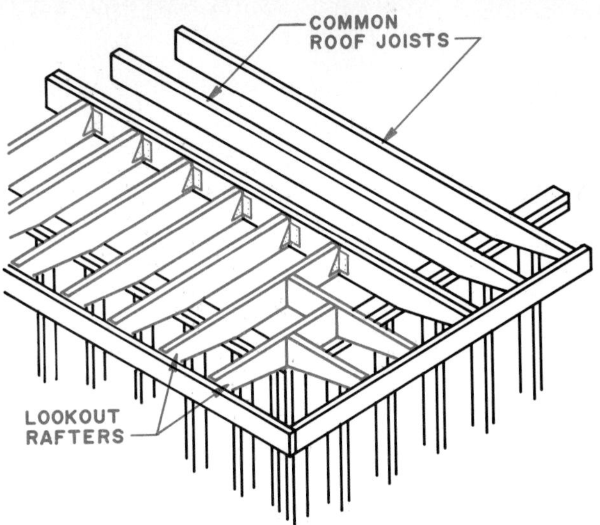

FIGURE 13-83

Lookout rafters are needed on two sides of a flat roof to provide an overhang.

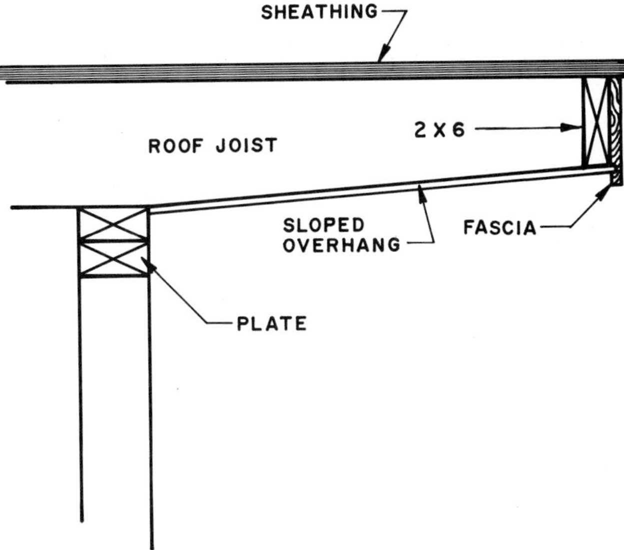

FIGURE 13-84 *(above)*

The overhang on a flat roof can be sloped.

GAMBREL ROOFS

A *gambrel roof* is actually two gable roofs with different slopes (Fig. 13-85). It makes more of the area below it useful for living space. The framing of this roof involves laying out two different rafters. The lower slope is steeper than the upper slope (Fig. 13-86).

The lower slope rafters have a bird's-mouth that rests on the plate. The top end rests against a *purlin*. The rafter is laid out like a common rafter. The plumb cut is made with a seat for the purlin. Installation of lower slope rafters is shown in Fig. 13-87.

The upper slope rafters rest on the purlin. They are notched so that they meet the lower slope rafter at the purlin. The other end rests against a ridge.

A typical framing plan is in Fig. 13-88. Notice that the overhang is framed using blocking running to the wall studs.

Most generally, the gambrel roof follows a semicircular shape. A carpenter can make a full-size layout on the subfloor. This will provide rafter patterns (Fig. 13-89).

Using the run as the radius, draw a quarter circle. Draw a perpendicular to the run line. This locates the center of the ridge. On the house plan find the design height of the purlin. Draw this perpendicular to the run line. Where it touches, the circle locates the ends of the rafters. Draw the rafter patterns. Draw the seat for the purlin. Get its size from the plans. Subtract half the thickness of the ridge from the upper rafter.

FIGURE 13-85 *(below)*

A house with a gambrel roof. *(Courtesy of Home Planners, Inc.)*

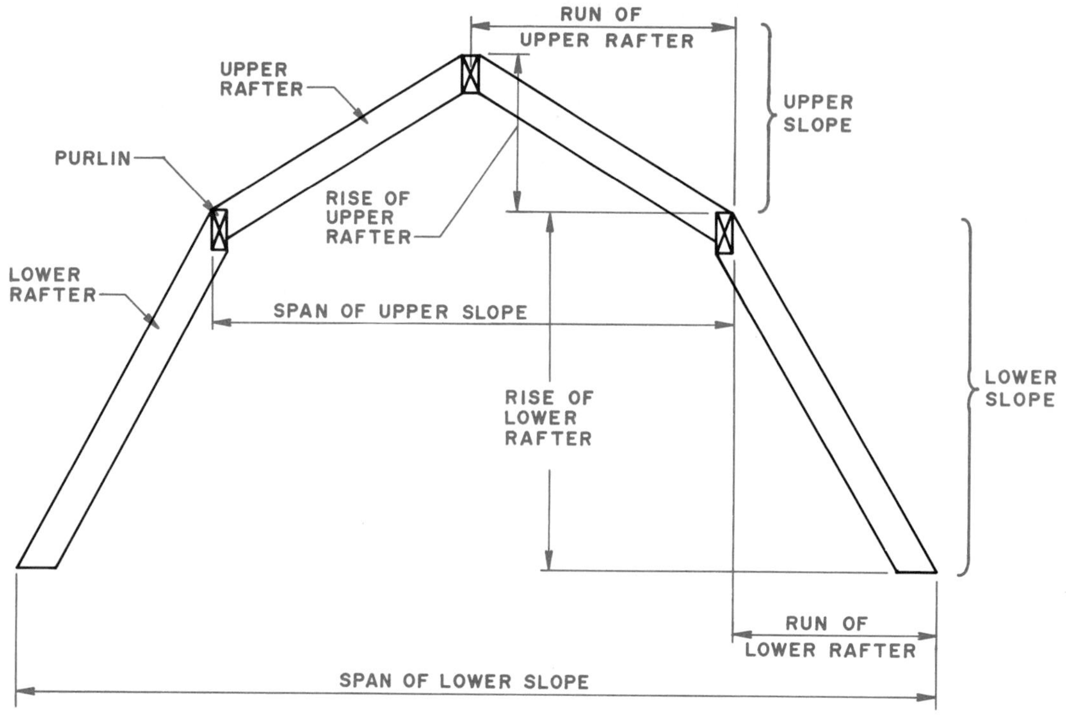

FIGURE 13-86

The framing plan for a gambrel roof.

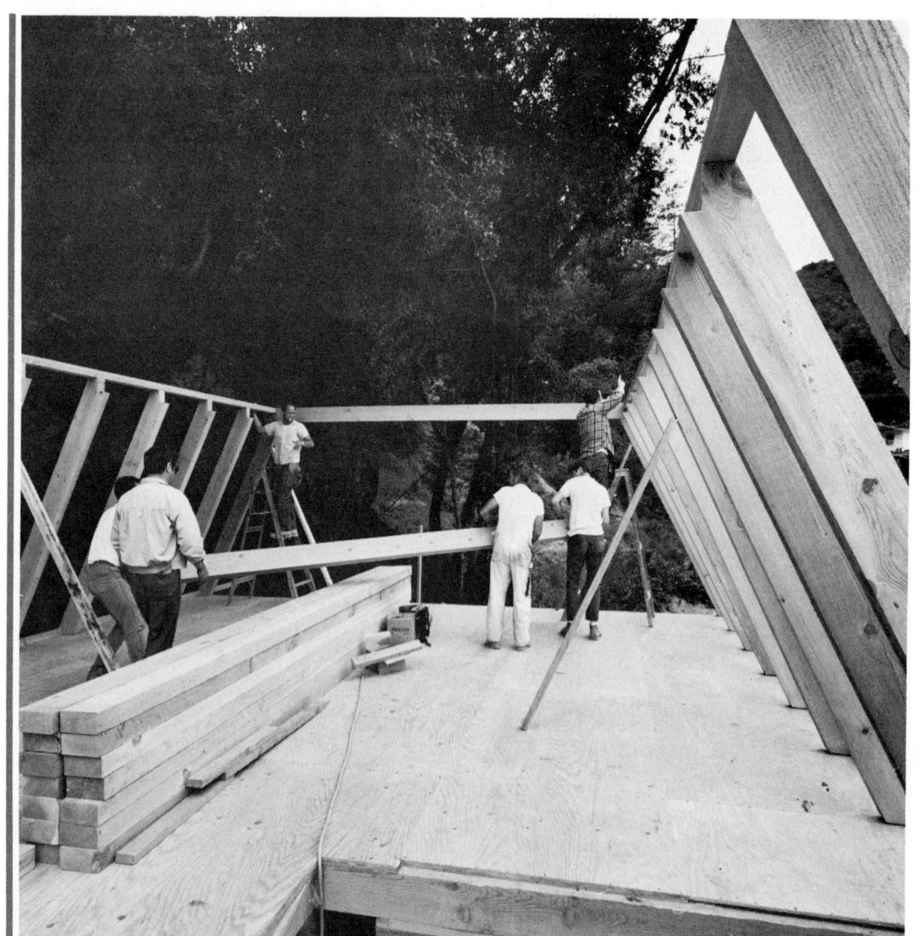

FIGURE 13-87

The lower-sloped rafters for a gambrel roof have been installed. The ceiling joists are now being nailed in place. *(Courtesy of American Plywood Association)*

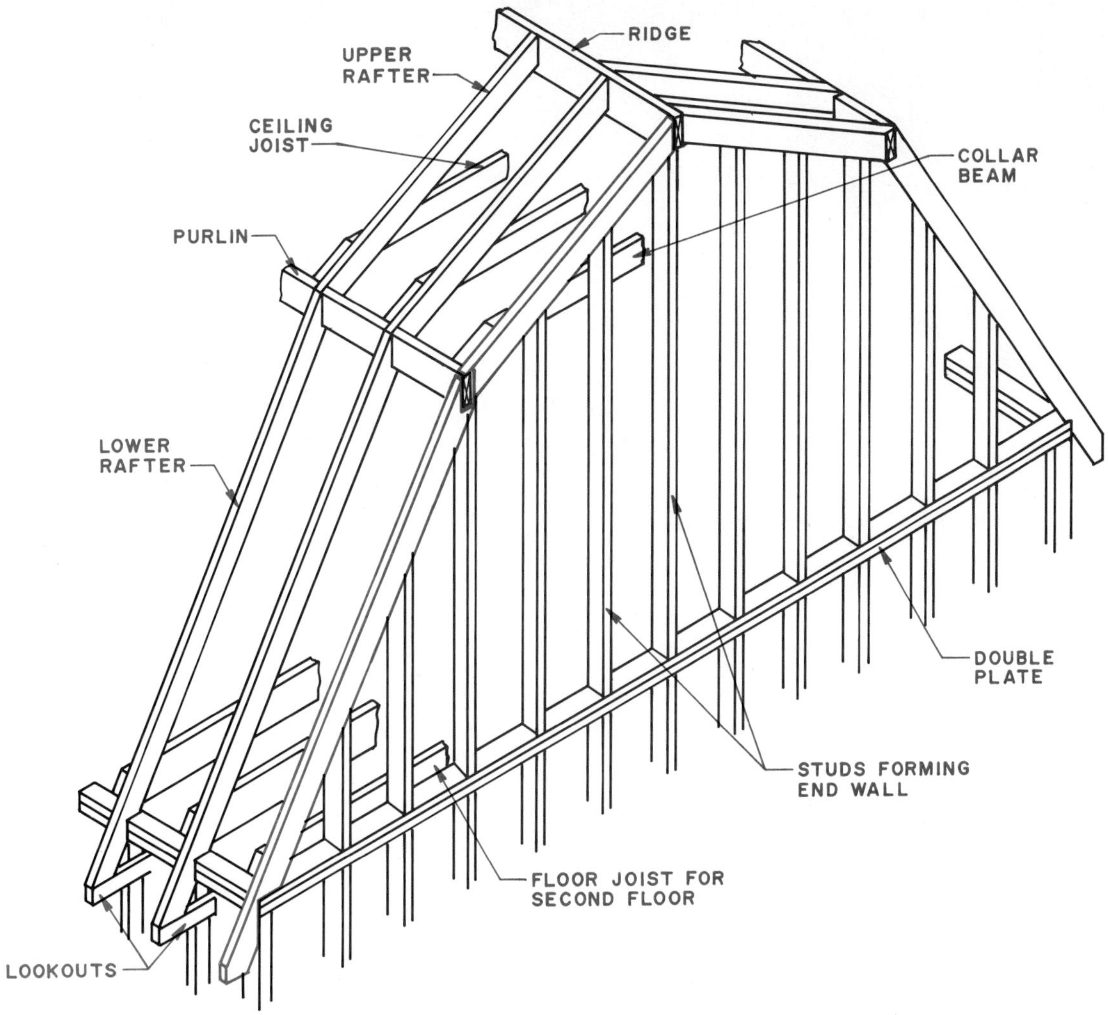

FIGURE 13-88

A complete framing plan for a gambrel roof.

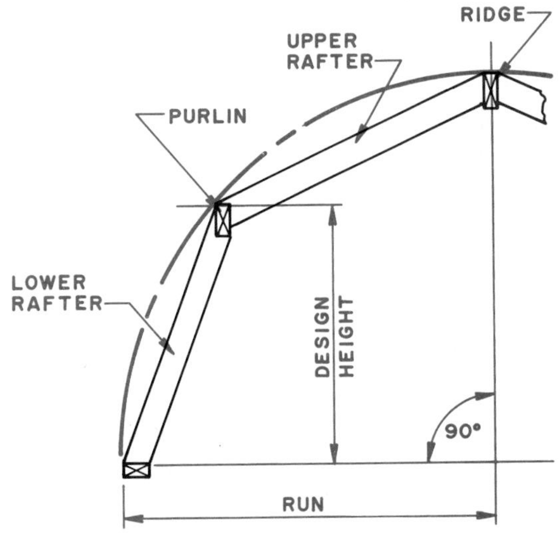

FIGURE 13-89

A full-size rafter layout for a gambrel roof.

MANSARD ROOFS

The *mansard roof* provides a full second floor. The lower slope is very steep. The upper slope is a hip roof (Fig. 13-90). The lower slope actually covers the second-floor wall. One advantage is that the second-floor exterior wall never needs painting.

Typical construction details are shown in Fig. 13-91. The second-floor exterior wall is framed in the conventional way. The floor joists are cantilevered beyond the plate. The lower slope rafters run from the upper slope rafter to a plate on this cantilevered section (Fig. 13-92).

A dormer-type window can be used in this roof. After roof sheathing is applied, it must be carefully flashed on all sides (Fig. 13-93). The window area can also be recessed (Fig. 13-94).

FIGURE 13-90

A house with a mansard roof. *(Courtesy of Home Planners, Inc.)*

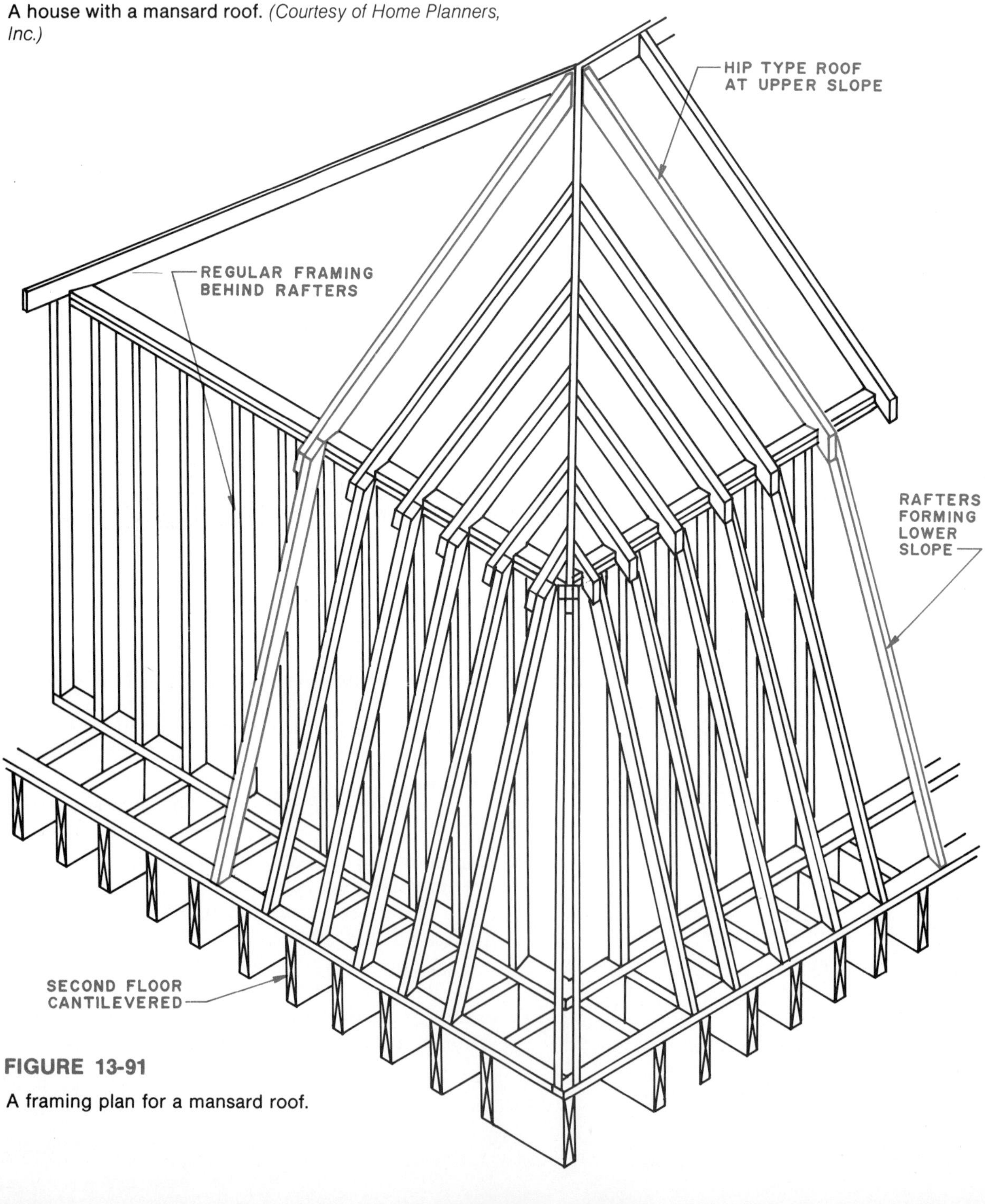

FIGURE 13-91

A framing plan for a mansard roof.

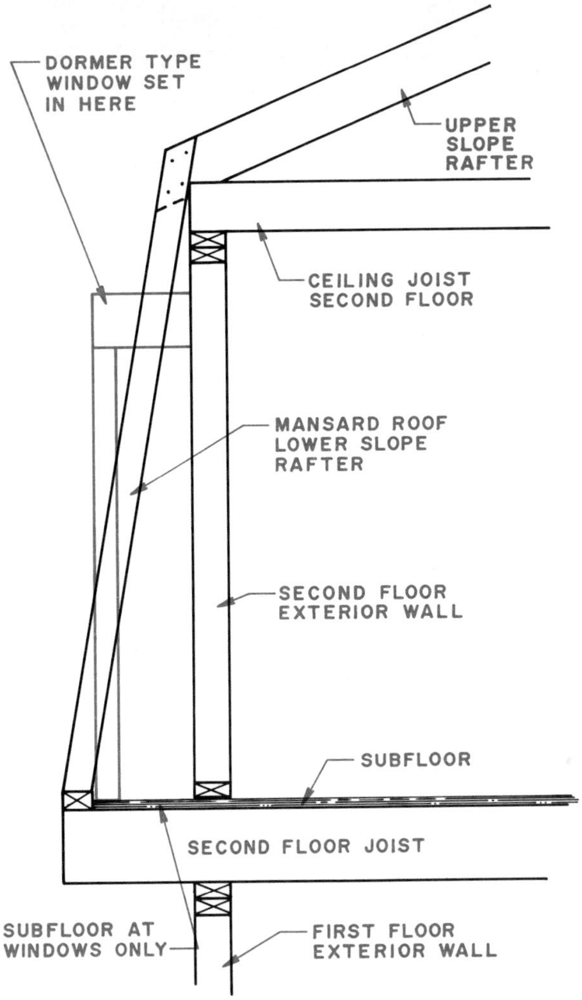

FIGURE 13-92

Dormer windows framing in a mansard roof.

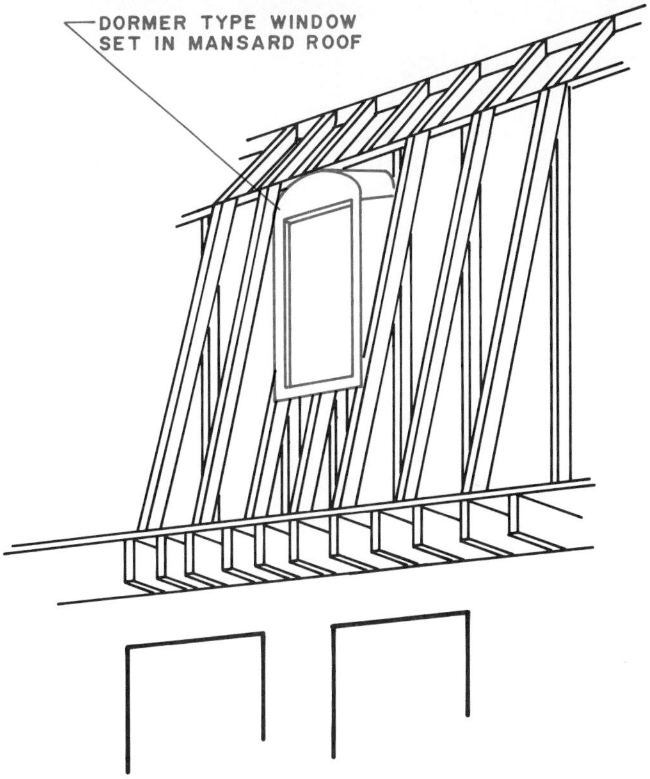

FIGURE 13-93

One type of dormer in mansard roof.

FIGURE 13-94

A window recessed into a mansard roof.

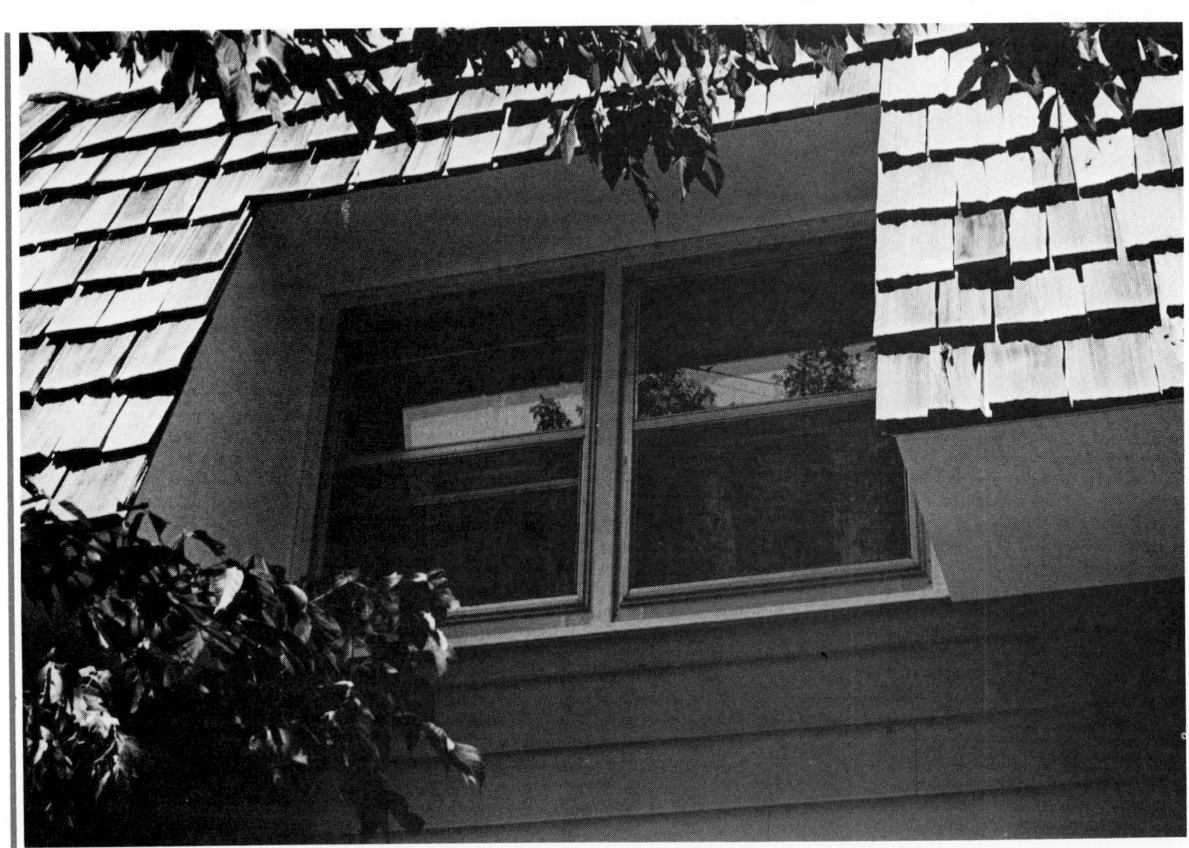

ROOF OPENINGS

Roof openings are used for chimneys and skylights. They are framed in the same way as openings in floors (Fig. 13-95). A double header is placed on each side. The headers should be in a vertical position (Fig. 13-96). Openings should clear masonry by at least 2 in.

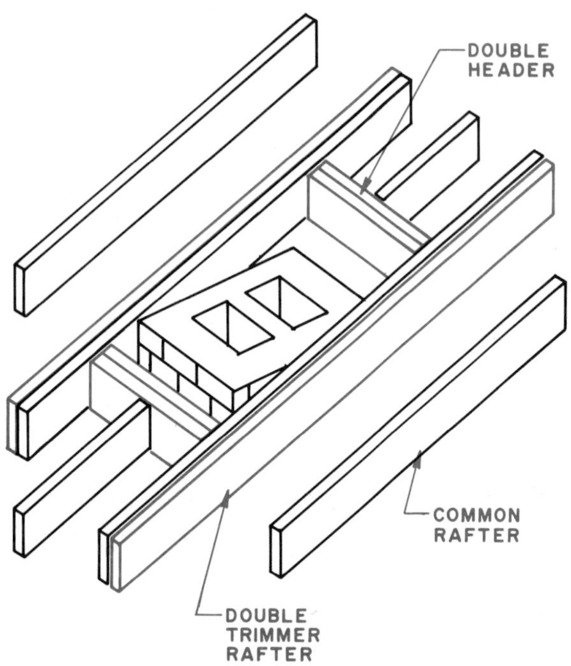

FIGURE 13-95

Nail double leaders around openings in a roof.

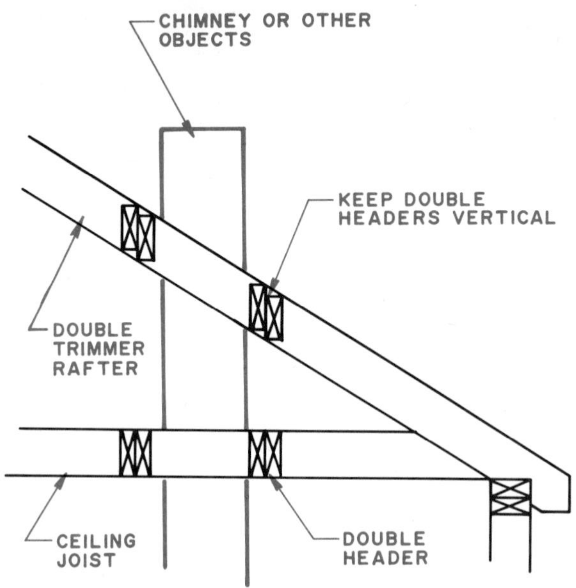

FIGURE 13-96

Roof opening leaders are kept in a vertical position.

WOOD ROOF TRUSSES

A *wood roof truss* is a rigid framework of structural members designed to carry roof loads over long spans with no intermediate support. They are formed by joining the structural members in triangular shapes (Fig. 13-97).

Most trusses are mass-produced using jigs in a factory. The assembled trusses are shipped to the construction site and set in place on the

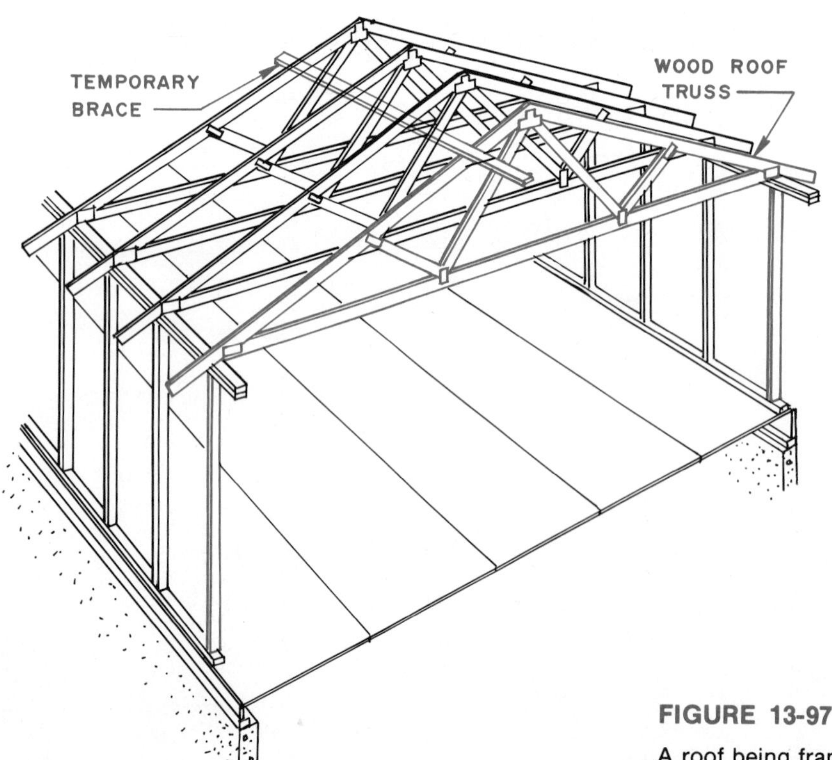

FIGURE 13-97

A roof being framed with wood trusses.

walls. Carpenters are occasionally expected to build trusses on the construction site.

Wood trusses most commonly used for house construction are the W-type, the King-post, and the scissors (Fig. 13-98). These are used on rectangular houses that have a constant width requiring only one type of truss. Trusses can be designed for use on L-shape plans and hip roofs. Special hip trusses are needed for each hip end and valley.

Most truss designs are for a 24-in. on-center spacing. This requires thicker roof sheathing and interior finish material than conventional framing on 16-in. centers.

W-Type Truss The *W-type truss* is most commonly used. It has four web members which reduce the distance spanned by the upper chord. This means that smaller structural members can be used than for conventional rafters. It also permits the use of a lower grade of lumber.

King-Post Truss The *King-post truss* has the simplest design. It has only one vertical center post connecting the upper and lower chords. It will not span the same distance as a W-type truss using the same-size structural members. It is more economical for short spans because it has fewer parts. For example, if the King-post and W-type trusses are made from 2 × 4-in. members and assembled in the same manner, the King-post will span 20 ft while the W-type will span 32 ft.

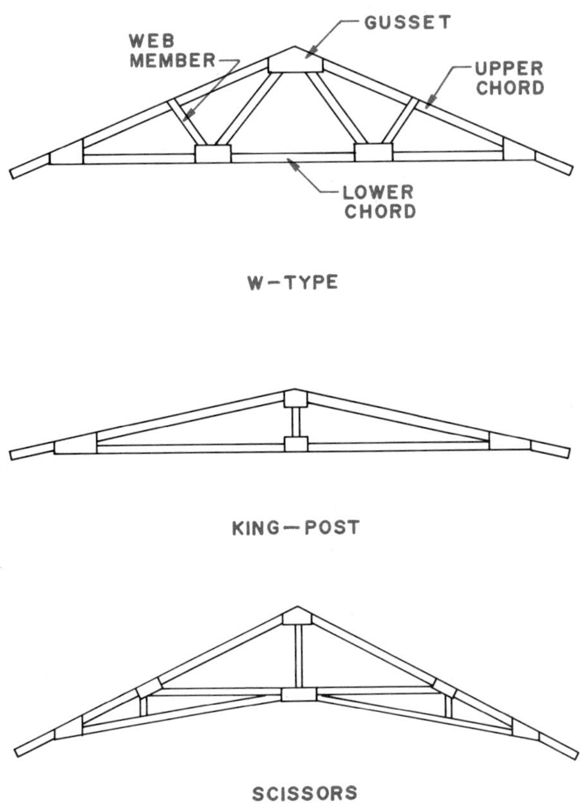

FIGURE 13-98

Commonly used types of trusses.

Scissors Truss The *scissors truss* is a special type used for houses requiring sloped ceilings. These are referred to as cathedral ceilings. They are more complex than W-type trusses. Scissors trusses are more economical for framing cathedral ceilings than is conventional construction.

TRUSS DESIGN

A roof truss is subject to snow and loads. It must also carry the weight of the sheathing and roofing material. The lower the slope of the roof, the greater are the stresses on it.

The design of a truss is the work of an engineer. There are many properly engineered designs available. If the carpenter is to build the trusses, preengineered plans should be obtained and followed carefully. Two sources are TECO Products and Testing Corporation, 5530 Wisconsin Ave., Washington, D.C. 20015, and Midwest Plan Service, Iowa State University, Ames, Iowa 50010.

Trusses are usually fabricated using plywood gussets, metal gussets, or split-ring connectors and bolts.

A typical truss with plywood gussets is shown in Fig. 13-99. The plywood gussets have the grain of the outer ply running horizontally. They are from standard plywood with exterior glue or exterior-sheathing-grade plywood. The recommended sizes for the gussets are shown in Fig. 13-99. In areas having high humidity, gussets are glued in place with a resorcinol glue. In dry climates a casein glue is used. Gluing should be done under careful temperature control. Use the glue manufacturer's recommendations.

The plywood gussets are nailed or stapled in place. Use 4d nails for ⅜-in.-thick gussets. 6d nails are used for ½- to ⅞-in.-thick gussets. Nails are spaced 3 in. apart on ⅜-in. plywood. They are spaced 4 in. apart on ½-in. or thicker plywood.

If the structural members are 4 in. wide, set the nails in ¾ in. from the edge in two rows. For 6-in.-wide material, use three rows of nails.

There are a variety of metal gussets. One type is shown in Fig. 13-100. Before using these gussets, careful study of proper joint design is necessary. A truss using metal gussets is in Fig. 13-101.

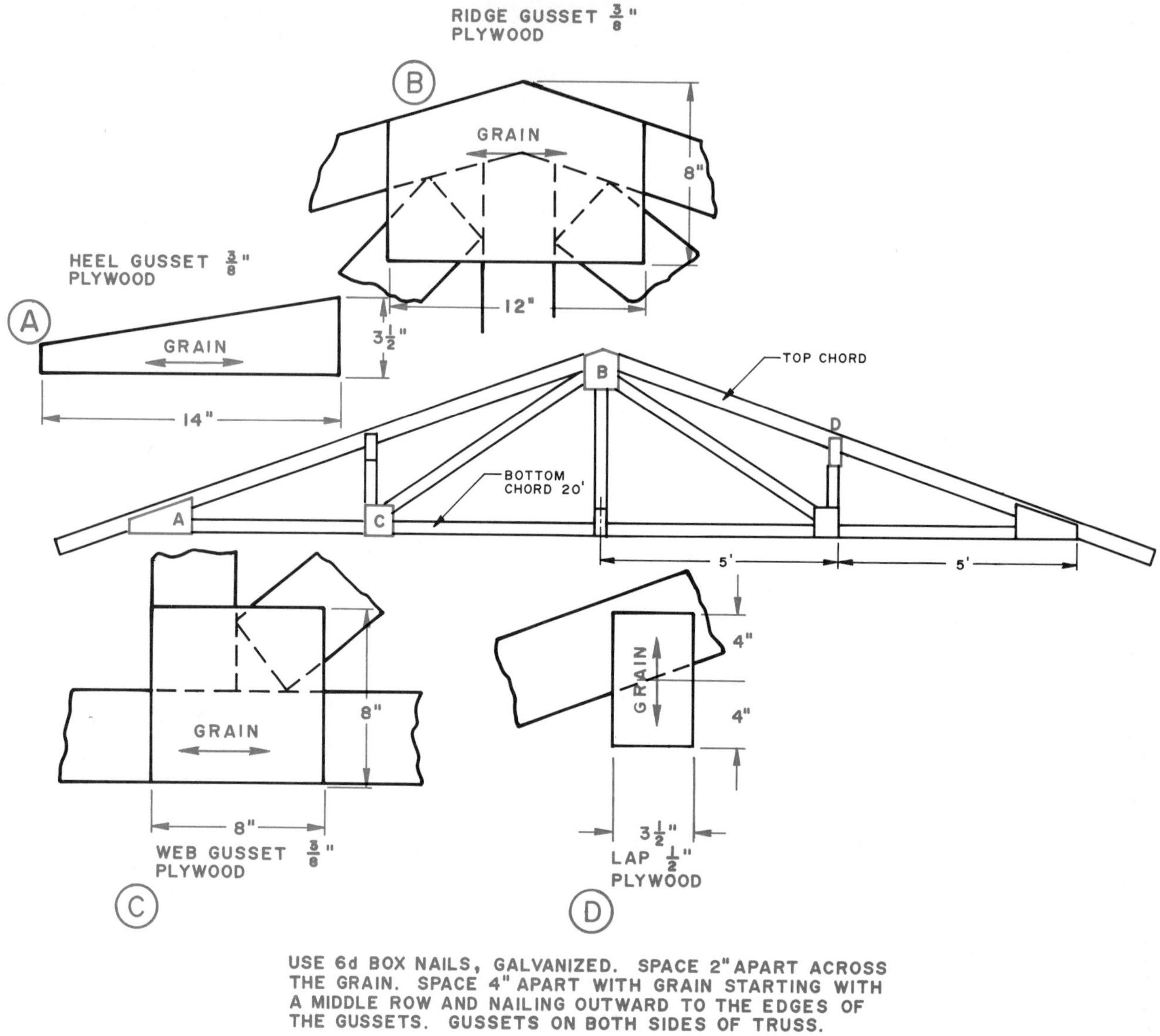

RIDGE GUSSET $\frac{3}{8}$"
PLYWOOD
B
GRAIN
8"
12"
HEEL GUSSET $\frac{3}{8}$"
PLYWOOD
A
GRAIN
$3\frac{1}{2}$"
14"
TOP CHORD
B
D
BOTTOM
CHORD 20'
A
C
5'
5'
GRAIN
8"
8"
WEB GUSSET $\frac{3}{8}$"
PLYWOOD
C
GRAIN
4"
4"
$3\frac{1}{2}$"
LAP $\frac{1}{2}$"
PLYWOOD
D
USE 6d BOX NAILS, GALVANIZED. SPACE 2" APART ACROSS THE GRAIN. SPACE 4" APART WITH GRAIN STARTING WITH A MIDDLE ROW AND NAILING OUTWARD TO THE EDGES OF THE GUSSETS. GUSSETS ON BOTH SIDES OF TRUSS.

FIGURE 13-99

A typical King-post truss with plywood gussets.

TOP
CHORD
WEB
TYPICAL NAIL—ON TRUSS PLATE

FIGURE 13-100

Metal gussets are nailed to the wood truss members.

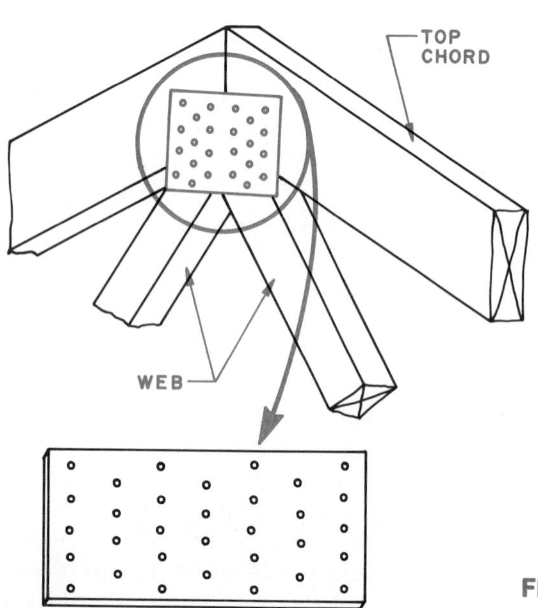

The *split-ring roof truss* has been in use for many years. It is easy to assemble and ensures accurate fabrication. Many engineered and tested designs are available. The system involves cutting a circular recess in the mating structural members (Fig. 13-102). The split-ring is inserted in the recess. When the two halves are bolted together, the ring keeps them in alignment and strengthens the joint. Split-ring trusses are used primarily for specialized situations. These include churches, heavy construction, and commercial buildings. Some small-volume builders use them because they

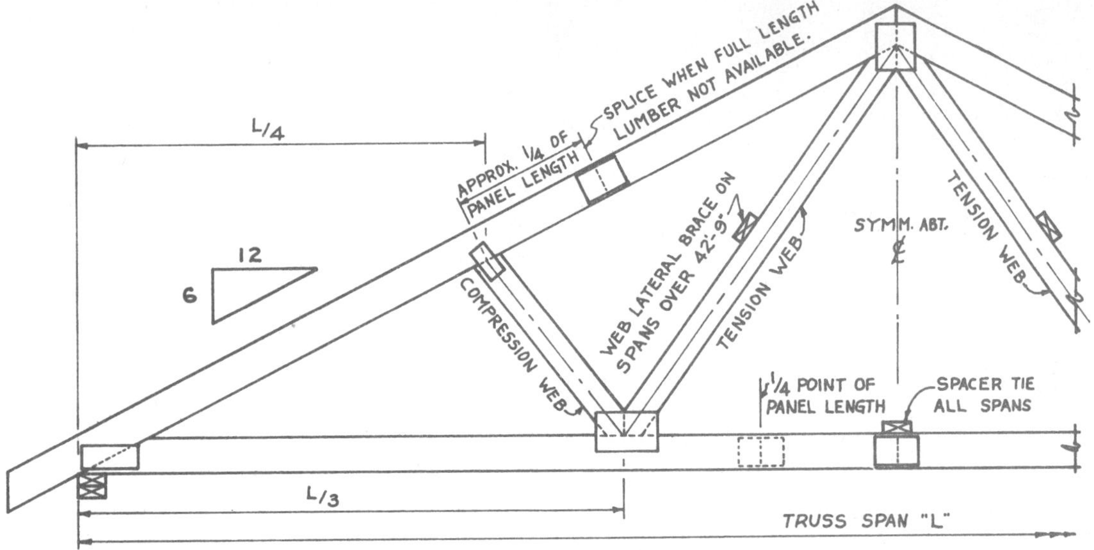

FIGURE 13-101

A W-type truss with metal gussets.

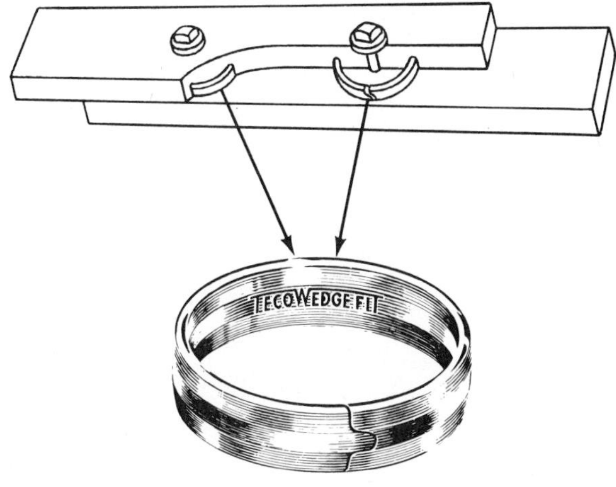

FIGURE 13-102

A split-ring truss connector. *(Courtesy of Teco Products and Testing Corporation)*

can be easily built by on-site carpenters. A typical preengineered split-ring truss is shown in Fig. 13-103.

HANDLING TRUSSES

Trusses are designed to carry a load when in a vertical position. They should be lifted and stored in a vertical position. If they must be handled in a flat position, they should be placed on supports running their entire length.

Erecting Trusses

The location of each truss is marked on the top wall plate. Trusses are then set in place. Large heavy trusses are best handled with a crane. It lifts them into position and holds them there as they are fastened to the wall and braced. Smaller trusses can be set on each wall with the point down (Fig. 13-104). A carpenter then rotates the point upright while two other carpenters fasten the truss to the wall and brace it. Normally, three workers are needed to set trusses in place (Fig. 13-105).

The best way to join the truss to the plate is with a metal anchor. This provides good resistance to upward stresses and horizontal thrust (Fig. 13-106).

Trusses can be toenailed to the plate. The plywood gusset makes toenailing difficult. If this is done, use two 10d nails on each side of the lower chord (Fig. 13-106). Since this may split the plywood gusset, it is desirable to predrill the nail holes. It is best to supplement this with a metal strap.

The trusses are braced near the ridge with a 2 × 4-in. board. It holds them in position until the sheathing is in place (Fig. 13-107).

FIELD-BUILT TRUSSES

Generally, builders buy factory-assembled trusses. These are assembled on jigs and delivered to the site ready to erect. Often these cost less than those built on the site by carpenters. A typical truss assembly table is given in Fig. 13-108.

Site-built trusses must be built according to an accepted engineering design. They are assembled on a jig in the same manner as factory-built trusses. If glued gussets are used, the trusses should be built in a building where temperatures can be controlled.

DEAD LOAD + LIVE LOAD ON ROOF 35#/□'
DEAD LOAD CEILING FRAMING 10#/□'
TRUSSED RAFTERS SPACED 2'-0" o.c.

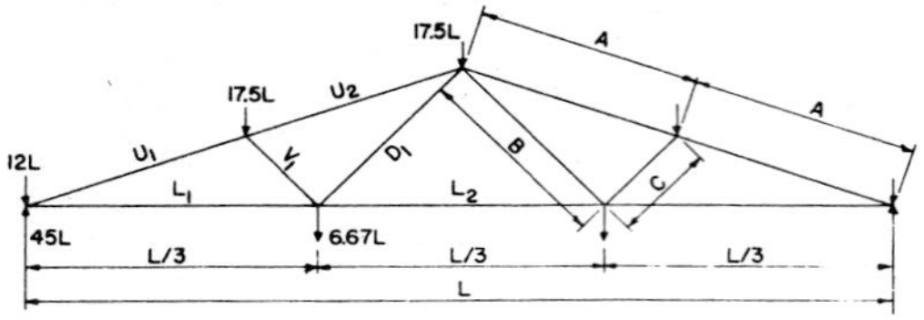

LUMBER														
SPANS	2"X8"		2"X6"				2"X4"						F.B.M. DRYWALL CEILING	F.B.M. PLASTER CEILING
	18'	16'	18'	16'	14	12'	18'	16'	14'	12'	10'	8'		
20'-0"						2				2		2	51	51
22'-0"					2				2		2		61	61
24'-0"					2				2		2		61	61
26'-0"				2				2			2		68	68
28'-0"		2		(2)				2		2			80	91
30'-0"	2		(2)				2			2			88	100
32'-0"	2		(2)				2			2			88	100

NOTE: PIECES IN PARENTHESES ( ) ARE USED IN PLACE OF EQUAL LENGTH 2X4'S WHEN PLASTER CEILING IS USED.

HARDWARE			
SPAN		ITEM	SIZE
20,22 24,26	28,30 32		
11	13	TECO WEDGE FIT SPLIT RINGS	2 1/2" DIA.
2	2	TECO TRIP-L-GRIP	TYPE-A
4	6	BOLTS	1/2"X4"
2	2	BOLTS	1/2"X6"
1	1	BOLT	1/2"X7 1/2"
14	18	WASHERS	1/2"

NOTES

THIS DESIGN SUGGESTION IS TO BE USED AS A GUIDE ONLY BY QUALIFIED DESIGNERS IN THE PREPARATION OF THEIR FINAL DESIGN.

LUMBER SHALL BE OF A GROUP "B" CONNECTOR LOAD SPECIES AND OF SUFFICIENT QUALITY TO PERMIT ALLOWABLE UNIT STRESSES OF:
f = 900 p.s.i EXTREME FIBER IN BENDING
c = 900 p.s.i COMPRESSION PARALLEL TO GRAIN
E = 1,760,000 p.s.i MODULUS OF ELASTICITY

THIS DESIGN SUGGESTION IS PREDICATED ON THE USE OF TECO WEDGE FIT 2 1/2" DIAMETER DOUBLE-BEVEL SPLIT RINGS AS MANUFACTURED BY THE TIMBER ENGINEERING CO., WASH., D.C. (SEE CROSS SECTION DETAIL)

BOLTS SHALL BE 1/2" DIAMETER MACHINE BOLTS, WITH WASHERS, AND SHALL BE TIGHT AT TIME OF ERECTION.

END DISTANCES SHALL BE 5 1/2" UNLESS OTHERWISE NOTED.

THE DIMENSIONS SHOWN WILL PROVIDE APPROXIMATELY 1/2" CAMBER AS INDICATED.

TRUSSED RAFTERS SHALL BE ANCHORED TO WALL PLATES WITH TRIP-L-GRIP FRAMING ANCHORS AS MANUFACTURED BY THE TIMBER ENGINEERING COMPANY, WASHINGTON, D.C. (SEE ANCHORAGE DETAIL)

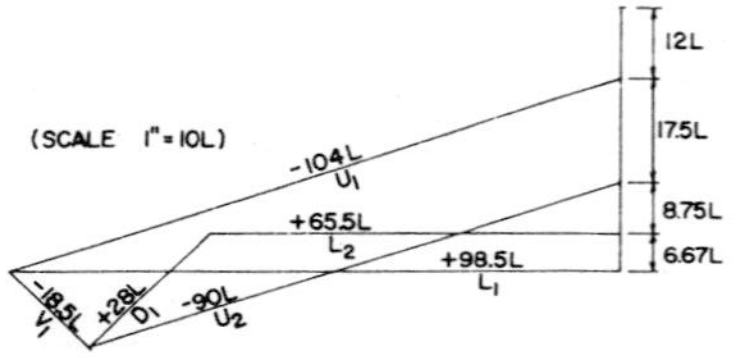

SPAN L	DIMENSIONS			DESIGN STRESSES					
	A	B	C	U_1	U_2	L_1	L_2	V_1	D_1
20'-0"	5'-3 1/4"	4'-8 3/16"	2'-3 15/16"	2080	1800	1970	1310	370	560
22'-0"	5'-9 9/16"	5'-1 7/8"	2'-6 3/4"	2290	1980	2170	1440	407	616
24'-0"	6'-3 7/8"	5'-7 1/2"	2'-9 9/16"	2500	2160	2360	1570	444	672
26'-0"	6'-10 3/16"	6'-1 3/16"	3'-0 7/16"	2700	2340	2560	1700	481	728
28'-0"	7'-4 9/16"	6'-6 13/16"	3'-3 1/4"	2910	2520	2760	1830	518	784
30'-0"	7'-10 7/8"	7'-0 1/2"	3'-6 1/16"	3120	2700	2960	1960	555	840
32'-0"	8'-5 3/16"	7'-6 3/16"	3'-8 7/8"	3330	2880	3150	2100	592	896

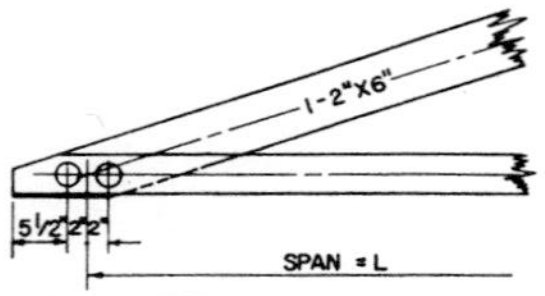

ALTERNATE HEEL DETAIL

TOP CHORD SIZE MAY BE 2"X6" FOR ALL SPANS IF A MATERIAL IS USED WHICH WILL PROVIDE f = 1200 p.s.i., c = 900 p.s.i., AND E = 1,760,000 p.s.i. THE HEEL DETAIL SHOWN ABOVE APPLIES TO SPANS 28' TO 32' WITH THE 2"X6" TOP CHORD.

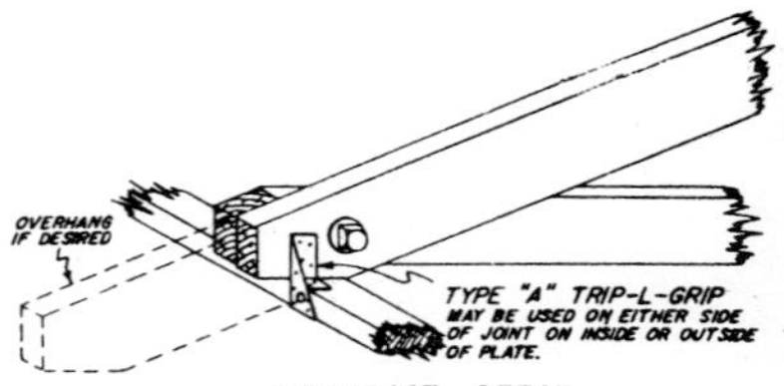

ANCHORAGE DETAIL

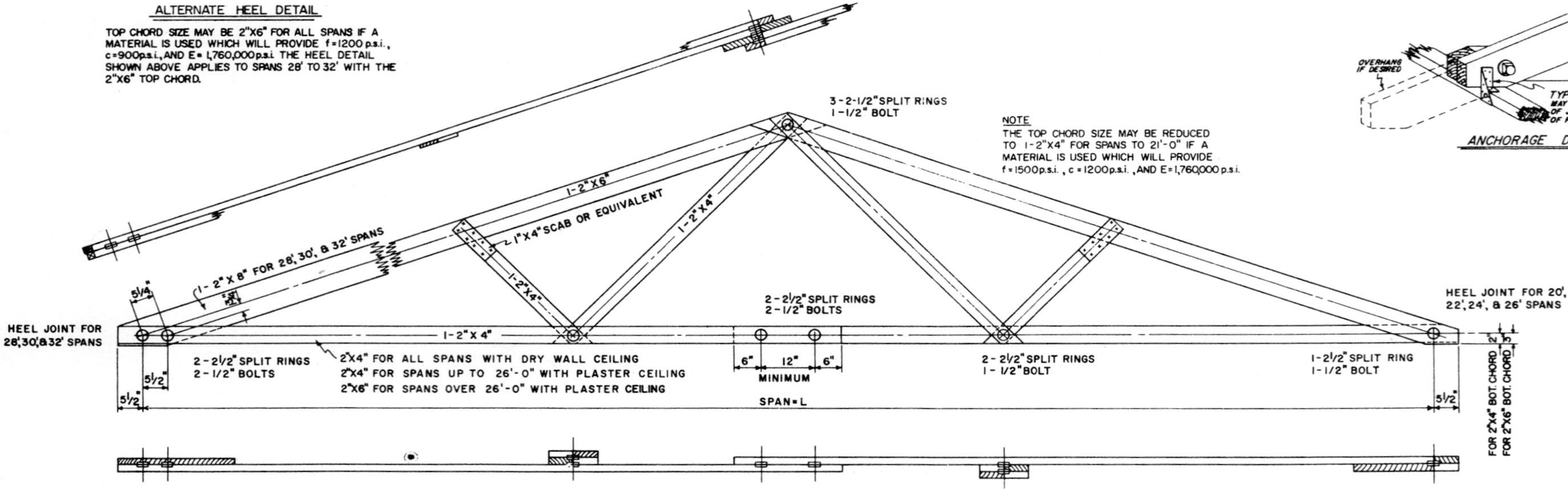

FIGURE 13-103

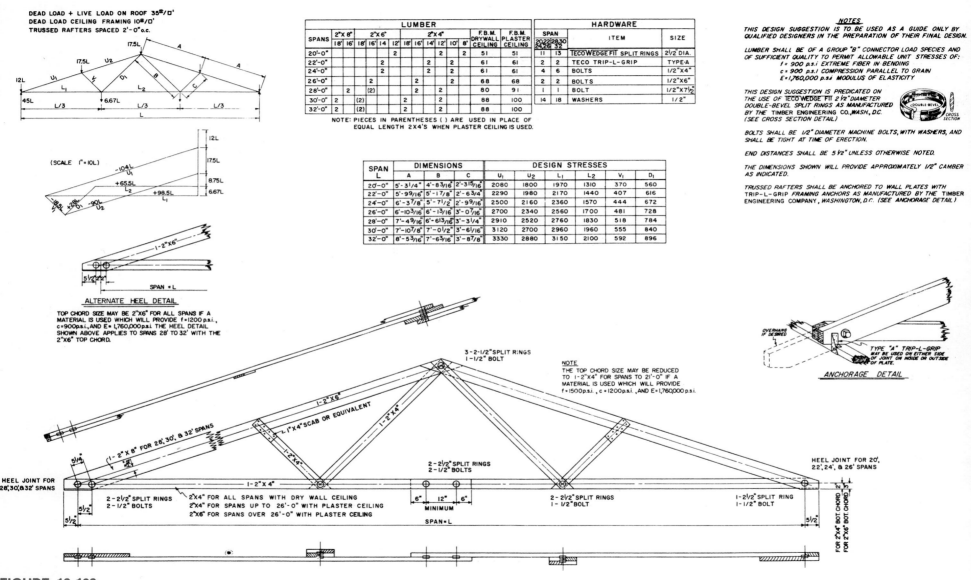

Design data for a W-truss using split-ring connectors. *(Courtesy of Teco Products and Testing Corporation)*

FIGURE 13-104

Trusses are set on each wall with the point down. *(Courtesy of Lumber Dealers Research Council)*

(A)

(B)

FIGURE 13-105

(A) A notched 2 × 4 in. is used to raise the truss. (B) The truss is rotated to a vertical position.

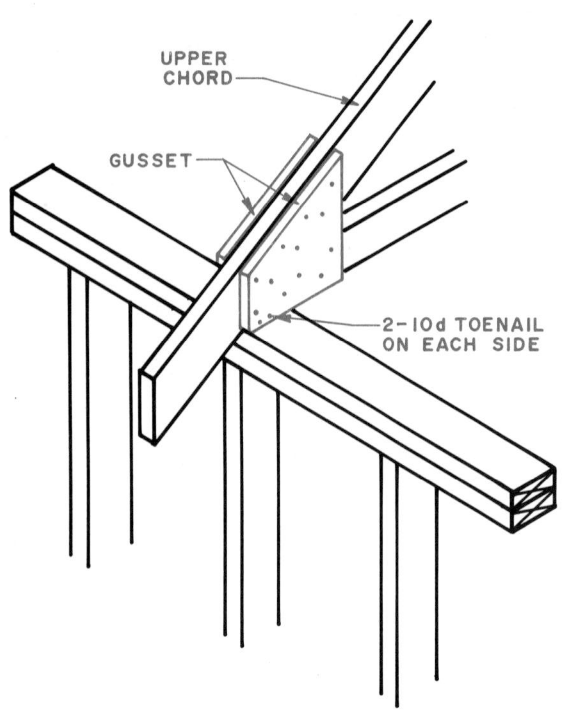

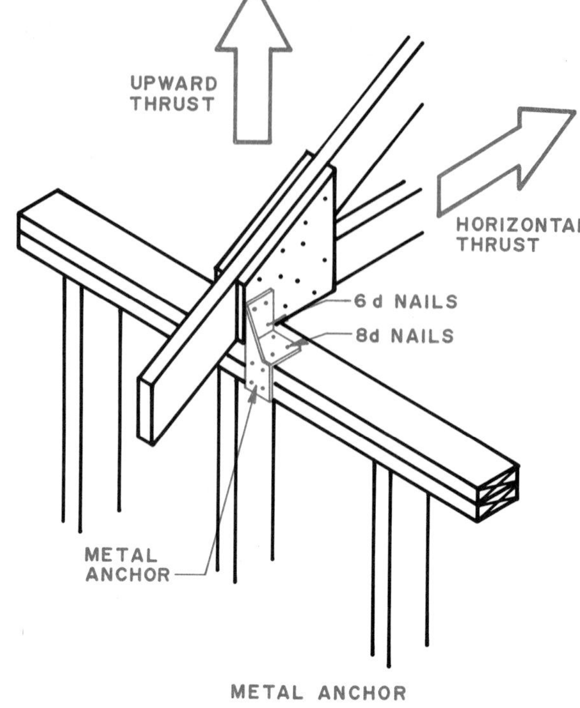

FIGURE 13-106

Trusses are joined to the wall by toenailing or with metal anchors.

FIGURE 13-107

Nail a horizontal brace near the top of the truss to hold it in position.

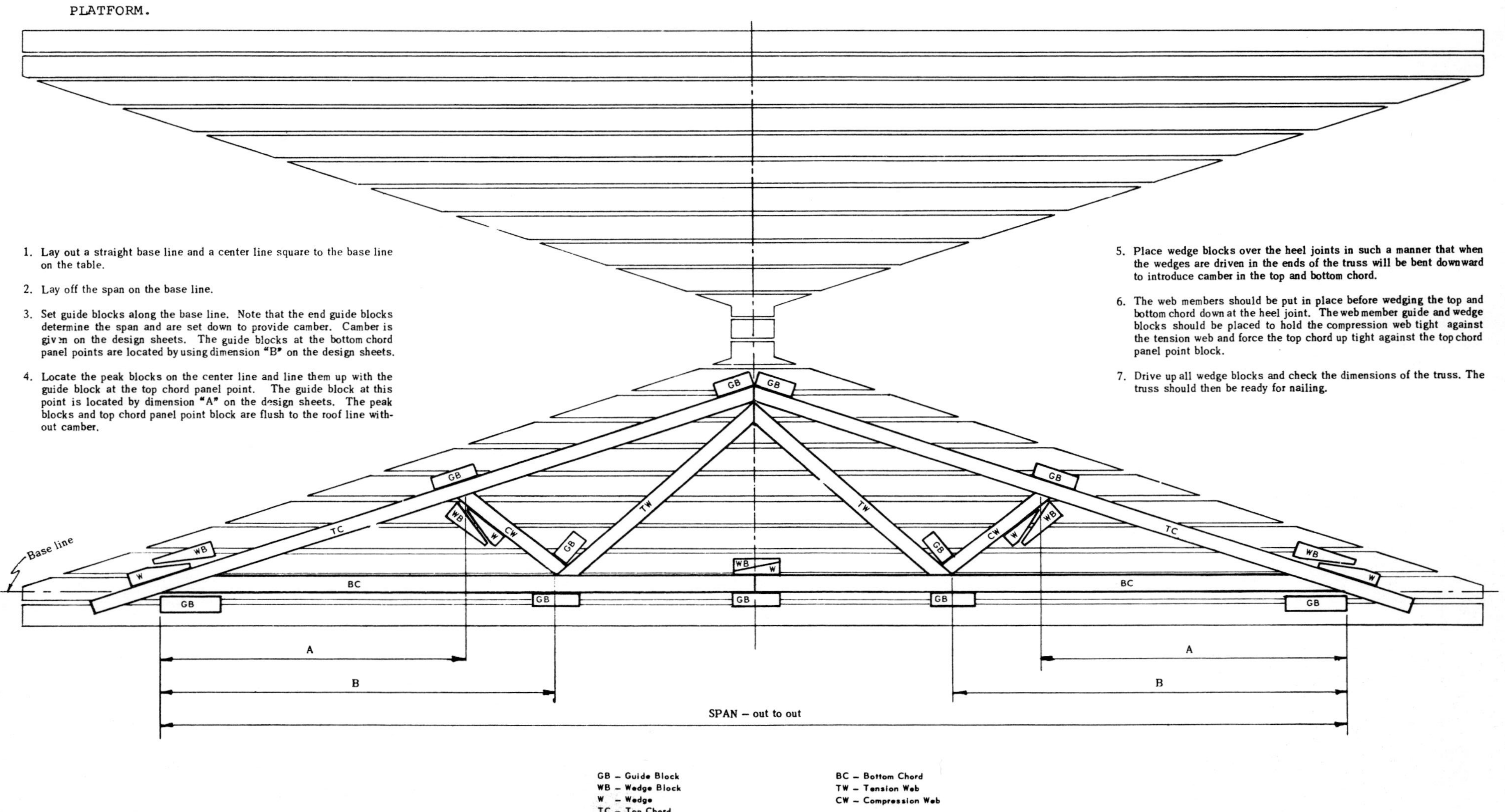

Suggested Assembly Table
THE FLOOR OF THE BUILDING USING THE TRUSSES CAN ALSO BE USED AS AN ON-SITE FABRICATING PLATFORM.
1. Lay out a straight base line and a center line square to the base line on the table.
2. Lay off the span on the base line.
3. Set guide blocks along the base line. Note that the end guide blocks determine the span and are set down to provide camber. Camber is given on the design sheets. The guide blocks at the bottom chord panel points are located by using dimension "B" on the design sheets.
4. Locate the peak blocks on the center line and line them up with the guide block at the top chord panel point. The guide block at this point is located by dimension "A" on the design sheets. The peak blocks and top chord panel point block are flush to the roof line without camber.
5. Place wedge blocks over the heel joints in such a manner that when the wedges are driven in the ends of the truss will be bent downward to introduce camber in the top and bottom chord.
6. The web members should be put in place before wedging the top and bottom chord down at the heel joint. The web member guide and wedge blocks should be placed to hold the compression web tight against the tension web and force the top chord up tight against the top chord panel point block.
7. Drive up all wedge blocks and check the dimensions of the truss. The truss should then be ready for nailing.
Base line
GB
WB
W
TC
CW
TW
BC
A
B
SPAN – out to out
GB – Guide Block
WB – Wedge Block
W – Wedge
TC – Top Chord
BC – Bottom Chord
TW – Tension Web
CW – Compression Web

FIGURE 13-108

A typical truss assembly jig table layout. *(Courtesy of Teco Products and Testing Corporation)*

STUDY QUESTIONS

After reading the chapter, answer each of the following questions. If you do not know the answer, review the chapter.

1. What are the common types of roofs?
2. How does the roof span differ from the run?
3. How does the roof slope differ from the pitch?
4. List the types of rafters used in roof framing.
5. What are the three methods for laying out a rafter?
6. What are the two types of dormers?
7. What types of trusses are used in residential construction?
8. What types of gussets and connectors are used to manufacture trusses?

IMPORTANT TECHNICAL TERMS

Following are technical terms that you should be able to use as part of your working vocabulary. Write a brief description of the meaning of each term.

Flat roof
Shed roof
Butterfly roof
Gable roof
Hip roof
Gambrel roof
Mansard roof
Span
Run
Measuring line
Rise
Unit of rise
Unit of run
Slope
Pitch
Common rafter
Hip rafter
Valley rafter
Hip jack rafter
Valley jack rafter
Cripple jack rafter
Plumb cut
Bird's-mouth
Tail
Tail cut
Ridge board
Collar beams
Louver
Open soffit
Boxed soffit
Gable end
Lookouts
Backing off a hip rafter
Common differences of hip jack rafters
Dormer
Purlin
Truss

14 FINISHING THE ROOF

After the roof is framed, it is covered with sheathing. This is solid wood, plywood, or wood fiber materials. The sheathing is covered with an underlayment such as asphalt-saturated felt. Flashing is placed at the eaves, around openings in the roof, and where it meets vertical walls. The final finish is some type of shingle. Commonly used are asphalt, mineral fiber, and wood shingles.

ROOF SHEATHING

The common types of roof sheathing include lumber, plywood, planking, and wood fiber decking. These materials are nailed to the rafters. They support the finished roofing material. Plywood or solid wood sheathing is used for pitched roofs. Planking and wood fiber decking is used for homes in which the underside of the material forms the exposed ceiling.

Lumber Sheathing *Lumber sheathing* can be laid closed or spaced (Fig. 14-1). The boards should be 6 to 8-in. wide. If placed on rafters spaced 16 to 24 in. on center, they should be ¾-in. thick. Use two 8d common nails at every rafter. If square-edge boards are used, their ends must meet on a rafter. Not more than two adjacent boards should have joints on the same rafter. If end-matched tongue-and-groove boards are used, the end joints can fall between rafters. In no case should the joints of two adjoining boards be in the same rafter space. Each board must be supported by at least two rafters.

When *wood shingles* or *shakes* are used in damp climates, the boards should be spaced. This helps in drying the roof. The wood

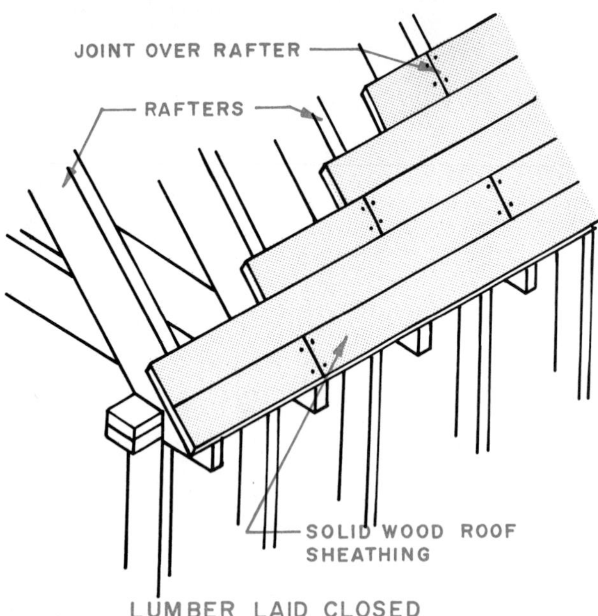

FIGURE 14-1

Solid lumber sheathing can be laid with no space between the boards.

strips are 1 × 3 or 1 × 4-in. stock. They are spaced on center the same as the shingles are laid to the weather. Laid to the weather means the amount of shingle actually exposed. For example, a 10-in. shingle lapped 5 in. by the one above it is laid 5 in. to the weather. For shingles laid 5 in. to the weather over 1 × 4-in. lumber, there will be about 1½ in. of space between the boards. Two 8d common nails should be placed in each board at each rafter (Fig. 14-2).

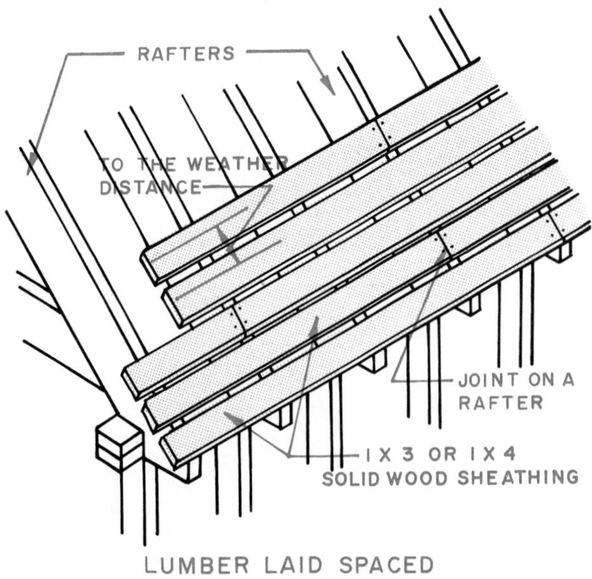

FIGURE 14-2

For certain roofing materials, lumber sheathing is best laid with a space between each piece.

Plywood Sheathing *Plywood sheathing* is laid with the face grain perpendicular to the rafters (Fig. 14-3). The sheathing grades of plywood include C-D Interior, C-D Interior with Exterior Glue, C-C Exterior, Structural I C-D Interior, Structural II C-D Interior, Structural I C-C Exterior, and Structural II C-C Exterior. These are all unsanded engineered grades. The C-D Interior with Exterior Glue is the most frequently used for roof and floor sheathing. Carpenters often refer to this as CDX. (Review Chapter 7 for details on plywood grades.)

The thickness of the plywood used depends upon the distance between rafters, the length of the unsupported edge of the panel, and the roof load. Design data are given in Fig. 14-4. The panel identification index is a set of two numbers separated by a slash, such as 32/16. The first number, 32, is the recommended center-to-center spacing of rafters when the panel is used as roof sheathing. The other number, 16, is the center-to-center spacing of floor joists when the panel is used for subflooring.

Under normal conditions 5/16-in. plywood is used for 16-in. rafter spacing when wood or asphalt shingles are used. When the span is 24 in., ⅜-in. plywood is used. Heavier roofs, such as slate, tile, and asbestos cement, require ½-in. plywood for 16-in. rafter spacing and ⅝-in. for 24-in. spacing.

If interior-type sheathing is used, no edges

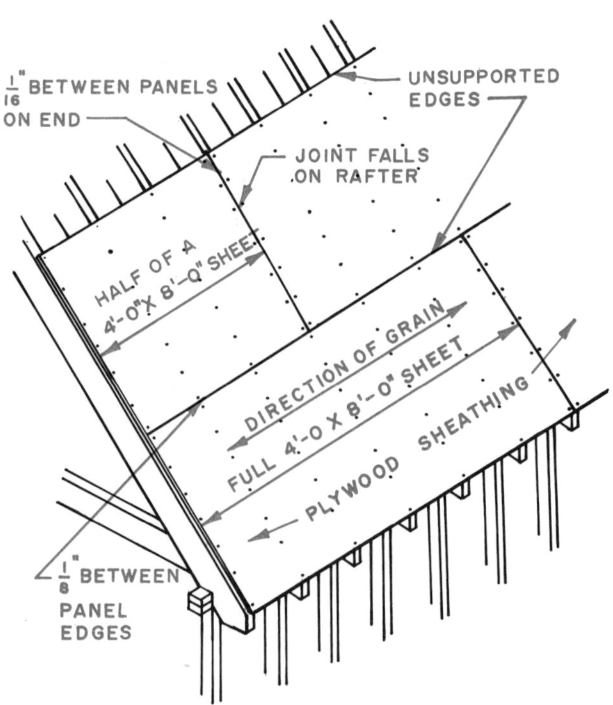

FIGURE 14-3

Plywood sheathing is laid with the face grain perpendicular to the rafters.

C-D
16 / 0 APA
INTERIOR
PS 1-74 000

IDENTIFICATION INDEX	PLYWOOD THICKNESS (in.)	MAXIMUM SPAN (in.)	UNSUPPORTED EDGE, MAX. LENGTH (in.)	ALLOWABLE LIVE LOADS (PSF)[a] Spacing of Supports Center to Center (in.)									
				12	16	20	24	30	32	36	42	48	60
12/0	5/16	12	12	150									
16/0	5/16, 3/8	16	16	160	75								
20/0	5/16, 3/8	20	20	190	105	65							
24/0	3/8	24	20	250	140	95	50						
	1/2		24										
32/16	1/2, 5/8	32	28	385	215	150	95	50	40				
42/20	5/8, 3/4, 7/8	42	32		330	230	145	90	75	50	35		
48/24	3/4, 7/8	48	36			300	190	120	105	65	45	35	
48/24	3/4, 7/8	48	36				225	125	105	75	55	40	
2-4-1	1 1/8	72	48				390	245	215	135	100	75	45
1 1/8 in. Grp. 1 and 2	1 1/8	72	48				305	195	170	105	75	55	35
1 1/4 in. Grp. 3 and 4	1 1/4	72	48				355	225	195	125	90	65	40

[a]Loads apply only to C-C EXT-APA, Structural IC-D, INT-APA, and Structural IC-C EXT-APA.

FIGURE 14-4

Design data for plywood roof sheathing. *(Courtesy of American Plywood Association)*

of the plywood panel should be exposed to the weather.

The carpenter should make a roof sheathing layout. This can be a freehand sketch to approximate scale. Draw a rectangle representing half the gable roof. One side is the length of the roof. The other is the rafter length plus overhang.

Start the layout at one corner with full 4 × 8-ft sheets. The second row of sheets starts with a half panel. In this way the joints are staggered. The top row will most likely have panels that must be cut to fit the remaining width. This layout can also be used to figure the number of sheets needed. Remember, it represents only half the roof. The number of sheets will have to be doubled (Fig. 14-5).

If a roof is to have an open soffit and sheathing with interior glue is used, the plywood covering the soffit will have to be panels with exterior glue. These are marked with an × on the layout. Generally, the soffit panels used will be thicker than the others. This is so that the roofing nails will not break through and be visible from below. The carpenter will have to shim up the thinner panel where the two types meet. The panels not over the soffit could be the same thickness, but this costs more.

Panels 5/16 to 1/2 in. thick are fastened with

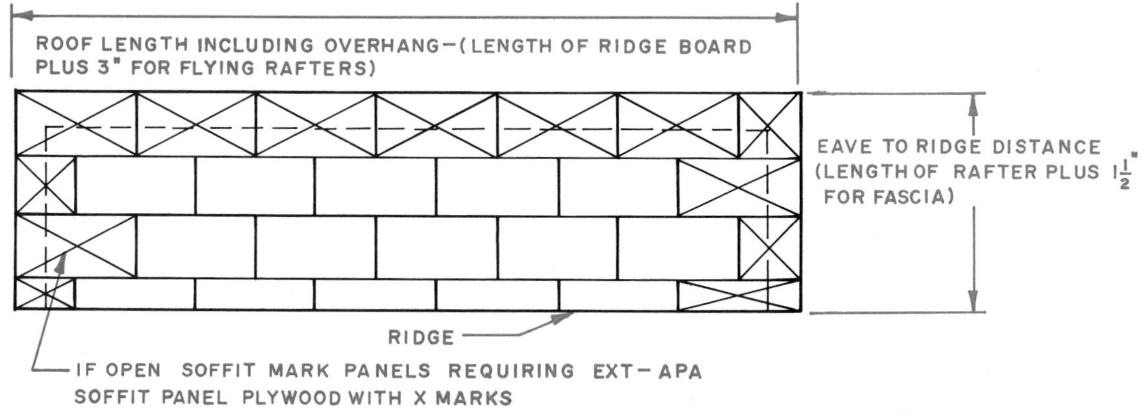

FIGURE 14-5

A roof sheathing layout for one side of a gable roof. *(Courtesy of American Plywood Association)*

6d common, ring shank, or spiral thread nails. Panels up to 1 in. thick require 8d common nails. They are spaced 6 in. on center on panel edges (Fig. 14-6). They are spaced 12 in. on center inside the panel. Always allow 1/16 in. between panel ends and 1/8 in. between panel edges.

Sometimes the unsupported edges of sheathing panels require blocking to increase stiffness. Instead of wood blocking, aluminum clips can be used. They are made in thicknesses from 5/16 to 13/16 in. They are tapered to their edges. They are installed by slipping them over the edges of the plywood panels (Fig. 14-7).

A fast way to move plywood panels to the roof is with a conveyor or crane (Figs. 14-8 and 14-9). On low-pitched roofs, the sheets can be stacked on the rafters. Steep-pitched roofs require that a few pieces of bracing be nailed to the rafters to keep the sheet from sliding off the roof.

On small jobs a carpenter on the ground can raise the sheets to a person on the roof. This person pulls them onto the roof and stacks them. It helps if they can be moved from the delivery truck directly onto the roof.

FIGURE 14-6

Plywood roof sheathing is nailed to the rafters. The ends of the sheet must rest on a rafter. *(Courtesy of American Plywood Association)*

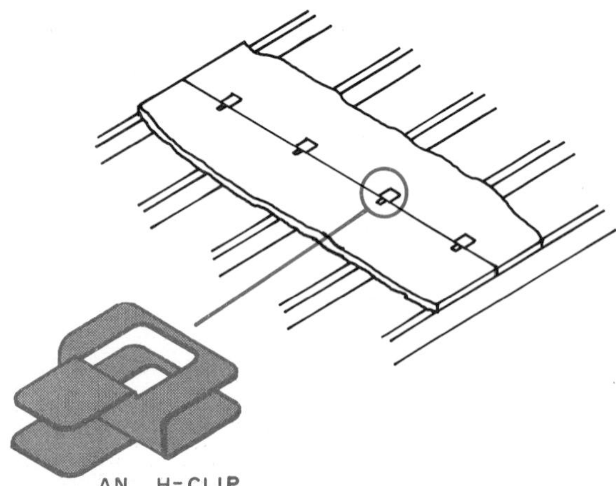

FIGURE 14-7

When unsupported edges of plywood sheathing require blocking, aluminum H-clips can be used.

FIGURE 14-8

Conveyors can be used to move sheathing to the roof. *(Courtesy of American Plywood Association)*

FIGURE 14-9

Large stacks of sheathing can be lifted onto the roof with a crane. *(Courtesy of American Plywood Association)*

Plank Roof Decking

Plank decking is tongue-and-groove wood 2 in. and thicker. It is used in flat and low-pitched roofs and in post-and-beam construction. It will span distances of 6 to 8 ft between supports. This material is covered in detail in Chapter 23.

Fiberboard Roof Decking *Fiberboard decking* is available in sheets 2 by 8 ft. The edges are tongue-and-grooved and the ends are square. The spacing between supports varies with the thickness. Typical spans are given in Fig. 14-10. On some types one surface is finished and serves as the finished interior ceiling (Fig. 14-11).

Fiberboard decking is usually nailed with corrosion-resistant nails. They are spaced 4 to 5 in. apart. They should penetrate the rafter at least 1½ in.

SHEATHING AT CHIMNEY OPENINGS

Rafters and headers should clear chimneys at least 2 in. Sheathing should overhang these but clear the chimney ¾ in. (Fig. 14-12).

THICKNESS (in.)	MAXIMUM JOIST SPACING (in.)
1½	24
2	32
3	48

FIGURE 14-10

Typical spans for fiberboard-type roof decking.

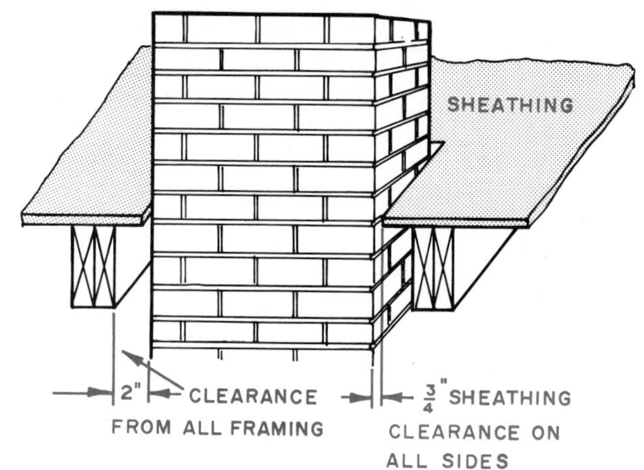

FIGURE 14-12

Wood framing and sheathing must clear chimneys.

SHEATHING AT THE GABLE END

When gable ends have little or no projection, the sheathing is set flush with the outside of the wall sheathing (Fig. 14-13). If it extends beyond the end wall, the sheathing should span at least three rafter spaces. This provides necessary strength and prevents sagging. Projections beyond 16 in. require special ladder framing.

FINISH ROOF COVERINGS

Shingles of various kinds are used on sloped roofs. Shingles shed water. Flat and low-pitched

FIGURE 14-11

Fiberboard roof decking produces a finished interior ceiling. *(Courtesy of Timber Structures, Inc.)*

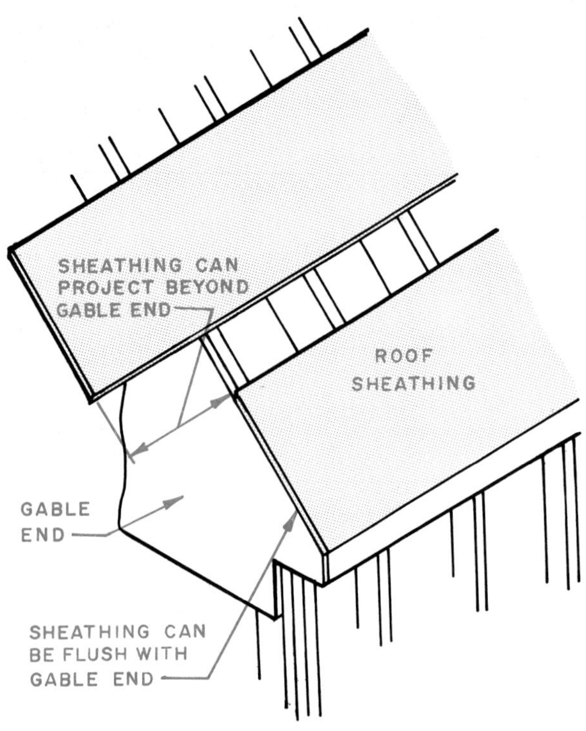

FIGURE 14-13

Sheathing can extend beyond the gable end or be set flush with it.

roofs use a builtup roof that provides a waterproof layer. Shingles contribute to the overall appearance because they are highly visible. They add color, texture, and pattern to the roof surface.

UNDERLAYMENT

The sheathing is covered with an *underlayment.* The material used varies with the type of finished roofing. The layers are edge-lapped 2 in. and end-lapped 4 in. (Fig. 14-14). Underlayment should be lapped 6 in. from both sides over all hips and ridges.

FIGURE 14-14

Underlayment is lapped on the edges and ends. It is placed below the rake drip edge.

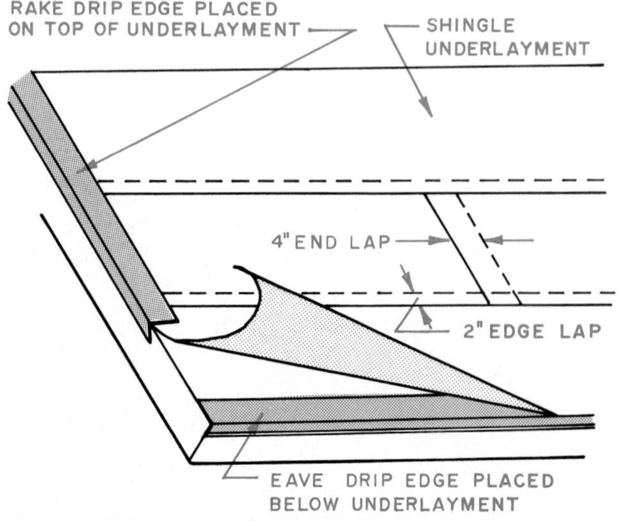

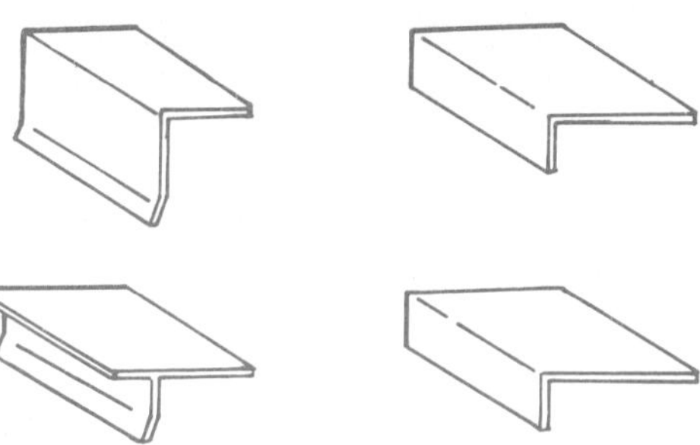

FIGURE 14-15

Typical types of metal drip edges.

DRIP EDGE

The edge and rake of the roof should have a metal *drip edge* applied. It is usually galvanized steel. It is designed to protect the edges of the sheathing and prevent leaks. At the eave the underlayment goes over the drip edge. At the rake of the roof it goes under the drip edge (Fig. 14-15).

ROOF FLASHING

Flashing is metal or some other material placed where special protection is needed to prevent water from entering. Anything that pierces the roof requires flashing. Examples include pipes, a chimney, or dormers. When two roofs intersect, as at a valley, flashing is necessary.

Flashing is generally galvanized metal, aluminum, copper, or stainless steel. The nails used to hold the flashing in place should be of the same material as the flashing. Mineral-surfaced roll roofing is also used.

Eave Flashing The eaves must be flashed in areas where the outside design temperature is 0°F or colder or where there is the possibility of ice forming along the eaves. This keeps the moisture backed up from penetrating the shingles and underlayment. The exact procedure varies with the type of finish roofing material.

Valley Flashing After the underlayment is in place and the eaves are flashed, the valleys and intersections with walls are flashed. As the underlayment is nailed in place a 36-in.-wide strip is nailed in the valley. The underlayment overlaps this 6 in. (Fig. 14-16). Use just enough roofing nails to hold the felt in place. If an open valley is wanted, two layers of mineral-surfaced roll roofing are added. An open valley is one

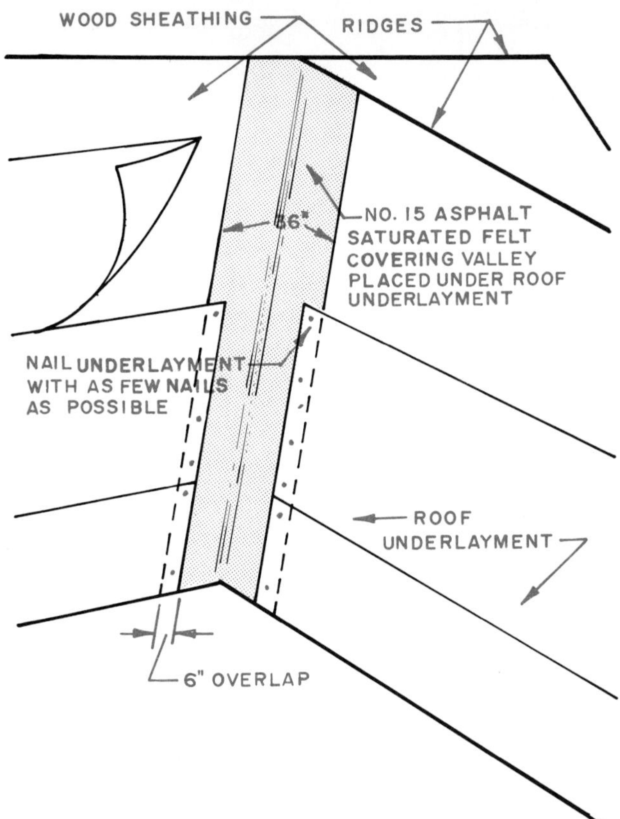

FIGURE 14-16

The valley underlayment is laid below the roof underlayment.

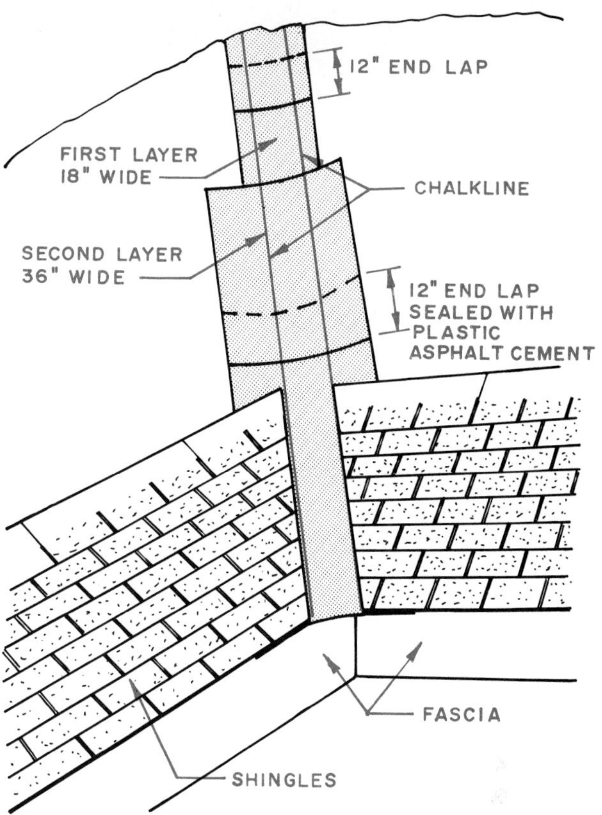

FIGURE 14-17

An open valley is flashed with two layers of mineral-surfaced roll roofing.

where the shingles on the intersecting roof do not meet. The layers are applied as shown in Fig. 14-17. The first layer is 18 in. wide. The second layer is 36 in. wide. The roof shingles are cemented to the flashing. Notice that the flashing is nailed only along the edges.

If a closed valley is wanted, one layer of 55-lb roll roofing is laid in the valley (Fig. 14-18). A closed valley is one in which the shingles overlap, covering the valley flashing.

Valleys can also be flashed with metal flashing. This is usually aluminum or galvanized steel. It helps if the metal has a seam folded in the center (Fig. 14-19). This helps keep the water from one roof from running under the shingles.

The line of the shingles on open valley flashing is located with a chalk line. Snap a chalk line from the ridge to the eave. Space 6 in. apart at the ridge and slope about ⅛ in. per foot of valley to the eave. The shingles are cut to the chalk-line mark, forming a straight edge (Fig. 14-20).

As the shingles are laid to the chalk line, their upper corner is cut on about a 45° angle. This helps prevent water from running under the shingles. The shingles are cemented to the valley lining with asphalt cement. No nails should be exposed along the valley.

Closed valley shingles are overlapped at the valley. This provides a double cover of shingles over the valley flashing. Each row of shingles is laid alternately over the other (Fig. 14-18). No nails are put closer than 6 in. from the

FIGURE 14-18

A closed valley is flashed with a layer of 55-lb roll roofing.

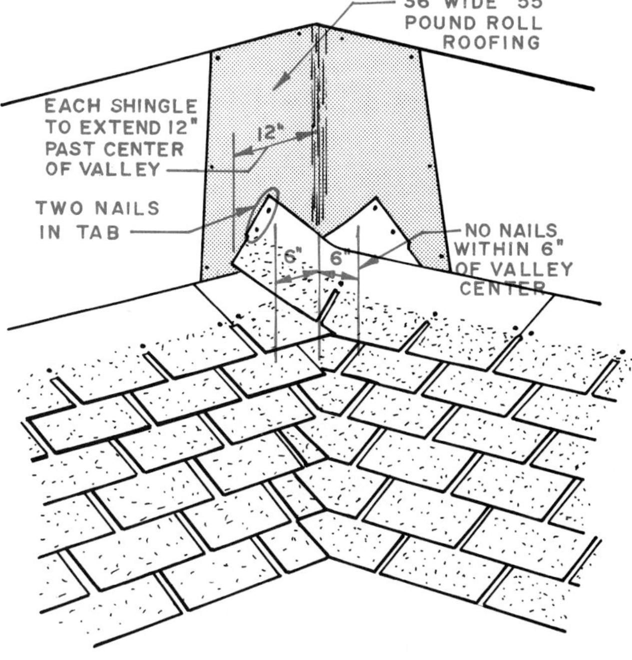

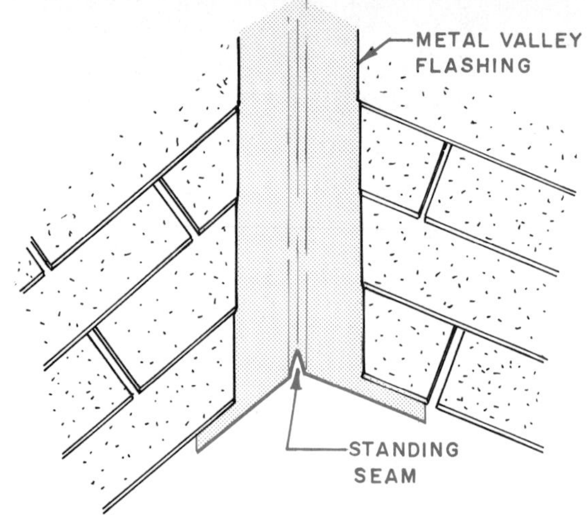

FIGURE 14-19

An open valley can be flashed with metal flashing.

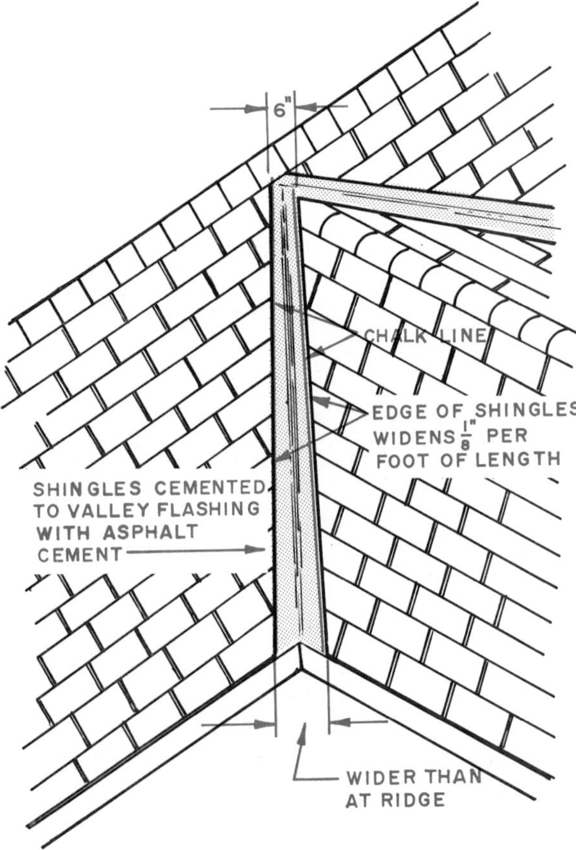

FIGURE 14-20

Shingles are cut to a chalk line along the valley flashing.

center of the valley. Use two nails at the end of each shingle.

Vertical Wall Flashing When the roof meets a wall, the intersection must be flashed. Generally, this is done with a stepped metal flashing applied over the end of each course of shingles (Fig. 14-21). The flashing is 2 in. wider than the exposed face of the shingle. It is laid so that

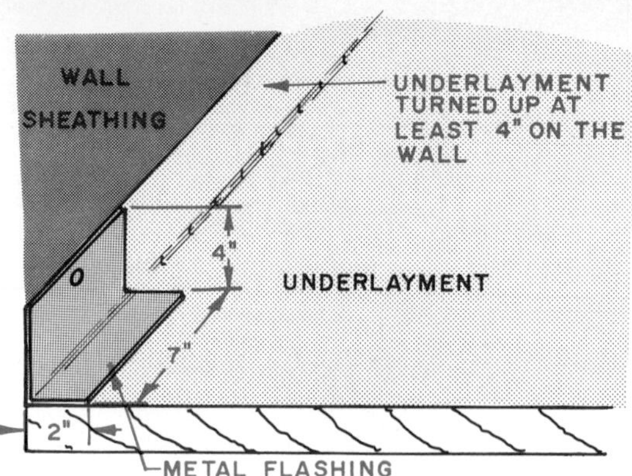

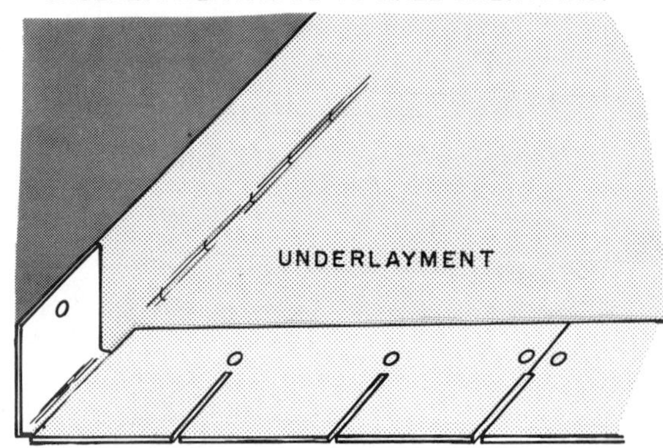

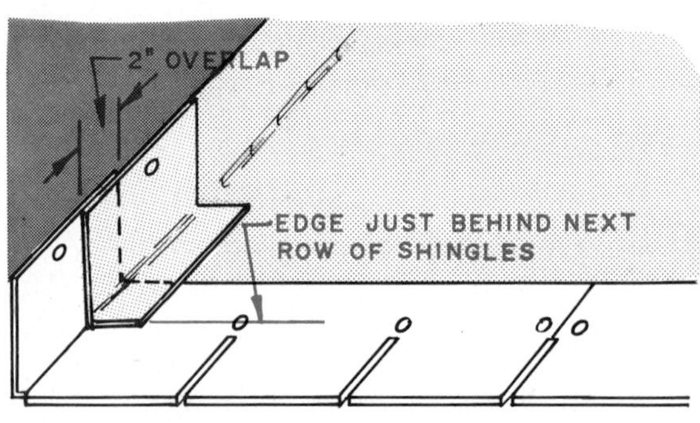

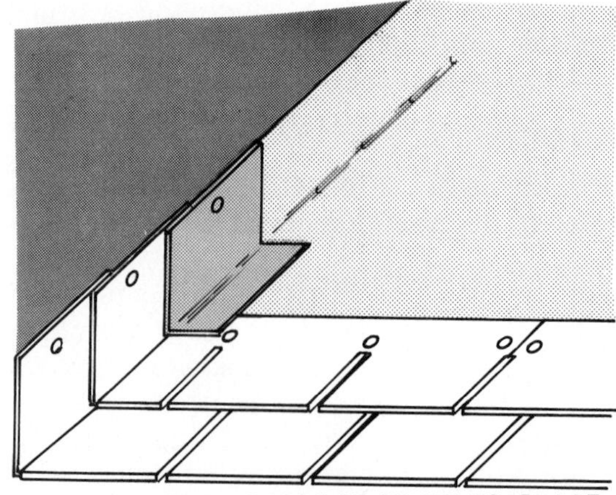

4. NAIL SECOND COURSE OF SHINGLES. PLACE
METAL FLASHING ON TOP, REPEAT THIS TO
THE RIDGE.

FIGURE 14-21

The intersection of a roof and a wall is flashed with stepped metal flashing.

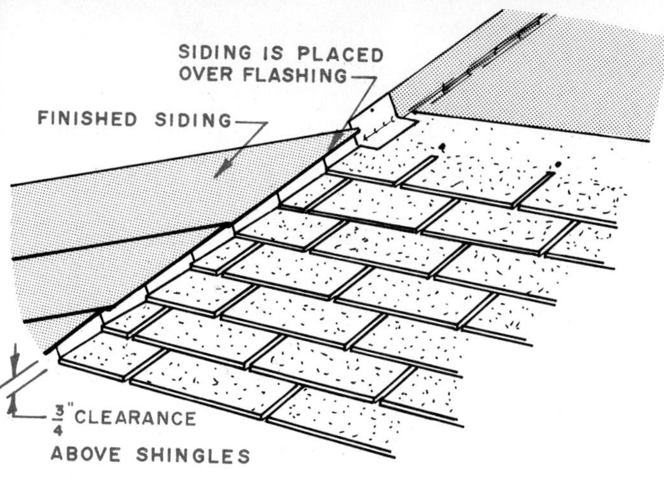

FIGURE 14-22

Siding is placed over the stepped roof flashing.

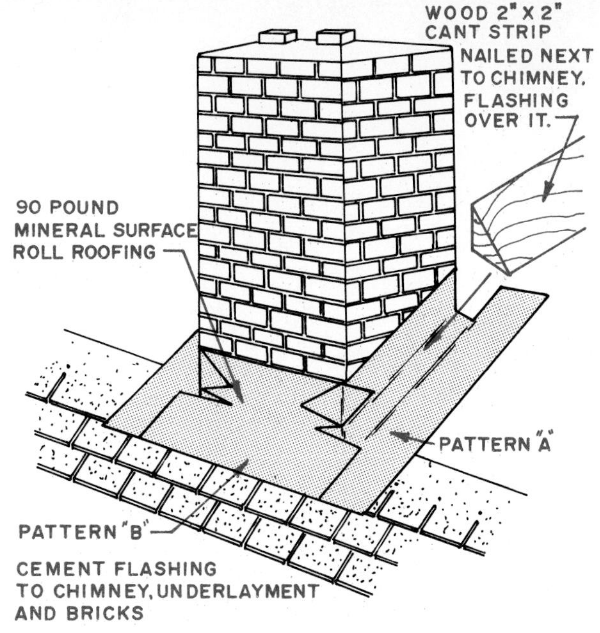

each piece overlaps the next 2 in. It is bent so that 4 in. is against the wall and 2 in. extends on to the roof. Each flashing piece is placed slightly up the roof from the exposed edge of the shingle. In this way it will not be visible. Nail each metal piece to the wall sheathing with one nail. After the flashing is in place the siding is nailed over it. The siding should clear the roof at least 1 in. (Fig. 14-22). Notice that the asphalt felt underlayment is turned up about 4 in. along the vertical wall.

Flashing the Chimney First run the shingles up to the front face of the chimney. Then apply a coat of asphalt primer to the bricks in the area to which the flashing is to be applied. Cut and nail 2 × 2-in. wood cant strips around the chimney (Fig. 14-23).

Now cut to size 90-lb mineral-surfaced roll roofing. Apply a bed of asphalt plastic cement to the shingles around the chimney. Stick the mineral felt to the roof and roll it up the sides of the chimney. Then stick it to the chimney the same way. If necessary a few roofers' nails can be driven through the top edge of the roofing material into the mortar.

Stepped metal flashing is then set into the mortar joints just above the edge of the roll roofing. This is called *counterflashing*. The mortar must be cut out 1½ in. deep. The metal is placed in the groove and fresh mortar is inserted. The flashing is then bent down over the roll roofing flashing. The stepped counter flashing should overlap 3 in. The counterflashing on the face of the chimney can be one long piece (Fig. 14-24). Now lay the shingles up to the chimney. Cement them to the roll flashing on the roof.

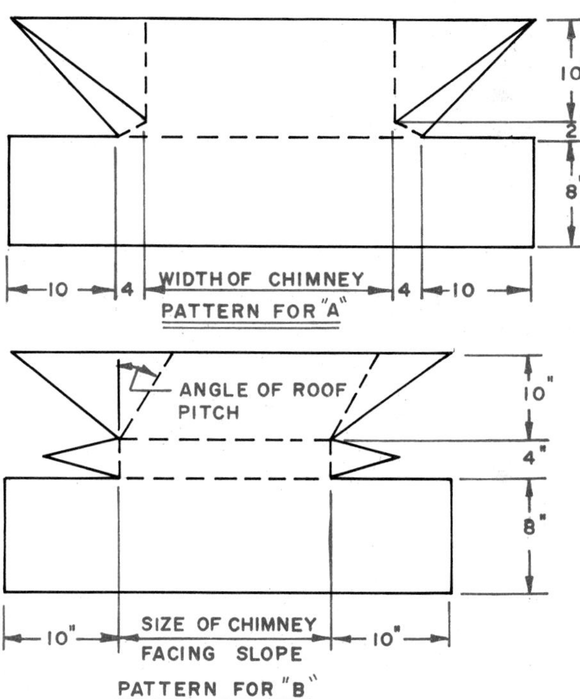

FIGURE 14-23

Patterns for cutting the mineral-surfaced roll roofing flashing at a chimney.

Stack Flashing Vent pipes from the sewer system project through the roof. These are flashed by laying the shingles up to the pipe. Cut a hole in one shingle and place over the pipe. Slide a metal flashing unit over the stock. The base will bend to fit the slope of the roof. Nail the metal base to the roof in the area above the shingles. Continue shingling the roof. Cut each shingle to fit around the pipe. The shingles go on top of the metal flashing unit (Fig. 14-25). Apply plastic asphalt cement around the base of the pipe. This will stick the shingles to the metal flashing unit.

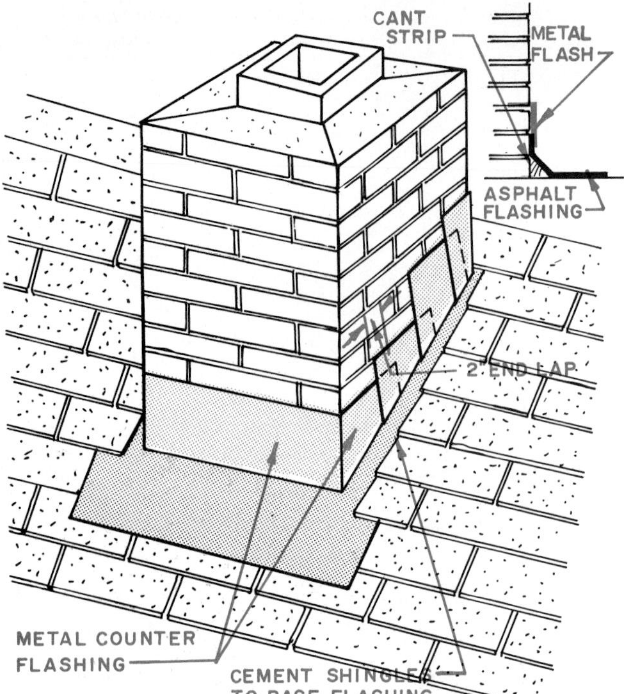

FIGURE 14-24

Applying the metal counterflashing on a chimney.

FIGURE 14-25

Installing a metal flashing sleeve on pipes that penetrate the roof.

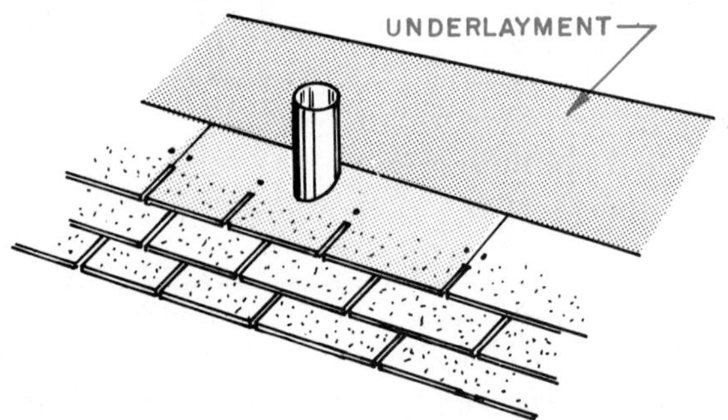

1. LAY SHINGLES TO THE PIPE. CUT ONE TO FIT AROUND THE PIPE.

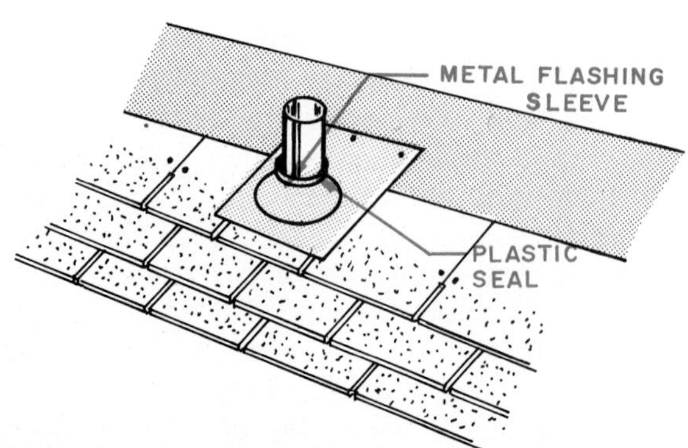

2. SLIDE A METAL FLASHING SLEEVE OVER THE PIPE. THE PLASTIC SEAL FITS TIGHTLY AROUND THE PIPE MAKING A WATER TIGHT SEAL.

SHINGLES

Shingles commonly in use are asphalt, mineral fiber, wood shakes, and wood shingles.

Asphalt Shingles

Asphalt shingles should provide double coverage on the roof. They should be at least 235 lb per square of shingles. A square covers 100 square feet of roof. In areas subject to winds in excess of 75 miles per hour, wind-resistance shingles should be used.

Nails should be corrosion-resistant with sharp points and large flat heads. Threaded nails must be used in plywood sheathing. Nails should be at least 1¼ in. long for new construction (Fig. 14-26). Power staplers and nailers are often used to fasten shingles to sheathing (Fig. 14-27).

Underlayment for asphalt shingles is No. 15 asphalt-saturated felt.

The types of shingles are tab square butt, tab hexagonal, and individual. Strip shingles are nailed with at least four nails per shingle. Individual shingles need two nails per shingle (Fig. 14-28). The tabs are cemented down to prevent wind damage (Fig. 14-29).

FIGURE 14-26

Typical nails used to install asphalt shingles.

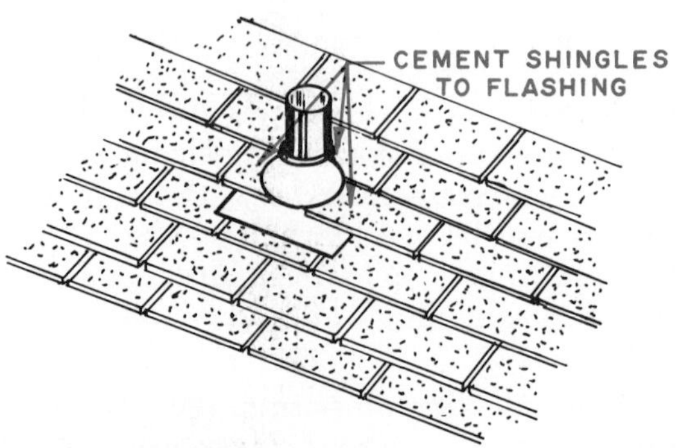

3. LAY SHINGLES IN NORMAL MANNER CUTTING TO FIT AROUND PIPE.

FIGURE 14-27

Asphalt shingles can be installed with staples. *(Courtesy of Bostitch Textron)*

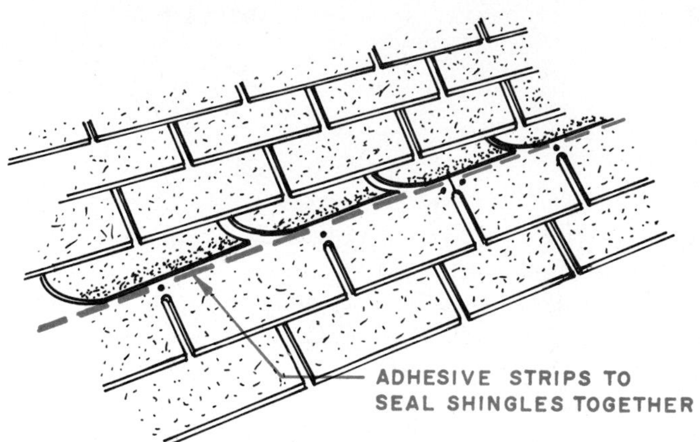

FIGURE 14-29

Adhesive strips on the shingles seal them together after they are laid.

Eave Flashing for Asphalt Shingles A layer of 90-lb mineral-surfaced roll roofing or 55-lb smooth roll roofing is used over the underlayment. It lays over the drip edge about 3/8 in.

For roofs sloped 4 in 12 or steeper, the flashing is nailed in place up the roof to a point that is 12 in. inside the interior (Fig. 14-30). If the roof has a large overhang so that a single 36-in.-wide roll will not reach the joint between it and the next layer, it must be cemented. The joint should be located so that it is not over the interior of the building.

For roofs sloped from 2 in 12 to 4 in 12, the flashing is *cemented* in place up the roof to a point that is 24 in. inside the interior (Fig. 14-31).

FIGURE 14-28

Nailing patterns for asphalt shingles.

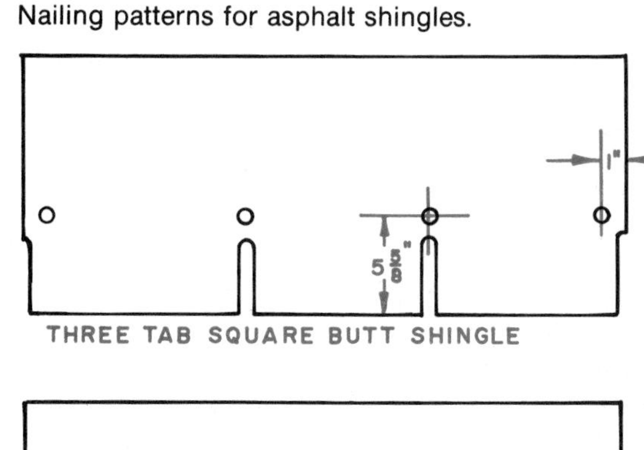

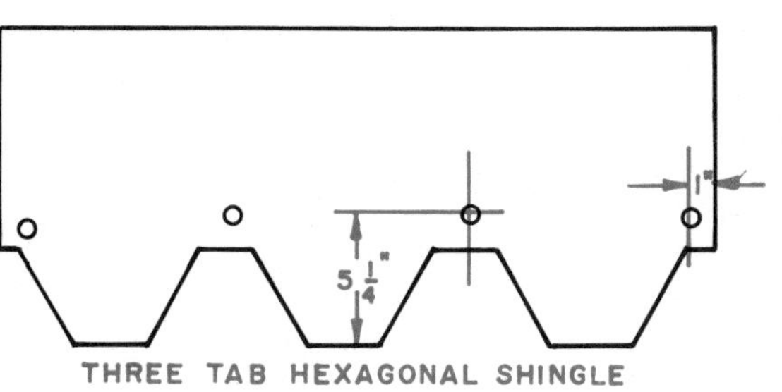

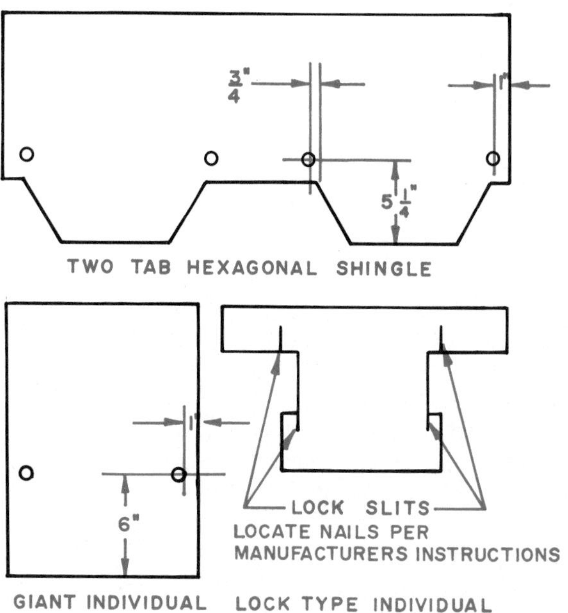

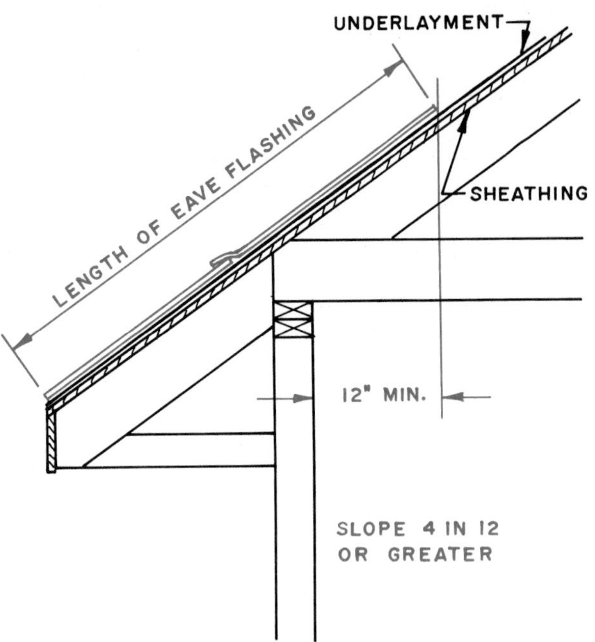

FIGURE 14-30

Recommended eave flashing for asphalt shingles on roofs with a pitch of 4 in 12 and steeper.

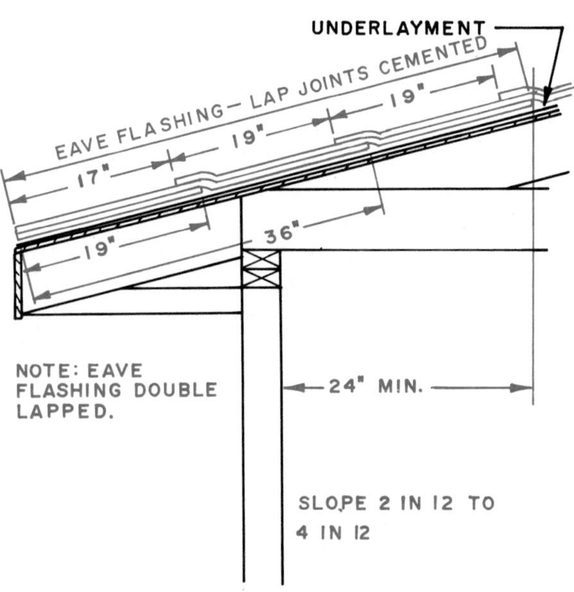

FIGURE 14-31

Recommended eave flashing for asphalt shingles on roofs with a pitch 2 in 12 to 4 in 12.

Applying Asphalt Shingles Once the roof has its underlayment, drip edge, and flashing in place, shingles can be applied. Asphalt shingles can be laid in several ways. The way selected depends upon the pattern desired. The simplest way, which requires the least cutting, is shown in Fig. 14-32. This pattern has the break joints in halves.

The starter course is a row of shingles laid with the tabs facing up the roof. This provides

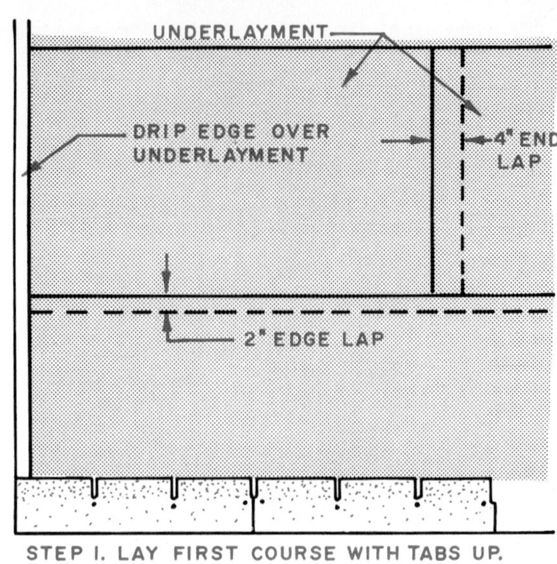

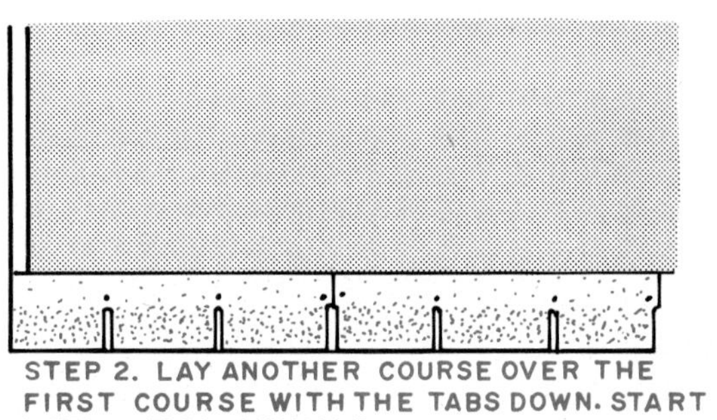

STEP 2. LAY ANOTHER COURSE OVER THE
FIRST COURSE WITH THE TABS DOWN. START
WITH A FULL SHINGLE.

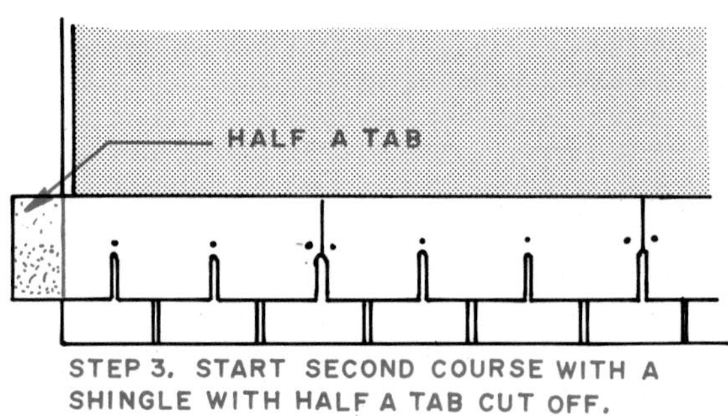

HALF A TAB
STEP 3. START SECOND COURSE WITH A
SHINGLE WITH HALF A TAB CUT OFF.

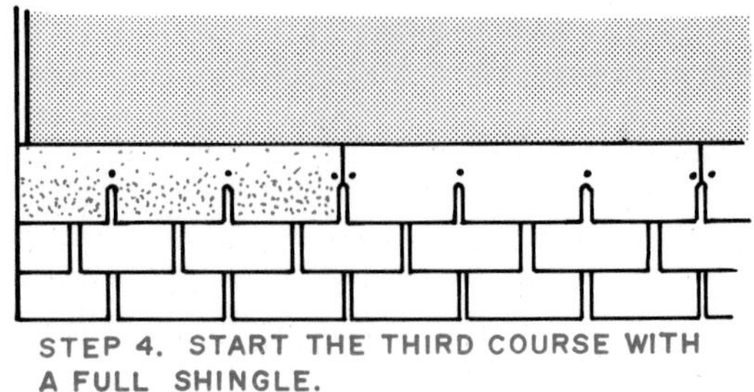

STEP 4. START THE THIRD COURSE WITH
A FULL SHINGLE.

FIGURE 14-32

Steps for laying asphalt shingles.

a backup for the first regular course and fills in the spaces between the shingles. The nails should be located so that they are covered by the first regular layer of shingles.

The first regular layer is started with a full three-tab shingle. The second layer is started with a three-tab shingle having one-half the end tab cut off. This staggers the cutouts on the center of the full tab below it. The third layer can be started with a full shingle. The system repeats itself.

When a roof is a large unbroken surface, the shingles can be started on the rake end that is most visible. If both rakes are equally visible, start laying the shingles in the center and work toward both rakes. Run a chalkline from the ridge to the eave to get an accurate alignment.

If a roof has a projection such as a dormer or a valley, start the shingles from a rake and lay toward the dormer or valley. The shingles should overhang the drip edge on the rake by ¾ in.

Another pattern allows the tab cutouts to break on thirds (Fig. 14-33). A random pattern can be laid by removing different amounts from the starting tab (Fig. 14-34).

The shingles are laid from each eave toward the ridge or hip (Fig. 14-35). The hip or ridge is finished by using hip or ridge shingles or by cutting 9 × 12-in. pieces from the square-butt shingle strips. They are applied by bending them lengthwise in their center. They are nailed over the ridge or hip. A 5-in. exposure is normally used (Fig. 14-36). Each shingle is fastened with one nail on each side of the hip or ridge. The nails are set 1 in. in from the edge and 5½ in. from the exposed end of the shingle.

Interlocking Asphalt Shingles ***Interlocking shingles*** are designed to hold the tabs down during high winds. Each shingle is locked to

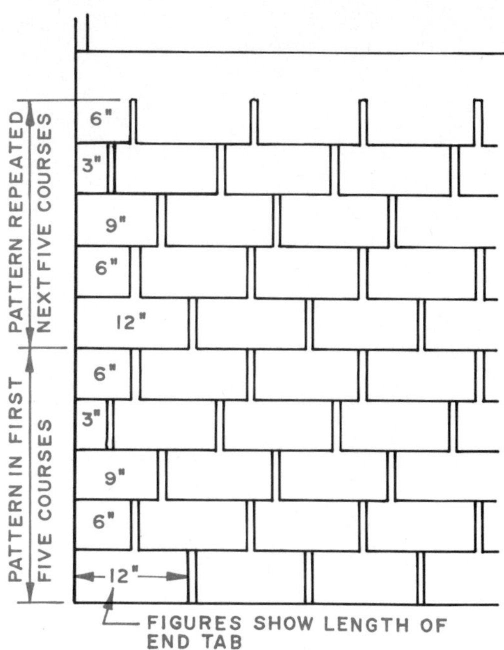

FIGURE 14-34

A plan for laying asphalt shingles in a random pattern.

FIGURE 14-35

Shingles are laid from the eave toward the ridge. *(Courtesy of CertainTeed Corp.)*

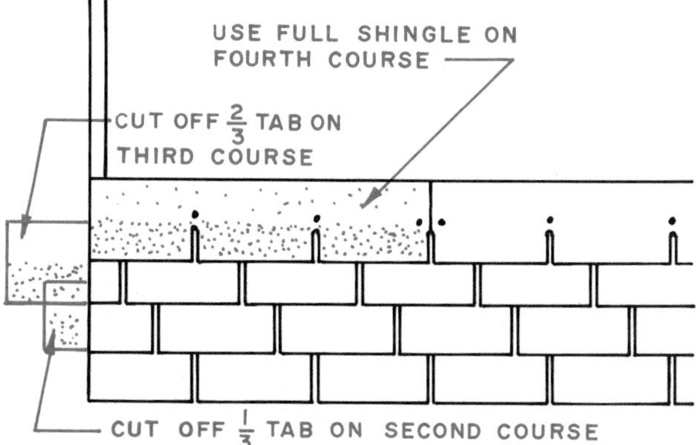

FIGURE 14-33

A pattern for laying asphalt shingles which alternates the tab cutouts by thirds.

FIGURE 14-36

Hip shingles are nailed over the roof hip rafter. *(Courtesy of CertainTeed Corporation)*

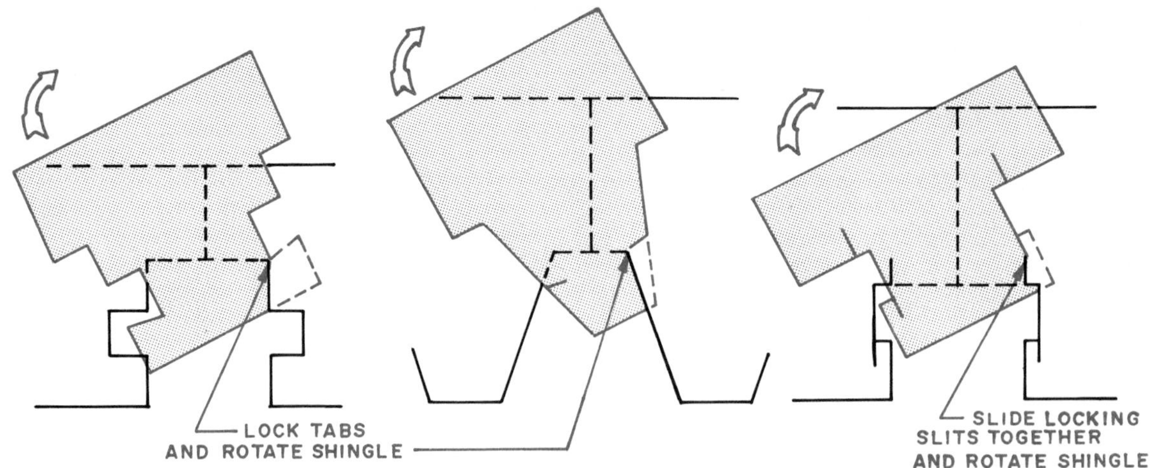

FIGURE 14-37

Typical interlocking individual shingles.

the one below it in some way. There are a variety of designs available. Three commonly found types are shown in Fig. 14-37.

Mineral Fiber Shingles *Mineral fiber shingles* are made in four types. These are individual units, multiple units, Dutch lap units, and ranch-style units. Each is applied in a different way, giving a different appearance to the roof.

The shingles are made with prepunched nail holes. Dutch lap and ranch-style shingles have additional holes for storm anchors (Fig. 14-38). The storm anchor holds the end of one shingle to the face of the one below it.

Packages of shingles should be stored in a clean, dry area prior to installation. Moisture and dampness can discolor the shingles when they are in the packages.

Nails used should be corrosion-resistant. Usually, galvanized or aluminum nails are used. They should have large heads about 3/8 in. in diameter. They should be 1¼ in. long with a shank diameter of 0.120 in. Smooth nails can be used with solid wood sheathing, but threaded nails are preferred. Threaded nails are required with plywood sheathing.

Underlayment is No. 15 asphalt-saturated asbestos roofing felt or No. 30 asphalt-saturated organic felt. For roofs with slopes of 3 in 12 to 5 in 12, two layers of underlayment are required. Only one layer is used for roofs 5 in 12 or steeper. Drip edges are used in the same manner as described earlier.

Eave Flashing for Mineral Fiber Shingles Eaves are flashed by a second layer of underlayment over the first layer. A continuous layer of plastic asphalt cement is applied with a comb trowel to the first underlayment. The second layer is placed over the cement. For roofs with slopes of 5 in 12 or greater, this flashing must reach a point 12 in. inside the interior. For roofs 3 in 12 to 5 in 12 the cement eave flashing must extend up the roof to a point at least 24 in. inside the interior of the building.

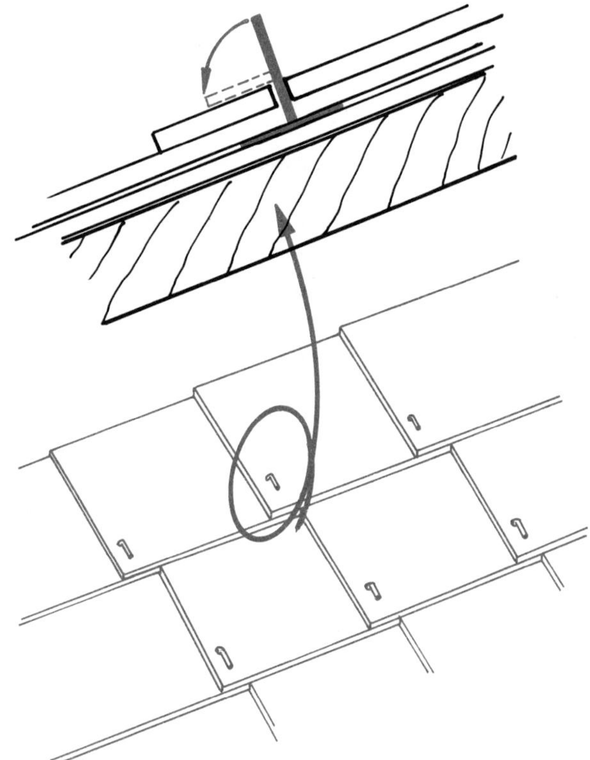

FIGURE 14-38

Mineral fiber shingles use metal storm anchors to help them resist high winds.

Applying Mineral Fiber Shingles There are several ways in which these shingles can be applied. The following paragraphs describe applying individual-type shingles (Fig. 14-39). A ¼ × 2-in. wood cant strip is nailed along the eave. This gives the proper pitch to the starter shingles.

Apply a row of starter shingles at the eave. Roll back the underlayment and eave flashing and apply the shingles directly over the cant strip and the sheathing. The shingles overlap the drip edge and rake by ¾ in. Replace the underlayment over the starter shingles. Start the first course of shingles at the rake with a half shingle. It overhangs the rake by ¾ in. and lines up with the edge of the starter shingles. Continue the first row with full shingles. Start the second row with a full-width shingle. The third row starts with a half shingle. Alternate all rows using this pattern.

To keep the rows straight, run a chalk line for each row.

A different pattern can be developed by staggering the butt line. The shingles have two sets of holes punched to permit staggered application. The first course is applied in a straight line. The second course is staggered (Fig. 14-40).

Hips and ridges are covered as shown in Fig. 14-41. Wood nailing strips equal in thickness to two roof shingles are nailed at the hip or ridge. This provides a flat surface. The wood strips are covered with two thicknesses of underlayment.

The first hip and ridge shingle laid is half length. It is covered with a full-length shingle. The rest are applied using normal exposure to the weather. Each shingle is held with two nails. Alternating shingles are lapped on either side of the ridge or hip. The ridge joint is sealed with plastic asphalt roofing cement.

Wood Shakes

Wood shakes are available in three types. These are hand-split-and-resawn, taper-split, and straight-split. They are sold in 18-, 24-, and 32-in. lengths. They are 100% clear wood and 100% heartwood. The taper-split and straight-split are made by hand. The hand-split-and-resawn shakes are first split by hand and have the back side sawed smooth (Fig. 14-42). The sizes commonly available are listed in Fig. 14-43.

The nails used are hot-dipped galvanized steel or aluminum. The shanks may be smooth or threaded. Threaded nails provide greater

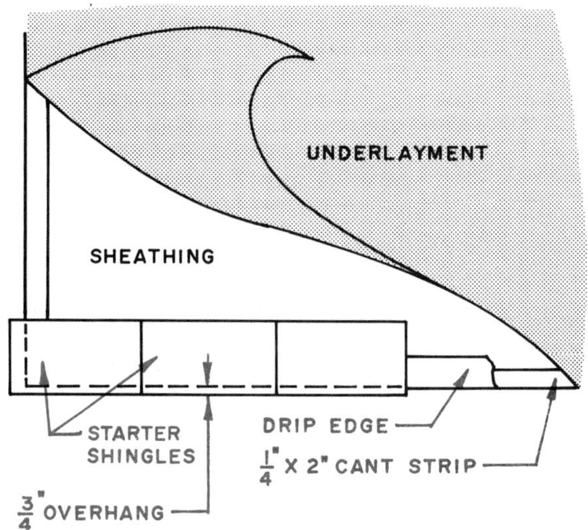

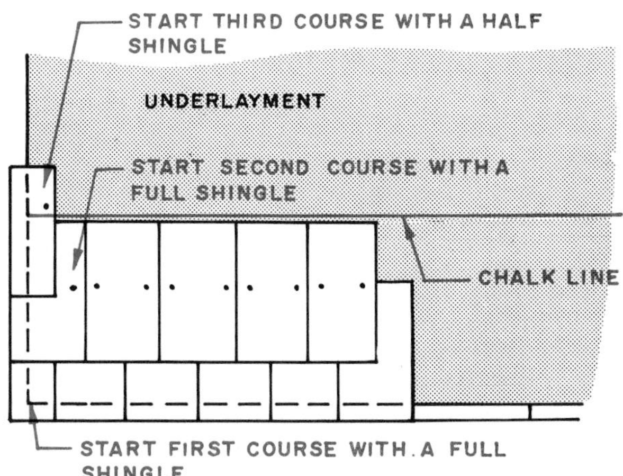

FIGURE 14-39

How to install mineral fiber shingles with a straight butt line.

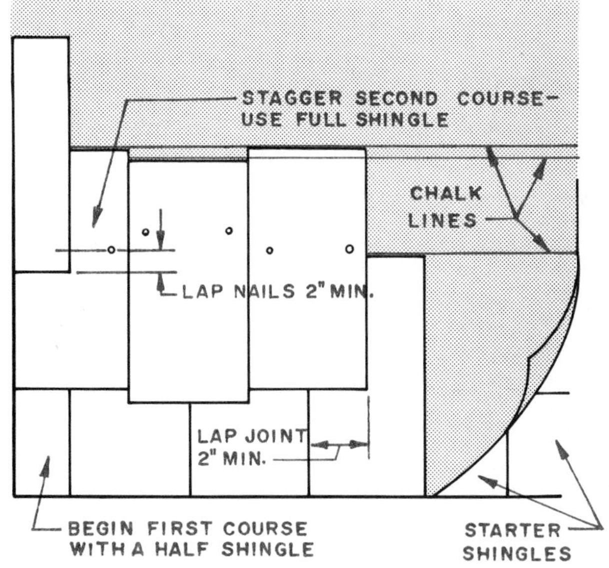

FIGURE 14-40

A plan for laying mineral fiber shingles with a staggered butt line.

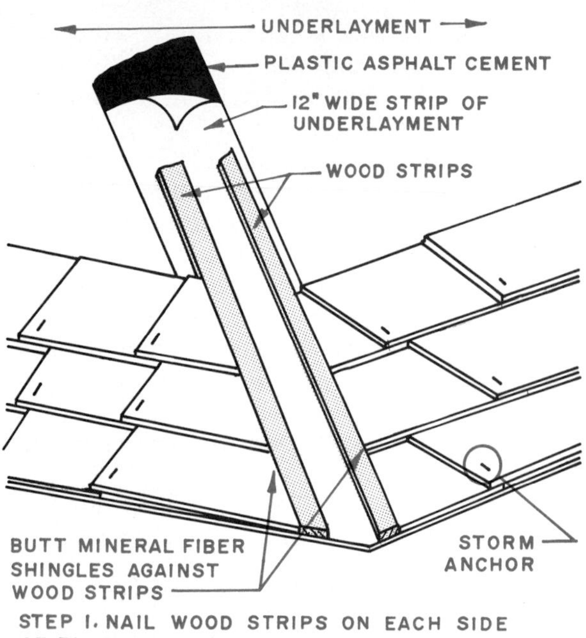

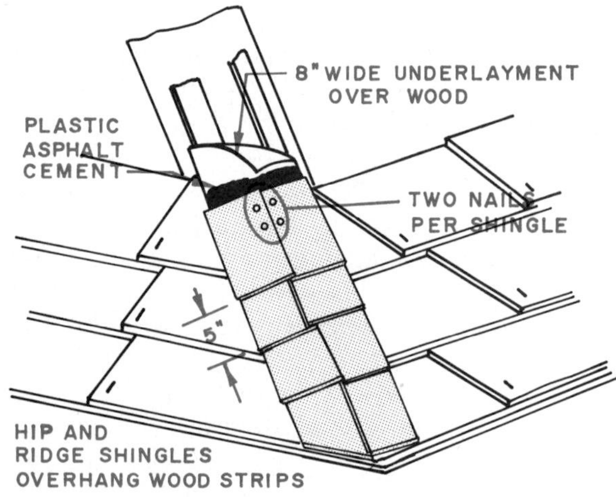

FIGURE 14-41

How to shingle a ridge or hip with mineral fiber shingles.

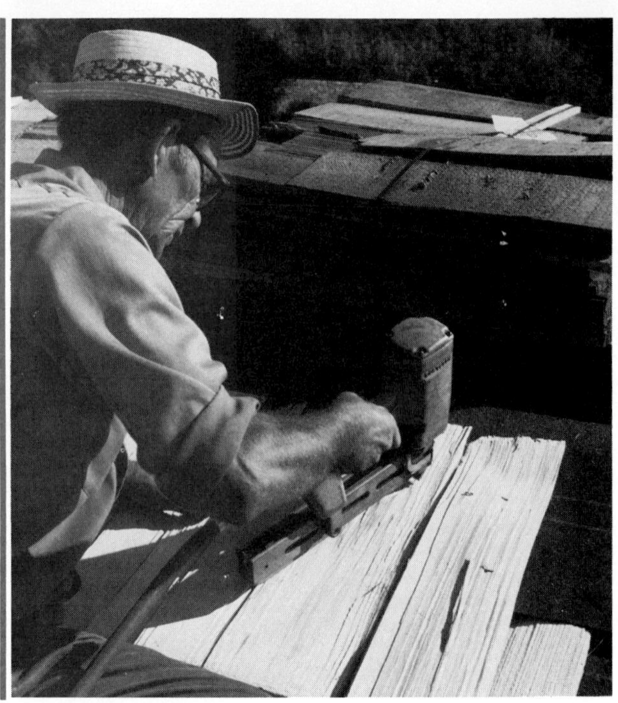

FIGURE 14-42

Hand-split wood shakes being installed using a power stapler. *(Courtesy of Bostitch Textron)*

holding power. Shakes ¾ to 1¼ in. thick require 6d nails. Those ½ to ¾ in. thick require 5d nails. Nails used on the hip and ridge shakes should be 2d longer than those for the roof.

Wood shakes are recommended for roofs sloping 4 in 12 or steeper. If used on a 3 in 12 roof, solid sheathing plus additional underlayment is needed.

Shakes can be applied to double or triple cover the roof. The to-the-weather exposure for double coverage is 13 in. for 32-in. shakes, 10 in. for 24-in. shakes, and 7½ in. for 18-in. shakes. For triple coverage use 10 in. exposure for 32-in. shakes, 7½ in. for 24-in. shakes, and 5½ in. for 18-in. shakes (Fig. 14-44).

GRADE	LENGTH × THICKNESS (in.)	DESCRIPTION
No. 1 hand-split and resawn	18 × ½ to ¾ 18 × ¾ to 1¼ 24 × ½ to ¾ 24 × ¾ to 1¼ 32 × ¾ to 1¼	These shakes have split faces and sawn backs. Cedar blanks or boards are split from logs and then run diagonally through a bandsaw to produce two tapered shakes from each.
No. 1 taper-split	24 × ½ to ⅝	Produced largely by hand, using a sharp-bladed steel froe and a wooden mallet. The natural shingle-like taper is achieved by reversing the block, end for end, with each split.
No 1. straight-split (barn)	18 × ⅜ 24 × ⅜	Produced in the same manner as taper-split shakes except that by splitting from the same end of the block, the shapes acquire the same thickness throughout.

FIGURE 14-43

Typical types and sizes of red cedar shakes.

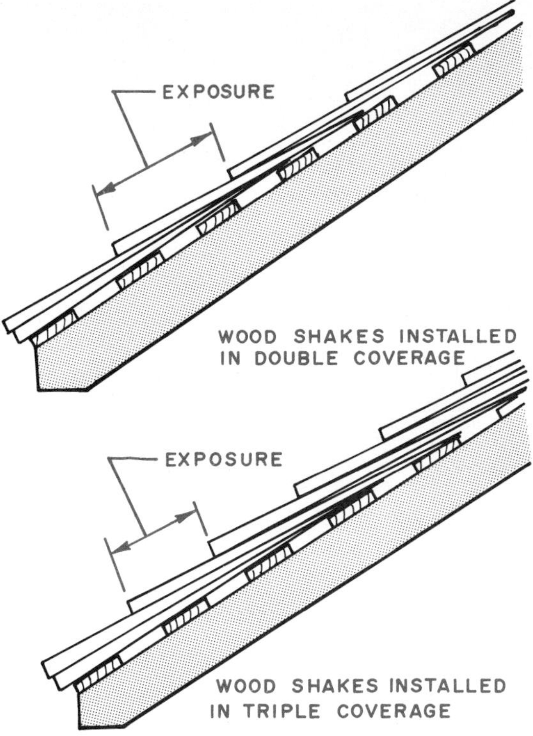

FIGURE 14-44

Wood shakes can be applied producing double or triple coverage.

Shakes can be applied over solid or spaced roof sheathing. Solid sheathing is recommended in areas having outside design temperatures of 0°F and wind-driven snow.

The underlayment is No. 30 asphalt-saturated felt. One 36-in.-wide strip is placed at the eave.

Eave Flashing for Wood Shakes For roofs with slopes of 4 in 12 or steeper, a second layer of underlayment is laid over the first layer. It should extend up the roof to a point 24 in. inside the building. If the flashing required is wider than 36 in., the joint between layers should fall outside the interior wall line. The joint should be cemented.

For roofs with slopes of 3 in 12 to 4 in 12 or areas with severe icing, the eave flashing is the same as just described except that the two layers are cemented together with plastic asphalt cement. It is applied with a comb trowel.

The shakes should extend 1½ in. beyond the edge of the sheathing at the rake and eave. A metal drip edge is not required.

It is recommended that copper flashing not be used with wood shakes. Contact between the copper and wood causes premature deterioration of the copper.

The open valley flashing method is recommended. A 20-in.-wide No. 30 asphalt saturated-felt strip is nailed in place. It is covered with a 20-in.-wide metal flashing strip. The metal should have the edges turned back to form a bead on each edge.

Applying Wood Shakes First lay the underlayment and eave flashing. The starter course of shakes should be doubled. The first layer is a 15-in.-long shake made especially for this purpose (Fig. 14-45). The second layer is nailed over the starter layer (Fig. 14-46).

FIGURE 14-45

Nailing the starter course of wood shakes. *(Courtesy of Red Cedar Shingle and Handsplit Shake Bureau)*

FIGURE 14-46

The second course of wood shakes is nailed over the starter course. *(Courtesy of Red Cedar Shingle and Handsplit Shake Bureau)*

After each course is laid, an 18-in.-wide strip of No. 30 asphalt-saturated felt underlayment is placed over the top part of the shake. It extends up on to the sheathing (Fig. 14-47). The underlayment should extend over the top of the shake a distance equal to twice the exposure. For example, an 18-in. shake is laid 7½ in. to the weather. The felt will extend 15 in. down on the shake. Since the felt is 36 in. wide, this allows 21 in. to extend up on the sheathing.

Shakes are nailed ¼ in. apart on their edges. This allows for expansion. The joints between shakes should be at least 1½ in. from any joints in the adjoining row.

Each shake requires two nails. They are placed 1 in. in from each edge and 1 to 2 in. above the butt line of the course to be nailed over it (Fig. 14-48). Nails should be driven so that the heads rest on the surface. Do not drive the heads into the wood (Fig. 14-49).

The last row of shakes at a hip or ridge requires extra nails. It is recommended that the smoother shakes be saved for this row. Lay a strip of No. 30 asphalt-saturated felt 12 in. wide over the hip or ridge. Shakes nearly 6 in. wide should be sorted out and saved for the hip and ridge.

To start laying the shakes on the hip or ridge, nail a wood strip 6 in. from the centerline. Put one on each side. These are used to line up the shakes as they are nailed in place. Start each with a double starter course. Then alternately overlap the shakes as they are laid (Fig. 14-50).

To lay shakes into a valley, cut them to be an equal distance from the ridge formed in the valley. A piece of wood cut to the desired distance can be used as a spacer (Fig. 14-51). There are many variations in roof appearance that can be had by varying the exposure.

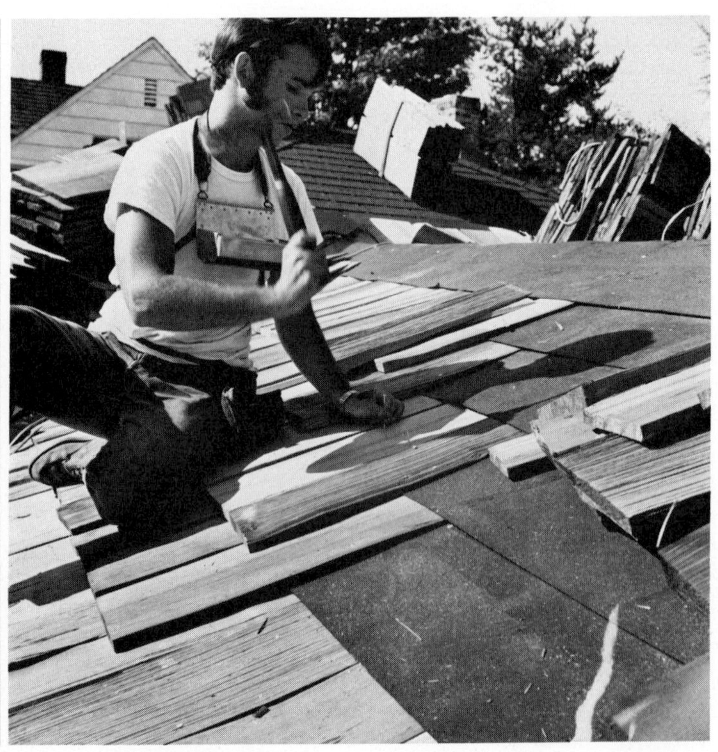

FIGURE 14-47

Overlap each course of wood shakes with underlayment. *(Courtesy of Red Cedar Shingle and Handsplit Shake Bureau)*

FIGURE 14-48

Shakes can be secured with power-driven nails. *(Courtesy of Bostitch Textron)*

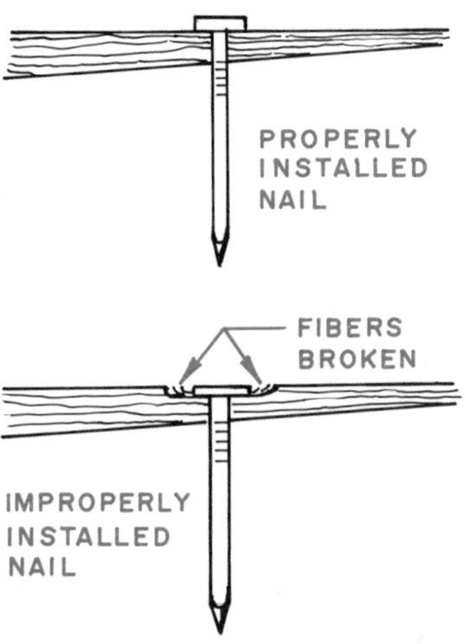

FIGURE 14-49

Do not drive the nails or staples into the shingle so that the wood fibers are broken.

Wood Shingles

Wood shingles are available in three grades: No. 1 (Blue Label), No. 2 (Red Label), and No. 3 (Black Label). They are sold in 16-, 18-, and 24-in. lengths. The No. 1 grade is the premium grade for roofs and sidewalls.

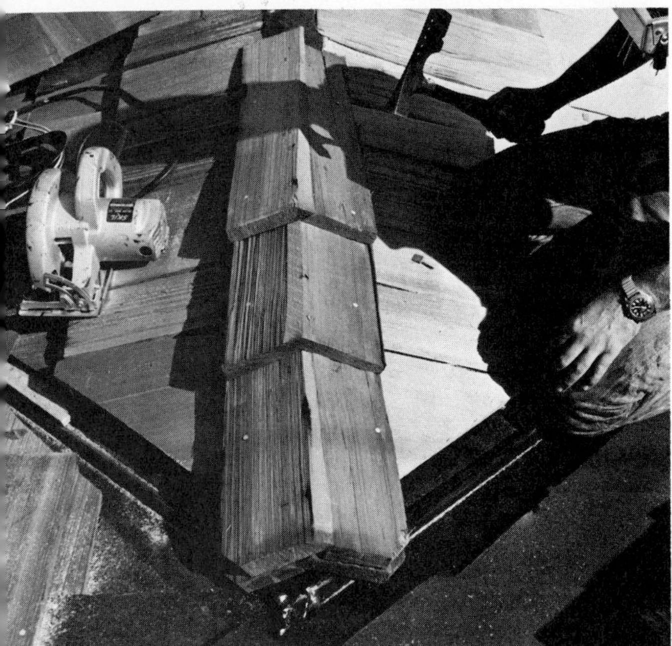

FIGURE 14-50

Wood shakes alternately overlapped when capping a ridge or hip. *(Courtesy of Red Cedar Shingle and Handsplit Shake Bureau)*

FIGURE 14-51

Wood shakes are spaced evenly from the ridge of metal valley flashing. *(Courtesy of Red Cedar Shingle and Handsplit Shake Bureau)*

They are 100% heartwood, 100% clear, and 100% edge grain. The No. 2 grade is a good grade for all applications. It has 10 in. clear on 16-in. shingles, 11 in. clear on 18-in. shingles, and 16 in. clear on 24-in. shingles. Flat grain and limited sapwood are permitted. The No. 3 grade is a utility grade. It is used on economy applications. It has less clear wood than does the No. 2 grade.

Wood shingles are sawed smooth on both sides.

The nails used must be hot-dipped galvanized steel or aluminum. They follow the same requirements as nails for wood shakes. Shingles 16 and 18 in. long require 3d nails. Shingles 24 in. long require 4d nails. Nails 2d longer are used to apply hip and ridge shingles.

The lower the slope of the roof, the smaller the shingle exposure used. Recommended exposures are shown in Fig. 14-52. This provides triple coverage. Wood shingles are not recommended for roofs below a 3 in 12 slope.

Either solid or spaced roof sheathing can be used.

Underlayment between the shingles and sheathing is not required. However, it is considered good practice to use it. A No. 15 asphalt-saturated felt is recommended.

The wood shingles should extend 1½ in. beyond the edge of the sheathing at the rake and eave. A metal drip edge is not required.

Applying Wood Shingles The first course should be doubled or tripled and extend 1½ in. beyond the sheathing. Leave ¼ in. between shingles to allow for expansion. Joints between shingles in any course should be at least 1½ in. from the joints in the courses above and below them (Fig. 14-53).

Shingles laid to a valley or on a hip or ridge are handled as explained for wood shakes.

Each shingle should have only two nails. These are not more than ¾ in. from the side edge of the shingle. They should be not more than 1 in. above the exposure line (Fig. 14-53). Wood shingles can be laid in a variety of ways. These produce different patterns (Fig. 14-54). One is a Dutch weave pattern. Some shingles

SHINGLE LENGTH (in.)	SINGLE EXPOSURE (in.)		
	5 in 12 Slope	4 in 12 Slope	3 in 12 Slope
16	5	4½	3¾
18	5½	5	4¼
24	7½	6¾	5¾

FIGURE 14-52

Recommended exposures for wood shingles.

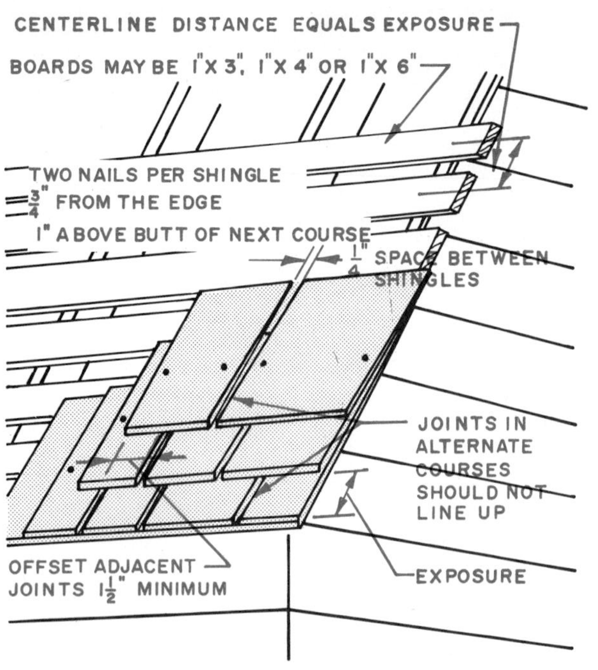

FIGURE 14-53

How to install wood shingles using spaced wood sheathing.

are doubled at random over the roof area. A thatch pattern has some shingles placed 1 in. above and below the straight line of shingles. The serrated pattern has one course laid double every three, four, or five courses. The pyramid pattern adds two extra shingles, one wide covered by one narrow, located at random over the roof.

It is always a problem to move bundles of shingles to a roof. One way is to use a lift truck (Fig. 14-55). Notice the use of wood boards nailed on this steep roof to give the roofer a foothold.

BUILT-UP ROOFS

Built-up roofs are made of a series of layers of felt, asphalt, and gravel. It is applied by roofing companies specializing in this type of construction (Fig. 14-56).

Dutch Weave

Thatch

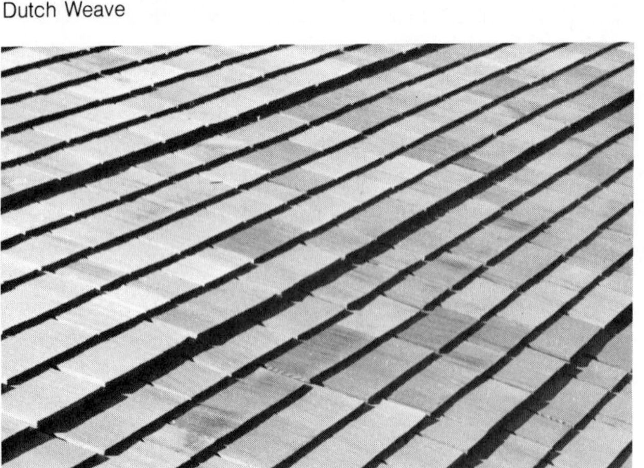

Serrated

Pyramid

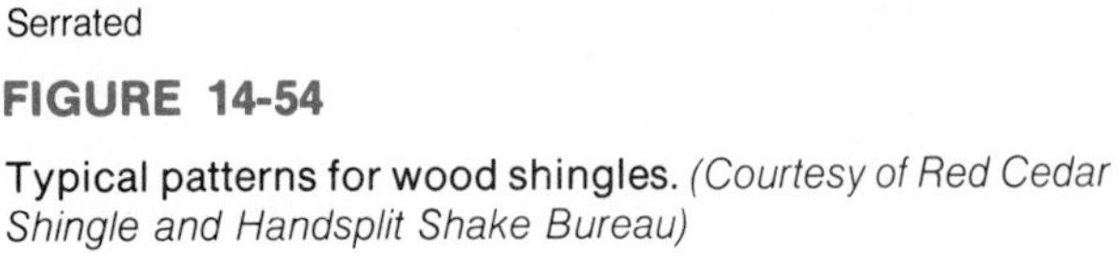

FIGURE 14-54

Typical patterns for wood shingles. *(Courtesy of Red Cedar Shingle and Handsplit Shake Bureau)*

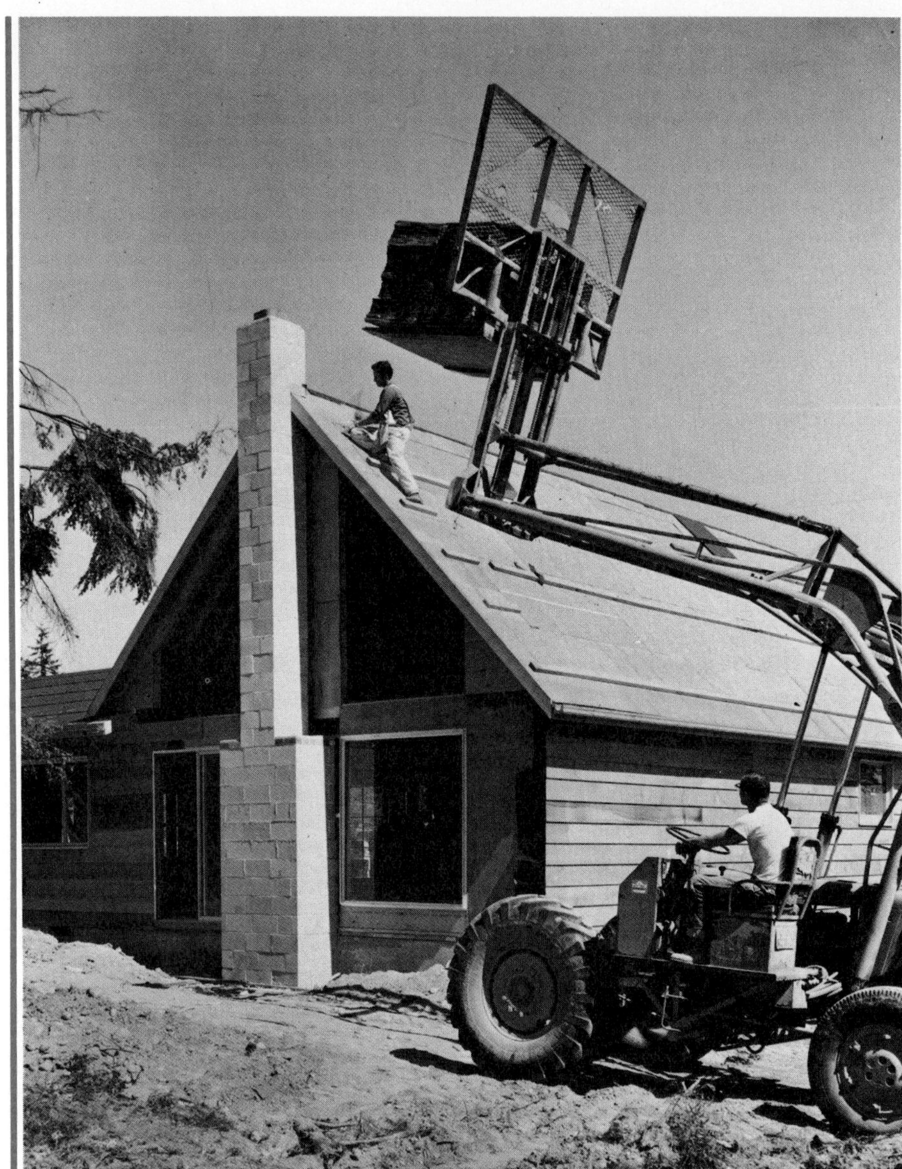

FIGURE 14-55

A lift truck is used to move wood shakes and shingles to the roof. Notice the wood strips nailed to the steep roof to provide a foothold for the roofer. *(Courtesy of American Plywood Association)*

FIGURE 14-56

Roofing companies apply built-up roofs. *(Courtesy of National Roofing Contractors, Assoc.)*

FIGURE 14-57

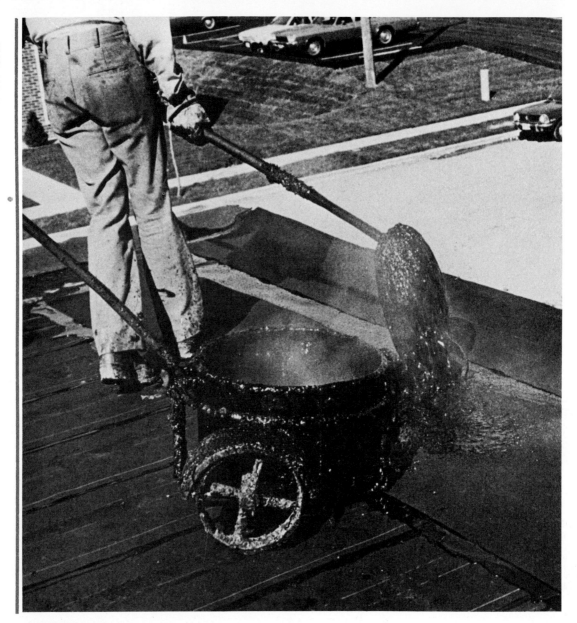

The tar or asphalt is hot-mopped on the layers of felt.
(Courtesy of National Roofing Contractors, Assoc.)

Built-up roofs are used on flat roofs or sloped roofs not steeper than 2 in 12 pitch.

Built-up roofs have three to five layers of roofers' felt which are mopped with tar or asphalt (See Fig. 14-57). The final layer is also mopped with asphalt and covered with gravel. The tar holds the gravel in place; a typical example is in Fig. 14-58. Sometimes rigid insulation is glued to the roof deck and the built-up roof is applied over this.

The edge of the roof has a metal gravel stop which serves as the drip edge. It is placed

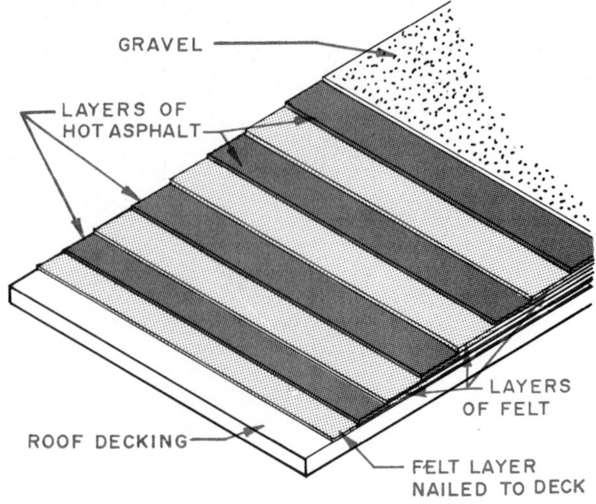

FIGURE 14-58

The layers of a typical built-up roof.

over the layers of felt and asphalt and is flashed with several layers of felt and asphalt (Fig. 14-59).

When a built-up roof meets a wall a cant strip is used. The layers of felt are extended 4 inches up the wall. See Fig. 14-60.

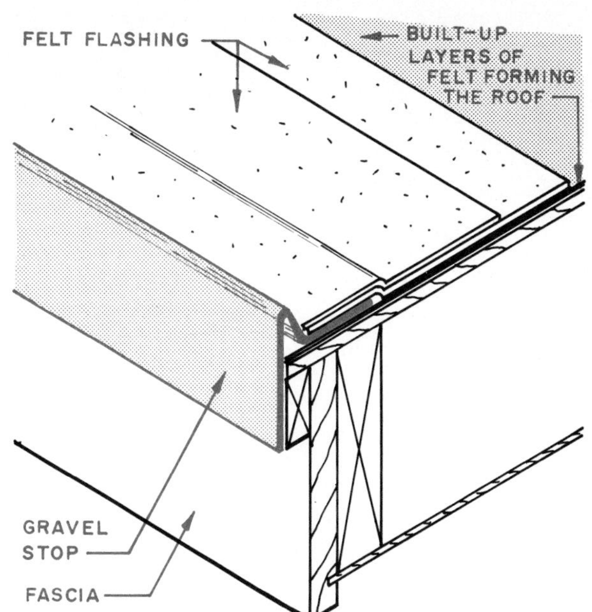

FIGURE 14-59

A gravel stop is used on the edge of the roof. It is flashed with several layers of felt.

Metal Roofing Corrugated metal roofing is used on garages, storage buildings and farm buildings. It is available in widths up to 4 ft 0 in. and lengths up to 24 ft 0 in. Normally it is used on roofs sloping 4 in 12 or steeper. It can be used on 3 in 12 roofs if a single panel reaches from the ridge to the eave.

The panels are fastened to purlins. A purlin is a structural member running perpendicularly to the rafters (Fig. 14-61). Usually 2-in × 4-in. stock is used. The spacing should follow the directions of the manufacturer. Filler strips are sold with the panels. They are set at the eave and ridge. Usually the panel is cut so it overhangs 2 in. to 3 in. at the eave.

The panels are nailed to the purlins through the top of the ribs. Special 1¾-in. screw-type nails are used. They have a neoprene washer below the head to prevent leaks. A metal ridge cap is used at the ridge. The rake edge is sealed with a metal drip strip.

ESTIMATING MATERIALS

To ascertain the number of squares of shingles needed, figure the area of the roof to be covered. This is figured in square feet. Subtract the area of openings such as corners. Add 10% of the total for cutting and waste. The final total area is divided by 100 square feet (a square of shingles) to find the number of squares needed. Measurements used are shown in Fig. 14-62.

A simple gable roof requires only that the length of the ridge be multiplied by the rafter length. This will give the area for half the roof. Multiply by 2 to get the total area.

Intersecting gables have a triangular area to figure; figure it as a square. This total will then be used as the area for the triangular area on both sides of the roof. The rectangular area is found by multiplying the ridge length of the triangular area by the rafter length.

To figure the end area of a hip roof multiply the length at the eaves by ½ the length of a common rafter. Multiply by 2 to get the area on both ends of the building. The area of the sides of the roof is found by adding the lengths of the ridge and eave and dividing this by 2. Then multiply this by the length of a common rafter. Multiply this by 2 to get both sides of the roof.

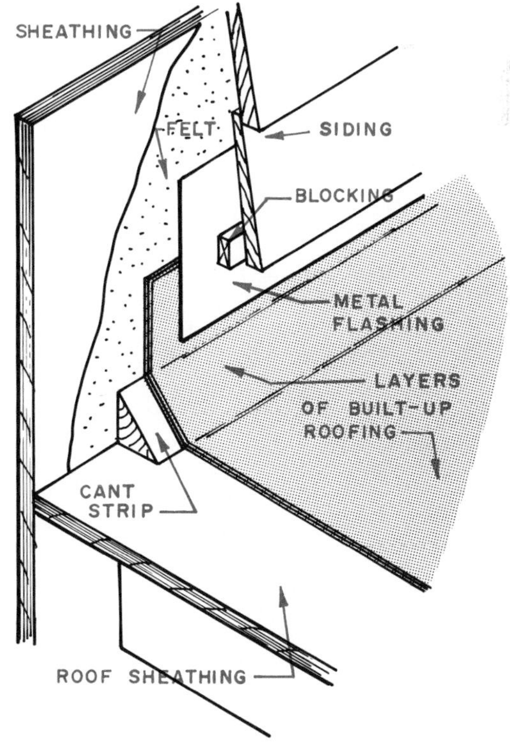

FIGURE 14-60

Built-up roofing is lapped up the side wall.

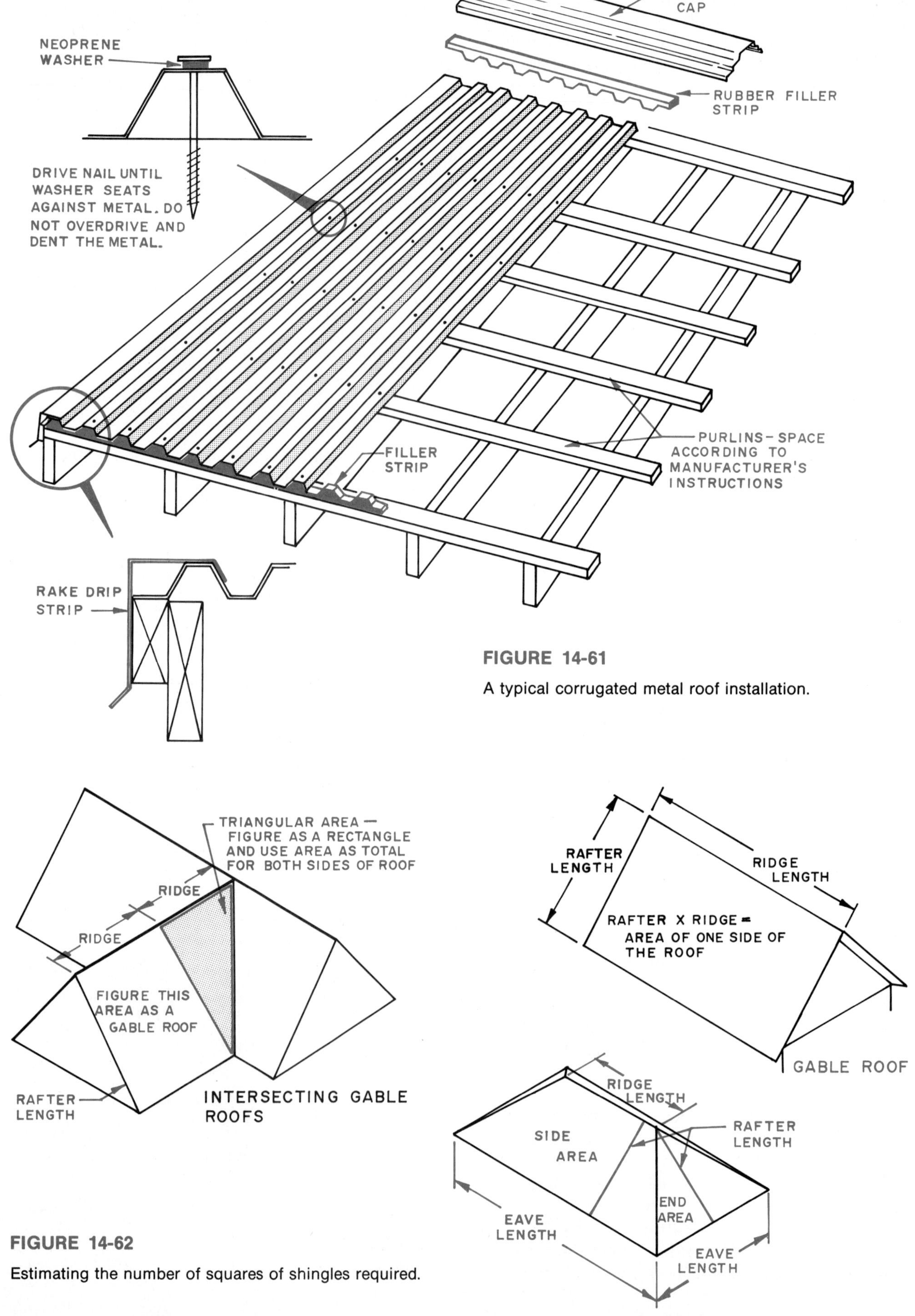

FIGURE 14-61

A typical corrugated metal roof installation.

FIGURE 14-62

Estimating the number of squares of shingles required.

STUDY QUESTIONS

After reading the chapter, answer each of the following questions. If you do not know the answer, review the chapter.

1. What are the various materials used to finish a roof?
2. What are the common roof sheathing materials?
3. What does "laid to the weather" mean?
4. What is the sheathing grade of plywood most commonly used for roof construction?
5. How can you tell the maximum distance between rafters plywood sheathing will span?
6. In what direction is the face grain of plywood sheathing placed in relation to the rafters?
7. How can a carpenter decide on how to start laying sheets of roof sheathing?
8. How are unsupported edges of plywood sheathing supported?
9. How is the underlayment placed in relation to the metal drip edge on the rake and eave?
10. What materials are used for flashing?
11. What are the commonly used types of shingles?
12. How are asphalt shingles prepared so that the wind does not lift them?
13. What are the three types of wood shakes?
14. How do wood shakes and wood shingles differ?

IMPORTANT TECHNICAL TERMS

Following are technical terms that you should be able to use as part of your working vocabulary. Write a brief description of the meaning of each term.

Plank decking
Fiberboard decking
Underlayment
Drip edge
Flashing
Shingles
Shakes

15 DOORS AND WINDOWS

After the exterior walls are sheathed and the roof is in place, many carpenters prefer to install the exterior doors and windows. Interior door frames are installed after the partitions are erected.

Windows and doors are manufactured in large quantities. They are accurately made and are generally assembled into a single unit. The windows have the sash and glass installed in the frame. The carpenter has to insert this unit into the rough opening and install it properly. Some door units come with the frame and door assembled in one unit. Others have the frame separate. After the frame is installed, the carpenter installs the door.

WOOD WINDOWS

Wood windows are manufactured using select woods. The moisture content is carefully controlled. They are available primed, unprimed, and covered with a rigid vinyl sheath. Vinyl-clad windows never need painting (Fig. 15-1).

Wood windows have better insulation qualities than do metal windows. The wood parts are generally 1⅜ in. thick. They also tend to sweat less than metal because wood is a good insulator.

METAL WINDOWS

Metal windows are available in steel and aluminum. Steel windows are covered with a baked-on primer and vinyl coverings. Aluminum windows are available in the natural material or anodized in black, brown, and gold. They are carefully built, so they have little air infil-

FIGURE 15-1

These double-hung windows add to the appearance of a home. These are covered with a rigid vinyl sheath. *(Courtesy of Andersen Corporation, Bayport, MN 55003)*

tration. Since metal is a good conductor of heat, the frames tend to cause moisture to condense on them. Brick or concrete sills are recommended for this reason (Fig. 15-2).

TYPES OF WINDOWS

Windows fall into three broad groups: *sliding*, *fixed*, and *swinging*. These are available in a variety of designs. For example, sliding windows move horizontally and vertically. Swinging windows open from the top, bottom, or side of the frame.

Double-Hung Windows

Double-hung windows are sliding units that move vertically. They have a top and a bottom sash. Both move in metal tracks. One type of window has springs behind the tracks pressing the track against the edge of the window. This holds the sash in the desired position. Double-hung windows are shown in Fig. 15-2.

Double-hung windows are available in many sizes. Screens and storm windows are installed on the outside of the unit.

Single-Hung Windows

A *single-hung window* is much like the double-hung. The upper sash is fixed. The lower sash moves vertically. In appearance it resembles the double-hung window.

FIGURE 15-2

Metal windows used in residential construction. *(Courtesy of Rusco Industries, Inc.)*

Horizontal Sliding Windows

Horizontal *sliding windows* generally have two or three *sashes* (Fig. 15-3). On two-sash windows, one sash moves and the other is fixed. On three-sash windows, the center sash is fixed and the two outside sashes slide (Fig. 15-4).

FIGURE 15-3

A two-sash horizontal sliding window. *(Courtesy of Andersen Corporation, Bayport, MN 55003)*

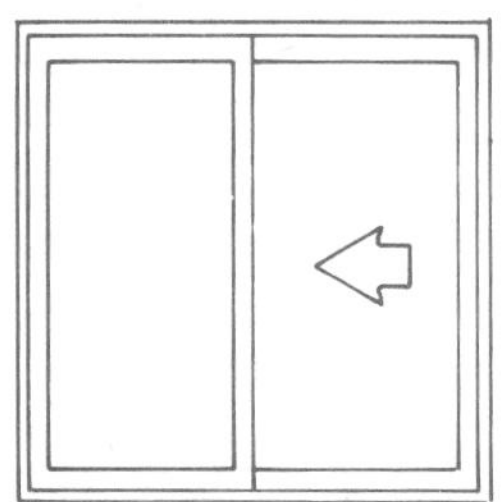

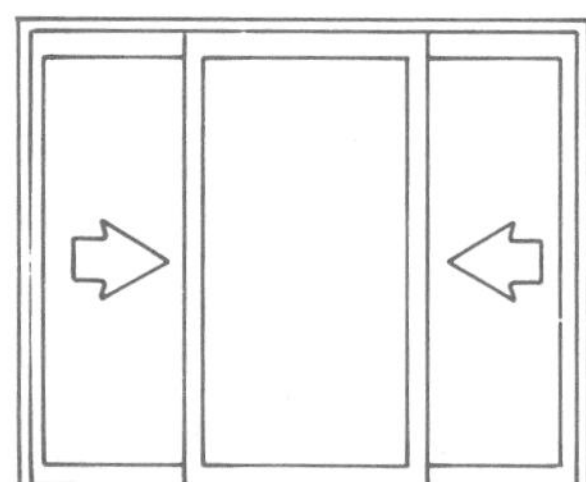

FIGURE 15-4

Horizontal sliding windows are available in two- and three-sash units.

Casement Windows The sash in a *casement window* is hinged on the side and swings outward (Fig. 15-5). It can be a single sash and hinge from right or left sides. Often two or more sashes are in a single unit (Fig. 15-6). Usually, the sashes hinge in opposite directions. Sometimes some of the sashes are fixed (Fig. 15-7). The windows are opened and closed with a crank. Since they open out, screens and storm sash are installed on the inside of the unit.

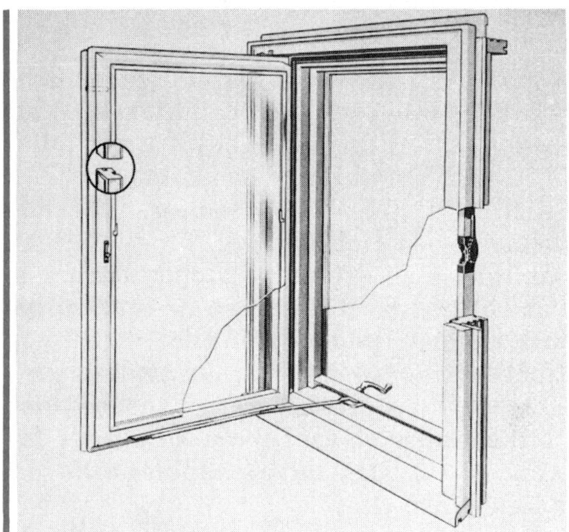

FIGURE 15-5

A single-sash casement window. *(Courtesy of Andersen Corporation, Bayport, MN 55003)*

FIGURE 15-6

These window units are made up of two double-sash casement windows. *(Courtesy of Andersen Corporation, Bayport, MN 55003)*

FIGURE 15-7

Casement windows are available in a variety of units combined in a single frame.

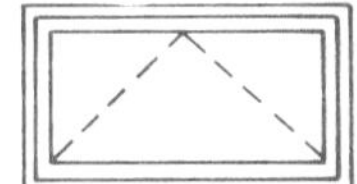
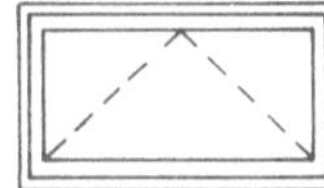
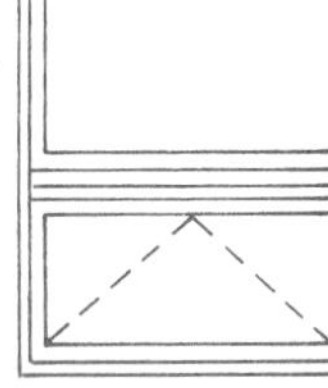

FIGURE 15-9

Awning windows can be installed as a single unit or combined with a fixed sash to form a larger window.

Awning and Hopper Windows *Awning windows* have the hinges at the top of the sash. The sash swings out (Fig. 15-8). They are available combined with several awning units or with a fixed sash (Fig. 15-9). They are operated using a cranking device. One advantage to these windows is that they can remain open during a rain. Since they open out, screens and storm windows are installed on the inside.

Hopper windows are just like awning windows except they are hinged at the bottom. They are installed so that the sash swings inward (Fig. 15-10). They have a handle at the top which locks them closed.

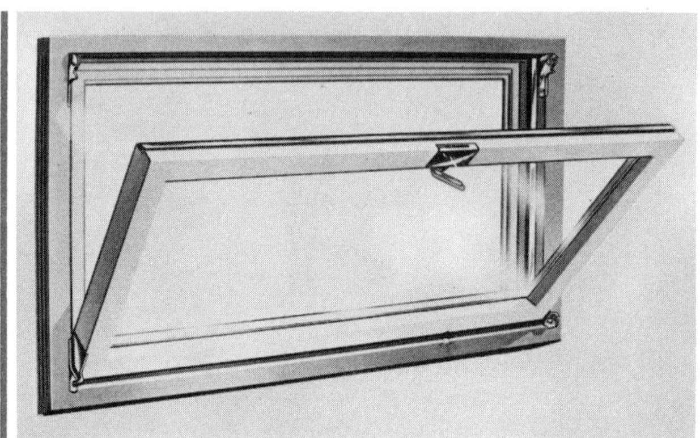

FIGURE 15-10

A hopper window hinges at the bottom. *(Courtesy of Andersen Corporation, Bayport, MN 55003)*

Jalousie Windows *Jalousie windows* are made of many narrow horizontal glass slats (Fig. 15-11). Each slat is held on each end by a metal clip. Each slat pivots much like a venetian blind. The slats are geared so one crank opens and closes them. This type of window permits the entire opening to provide ventilation. When closed, they are poor in retarding air leakage and are generally used on porches and breezeways.

FIGURE 15-8

A single awning window unit. *(Courtesy of Andersen Corporation, Bayport, MN 55003)*

FIGURE 15-11

Jalousie windows enable the total window opening to provide ventilation. *(Courtesy of Airmaster)*

Fixed Windows A *fixed window* is a sash set in a frame. It does not move. It provides light and a view but no ventilation. Fixed windows are often used in connection with other units which will open (Fig. 15-12).

Bay and Bow Windows A *bay window* has three window units. The two side units are set on an angle, such as 30 or 45° (Fig. 15-13). The units meet at specially designed angle mullion posts. The unit has head and seat boards to enclose the area.

A *bow window* is made up of several window units placed to form a curved surface (Fig. 15-14). It also uses specially designed mullions to join the individual window units. These usually contain four to seven window units. The casement-type window is popular for use in bow window units. It has head and seat boards to enclose the area.

Both the bay and bow windows are shipped completely assembled. The carpenter needs to build the proper rough opening and then install the entire unit in it.

Details for joining bay and bow units to the exterior wall are shown in Figs. 15-15 and 15-16.

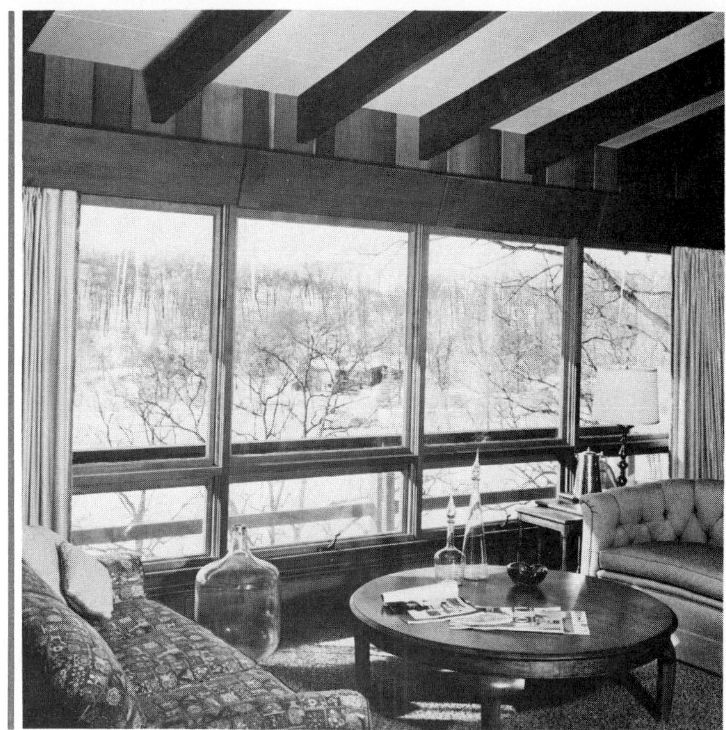

FIGURE 15-12

The large upper windows are fixed. The small windows below are crank-operated awning windows. *(Courtesy of Wood Conversion Co.)*

FIGURE 15-13

A bay window has three sashes. The side units are on a 30 or 45° angle.

FIGURE 15-14

A five-unit bow window. Some of the sashes are casement units, which provide ventilation. *(Courtesy of Andersen Corporation, Bayport, MN 55003)*

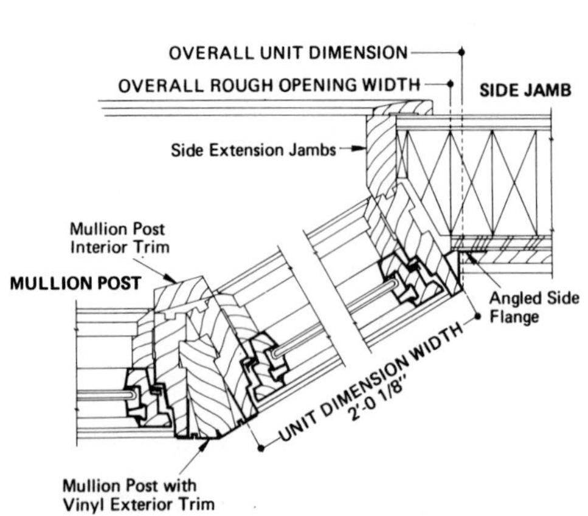

Horizontal section through the side jamb mullion.

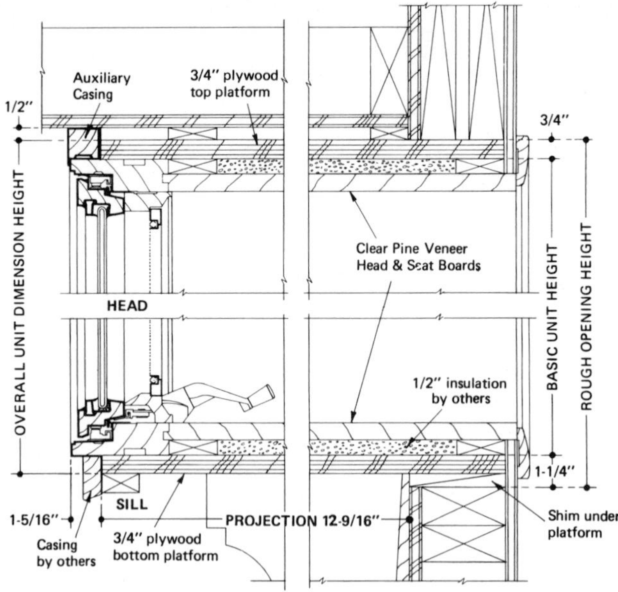

Vertical section through the window unit.

A BAY WINDOW.

FIGURE 15-15

Construction details for a bay window. *(Courtesy of Andersen Corporation, Bayport, MN 55003)*

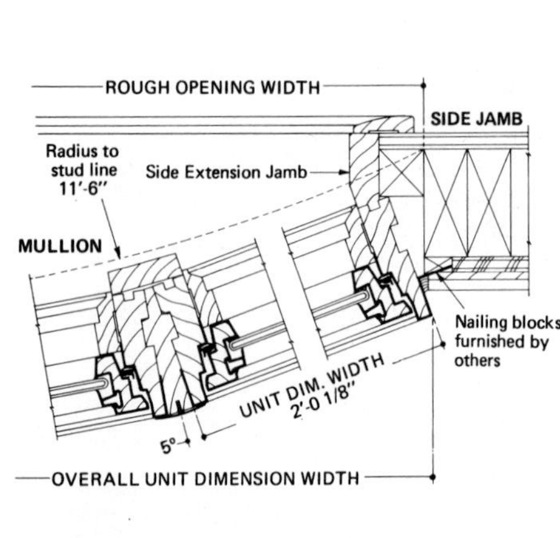

Horizontal section through the mullion.

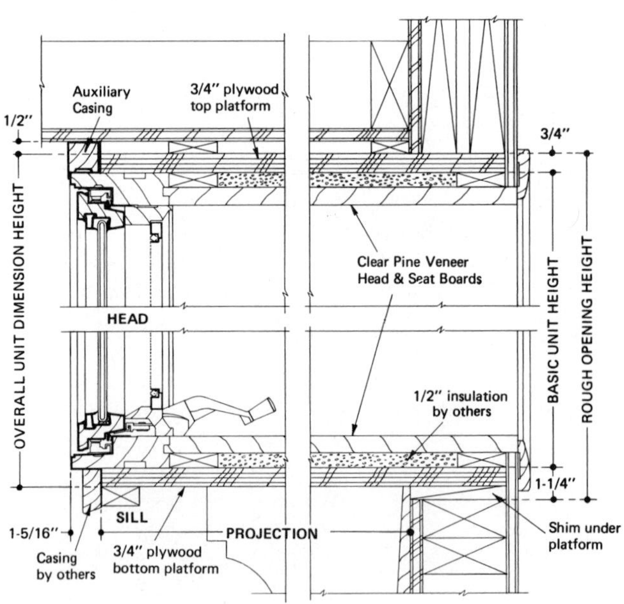

Vertical section through the window unit.

A BOW WINDOW.

FIGURE 15-16

Construction details for a bow window. *(Courtesy of Andersen Corporation, Bayport, MN 55003)*

WINDOW GLASS

Windows are a major source of heat loss in the winter and heat gain in the summer. If properly installed, there will be little seepage of air through the window unit itself. Considerable loss does occur through the glass. This can be greatly reduced by installing storm windows or purchasing windows with sealed double-glass panes (Fig. 15-17).

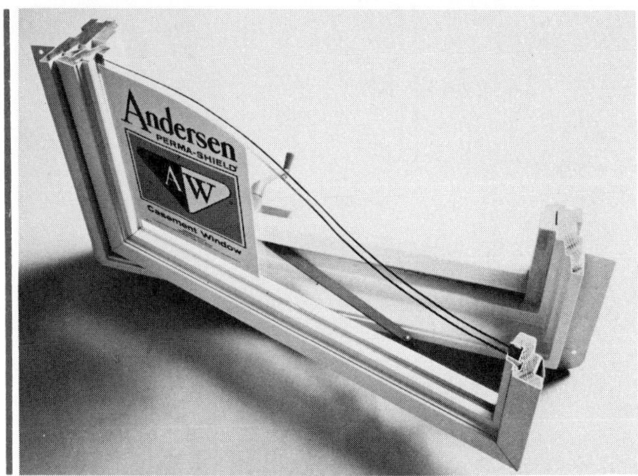

FIGURE 15-17

A section through a casement window showing the energy-saving double-panel glass. *(Courtesy of Andersen Corporation, Bayport, MN 55003)*

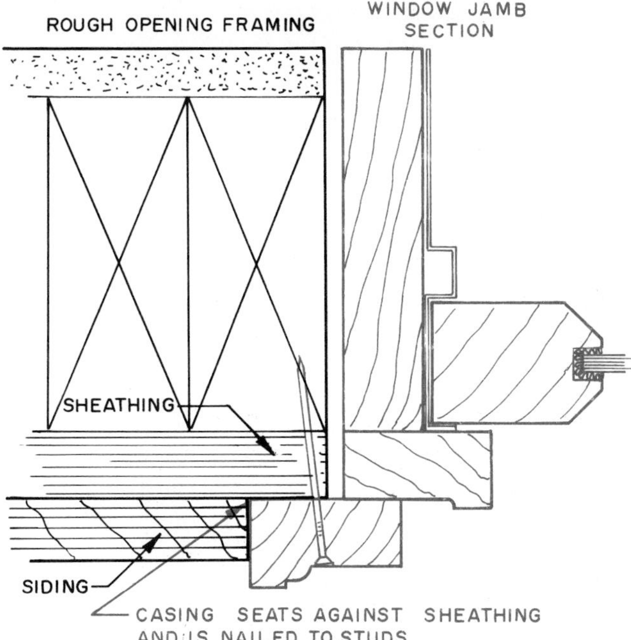

FIGURE 15-19

This type of window unit is installed by nailing the wood molding to the studs forming the rough opening.

SCREENS

Screens are needed to keep insects from entering the house. They are available in factory-built units sized to fit the various windows. Most use a light metal frame. They are available in sizes that cover the entire window opening or just half the opening. If storm windows are installed, screens will be a part of the unit and run in one of the metal tracks.

MUNTINS

Muntins are the vertical and horizontal bars that break up a large glass pane into smaller panes (Fig. 15-20). Windows used to be made with the muntin built into the sash. The glass panes were actually small. Today, it is too expensive to make windows this way. The sash is made with one large piece of glass. Wood and plastic strips are made which clip to the frame and present the appearance of muntins. These are supplied by companies manufacturing the windows and cost extra.

WINDOW CONSTRUCTION

Windows are manufactured and assembled in a factory. The carpenter receives the unit ready to install in the *rough opening* (Fig. 15-18). The exterior molding is nailed to the frame and seats against the sheathing as the unit is slid into the rough opening (Fig. 15-19). Another window system uses a metal flange to hold the window to the wall (Fig. 15-20).

FIGURE 15-18

Factory-assembled window units are carefully packaged for shipping to the construction site. *(Courtesy of Andersen Corporation, Bayport, MN 55003)*

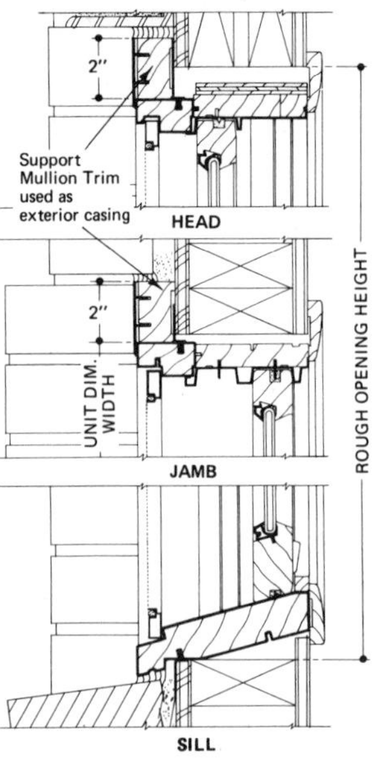

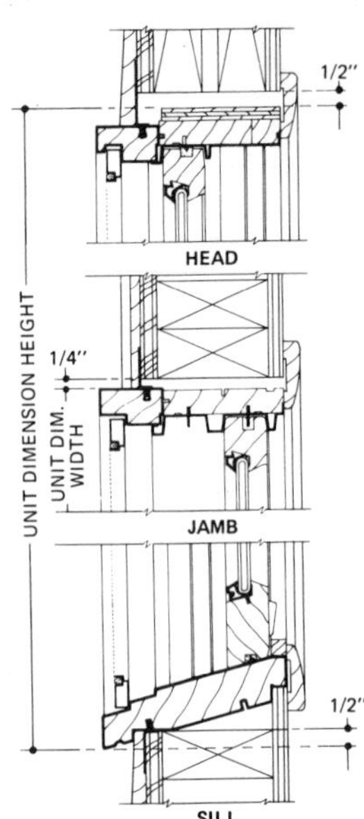

Double hung window in masonry construction.

Double hung window in frame construction.

FIGURE 15-21

Typical double-hung window construction and installation details. *(Courtesy of Andersen Corporation, Bayport, MN 55003)*

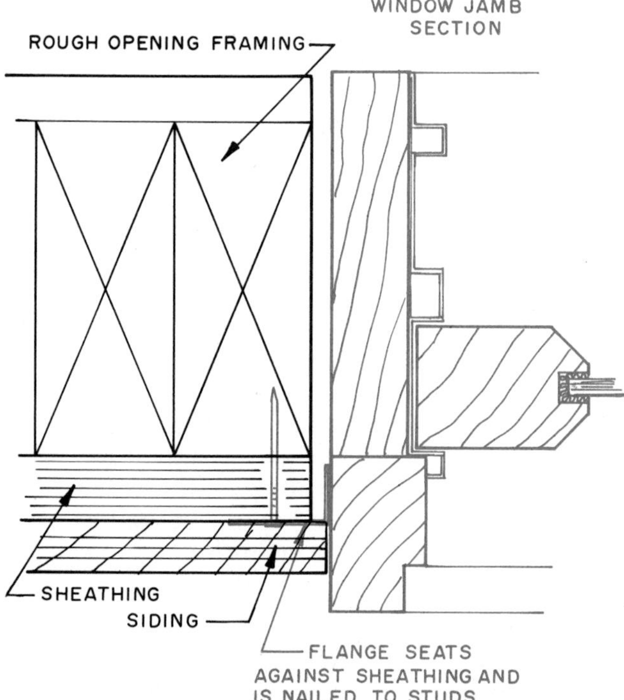

FIGURE 15-20

Some manufactures provide a metal nailing flange which is used to fasten the window to the framing. *(Courtesy of Andersen Corporation, Bayport, MN 55003)*

The *head* of the unit is the top assembly of parts. The *sill* describes the unit at the bottom. The *jamb* is the right and left vertical sides (Fig. 15-21). A *mullion* is a vertical member running between two windows. Its size depends upon the space wanted between windows (Fig. 15-22). If it is helping to support the ceiling or the roof, it could contain one or more studs (Fig. 15-23).

FIGURE 15-22

A mullion joining two window units. This carries no overhead load.

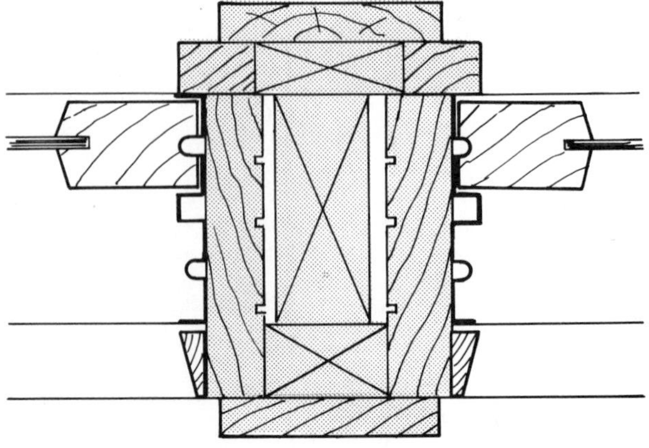

FIGURE 15-23

A mullion with a stud that carries an overhead load.

WINDOW INFORMATION

The carpenter should be able to secure all the information needed about windows from the house plan. The floor plan shows the location for each window (Fig. 15-24). Usually, windows are identified by a code. All windows of the same size and type use the same letter identification code.

The other details needed are in the window schedule. It identifies the windows by code, the number needed, the sash size, rough opening in frame and masonry walls, and often the specific brand (Fig. 15-25).

If this information is missing from the plan, the carpenter will need to contact the window supplier to get the needed information. If the window units are on the job, the carpenter can find the sizes by measuring them at these places.

JAMB EXTENSIONS

The width of the window jamb often needs to be increased because of the wall thickness. Most window manufacturers build their units to a standard width. They supply jamb extensions to make them fit walls that are thicker (Fig. 15-26).

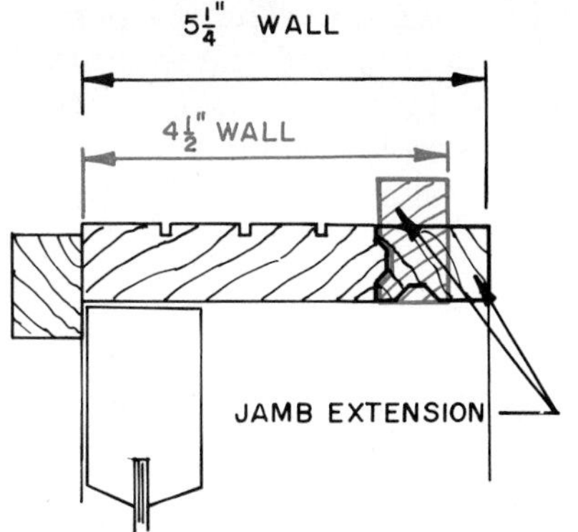

FIGURE 15-26

Jamb extension can be turned to accommodate two different wall thicknesses.

INSTALLATION DETAILS—WOOD WINDOWS

There are many companies manufacturing wood windows. Installation details for all of

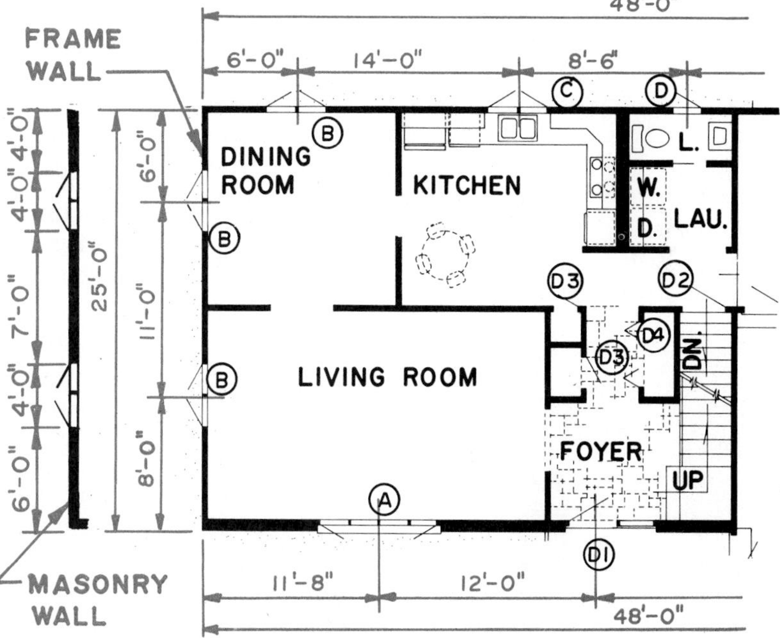

FIGURE 15-24

The floor plan for a frame building locates each door and window by its centerline. A masonry building gives the rough window opening from side to side of the masonry opening.

CODE	QUAN	NO. LTS	UNIT DIMENSION	ROUGH OPENING	DESCRIPTION
A	1	3	4'-11⅞" x 8'-0½"	5'-0⅜" x 8'-1"	CASEMENT ANDERSEN,C15,CP25,C15
B	3	2	4'-11⅞ x 4'-0"	5'-0⅜" x 4'-0½"	CASEMENT ANDERSEN C25
C	1	2	3'-4$\frac{13}{16}$" x 3'-4¾"	3'-5⅜" x 3'-5¼"	CASEMENT ANDERSEN CN 235
D	1	1	2'-11$\frac{15}{16}$" x 2'-0⅛"	3'-0½" x 2'-0⅝"	CASEMENT ANDERSEN C13

FIGURE 15-25

A typical window schedule as found on construction drawings.

them are somewhat similar. The carpenter should study the publications of the manufacturer before attempting to install windows with which he/she is not familiar. Some typical installation details are shown in Figs. 15-15, 15-16, and 15-21.

Installing Wood Windows A window unit must be handled very carefully on the job. It should be unloaded and stored so that it will not be damaged or twisted. It should be kept dry. Usually, each unit will have some type of bracing. This holds it square. Do not remove the bracing until the window is completely installed.

The rough opening should be about 1 in. larger than the window unit. This provides needed room to plumb and level the window. Nail a layer of builders' felt around the rough opening. This helps reduce air infiltration (Fig. 15-27).

Most window units are installed by sliding them into the rough opening from the outside. Be certain to hold or brace the window so that it does not fall out while being plumbed and leveled.

Now place wedge blocks under the sill (Fig. 15-28). Wood shakes make good wedges. Lightly tap the wedges under the sill until the frame is level. On small windows, place one wedge on each end. On large windows, place a wedge or two in the center to prevent sagging. Once the sill is level, put one nail through the casing into the studs on the lower end on each side. This will hold the sill level. On windows using a metal flange instead of a wood casing, nail the flange to the wall (Fig. 15-20).

FIGURE 15-27

A layer of builders' felt is nailed around the rough opening before the window unit is set in place.

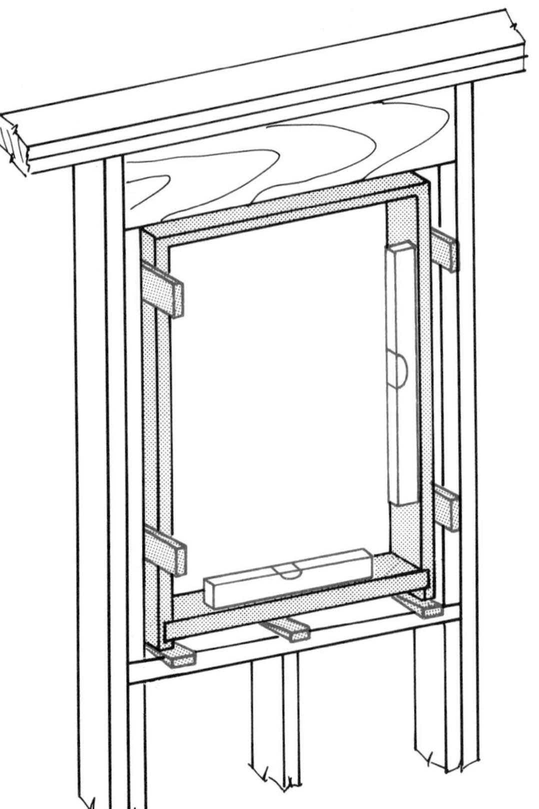

FIGURE 15-28

After sliding the window unit into the rough opening, level it with wood wedges.

Next, plumb the jambs. Place wedges on each side of the jamb. Tap them together until the jamb is plumb. Use a long level (Fig. 15-28). Drive a nail through the casing or flange on each side at the top of the window.

Now recheck to see if it is plumb and level. Try moving the sash to see if it works easily. If all is in order, nail the unit in place. This is done by nailing 16d casing nails about 16 in. apart through the casing.

Be especially careful when nailing the casing. It is visible when the job is finished and hammer marks are absolutely forbidden. Use a nail set to place the head of the nail below the surface of the casing.

INSTALLATION DETAILS—METAL WINDOWS

The carpenter should become familiar with the various ways in which metal windows are installed (Fig. 15-29).

Installing Metal Windows Be familiar with the manufacturer's recommendations for in-

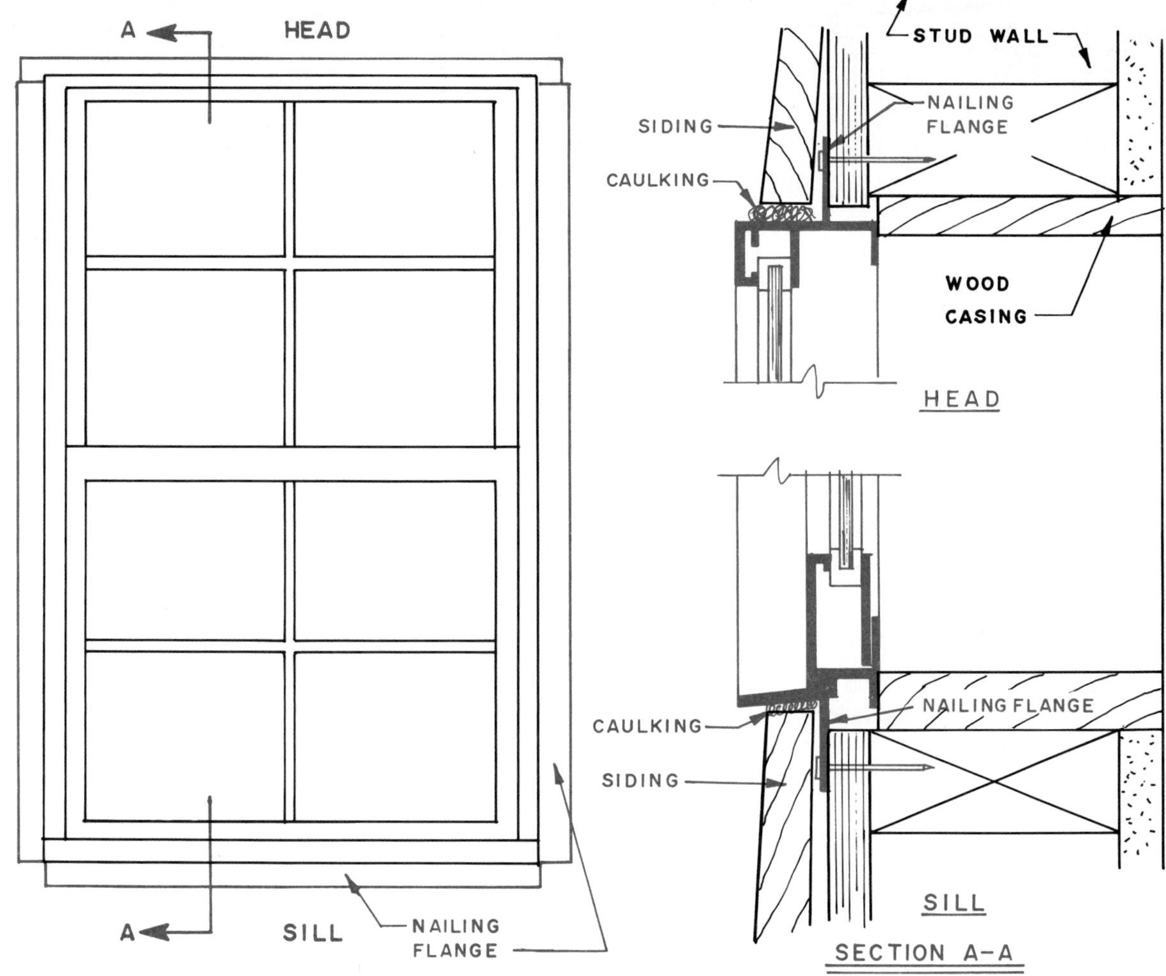

FIGURE 15-29

Typical installation details for a metal window.

stalling the window. Keep the unit completely assembled. Lock the sash together. Carefully lift it in place over the rough opening. Check it for levelness. When level, drive two nails through the holes in the upper corners of the flange. Check to see if it is plumb. Be certain that it is not twisted or distorted in any way. Nail the two lower corners of the flange. Now operate the windows to see if they move easily. If all is in order, finish nailing the flange to the wall. Use the nails recommended by the manufacturer.

WOOD DOORS

Wood doors may be solid- or hollow-core. They may be flush or panel types. The *flush-type door* can be solid- or hollow-core. The *solid-core door* has the inner layers made from small wood blocks glued together. Over this is glued several layers of veneer (Fig. 15-30). Solid-core doors are generally used as exterior doors. They are heavy, strong, and fire-resistant. They resist the weather better than do hollow-core doors.

The *hollow-core flush door* is generally used as an interior door. It is lighter and costs less than the solid-core. It is made with an internal honeycomb of interlocking strips with a solid wood frame. Layers of veneer are glued to the frame forming the surface (Fig. 15-31).

Panel doors with solid wood frames and solid wood panels are also used as exterior doors. They are made of rails, stiles, and panels. The *rails* are the main horizontal members. The *stiles* are the main vertical members (Fig. 15-32). These doors can have *lights* (glass panes) as well as wood panels. Some typical designs are in Fig. 15-33.

STEEL-FACED DOORS

Steel-faced doors have the desired design stamped into sheet steel. These are laminated

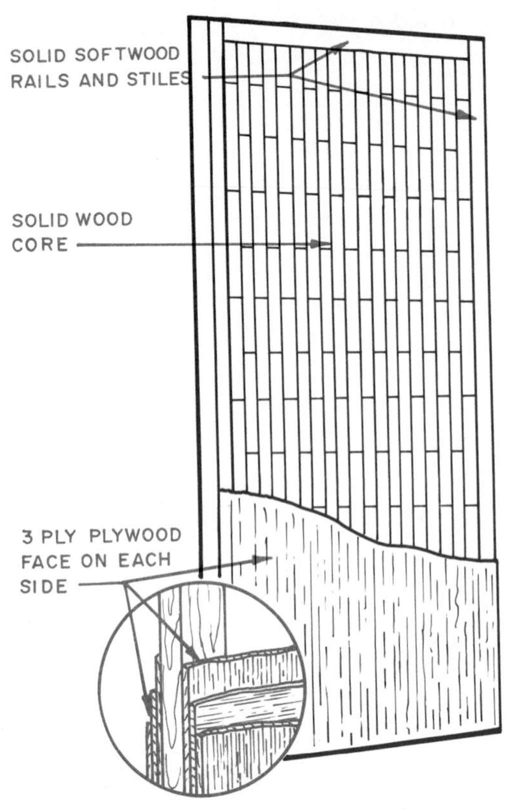

FIGURE 15-30

A solid-core flush door.

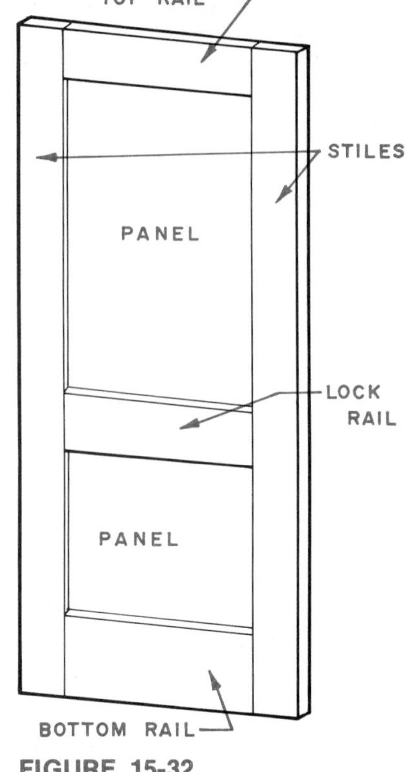

FIGURE 15-32

A panel door.

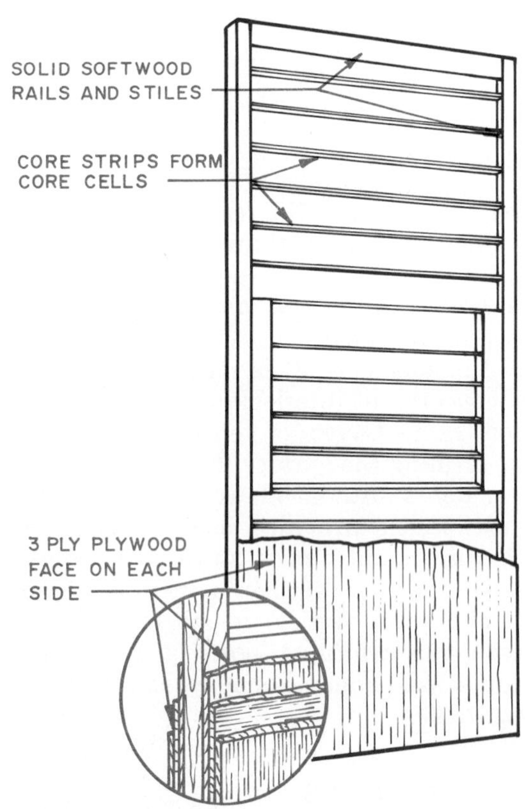

FIGURE 15-31

A hollow-core flush door.

FIGURE 15-33

Some typical wood exterior door designs.

to a *core* of expanded *polystyrene.* The metal sheets forming the inside and outside of the door do not touch. This prevents the passing of heat or cold from one face to the other. They are usually used as exterior doors.

WEATHERSTRIPPING AND THRESHOLDS

Exterior doors must have weatherstripping on all sides. This is usually done by nailing a vinyl or metal strip at the top and sides of the door. When the door closes, the strip is pressed against the door (Fig. 15-34).

Metal *thresholds* with a flexible vinyl strip are used on the sill. When the door closes, the vinyl strip is pressed against the bottom of the door (Fig. 15-35).

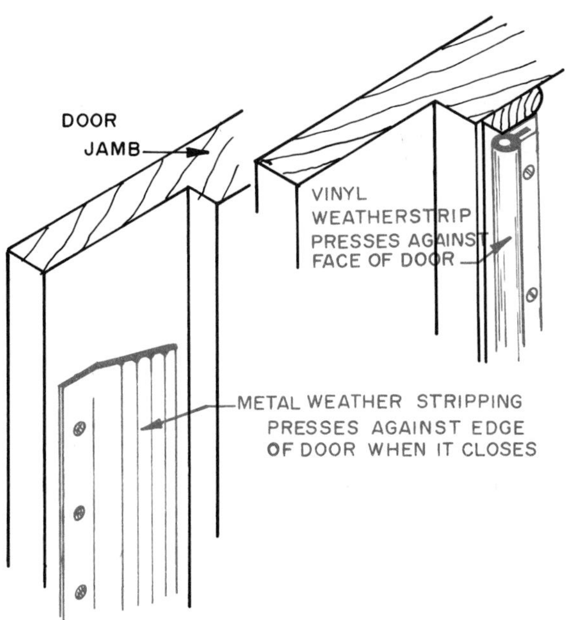

FIGURE 15-34

Typical types of weatherstripping used on exterior doors.

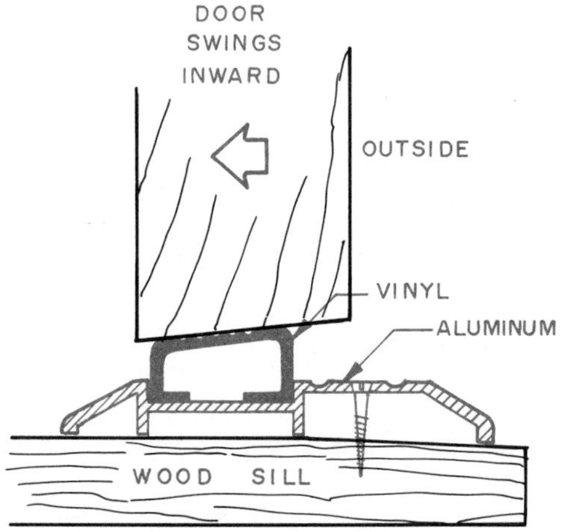

FIGURE 15-35

An aluminum threshold with a vinyl insert.

WOOD EXTERIOR DOOR FRAMES

Wood *door frames* are manufactured in large quantities. They are available either in knocked-down or assembled form. Since all the joints are cut, it is easy to assemble them on the job (Fig. 15-36). Some door frames are shipped with the door already hung. Exterior door frames have a wood sill. In masonry buildings the sill is often stone or concrete rather than wood.

The exterior door frame has a ½-in.-deep rabbet on the inside edge to receive the door. It also has a rabbet on the outside edge formed by

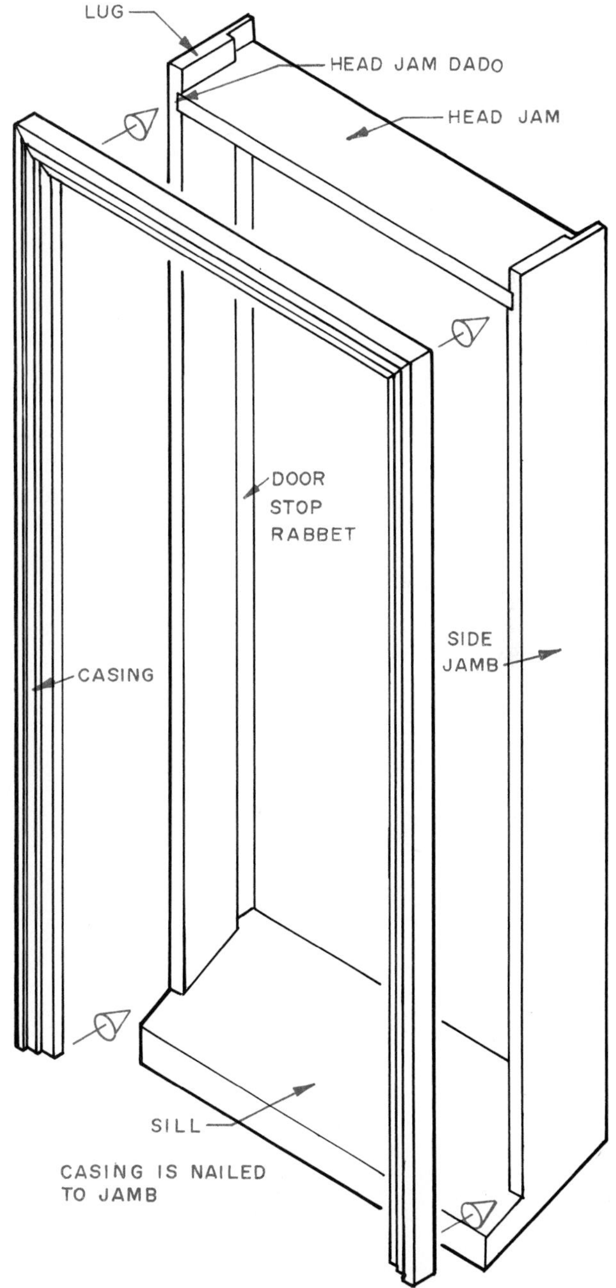

FIGURE 15-36

An exterior door frame and casing.

the casing. This is for a screen or storm door (Fig. 15-37). Exterior doors swing into the house, so this rabbeted edge must face in. The screen or storm door swings out.

Installing Exterior Wood Door Frames The door sill rests on the header and floor joists (Fig. 15-38). To get the top of the sill level with the finished floor, the subfloor must be cut out and the header and floor joists notched. If necessary, add additional support to hold the edge of the subfloor (Fig. 15-39).

Next, cover the edges of the rough opening with builder's paper. This reduces air infiltration.

Now insert the frame in the rough opening. Use blocking to get the sill level and seated firmly. Check the sill with a level. When the sill is level and centered in the rough opening, drive a nail through the casing on each side near the sill. Do not drive it all the way in because it may be necessary to remove it to make adjustments.

Now place wedges on each side. Check the sides to see if they are plumb. Adjust the wedges until the sides are plumb. Check with a long level or a level with a long, straight edge. Drive a casing nail through the casing on each side at the top (Fig. 15-40).

Now insert blocking between the studs and jamb where each hinge is to be located and behind the lock plate. Locate other blocks necessary to firm up the frame. Drive two 16d casing nails spaced ¾ in. apart into each block (Fig. 15-41). Be careful that the blocking does not bow the frame.

Then face-nail the casing using 16d casing nails spaced 16 in. apart. Remember, the door framing is visible. No hammer marks are allowed. Use a nail set to place the nail heads below the surface.

If a frame with a prehung door is to be installed, the same procedures are used. It is suggested that the door be removed from the frame while it is being installed.

FIGURE 15-37

The door side jamb and casing form rabbets to receive the door and screen or storm door.

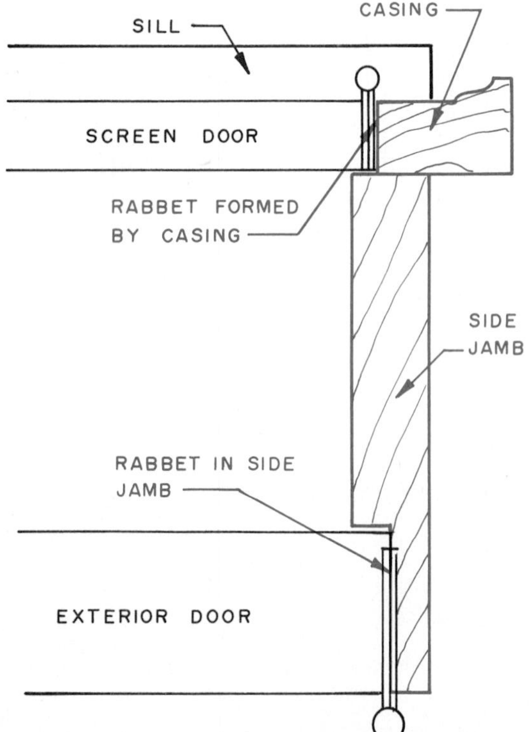

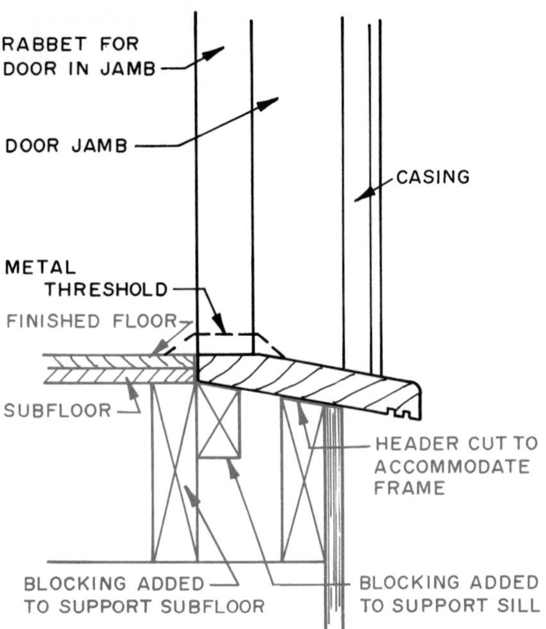

FIGURE 15-38

The exterior door sill rests on the header and floor joists.

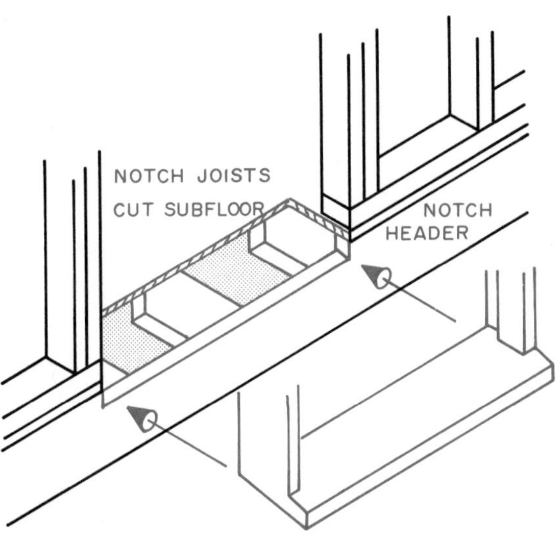

FIGURE 15-39

The floor joists must be cut to receive the door sill.

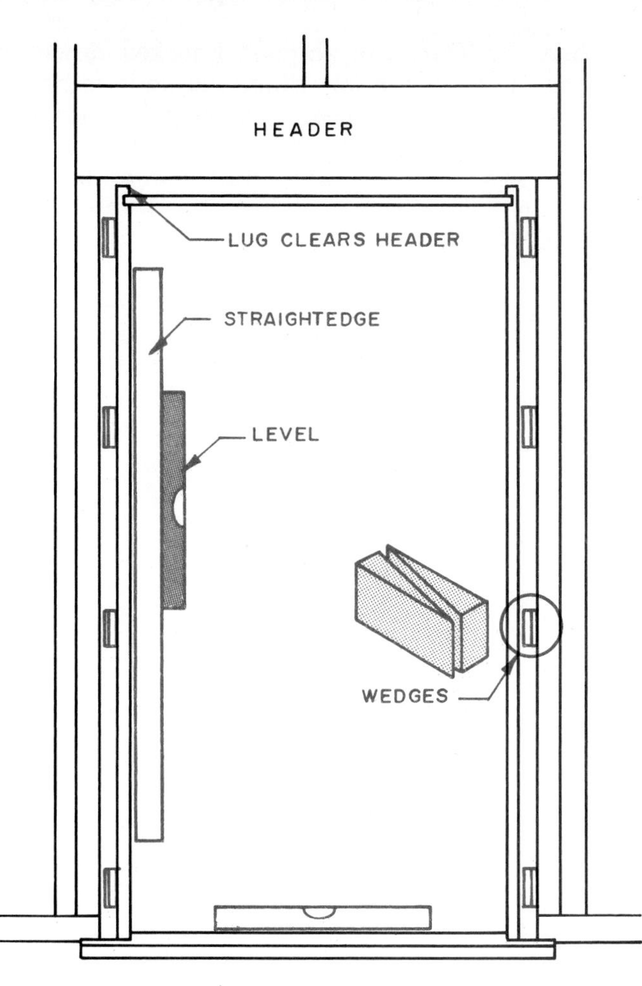

FIGURE 15-40 *(left)*

Use wedges to position the door frame. Check to see that it is plumb and level.

SETTING A DOOR FRAME IN A SOLID MASONRY WALL

In solid masonry construction the door frames are set in place after the subfloor is laid. The masons lay the wall to the frame.

The head of the door must be the proper height above the finished floor. The side jambs might have to be cut shorter.

Set the door frame on the masonry sill. The frame should have diagonal braces and a spacer at the bottom. These keep it square and the correct size. Brace the frame to the subfloor. Make certain that it is level and plumb and securely nailed in place (Fig. 15-42).

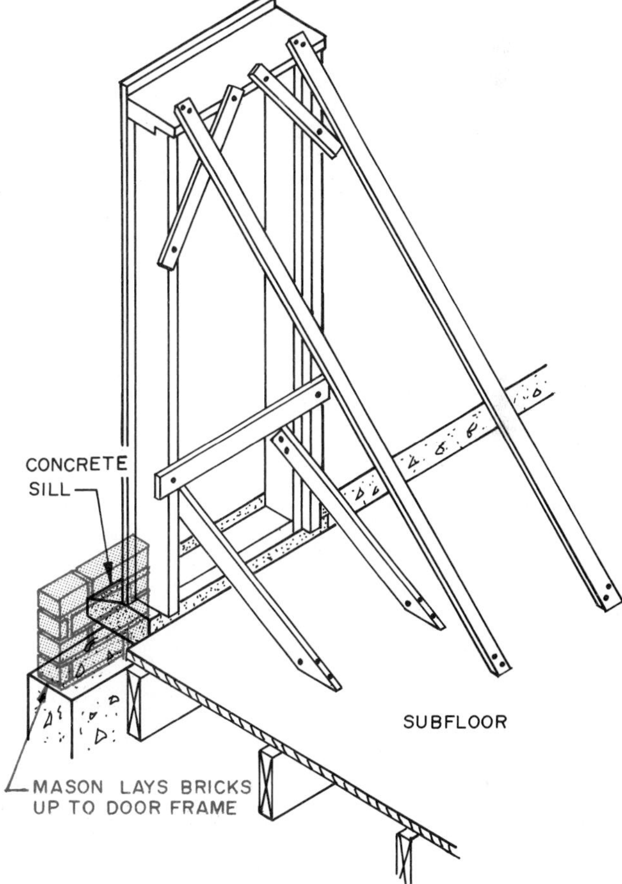

FIGURE 15-42 *(above)*

Set the door frame plumb and level and strongly brace it.

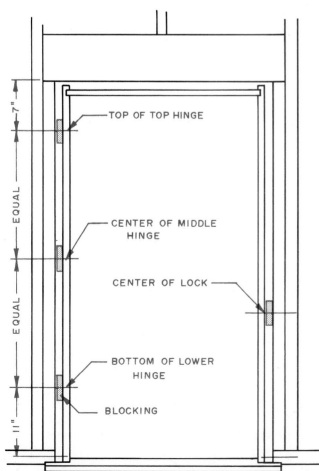

FIGURE 15-41 *(left)*

Insert blocking at the location of each hinge and the lock.

The inside edge of the jamb must line up with the inside surface of the finished wall material. The plan must show a typical wall section. This tells the carpenter the amount of furring to be used inside and the thickness of the finished wall (Fig. 15-43).

The bricklayer lays the brick up next to the casing on the door frame. The frame is anchored to the brick wall as it is laid. The crack between them is later caulked (Fig. 15-44).

WOOD INTERIOR DOOR FRAMES

Interior door frames are flat. They do not have a rabbet or a sill (Fig. 15-45). Most commonly used widths are 4½ in. for walls with ½-in. drywall and 5¼ in. for walls with plaster. The frames are cut to finished size and sanded in a factory. They are shipped knocked down for assembly on the job. Since the dado is cut for the head jamb, assembly is easy. Use three 8d casing or finishing nails.

Also available are adjustable jambs. They can be adjusted to fit different wall thicknesses (Fig. 15-46).

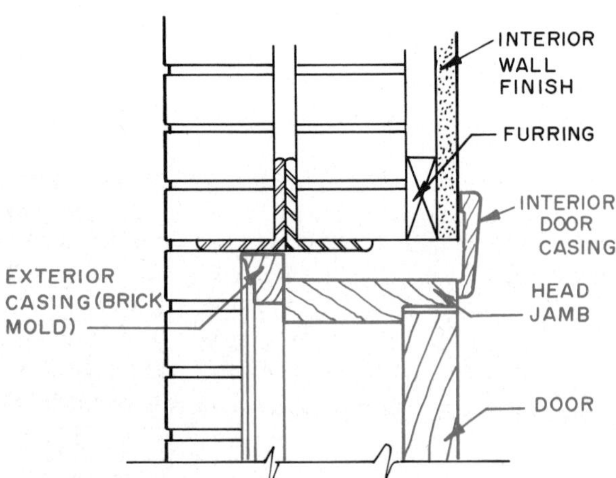

FIGURE 15-43

A typical wall detail found on construction drawings. It shows the thickness of the finished wall.

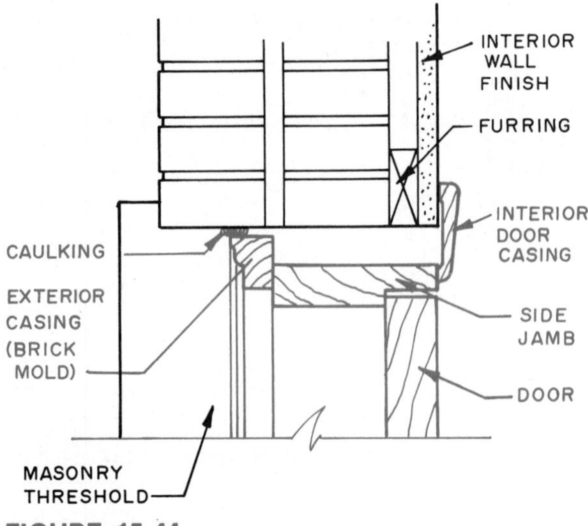

FIGURE 15-44

A horizontal section through a doorjamb as it meets a masonry wall. The joint is caulked after the masonry units are laid.

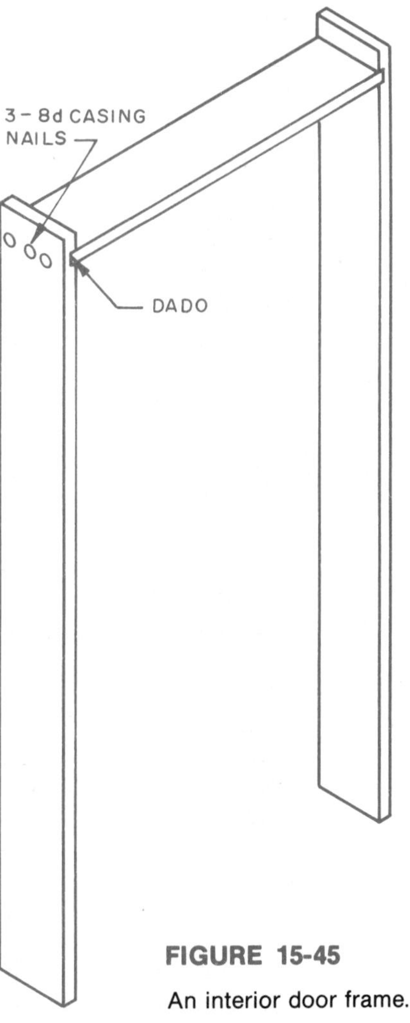

FIGURE 15-45

An interior door frame.

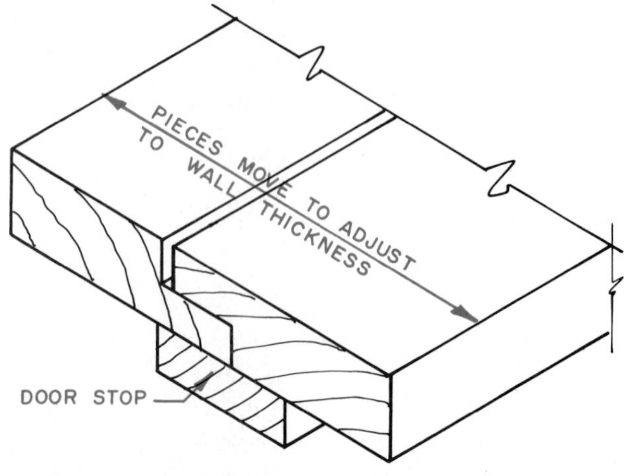

FIGURE 15-46

Some interior door frames have adjustable jambs.

Installing Wood Interior Door Frames Wood interior frames are installed much the same as exterior frames. Since the frame has no sill, it will need a brace at the floor to keep the side jambs the proper distance apart. It should be the same length as the distance between the side jambs at the head.

Slide the assembled frame into the rough opening. Part of the lug can be cut off if it does not clear the header. If the finish flooring has been laid, the side jambs can rest on it. If not, put blocks of wood under them to allow for this material. Place the brace between the side jambs at the floor.

Center the frame in the opening. Lightly tap double wedges between the frame and the studs (Fig. 15-47). Next, locate the nailing line (Fig. 15-48). All nails should be hidden by the door stop. Measure in from the edge of the jamb the thickness of the interior doors plus half the width of the door stop to be used. This will give the line for nailing the frame to the studs. The door stop will cover the heads of the nails.

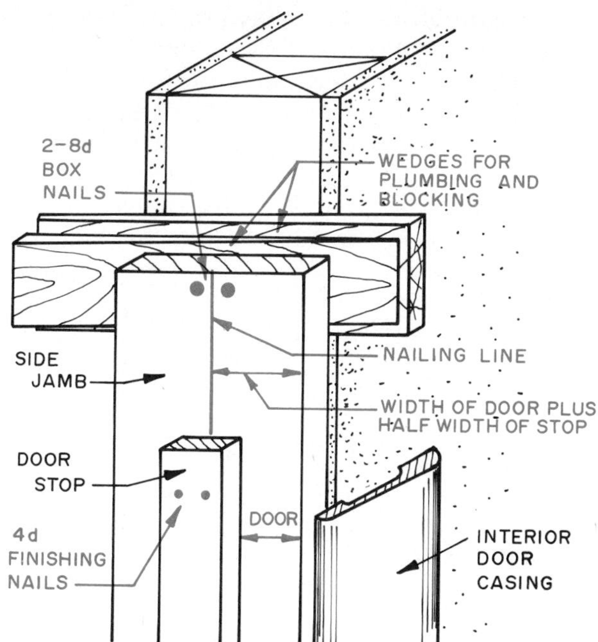

FIGURE 15-48

Locate the nailing line so that the doorstop covers the nail heads.

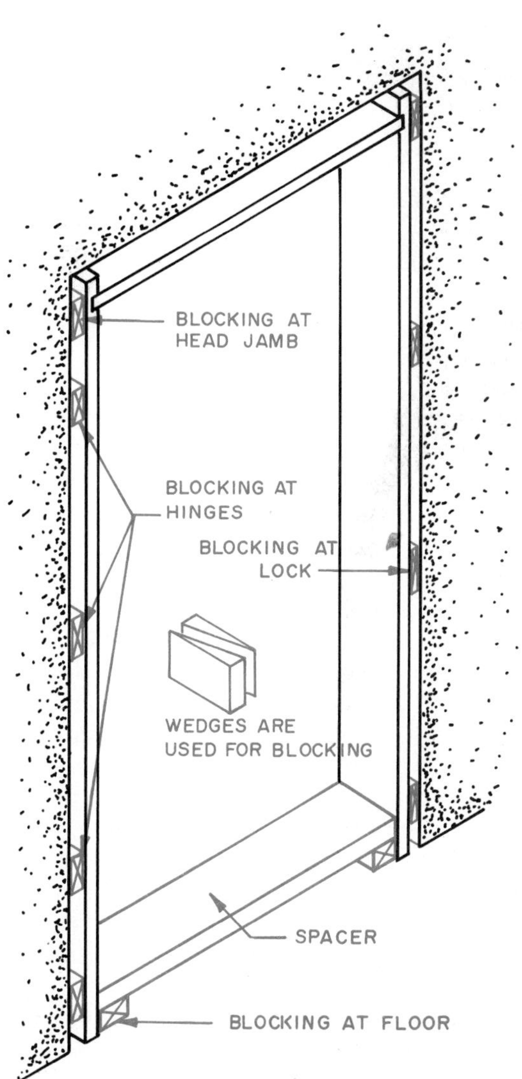

FIGURE 15-47

Use wedges to position the interior door frame in the rough opening.

Using a long level, check the side jambs for plumb. Adjust the wedges until the jambs are plumb and straight. Remember, the wedges can force the jamb to bow. Nail two 8d nails through the jamb and wedges into the studs at both sides at the top and bottom. After rechecking for plumb, nail through the other wedges. On the hinge side of the frame, locate blocking about where each hinge will go. This is usually 7 in. from the top and 11 in. from the bottom. Also put one in the center of the side jamb. On the lock side, locate wedges where the lock will hit. This is usually 36 in. from the floor. Put at least one more set of wedges in the space above this. Drive two 8d casing nails about ½ in. apart through each set of wedges.

Split door frames for interior use are available. Each half contains interior trim plus half the side and head jamb. The door is hung on one side.

HANGING A DOOR

In most buildings a variety of door types are used. Check the plans to be certain that you are installing the proper door. Notice how it is to swing. Put a mark on the hinge side so that no mistake can be made.

Doors are expensive and easily damaged.

Handle carefully so that they are not scratched or have edges damaged. Store flat and cover with plastic to keep dry and clean.

The door may require a little trimming to fit the door frame. Usually, very little need be removed from a door.

Never cut a great deal off any one edge. This will produce a thin stile and a weak door.

Trim the door so that it clears the door frame 1/16 in. on all sides. Usually 5/8 in. clearance on the bottom will be enough. If metal weather stripping is to be used, a side clearance of 1/8 in. will be necessary.

Since most doors require very little trimming, the amount required is removed with a plane. A hand or power plane can be used. Stand the door on edge. The best way to support it is with a door holder (Fig. 15-53). Remove some material from each side. If it must be trimmed with a power saw, use a fine-tooth blade to get a smooth cut. Cut a little large so that the edge can be planed to its final size. Once the door is to its final size, plane a bevel on the edge on the lock side of the door (Fig. 15-49). This is needed so that the edge will not hit the door frame as the door opens. Usually, a bevel about 1/8 in. will be enough.

Now sand the planed edges and corners so that they are slightly round and smooth. Always sand with the grain of the wood. Use a fine-grit abrasive paper.

The door is now ready to fit into the frame and have the hinges and lock installed.

Installing the Hinges Hinges are mounted in gains. A *gain* is where a portion of the wood has been removed, forming a depression (Fig. 15-50). The hinges used for doors are called *butt hinges* or sometimes butts. Some have loose pins. The door can be removed when the pin is pulled out. Others have fast pins. They cannot be removed. The leaves of the hinge meet in a *barrel*. The pin slides into the barrel (Fig. 15-51).

The hinge should be located about 1/4 in.

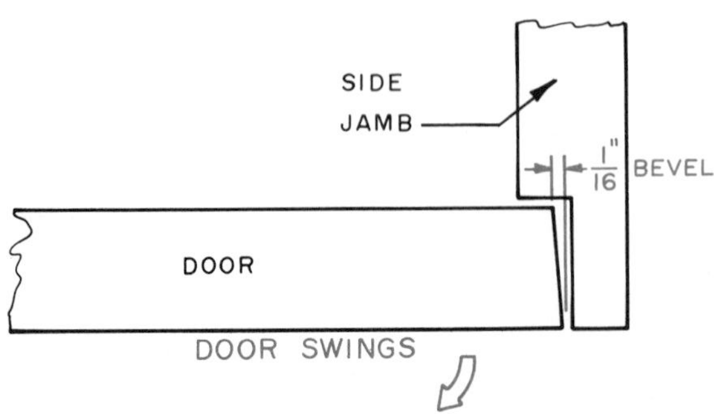

FIGURE 15-49

Put a bevel on the edge of the door that will have the lock.

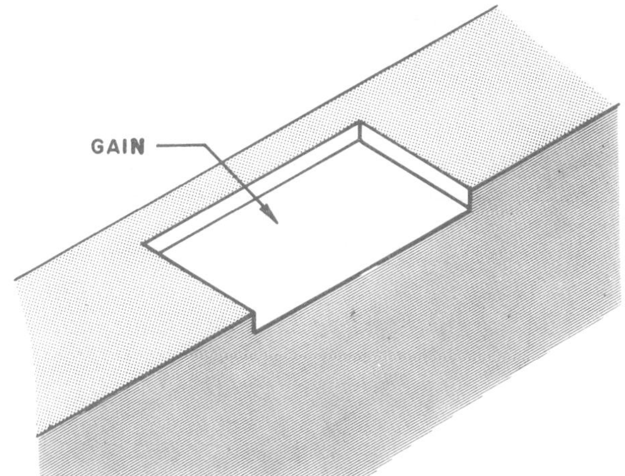

FIGURE 15-50

Gains are cut to hold the hinges.

FIGURE 15-51

A loose-pin door hinge. *(Courtesy of Stanley Hardware)*

from the face of the door. This usually places the barrel out from the wall far enough to permit the door to open fully without hitting the interior trim (Fig. 15-52).

Gains are cut in the edge of the door and in the jamb. A hinge mortising template is used. It is set so that the gains are located the desired distance from the top and bottom of the door. This is usually 7 in. from the top and 11 in. from the bottom. The template is placed on the edge of the door (Fig. 15-53). A router is placed on the template and is used to cut the gain. The depth of cut should equal the thickness of the hinge leaf. The router leaves a round corner for round corner hinges. The corners have a 5/8-in. radius and require a router cutter with a 1¼-in. diameter.

If a hinge with rectangular leaves is used, the rounded corner is cut square with a chisel.

If a hinge butt template is not available, the gains can be cut other ways. First mark the gain on the door and jamb. To do this, place the door in the frame. Using wedges on the bottom

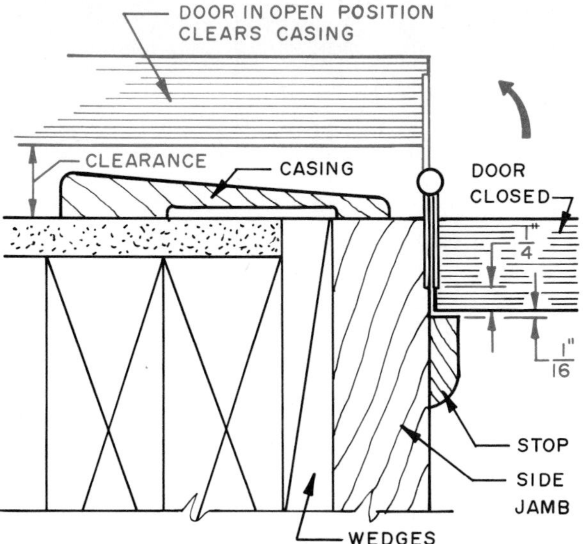

FIGURE 15-52

The hinge is set so that the door will clear the trim when it is opened.

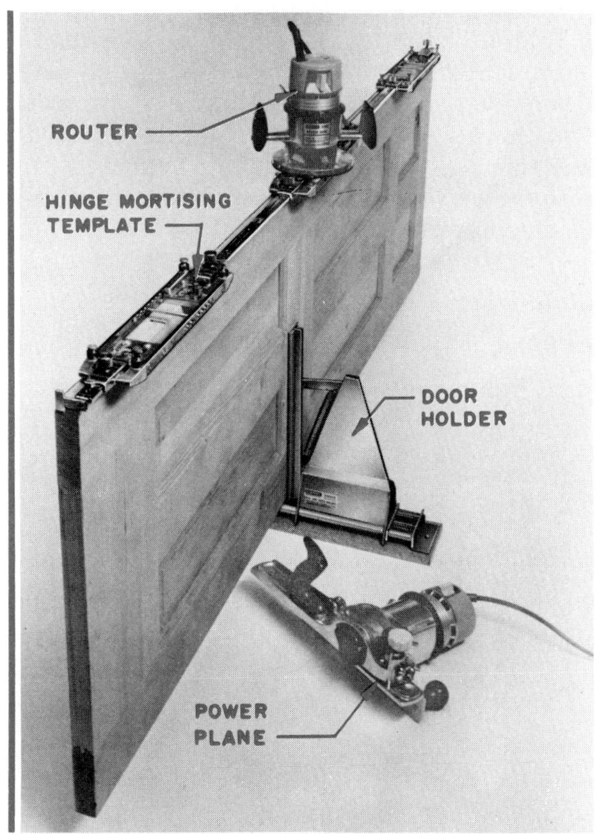

FIGURE 15-53

Gains in door edges are cut using a hinge mortising template and a router. The door is held with a door holder. *(Courtesy of The Stanley Works)*

and sides, locate it in the desired position. Measure the location of the hinges from the top and bottom of the door. Mark on the door and the jamb (Fig. 15-54). Remove the door and lay out the gain as shown in Fig. 15-55. Mark the jamb

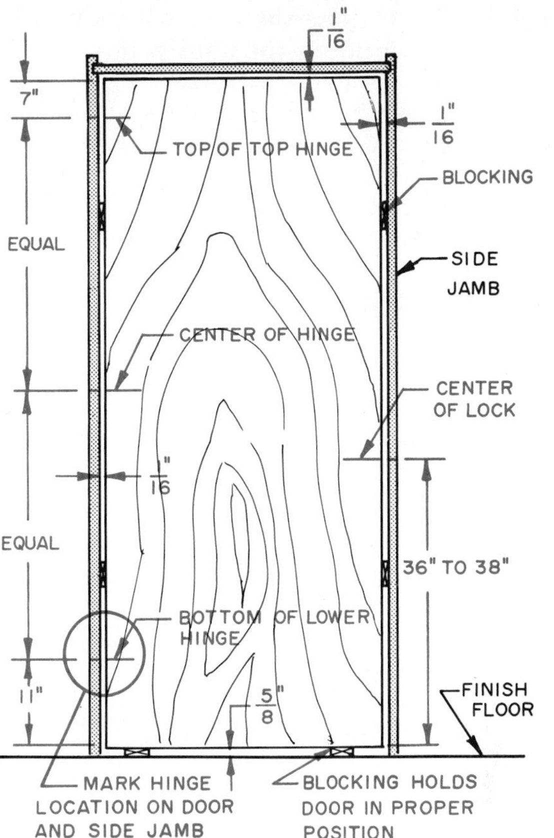

FIGURE 15-54

Use wedges to position the door in the frame. Measure and mark the location of hinges and the lock.

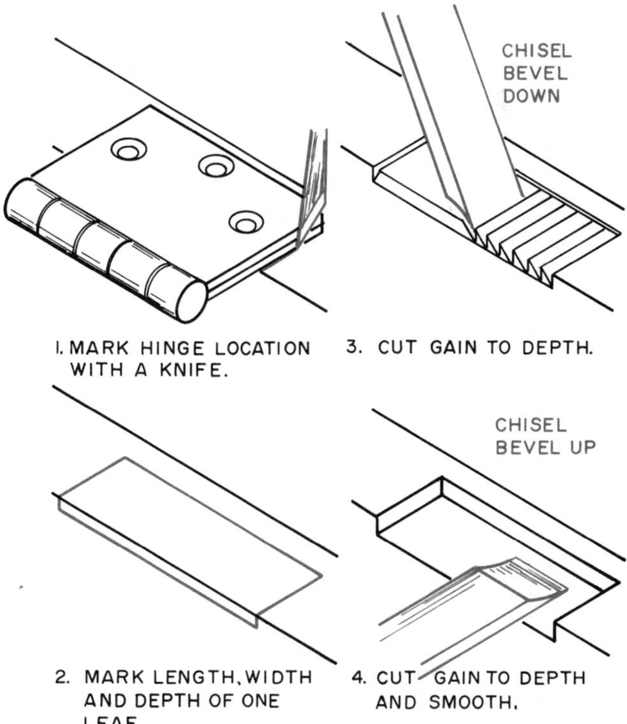

FIGURE 15-55

How to cut a gain with a wood chisel.

the same way. Outline the gain with a chisel. Remove the material in the gain with a router or a chisel.

Pull the pin on the hinge. Place one leaf in the gain on the door. Drill holes for the screws. Install the screws. Install the other leaf on the jamb. Set the door in the jamb, line up the barrels, and drop the pin in place (Fig. 15-56). Check to see that the door swings freely and has the required clearances.

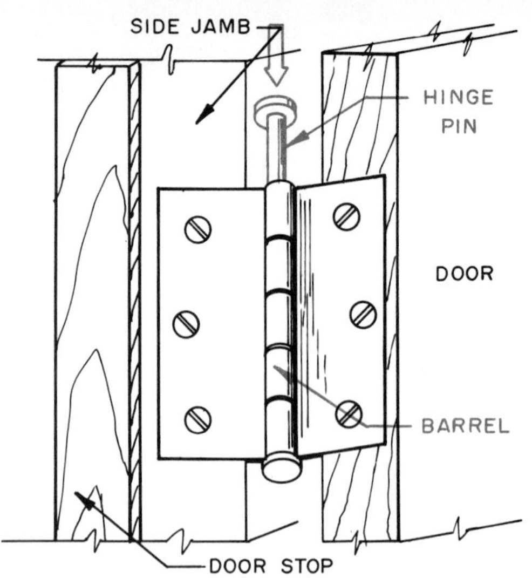

FIGURE 15-56

Separate the hinge leaves. Screw to the door and frame. Place the door in position and insert the hinge pin.

DOOR LOCKS

There are a variety of exterior door locks available. The cylindrical and tubular are most commonly used. The *cylindrical lock* has a large cylindrical case containing the locking mechanism. It requires that a large hole be bored in the face of the door. A smaller hole is required in the edge of the door for the latch unit (Fig. 15-57).

A *tubular lock* is much like the cylindrical lock. It is not as secure a mechanism. It requires a small hole in the door face and edge of the door (Fig. 15-58).

Dead bolts are another type of lock. They have a large solid plunger that enters a metal plate in the jamb. Some can only be opened and locked from inside the house. Others have an inside locking lever but can be opened with a key from the outside (Fig. 15-59).

Interior door locks are not key-operated. Some are used simply to hold the door closed. Neither side has a lock. Others have a button or knob on one side which is a lock. This is placed in the room so that the door can be locked inside the room (Fig. 15-60). They are installed in the same manner as exterior locks.

FIGURE 15-57

A typical cylindrical lock used on exterior doors.

FIGURE 15-59

A dead bolt increases security.

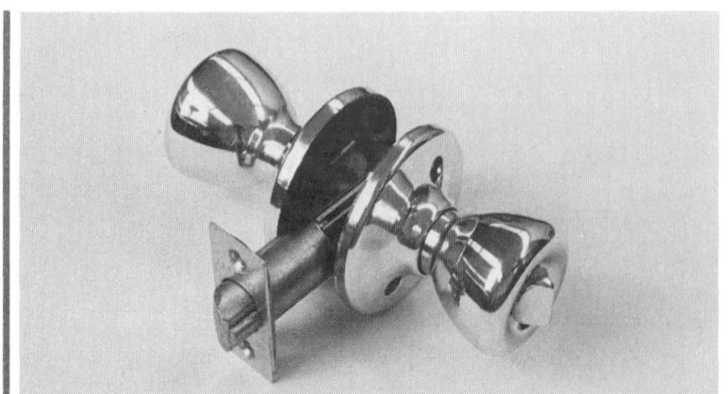

FIGURE 15-58

A tubular lock is used on exterior doors.

FIGURE 15-60

A typical interior door lock.

Installing Door Locks A set of instructions comes with each lock. The carpenter should study these carefully. Most locks follow the same basic installation procedures.

The first thing the carpenter must understand is the "hand" of the door. Some locks can be used on either left- or right-hand doors. Others cannot be reversed. When ordering locks the carpenter must specify the hand of the door. The hand of the door refers to which way the door swings. The hand is determined by facing the outside of the door. This is the exterior of the house for exterior doors. It is the hall side of interior doors. If the door is hinged on the left side and swings inward, it is a left-hand (LH) door. If it is hinged on the right side and swings inward, it is a right-hand (RH) door. If it is hinged on the left side and opens outward, it is a left-hand reverse (LHR). If it hinges on the right side and swings outward, it is a right-hand reverse (RHR) (Fig. 15-61).

To install the lock, open the door and hold it in place with a couple of wedges under it. Measure from the floor 38 in. to locate the height of the lock. Mark this with a very light line. Do not mark heavily because the lines will have to be removed when the door is finished. Place the template furnished with the lock on the face and edge of the door (Fig. 15-62). Mark the centers of the holes on the door face and edge. Bore the required holes the size and depth required. This can be done by hand with a brace and auger and expansion bits.

Next lay out and cut the mortise on the door edge for the front plate of the latch unit. This can be cut with a chisel (Fig. 15-62).

Install the lock as per manufacturer's directions.

The strike plate will have to be mortised into the jamb. Locate it so that the latch bolt is centered in the opening in the striker plate. Mark around the plate on the jamb. Cut the mortise for the plate. A deeper recess is needed for the latch bolts (Fig. 15-63).

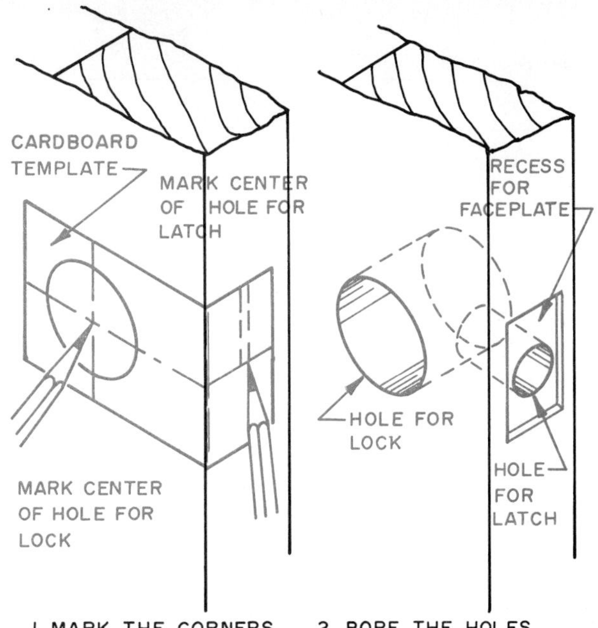

FIGURE 15-62

How to install a door lock.

METAL DOOR FRAMES

When metal doors are used, the frame comes as part of the total unit. The frame is in a knocked-down condition. It is assembled on the job. The corners are held together with interlocking tabs. The door and frame are prepared to receive the hinges and locks. Some are available completely assembled.

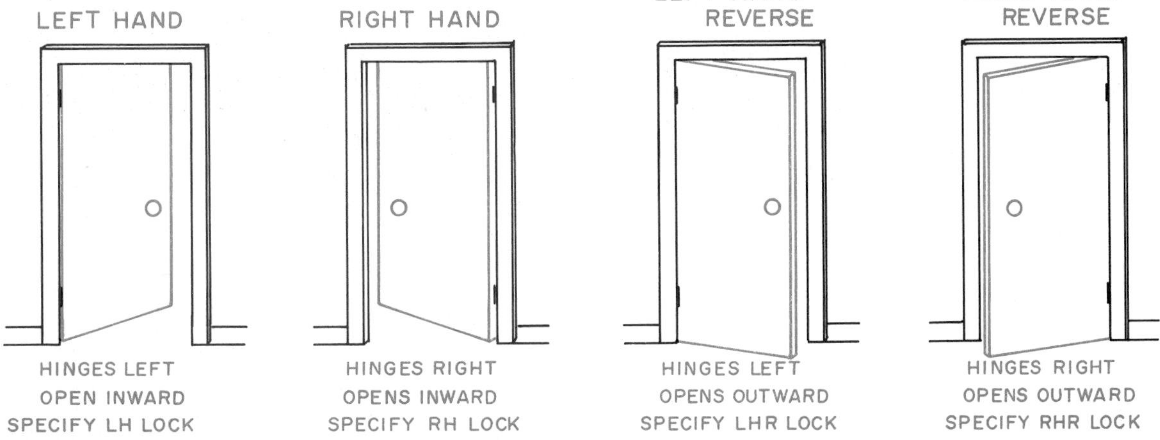

FIGURE 15-61

The "hand" of a door refers to which way it opens and the location of the lock.

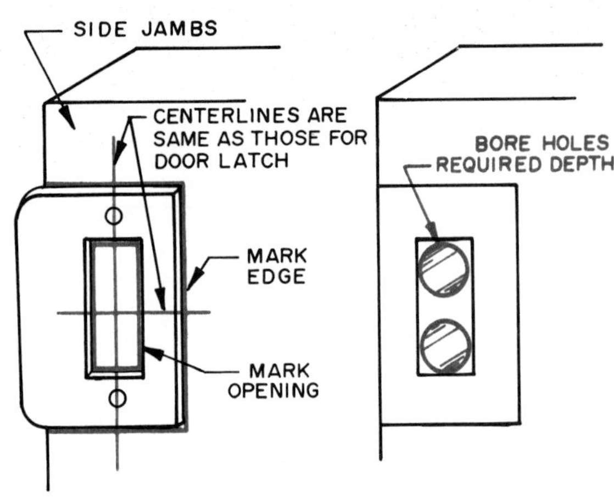

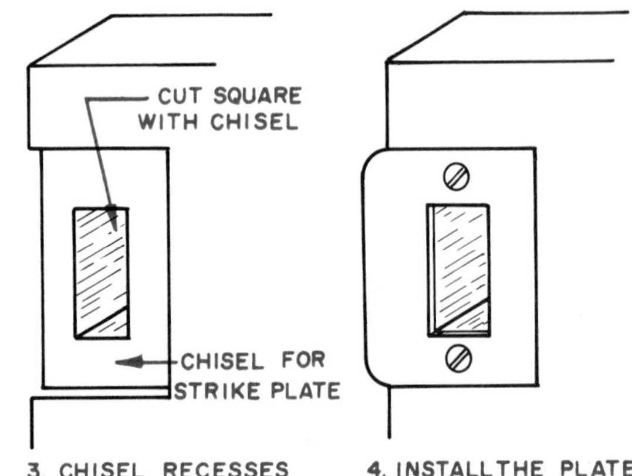

FIGURE 15-63

How to install the strike plate.

SLIDING GLASS DOORS

Sliding glass doors are available in a variety of sizes and with wood or metal frames (Fig. 15-64). They slide on nylon or stainless steel tracks. They are weatherstripped and use safety glass. They may be glazed with one thickness of glass or be double-glazed with an airspace between two panes of glass.

They usually come with one fixed door and one sliding door. Units with three doors are available (Fig. 15-65).

The frame is usually factory assembled. It is set in the rough wall opening the same as a window or door. Installation details are shown in Fig. 15-66. Metal door frames are fastened to the rough opening studs with wood screws. The frame has screw holes already drilled and countersunk for this purpose. The proper screws are supplied with the frame. Wood units are nailed to the studs.

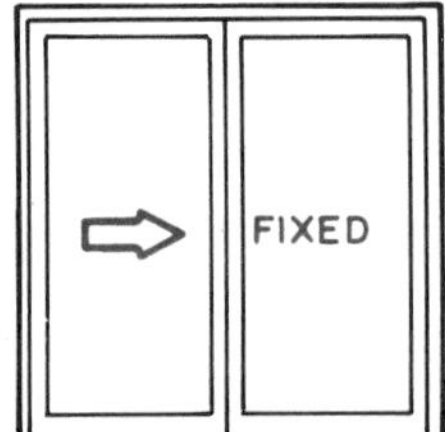

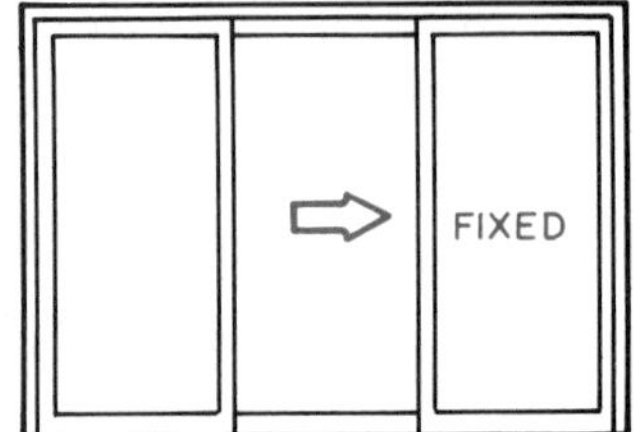

FIGURE 15-65

Typical sliding-glass-door units.

FIGURE 15-64

A sliding glass door. *(Courtesy of Andersen Corporation, Bayport, MN 55003)*

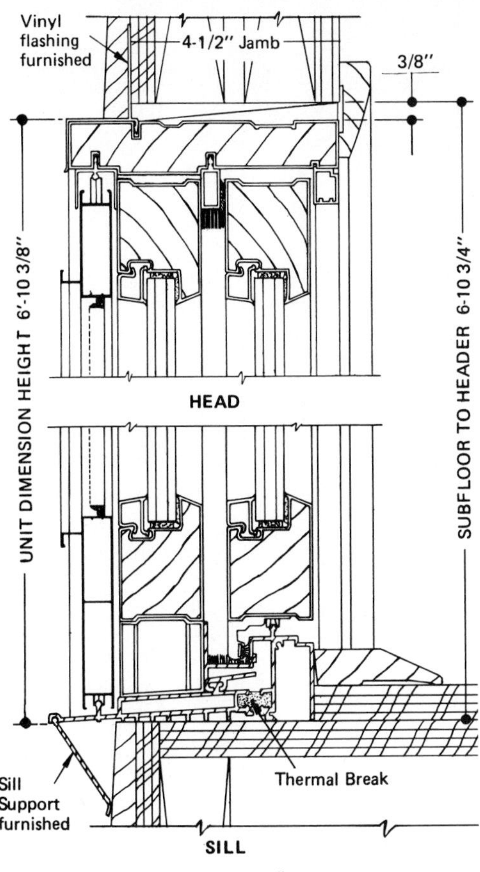

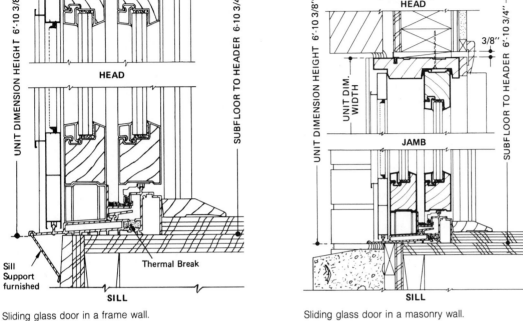

Sliding glass door in a frame wall.

Sliding glass door in a masonry wall.

FIGURE 15-66

Installation and design details of a sliding-glass-door unit. *(Courtesy of Andersen Corporation, Bayport, MN 55003)*

After the frame is firmly set in place, the fixed window is installed. The sliding door is then set on its track. Carpenters should follow the installation instructions that come with each door.

SLIDING POCKET DOORS

The *sliding pocket door* is used between rooms. It saves wall space because it slides into a frame built inside the wall. One type is shown in Fig. 15-67. The rough opening must be built to the design recommended by the manufacturer. The door framework is installed while the wall is being rough-framed.

BYPASS SLIDING DOORS

The *bypass sliding door* is often used on closets. It saves space because there are no doors swinging out into the room. It permits access to only half the closet at any one time. The door opening is framed the same as for swinging doors. The hardware for the sliding door consists of an

FIGURE 15-67

Hardware for a sliding pocket door installed in a framed opened in an interior wall. *(Courtesy of Stanley Hardware)*

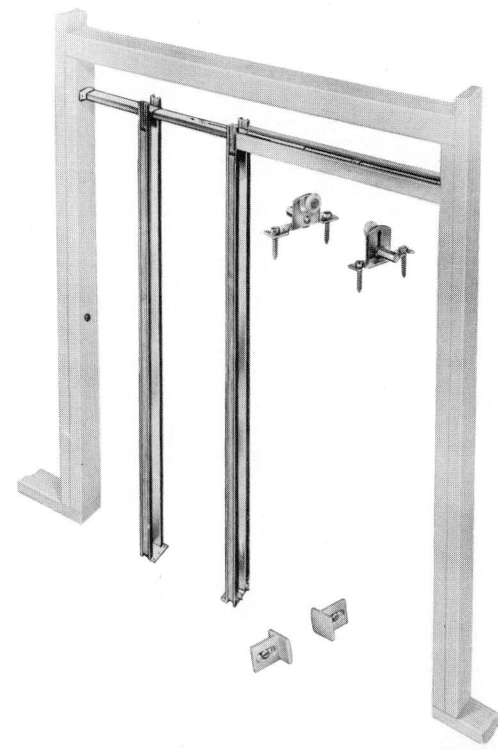

overhead track and a floor guide (Fig. 15-68). The track is usually fastened to the head jamb with screws. The floor track is screwed to the finished floor.

The rollers are screwed to the top edge of the door. Once this is done, slip the rollers in the track to hang the door. Place the floor track under the door, check the door with a level to make certain it is plumb, and fasten the bottom track in place.

FOLDING DOORS

The *folding door* is generally used on closets and between rooms. When open, it takes up little space. It also provides access to the entire opening (Fig. 15-69). The opening is framed in the same way as swinging doors. Be certain that the finished size is as recommended for the doors to be installed.

These units are sold complete with the doors and all hardware. The doors are joined with hinges. The edge of the door next to the jamb has a pivot on the top and bottom. This pivot carries the weight of the door and serves as the hinge at the jamb. Across the top of the opening, a track is installed. A pin or rollers are on the top of each door section. They run in the overhead track. The track serves as a guide and carries the weight of the door (Fig. 15-70).

Small four-door units for closets are available. They pivot on pins at the outer edge of each door. The center door has a pin that runs in a track. It serves as a guide. It carries no weight (Fig. 15-71).

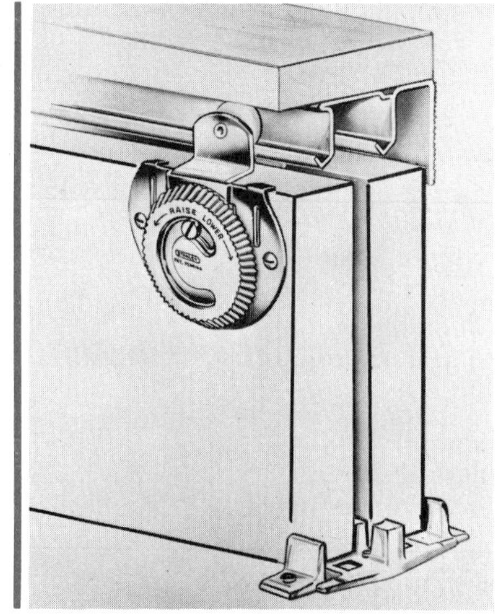

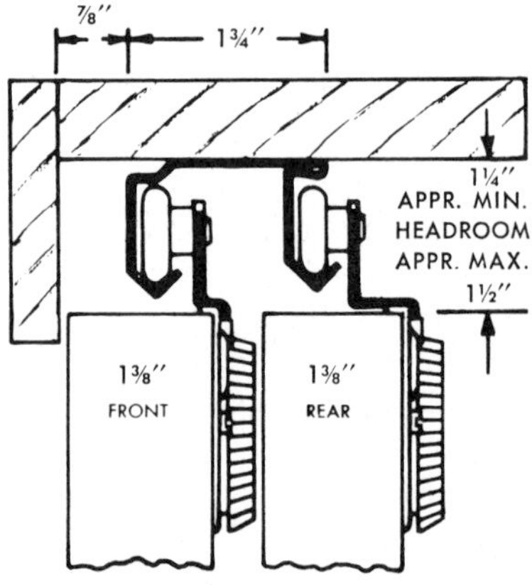

FIGURE 15-68

The track and floor guides for bypass sliding doors. *(Courtesy of Stanley Hardware)*

FIGURE 15-69

Folding doors are used to divide rooms into smaller areas. *(Courtesy of Rolscreen Co.)*

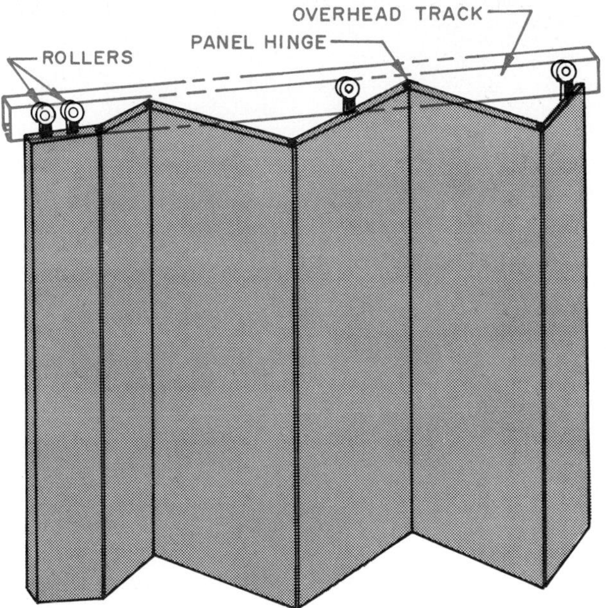

FIGURE 15-70

A typical track system used on folding doors.

FIGURE 15-71

A typical four-panel closet folding door.

GARAGE DOORS

The two commonly used *garage doors* are the one-piece *swing-up* and the *roll-up* door with hinged sections (Fig. 15-72). They are available in a variety of designs. They are made of wood, steel, aluminum, aluminum frame with fiberglass panels, and polyethylene (Fig. 15-73).

The size of the door is shown on the working drawing. A single garage door is usually 7 ft 0 in. high and 8 ft 0 in., 9 ft 0 in., or 10 ft 0 in.

FIGURE 15-72

A wood-framed hinged-section roll-up garage door. *(Courtesy of Overhead Door Corporation)*

FIGURE 15-73

A fiberglass hinged-section roll-up garage door. *(Courtesy of Overhead Door Corporation)*

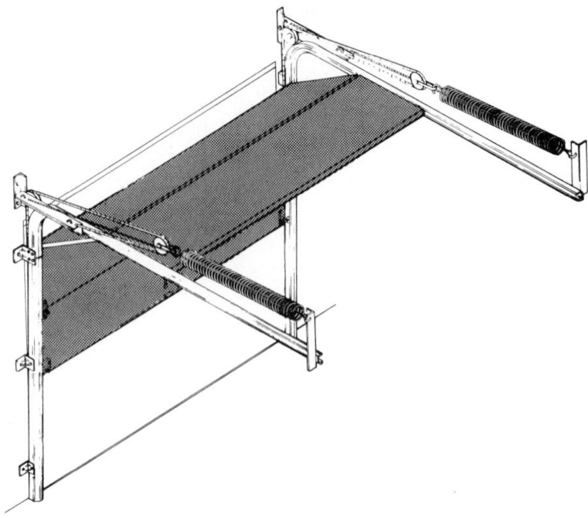

FIGURE 15-74

The track system for a hinged-section roll-up garage door. *(Courtesy of Overhead Door Corporation)*

wide. A double door is usually 7 ft 0 in. high by 16 ft 0 in. or 18 ft 0 in. wide. Other heights and widths are available.

The door with hinged sections opens using some form of track. Wheels on the sides of the door roll inside the track (Fig. 15-74). Electric door openers can be used. They are activated with a remote-control transmitter. The swing-up door uses several metal levers to hinge the door up against the ceiling of the garage.

Garage Door Frames Details for framing a garage door opening are shown in Fig. 15-75. The side jamb and head board are flat stock with no rabbets. They are installed as described for door frames. Each part is set plumb and level using wedges. When set, it is nailed through the jamb and wedge into the rough framing (Fig. 15-76). Generally, the inside dimensions of the finished door frame are the same as the

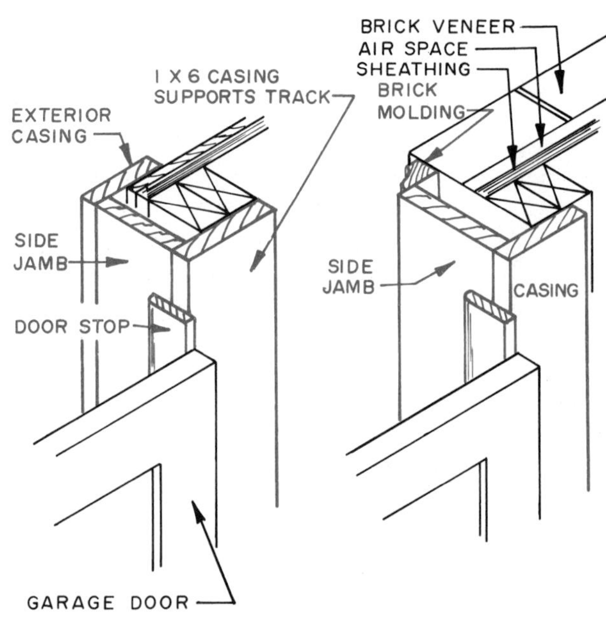

FIGURE 15-75

How to frame the jamb for a garage door.

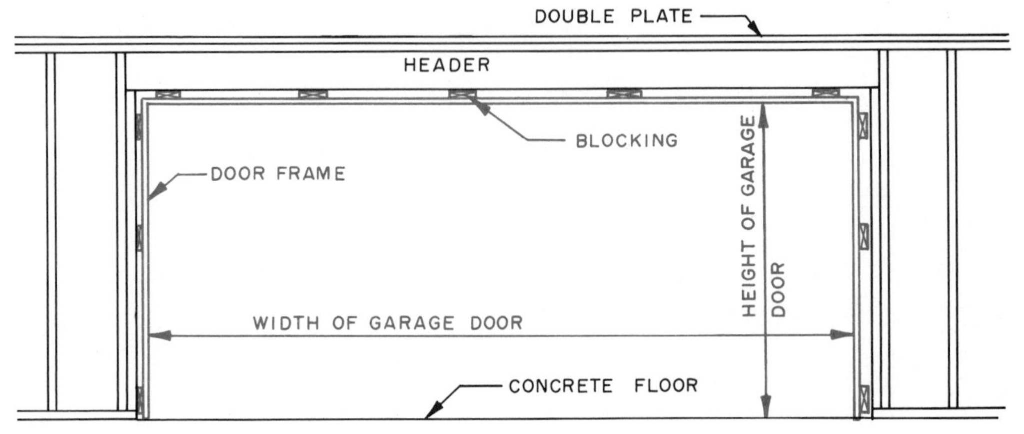

FIGURE 15-76

The proper way to frame the garage door opening.

size of the door. It is best to refer to the manufacturer's directions that come with the door.

After the door frame is installed, the hardware for the door is hung. The door is then mounted. The door stop trim can now be nailed to the door frame (Fig. 15-75). In this way it can be set to fit against the door, sealing any space between it and the door frame.

Installing the Door The garage doors are installed by companies specializing in this work. The carpenter prepares the finished door opening. The garage door contractor installs the door.

If a carpenter installs the door, the installation directions that come with it should be followed carefully.

STUDY QUESTIONS

After reading the chapter, answer each of the following questions. If you do not know the answer, review the chapter.

1. What materials are used to make metal windows?
2. What are the common groups of windows?
3. How do double-hung and single-hung windows differ?
4. How do casement windows differ from awning windows?
5. What is a jalousie?
6. What is the difference between a bay window and a bow window?
7. What purpose does a jamb extension serve?
8. How do carpenters check to see if a window is plumb and level?
9. What are the two types of wood doors?
10. How do interior and exterior door frames differ?
11. What tool is used to trim a door to fit inside the frame?
12. What are two commonly used exterior door locks?
13. What is the proper height off the floor for a door handle and lock?
14. Garage doors are made from what kinds of materials?

IMPORTANT TECHNICAL TERMS

Following are technical terms that you should be able to use as part of your working vocabulary. Write a brief description of the meaning of each term.

Sliding window
Fixed window
Swinging window
Sash
Jalousie window
Bay window
Bow window
Muntin
Rough opening
Window head
Window jamb
Solid-core door
Hollow-core door
Lights
Polystyrene core
Threshold
Door frame
Hinge gain
Hinge barrel
Cylindrical lock
Tubular lock
Dead bolt
Pocket door
Sliding door
Folding door
Garage door

16 FINISHING THE EXTERIOR

After the exterior walls have the sheathing applied and the doors and windows installed, they are flashed and the finish exterior material is applied. Also, the work on the cornice and rake on the gable end is completed. These steps finish the exterior of the house (Fig. 16-1).

CORNICE CONSTRUCTION

The design details of the *cornice* should be found on the working drawings. On some houses cornices are very simple. Mainly, they serve to close the end of the roof and seal the overhang. On other house designs they are an important part of the style. These use moldings of various types and can become very decorative.

The work on the cornice must be carefully done. It is one of the parts of the building that is exposed to view. Each part must be cut accurately and fit neatly. Hammer marks and split boards are not acceptable.

The parts of a typical cornice are in Fig. 16-2. The *fascia* header is nailed to the ends of the rafters. It is 2-in. material and is the main horizontal trim member. The fascia board is nailed to the header. The gutter is hung on it. The ledger is a 2 × 4-in. member nailed to the studs in the exterior wall. The *lookouts* are nailed to it. The lookouts are usually 2 × 4-in. material. They carry the *soffit*. The soffit is usually plywood, hardboard, solid wood, gypsum board, or specially manufactured metal sheets (Fig. 16-3).

The *frieze* covers any irregular openings between the soffit and the wall sheathing. It gives a straight edge to butt the finished wall material against.

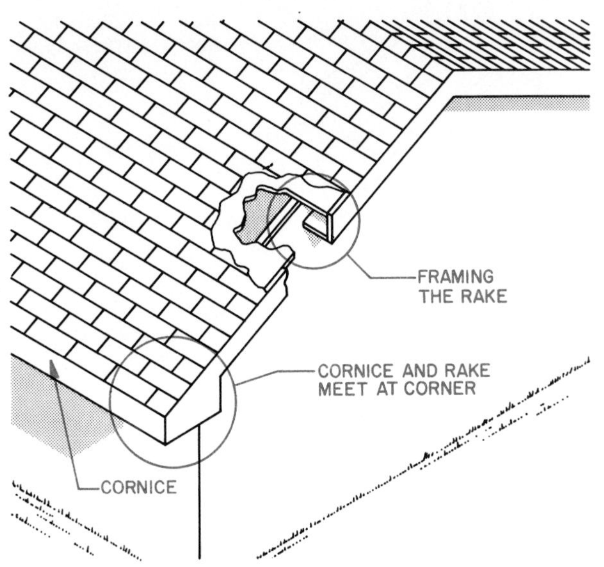

FIGURE 16-1

Cornice and rake construction are a part of the exterior finish.

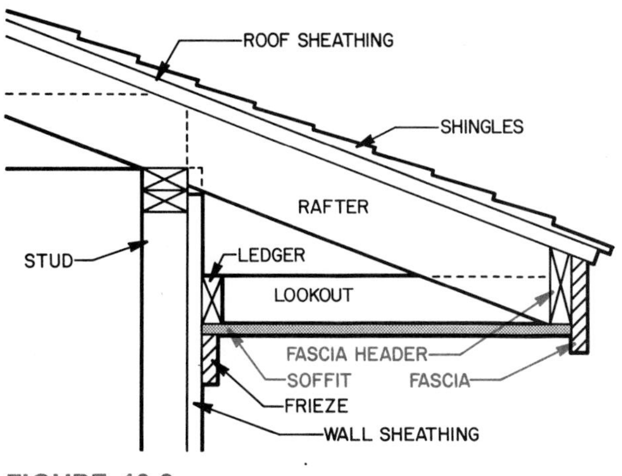

FIGURE 16-2

A typical horizontal cornice seals the area below the rafter overhang.

FIGURE 16-3

A plywood soffit being nailed to the lookout members. *(Courtesy of American Plywood Association)*

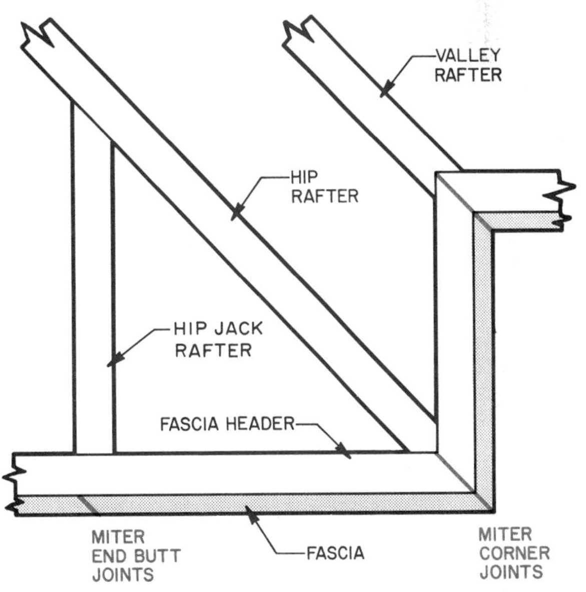

FIGURE 16-4

The inside and outside corners of the fascia board are mitered.

The material used in the cornice should be of good quality. It is exposed to the weather and view. It must hold paint and not warp or decay. Often decay-resistant wood such as redwood or cypress is used. Other woods can be treated to resist decay. When these are cut, the cut ends need to be painted with a decay-resistant solution. All joints between boards should be caulked to keep moisture from entering the cornice.

To get a neat joint and resist the entrance of water, joints in the fascia board should be mitered (Fig. 16-4).

The design of the cornice is drawn by the architect. The carpenter must build it according to the drawings. Some typical cornice framing plans follow. A cornice for a flat roof is shown in Fig. 16-5. The fascia header is usually 2-in.-thick stock. The soffit is made of two pieces. The space left between them is covered with screen forming a vent.

An open box cornice with a sloping soffit is shown in Fig. 16-6. The soffit is nailed to the bottom of the rafters. A wide box soffit with 2 × 4-in. lookouts is shown in Fig. 16-7. This example is with a frame wall. The soffit is nailed to the bottom of the lookouts. A frieze board seals the soffit to the sheathing. It is notched to

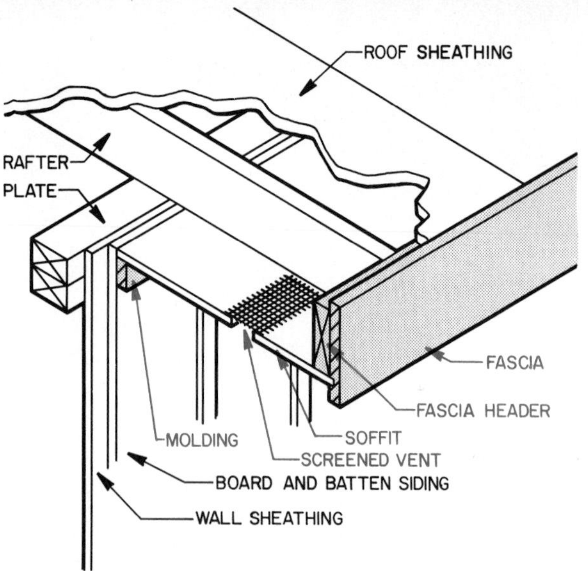

FIGURE 16-5

Typical cornice construction for a flat roof.

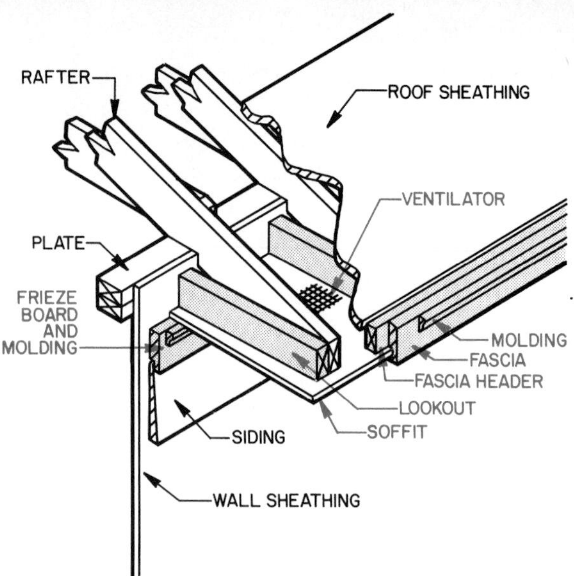

FIGURE 16-7

A wide boxed soffit on a frame exterior wall.

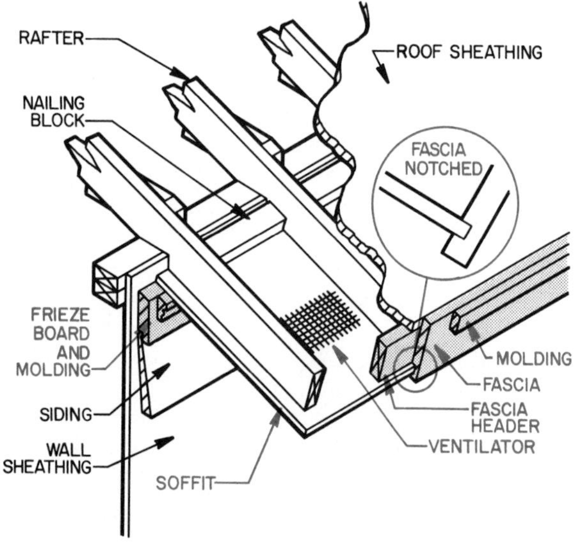

FIGURE 16-6

An open box cornice has a sloping soffit.

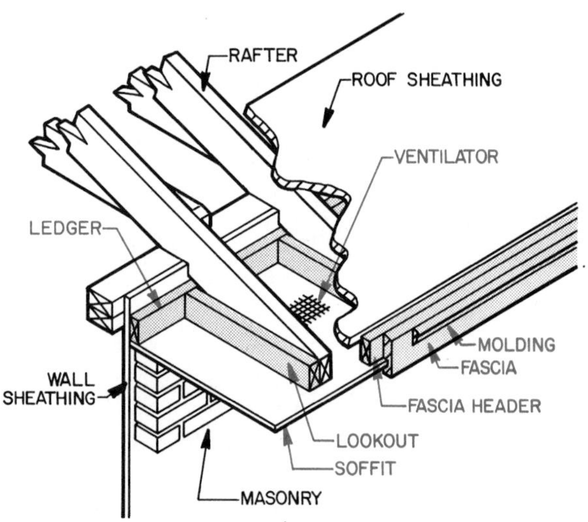

FIGURE 16-8

A wide boxed soffit on a brick-veneer exterior wall.

receive the wood siding. A design often used when a masonry veneer wall is built is shown in Fig. 16-8. A narrow box cornice does not use lookouts (Fig. 16-9). The soffit is nailed to the bottom of the rafter. A close cornice on a frame wall has the frieze board next to the sheathing. It is nailed into the ends of the rafters (Fig. 16-10). The roof sheathing overlaps the frieze board. A molding is used to seal the space between it and the sheathing.

A close cornice on a masonry veneer wall

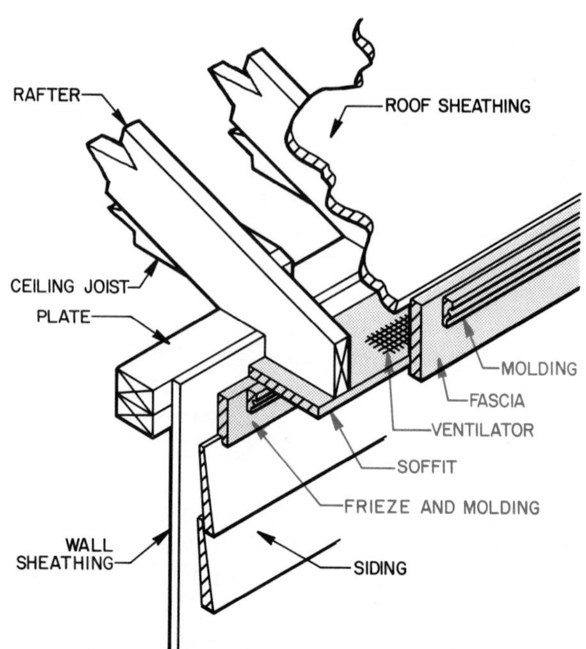

FIGURE 16-9

A narrow box cornice provides a little overhang.

is shown in Fig. 16-11. This design permits the use of a wide frieze board and dentil blocks. This is typical of some colonial house styles and is very decorative. An open cornice has no soffit. The space between the rafters has to be blocked. The rafter tails can be straight or curved (Fig. 16-12). For some types of classic houses the design calls for a gutter to be built in the edge of the roof. A typical framing detail is shown in Fig. 16-13.

Study Chapter 13 for additional details on framing cornices.

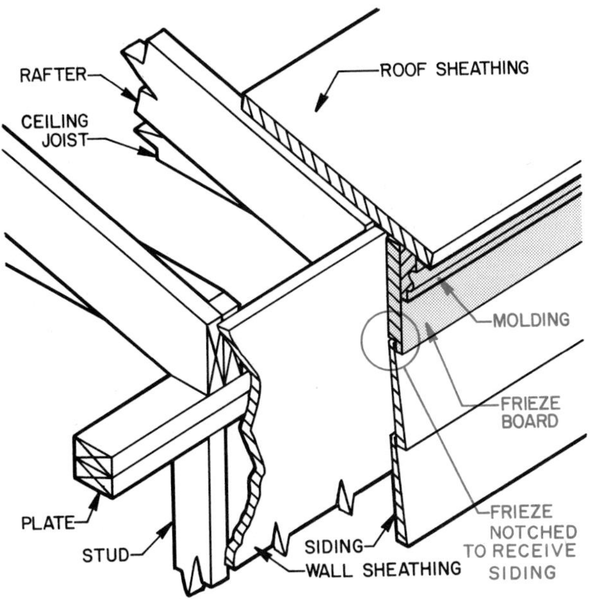

FIGURE 16-10

A close cornice uses a frieze board next to the sheathing.

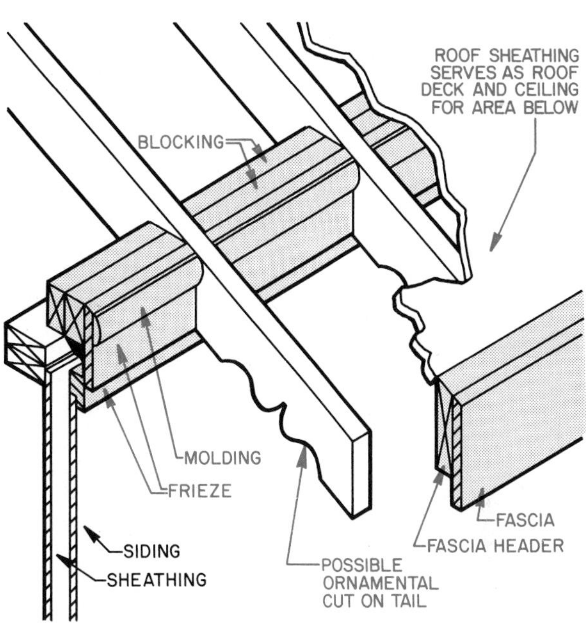

FIGURE 16-12

This open cornice has the wall area to the roof covered and does not use a soffit.

FIGURE 16-11

A close cornice on a masonry veneer wall uses blocking for nailing the frieze board.

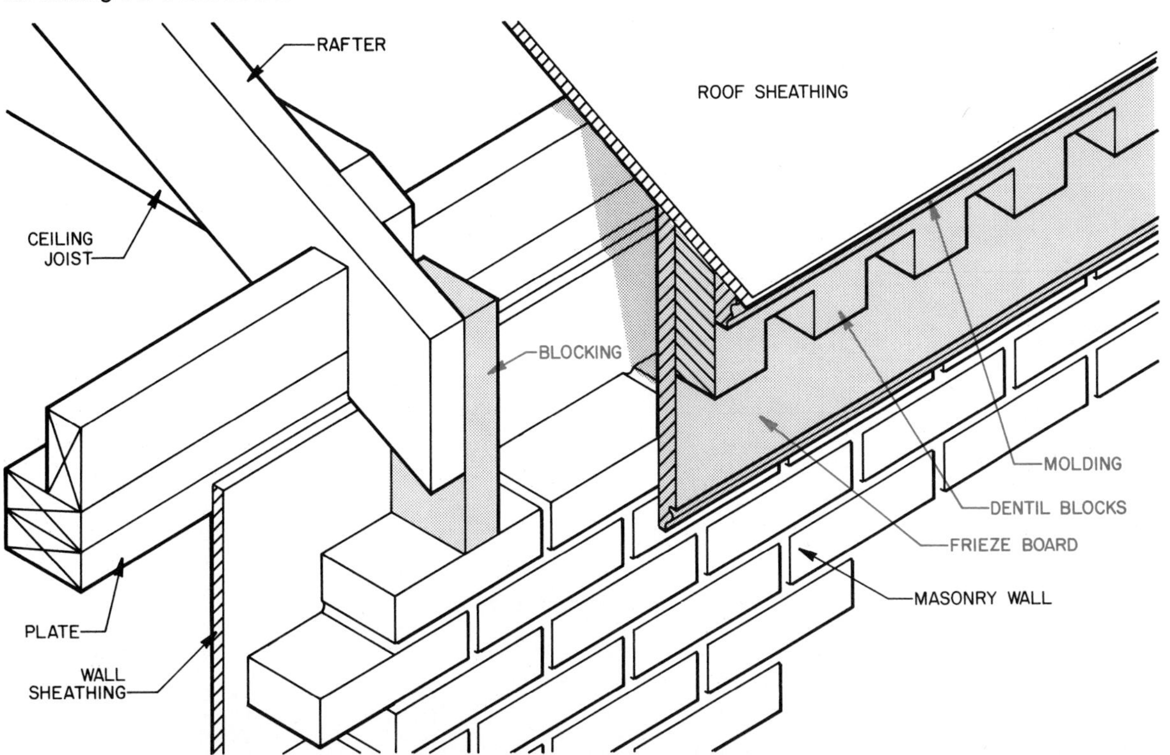

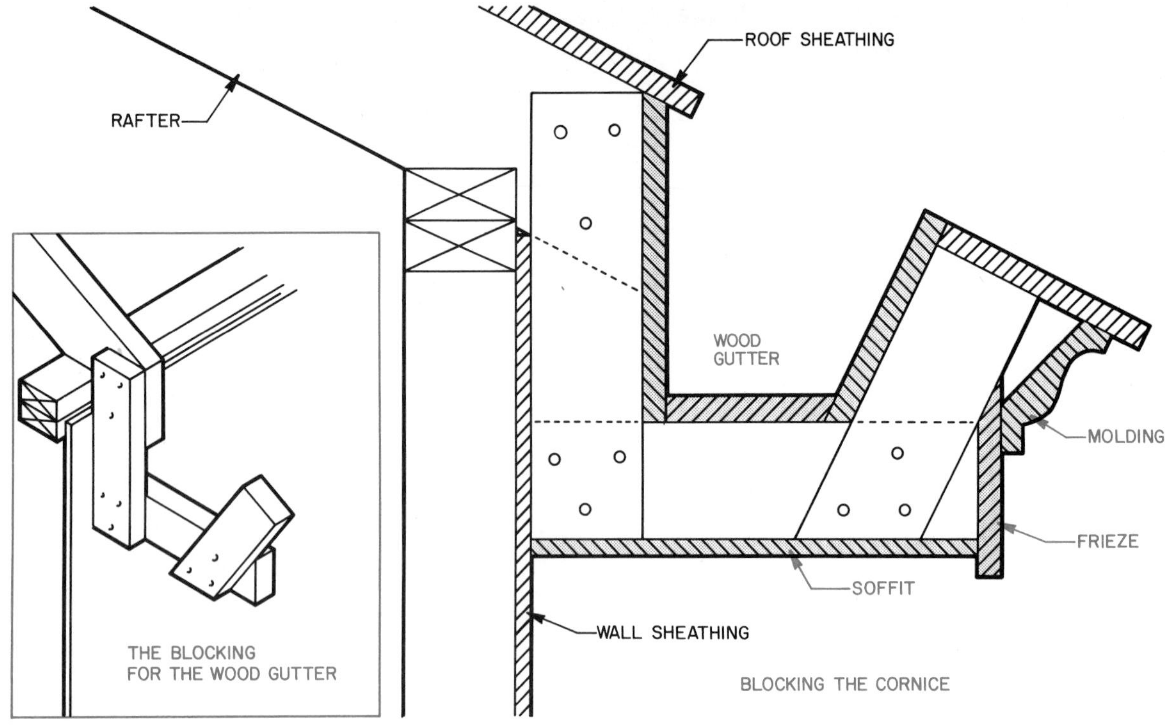

FIGURE 16-13

A wood gutter can be built in the edge of the roof.

RAKE CONSTRUCTION

The *rake* on the gable end must also be enclosed. A close rake is shown in Fig. 16-14. Notice that the fascia block is notched to receive the wood siding. The fascia is wider than the fascia block. A close gable end extension is shown in Fig. 16-15. The length of the lookout can vary. The architect decides on the amount of overhang.

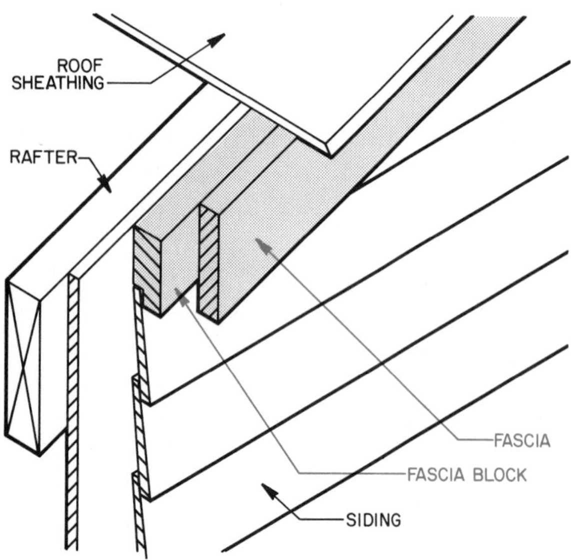

FIGURE 16-14

Framing for a close rake.

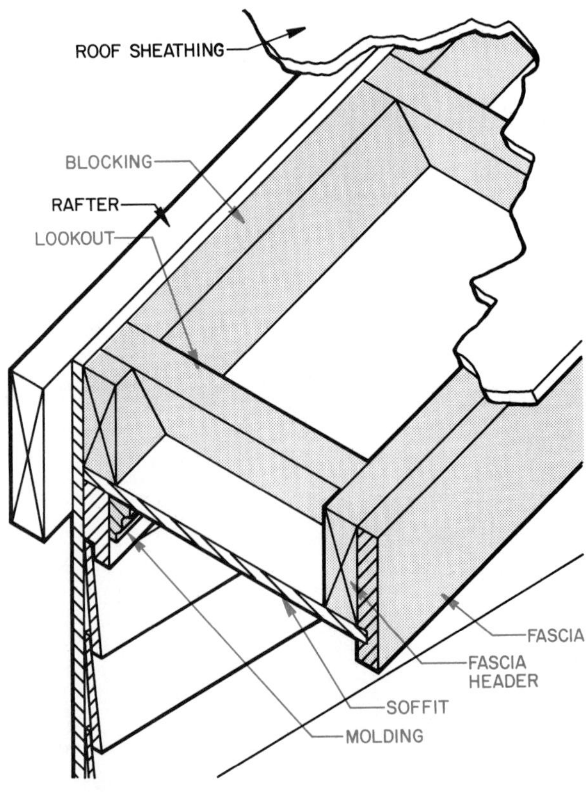

FIGURE 16-15

A close gable end extension is built using lookouts and has a soffit.

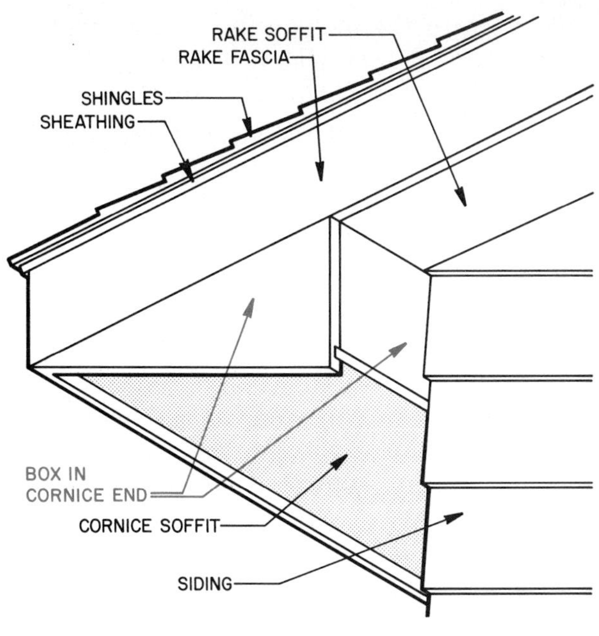

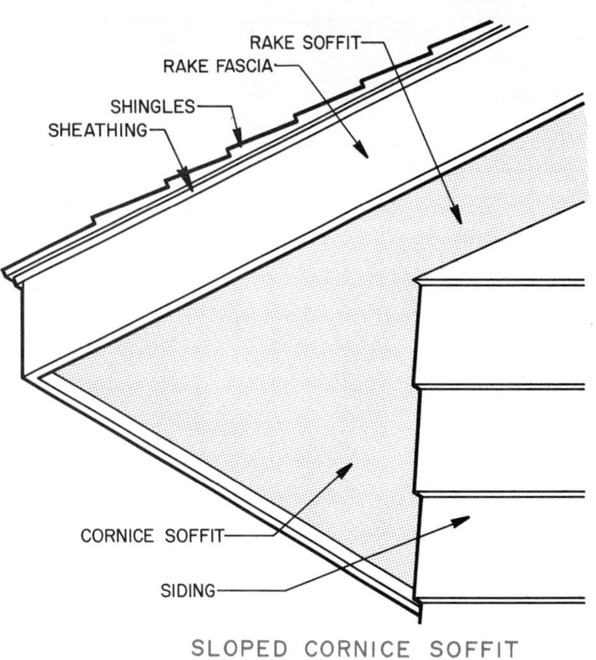

FIGURE 16-16

The cornice fascia and rake fascia meet at the corner of the roof.

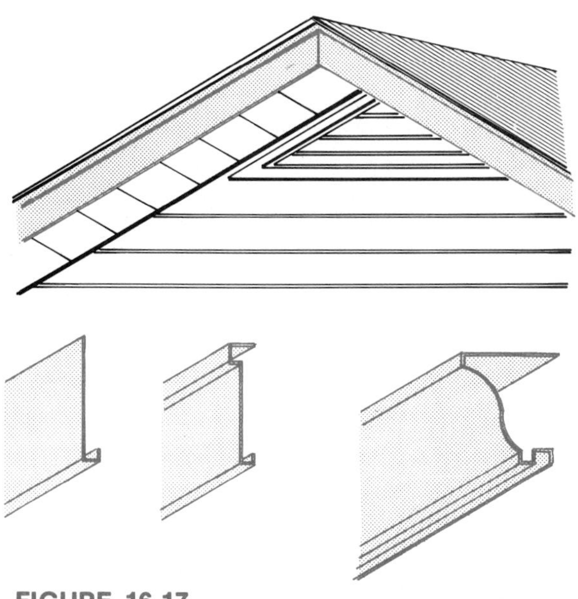

FIGURE 16-17

Metal and vinyl products are available to cover the fascia and the rake.

Usually, this type is used for overhangs up to 12 in. For overhangs larger than this, the lookouts extend into the roof. (See Fig. 13-40 for details.) The rake is trimmed the same as the close gable end extension. In all cases the fascia is grooved to receive the soffit. The groove is cut ⅜ in. from the edge of the fascia board.

The cornice fascia and the rake fascia meet at each corner of the house. How these are joined depends upon the design produced by the architect. Following are some of the ways in which this is done. If the cornice has a sloping soffit the corner can be built as shown in Fig. 16-16. The cornice and rake fascia boards meet in a mitered corner. The soffits are flush.

If the cornice soffit is horizontal, the corner can be boxed in as shown in Fig. 16-16. This produces a neat, weather-tight corner.

The fascia on the rake can be covered with metal or plastic preformed materials. This eliminates the need to paint it (Fig. 16-17).

SOFFIT MATERIALS

The commonly used soffit materials are plywood, hardboard, and gypsum board.

Plywood Soffits Plywood should be of the exterior type. The recommended thickness and spans for closed plywood soffits are listed in Fig. 16-18. The grain of the face panel runs perpendicular to the lookouts. Rustproof box or casing nails should be used. Use 6d for 5/16- and 7/16-in.-thick panels and 8d for ⅝-in. panels. Space nails 6 in. apart on the edges of the panel and 12 in. apart on intermediate supports. Leave 1/16 in. between the end and edge joints of the panels.

To install plywood panels, first cut them to the proper width. Push the edge of the panel

303 SIDING O/L MDO GROUP 3 16 oc SPAN EXTERIOR PS 1-74 000 APA	303 SIDING 6-S GROUP 1 24 oc SPAN EXTERIOR PS 1-74 000 APA	A-C GROUP 4 EXTERIOR PS 1-74 000 APA

NOMINAL PLYWOOD THICKNESS	GROUP	MAXIMUM SPAN (in.) ALL EDGES SUPPORTED
5/16-in. APA 303 Siding 3/8-in. APA Sanded		24
7/16-in. APA 303 Siding 1/2-in. APA Sanded	1, 2, 3, or 4	32
5/8-in. APA 303 Siding or APA Sanded		48

(A)

PANEL DESCRIPTIONS, MINIMUM RECOMMENDATIONS	GROUP	MAXIMUM SPAN (in.)
7/16-in. APA 303 Siding 1/2-in. APA Sanded	1, 2, 3, 4 1, 2, 3, 4	16
1/2-in. APA Sanded 5/8-in. APA 303 Siding 5/8-in. APA Sanded 3/4-in. APA 303 Siding	1, 2, 3	
5/8-in. APA Sanded 3/4-in. APA 303 Siding 3/4-in. APA Sanded	1 1 1, 2, 3, 4	32
1 1/8-in. APA Textured	1, 2, 3, 4	48

(B)

Note: Face grain across supports.

FIGURE 16-18

Design data for plywood soffits. (A) Exterior closed. (B) Exterior open, with combined ceiling/decking. *(Courtesy of American Plywood Association)*

into the groove in the fascia. Push the sheet up against the lookouts. When it is in the proper position, set a few nails to hold it up. Then proceed nailing it in place.

If the soffit is the open type, the plywood panel serves as roof decking and as the exposed ceiling. The recommended thickness and types of plywood are shown in Fig. 16-18. If the rafter spans are 32 or 48 in., the sheathing should have blocking below its edge joints which run between rafters unless it is tongue-and-groove sheathing or metal plyclips are used for support. For open soffits use smooth, ring-shank, or spiral thread nails. Use 6d for 1/2-in.-thick plywood and 8d for 5/8- to 1-in. panels. Use 8d ring-shank or spiral thread or 10d common smooth nails for 1 1/8-in. thick panels. Space nails 6 in. at panel edges, 12 in. at intermediate supports. On 48-in. spans, space nails on intermediate supports 6 in. apart. The panels should be placed with the face grain perpendicular to the rafters.

Hardboard Soffits

It is recommended that hardboard panels with factory-applied primer be used. If unprimed hardboard is used, prime it according to the manufacturer's instructions.

The panels must have support on all edges and ends. If the soffit is over 16 in. wide, a row of blocking is needed in the center of the panel (Fig. 16-19).

Use corrosion-resistant 5d or 6d box nails. Space them about 4 in. apart on all edges and ends. Space about 6 in. apart on intermediate supports. Keep the nails at least 3/8 in. from the edge of the panel.

Hardboard panels can be installed on soffits under 24 in. wide with metal channels. The channels are nailed in place. The hardboard is placed in the channels. The ends are also supported with metal channels (Fig. 16-20).

Gypsum Board Soffits

Soffits made from exterior gypsum ceiling board are used where there is no direct exposure to the weather. It can be used for ceilings in carports and porches. This material is in panels 1/2 and 5/8 in. thick, 4 ft wide, and 8 and 12 ft long. It has a water-resistant noncombustible core. The panels have a brown back paper covering and gray water-repellent face paper.

They are installed by nailing to wood framed soffits or by screwing to metal frames.

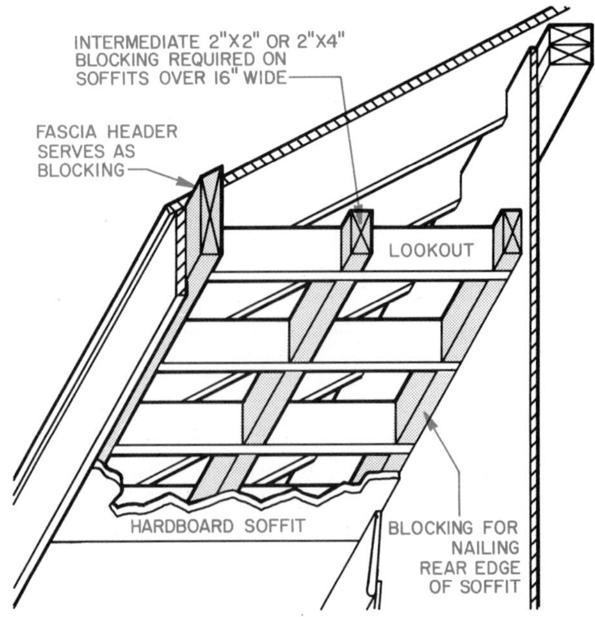

FIGURE 16-19

Construction details for installing hardboard soffits.

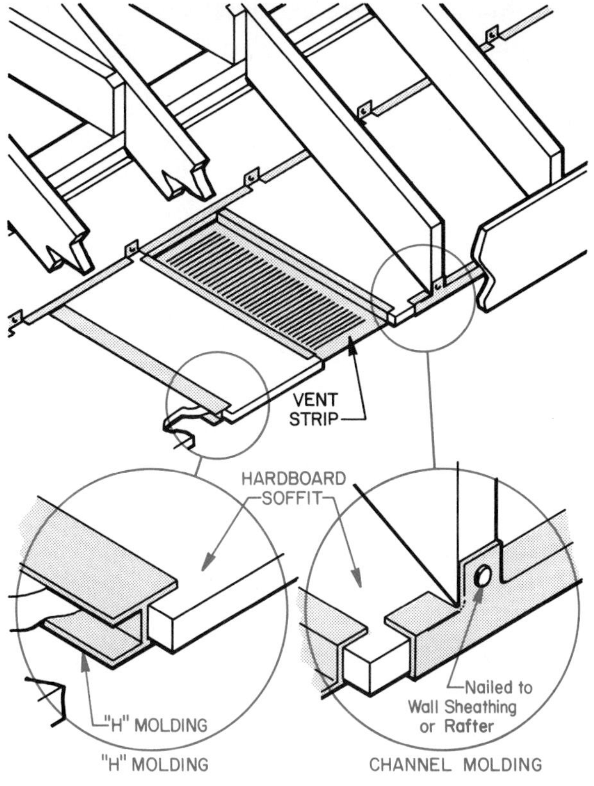

FIGURE 16-20

Hardboard soffits can be installed with metal moldings.

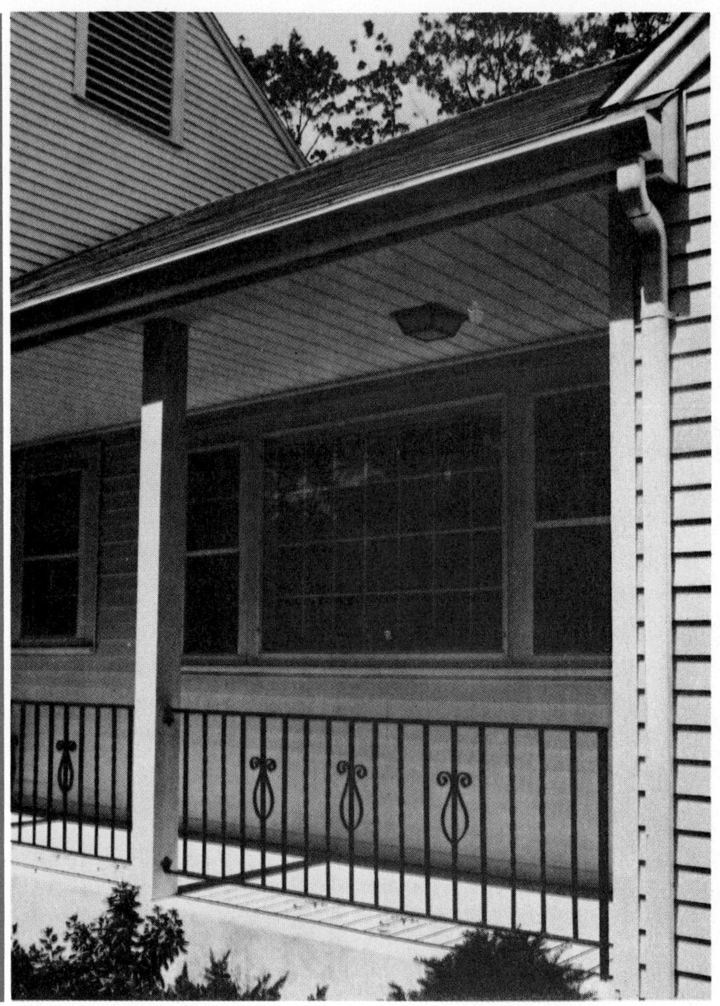

FIGURE 16-21

This house uses prefabricated vinyl soffits, siding, and and guttering. *(Courtesy of Bird and Son)*

Supporting members, such as soffit lookouts, should not be spaced more than 24 in. apart. Rust-resistant 6d nails with ¼-in. flat heads are used. They should be placed about ⅜ in. in from the edge of the panel and 6 in. apart on all edges. Joints between ends of the panels are taped and finished the same as for interior drywall construction. This is explained in Chapter 18. The panels can be painted.

PREFABRICATED CORNICE MATERIALS

There are a variety of prefabricated cornice soffit materials available.

Vinyl soffits are available in solid and perforated panels (Fig. 16-21). They run from the fascia to the wall and fit into channels nailed to these. If the soffit is wider than 18 in., an intermediate nailing strip is required (Fig. 16-22). The soffit panels should be cut ¼ in. shorter than the distance between the channels. This allows for expansion. Fit the panels into the channels and push them together.

Metal soffit systems using similar installation procedures are available (Fig. 16-23). The soffit material is held by channels nailed along the overhang (Fig. 16-24).

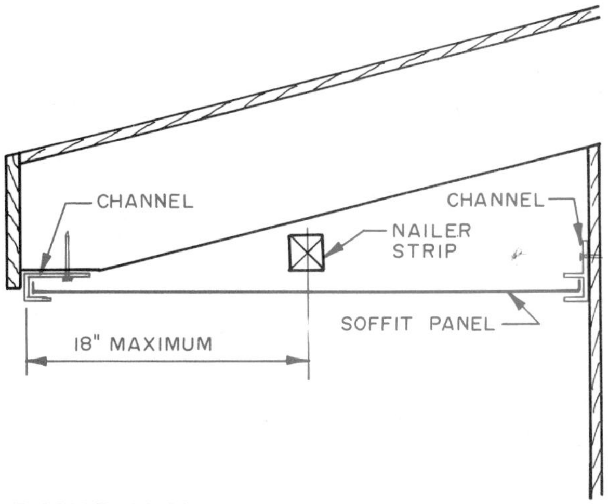

FIGURE 16-22

Vinyl soffits are held in place with channels.

FIGURE 16-23 *(above)*

Metal soffits are mitered at the corner and the joint is covered with a metal strip. *(Courtesy of Kaiser Aluminum and Chemical Corporation)*

FLASHING DOOR AND WINDOW OPENINGS

Door and window openings must be carefully flashed before the finished siding is installed. This step is essential to providing a weather-tight wall.

Aluminum flashing is generally used. It is easy to cut and bend. It will never rust. Always use aluminum nails with aluminum flashing. Metal flashing is required over the drip caps on doors and windows. Often, a drip cap is not used. The flashing fits over the window casing (Fig. 16-25). A strip about 6 in. wide is needed. Bend it to the shape of the drip cap. Some persons prefer to lightly nail the flashing to the edge of the drip cap. Use three or four nails to hold it to the sheathing above the window (Fig. 16-26). Caulk between the window casing and the sheathing before installing the flashing.

Some carpenters prefer to flash the openings on the sides as well. This is not mandatory but is a good practice (Fig. 16-27).

After the flashing is in place, the finished siding can be installed.

FIGURE 16-24 *(below)*

Typical installation details for aluminum soffits and fascias. *(Courtesy of Kaiser Aluminum and Chemical Corporation)*

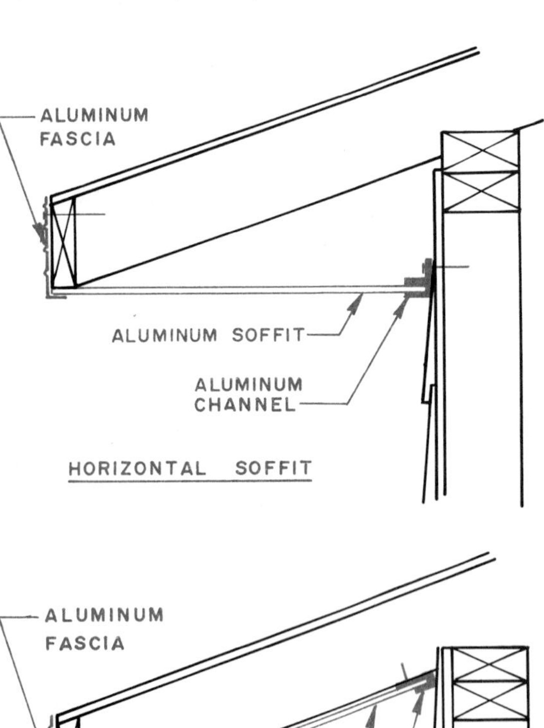

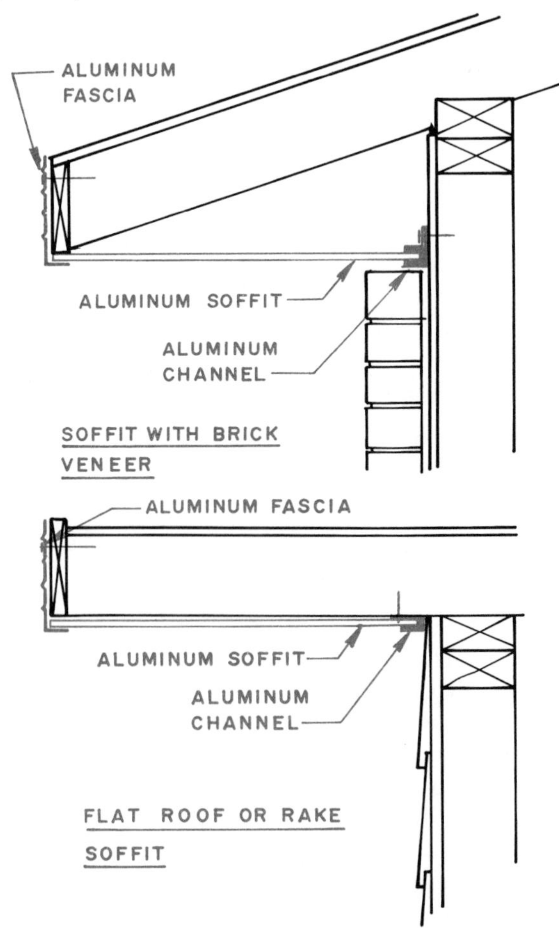

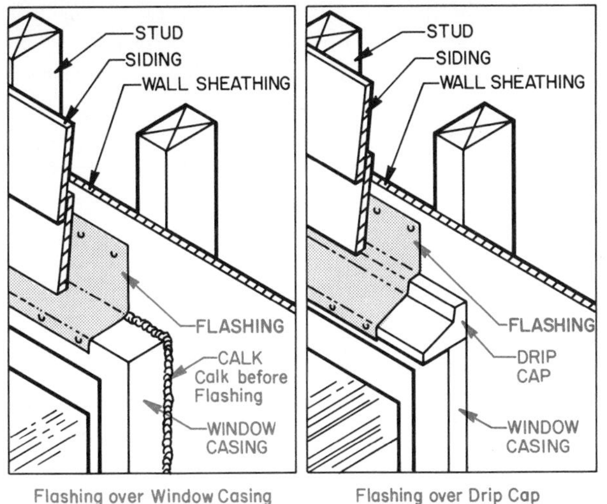

FIGURE 16-25

Flashing is placed above each window and the siding overlaps it.

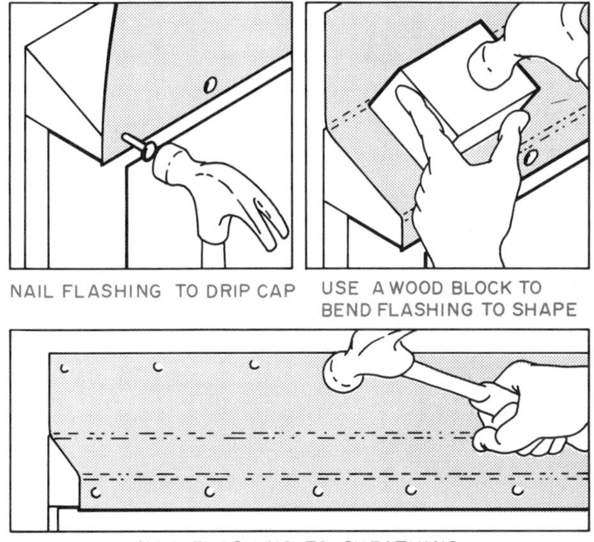

FIGURE 16-26

How to bend aluminum flashing to fit the drip cap.

FIGURE 16-27

It is advisable to flash the sides of windows.

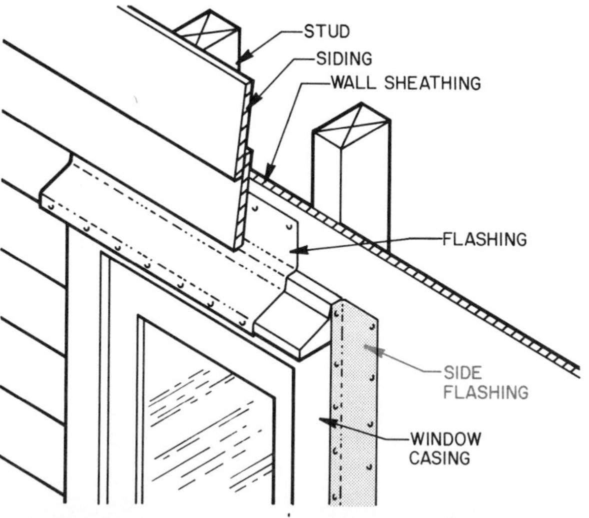

After the cornice and rake of the roof are finished and the door and window openings are flashed, the exterior wall siding is installed.

There are a wide variety of exterior wall finishing materials available. These include solid wood, plywood, mineral fiber, hardboard, steel and aluminum, vinyl plastic, wood shingles, stucco, and masonry. These are available in a wide variety of sizes, colors, and textures.

Solid Wood Siding

Solid wood siding is made in a wide variety of sizes and shapes. It is also available in different species of wood, such as redwood, cypress, white pine, sugar pine, yellow pine, hemlock, spruce, yellow poplar, Douglas fir, and western larch. These materials are easy to work and have good paint-retention qualities.

Siding should be of good quality. It should be free of knots, pitch, and wane. In most localities a 12% moisture content is recommended. In the southwest it is drier and a 9% moisture content is recommended.

Some of the common types of solid wood siding are listed in Fig. 16-28.

Nails

Several kinds of nails are used with wood siding. These are finishing, casing, sinker head, and sinker head ring-shank nails (Fig. 16-29). These nails can be set below the surface of the siding with a nail set. After the siding is primed, the hole is filled with putty.

These nails must be of aluminum or be galvanized. Steel nails will rust even if set and puttied.

The diamond point is the most commonly used type of nail point. If there is danger of the wood splitting, the nail can be blunted. Blunt-pointed nails are available. The threaded shank nails have greater holding power.

Installing Wood Siding

If the sheathing is plywood or solid wood, the siding is nailed directly into it. Space the nails every 24 in. Solid wood sheathing must be covered with a layer of builder's felt lapped 4 in. on the edges. Gypsum, plastic foam, and fiberboard sheathing will not hold nails. The siding must be nailed through the sheathing into the studs. The nail must penetrate the sheathing and the stud at least 1½ in.

Horizontal Siding

Some wood siding is designed to be applied horizontally. Others are applied vertically. *Horizontal siding* may be installed by starting at the bottom course. This overlaps the foundation about 1 in. The bottom

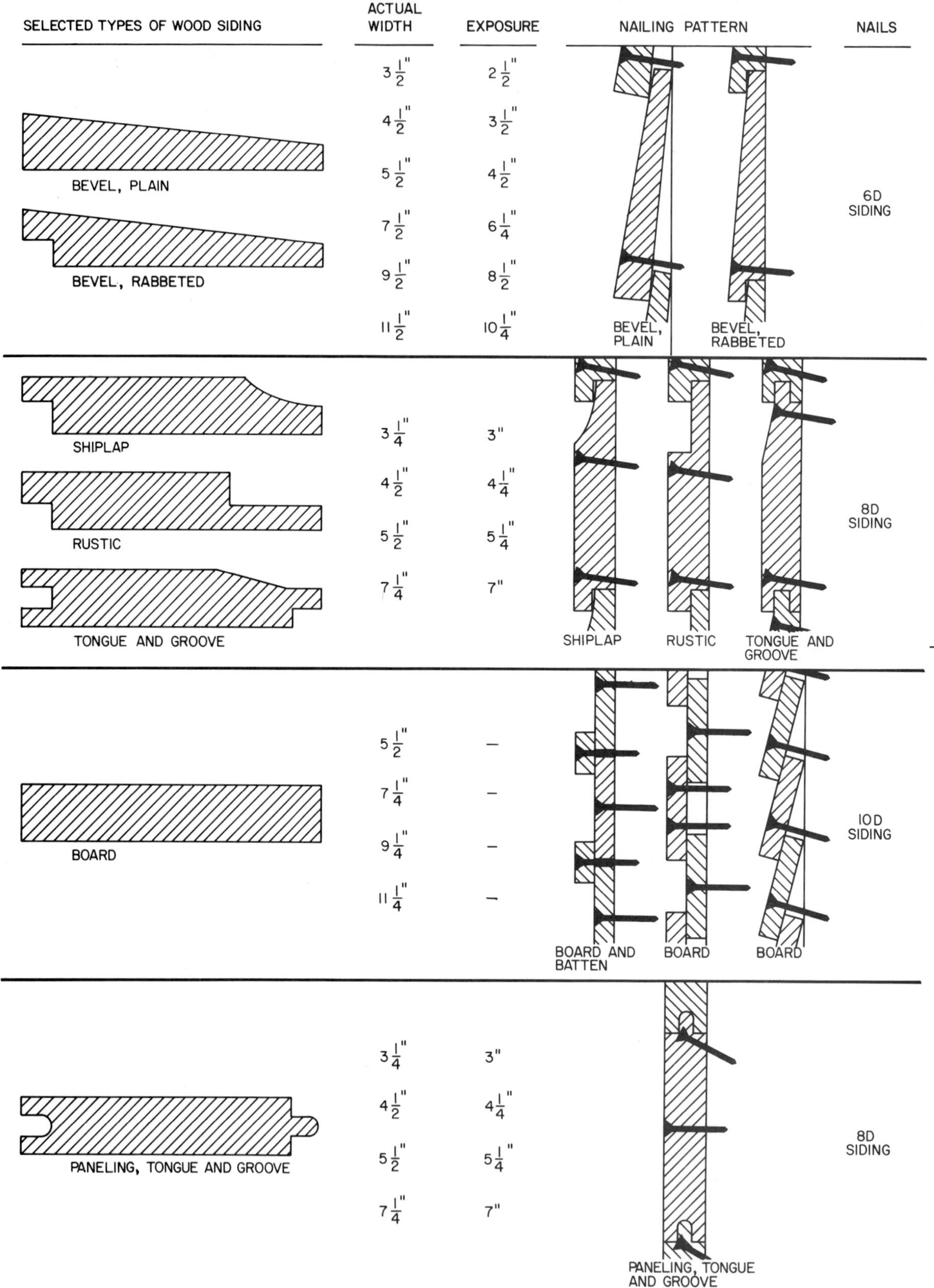

FIGURE 16-28

The common types of wood siding.

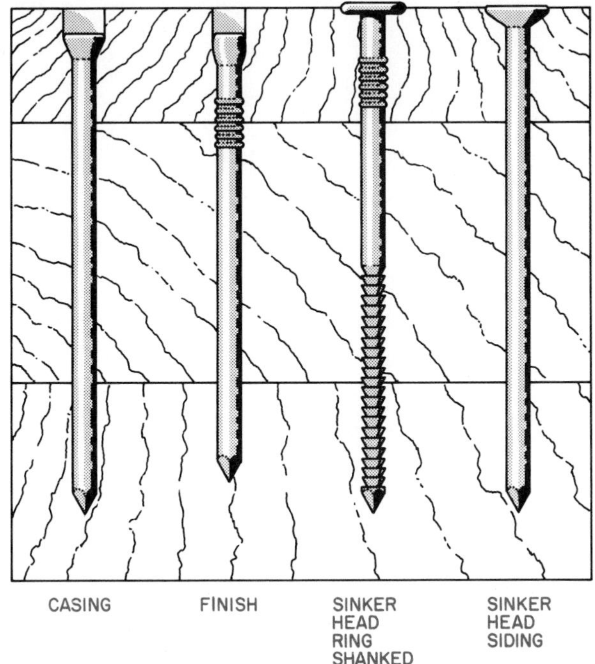

FIGURE 16-29

Nails used in installing exterior wood siding.

course should be at least 8 in. above any soil. A wood starting strip is nailed to the sill. It overlaps the foundation ½ in. It is the same thickness as the thin edge of the siding. This gives the first course the correct angle (Fig. 16-30).

Next locate the courses of siding. Siding starts 1 in. below the top of the foundation. Usually, it runs to the soffit. Mark this distance on a story pole. A story pole is a 1 × 2-in. wood member on which the location of the courses of siding are marked (Fig. 16-31). Divide the total vertical distance into equal parts which represent the exposure. Mark these on the story pole. Sometimes it is necessary to vary the exposure a little so that the siding aligns with the top and bottom of the window without being notched.

Transfer the marks from the story pole to each corner and door and window casing (Fig. 16-32). This locates the bottom edge of each course.

Next locate the top edge of the first course. Snap a chalk line on the sheathing to mark it (Fig. 16-33). Nail a starter strip that overlaps the top edge of the foundation about ½ in. The starter strip should be the same thickness as the thin edge of the siding. Nail the first course of siding to the sill. Some carpenters locate these nails in line with the wall studs. They serve as guides for locating the studs as the siding is set in place.

After the first course is in place, snap a chalk line from the first mark of the story pole. This locates the bottom edge of the second course (Fig. 16-34). Another way to set the exposure is to make a notched board which is used to measure the exposure (Fig. 16-35).

Sometimes, it is not possible to adjust the

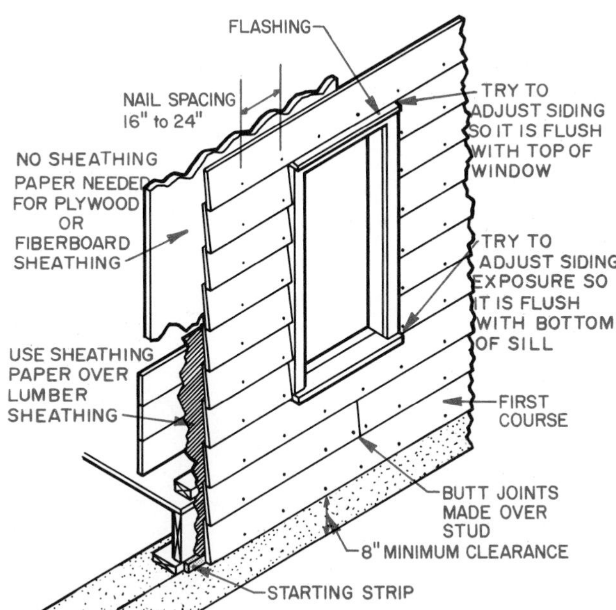

FIGURE 16-30

Proper installation of wood siding.

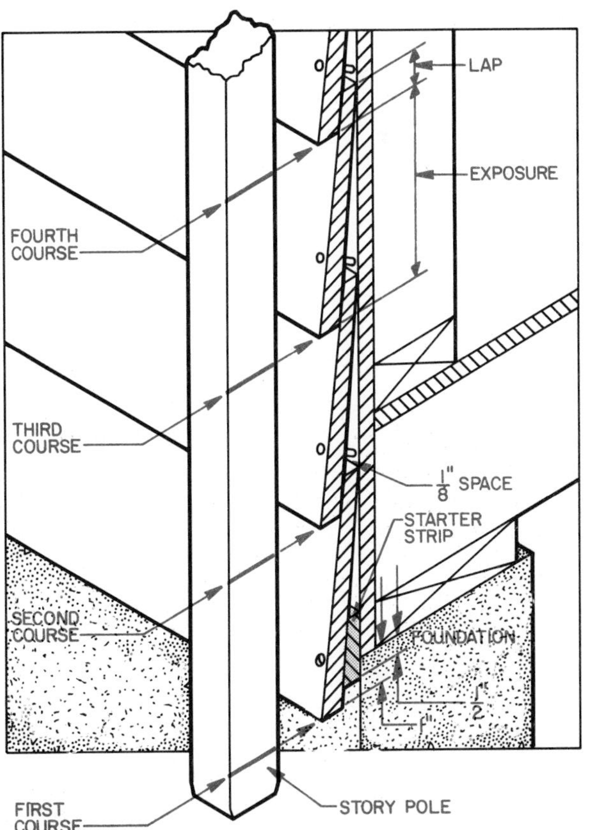

FIGURE 16-31

The courses of siding are located with a story pole.

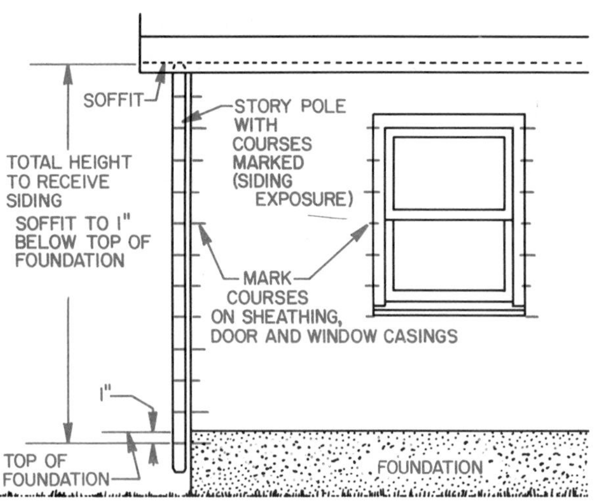

FIGURE 16-32

Siding courses are marked on windows and corners using a story pole.

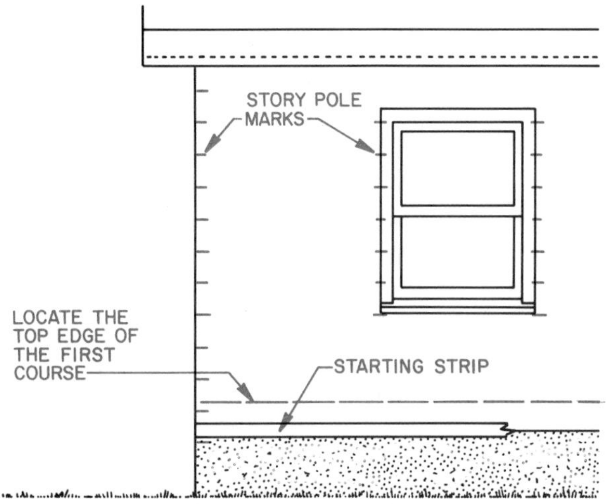

FIGURE 16-33

To locate the first course of siding, snap a chalk line marking the top edge on the sheathing.

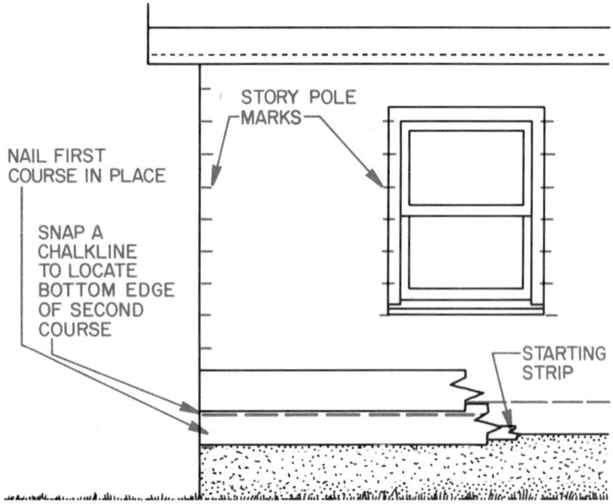

FIGURE 16-34

To locate the second course of siding, snap a chalkline from the first mark from the story pole. This is the bottom of the siding.

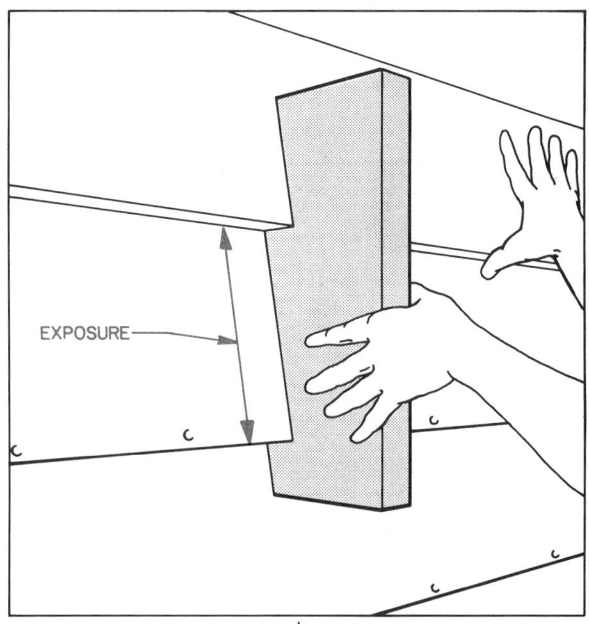

FIGURE 16-35

Siding exposure can be set with a carpenter-made notched device.

siding exposure so that the boards come out full size above and below the windows (Fig. 16-36). The siding must be notched to fit around the window (Fig. 16-37).

Beveled wood siding can also be started on a *water table*. A 1 × 2-in. member is nailed

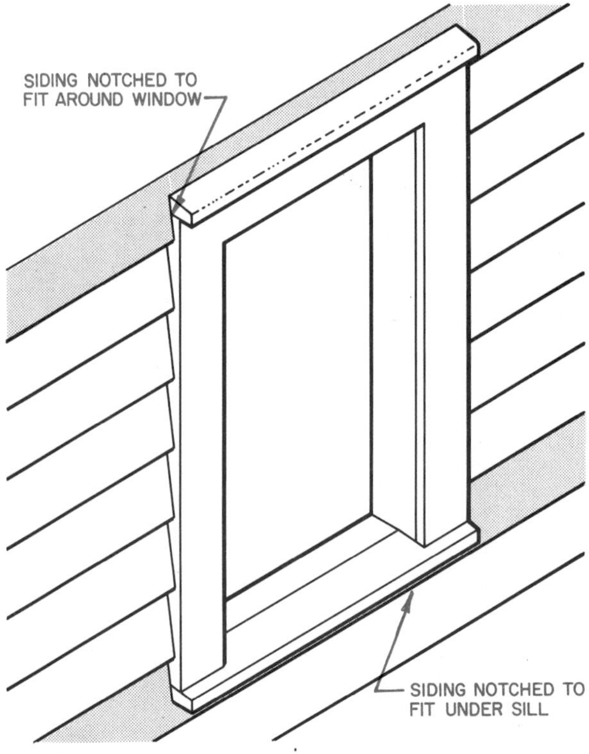

FIGURE 16-36

Sometimes, it is necessary to notch a piece of siding around a window or door.

FIGURE 16-37

A carpenter fitting notched siding between windows. *(Courtesy of California Redwood Association)*

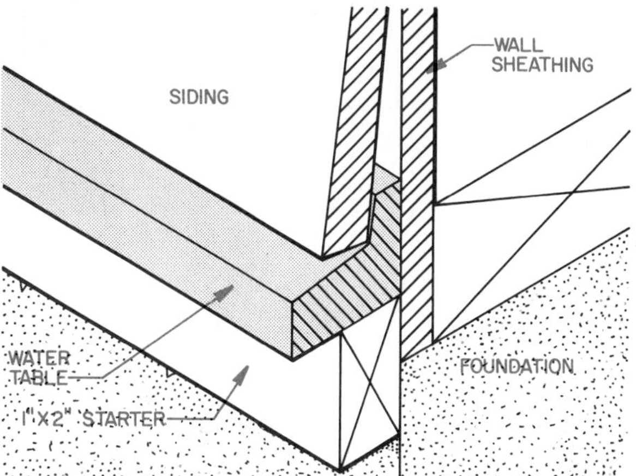

FIGURE 16-38

Wood siding can be started with a water table.

over the joint between the foundation and sheathing. The water table is nailed above this. The siding begins on top of the water table (Fig. 16-38). When this construction is used, the distance to be divided into equal parts for finding siding exposure begins at the top of the water table.

Following is an example of how siding spacing could be figured. Assume an 11¼-in. bevel siding with a 10¼-in. exposure is used. The distance from 1 in. below the foundation to the bottom of the window is 30 in. The length of the window from the bottom of the sill to the top of the drip cap is 49 in. Dividing these by 10¼ in. reveals that the lower area needs three courses. The window needs five courses. The exposure on the window area is 10 in. The exposure on the lower wall is 9 15/16 in. These are both very close to the required 1-in. overlap. They are different exposures, but the difference is so small that it will not be noticed. If it becomes necessary to go much below the 1-in. overlap, it would be best to notch the siding.

Many horizontal wood sidings have rabbeted edges. The amount of overlap cannot be changed much. The location of the first course below the foundation can be varied some, but this is the only possibility for adjustment.

When nailing wood siding, the nail should miss the piece below. This will allow the siding to expand without interference from the course above or below it (Fig. 16-31).

Avoid butt joints whenever possible. Use the long pieces for the long unbroken runs. Use the shorter pieces between windows or between corners and windows. If a butt is necessary, it should occur over a stud unless plywood or solid wood sheathing is being used.

A carpenter-made gauge can be used to mark the length of siding which fits between two windows or a window and a corner board (Fig. 16-39). To mark the length, place one end of the siding against one of the casings. Be certain that the end is cut clean and square. Lay the other end across the second casing. Place the gauge on the siding and slide it against the casing. Mark the length along the side of the gauge. Cut to length and nail in place (Fig. 16-40).

The piece should not fit too tightly. A slight space is needed at each end. This allows the siding to expand without buckling. The space is filled with a soft caulking. Remember to coat the cut ends of siding with a water repellent before nailing them in place.

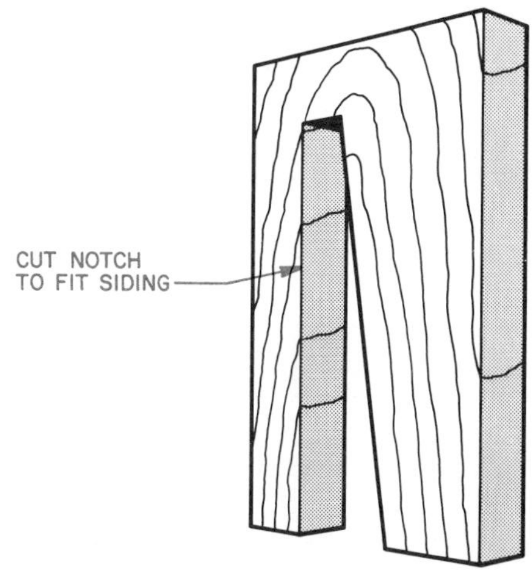

FIGURE 16-39

This carpenter-made gauge is used to mark siding to length between windows or doors and windows.

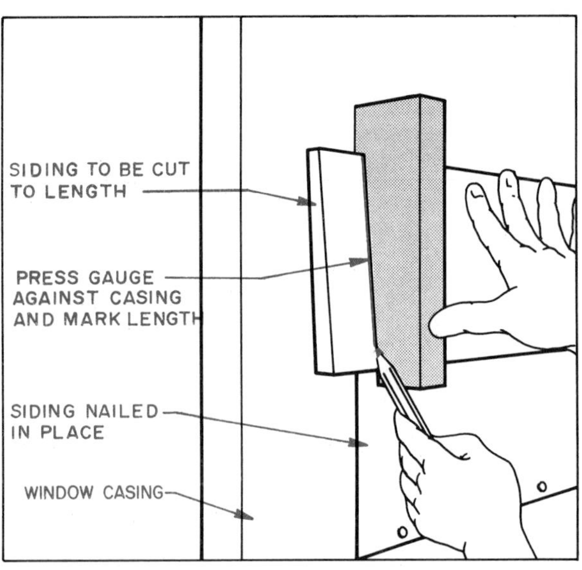

FIGURE 16-40

How to mark siding to length.

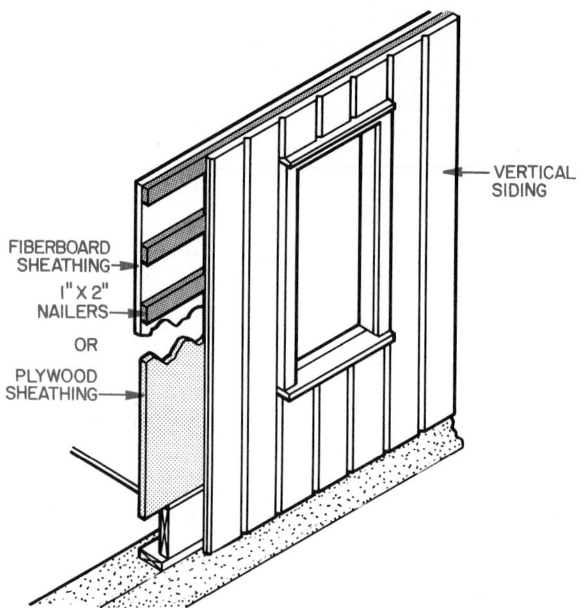

FIGURE 16-42

How to install vertical siding over plywood and fiberboard and other sheathing that will not hold nails.

Siding that is rabbeted or tongue-and-grooved is applied in the same manner as bevel siding. The nailing patterns are shown in Fig. 16-28.

If tongue-and-groove paneling is used as horizontal siding, it is blind-nailed through the tongue. It is face-nailed in the center of each piece if it is 6 in. wide or wider.

Vertical Wood Siding *Vertically applied siding* has interlapping joints (Fig. 16-41). It is nailed the same as if it is used as horizontal siding. It is best if applied over plywood or solid wood sheathing. If fiberboard, gypsum, or plastic foam-type sheathing is used, 1 × 2 or 1 × 4-in. wood blocking strips must be nailed horizontally along the wall. Space them 16 to 24 in. apart. Nail the siding to them. Space the nails about 16 in. apart (Fig. 16-42). Areas commonly using vertical siding are gable ends and the area surrounding the front entrance.

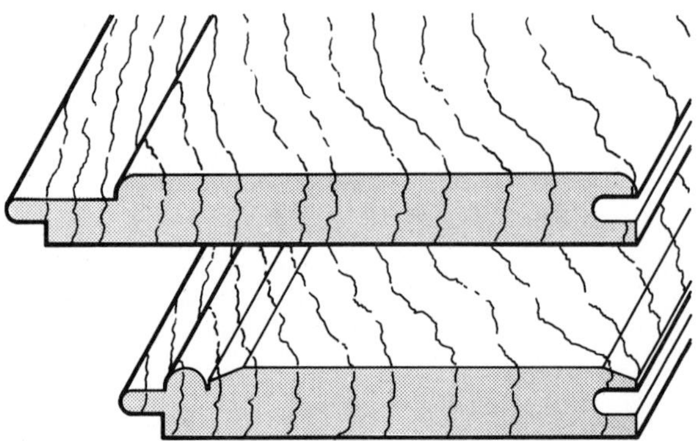

FIGURE 16-41

Vertically applied siding has tongue-and-groove joints.

Square-edge boards are also used as vertical siding (Fig. 16-43). They require the same sheathing preparation as just discussed. The layer of boards or battens next to the sheathing are nailed with 8d nails at each blocking strip. Wide under boards require two nails. Battens require one nail. The nails in the top boards should miss the under boards. Use 12d nails through the top boards. The nailing patterns and board arrangements are in Fig. 16-44.

The bottom edges of the boards are cut on an angle to form a *drip edge* (Fig. 16-45).

Siding at the Corners Siding meets at external and internal corners. How this is handled depends somewhat upon the architect's design. In exterior corners the corner boards can be mitered (Fig. 16-46). This is a difficult cut because the siding is usually tapered and nailed sloping. The mitered corners must fit very tightly so that moisture does not penetrate.

Often, specially manufactured metal corners are used (Fig. 16-46). This gives the same appearance as a mitered corner. The metal corners are nailed in place as each course of siding is installed.

Corner boards are also used (Fig. 16-46). They are usually 1⅛ to 1⅜ in. thick. The width varies depending upon the instructions of the designer. Usually, a rather narrow board is used. The corner boards are butted and nailed through the sheathing into the corner stud. Notice that one board is narrower than the other. They are set in place before the siding is

FIGURE 16-43

Square-edge boards are used as a vertical siding. *(Courtesy of The Garlinghouse Co., Inc.)*

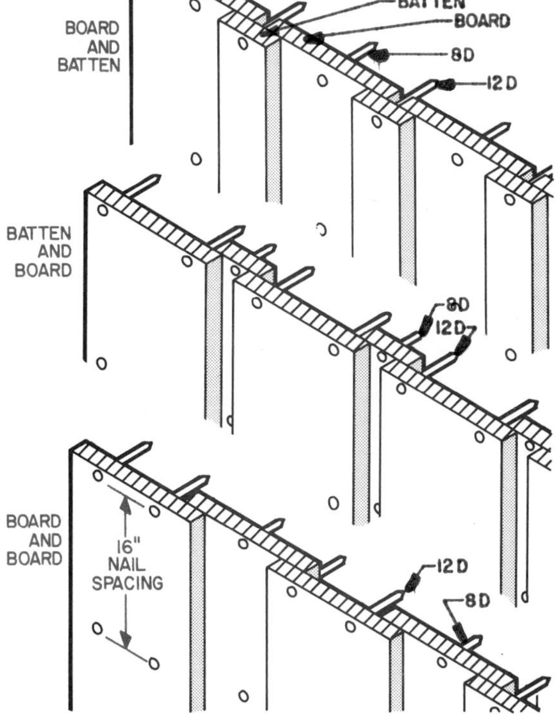

FIGURE 16-44

Installation patterns and details for square-edge boards used as vertical siding.

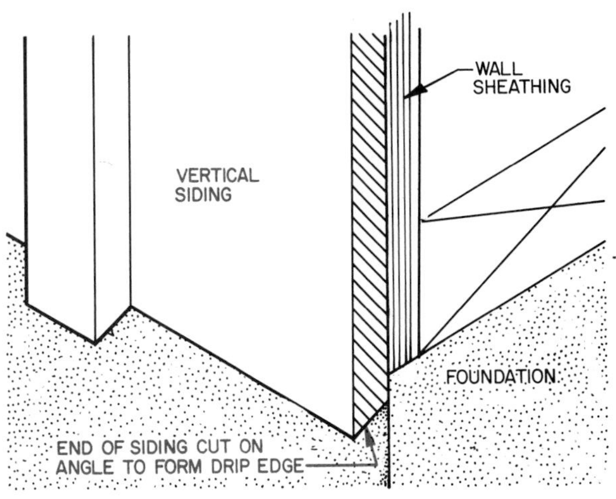

FIGURE 16-45

Cut the bottom edge of the siding on an angle to form a drip edge.

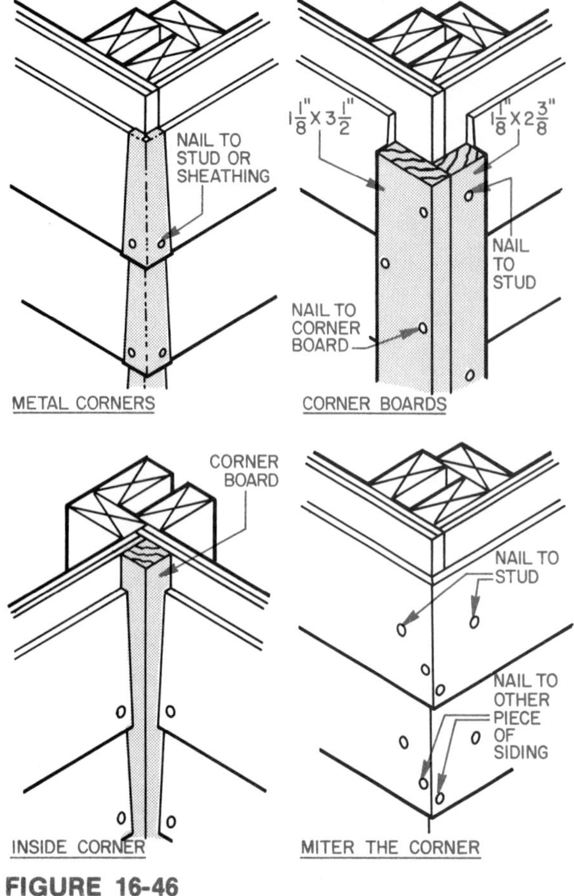

FIGURE 16-46

How to form internal and external corners when using horizontally applied wood siding.

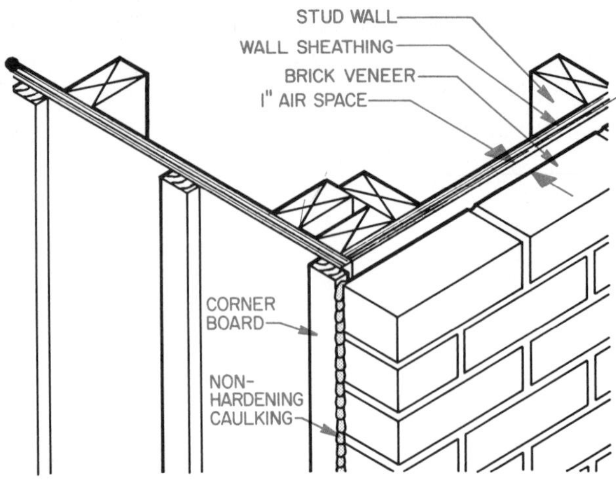

FIGURE 16-47

How to form an exterior corner between wood siding and a masonry wall.

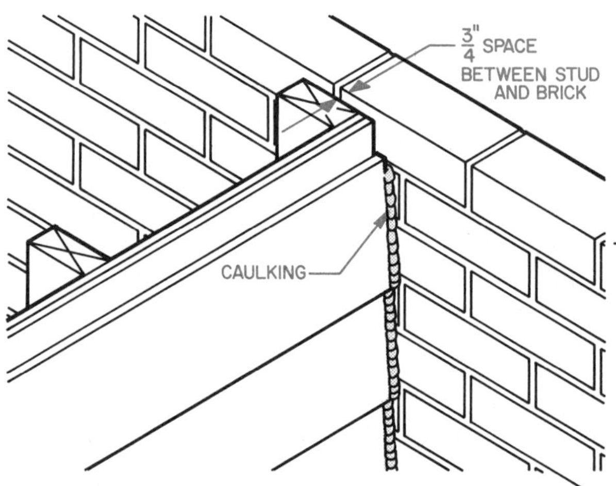

FIGURE 16-48

How to form an interior corner between wood siding and a masonry wall.

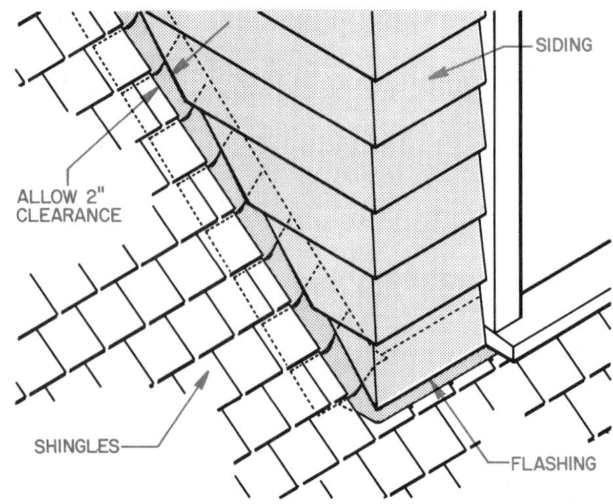

FIGURE 16-49

Siding is cut to parallel a roof and clears it about 2 in.

installed. The siding must be cut to length carefully so that it fits against the corner board. The crack between them is caulked after they are primed.

Interior corners are butted against a 1⅛- or 1⅜-in. square corner board. The board is nailed to the stud before the siding is installed. The joint between the siding and the corner board is caulked after the siding is primed (Fig. 16-46).

When siding meets a masonry wall, a watertight joint is needed. One way to frame an exterior corner is shown in Fig. 16-47. A corner board is nailed to the corner studs. The space between it and the masonry is caulked. It is recommended that a layer of roofing felt be nailed around the corner before the siding and masonry units are set in place.

The siding on an interior corner butts the masonry wall. The space between them is caulked (Fig. 16-48).

When siding ends against a roof surface, the edge is cut parallel with the angle of the roof (Fig. 16-49). It should clear the roof surface about 2 in. The siding is on top of the flashing. The cut end should be coated with a water-repellent preservative.

Gable-End Treatment Often, the designer uses a different siding or treatment on the *gable end.* This requires the carpenter to prepare the joint between these so that it is neat and waterproof. One way to do this is with a *drip cap* (Fig. 16-50). The drip cap is nailed to the plate. It is then flashed the same as discussed for windows. The gable siding is applied over the flashing.

Another technique is to apply 1 × 2-in. blocking over the sheathing on the gable end (Fig. 16-51). This places the gable siding out over the lower siding. The edge of the gable siding is undercut to form a drip edge.

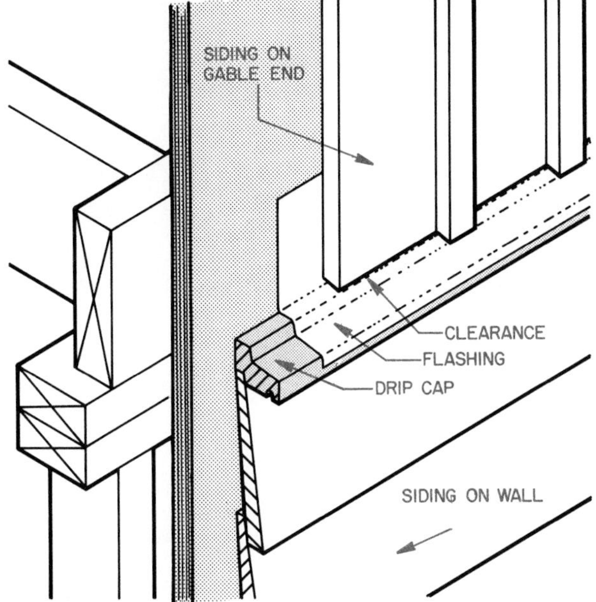

FIGURE 16-50

The siding on a gable end can join the siding below by using a drip cap.

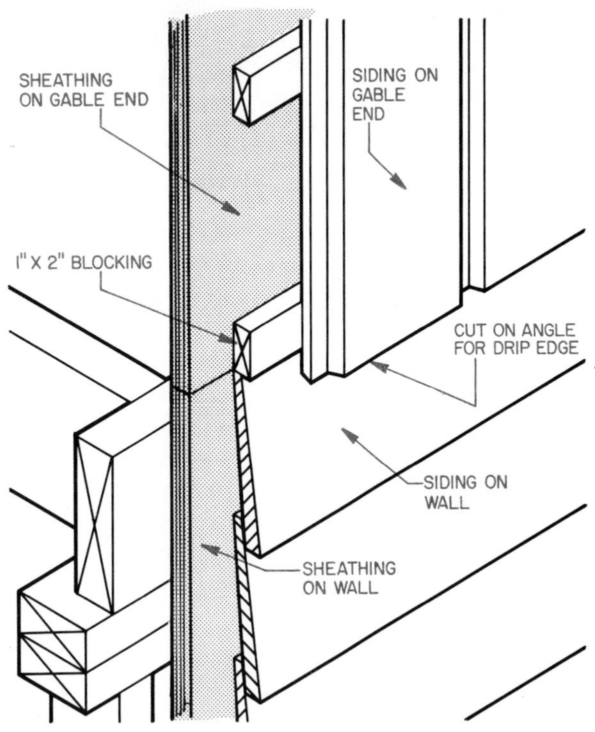

FIGURE 16-51

Siding on a gable can overlap the siding below.

Plywood Siding *Plywood siding* is available in panels and lap siding. The standard sizes are shown in Figs. 16-52 and 16-53. The panels are available with a variety of textures (Fig. 16-54).

Plywood siding is very strong and produces a rigid wall. A plywood-sheathed or -sided wall is many times more rigid than one using solid wood sheathing and siding.

Plywood siding panels can be applied directly to the studs (Fig. 16-55). The panels may be set horizontally or vertically (Figs. 16-56

303 SIDING 6-S/W
T 1-11
19/32 INCH
GROUP 2
24 oc SPAN
EXTERIOR
PS 1-74 000
APA

303 SIDING O/L
MDO
GROUP 3
16 oc SPAN
EXTERIOR
PS 1-74 000
APA

PLYWOOD SIDING[a]		MAX. STUD SPACING (in.)		NAIL SIZE (Use noncorrosive box, siding, or casing nails)	NAIL SPACING (in.)	
Description (All Species Groups)	Nominal Thickness (in.)	Face Grain Vertical	Face Grain Horizontal		Panel Edges	Intermediate
MDO EXT-APA	11/32 and 3/8	16	24	6d for panels ½ in. thick or less;		
	½ and thicker	24	24			
303-16 o.c. Siding EXT-APA	5/16 and thicker	16	24	8d for thicker panels[b]	6	12
303-24 o.c. Siding EXT-APA	7/16 and thicker	24	24			

[a]Panel sizes 4 × 8 ft, 4 × 12 ft, 4 × 14 ft, and 4 × 16 ft.

[b]If applied over sheathing thicker than ½ in., use the next regular nail size.

FIGURE 16-52

Standard sizes of plywood panel siding. *(Courtesy of American Plywood Association)*

TYPICAL WIDTH (in.)	MIN. LAP SIDING THICKNESS (in.)	MIN. BEVEL BUTT THICKNESS (in.)	WIDTH (in.)			NAILING	
12, 16, or 24	3/8	9/16	16	6d	noncorrosive	One nail per	4 in. at vertical joint;
	1/2		20	8d	siding or casing	stud along	8 in. at studs if siding
	5/8		24	8d	(galv. or alum.)	bottom edge	wider than 12 in.

Note: Minimum head lap, 1½ in.

FIGURE 16-53

Standard sizes of plywood lap and bevel siding. *(Courtesy of American Plywood Association)*

Fine Line

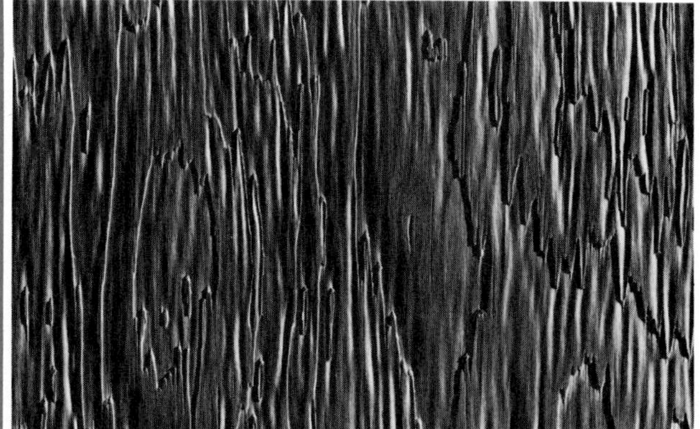

Brushed

Texture I-II

Kerfed

Rough Sawn

Channel Groove

FIGURE 16-54

Surface texture and patterns of selected plywood siding panels. *(Courtesy of American Plywood Association)*

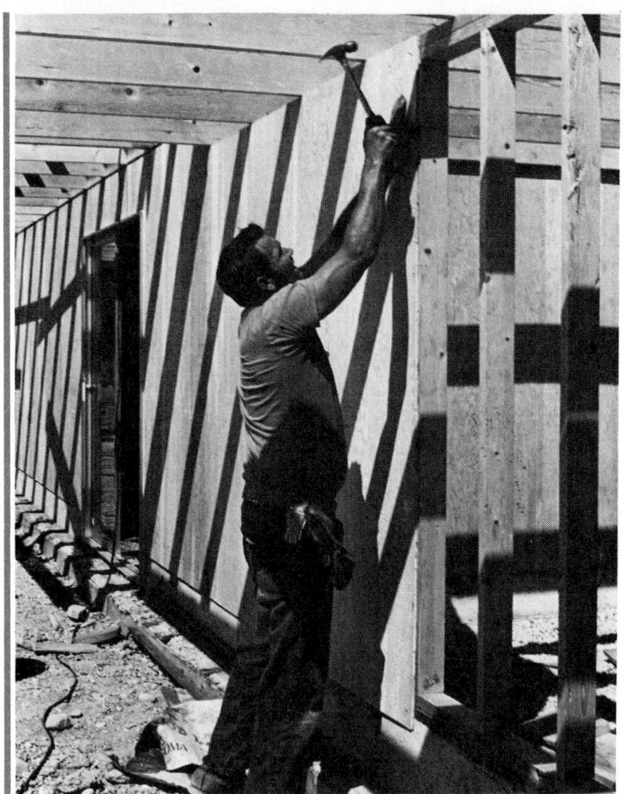

FIGURE 16-55

Plywood siding panels can be applied directly to the studs. *(Courtesy of American Plywood Association)*

FIGURE 16-57

Installing plywood siding panels in a horizontal position. *(Courtesy of American Plywood Association)*

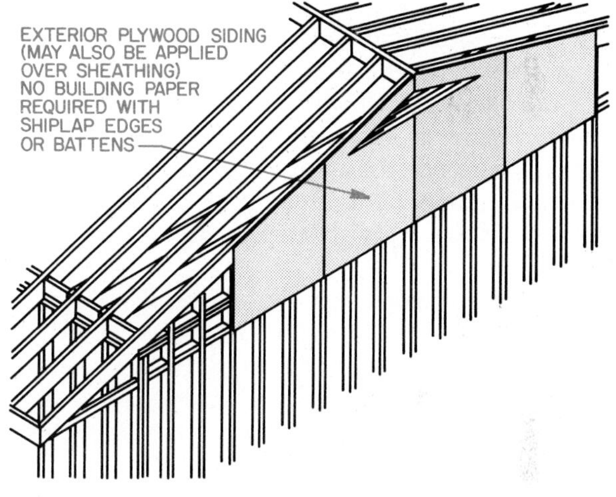

FIGURE 16-58

Plywood siding panels can be used on gables.

and 16-57). Plywood siding is also used on gables (Fig. 16-58). No diagonal bracing is necessary under these conditions. The joints between panels must fall on a stud or horizontal blocking. If the vertical joint is shiplapped, backed with building paper, or covered with battens, it need not be caulked (Fig. 16-59). If it is to be caulked, the edges should be treated with a water repellent.

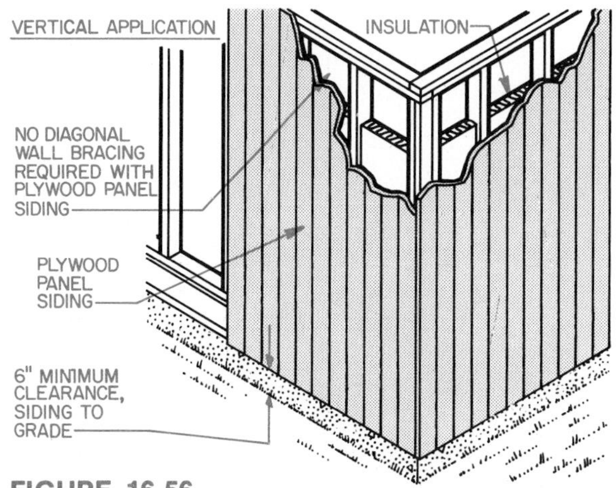

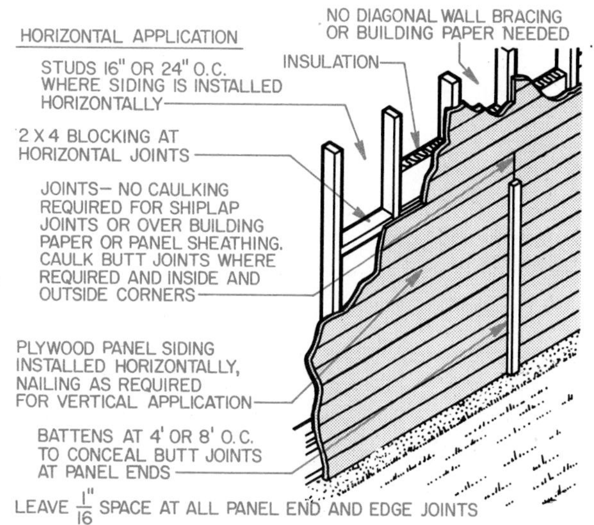

FIGURE 16-56

How to apply plywood siding panels in vertical and horizontal positions.

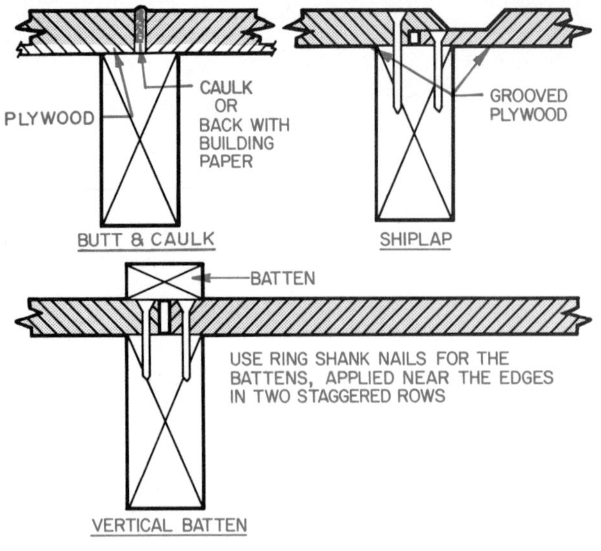

FIGURE 16-59

Joints used in installing plywood siding panels in a vertical position.

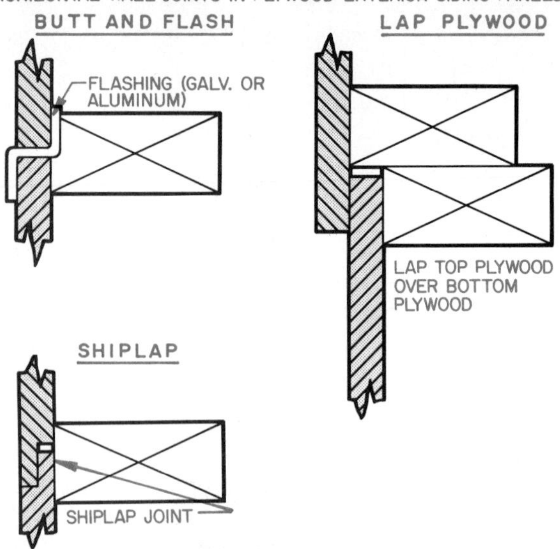

FIGURE 16-60

Types of horizontal joints that are used with plywood siding panels.

Horizontal wall joints have either a shiplap joint or butt and flashing (Fig. 16-60).

Plywood panels are joined at inside and outside corners as shown in Fig. 16-61. Inside corners are formed by butting and caulking the corner. Outside corners can be rabbeted and caulked or butted with corner boards over the corner.

When plywood panels are used on multistory buildings a horizontal joint will occur. This must be waterproof and allow for settling and shrinkage of the wood structural frame. Recommendations for making horizontal beltline joints are given in Fig. 16-62.

Framing at windows and doors is much like that for solid wood. Details are presented in Fig. 16-63.

Plywood panels can be applied over any type of sheathing. This produces an even more rigid wall and increases its *R*-factor (resistance to heat loss) (Fig. 16-64).

Plywood lap siding is applied as described for horizontal solid wood siding. It can be applied directly to the studs if they are 16 in. on center. If this is done, the wall needs let-in bracing. Building paper is applied over the studs (Fig. 16-65). Generally, lap siding is applied over sheathing. Studs can be spaced 24 in. on center. Use 6d nails for ½-in.-thick siding. Use 8d nails for thicker siding. Space the nails 6 in. apart ½ in. from the bottom edge. Space nails 12 in. apart on intermediate studs. Under each butt joint and at each corner, install wedges (Fig. 16-65). Coat the edges cut off the siding with water repellent.

Start the first course with a ⅜ × 1½-in. starter strip. Siding up to 12 in. wide should be overlapped 1 in. Widths over 12 in. should be lapped 1½ in.

Vertical butt joints should be staggered. They should be centered on a stud.

Plywood lap siding is joined at inside and outside corners in the same manner as described for solid wood siding.

FIGURE 16-61

How to form inside and outside corners when using plywood siding panels.

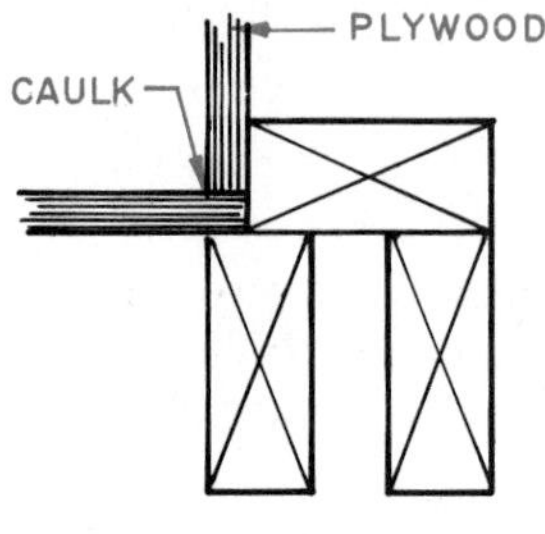

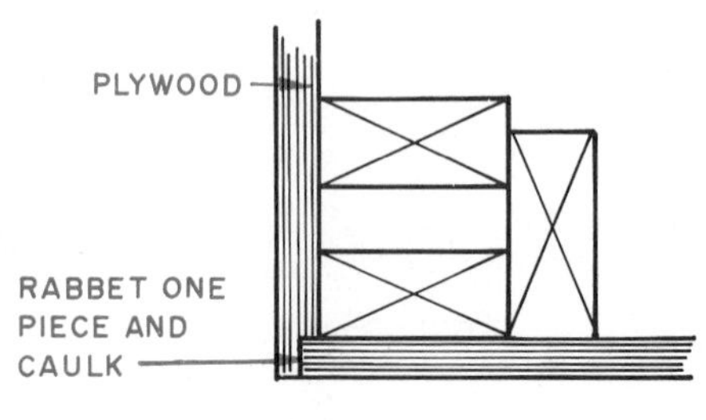

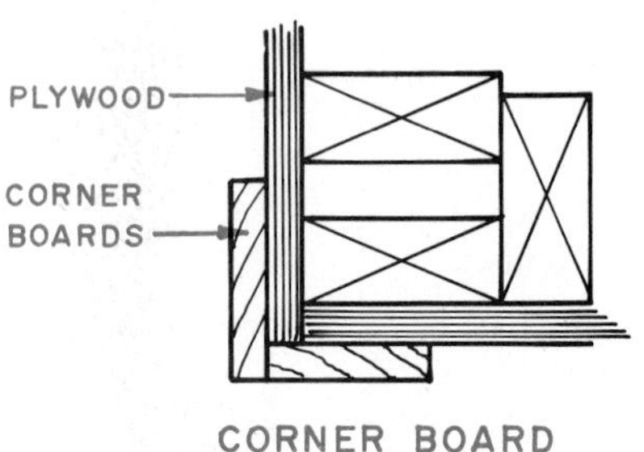

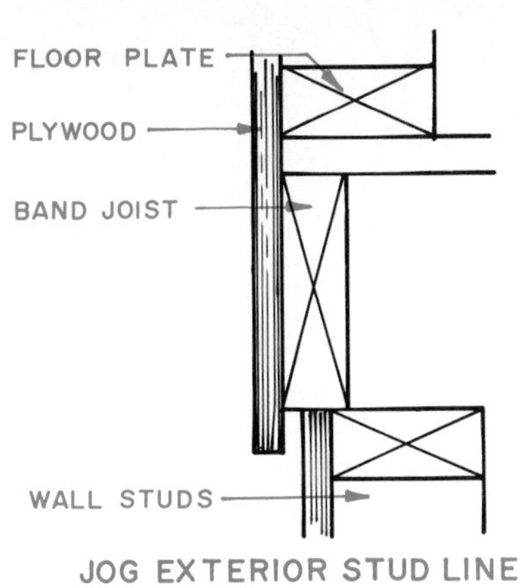

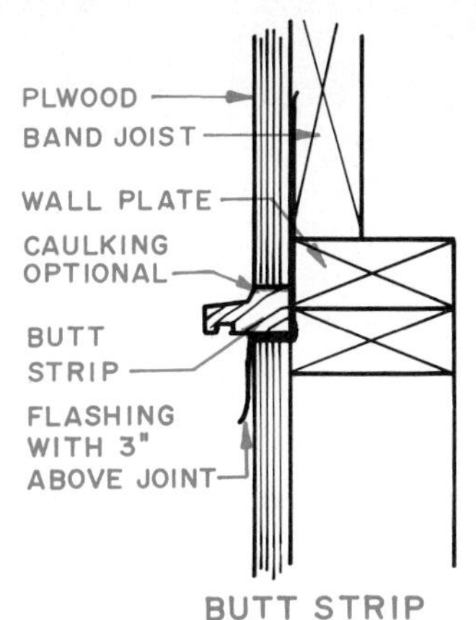

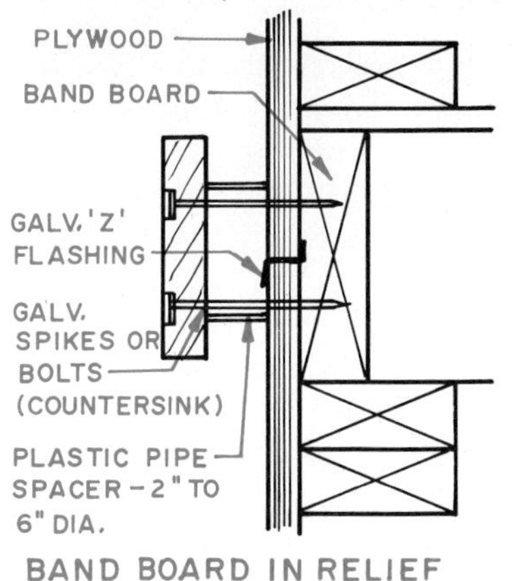

FIGURE 16-62 *(above)*

How to form horizontal beltline joints between plywood siding panels.

FIGURE 16-63 *(below)*

How to install plywood siding panels at windows and doors.

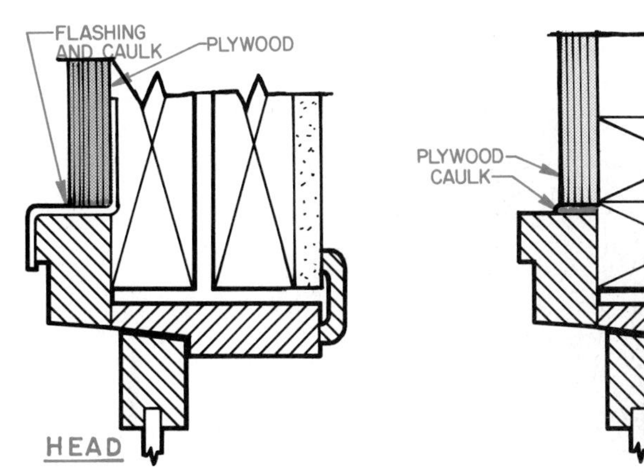

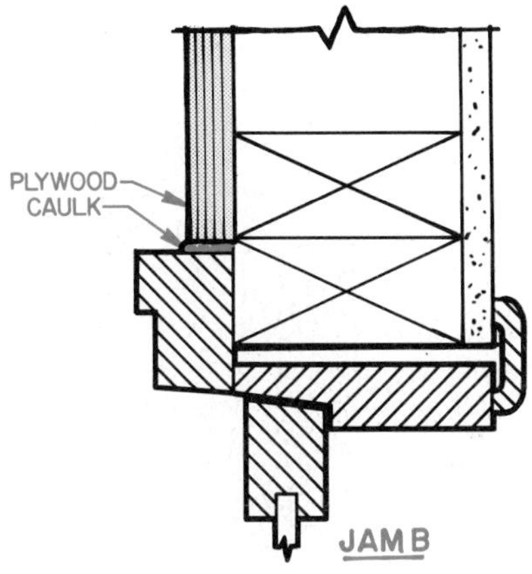

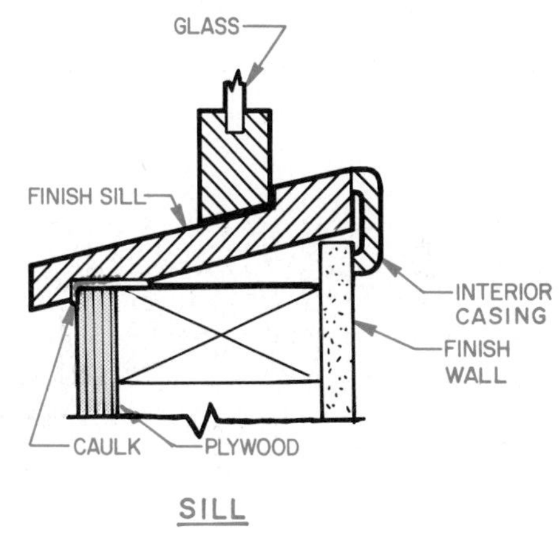

FIGURE 16-64

Plywood siding panels can be nailed over sheathing materials of all kinds. *(Courtesy of American Plywood Association)*

FIGURE 16-65

When plywood lap siding is nailed directly to the studs, a layer of builder's felt is needed. The studs are braced with a let-in diagonal brace.

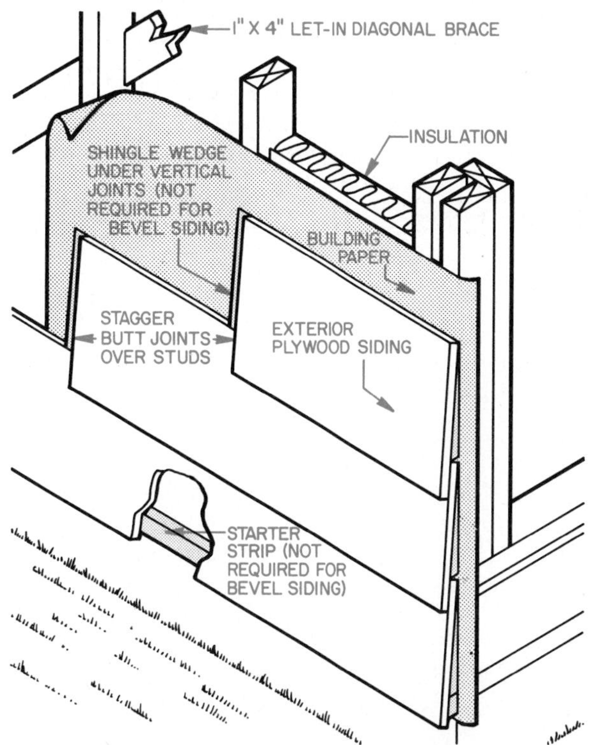

Hardboard Siding *Hardboard siding* is used much like plywood siding. It is available in panels and lap siding. Sizes are shown in Fig. 16-66. This material has a tough, dense surface. It will not split or splinter and resists denting. Since they are manufactured panels, the surfaces are free of imperfections. Panels are available in a variety of surface patterns (Fig. 16-67).

They are cut with hand and power woodworking tools. Siding is available with a factory-primed surface. This gives a superior surface for the finished coats of paint. Panels that are vinyl-coated are also available.

Installing Hardboard Siding Hardboard siding can be applied directly to studs if spaced 16 in. on center. If placed over sheathing, the studs can be spaced 24 in. on center. The lowest edge should be at least 8 in. from the ground (Fig. 16-68).

The panels should be applied with 8d corrosion-resistant nails spaced 4 in. apart on the edges. They should be in ½ in. from the edge of the panel. Nails should be 8 in. apart on intermediate members.

Panels	48 × 96 in., 48 × 108 in., 48 × 120 in.
Lap siding	widths: 8 in., 12 in.; lengths: 12 ft, 16 ft

FIGURE 16-66

Sizes of hardboard siding.

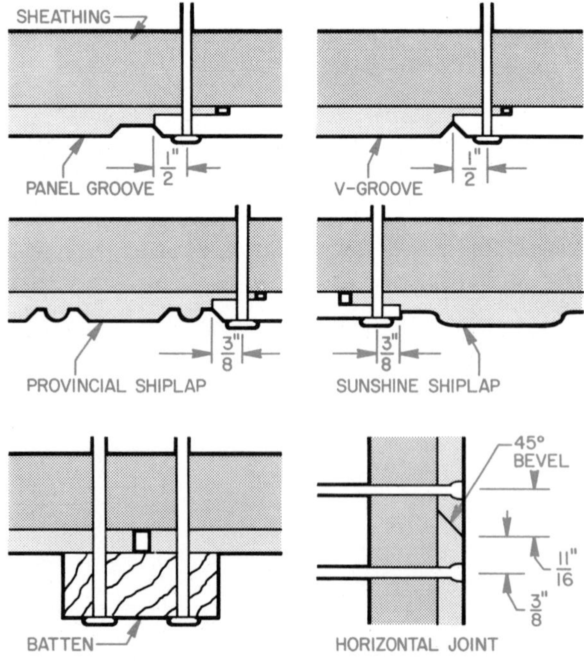

FIGURE 16-67

Typical surface patterns and joints available in hardboard siding panels.

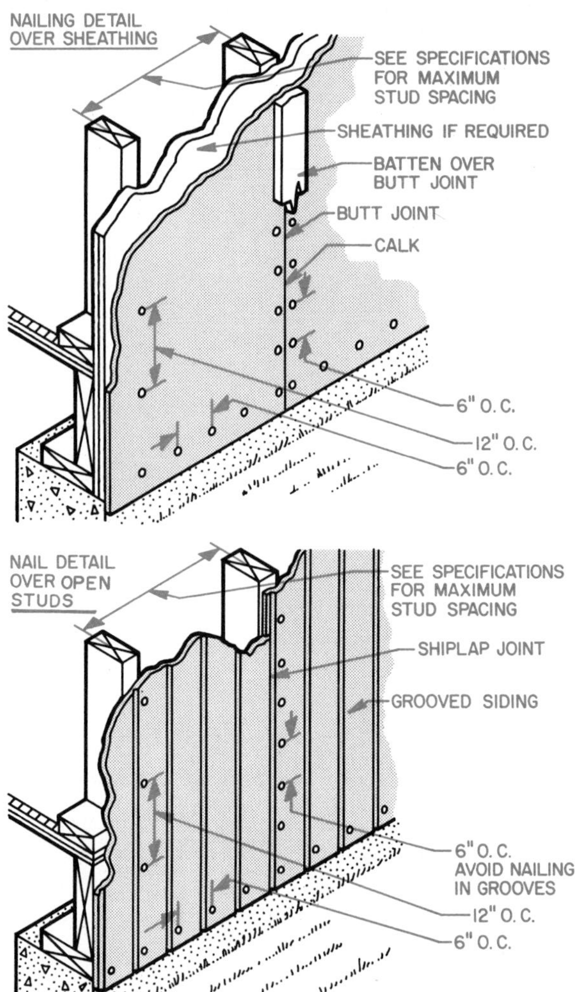

FIGURE 16-68

How to install hardboard siding panels.

Hardboard panels are installed in the same manner as described for plywood panels. One type of horizontal joint that can be used is a bevel cut. This is caulked before the two parts are set together. Some carpenters prefer to use wood corner boards on inside corners. This helps seal out moisture (Fig. 16-69).

Vinyl-covered hardboard panels used as board-and-batten siding have snap-on prefinished batten strips. The hardboard batten strip is notched to receive the prefinished cover (Fig. 16-70). The batten covers are finished in a color that matches the vinyl siding.

Hardboard lap siding is installed as explained for plywood lap siding. It is applied with 8d galvanized nails driven about ½ in. from the top edge. The lower edge is nailed through both layers of siding (Fig. 16-71). Vinyl-covered lap siding is nailed on the top edge and bonded to the course below with adhesive. The adhesive is applied at intervals. It is not applied solid because the wall needs ventilation spaces. Some types of hardboard siding have mounting

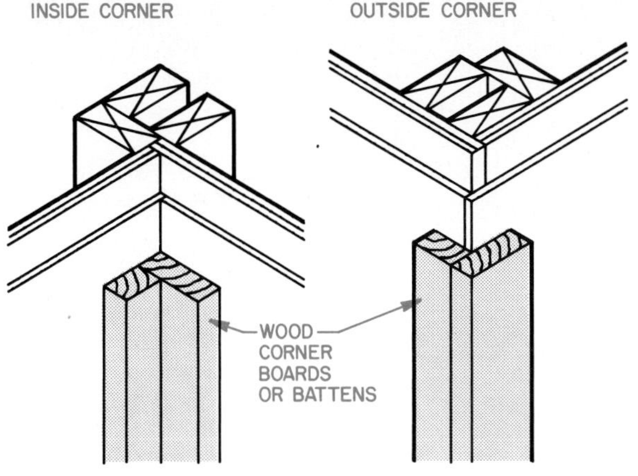

FIGURE 16-69

How to frame interior and exterior corners when installing hardboard siding panels.

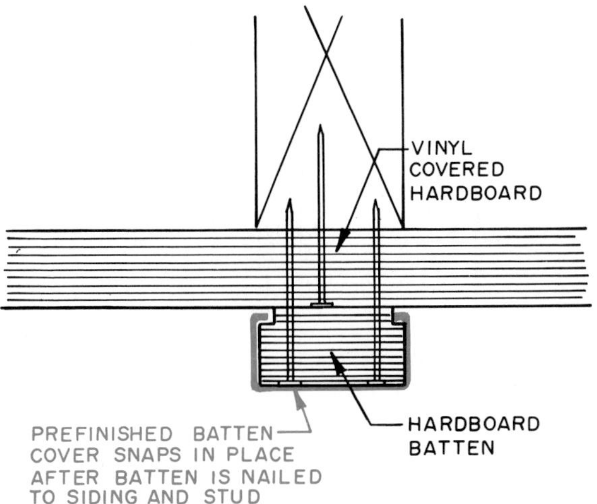

FIGURE 16-70

Hardboard batten strips are covered with a vinyl coating.

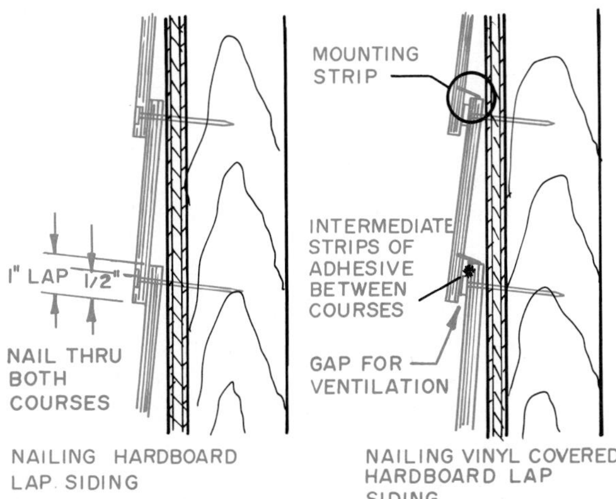

FIGURE 16-71

How to install hardboard lap siding.

strips. These strips rest on the top of the row below. This automatically spaces the courses of siding (Fig. 16-71).

Metal and Plastic Siding There are a variety of metal and plastic siding materials available. Metal is usually *steel* or *aluminum* with a baked-on finish. This surface requires no painting. Care must be exercised when installing it so as not to damage the surface. Solid vinyl plastic siding is also available in a variety of colors. To properly install these materials, follow the instructions of the manufacturer.

Following is a description of one product.

1. The courses are laid out in the same manner as described for wood siding.
2. A chalkline is run around the bottom of the house to locate the first course. It is important that it be accurately located (Fig. 16-72).
3. Nail the starter strip to the sheathing (Fig. 16-73).
4. Next nail the inside and outside corner

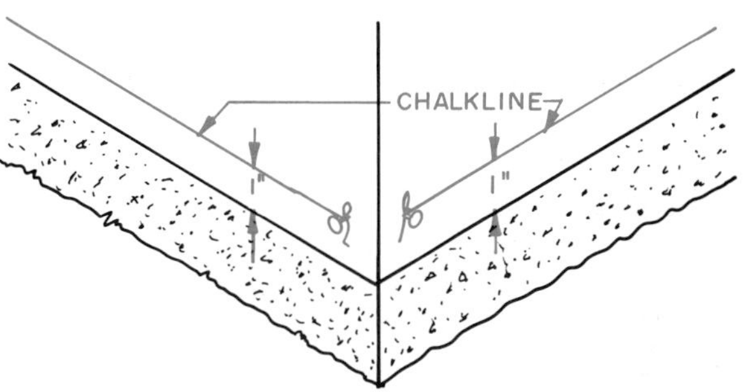

FIGURE 16-72

Using a chalkline, locate a line around the lower edge of the building 1 in. from where the wall meets the foundation. *(Courtesy of GAF Corporation)*

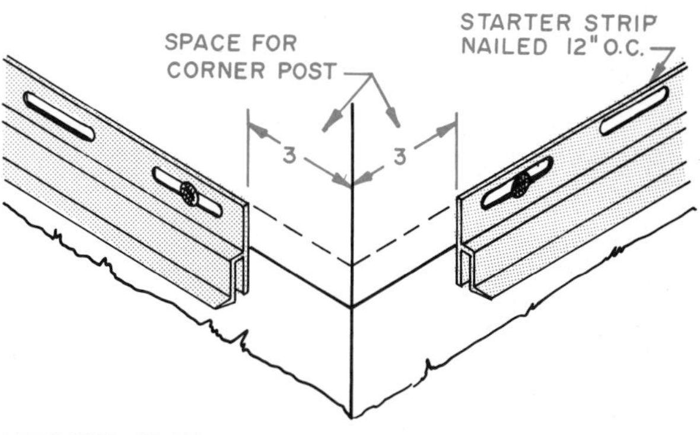

FIGURE 16-73

Nail a starter strip with the upper edge on the chalkline. Leave a ¼-in. gap between the starter strip and the corner posts. Drive the nails in the center of the slot provided. *(Courtesy of GAF Corporation)*

posts (Fig. 16-74). These are made from the same material as the siding. They are formed with flanges that are nailed to the sheathing. Drive the nails so that they are about 1/32 in. from touching the material (Fig. 16-75). Usually, nails are placed every 12 in. Center the nails in the slot.

5. Attach the channel to receive the siding around doors and windows (Fig. 16-76).
6. Place the first course of siding in the starter strip (Fig. 16-77). If it requires backers, these should be in place at this time. Nail the panel to the sheathing per the manufacturer's directions. Overlap as recommended by the manufacturer. Continue with each course until the wall is finished (Fig. 16-78).
7. Pieces will have to be cut to length to fit between windows and between corners and windows. Leave space for expansion. Caulk all corners (Fig. 16-79).

Wood Shingles and Shakes Review the material on wood shingles and wood shakes in Chapter 14. These are applied to walls in much the same way as described for roofs.

Wood shingles and wood shakes are usually applied in single-course or double-course coverage. It is best if plywood or solid wood sheathing is used. Builder's felt must be placed over solid wood sheathing. If nonwood sheathing is used, 1 × 4-in. nailing strips applied horizontally on the wall will be needed.

Single-course shingles are laid as explained for bevel lap siding. The starter course is doubled (Fig. 16-80). Shingles up to 8 in. wide are held with two nails. Over this use three nails. They are nailed about 1 in. above the butt of the next course and 3/4 in. from each edge. Threaded nails are recommended if plywood sheathing is used. Use 4d nails.

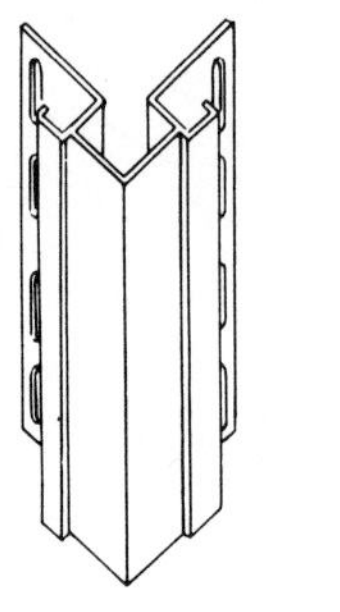

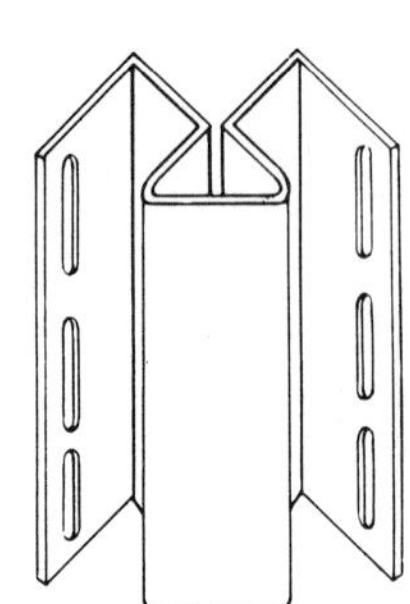

FIGURE 16-74

Inside and outside corners used when installing metal and plastic lap siding. *(Courtesy of GAF Corporation)*

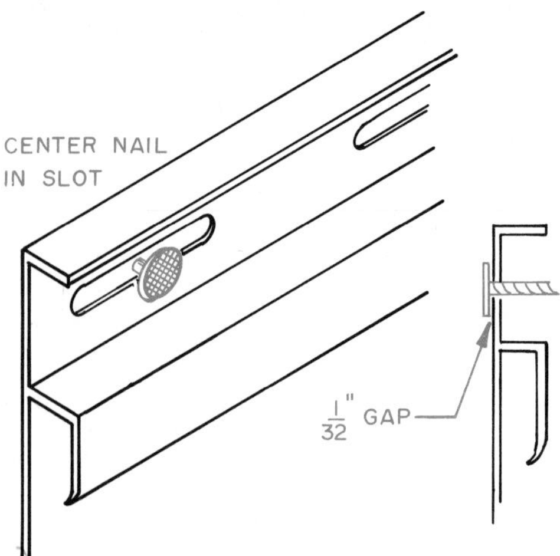

FIGURE 16-75

Set the corner post nails so that they almost touch the post. *(Courtesy of GAF Corporation)*

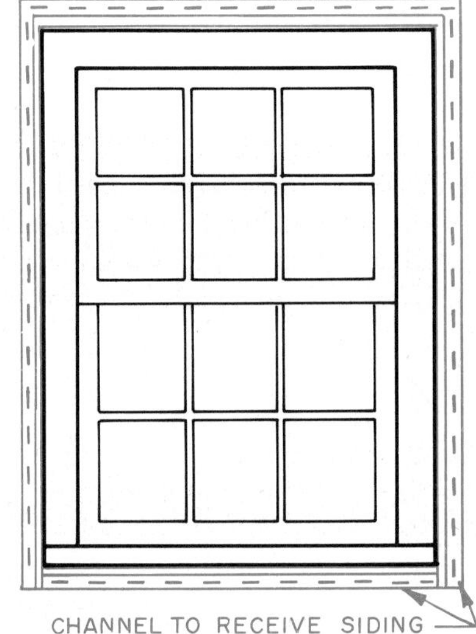

FIGURE 16-76

Place channel strips around doors and windows. *(Courtesy of GAF Corporation)*

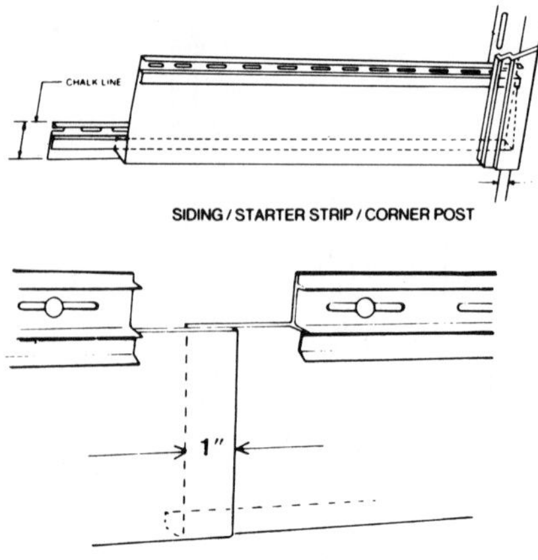

FIGURE 16-77

Place the first piece of siding on the starter strip. Start at the corner post. Overlap as required. *(Courtesy of GAF Corporation)*

FIGURE 16-78

Place the pieces of siding in the strips. *(Courtesy of GAF Corporation)*

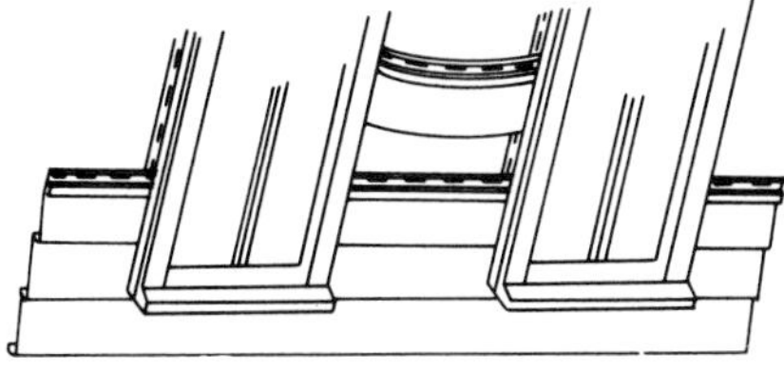

FIGURE 16-79

Fit the siding between openings. *(Courtesy of GAF Corporation)*

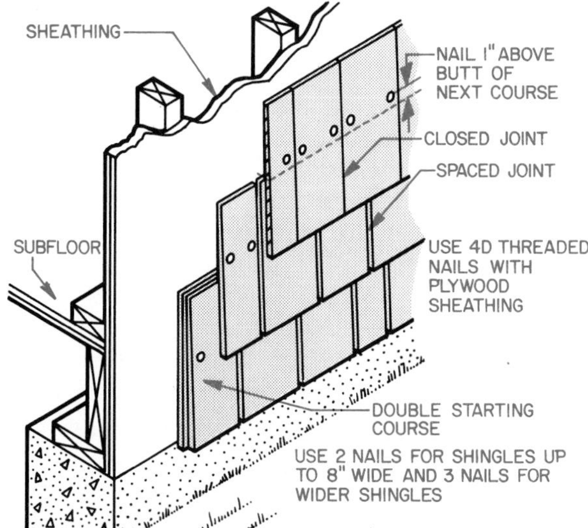

FIGURE 16-80

How to lay single-course wood shingles.

Usually, only one-half or less of the shingle is exposed. The amount of exposure depends upon the desired appearance. They should be laid with a ⅛-in. space on each side to allow for expansion. Shingles expand when they get wet (Fig. 16-81).

When nailing a double course, the under course is nailed to the sheathing. The top course drops below the undercourse about ¼ to ½ in. (Fig. 16-82). The undercourse has only enough

FIGURE 16-81

Nail with half or less of the shingle exposed. *(Courtesy of Red Cedar Shingle and Handsplit Shake Bureau)*

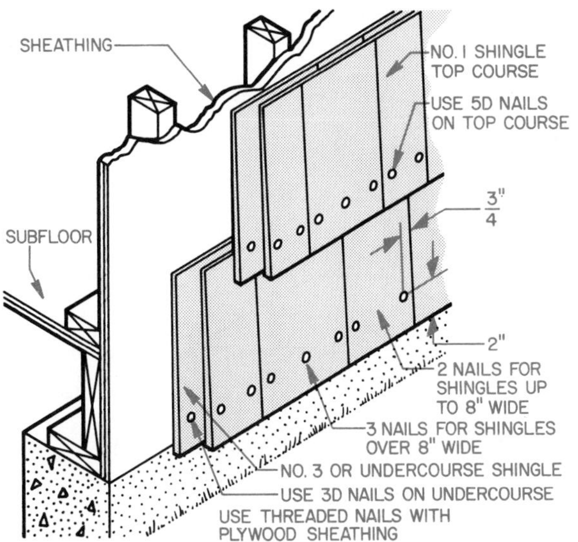

FIGURE 16-82

How to lay double-course wood shingles.

nails to hold it in place. The top course is nailed about 2 in. above the end and ¾ in. from each edge. Use zinc-coated nails with a small, flat head. Use 3d nails in the undercourse. Use 5d nails in the top course.

The undercourse can be low-quality shingles because they are not seen. The top course should be first-grade shingles.

The side joints between shingles should be spaced so they are at least 1½ in. from joints in the under shingles.

Closed or open joints can be used. Closed joints make the siding appear more like wood lap siding. Open joints emphasize the individual shingles.

Shakes and shingles can be mitered at inside and outside corners. Inside corners can also be made by placing shingles againt a corner board. Outside corners can be formed by lapping the shingles on alternate rows (Fig. 16-83). In all cases the corners should be flashed.

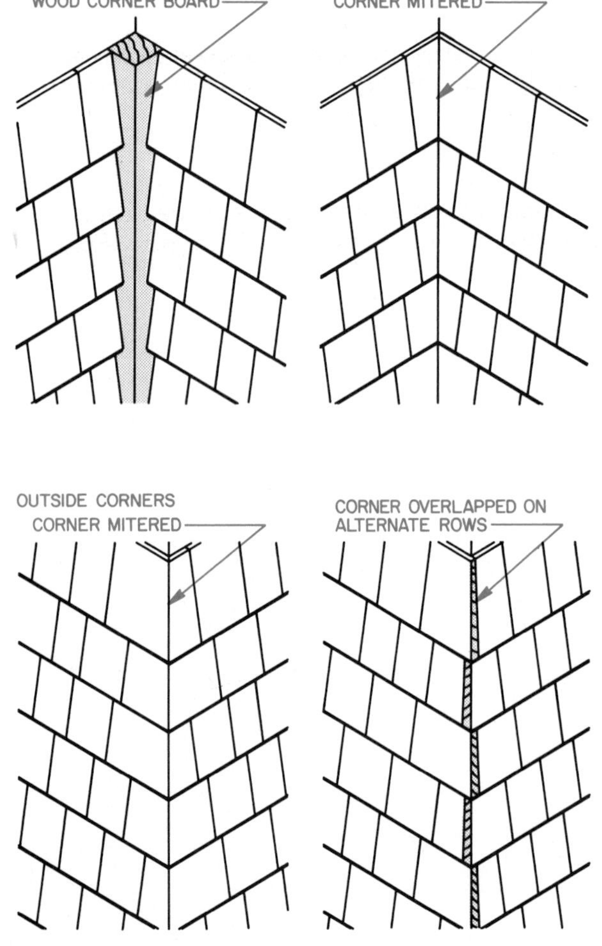

FIGURE 16-83

How to form inside and outside corners when installing wood shingles and shakes.

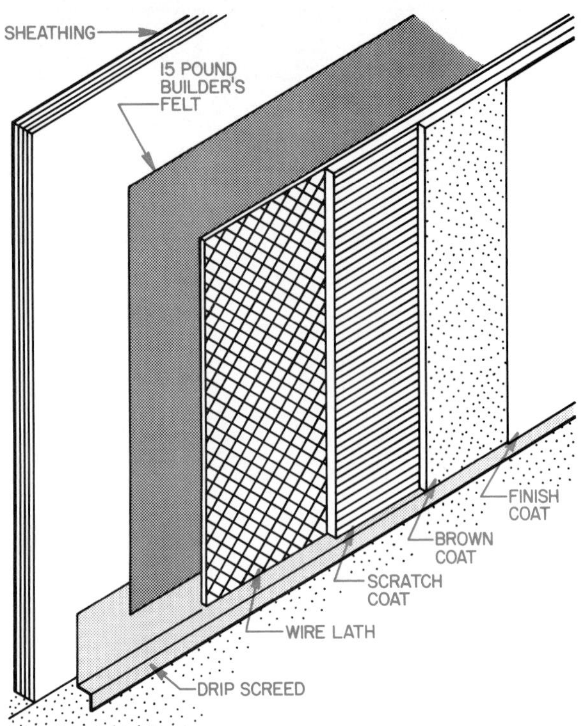

FIGURE 16-84

The layers forming a stucco wall when sheathing is used.

Stucco Wall Finish *Stucco* is a cement mortar finish. It is used a great deal in the Southwest. The cement mortar is applied over a coated, expanded metal lath.

The wall may or may not be sheathed. If the sheathing is used, it may be any typical type. The sheathing is covered with 15-lb builder's felt. When using wood or plywood sheathing, the metal lath can be nailed to the sheathing. Other kinds require that it be nailed to the studs. The nails should be corrosion-resistant and long enough to penetrate at least ¾ in. (Fig. 16-84).

If no sheathing is used, the wall framing must have diagonal bracing let into it. Rows of soft 18-gauge steel line wire are stretched horizontally across the studs. They are spaced 6 in. apart and stapled every fifth stud. The builder's felt is nailed over the wire. It is lapped 3 in. on the edges. The metal lath is nailed over the felt (Fig. 16-85).

If a two-story building is to be built, it is recommended that balloon framing be used. It will shrink less than platform framing. Shrinkage will cause the stucco coating to bulge and crack.

The lath should be at least ¼ in. away from the sheathing. This space is needed so that the stucco can flow behind the lath. Galvanized furring strips or self-furring metal lath provide this spacing.

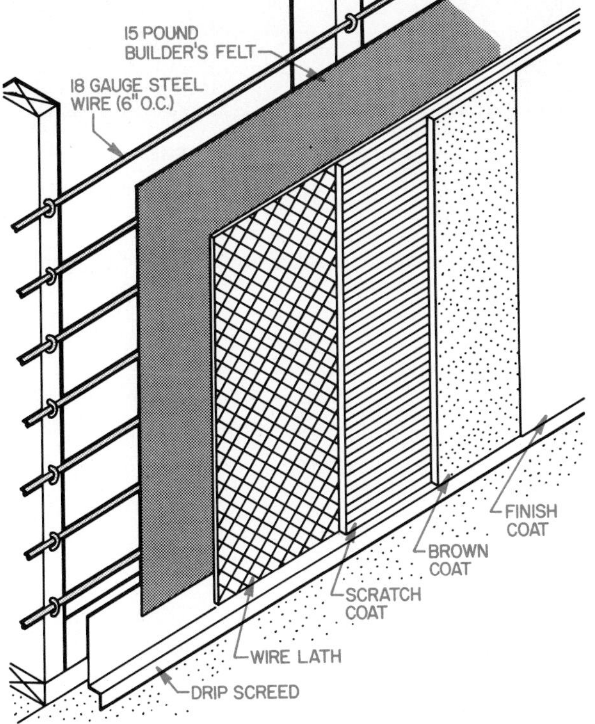

FIGURE 16-85

Forming a stucco wall without sheathing.

Usually, three coats are applied. The first coat is the scratch coat. It is worked into the metal lath. The second coat is the brown coat. It builds up the thickness of the layer and smooths irregularities in the scratch coat. The final coat is the finish coat. This can vary depending upon the texture desired. It can be troweled smooth or brushed to a rough texture. Coloring can be added to the finish coat. The wall can be painted.

Usually, a metal molding is applied on the edges and around openings in the wall.

Masonry Veneer A wood-framed wall can use brick or stone as the finish material. The foundation provides a 5-in. ledge for the masonry units. A 1-in. air space is left between the sheathing and masonry. When using plywood or fiberboard sheathing, it is not necessary to cover it with builder's felt (Fig. 16-86).

Flashing at the base of the wall is needed. The flashing should extend below the bottom of the sheathing. It should extend at least 6 in. up on the sheathing. If the sheathing is covered with builder's felt, it should lap over the flashing.

Corrosion-resistant ties are used to join the masonry to the sheathing. They are spaced in rows 16 in. apart and spaced 32 in. apart vertically. If the sheathing will not hold nails, the ties must be nailed into the studs.

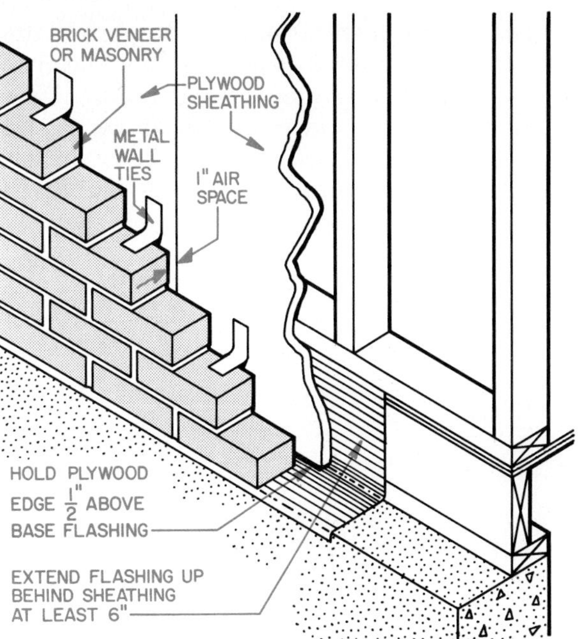

FIGURE 16-86

How to install a masonry veneer over a frame wall.

LOUVERED VENTILATORS

As the wall is framed, *louvered ventilator* openings must be framed. They are framed the same as rough openings for windows.

Louvered ventilators are available in a variety of sizes and materials. Most common are wood and aluminum. The triangular and square shapes are most commonly used (Fig. 16-87).

FIGURE 16-87

Typical louvers used for attic ventilation.

The rough opening is sized to permit the louver to slide inside the opening. The louver is installed from the outside of the building. After the exterior siding is in place, the louver is slid into the opening. It is nailed through its flanges through the siding into the framing. The edges are caulked (Fig. 16-88).

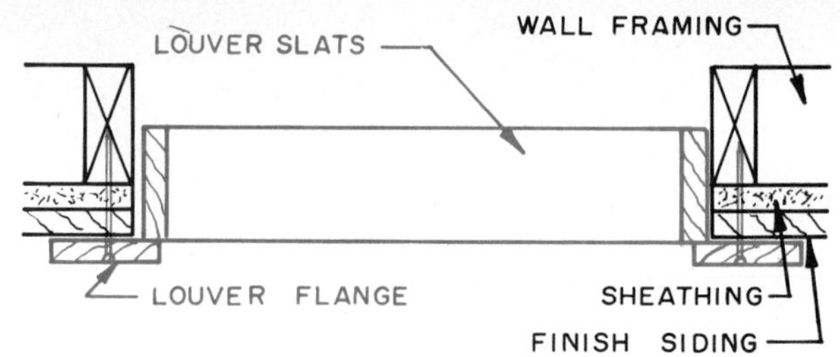

FIGURE 16-88

Louvers can be set in the gable end and nailed to the studs.

STUDY QUESTIONS

After reading the chapter, answer each of the following questions. If you do not know the answer, review the chapter.

1. What are the parts of a cornice?
2. How is the rake finished?
3. What materials are used for soffits?
4. What types of materials are used as the finished exterior siding?
5. Which types of sheathing will not hold nails?
6. How is each course of horizontal siding located?
7. What is the purpose of a corner board when installing horizontal wood siding?
8. What is a stucco finish?
9. How should panels of plywood siding be joined when forming vertical and horizontal joints?

IMPORTANT TECHNICAL TERMS

Following are technical terms that you should be able to use as part of your working vocabulary. Write a brief description of the meaning of each term.

Cornice
Fascia
Lookouts
Soffit
Frieze
Rake
Solid wood siding
Horizontal siding
Vertical siding
Drip edge
Corner boards
Gable end
Plywood siding
Hardboard siding
Vinyl siding
Steel siding
Aluminum siding
Wood-shake siding
Stucco
Masonry siding
Louvered ventilators

17 THERMAL INSULATION AND SOUND CONTROL

Thermal insulation is used to reduce the transmission of heat through ceilings, walls, and floors. In the summer it retards exterior heat from entering, thus reducing air-conditioning costs. In the winter it helps hold the furnace heat inside the building, thus reducing heating costs. In both cases it enables the building to maintain a rather stable temperature which increases the comfort of the occupants.

INSULATING VALUES

Some materials have greater ability to retard the transmission of heat than others. They are good insulators. For example, wood is a much more effective insulator than is brick or metal.

The effectiveness of material to retard heat is called their *R-value.* The *R*-value is a measure of the resistance the material has to the flow of heat. The higher the *R*-value, the greater the resistance to the flow of heat or the better it is as an insulator.

The *R*-values for selected construction materials are listed in Fig. 17-1. *R*-values for insulation are given in Fig. 17-8.

The *R*-value is the total resistance value of a particular material. It is calculated from the *k-value.* The *k*-value represents the heat loss. It is defined as the amount of heat that will pass in 1 hr through 1 square foot of material 1 inch thick per 1 degree Fahrenheit of temperature difference between the two faces of the material. This is measured in *British thermal units* (Btu). The lower the *k*-value, the better are the insulating qualities. A British thermal unit is the amount of heat needed to raise the temperature of 1 pound of water 1 degree Fahrenheit.

STRUCTURAL AND FINISH MATERIALS		*R*-VALUE
Wood bevel siding, ½ × 8 in., lapped		R 0.81
Wood siding shingles, 16 in., 7½ in. exposure		R 0.87
Asbestos-cement shingles		R 0.03
Stucco, per inch		R 0.20
Building paper		R 0.06
½-in. nail-base insul. board sheathing		R 1.14
½-in. insul. board sheathing, regular density		R 1.32
25/32-in. insul. board sheathing, regular density		R 2.04
¼-in. plywood		R 0.31
⅜-in. plywood		R 0.47
½-in. plywood		R 0.62
⅝-in. plywood		R 0.78
¼-in. hardboard		R 0.18
Softwood, per inch		R 1.25
Softwood board, ¾ in. thick.		R 0.94
Concrete blocks, three oval cores		
Cinder aggregate, 4 in. thick		R 1.11
Cinder aggregate, 12 in. thick		R 1.89
Cinder aggregate, 8 in. thick		R 1.72
Sand and gravel aggregate, 8 in. thick		R 1.11
Lightweight aggregate (expanded clay, shale, slag, pumice, etc.), 8 in. thick		R 2.00
Concrete blocks, two rectangular cores		
Sand and gravel aggregate, 8 in. thick		R 1.04
Lightweight aggregate, 8 in. thick		R 2.18
Common brick, per inch		R 0.20
Face brick, per inch		R 0.11
Sand-and-gravel concrete, per inch		R 0.08
Sand-and-gravel concrete, 8 in. thick		R 0.64
½-in. gypsumboard		R 0.45
⅝-in. gypsumboard		R 0.56
½-in. lightweight-aggregate gypsum plaster		R 0.32
25/32-in. hardwood finish flooring		R 0.68
Asphalt, linoleum, vinyl, or rubber floor tile		R 0.05
Carpet and fibrous pad		R 2.08
Carpet and foam rubber pad		R 1.23
Asphalt roof shingles		R 0.44
Wood roof shingles		R 0.94
⅜-in. built-up roof		R 0.33
Glass		
Single glass (winter)	U = 1.13	
Single glass (summer)	U = 1.06	
Insulating glass (double)		
¼ in. air space (winter)	U = 0.65	
¼ in. air space (summer)	U = 0.61	
½ in. air space (winter)	U = 0.58	
½ in. air space (summer)	U = 0.56	
Storm windows		
1–4 in. air space (winter)	U = 0.56	
1–4 in. air space (summer)	U = 0.54	

FIGURE 17-1

R-values of construction materials.

The *R*-value is equal to $1/k$. For example, the *k*-value of brick is 5. The *R*-value is 1/5 or 0.2 per inch of brick. A 3-in.-thick brick has an *R*-value of 0.6.

The combined value of all the materials forming a part of the building is given in terms of *U*. *U* is the *overall coefficient of transmission.* It includes the materials in the building section, air spaces, and surface air films. Surface air films provide a small degree of heat retardation. See the *R*-values in Fig. 17-1. The *U* is the reciprocal of *R*. $U = 1/R$. If the *R*-value of a wall is 13.57, the *U* of the wall is 1/13.57 of 0.07. The smaller the *U*, the higher the insulation value (Fig. 17-2).

	R-VALUE
Outside film surface	0.17
Wood siding shingles	0.87
⅜-in. plywood sheathing	0.47
4-in. glass fiber insulation	11.00
½-in. gypsum wallboard	0.45
Inside air film	0.61
R-value	13.57

$$U = \frac{1}{R} = \frac{1}{13.57} = 0.07$$

FIGURE 17-2

How to calculate the *U*-value of a wall.

TYPES OF INSULATION

Several types of insulation are currently in use. *Batt-type insulation* is usually glass fiber or rock wool. They are available in 15- and 23-in. widths. They are available in thicknesses from 2 to 6 in. and in lengths of 4 and 8 ft. They are available with and without a vapor barrier (Fig. 17-3). *Insulation blankets* are the same as described for batts and are available in lengths of 40 to 60 ft (Fig. 17-4).

Rigid board insulation is available in extruded polystyrene, urethane, expanded polystyrene, and glass fiber. Polystyrene and urethane boards serve as vapor barriers. They must be covered with ½-in. gypsum wallboard to meet

FIGURE 17-3

Stapling batt-type insulation between ceiling joists. The vapor barrier faces the heated room. *(Courtesy of Owens Corning Fiberglas)*

FIGURE 17-4

Stapling insulation blankets to a stud wall. *(Courtesy of Bostitch Textron)*

FIGURE 17-5

Rigid board insulation can be applied over the foundation. *(Courtesy of Dow Chemical U.S.A.)*

fire safety requirements. The sheets are available from ¾ to 4 in. thick, 24 to 48 in. wide, and generally 8 ft long (Fig. 17-5).

Loose-fill insulation is available in glass fiber, rock wool, cellulosic fiber, vermiculite, and perlite. It is installed by pouring or blowing in wall, ceiling, and floor cavities (Fig. 17-6).

Reflective insulation is usually a metal foil or foil-surfaced material. Its thickness is not as important as the number of layers of reflective material. The spaces between layers of foil should be at least ¾ in. (Fig. 17-7). Gypsum wallboard is available with a single foil layer on the back.

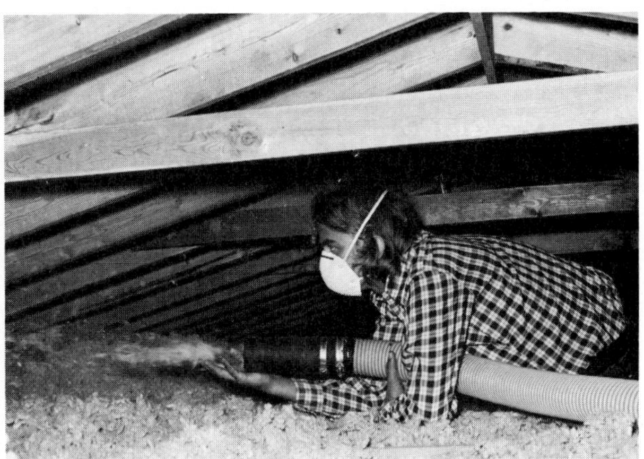

FIGURE 17-6

Loose-fiber-type insulation being blown into an attic. Notice that the worker is wearing a respirator. *(Courtesy of 3M Co.)*

FIGURE 17-7

Reflective insulation is installed with staples. *(Courtesy of Bostitch Textron)*

MATERIAL	THICKNESS (in.)		R-VALUE
Rock wool and cellulose	4	Blown	R 11
	6½	Blown	R 19
	13	Blown	R 38
Extruded polystyrene sheets	1		R 5
	1½		R 7.5
Glass fiber	2	Batt	R 7
	4	Batt	R 11
	6	Batt	R 19
	6	Blown	R 13
	12	Blown	R 26

FIGURE 17-8

Typical *R*-values for selected types of insulation.

Insulating board is usually used as sheathing on exterior walls. It is available in ½- and 25/32-in. thicknesses.

The *R*-values for various types of insulation are given in Fig. 17-8.

VAPOR BARRIERS

A *vapor barrier* is a material through which moisture will not pass. It is applied on walls, ceilings, and floors facing the heated area (Fig. 17-9). Moisture is generated inside of the house from activities such as cooking and laundry. If this is permitted to pass through the wall to the cold exterior sheathing, it will condense. This will wet the insulation, wood framing members, sheathing, and wood siding. Paint will peel off wood siding if there is no vapor barrier.

If it is necessary to cut a hole in a vapor barrier for a pipe or wiring, be certain to carefully fit the vapor barrier closely around it. Materials forming effective vapor barriers include asphalt-laminated paper, aluminum foil, and plastic films. Most blanket and batt-type insulation has a vapor barrier on one side.

FIGURE 17-9

Applying a polyethylene vapor barrier over insulation blankets in a frame wall. *(Courtesy of Owens Corning Fiberglas)*

WHEN TO INSULATE

After the building is framed, the carpenters begin to finish the exterior. At this time the plumber, electrician, and heating and air-conditioning contractors move in to do their part. The electrician can run all the circuits and install the boxes for switches, lights, and duplex outlets (Fig. 17-10). The finish fixtures are not installed at this time. The plumber can put in all the pipes that carry water to the fixtures. All sewer lines and vents can be installed. The fixtures are not installed until later. The furnace, air conditioner, and ducts are installed. After this is finished, the building is ready to have the insulation installed (Fig. 17-11).

WHERE TO INSULATE

All heated areas should be surrounded by insulation (Fig. 17-12). This includes the walls, ceil-

FIGURE 17-10

The electricians install the wiring before insulation is installed. *(Courtesy of International Brotherhood of Electrical Workers)*

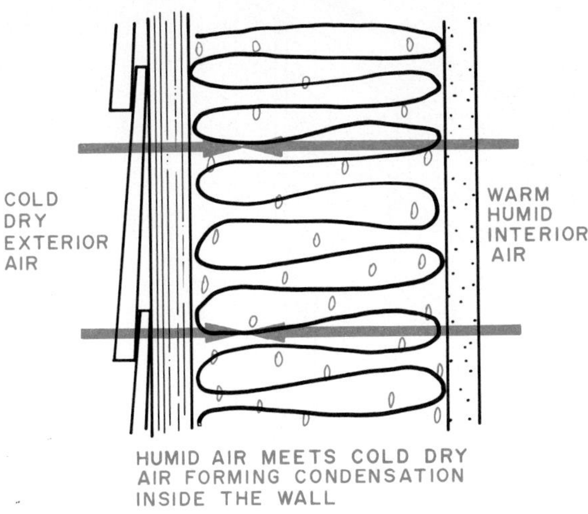

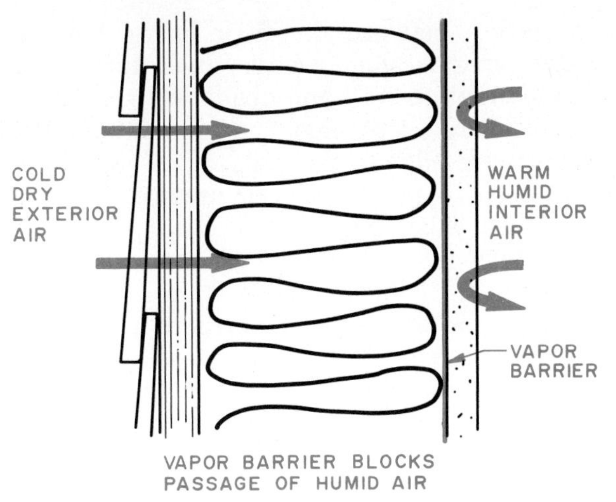

FIGURE 17-11

The vapor barrier blocks the passage of interior moisture through the wall.

ing, and floor. If the space below a room is heated, insulation is not required in the floor. It can be placed there, however, as a sound barrier. It should not have a vapor barrier. If an attic is not heated, place insulation in the attic floor but not between the rafters. Ventilate the attic space so that moisture does not collect.

Ventilated crawl spaces should have the

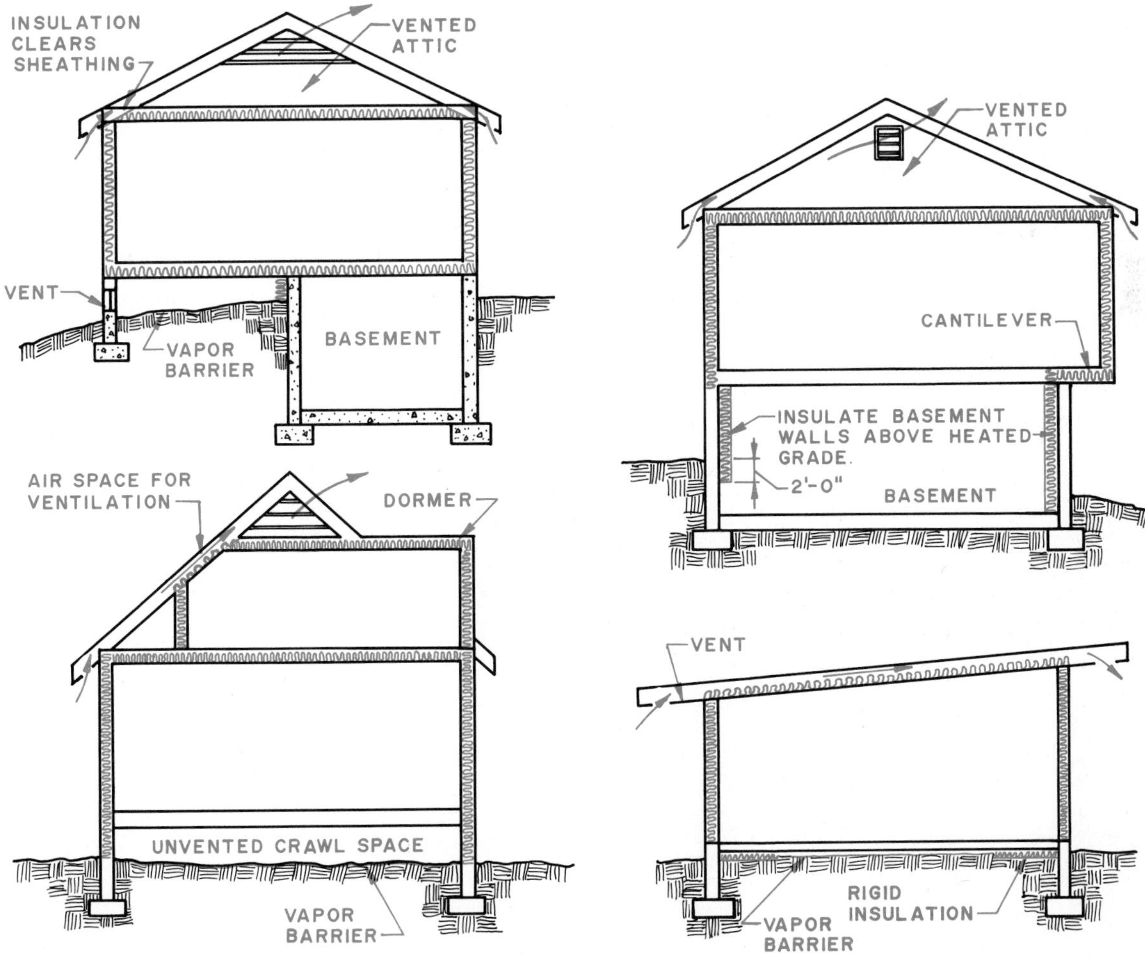

FIGURE 17-12

Recommended places to install insulation.

insulation between the floor joists. Unvented crawl spaces should have insulation around the foundation wall and header (Fig. 17-12).

Concrete slab floors are insulated around the outside edge of the building. A rigid insulation such as polystyrene or urethane in a 2-in. thickness is commonly used (Fig. 17-12).

Basement walls above grade should be insulated to 2 ft below grade. This is not recommended for the extreme northern parts of the United States. The extreme frost penetration may cause heaving of the foundation.

Insulation should be placed in all openings around doors and windows (Fig. 17-13). It should not be packed in too solidly. The insulation value comes from the air spaces that exist between the fibers. A vapor barrier is installed over the insulation.

Windows should have double-pane insulating glass or storm windows. This is vital to a total insulation job. Weather stripping at all openings is required.

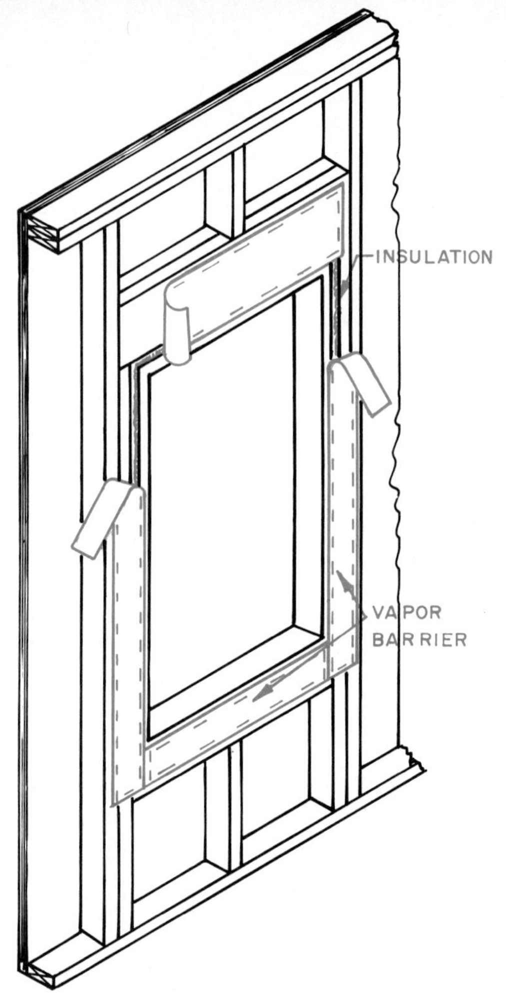

FIGURE 17-13

All openings around doors and windows should be filled with insulation and covered with a vapor barrier.

HOW MUCH INSULATION

The amount of insulation depends upon such things as the cost of fuel, climate, desired comfort, and type of construction. Other factors are the orientation of the building to use the sun for heating, weatherstripping, and thermal properties of doors and windows.

One recommendation for insulation is shown in Fig. 17-14 (map). Notice that even in the warm climates a ceiling should have an *R*-26 rating to handle air-conditioning requirements. In the most northern states the following *R*-values are recommended: ceiling, 38; walls, 19; and floor, 22. In the southern states the recommendations are ceiling, 26; walls, 13; and floor, 11.

Following are examples of typical construction that will satisfy the central band of states.

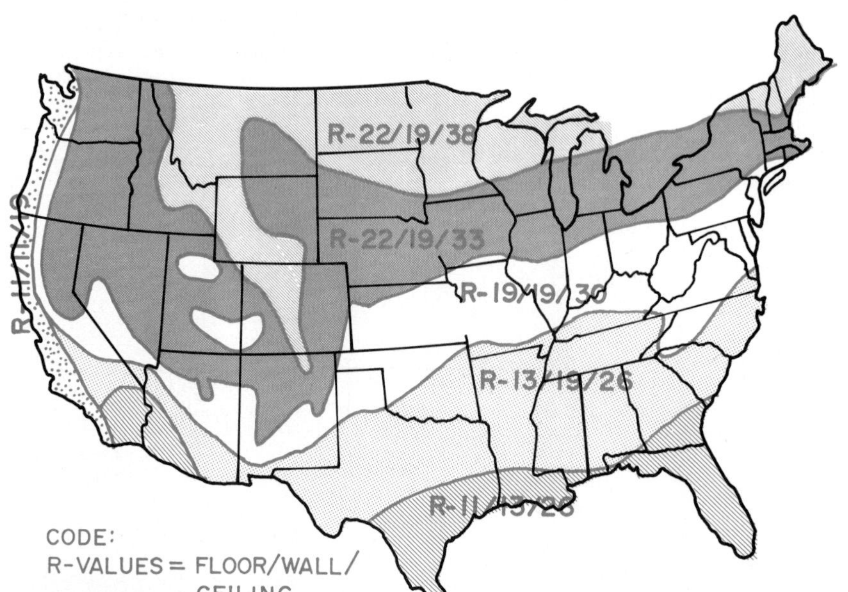

FIGURE 17-14

Recommended *R*-values for various parts of the United States.

CEILING		EXTERIOR WALL		FLOOR	
Surface air film	0.61	Surface air film	0.61	⅝-in. plywood subfloor	0.78
6-in. glass fiber batt	19.0	½-in. insulation board sheathing	1.14	¼-in. hardboard	0.18
4-in. glass fiber batt	11.0	Wood bevel siding	0.81	Asphalt floor tile	0.05
½-in. gypsum board	0.45	6-in. glass fiber batt	19.00	6-in. glass fiber batt	19.0
Inside air film	0.61	½-in. gypsum board	0.45	*R*-value	20.01
R-value	31.67	Inside air film	0.61		
		R-value	22.62		

HOW TO INSTALL INSULATION

Blanket and batt-type insulation have tabs that are stapled to the wood framing members. The vapor-barrier side is placed facing the room. This prevents moisture from getting to the insulation. Some prefer to staple the tabs to the face of the studs. The vapor barrier protects them from moisture. Strips of vapor-barrier material could be stapled to the sole and plate as well (Fig. 17-15).

Others prefer to staple the tabs on the inside of the studs (Fig. 17-15). While this does not protect the studs from moisture, it does make it easier to install the interior finish material, such as gypsum board. The blankets are cut long enough so they fit tightly against the plate and sole. Some prefer to cut them long enough to form a tab for stapling to the plate and sole (Fig. 17-15). Staples are placed every 8 in. on each tab.

When installing insulation in walls, be certain to push it into the wall until it touches the sheathing.

Insulation is made in standard widths to fit between studs. Because of windows and doors, some spaces will be smaller than standard. To insulate these, cut the insulation 1½ in. wider than the space. Pull the vapor barrier loose on each edge to form a tab for stapling (Fig. 17-16). Rigid insulation is cut slightly oversize and pressed into place.

Insulation should be cut around electrical boxes and should fit tightly against them. Be certain to fill the space between the box and the sheathing (Fig. 17-17).

Generally, water lines are not placed in exterior walls because they are likely to freeze. If they are, be certain to place insulation between them and the sheathing (Fig. 17-18).

Rigid insulation is also used on walls. In a stud wall it is pressed in between the studs. Since it does not have a vapor barrier, a 2-mil polyethylene film vapor barrier is stapled to the entire wall. It must cover the sole and top plates. Usually, door and window openings are covered and cut out after the finished interior wall material is in place (Fig. 17-19). The wall can also be covered with foil-backed gypsum board. The foil serves as a vapor barrier.

Masonry walls are insulated on the inside surface by fastening furring strips to them. These are usually 1 × 2 or 2 × 2-in. strips. They are spaced 16 or 24 in. on center. When using 1 × 2 in. furring, a 1-in.-thick rigid-type insulation is used. Since it does not have a vapor barrier, a 2-mil polyethylene film or foil-backed gypsum board can be placed over it (Fig. 17-20).

FIGURE 17-15

Insulation can be stapled to the face of the stud or to the inside. Do not compress the fibers.

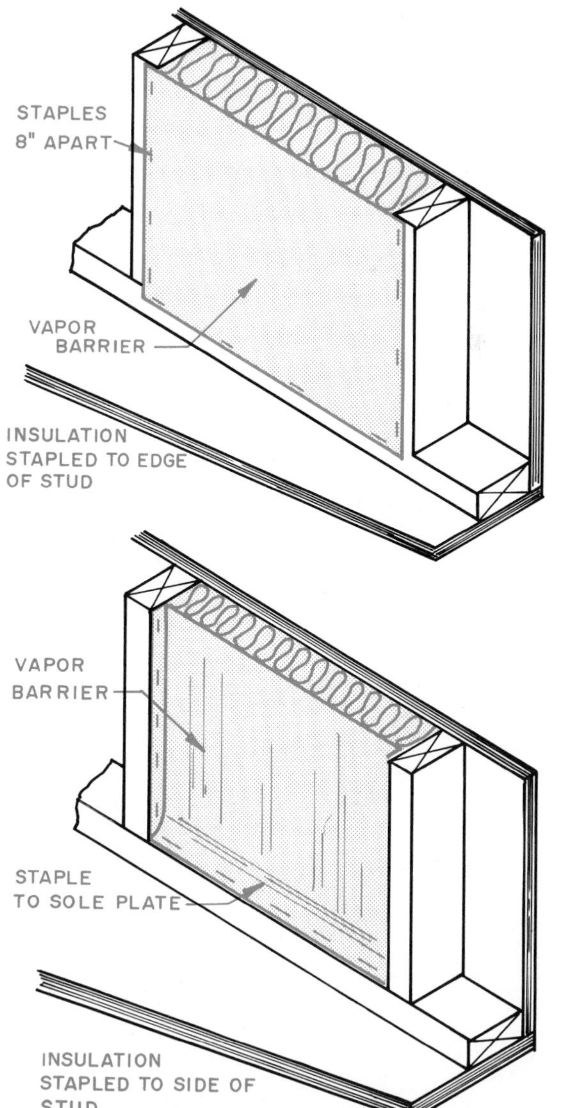

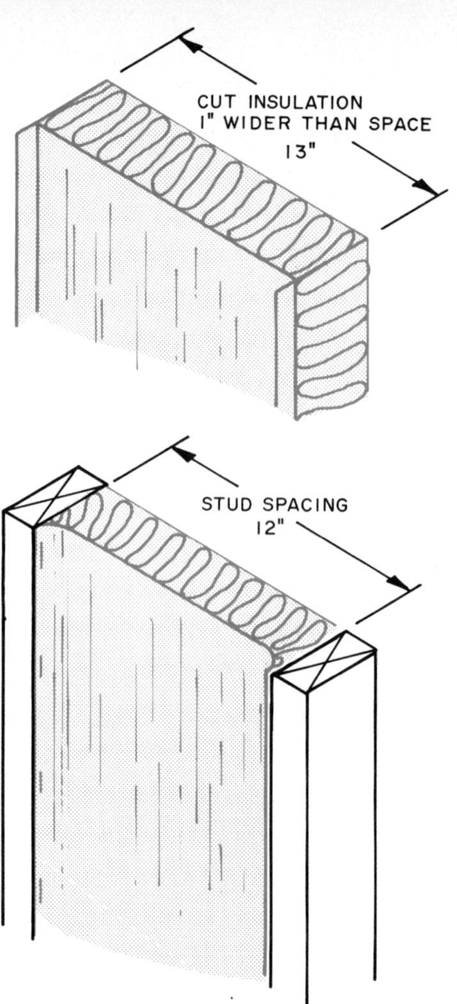

FIGURE 17-16

How to fit insulation between studs spaced closer than standard.

FIGURE 17-17

Fit insulation around electrical boxes and other openings.

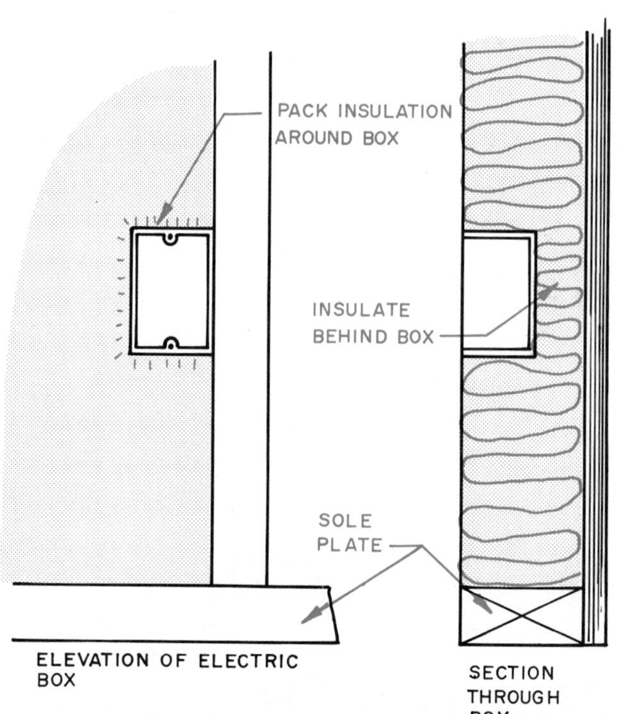

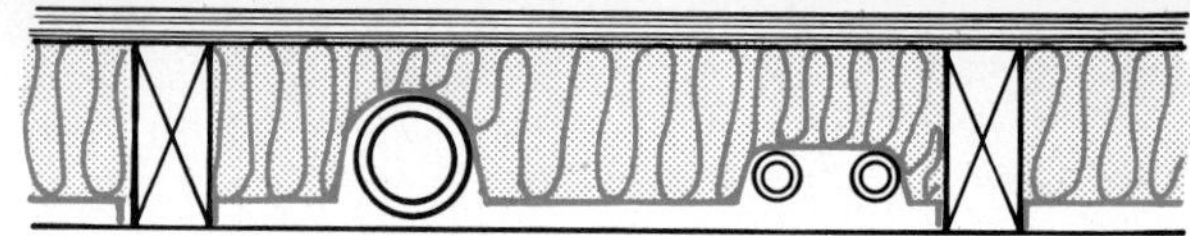

FIGURE 17-18

Place insulation between the sheathing and water and sewer pipes.

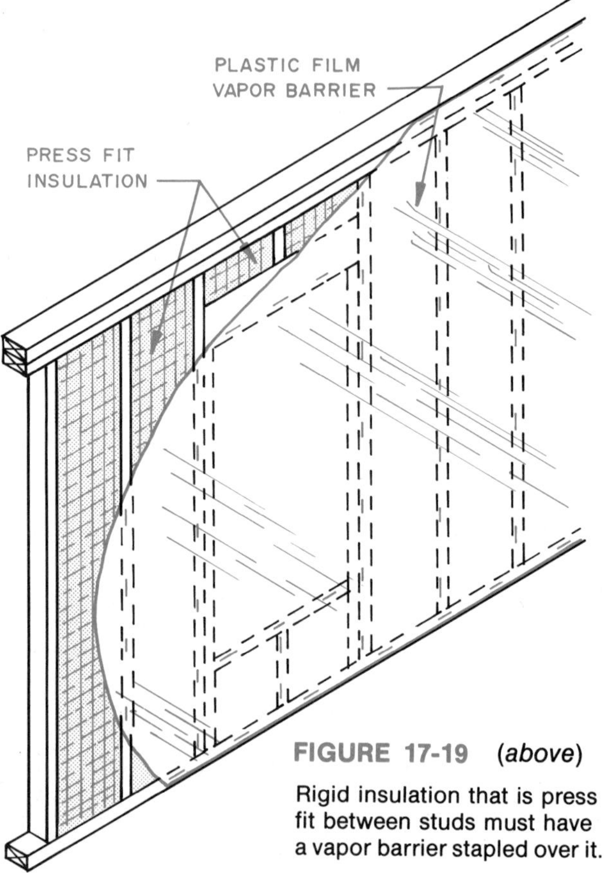

FIGURE 17-19 (*above*)

Rigid insulation that is press fit between studs must have a vapor barrier stapled over it.

FIGURE 17-20 (*below*)

Masonry walls can be insulated with rigid insulation panels. Furring strips are used for nailing paneling or gypsum board finished wall.

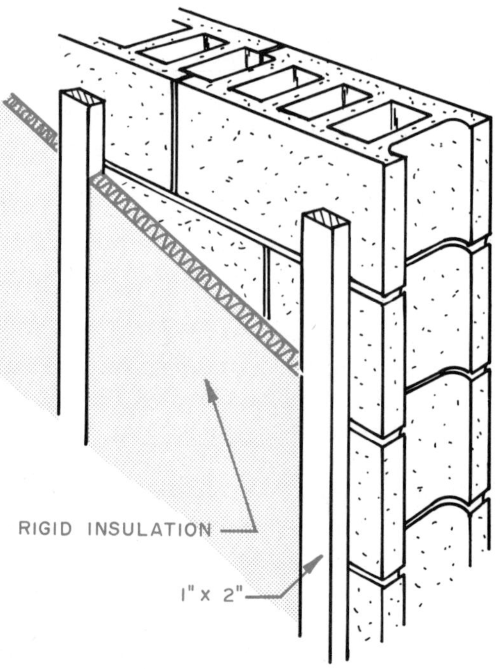

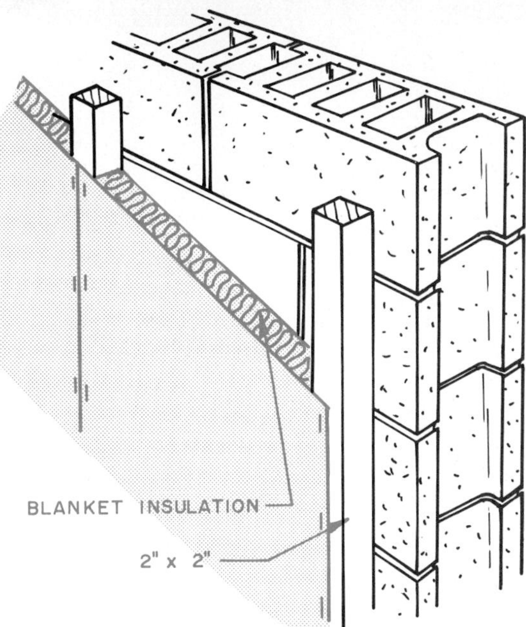

FIGURE 17-21

Batts or blankets can be used to insulate masonry walls.

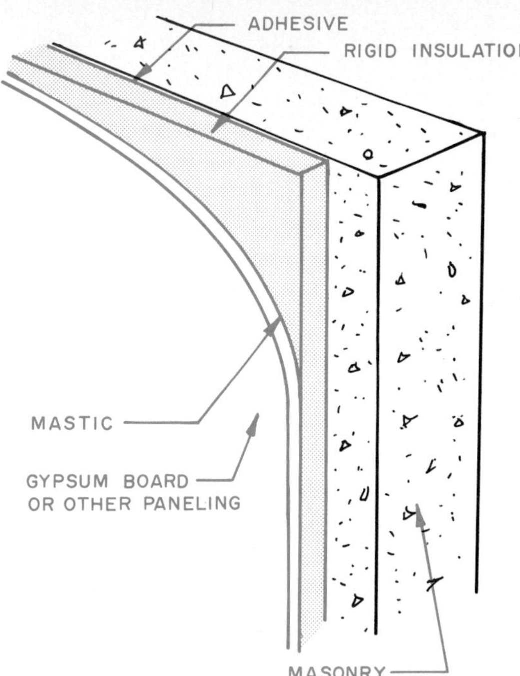

FIGURE 17-22

Rigid insulation can be glued to the masonry wall and finish wall material is glued to the insulation.

The 2 × 2-in. furring can use 2-in. batts or blankets which are stapled to the furring (Fig. 17-21). The tabs are stapled to the face of the furring strip. Rigid insulation is fastened to masonry with an adhesive. The finish interior wall is then glued to the insulation (Fig. 17-22).

Basement walls can be insulated by gluing rigid insulation to them or by framing a 2 × 4 wall and installing blankets (Fig. 17-23). Run the insulation 2 ft below the ground line (Fig. 17-24). Be certain to insulate the header (Fig. 17-25).

Basement walls can also be insulated by gluing rigid insulation to the exterior of the wall. The part that is above the surface of the earth must be protected against ultraviolet light and physical damage. Common protection materials include a layer of cement—asbestos board, pressure-treated wood, stucco, rigid vinyl sheet, portland cement coatings with a liquid latex addition, or vinyl-acrylic coatings (Fig. 17-26).

FIGURE 17-23

Basement walls are often insulated by building a 2 × 4 wall inside and insulating it in the same manner as an exterior stud wall.

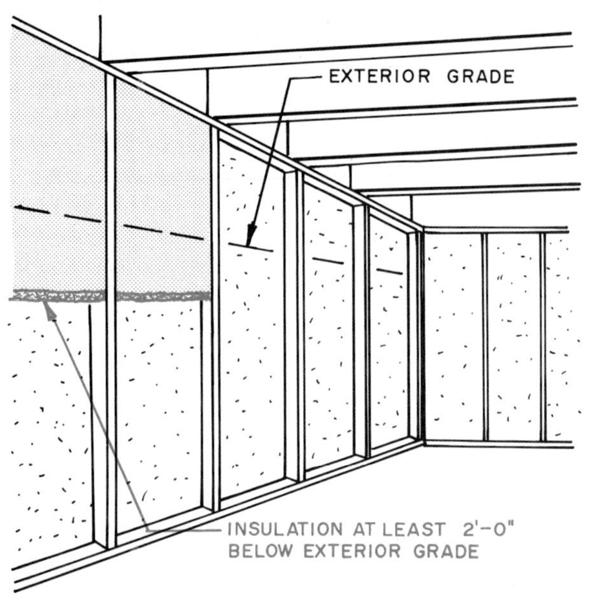

FIGURE 17-24

Basement insulation should extend at least 2 ft below the exterior grade.

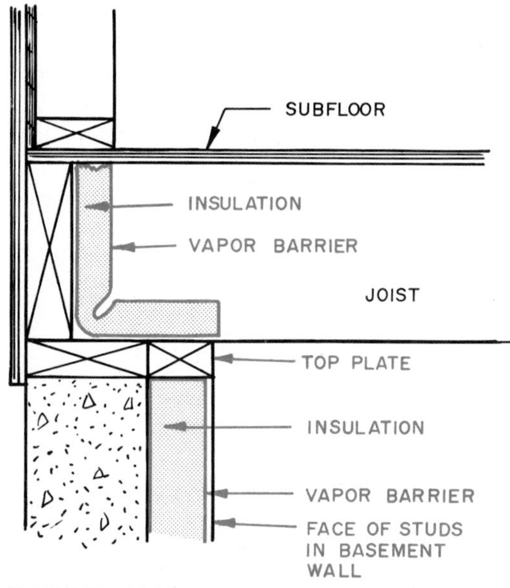

FIGURE 17-25

Headers must be insulated.

Blanket insulation may be installed in the ceiling by stapling to the joists, placing pressure fit blankets between studs, or laying the blankets in place from above after the finish ceiling is in place (Figs. 17-27 and 17-28).

When installing by stapling, keep the insulation as close to the plate as possible. Place the vapor barrier facing the room. If a gap exists at the plate, fill it with loose insulation (Fig. 17-29).

If the building uses soffit vents, the insulation should not block the flow of air into the attic (Fig. 17-30). This type of installation is

FIGURE 17-26

Foundations can be insulated with rigid insulation on the outside. This material is being given a protective coating of portland cement having a liquid latex addition. *(Courtesy of Dow Chemical U.S.A.)*

FIGURE 17-27

Blanket insulation can be installed in the ceiling before the finish ceiling is nailed to the joists. *(Courtesy of Owens Corning Fiberglas)*

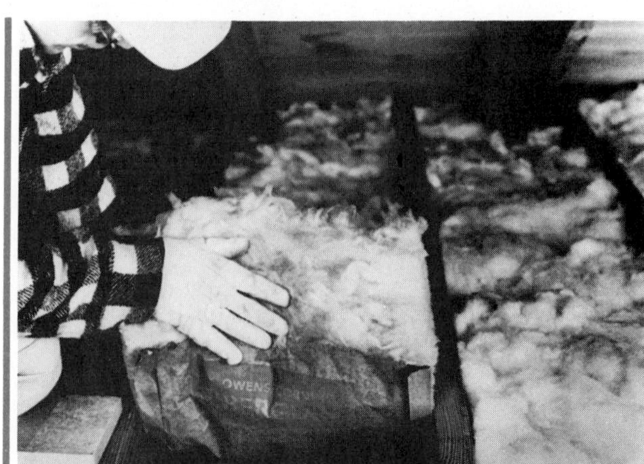

FIGURE 17-28

Blanket insulation can be installed in the ceiling after the finish ceiling is in place. *(Courtesy of Owens Corning Fiberglas)*

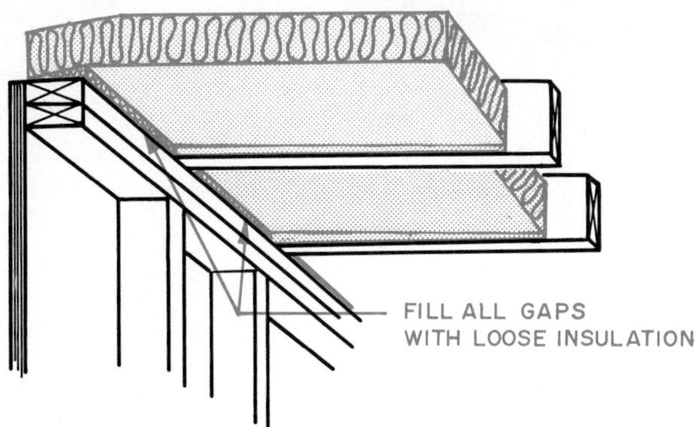

FIGURE 17-29

Fill any gaps at the plate with loose insulation.

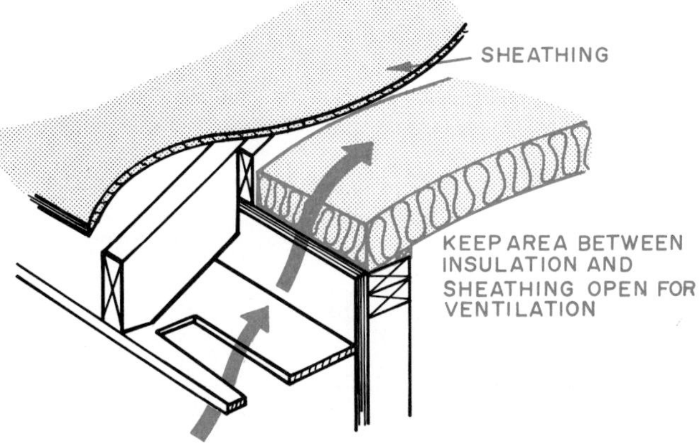

FIGURE 17-30

If soffit vents are used, do not block the flow of air with insulation.

done from below before the finish ceiling is in place.

Pressure-fit blankets are also installed from below. While they should overlap the top plate, they should not block attic ventilation. This type of insulation does not have a vapor barrier. If the attic is properly ventilated, no vapor barrier is needed (Fig. 17-31). Requirements call for the total net free ventilating area to the area of the ceiling to be 1/300 or more. Half of the ventilation area must be provided by ventilators located in the upper half of the attic space. The balance of ventilation can come from eave vents.

The blankets can be laid on top of the finish ceiling. It is placed with the vapor barrier toward the room. It is not stapled.

Insulation should be cut to go around pipes and wiring in the ceiling. It should be carefully placed on all sides so that no voids are created.

Insulation should be placed 3 in. away from and not on top of recessed light fixtures. The exposed fixture is needed so that the excess heat from the bulbs can escape. Stuff loose noncombustible insulation with no vapor barrier between joists and chimney (Fig. 17-32).

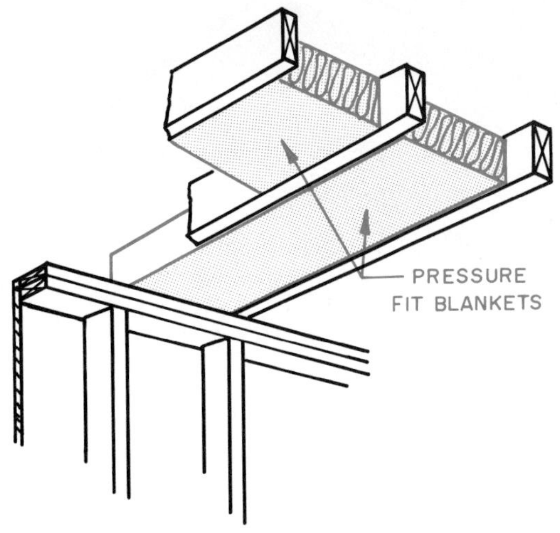

FIGURE 17-31

Pressure-fit insulation blankets can be installed between ceiling joists before the ceiling material is nailed in place.

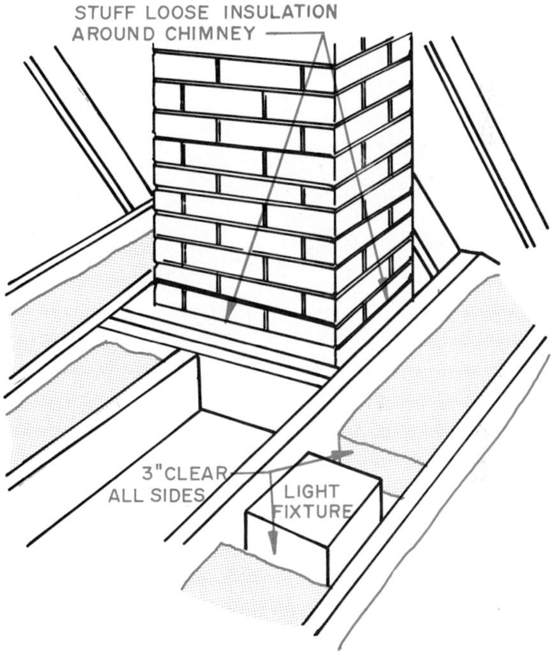

FIGURE 17-32

Fill gaps around the chimney with loose insulation. Keep insulation clear of all recessed light fixtures.

Vented crawl spaces are insulated by placing blankets between the floor joists (Fig. 17-33). The vapor barrier should face the floor. Usually, the insulation is held in place with chicken wire or bow wire (Fig. 17-34). It is recommended that the soil under the building be covered with a vapor barrier. This reduces the moisture from the ground.

Unvented crawl spaces have the foundation wall insulated. First place a vapor barrier on the ground below the building. Join it to the foundation with tape or adhesive. The blanket

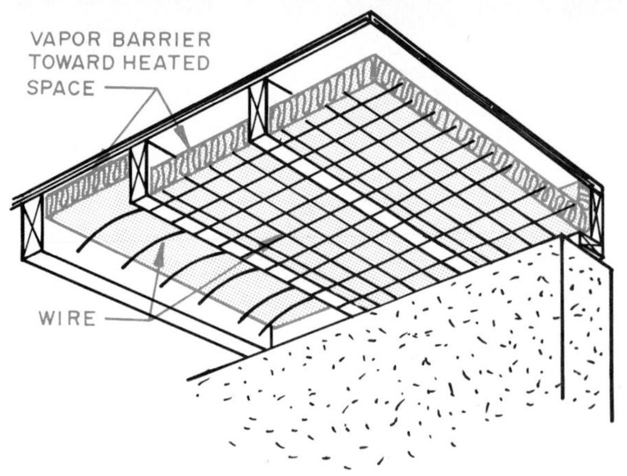

FIGURE 17-33

Vented crawl spaces are insulated with blankets between the joists.

FIGURE 17-34

Insulation blankets below the floor are held in place with chicken wire. *(Courtesy of Owens Corning Fiberglas)*

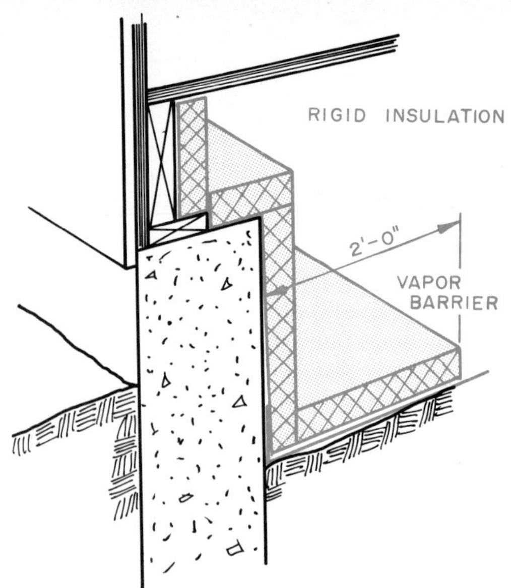

FIGURE 17-36

Rigid insulation can be used to insulate the walls of unvented crawl spaces.

insulation is placed against the header and must touch the subfloor. It then drapes over the foundation to the ground and extends 2 ft over the ground. It is easiest to install before the subfloor is nailed into place (Fig. 17-35). Rigid insulation is also excellent for this purpose. It can be fastened to the wall with an adhesive (Fig. 17-36).

The vapor barrier on unvented crawl spaces should face the crawl space. It will be the warm side.

The header area on two-story houses must be insulated (Fig. 17-37). The pieces are stapled

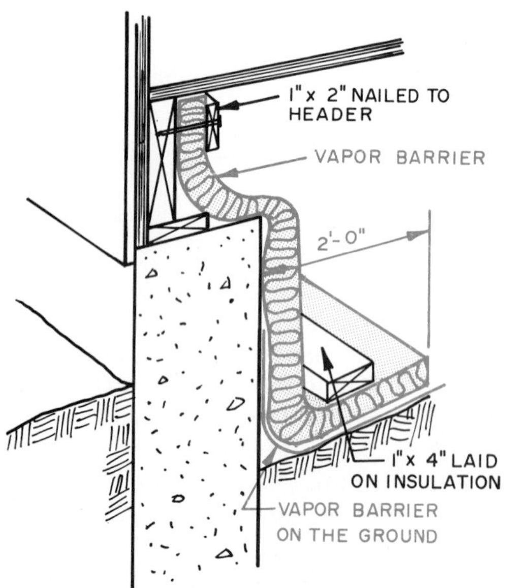

FIGURE 17-35

Unvented crawl spaces should have the walls insulated.

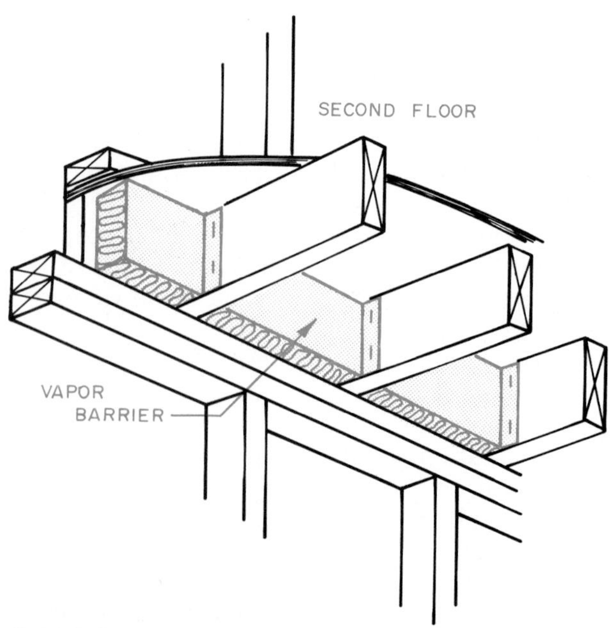

FIGURE 17-37

The header area between levels of two-story houses must be insulated.

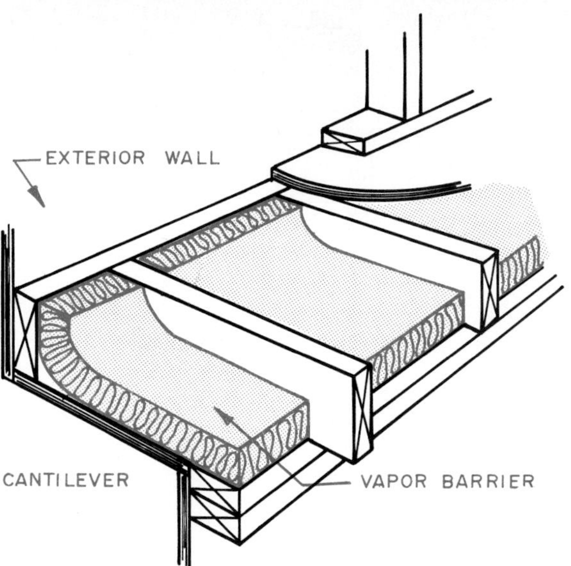

FIGURE 17-38

Cantilevered areas must be fully insulated.

to the floor joists. If the floor cantilevers over the exterior wall, the header and floor area should be insulated (Fig. 17-38). This insulation can be wedged in place without stapling (Fig. 17-39).

Concrete slab construction requires insulation around the exterior edge of the floor. This is rigid insulation that is placed before the floor is poured. Some typical construction details are shown in Fig. 17-40. A polyethylene vapor barrier is placed over the gravel fill under the slab. It should lap under the rigid insulation.

If a flat or shed roof is to be insulated, space must be left between the insulation and the sheathing. This space is needed for ventilation. Ventilation will keep the space from sweating, thus damaging the insulation (Fig. 17-41).

Rigid insulation can be installed on top of the sheathing. A 1 × 2-in. wood nailing strip is laid around the edge of the roof. Then a second layer of ½-in. plywood sheathing is nailed over the insulation to serve as a nailing base for the shingles. This is especially valuable when an exposed beam ceiling is used (Fig. 17-42).

Additional roof insulation for flat or shed roofs can be had by nailing rigid insulation over the vapor barrier on the inside of the roof. The finished ceiling material is nailed to the rafters through the insulation (Fig. 17-43).

VENTILATION

Attic spaces and the area under flat roofs must be *ventilated* to prevent condensation of moisture. Asphalt shingle and built-up roofs do not permit moisture to pass through them to the outside. Wood shingles and wood shakes do permit some moisture to pass.

The attic should be as cold as the exterior air. A warm attic will permit snow to melt and run to the eaves. Here it is cold and it freezes, forming an ice dam. This causes water to build up at the eaves and sometimes seep into the house. If the ceiling is properly insulated, the inside heat will not flow into the attic. If the attic is properly ventilated, the snow on the roof will not melt.

In the summer attic ventilation is needed to lower the air temperature. The ceiling insulation retards the flow of summer heat into the house.

FIGURE 17-39

Insulating a floor and the header over the foundation. *(Courtesy of Owens Corning Fiberglas)*

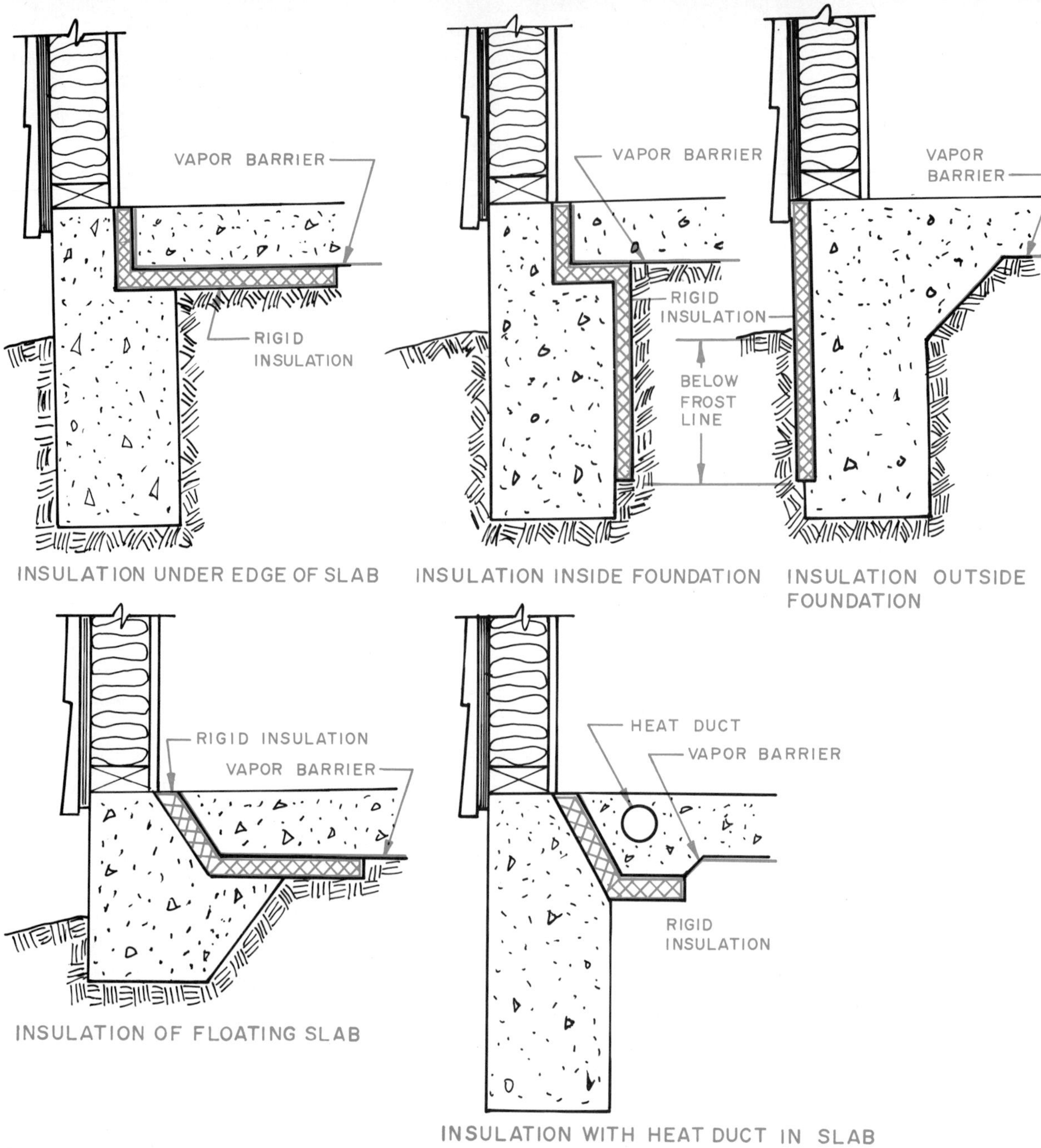

VAPOR BARRIER
RIGID INSULATION
INSULATION UNDER EDGE OF SLAB
VAPOR BARRIER
RIGID INSULATION
BELOW FROST LINE
INSULATION INSIDE FOUNDATION
VAPOR BARRIER
INSULATION OUTSIDE FOUNDATION
RIGID INSULATION
VAPOR BARRIER
INSULATION OF FLOATING SLAB
HEAT DUCT
VAPOR BARRIER
RIGID INSULATION
INSULATION WITH HEAT DUCT IN SLAB

FIGURE 17-40

How to place insulation when building concrete slab floors.

VENTILATE SPACE ABOVE INSULATION

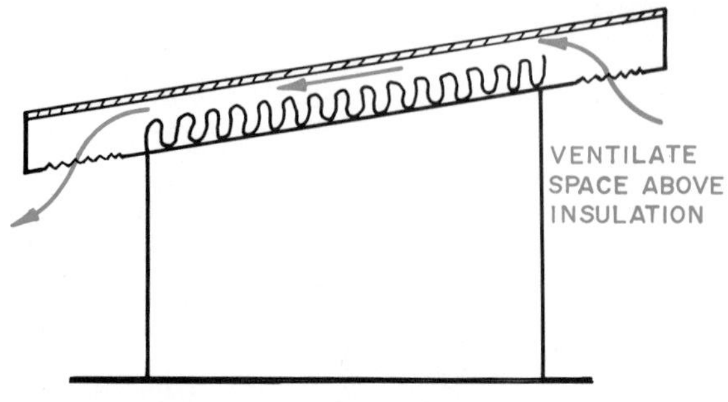

FIGURE 17-41

Shed and flat roofs must be ventilated to prevent moisture from condensing beneath the roof.

Gable-roofed houses usually have louvers in the end walls. These are especially effective if the prevailing winds strike the building in such a way as to move through the attic (Fig. 17-44). Hip-roof houses usually use inlet ventilators in the eaves and outlet ventilators at the ridge (Fig. 17-45). There are other types of ventilators available. One is placed near the ridge. It has a rotating fan-like top. The wind causes it to rotate, thus creating a draft through the attic. Another type is an electric fan. It is operated with a thermostat. When the attic temperature reaches a certain level, the fan comes on and pulls the air through the attic.

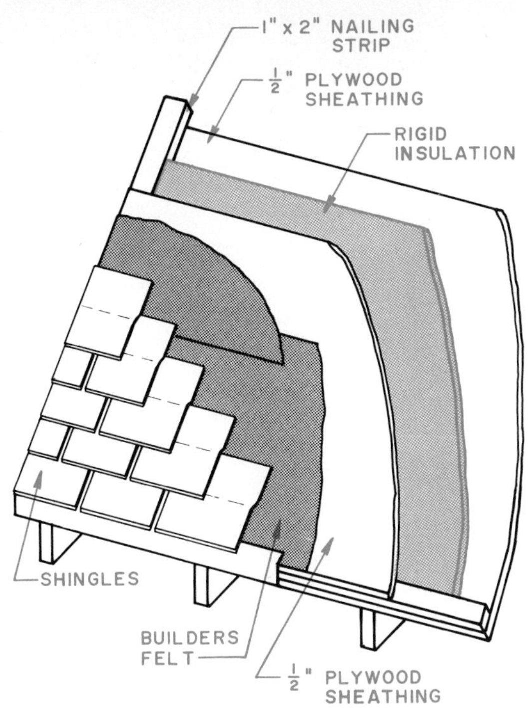

FIGURE 17-42

Open beam ceilings can be insulated by applying rigid insulation to the outside of the sheathing.

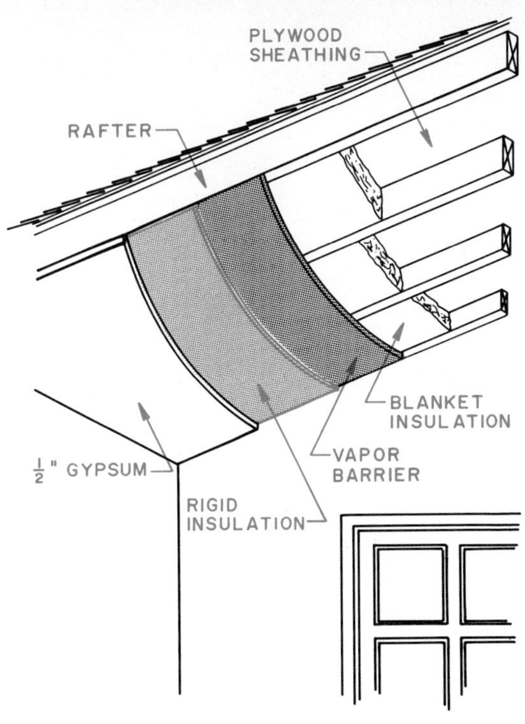

FIGURE 17-43

Flat and shed roofs can have rigid insulation installed on the bottom side of the rafters. *(Courtesy of Dow Chemical U.S.A.)*

FIGURE 17-44

Various ways to ventilate gable roofs and the recommended inlet/outlet ratios.

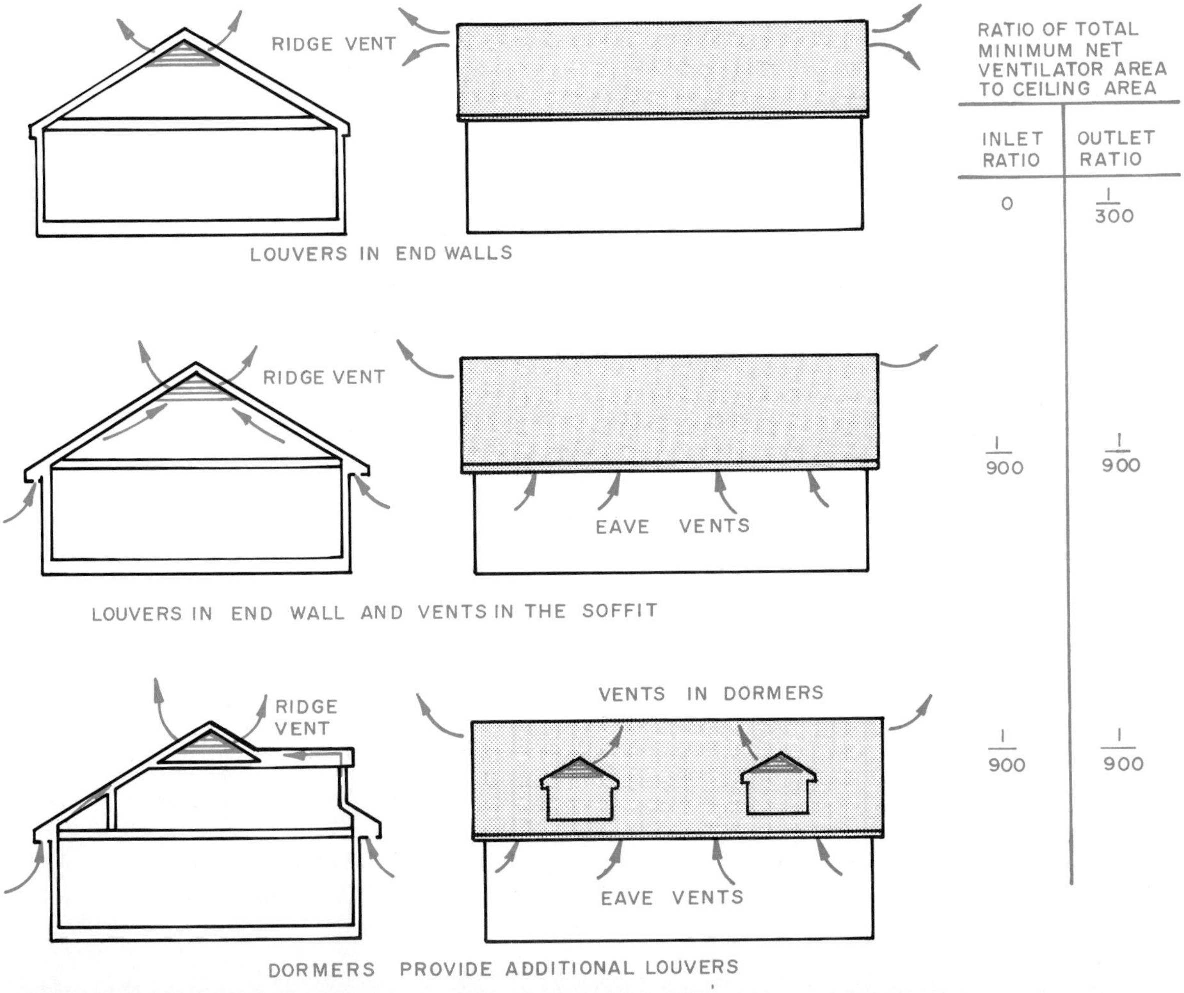

RATIO OF TOTAL MINIMUM NET VENTILATOR AREA TO CEILING AREA	
INLET RATIO	OUTLET RATIO
0	$\frac{1}{300}$
$\frac{1}{900}$	$\frac{1}{900}$
$\frac{1}{900}$	$\frac{1}{900}$

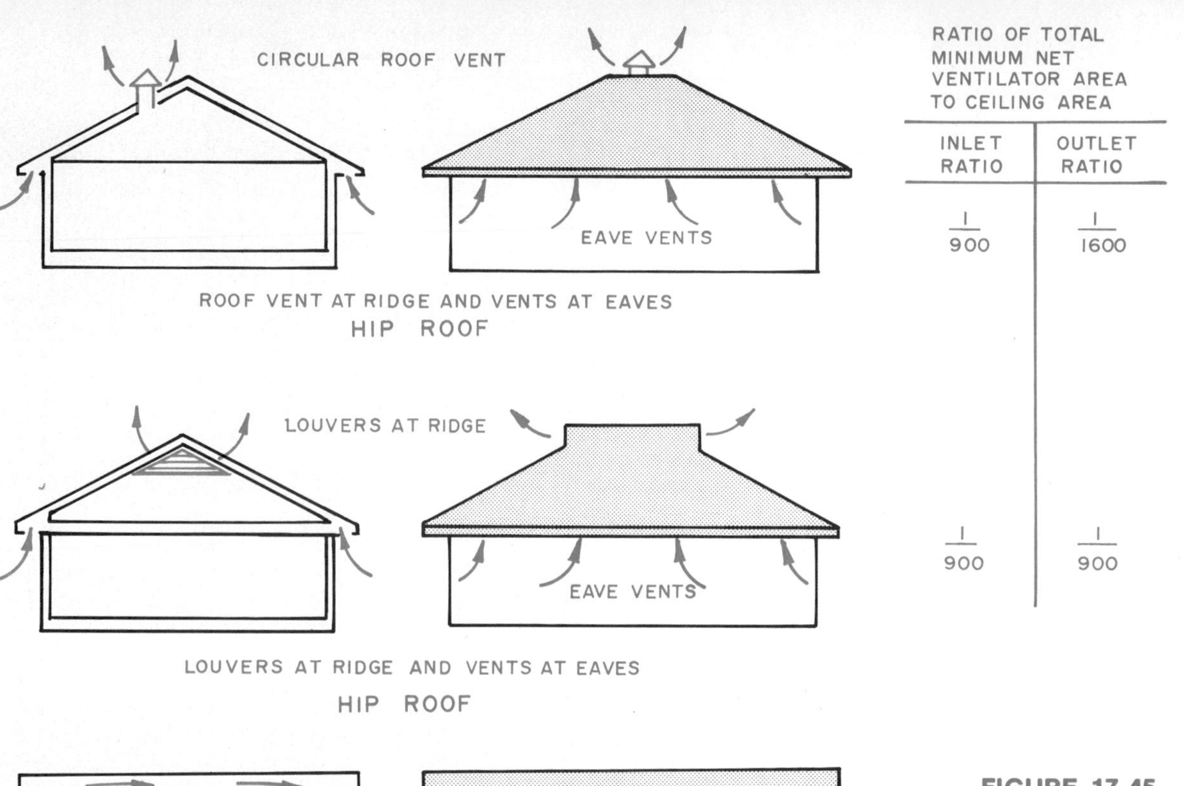

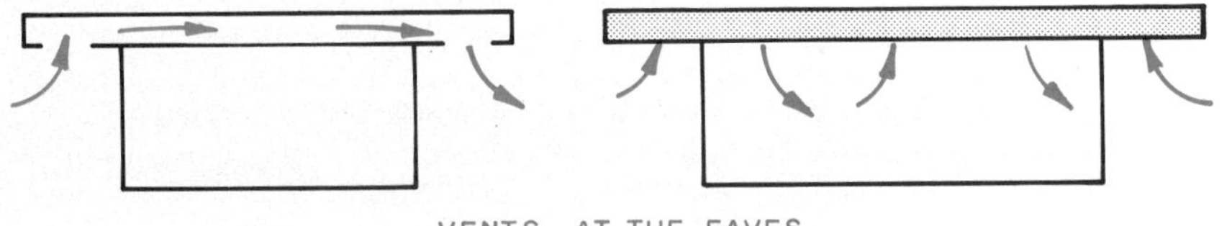

FIGURE 17-45

Ways to ventilate hip, flat, and shed roofs and the recommended inlet/outlet ratios.

Area of Ventilators The amount of ventilation needed is based on the ceiling area of the rooms below the attic. The vents are covered with screen to keep out insects and birds. The screen reduces the area available for ventilation; therefore, use as large a mesh screen as is practical. A very fine screen will clog with lint and dirt.

Gable roofs with louvers in the gable end require a ventilation area of 1/300 of the ceiling area. A building with 1500 square feet of ceiling area should have 5.0 square feet of ventilator space (Fig. 17-44). If vents are placed in the eaves, the vents should have an area of 1/900 and the louvers, 1/900. Since the screen reduces the vent area, they should be enlarged about double to get adequate ventilation.

A hip roof house requires a 1/900 ratio on the eave vents and a 1/600 area at the ridge. If the roof has louvers at the ridge, both inlet and outlet sizes are 1/900 (Fig. 17-45). It is more difficult to get air circulating between the insulation and sheathing in a flat or shed roof. Most flat or shed roofs have the vents in the overhang. They should be distributed evenly around the building. A ratio of 1/250 is recommended. A 1500-square-foot building would require 6 square feet of clear vent opening (Fig. 17-45).

SOUND INSULATION

It is desirable to have the bedroom areas insulated so that noise from the living area is blocked. It is also necessary to insulate the bathrooms.

Sound travels in waves that radiate from the source through the air until they strike a wall, floor, or ceiling. These surfaces are set in vibration by the fluctuating pressure of the sound wave. As the material vibrates, some of the sound is conducted to the other side. The resistance of a material to the passage of airborne sound is its *sound transmission class* (STC). The higher the STC, the better the sound barrier. Surfaces also reflect sound back into a room.

Following shows typical STC sound levels:

STC NUMBER	
25	Normal speech which can be understood easily
35	Loud speech audible but not intelligible
45	You must strain to hear loud speech
50	Loud speech not audible

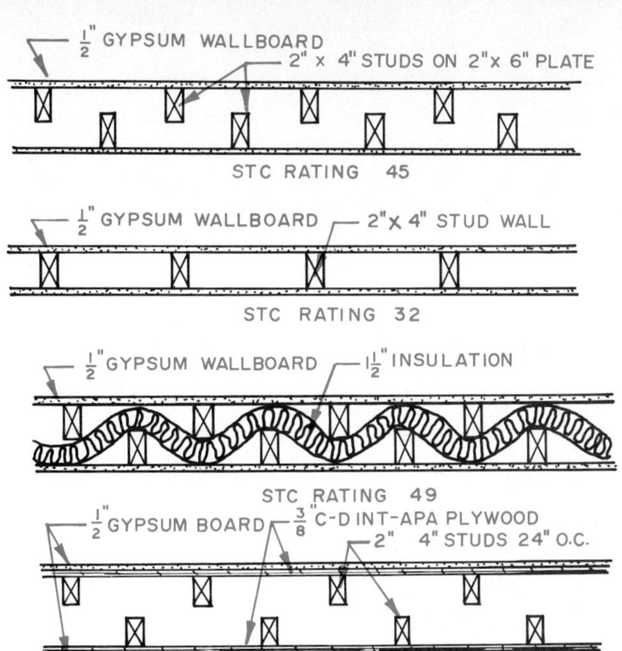

As sound hits a wall, the studs conduct the sound to the next room. An effective way to reduce this is to stagger the studs.

Other wall construction details are in Fig. 17-46.

Sound also passes through floors and ceilings. They must resist airborne noises as well as impact noises such as someone walking. Impact noise is expressed as *impact insulation class* (IIC). The greater the IIC value, the more resistant a material is to impact noise.

Typical floor-ceiling construction details are shown in Fig. 17-47. Notice how the use of

1/2" GYPSUM WALLBOARD
2" x 4" STUDS ON 2" x 6" PLATE
STC RATING 45
1/2" GYPSUM WALLBOARD
2" x 4" STUD WALL
STC RATING 32
1/2" GYPSUM WALLBOARD
1 1/2" INSULATION
STC RATING 49
1/2" GYPSUM BOARD
3/8" C-D INT-APA PLYWOOD
2" 4" STUDS 24" O.C.
STC RATING 46

FIGURE 17-46 (*left*)

Sound transmission class (STC) ratings for selected wall constructions.

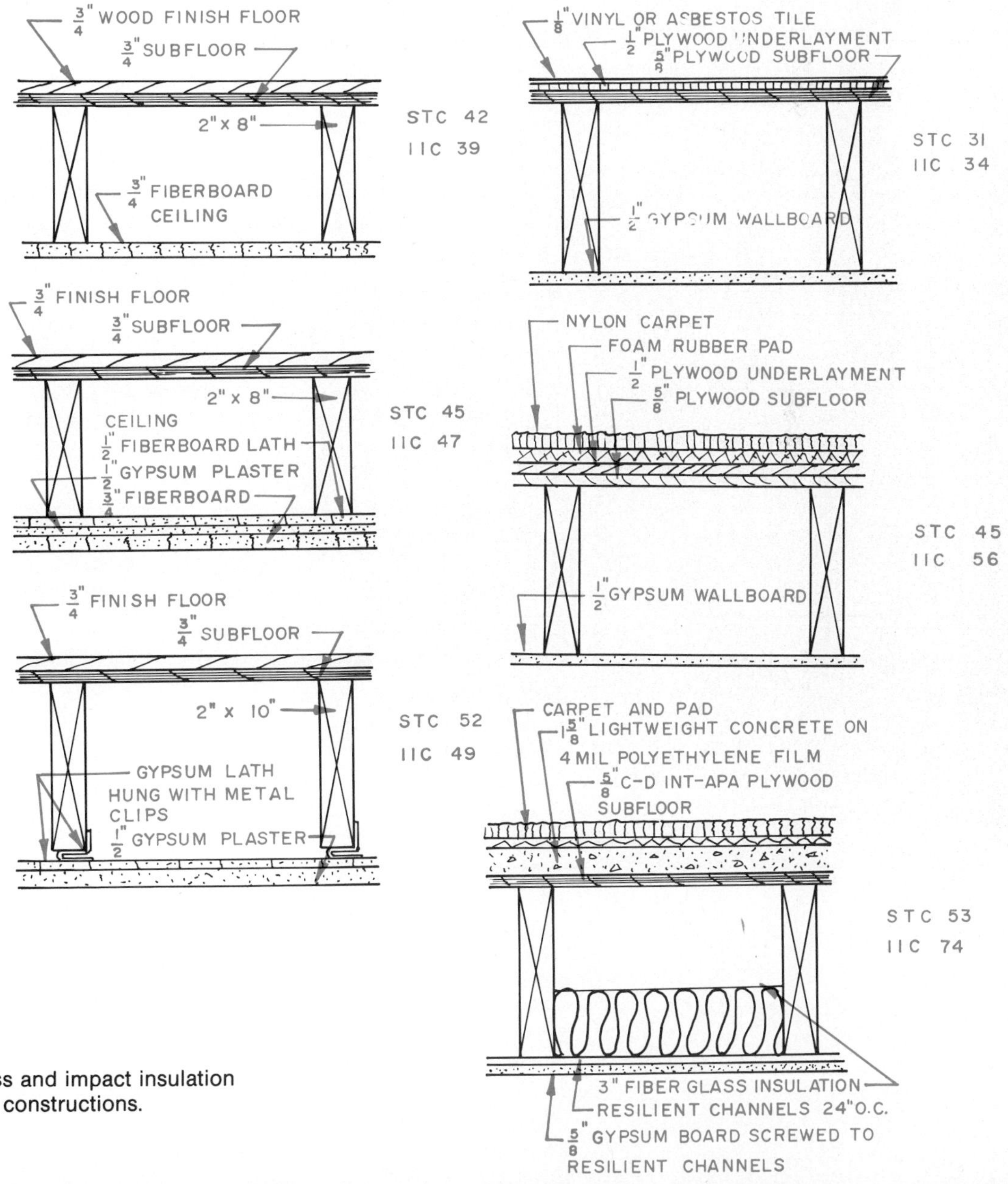

3/4" WOOD FINISH FLOOR
3/4" SUBFLOOR
2" x 8"
3/4" FIBERBOARD CEILING
STC 42
IIC 39
1/8" VINYL OR ASBESTOS TILE
1/2" PLYWOOD UNDERLAYMENT
5/8" PLYWOOD SUBFLOOR
1/2" GYPSUM WALLBOARD
STC 31
IIC 34
3/4" FINISH FLOOR
3/4" SUBFLOOR
2" x 8"
CEILING
1/2" FIBERBOARD LATH
1/2" GYPSUM PLASTER
3/4" FIBERBOARD
STC 45
IIC 47
NYLON CARPET
FOAM RUBBER PAD
1/2" PLYWOOD UNDERLAYMENT
5/8" PLYWOOD SUBFLOOR
1/2" GYPSUM WALLBOARD
STC 45
IIC 56
3/4" FINISH FLOOR
3/4" SUBFLOOR
2" x 10"
GYPSUM LATH HUNG WITH METAL CLIPS
1/2" GYPSUM PLASTER
STC 52
IIC 49
CARPET AND PAD
1 5/8" LIGHTWEIGHT CONCRETE ON 4 MIL POLYETHYLENE FILM
5/8" C-D INT-APA PLYWOOD SUBFLOOR
3" FIBER GLASS INSULATION
RESILIENT CHANNELS 24" O.C.
5/8" GYPSUM BOARD SCREWED TO RESILIENT CHANNELS
STC 53
IIC 74

FIGURE 17-47

The sound transmission class and impact insulation class (IIC) for selected floor constructions.

carpet increases the INR rating of a floor. A dropped ceiling or ceiling hung on metal hangers has a high STC rating.

Sound can also be handled by absorption. *Sound-absorbing materials* minimize the amount of noise by stopping the reflection of sound back into a room. Acoustic tile is an example. The porous material traps much of the sound. Fabric, such as curtains and carpets, are also good sound absorbers.

STUDY QUESTIONS

After reading the chapter, answer each of the following questions. If you do not know the answer, review the chapter.

1. How does insulation keep a house cool in the summer and warm in the winter?
2. What is the *R*-value of a material?
3. What is the *U*-value of a wall?
4. What are the commonly used types of insulation?
5. Why are vapor barriers used?
6. Where should insulation be placed?
7. How can basement walls be insulated?
8. How are vented crawl spaces insulated?
9. How are concrete slab floors insulated?
10. Why are unheated attics ventilated?
11. How do sound waves go through walls?
12. In what terms is impact noise expressed?
13. What are the recommended insulation *R*-values for northern states?

IMPORTANT TECHNICAL TERMS

Following are technical terms that you should be able to use as part of your working vocabulary. Write a brief description of the meaning of each term.

R-value
British thermal units
U-value
Batt insulation
Blanket insulation
Rigid insulation
Loose-fill insulation
Reflective insulation
Vapor barrier
Ventilation
Sound transmission class
Impact insulation class
Sound absorption

18 FINISHING THE INTERIOR

Interior finish includes the application of finish materials to walls and ceilings and installing trim where required. This is done after all electrical, plumbing, and heating systems are installed and the insulation is in place.

PLASTER

Plaster is made from gypsum. It is fire-resistant and provides a wall that will retard the spread of a fire.

The carpenter generally applies the plaster base and grounds. The grounds are wood strips that give the plasterer a guide as the plaster coats are installed.

Plaster-Base Materials

Plaster-base materials include gypsum lath, perforated gypsum lath, insulating fiberboard, and metal lath.

Gypsum lath is nailed or stapled to wood framing members and nailable steel framing. It is available with an aluminum foil laminated to the back, which creates a vapor barrier (Fig. 18-1).

The sizes available are listed in Fig. 18-2. The ⅜-in. thickness is used on studs spaced 16 in. on center. The ½-in. lath has insulating value and is used on walls facing an unheated area.

Installing Lath

Before fastening the lath to studs, the carpenter must be certain they are straight. They can be checked by running a chalk line the length of the wall. Bowed or warped studs must be straightened (Fig. 18-3). This is done by sawing into the hollow side

FIGURE 18-1

Gypsum lath plaster base is nailed to the studs. *(Courtesy of U.S. Gypsum)*

PRODUCT	THICKNESS in.	THICKNESS mm	WIDTH in.	WIDTH mm	LENGTH in.	LENGTH m	Pieces per Bundle	APPROX. WT. lb/ft²	APPROX. WT. kg/m²
Regular	⅜	5.9	16	406	48	1.2	6	1.4	7.1
Regular	½	12.7	16	406	48	1.2	4	1.8	8.9
Regular	½	12.7	24	610	96	2.4	2	1.8	8.9
Special fire resistant	⅜	5.9	16	406	48	1.2	6	1.4	7.1

TYPE	THICKNESS (in.)	WIDTH (in.)	LENGTH (in.)
Plain	⅜	16	48 or 96
	½	16	48
Perforated	⅜	16	48 or 96
	½	16	48
Insulating	⅜	16	48, 96, 144
	⅜ or ½	24	

FIGURE 18-2

Specifications for gypsum board materials. *(Courtesy of U.S. Gypsum)*

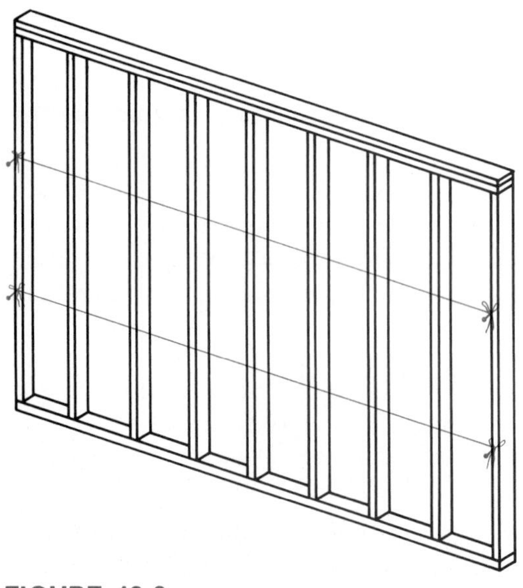

FIGURE 18-3

Walls can be checked for straightness with a chalk line.

at the middle of a bow and driving a wedge into the saw kerf until the stud is straight. Then nail a 1 × 4 in. wood scab on each side of the stud to hold it straight (Fig. 18-4).

If ceiling joists are out of alignment, they can be brought into alignment by installing 2 × 6 in. leveling plates perpendicular to them in the attic. The ceiling joists are pulled to the leveling plate and toenailed to it. Be certain that the 2 × 6 in. material is straight. Some carpenters prefer to use an L-shaped plate. This makes it easier to nail the ceiling joists to the ceiling plate and provides a stiffer plate (Fig. 18-5).

Gypsum lath is applied with the face out and the long dimension perpendicular to the framing members. The end joints should be staggered. Butt all joints together. Cut the lath to fit around electrical outlets and other openings.

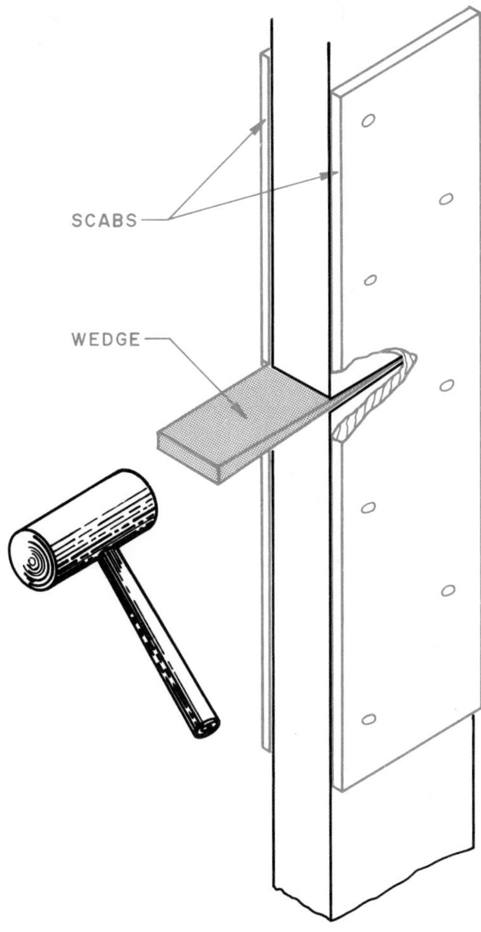

FIGURE 18-4

Studs that are bowed can be straightened by cutting into the middle of the bow and driving in a wedge.

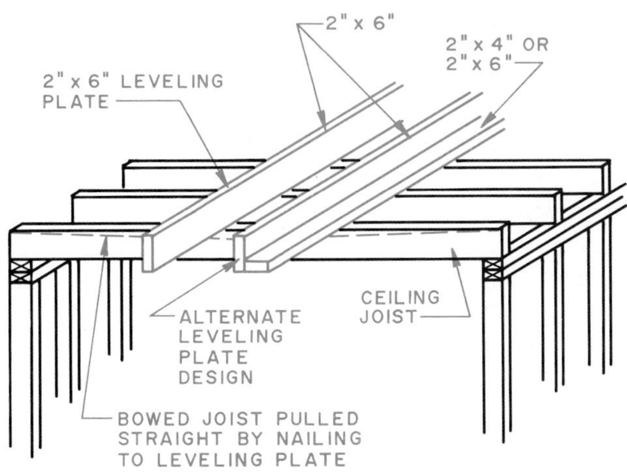

FIGURE 18-5

Leveling plates are used to bring ceiling joists into line.

Use four fasteners, 5 in. on center per 16-in. width of ⅜-in.-thick lath and five fasteners 4 in. on center for ½-in.-thick lath. Place the fasteners ⅜ in. from the edges and ends of the lath. Begin nailing in the center of the lath and proceed to the outer edges. When nailing, apply pressure adjacent to the nail being driven to make certain that the lath is tight against the framing member. The nailheads should be driven flush with the paper surface but not break the paper (Fig. 18-6). The size of fasteners recommended are shown in Fig. 18-7. Gypsum lath can also be fastened to steel studs with clips (Fig. 18-8).

Metal lath is available in a variety of types. The junior diamond mesh is a general-purpose lath and is especially good for ornamental, contour plastering. The diamond self-furring lath has projections that hold it away from the surface below. It is used for exterior stucco, column fireproofing, and over old surfaces. The four-mesh Z-riblath is more rigid than the others and is also used as a tie-on lath for ceilings. The ⅜-in. riblath is used when supports are spaced more than 16 in. on center but not over 24 in. on center. The stuccomesh is used as a base for stucco on the exterior of buildings (Fig. 18-9). Metal lath is applied to wood framing with its long dimension perpendicular to the framing members. Fasteners must be 1½-in. galvanized nails which engage two strands or ribs and penetrate the framing ¾ in. The ends of the lath must overlap 1 in. If the end joint occurs between supports, tie the ends together with 18-gauge wire. Stagger the end joints. The edges of the lower sheets should overlap the sheet above it.

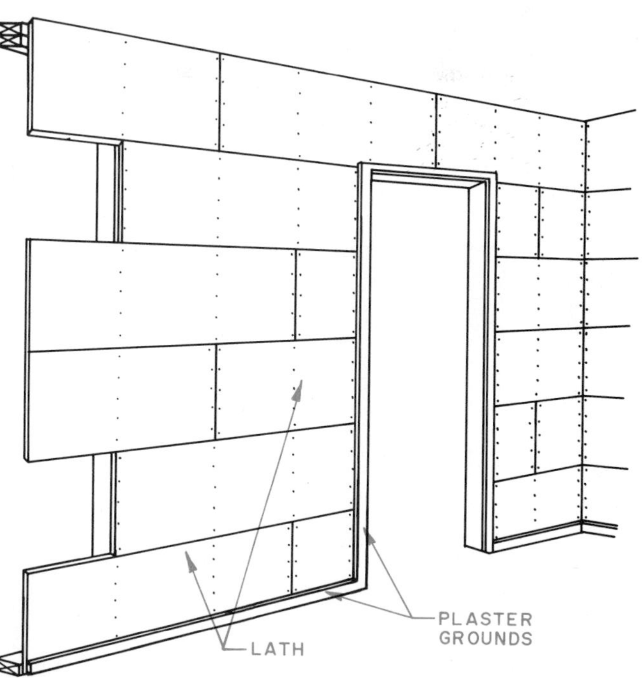

FIGURE 18-6

Gypsum lath is applied with its long dimension perpendicular to the studs.

FASTENER	LATH THICKNESS (in.)	SPACING
Nail 1⅛ in. 13 gauge, 9/16-in.-diameter flat head, blued	⅜	16-in.-wide lath, 4 nails for support, spaced 5 in. o.c., ⅜ in. from ends and edges. 24-in.-wide lath, 5 nails per support.
Nail 1¼ in. 13 gauge, 9/16-in.-diameter flat head, blued	½	16-in.-wide lath, 5 nails per support, spaced 4 in. o.c., ⅜ in. from ends and edges. 24-in.-wide lath, 6 nails per support.
Staple ⅞ in. long, 7/16 in. wide, 16 gauge flattened, galvanized	⅜	Crown parallel with long dimension of stud. 16-in.-wide lath, 4 nails per member, 5 in. o.c., ⅜ in. from ends and edges. 24-in.-width lath, 5 staples per support.
Staple same as above but 1 in. long	½	Same as above. If studs 24 in. o.c., use 5 staples per support at 4 in. o.c. 24-in.-wide lath, 6 staples per support.
Screws to steel studs	½ and ⅝	1-in. type S Bugle Head screws
Screws to wood framing	⅜, ½, and ⅝	1¼-in. type W Bugle Head

GYPSUM LATHING NAIL

TYPE S BUGLE HEAD SCREW

TYPE W BUGLE HEAD SCREW

FIGURE 18-7

Recommended types and sizes of fasteners.

USG Junior Diamond Mesh Lath

USG 4-Mesh Z-Riblath

USG ⅜" Riblath

USG Self-Furring Diamond Mesh Lath

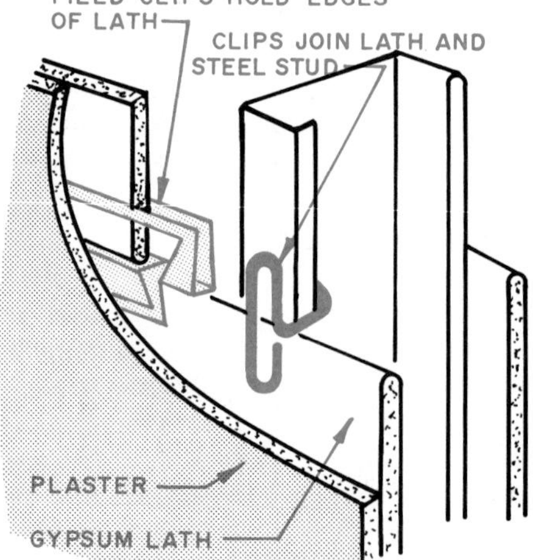

FIGURE 18-8

Gypsum lath is fastened to steel studs with metal clips.

FIGURE 18-9

Examples of metal lath. *(Courtesy of U.S. Gypsum)*

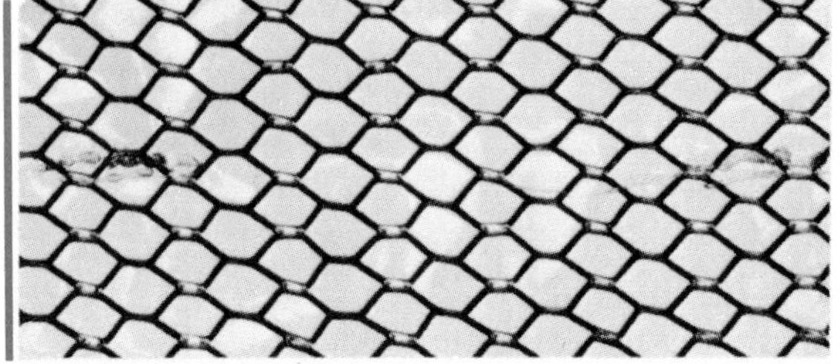
USG Poly-Backed Metal Lath

Reinforcing the Base As a building ages, the wood members will dry and shrink. The building might also settle. These can cause cracks in the plaster. Cracks occur most frequently around openings. These areas need to have an 8-in.-wide piece of expanded metal lath placed diagonally at the corners of doors and windows (Fig. 18-10). They are lightly secured to the gypsum lath with staples or nails. They are not nailed into the wood or metal frames. If they were, the twisting of the frame would cause the lath to bend and crack the plaster.

If a flush wood beam is used, the area below it should be reinforced with metal lath. The reinforcement should extend several inches beyond the width of the beam (Fig. 18-10).

Corners also must be reinforced. Special metal lath units are made for interior and exterior corners (Fig. 18-11). When installing corner reinforcement, check it with a level to be certain it is plumb.

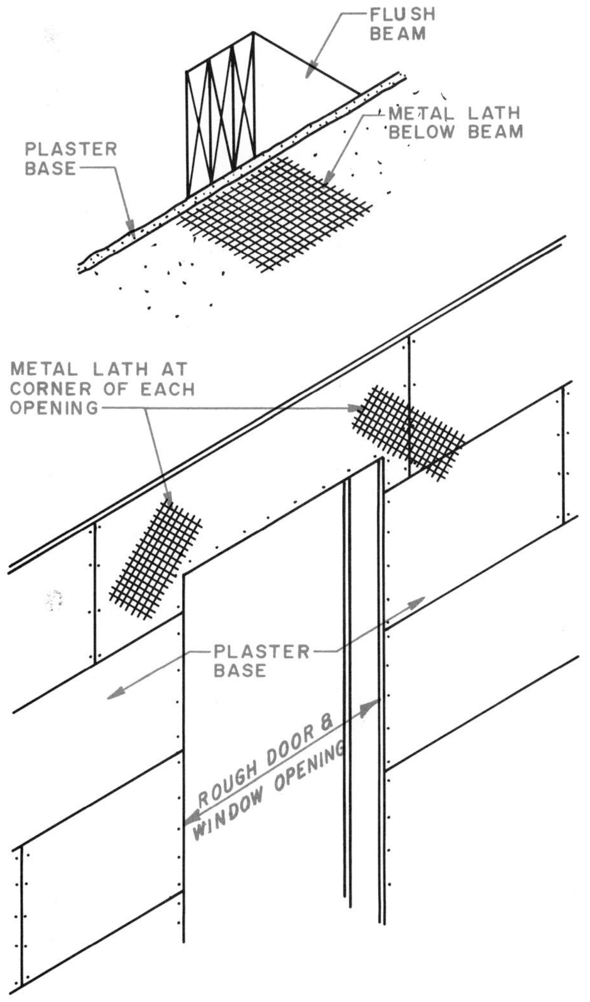

FIGURE 18-10

Metal lath is used at the corners of wall openings to help keep the plaster from cracking.

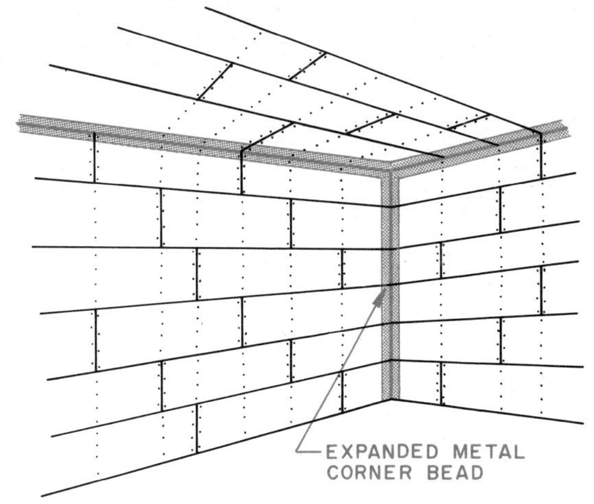

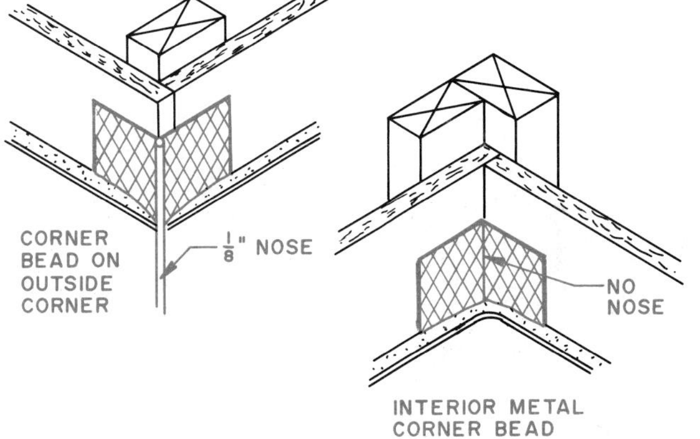

FIGURE 18-11

Expanded metal corner beads are used to reinforce plaster in the corners.

Installing Plaster Grounds *Plaster grounds* are wood strips used to assist the plasterer in getting the plaster the proper thickness. Plaster on gypsum lath is ½ in. thick and ⅝ in. thick on metal lath. The thickness of the ground is the thickness of the plaster plus the thickness of the lath. In a typical residence ⅜-in. lath plus ½-in.-thick plaster is used. This requires a ⅞-in.-thick ground. Grounds are usually 1 to 2 in. wide.

Plaster grounds are nailed around door and window openings and next to the floor (Fig. 18-12). Raise the floor ground so that it is above the top of the finished floor. Grounds are usually left in place after the plaster is applied and used for nailing the finished trim. If metal lath is used, metal grounds are provided.

If the windows have jamb extensions, these can be used as plaster grounds (Fig. 18-13).

Temporary ground strips can be used (Fig. 18-14). These are removed when the plaster is dry. Be certain that the grounds are level and plumb.

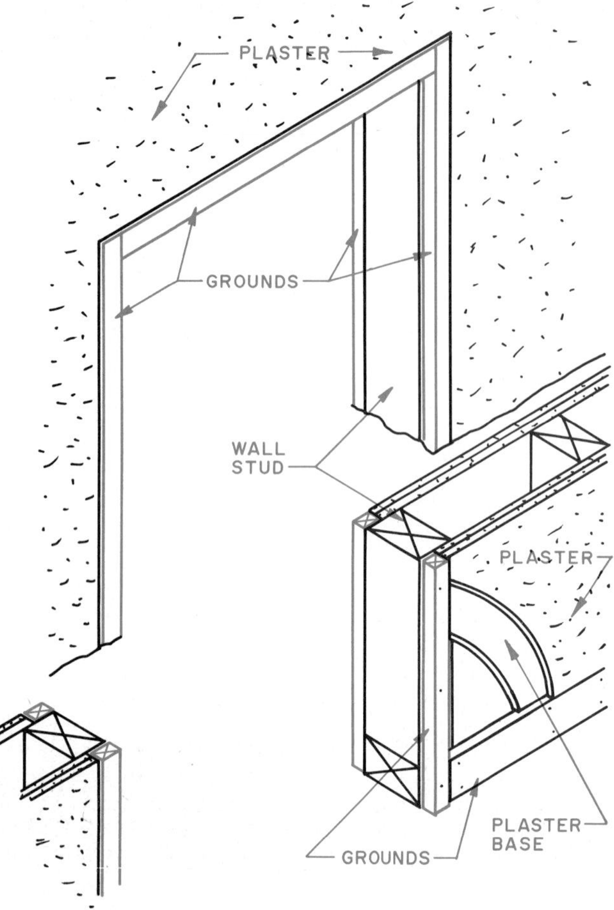

FIGURE 18-12

Plaster grounds are installed around door and window openings and next to the floor.

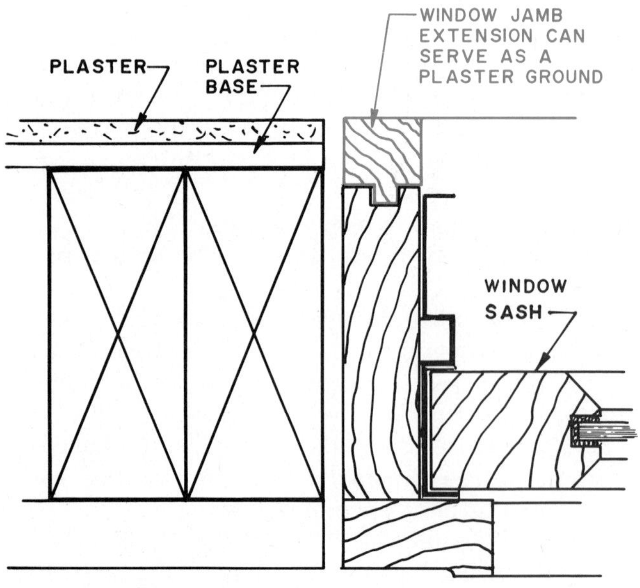

FIGURE 18-13

Jamb extensions can be used as plaster grounds.

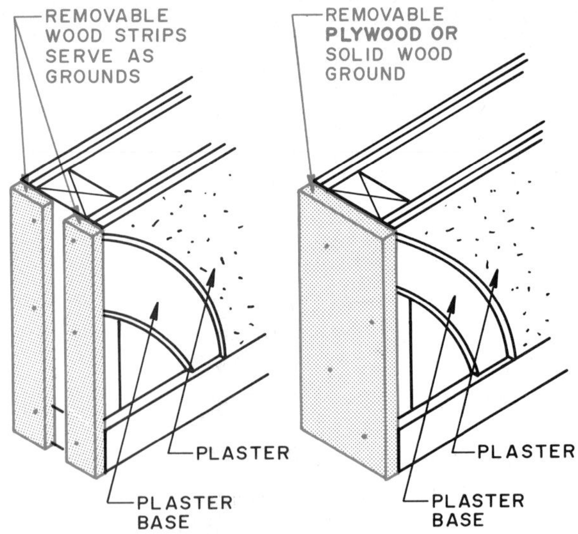

FIGURE 18-14

Temporary ground strips are removed after the plaster is hard.

Plastering Solid Masonry Walls Plaster can be applied directly to a masonry surface. Wood plaster grounds are used around openings and at the floor and ceiling. If the wall is to be insulated, it must be furred out, insulation installed, and the gypsum or metal lath fastened to the furring (Fig. 18-15).

Applying the Plaster The coats of plaster are applied to the lath by a plasterer. Carpenters should understand this process even though they do not do the work.

If gypsum lath is used, a two-coat plaster is used. Three-coat plaster is required for metal lath.

The two-coat systems have a basecoat applied to the lath. This should be applied with sufficient material and pressure to form a good bond to the base and to cover well (Fig. 18-16). Then a second layer of the base coat is applied to bring the plaster out to the grounds and straighten the wall or ceiling surface. This is called doubling back. It is done before the first layer is dry. It is left rough to receive the finish coat. The finish coat is a thin coat, about 1/16 in., and can be given many finishes. It can be troweled smooth, float-finished, sprayed, or have a rough texture finish.

The three-coat has a scratch coat applied to the metal lath (Fig. 18-17). It is applied with sufficient pressure to key into the openings in the metal lath. The second coat is the brown coat. It is applied after the scratch coat has hardened. It brings the surface out to the grounds and straightens the surface. It is left rough to receive the finish coat. The third coat is the finish coat.

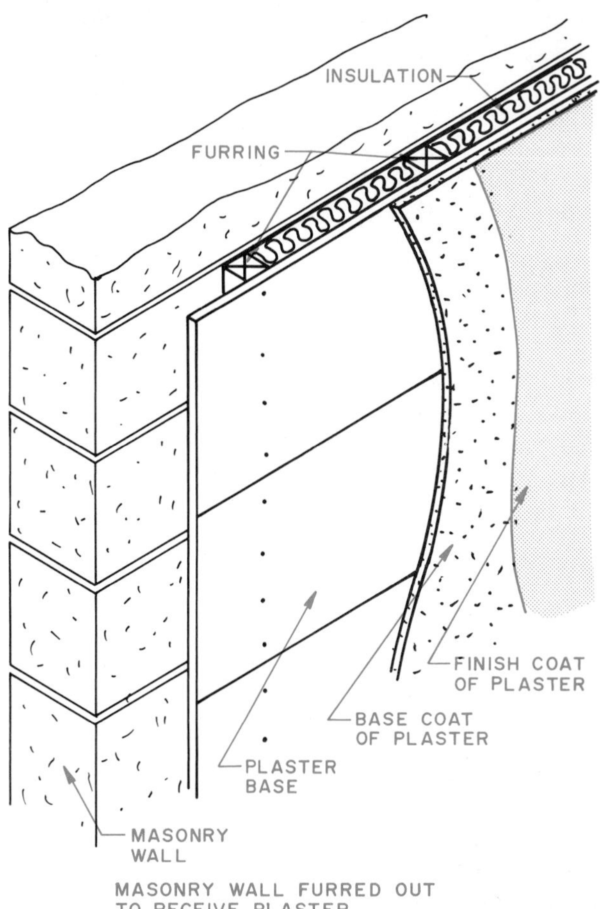

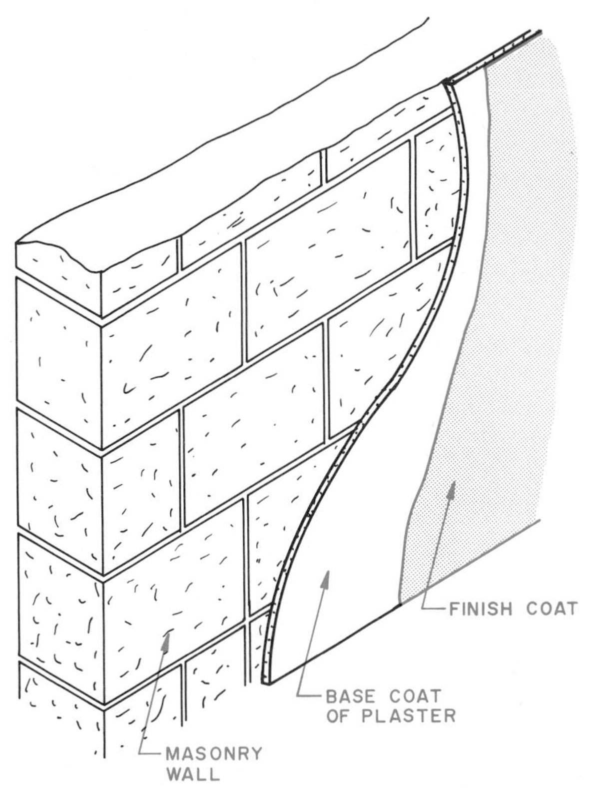

FIGURE 18-15

Plaster can be applied directly to a masonry wall or the wall can be furred out and plaster base used.

FIGURE 18-16

The two workers on the left are applying the second coat of plaster. The worker on the right is applying a machine-applied base coat. *(Courtesy of U.S. Gypsum)*

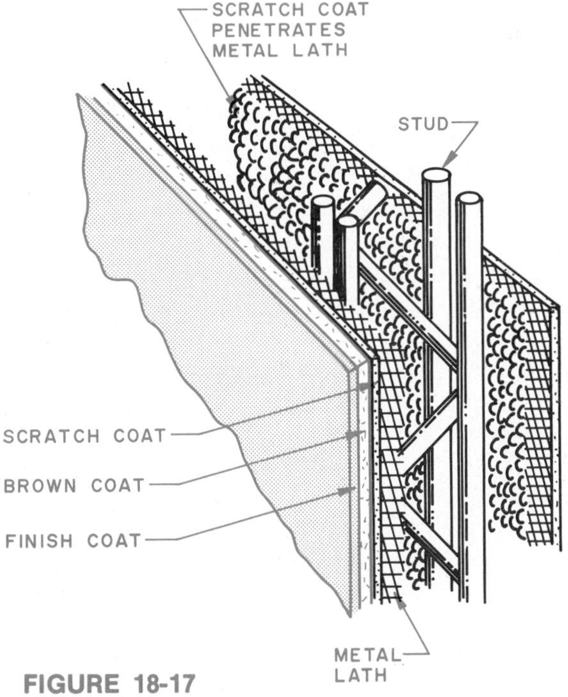

FIGURE 18-17

A typical three-coat plaster wall applied over a wall using metal studs.

Drywall finish includes materials such as gypsum board, plywood, fiberboard, and solid wood paneling which are fabricated in standard sizes and applied directly to the wall or ceiling framing.

In all cases the studs and ceiling joists must be straight and plumb. Procedures for checking and correcting these are explained in the section on plaster.

GYPSUM BOARD

Gypsum board is a laminate composed of a gypsum core with outer layers of paper. It is available in sheets 4 ft wide and 8 ft long. Lengths up to 16 ft can be ordered.

Gypsum board is made in four thicknesses, 5/8, 1/2, 3/8, and 1/4 in. The 5/8-in. thickness is recommended for the finest-quality single-layer construction. The 1/2-in. thickness is the most commonly used size for single-layer application in new construction. The 3/8-in. thickness is a lightweight panel and is used mainly in a double-layer application or in remodeling work where it is applied over existing surfaces. The 1/4-in. panel is lightweight and used as a base panel for improving sound control in multilayer partitions and covering old wall and ceiling surfaces.

Gypsum panels are used under wood and hardboard panels to provide fire protection. A special water-resistant panel is available for use in bathrooms and in areas of high moisture.

Gypsum panels are available with factory-applied vinyl facings in a wide range of coordinated decorator colors. The vinyl covering is tough and easily cleaned. The panels have beveled edges that form a shallow V-groove joint similar to that used in plywood paneling. Special moldings are available with the panels. They are installed with adhesives. They are not recommended for ceilings because the end joints are difficult to conceal. The vinyl patterns available include stipples, linens, embossed wood grain, textiles, and corks.

Square-edge gypsum panels are used as bases for walls and ceilings requiring a veneer finish. A veneer of 1/16 to 1/32 in. of specially formulated gypsum finish is applied in one or two coats over the base. The surface can be smooth or textured. They are durable and resist abrasion.

Base sheets that are used for veneer finishes have the gypsum core faced with a specially treated, multilayered paper designed to provide a maximum bond to veneer finishes. The paper's absorbent outer layers quickly and uniformly draw moisture from the veneer finish for proper application and finishing. The moisture-resistant layers keep the gypsum core dry and rigid. The face paper is folded around the long edges. The ends are square cut and finished smooth.

Fasteners

Special screws and nails are used to fasten gypsum panels to wood and steel studs and joists. There are a wide variety of self-drilling, self-tapping steel screws. The screw used to join 1/2-in. single-layer panels to steel or wood studs is shown in Fig. 18-7. This has a type S Bugle head, has a Phillips head recess, and is prepared to be driven with a power-driven screwdriver (Fig. 18-18). A 7/8- or 1-in. screw is used with steel studs. A 1¼-in.-length is used to fasten 3/8, 1/2, and 5/8-in.-thick panels to wood.

The proper type and size of nail is important to a successful application. Since wood shrinks as it dries, the nails tend to become loose. The use of the annular-ring-type nail provides 20% greater holding power than the smooth-shank nail. Longer nails tend to pop loose less frequently. A guide for the selection of nails is shown in Fig. 18-19.

Adhesives especially designed for applying gypsum panels to studs or other panels are available. They are applied to the base material

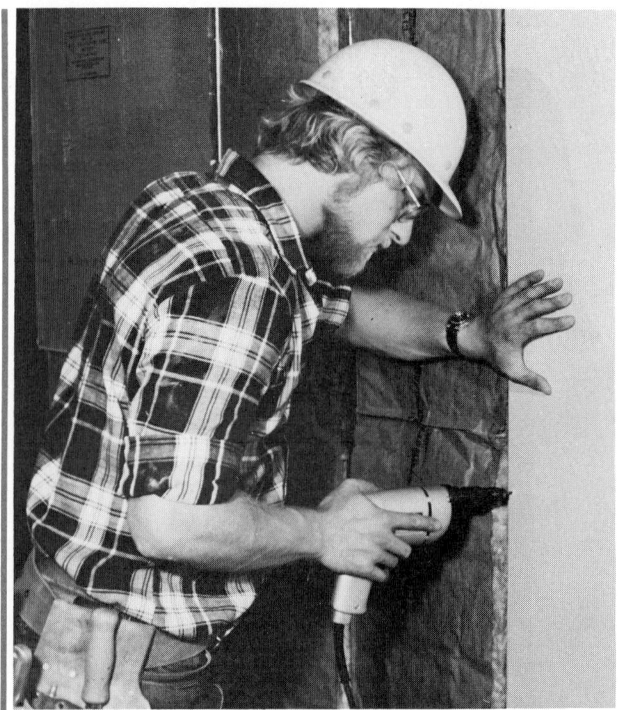

FIGURE 18-18

Gypsum panels can be installed with screw-type fasteners. *(Courtesy of U.S. Gypsum)*

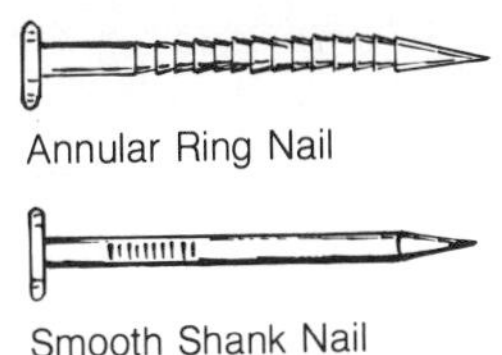

Material thickness (in.)	¼	⅜	½	⅝	¾	⅞
Nail length (in.)	1¼	1¼	1¼	1⅜	1½	1⅝

FIGURE 18-19

Typical nails used to install gypsum panels to wood studs. *(Courtesy of U.S. Gypsum)*

from a cartridge gun. A ⅜-in. bead of adhesive is recommended.

Joint Compounds and Tapes The joints between the gypsum panels are covered with a joint compound and tape. There are several *joint compounds* available. Some are best for embedding the tape and applying the coat over it. Others are best for the finish coat and texturing.

Joint tape is usually a high-strength cross-fibered paper tape. It is tapered to feather edges to aid in finishing the joint. The center is creased so that the tape can be easily folded to form angles. One type is stapled over the joint. The other has an adhesive backing and is pressed in place (Fig. 18-20).

Interior Texture Finishes The texture finish compound is used to produce a variety of finishes over the taped gypsum panels (Fig. 18-21).

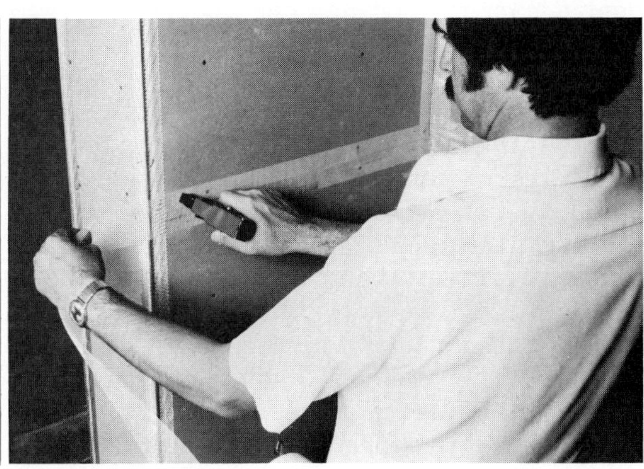

FIGURE 18-20

Joints between panels can be covered with a special tape that is stapled in place. *(Courtesy of U.S. Gypsum)*

Texture XII

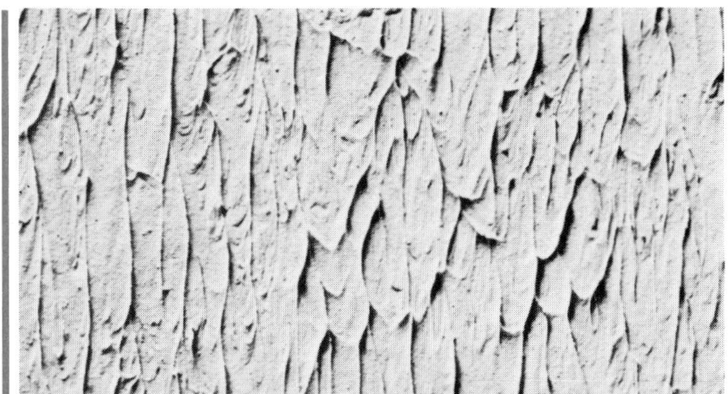

Roller

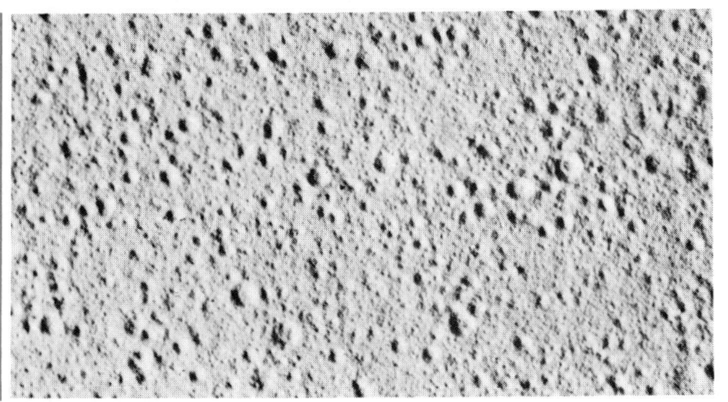

Medium light spray.

Lightly stippled with brush or roller.

FIGURE 18-21

Examples of textured finishes that can be applied over gypsum wall panels. *(Courtesy of U.S. Gypsum)*

The texture effect is obtained by brush, roller, or spray application. It helps conceal minor surface defects. It dries to a white finish and can be painted if desired (Fig. 18-22).

Installing Gypsum Panels Two methods for installing gypsum panels are commonly used: the single layer and the double layer. The single-layer method has one thickness of gypsum board applied directly to the studs. The double-layer method has one layer fastened to the studs and a second layer placed over the first and joined with adhesive plus a few nails or screws. This produces a strong wall that is resistant to cracking.

Gypsum panels can be applied with the long edge parallel to the studs or perpendicular to the studs (Fig. 18-23). Perpendicular application is preferred because it reduces the footage of joints 25%. It also has the strongest dimension of the panel running across the framing members. For ceiling application, use the method that produces the fewest joints.

The ceiling panels should be installed first. All joints between panels should be butted loosely. The edges should be placed next to each other. Never place a square butt end or a cut edge next to a rounded edge. Stagger butt joints when they occur and locate them as far from the center of walls and ceilings as possible.

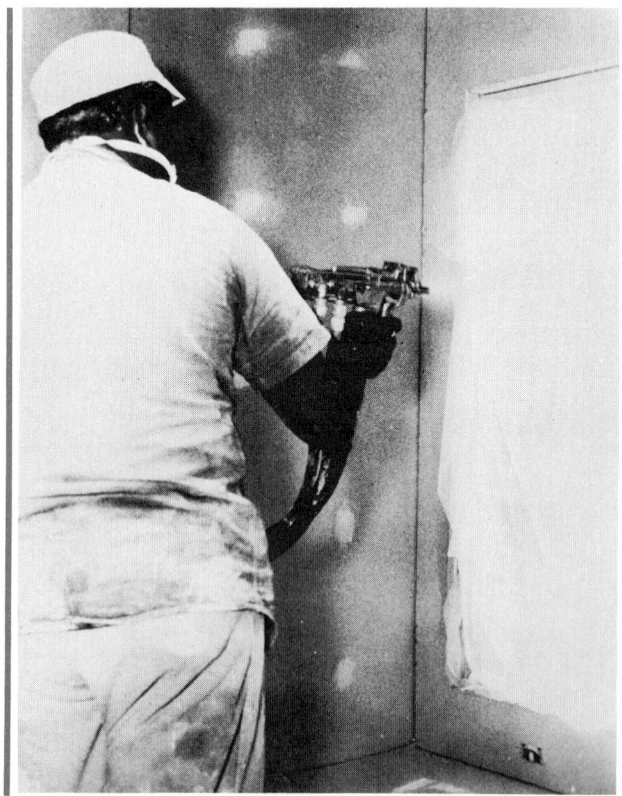

FIGURE 18-22

Applying a finish over a gypsum wall. *(Courtesy of Guerdon Industries)*

FIGURE 18-23

Most residential installations are made with the gypsum panel having its long edge parallel with the studs. *(Courtesy of U.S. Gypsum)*

All edges and ends of panels must rest on the framing. Exceptions to this can be made if the end joints are back-blocked with gypsum panels.

If metal trim is installed around doors, windows, and other edges install it before the panels are set in place (Fig. 18-24). Metal trim is stapled or nailed every 12 in. on center.

To cut gypsum panels, lay a straightedge where the cut is to occur. Score the panel along the straightedge with a utility knife. Bend the panel forward toward the cut. This will snap the core. Then cut the paper layer on the other side (Fig. 18-25). Smooth the edge with a rasp or coarse sandpaper.

Carefully measure the location of electrical boxes and other openings and cut out before installing the panel. A special electrical-outlet-box cutting tool is available. Other cuts can be made with a drywall saw.

To nail a panel, drive nails ⅜ in. from the

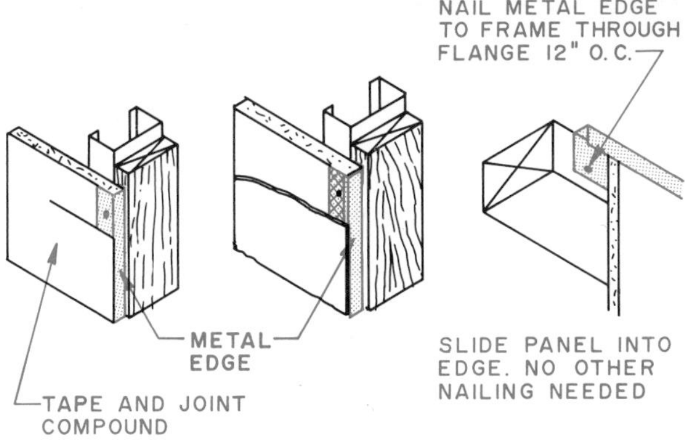
NAIL METAL EDGE TO FRAME THROUGH FLANGE 12" O. C.
METAL EDGE
TAPE AND JOINT COMPOUND
SLIDE PANEL INTO EDGE. NO OTHER NAILING NEEDED
SLIDE BOARD INTO METAL EDGE. NAIL THROUGH EDGE AND BOARD TO FRAME WITH REGULAR GYPSUM BOARD NAILS SPACED 9" O. C.

FIGURE 18-24

If metal trim is to be used on the edges of gypsum panels at door and window openings, put it in place before nailing the panels to the studs.

edges of the board (Fig. 18-26). Position nails on adjacent ends or edges opposite each other. Nails on ceilings are spaced 7 in. on center. On walls they are spaced 8 in. on center. Other spacing information is given in Fig. 18-27. Begin by nailing in the center of the board and proceed toward the outer edges. Keep the board pressed tightly against the frame while nailing. Drive the nails perpendicular to the face of the board. Drive them so that the head of the nail is in a shallow, uniform dimple which is formed by the last blow of the hammer (Fig. 18-28). Use a drywall hammer because it has a crowned head for setting nails (Fig. 18-29). Do not break the paper when setting the nail.

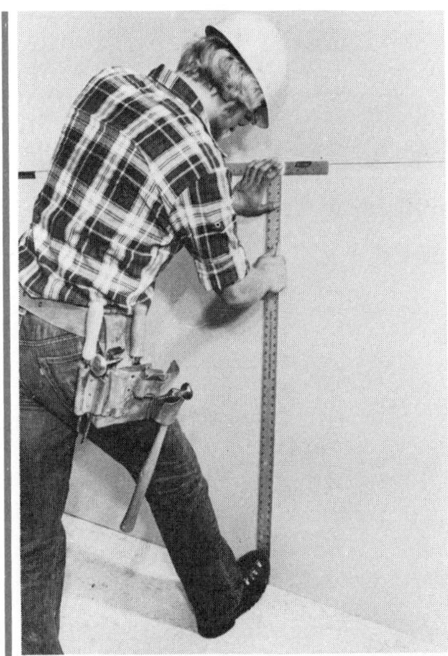

1. Score board with utility knife.

2. Snap forward breaking the gypsum line.

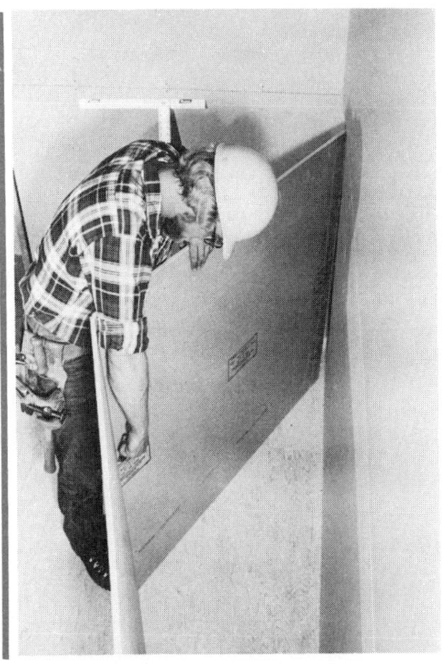

3. Cut paper on back of cut.

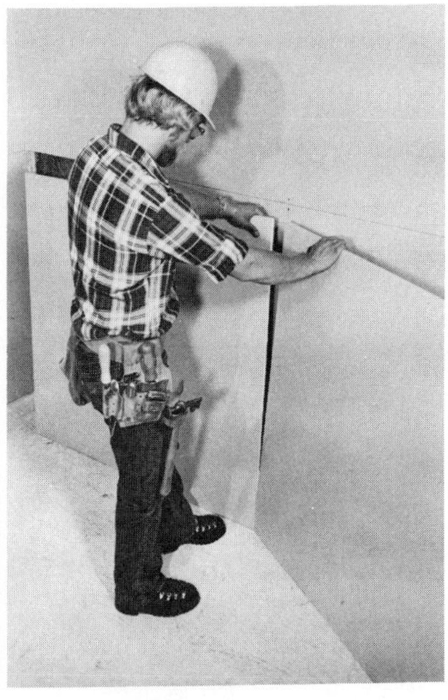

4. Separate the parts of the panel.

FIGURE 18-25

The steps to cut a gypsum panel. *(Courtesy of U.S. Gypsum)*

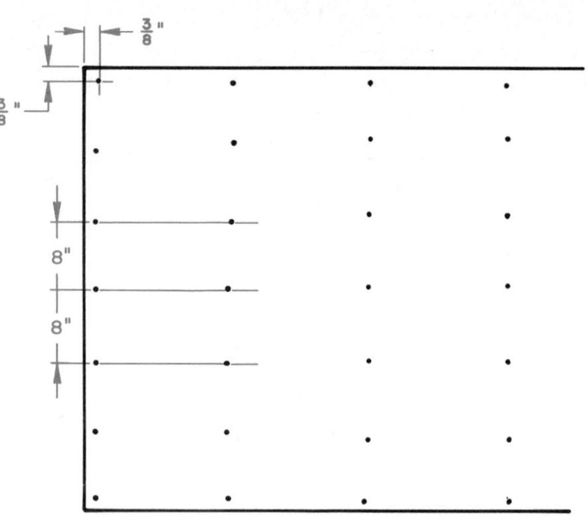

FIGURE 18-26 *(above)*

Recommended spacing for nails on gypsum board panels.

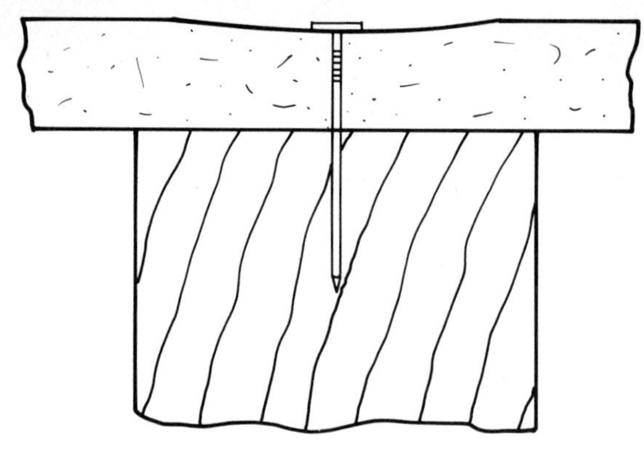

FIGURE 18-28

When nailing gypsum panels, drive the nail so that the surface has a shallow dimple.

TYPE OF CONSTRUCTION	TYPE OF FASTENER	LOCATION	FASTENER SPACING (in.)
Single layer	Nails	Ceiling	7
		Sidewall	8
	Screws	Ceiling	12
		Sidewall	16
Base layer of double layer with both layers mechanically attached	Nails	Ceiling	24
		Sidewall	24
	Screws	Ceiling	24
		Sidewall	24
Face layer of double layer with both layers mechanically attached	Nails	Ceiling	7
		Sidewall	8
	Screws	Ceiling	12
		Sidewall	16
Base layer of double layer with face layer adhesively attached	Nails	Ceiling	7
		Sidewall	8
	Screws	Ceiling	12
		Sidewall	16
Face layer of double layer with face layer adhesively attached	Nails	Ceiling	16 in. o.c. at edges and ends; one fastener per frame member at midwidth of board
		Sidewall	Fasten at top and bottom as required

(a)

TYPE OF CONSTRUCTION	TYPE OF FASTENER	LOCATION	FASTENER SPACING (in.)
Single layer	Screws	Ceiling	12
		Sidewall	16
Base layer of double layer with both layers mechanically attached	Screws	Ceiling	16
		Sidewall	24
Face layer and base layer of double layer with both layers mechanically attached	Screws	Ceiling	12
		Sidewall	16
Base layer and face layer of double layer with face layer adhesively attached	Same spacing as required for nails		

(b)

FIGURE 18-27

Recommended types of fasteners and spacing for various types of gypsum board installations. (A) Wood framing. (B) Steel framing. *(Courtesy of U.S. Gypsum)*

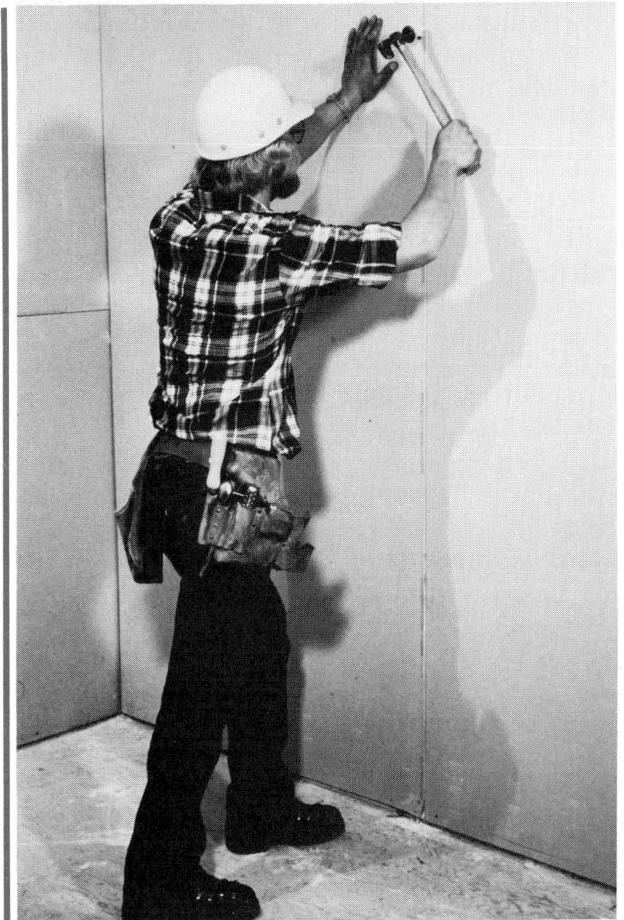

FIGURE 18-29

Installing gypsum panels with a drywall hammer.

The popping of nails can be reduced by double nailing. Space the first nails 12 in. on center and place a second nail 2 in. from the first nail on the framing members inside the panel (Fig. 18-30).

When fastening with screws, the application is much the same as nailing. Spacing is shown in Fig. 18-26.

FIGURE 18-30

Recommendations for double-nailing gypsum board panels.

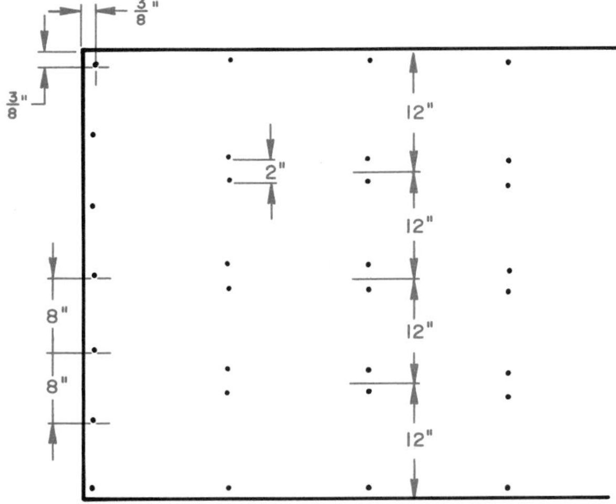

The screw gun must have a Phillips tip. It is then set to the length of the screw. The screw head should be driven slightly below the face of the panel. It should not break the paper.

Place the screw onto the Phillips tip. Press it perpendicular to the panel. Drive the screw against the panel (Fig. 18-31). Keep the screws at least $\frac{3}{8}$ in. from the edge of the panel.

To fasten panels with adhesive, cut the plastic nozzle on the cartridge as shown in Fig. 18-32. Lay a continuous $\frac{3}{8}$-in.-diameter bead to within 6 in. of the end of each framing member (Fig. 18-33). When two panels meet on a single framing member, each requires a $\frac{3}{8}$-in. bead. Do not apply adhesive to members such as bridging or bracing. Only use the wall or ceiling framing members. Adhesive is not required at inside corners, top and bottom plates, bracing, and fire stops, and is not used in closets.

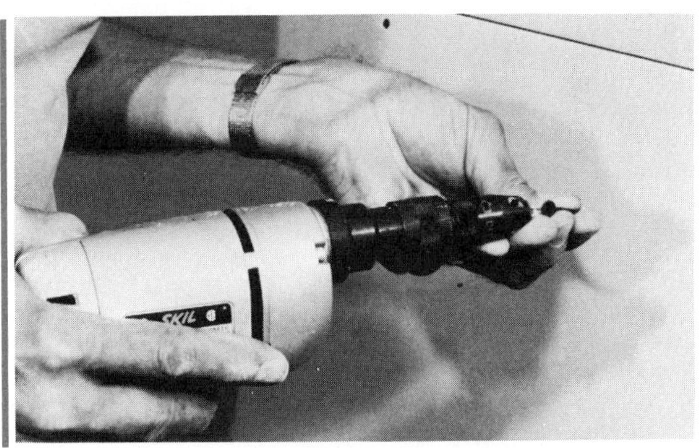

Press the screw onto the Phillips tip.

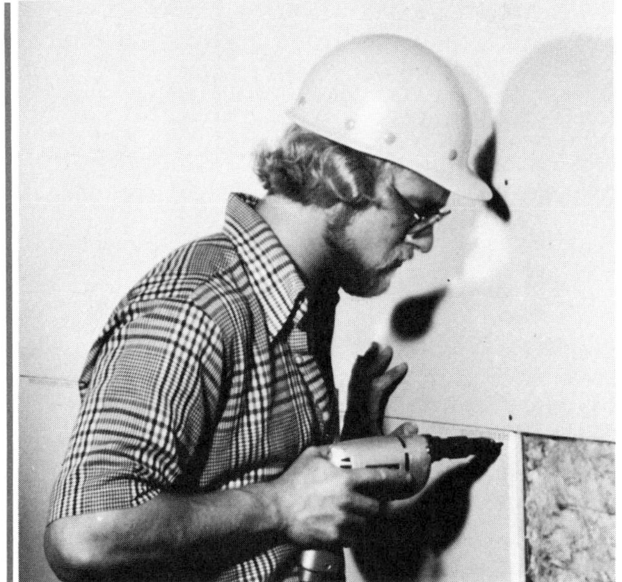

Drive the screw perpendicular to the panel.

FIGURE 18-31

Steps for installing gypsum panels with screws. *(Courtesy of U.S. Gypsum)*

FIGURE 18-32

How to cut the nozzle on adhesive tubes when used to fasten gypsum panels to studs.

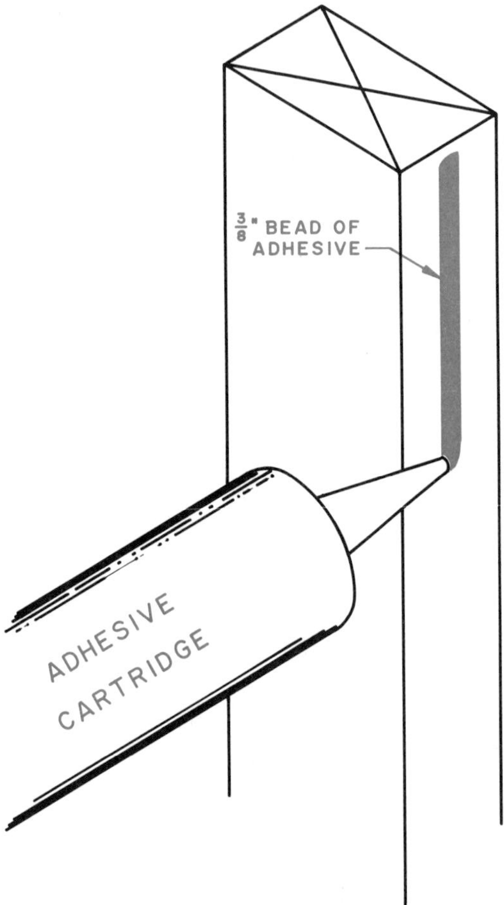

FIGURE 18-33

A continuous ⅜-in. bead of adhesive is recommended.

Place the gypsum panels shortly after the adhesive bead is applied and fasten immediately with the required nails or screws. Press the board firmly against each framing member to be certain full contact has been made.

When using adhesives, nail perpendicular ceiling panels at each framing intersection and 16 in. on center at each end. Place one temporary double-headed nail per framing member in the center of the panel until it is dry. This is usually 48 hours. If the ceiling panels are parallel (long edges parallel to framing), space fasteners 16 in. on center along board edges and at each framing intersection on the ends. Place temporary fasteners 24 in. on center on intermediate framing members and remove after 48 hours. Application to walls is the same manner as described for ceilings except that no intermediate fasteners are used.

If a molding is to be applied at the ceiling, the corner need not be taped (Fig. 18-34).

If a double-layer gypsum wall is to be built, the first layer is installed as just described for a single layer. The second panels are precut to fit in a particular place on the wall or ceiling. Adhesive is applied with a notched spreader (Fig. 18-35). It should produce four ⅜- by ½-in.-wide adhesive strips spaced 2 in. on center. The board is placed against the wall and pressed into place. Another technique is to apply the adhesive to the base surface and press the second panel against this.

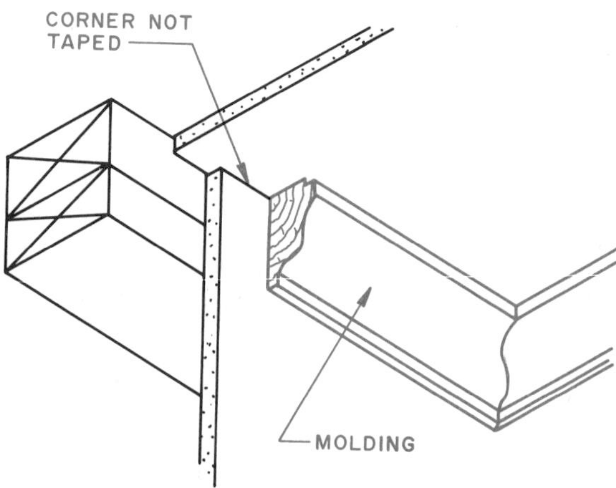

FIGURE 18-34

Molding can be used to cover the joint at the ceiling.

FIGURE 18-35

Adhesive being applied to a panel with a notched spreader. *(Courtesy of U.S. Gypsum)*

When laminating a second layer on a ceiling, the second layer must be nailed or screwed to the framing with fasteners placed 24 in. on center. On walls, nail or screw the entire outer edge of the board to the frame with fasteners spaced 24 in. on center.

Outside corners are reinforced using a metal mesh (Fig. 18-36). The corner bead is nailed over the gypsum panel and is covered with tape and compound (Fig. 18-37).

Inside corners are finished by folding the paper tape along the center crease. Apply joint cement to the gypsum board on both sides of the corner. Apply the tape over the cement in the same manner as for flat joints. Apply second and third coats of joint cement (Fig. 18-38).

Taping the Joints To tape a joint with tapered edges, first check to see that all nails are in place. Do this by drawing the finishing knife over the joint. Set any nails that protrude. Then trowel a layer of joint compound into the joint (Fig. 18-39). Center the tape over the joint and press it into the compound with the finishing knife (Fig. 18-40). Hold the knife at a 45° angle. Press hard enough to firmly seat the tape and remove excess compound. Next, immediately apply a skim coat of joint compound over the tape. Allow the joint to dry. Apply the first coat of compound over all nail heads. Use the finishing knife to remove excess compound, leaving the compound flush with the board surface.

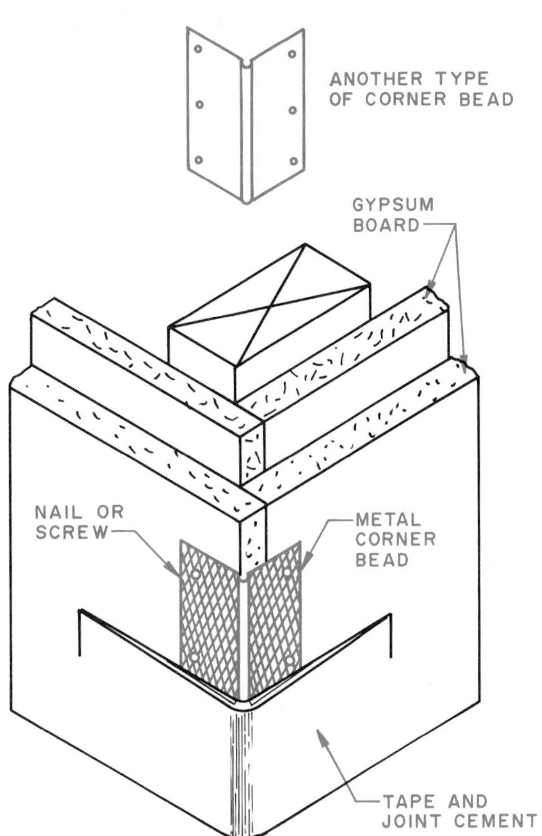

FIGURE 18-37

A finished outside corner.

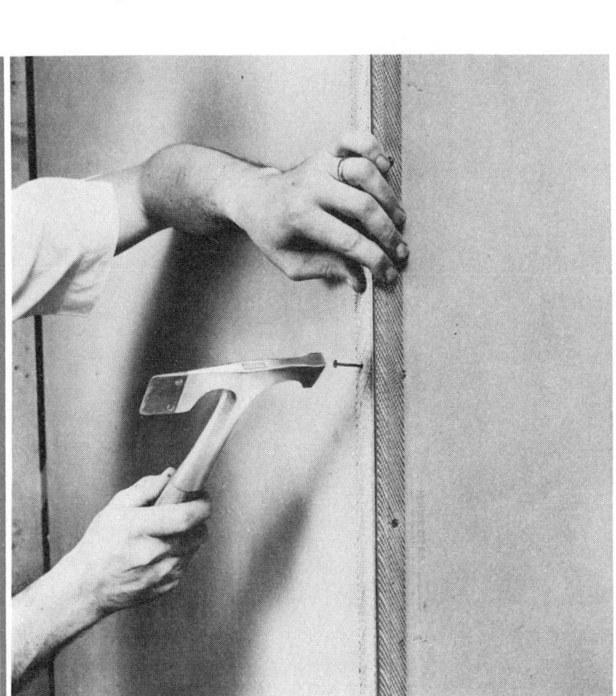

FIGURE 18-36

How to install metal mesh outside corner reinforcing. *(Courtesy of U.S. Gypsum)*

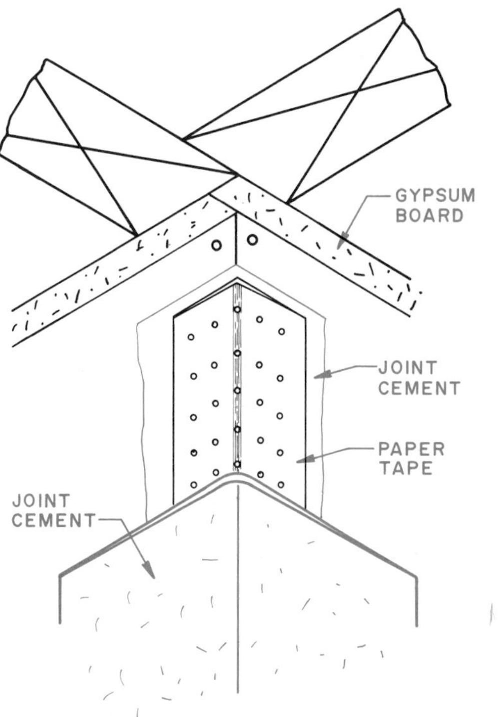

FIGURE 18-38

How to finish an inside corner.

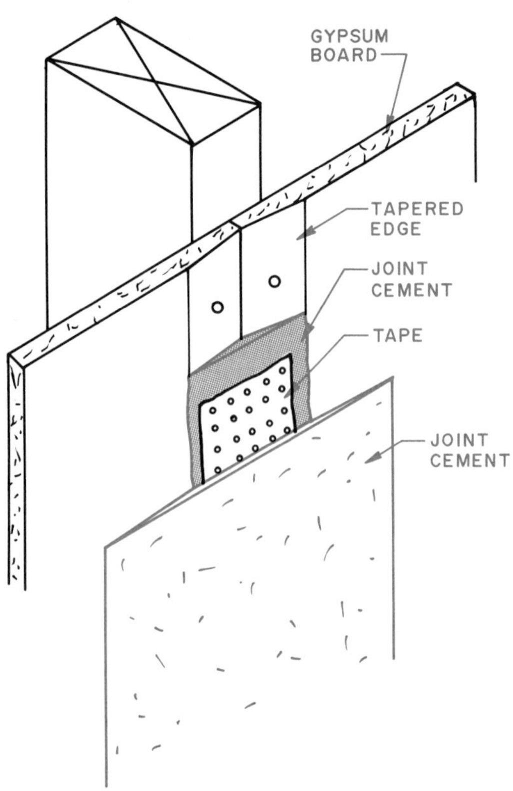

FIGURE 18-39

The layers required to finish joints between gypsum board panels.

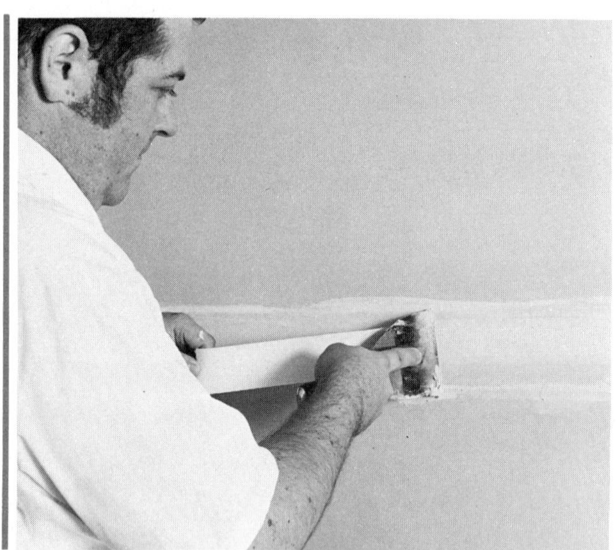

FIGURE 18-40

The tape is centered on the joint and pressed into the joint compound. *(Courtesy of U.S. Gypsum)*

Apply a first coat over all corner beads (Fig. 18-41). Allow to dry.

After all has dried about 24 hours, apply the next coat. It should be feathered on the edges but extend 2 in. beyond the edge of the first coat (Fig. 18-42). Spot over nailheads and corner beads.

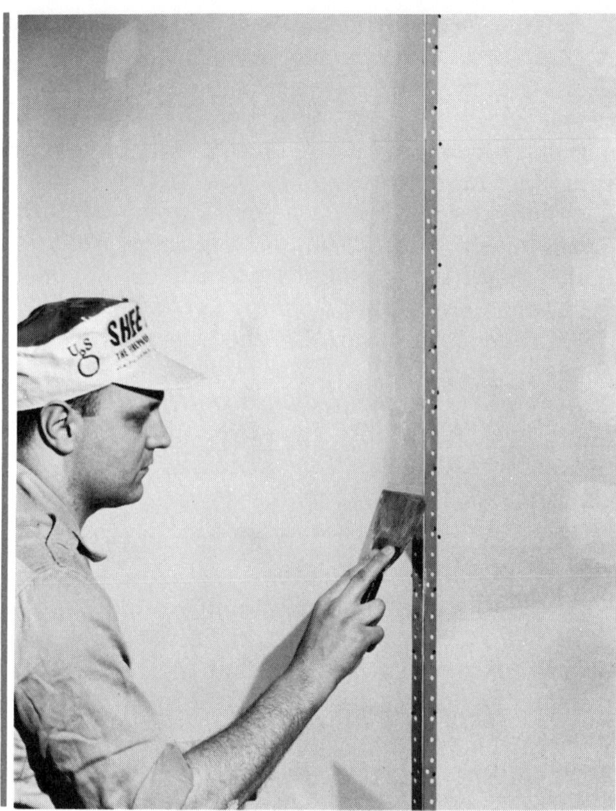

FIGURE 18-41

Applying the first coat of compound over a metal corner bead. *(Courtesy of U.S. Gypsum)*

Second coat applications.

Third coat application.

FIGURE 18-42

Applying the second and third coats of compound. *(Courtesy of U.S. Gypsum)*

When the second coat is dry, sand it lightly. Apply a thin finishing coat. Feather the edges about 2 in. beyond the edge of the second coat. Allow to dry and sand lightly. Use a very fine sanding cloth (about 220 grit). Be careful not to oversand and destroy the layers (Fig. 18-43). The worker should wear a respirator when standing.

Tape can be applied using an Ames mechanical tape applicator (Fig. 18-44). The compound can also be applied with an Ames compound applicator.

If the gypsum panel has a tapered edge, prefill the V-groove. Apply the compound directly over the V-groove with a 5 or 6 in. finishing knife. Wipe off compound beyond the groove. Allow to harden (Fig. 18-45). After this hardens, apply the tape and finishing compound as previously explained.

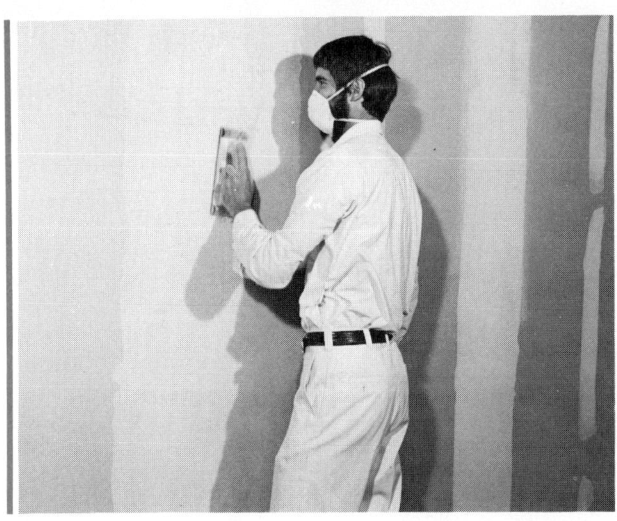

FIGURE 18-43

Each coat of compound is sanded lightly. A respirator protects the worker from breathing the heavy dust. *(Courtesy of 3M Co.)*

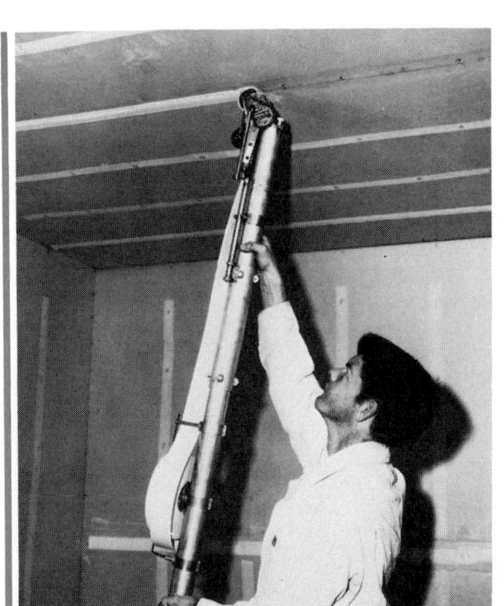

1. Mechanically tape joints.

2. Wipe down with broad knife.

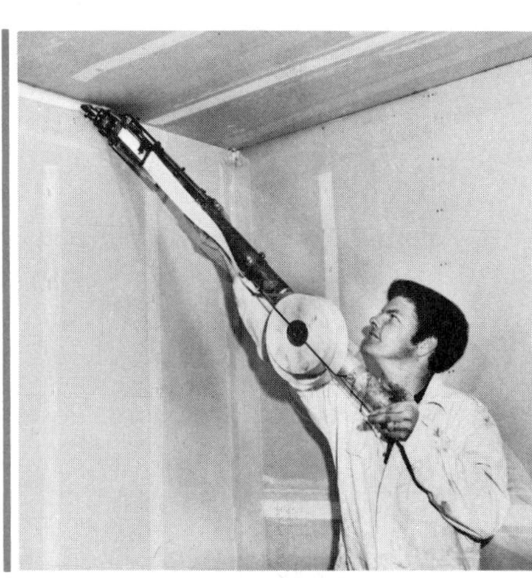

3. Mechanically tape interior corners.

4. Finish sides of angles with corner rollers and corner finisher.

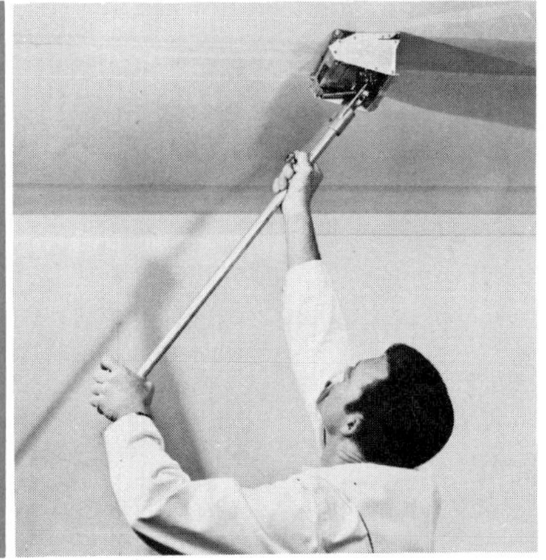

5. Apply second coat of compound over tape using a hand finisher.

6. Spot fastener heads and apply second coat to metal reinforcement.

FIGURE 18-44

Taping joints using an Ames tape applicator and an Ames compound applicator. *(Courtesy of U.S. Gypsum)*

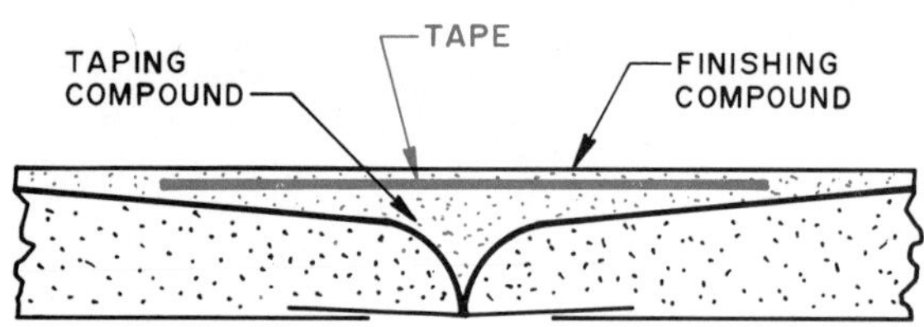

FIGURE 18-45

Section through a finished joint.

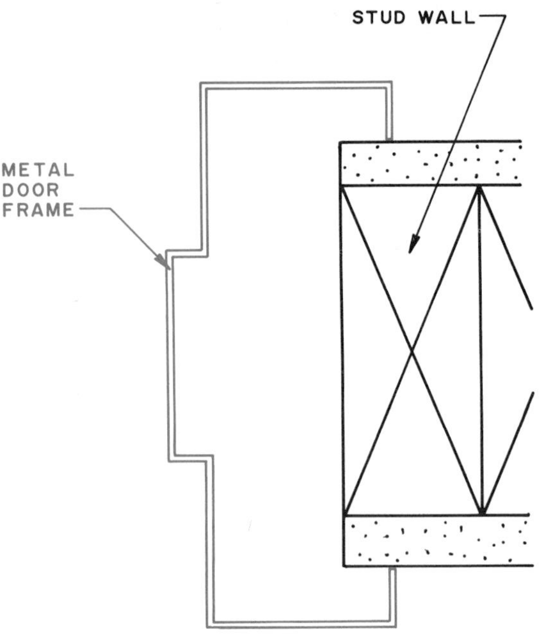

FIGURE 18-47

Metal door frames are installed after the gypsum panels are secured to the wall.

Metal Door Frames The frames are in three pieces. Each piece is installed separately (Fig. 18-46). The corners are diecut to form precise miters.

Prefabricated metal door frames can be installed after the drywall materials are installed. They can be installed over wood-framed walls or walls using metal studs (Fig. 18-47).

PLYWOOD WALL PANELING

Most *plywood paneling* is available with a prefinished surface. This is a tough factory-applied

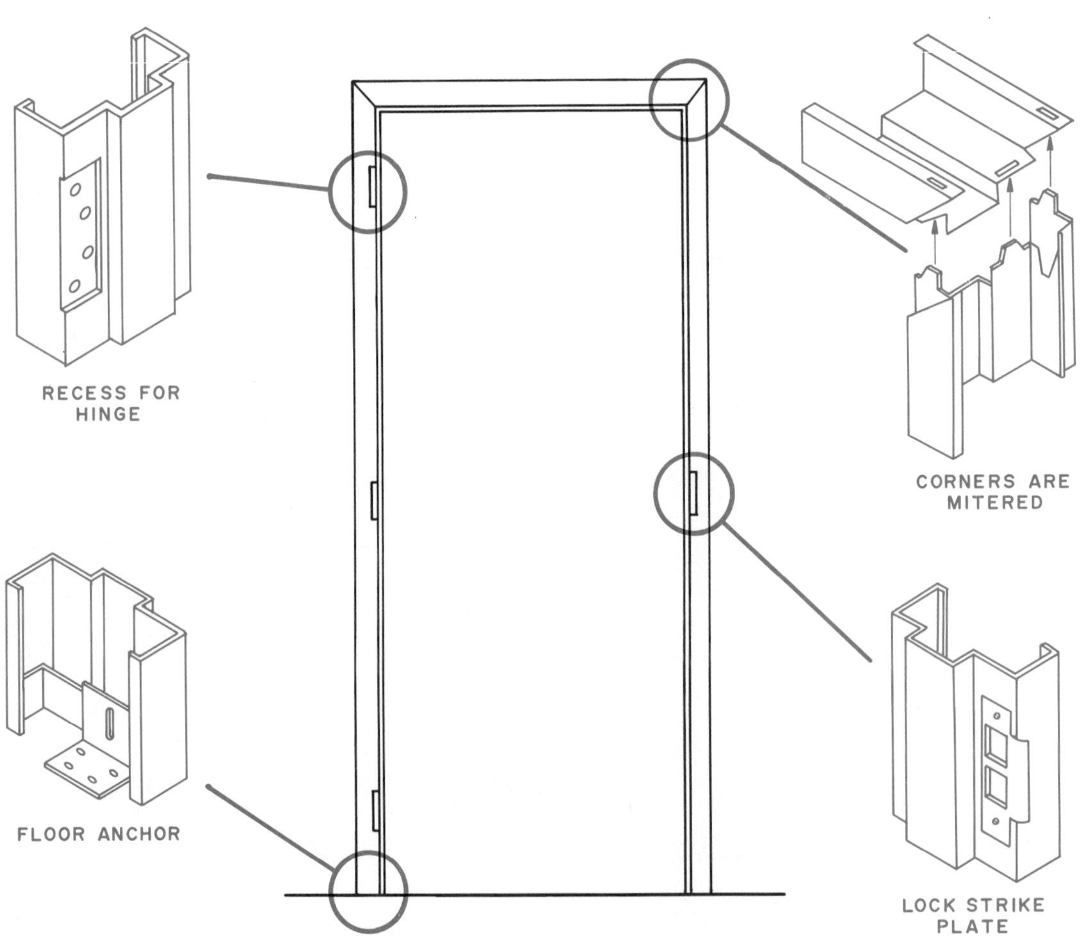

FIGURE 18-46

A typical metal door frame.

finish that resists damage and can be easily cleaned.

Plywood panels are available with hardwood and softwood species on the finished surface. The grain runs vertically. These range in thickness from ¼ to 7/16 in. Panels are made with the surface embossed, antiqued, or color-toned to achieve a wood grain or other decorative appearance. Some have a paper overlay with a woodgrain appearance laminated to the surface.

Other panels are made from particleboard or hardboard. They have wood-grain printed-paper overlays or a decorative appearance printed on the surface. They are available in thicknesses of 5/32, 3/16, and ¼ in.

Figuring Panels Needed To figure how many 4 ft 0 in. wide panels to buy, measure the length of each of the walls in the room. Divide this total number of feet by 4. This gives the number of panels. From this deduct the following, ½ a panel for each door, ¼ a panel for each window, and ½ a panel for a normal fireplace. Buy a full panel for any fractional-size panel that may be left after the deductions are taken.

Conditioning Panels Panels should be stored in a dry location. Do not store in new construction where moisture is high because of plastering or drywall construction. After all is dry, store the panels for 48 hours in the room in which they will be installed. This will permit them to adjust to the room temperature and humidity. Store the panels flat with spaces to allow air to circulate. They may be stored vertically on their 8 ft 0 in. side. Spacers are required between the sheets (Fig. 18-48).

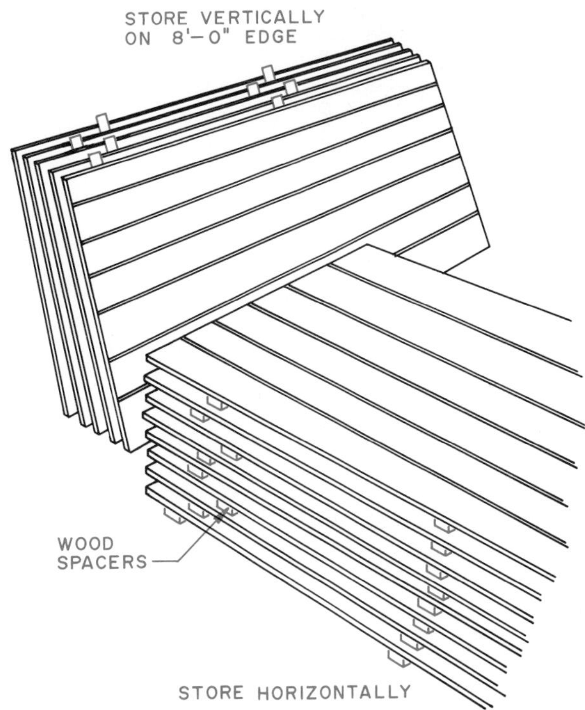

FIGURE 18-48

Plywood paneling should be stored vertically or horizontally.

Installing Plywood Paneling Plywood paneling is available in sheets 4 ft wide and 8 ft long. Longer sizes are available. It can be installed horizontally or vertically (Fig. 18-49). In most

FIGURE 18-49

This plywood paneling is installed with the grain running vertically. *(Courtesy of Georgia-Pacific)*

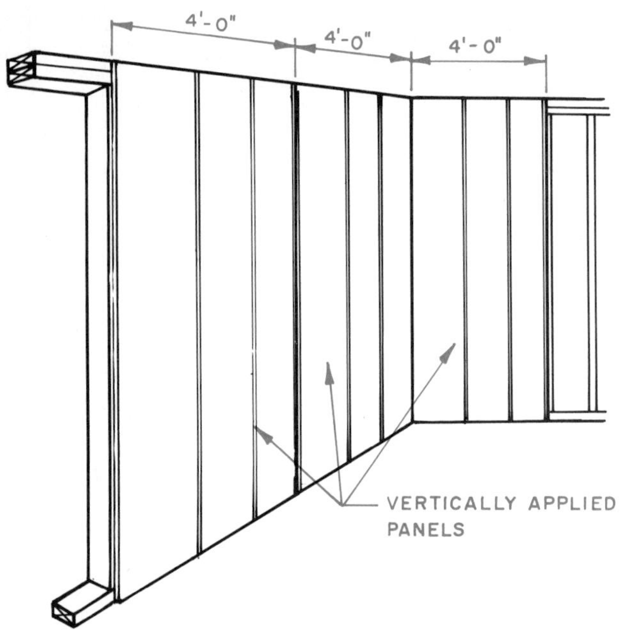

FIGURE 18-50

Vertically applied plywood panels.

cases it is installed vertically (Fig. 18-50). The 4-ft panel width enables the edges of the panel to rest on studs spaced 16 or 24 in. on center with no cutting (Fig. 18-51).

Occasionally, paneling is applied horizontally (Fig. 18-52). This makes the ceiling appear lower and the walls longer. Since all panel

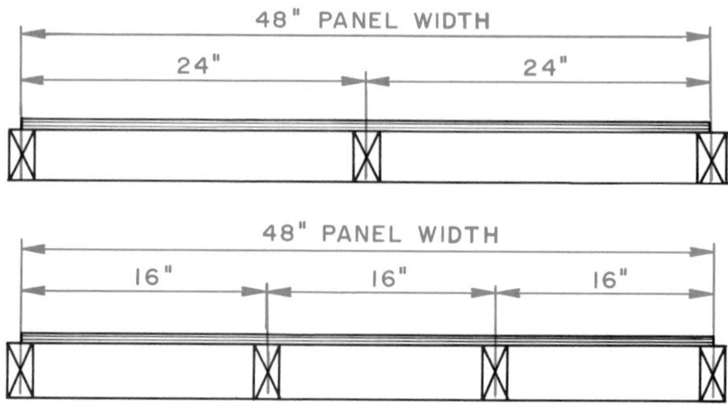

FIGURE 18-51

Plywood paneling will install without cutting on studs spaces 16 and 24 in. on center.

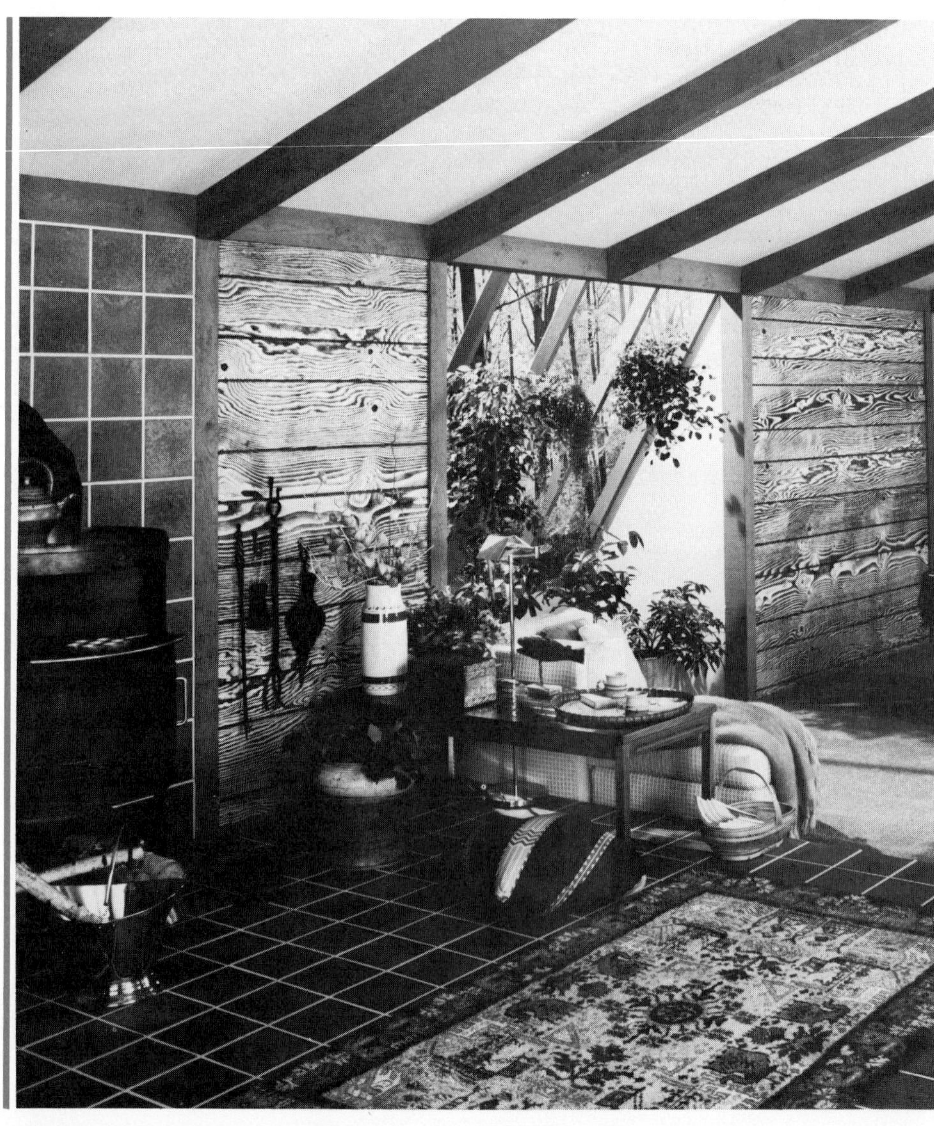

FIGURE 18-52

This paneling is installed horizontally. *(Courtesy of Georgia-Pacific)*

edges must rest on 2-in. blocking, horizontal members are needed when installing the sheets horizontally (Fig. 18-53). If the plywood paneling has tongue-and-grooved edges, the horizontal blocking can be omitted.

If the studs are spaced 16 in. on center, the paneling should be at least ¼ in. thick if no backing is used. If spaced 24 in. on center, ⅜-in. paneling should be used. Plywood paneling less than ¼ in. thick must have a backing material. Often ⅜-in. gypsum board or 5/16-in. plywood sheathing is nailed to the framing and the thin plywood paneling is fastened over this (Fig. 18-54). It is best if all plywood paneling is applied over a backing material.

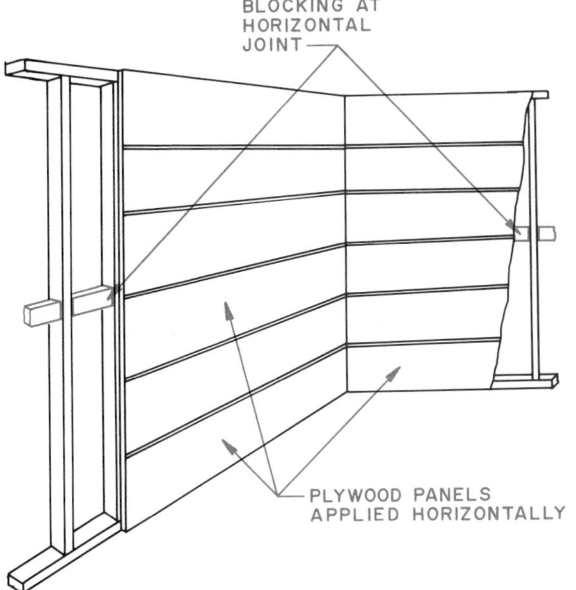

FIGURE 18-53

When installing panels horizontally, blocking is needed to support the joint in the center of the wall.

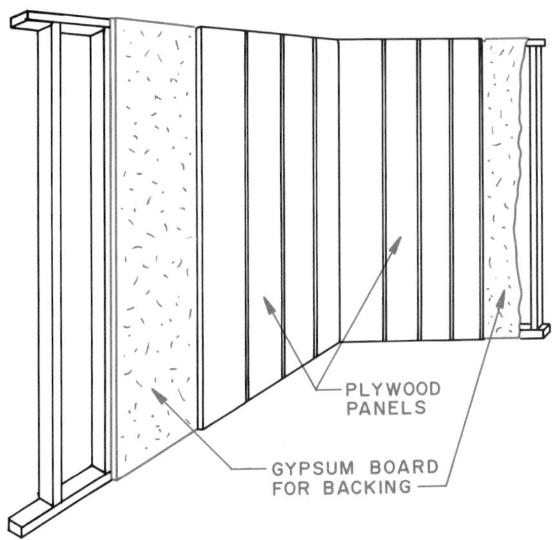

FIGURE 18-54

Plywood paneling should be installed over a gypsum board backing.

Installing Panels with Nails

Be certain that the insulation is covered with a vapor barrier.

Paneling nailed directly to studs requires 3d (1¼-in.) finishing or casing nails. When nailed to studs or furring strips the nails are spaced 6 in. apart on the panel edges and 12 in. apart on the panel interior (Fig. 18-55). When nailing over gypsum board or other backing, use 6d (2-in.) nails. Space the nails 6 in. apart on the edges and 2 in. apart on the panel interior.

Most panels have some type of groove running the length of the panel. These are spaced so that there is a groove every 16 in. Whenever possible, drive the nails in a groove. If it is necessary to nail in the face of the panel, set the head and cover it with material from a color putty stick. Most panels have lines spaced 16 in. on center printed on the back. These are used as a guide for applying beads of adhesive.

When installing plywood panels leave a hairline crack between their edges. This will allow for expansion of the panel and prevent buckling. Spray-paint the wall wherever two panels will meet. This will prevent a light background from showing through any crack between panels (Fig. 18-56).

Before installing the panel measure carefully and cut openings for electrical outlets

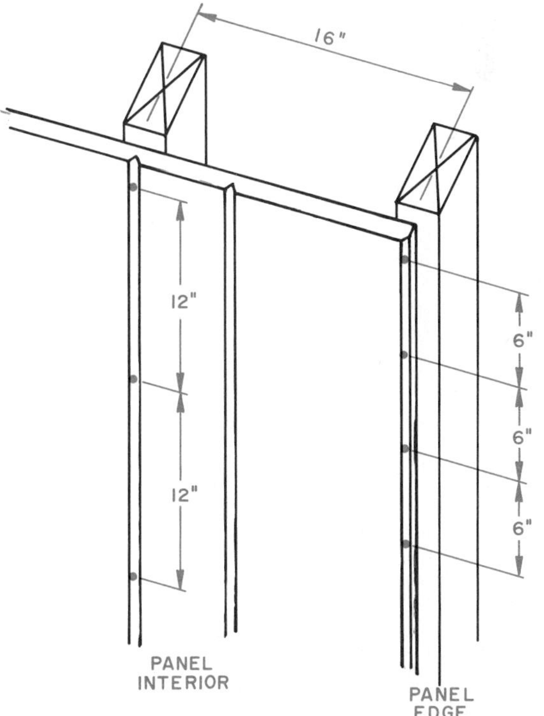

FIGURE 18-55

How to space nails on plywood paneling.

FIGURE 18-56

Paint the wall black where each crack between plywood panels will fall. *(Courtesy of Georgia-Pacific)*

(Fig. 18-57). A template the size of the electrical box can be used to mark the size of the opening. The box can be chalked and the panel set against it. The location of the box is marked on the back of the panel (Fig. 18-58). Bore holes at each corner and cut the opening with a compass or saber saw. Do not cut it more than ¼ in. oversize (Fig. 18-59).

When you come to a door or window, measure carefully and cut the panel around the opening. The panel usually butts the window frame and the trim goes over it. It can butt the trim if desired. This is sometimes done in remodeling jobs when it is not desirable to remove the trim (Fig. 18-60).

FIGURE 18-57

How to measure to locate cutouts for electrical outlets.

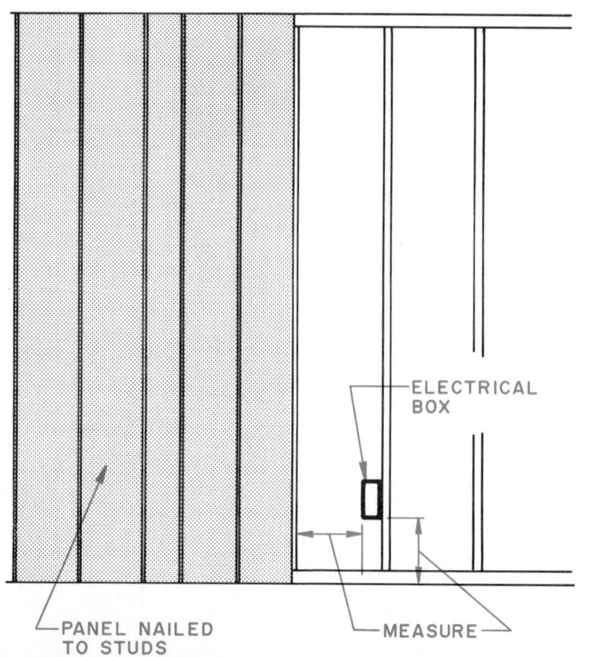

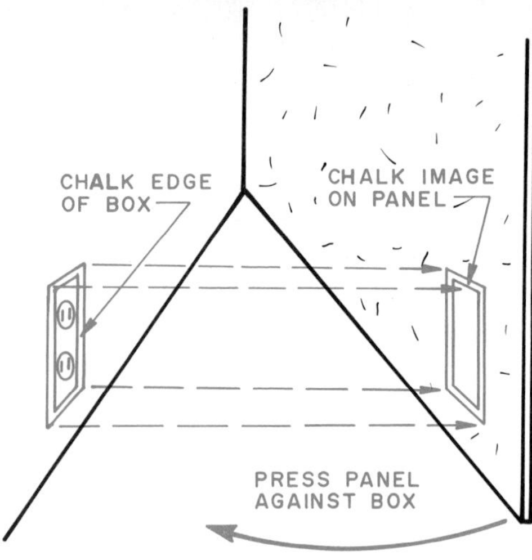

FIGURE 18-58

Chalk can be used to imprint on a panel the location of an electrical outlet.

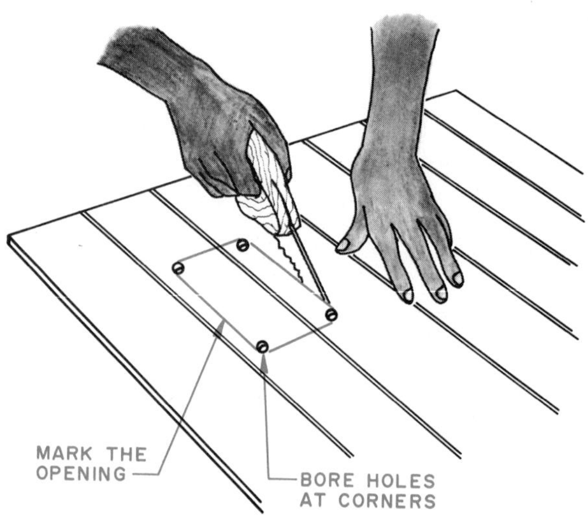

FIGURE 18-59

How to cut the opening for an electrical outlet.

FIGURE 18-60

Plywood paneling can be placed below or butt the window trim.

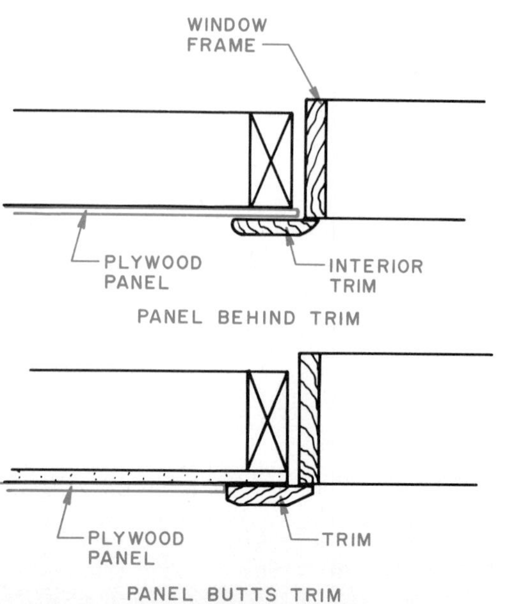

To install plywood paneling, start in one corner of the room. Using one or two nails, lightly tack the panel in place (Fig. 18-61). Then use a level or plumb line to get the panel plumb (Fig. 18-62). It may help if wedges are placed under the panel end next to the floor. This will help hold the panel plumb as it is nailed to the studs (Fig. 18-63). If the edge of the panel in the corner does not fit squarely against the wall, it will have to be cut to fit. The panel can be marked for cutting by drawing a line parallel to the intersecting wall. This can be marked with a compass (Fig. 18-64) or by placing a strip of wood the thickness of the amount to be removed against the edge. Cut off enough to get the panel edge parallel with the intersecting wall. It should clear the wall by about ⅛ in. The panel should clear the floor and ceiling ¼ to ½ in. to allow for expansion.

Now replace the panel on the wall, check it for plumb again, and nail in place. The following panels should be plumb because they are following the one just installed. It is a good idea to check them for plumb occasionally.

If the edge of the panel does not reach a stud, a 2 × 2 must be nailed to the inside of the stud. If the panel covers more than half the edge stud, there will not be enough room for the next

FIGURE 18-61

Start nailing a panel in one corner and secure with several nails. *(Courtesy of Georgia-Pacific)*

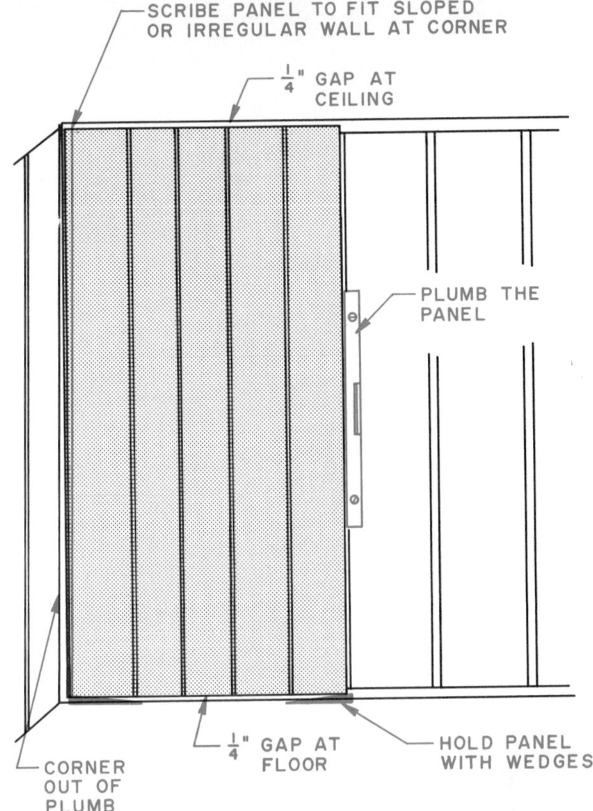

FIGURE 18-62

Check the panel to be certain that it is plumb before finishing nailing it to the studs.

FIGURE 18-63

Wedges below the panel at the floor will help hold it plumb as it is nailed.

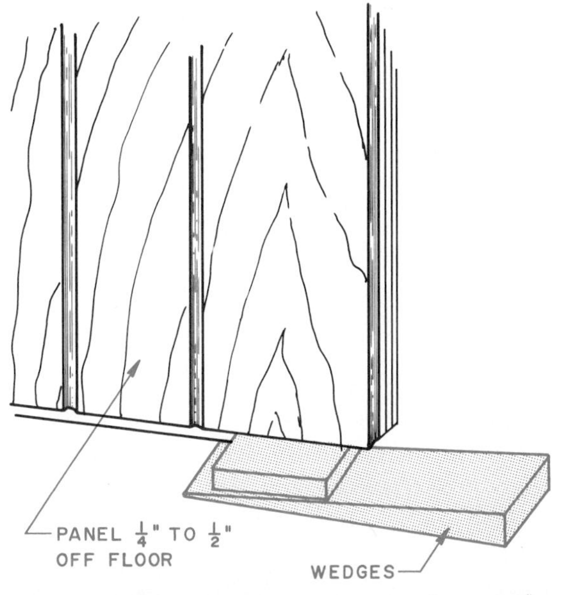

FIGURE 18-64

When the panel meets an irregular wall, scribe the edge to be cut.

FIGURE 18-66

A chalk line can be dropped to locate the line of the stud so that nails can be accurately placed.

panel. Nail a 1 × 2 or 2 × 2 to the outside face of the stud (Fig. 18-65).

It helps to mark stud locations on the floor and ceiling. Once the panel is in place, the studs are hidden. A chalk line can be dropped from these marks to help when nailing (Fig. 18-66). If it is difficult to find a stud, use a stud finder (Fig. 18-67).

Sometimes plywood panels are used as wainscoting. *Wainscoting* is paneling that goes from the floor partway up a wall. It is usually 32 to 36 in. high. It will have a *chair rail* on top. It is installed the same as just described. Horizontal blocking is recommended for the top edge of the panel and as a nailer for the chair rail (Fig. 18-68).

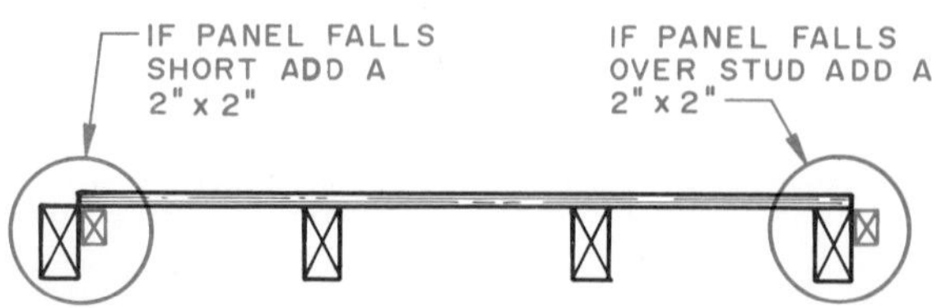

FIGURE 18-65

If a panel does not fall on the center of a stud, a 2 × 2 in. piece must be added.

FIGURE 18-67

A stud finder can be used to locate studs. *(Courtesy of Georgia-Pacific)*

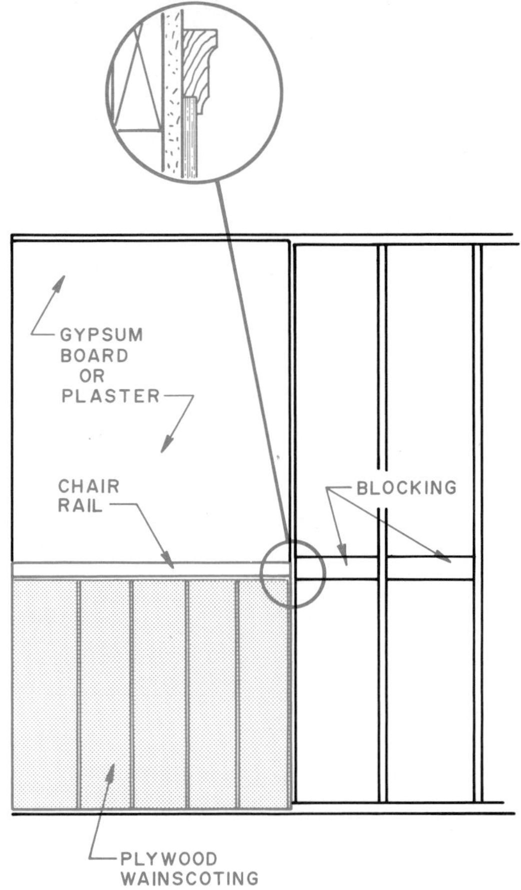

FIGURE 18-68

Horizontal blocking is used behind panels having horizontal joints.

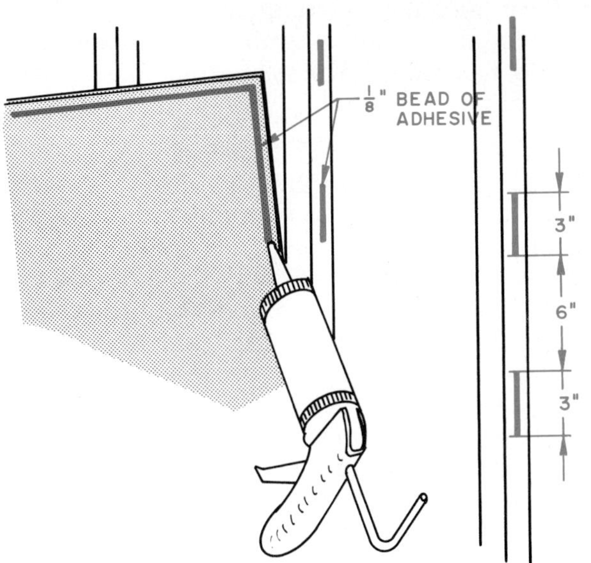

FIGURE 18-69

Plywood paneling can be fastened with beads of adhesive.

Fastening Plywood Panels with Adhesive

Paneling can be fastened to studs or backing with adhesive. When applied to studs, lay a continuous ⅛-in. bead on the top, bottom, and sides of the panel and also around all openings, such as doors and windows. Put 3-in.-long beads every 6 in. on intermediate studs. When gluing panel to backing, such as gypsum board, use the same procedure. Follow the instructions of the manufacturer of the adhesive (Fig. 18-69).

Place the panel against the studs or backing and lightly press in place. Plumb the panel. Drive several nails in the top edge of the panel to hold it in place. Then pull the bottom of the panel 6 to 8 in. from the wall. Block it there for about 10 minutes to allow the adhesive to set (Fig. 18-70). Then push the panel firmly against the wall. Tap or push the panel over each stud to get a firm bond. If necessary, drive a few finishing nails to hold the panel in place until the adhesive dries.

When installing paneling, arrange the sheets around the room to try to get the best

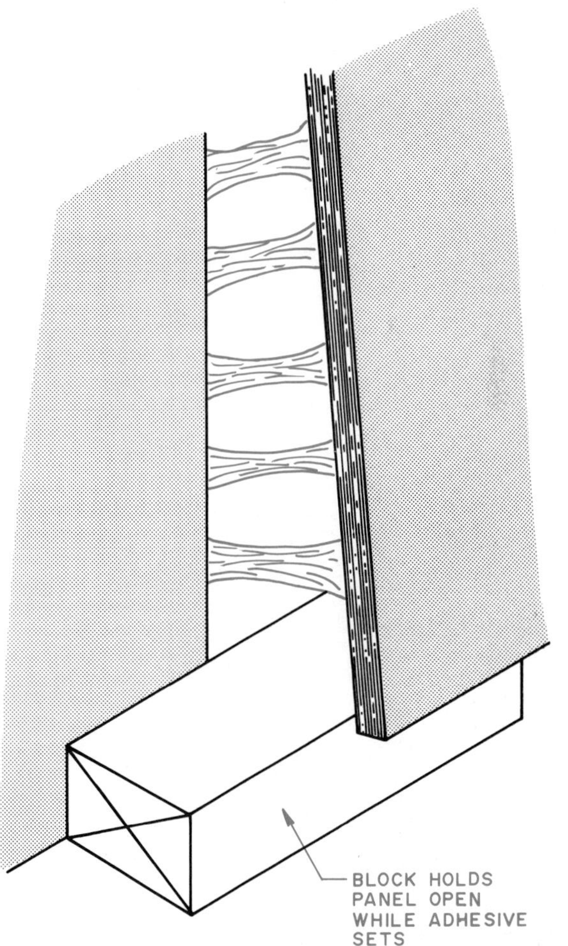

FIGURE 18-70

After pressing the panel into the adhesive on the wall, pull it out and block it for 10 minutes before pressing it permanently against the wall.

pattern and color arrangement. The sections with similar color or grain can be placed together to produce a more harmonious appearance.

Installing Plywood Paneling over Old Walls

To install paneling over an old wall, it is necessary to locate the studs. They are hidden from view and should be marked on the ceiling and floor. A chalk line can be run from the ceiling to the floor. This gives a guide for nailing into a stud.

If the old wall is irregular or in poor shape, nail a series of furring strips over it. Use wedges to bring the strips into plumb. This is similar to furring masonry walls as described in the following section on masonry walls.

Installing Plywood Paneling over Masonry

Masonry walls are usually furred out to receive paneling. A typical layout is shown in Fig. 18-71. If insulation is not needed, 1 × 2 in. strips are fastened to the wall. This is done with adhesives, a power-driven nailer, or by fastening into wood nailing plugs set in the wall during construction. Sometimes case-hardened nails are driven into the motar joint. Leave a ½-in. space between the ends of furring strips to allow for ventilation. Notice the use of wedges to get the furring straight. Furring must be placed around doors and windows (Fig. 18-72).

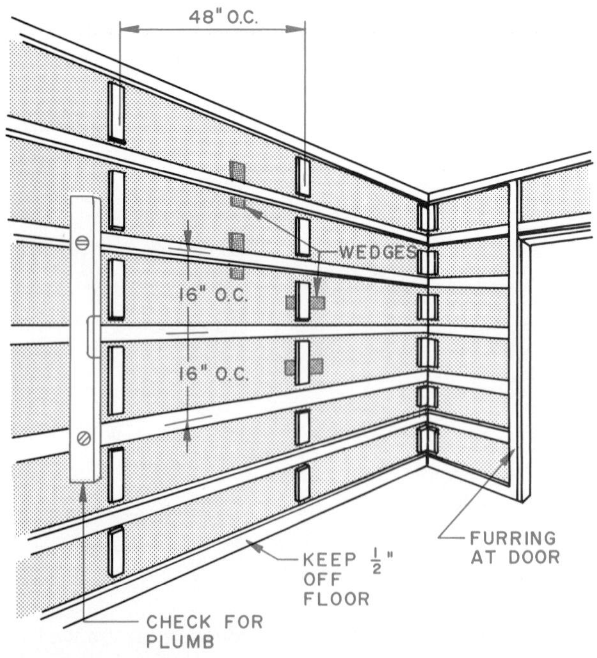

FIGURE 18-71

Masonry walls are furred out to receive paneling. Wedges are used to plumb the furring strips.

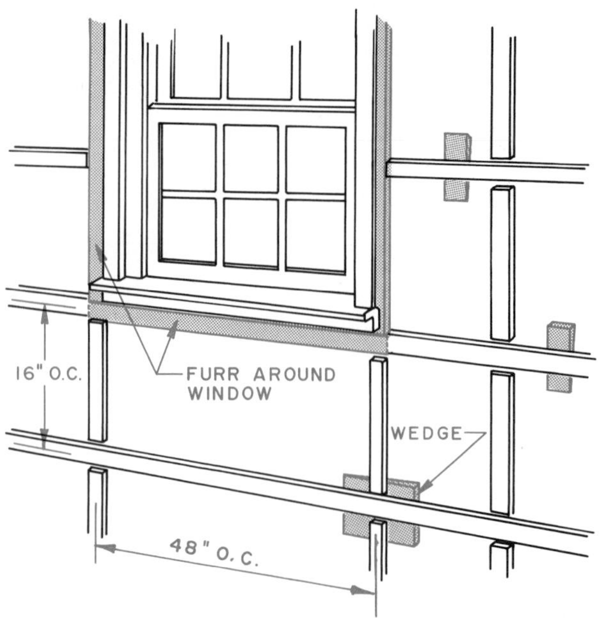

FIGURE 18-72

Furring is placed around windows and doors to provide a nailing base.

Special blocking is needed at the ceiling on basement walls running parallel with the floor joists (Fig. 18-73).

If the wall is to be insulated, a stud wall is built next to the masonry wall. 2 × 4 or 2 × 3 in. studs can be used. The wall is assembled on the floor and raised into place. It is wedged to the overhead floor joists and nailed to them (Fig. 18-74). The studs can be turned flatwise (the 4-in. dimension next to the wall). This saves space but reduces insulation. Since they may bow, they should be mechanically joined to the ma-

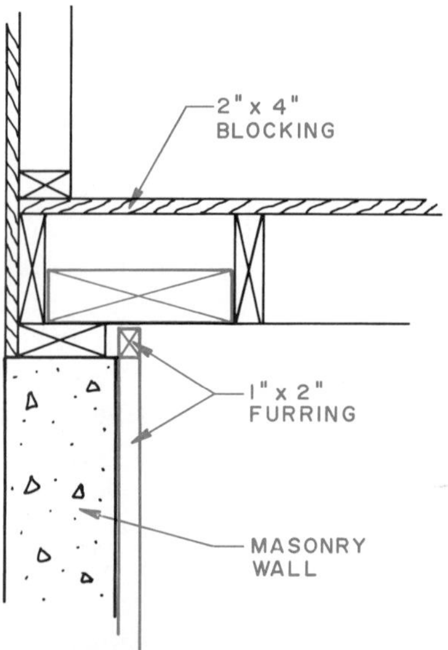

FIGURE 18-73

How to block at the ceiling when paneling basement walls.

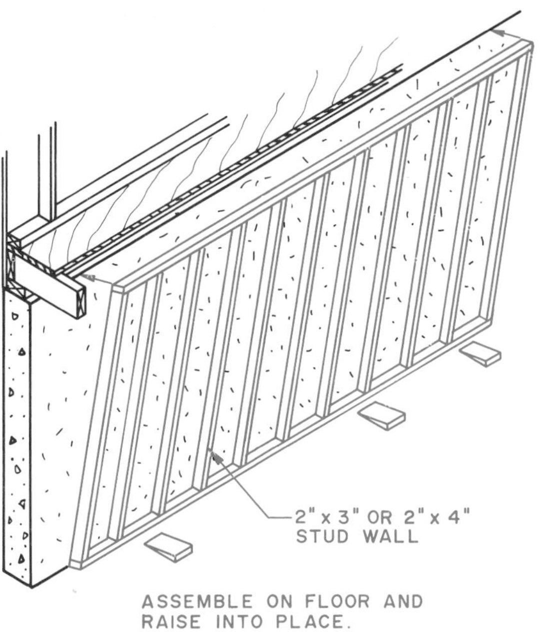

FIGURE 18-74

Stud walls are used next to masonry walls if insulation batts are to be used.

FIGURE 18-75

Paneling applied in a herringbone pattern. *(Courtesy of Georgia-Pacific)*

sonry wall. Paneling is installed as described for stud walls.

Be certain to apply a vapor barrier over the furring or studs before installing the plywood paneling.

How to Produce a Herringbone Pattern A herringbone pattern produces a dramatic appearance (Fig. 18-75). To produce this pattern, lay out the paneling cuts on a 45° angle (Fig. 18-76). Match up two sheets of paneling. Place

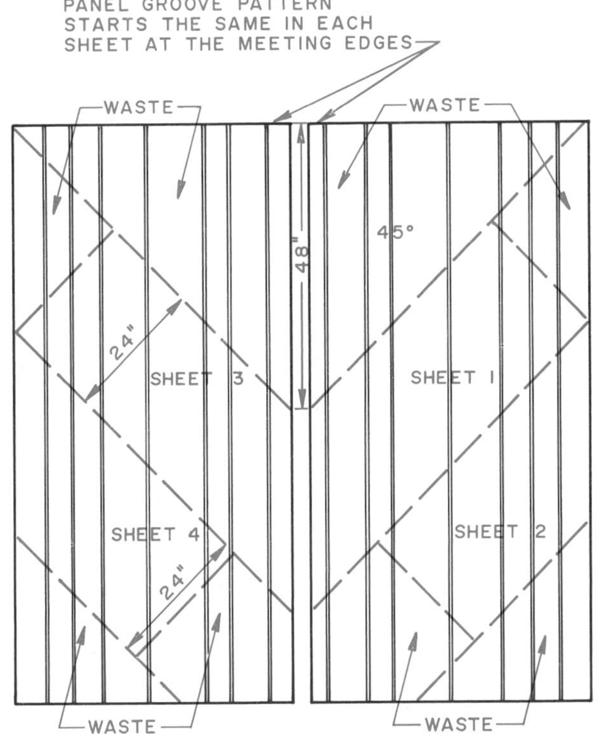

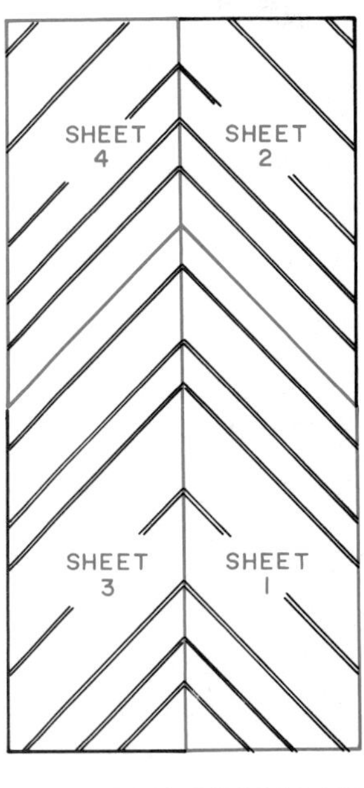

FIGURE 18-76

Cutting paneling to produce a herringbone pattern.

them so that the groove pattern is the same on the edge at which they meet. After cutting, pair up the opposite cuts where panels are nailed to the wall. The finished wall should have a batten strip vertically at the midpoint or a butt joint could be used.

Panels can be cut on 60° if a steeper angle is desired.

Installation Techniques Sometimes a ceiling is higher than 8 ft 0 in. This can be handled in several ways, depending upon the distance. If the distance beyond 8 ft 0 in. is small, the panel can be raised off the floor and a molding can be used at the ceiling. Higher ceilings require that a small panel be cut to fill the space above the 8 ft 0 in. sheet. Usually, some type of wood molding is placed over the end joint between the panels (Fig. 18-77).

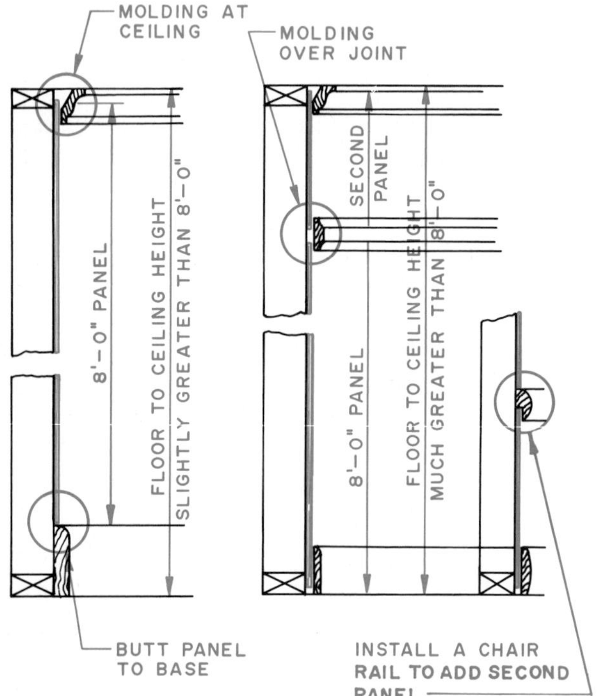

FIGURE 18-77

Ways to panel rooms with ceilings higher than the normal 8'-0".

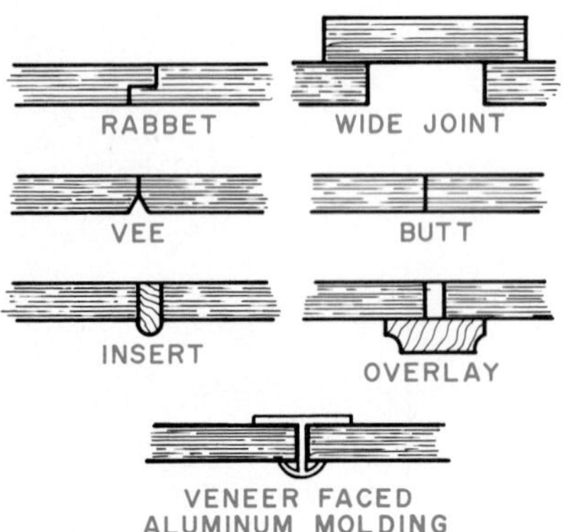

FIGURE 18-78

Types of joints used to join plywood panels.

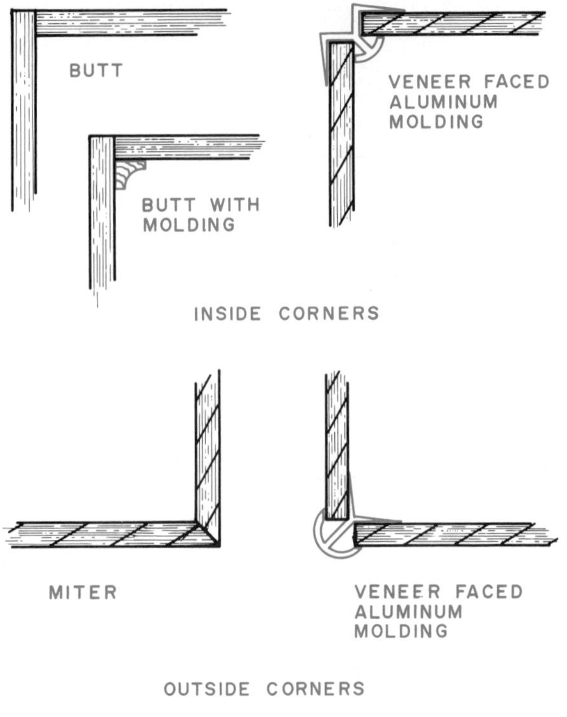

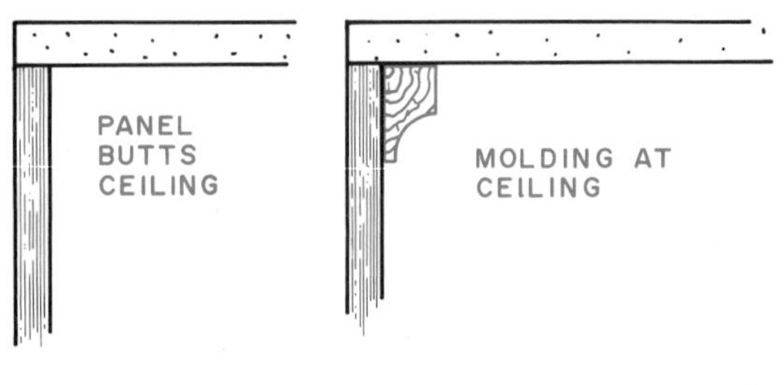

FIGURE 18-79

Ways to form inside and outside corners when using plywood paneling.

Plywood paneling generally has the long edge prepared as half a V-groove. When it matches another panel, a V-groove is made. Other types of joints can be made. Examples are shown in Fig. 18-78.

Inside corners are usually butted. A small molding can be nailed in place if desired (Fig. 18-79). Outside corners are usually mitered. A veneer-faced aluminum molding is available for forming outside corners. The panel can butt the ceiling squarely or a molding can be used.

FIGURE 18-80

When cutting plywood paneling with a portable circular saw, place the finished surface down. *(Courtesy of Georgia-Pacific)*

Cutting Plywood Paneling Plywood paneling usually has the final finish applied in a factory. The carpenter must handle the sheets carefully so that the surface does not get scratched. Also, when cutting the panels care must be taken to prevent splintering the edge. If the panel is cut on a table-type circular saw, the finished surface should face *up*. When cutting with a portable circular saw or saber saw, cut with the finished face down (Fig. 18-80). If cutting with a hand saw, use a finetooth saw and keep the finished face up.

INSTALLING HARDBOARD AND FIBERBOARD PANELING

Hardboard and *fiberboard paneling* are handled in the same way as described for plywood paneling. The sheets have a finished surface that must be protected from damage. If ⅛-in.-thick hardboard is used, a backing is necessary. Sheets ¼-in. thick can be used over studs spaced 16 in. on center. Sheets 7⁄16-in.-thick can be used with studs spaced 24 in. on center.

Fiberboard usually has tongue-and-groove edges. It must be ½ in. thick to be applied over studs 16 in. on center and ¾ in. thick for 24-in. spacing.

Nailing Hardboard Panels When nailed to studs or over a backing, the nails are spaced 4 in. apart on the outer edges and 8 in. at intermediate studs. The nail should be long enough to penetrate the stud ¾ in. (Fig. 18-81).

Gluing Hardboard Panels When applying hardboard panels with adhesive to studs, apply a continuous ⅛-in. bead to the stud at the outer edge of the panel. On intermittent studs apply 3-in.-long strips of adhesive with 6 in. between each. The same application is used on panels with backing (Fig. 18-69). Moldings and trim

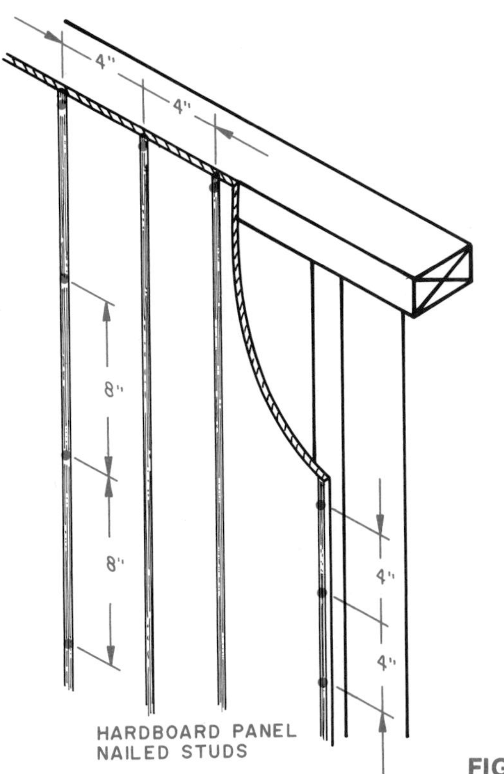

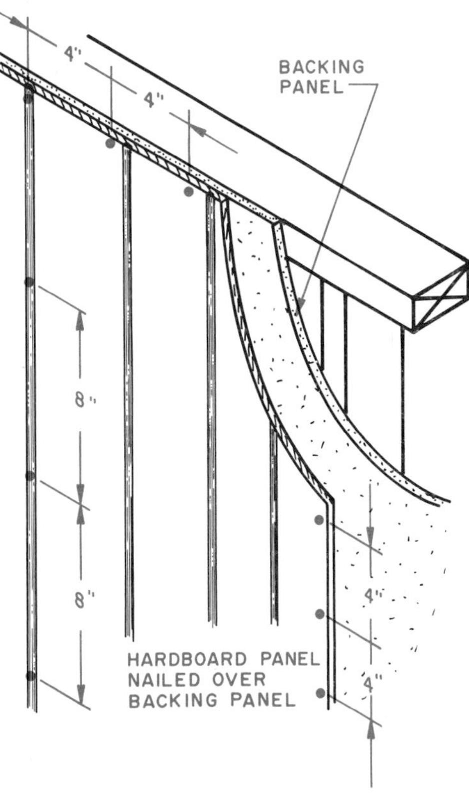

FIGURE 18-81

Nailing patterns for installing hardboard paneling.

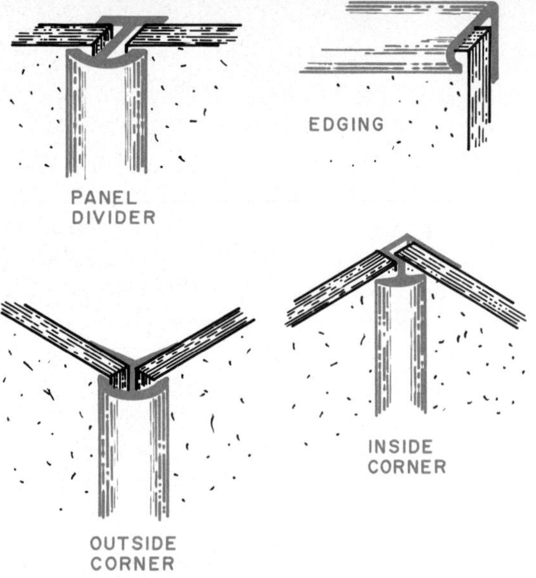

FIGURE 18-82

Aluminum corner strips are used to finish inside and outside corners when installing hardboard paneling.

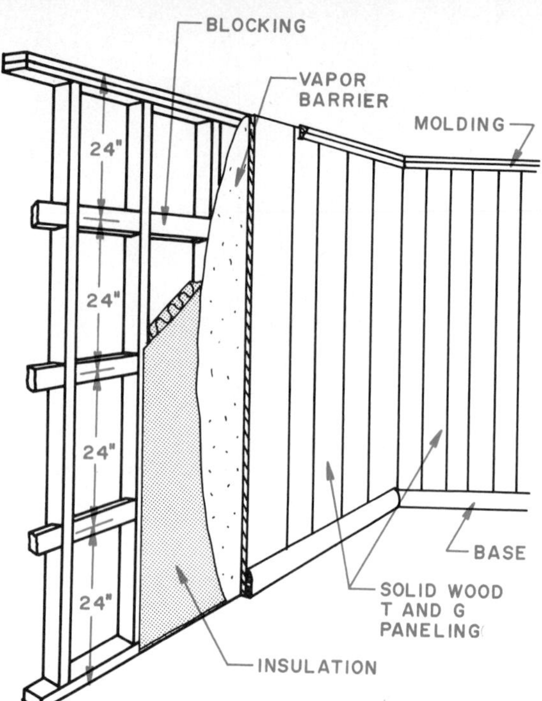

FIGURE 18-84

When installing solid wood paneling, nail horizontal blocking at 2′-0″ intervals.

are applied as described for plywood paneling. To trim edges and form corners, prefinished aluminum moldings are available (Fig. 18-82).

INSTALLING SOLID WOOD PANELING

There are many kinds of *solid wood paneling* available. Generally, they have a tongue-and-groove cut on the edges (Fig. 18-83). Most are 6 to 8 in. wide. Paneling should be ⅜ in. thick for 16-in. stud spacing, ½ in. thick for 20-in. stud spacing, and ⅝ in. thick for 24-in. stud spacing. The paneling may have a factory-applied finish or be finished after it is installed. The carpenter must be very careful to keep the finished surface clean and free from scratches or dents.

Wood paneling can be installed vertically or horizontally. It must be seasoned to the moisture content acceptable in your locality before it can be installed.

FIGURE 18-83

Typical patterns of solid wood paneling.

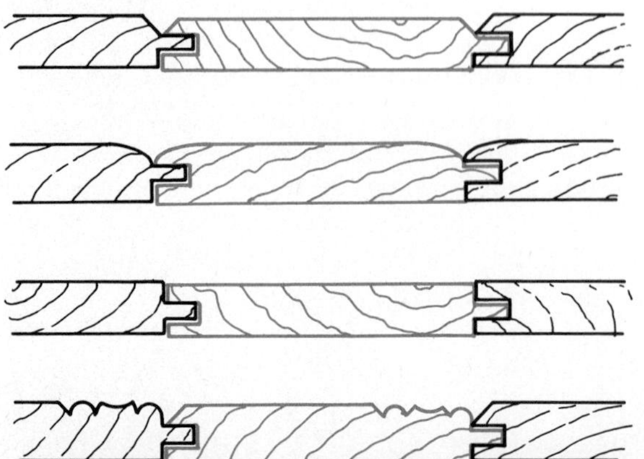

Installing Solid Wood Paneling Vertically

Be certain that the insulation is covered with a vapor barrier.

Install horizontal blocking or furring between studs at 24 in. on center (Fig. 18-84). Use 5d or 6d casing or finishing nails to fasten paneling to studs. Use one nail to blind-nail boards 6 in. wide or smaller to the stud. Use two nails for boards 8 in. or wider. One of these nails is blind-nailed in the tongue (Fig. 18-85). The other is face-nailed and is set and hidden with a colored putty stick.

Start the first piece of paneling in a corner. Make certain that it is plumb (Fig. 18-86). If the wall is out of plumb, trim the edge of the board to bring it parallel to the wall. Plane a 5° bevel on the edge. This helps produce a tight fit in the corner. Nail the other pieces, checking them for plumb. When the other side of the room is reached, scribe the last piece to fit. Cut it and plane a 5° bevel on the edge. Slip the tongue-and-grooved edges together and push the last piece into place.

FIGURE 18-85

How to nail solid wood paneling to blocking.

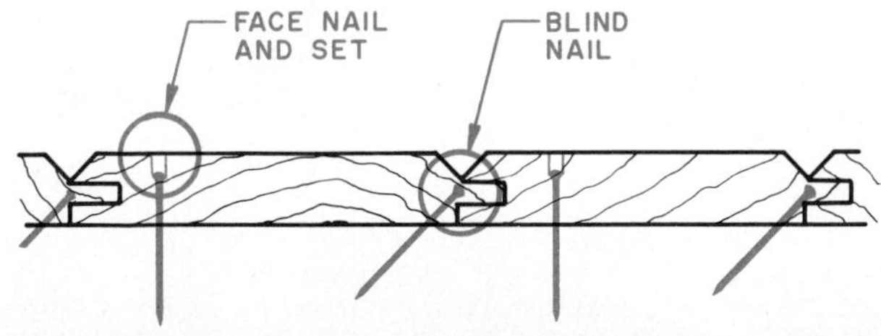

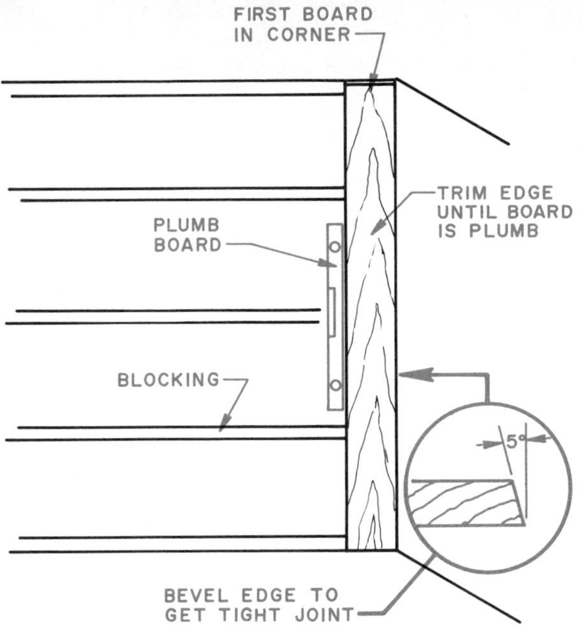

FIGURE 18-86

Start nailing solid wood paneling in the corner. Plumb the first piece.

Installing Solid Wood Paneling Horizontally

Horizontally applied paneling can be nailed directly to the studs. The span between studs and the thickness of paneling required is the same as for vertically applied paneling.

Start the first piece at the floor line. Using a level and wedges, be certain that it is level before nailing. Bevel the ends 5° so that they fit tightly to the wall. Outside corners are mitered. Blind-nail in the same manner as vertically applied siding (Fig. 18-87).

At the ceiling, bevel the edge of the board 5° or use a molding.

FIGURE 18-87

When installing solid wood paneling horizontally, locate the first piece at the floor. Be certain that it is level.

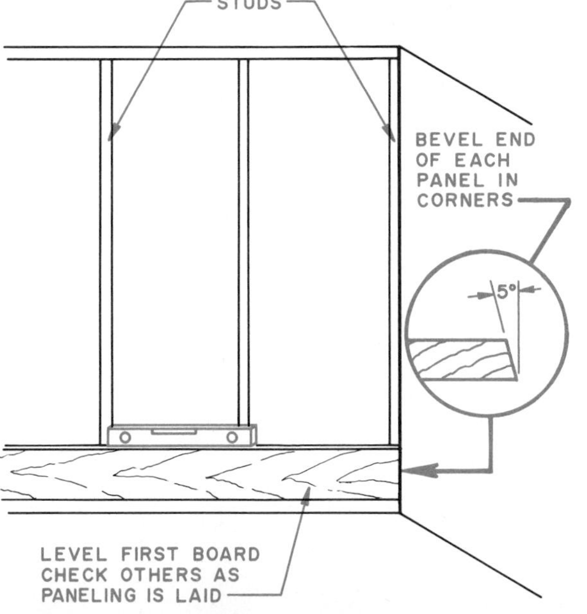

In the bath and kitchen the walls around the fixtures are usually covered with some water-resistant covering. Two commonly used are plastic-surfaced hardboards and ceramic, plastic, and metal tiles.

Plastic-surfaced hardboards are adhered to water-resistant gypsum board. Moldings are placed at the corners and edges. The adhesive is spread with a notched trowel over the entire wall. The sheets are pressed into the adhesive. The moldings are nailed to the wall before the adhesive is applied.

Plastic, ceramic, and metal tile is adhered to water-resistant gypsum board in the manner just described. After the adhesive has dried, grout is placed in the grooves between tiles. Grout is a paste-like cement which hardens, filling the grooves between the tiles. Some tiles are sold in sheets of 16 tiles with a flexible, silicone rubber grout already between them (Fig. 18-88). Silicone rubber is used to seal the seams, corners, and tub and pipe edges.

FIGURE 18-88

Ceramic tile is installed with 16 tiles together in a single unit. *(Courtesy of American Olean Tile Co.)*

Ceilings are most often covered with gypsum board or plaster. These were discussed earlier in this chapter. Ceiling tile and suspended ceilings also find extensive use.

CEILING TILE

Ceiling tile have acoustical value as well as being decorative. They are fiberboard being made from cane fibers and natural wood. They are completely decorated in the factory, so the carpenter must be careful to keep them clean and undamaged. A wide variety of surface textures are available. Some are sculptured while others have holes or slots to increase acoustical value.

Tiles are generally ½ in. thick and have tongue-and-groove edges. Common sizes are 12 × 12, 12 × 24, 16 × 16, and 16 × 32 in.

Planning the Tile Layout

Before installing tile, a plan for placing them must be developed. The tiles next to the walls are the same size on both sides of the room. Avoid having a small part of a tile, such as 3 or 4 in., next to the wall. Following are steps to make a plan:

1. Measure the size of the ceiling (Fig. 18-89). This was 9 ft 4 in. wide and 10 ft 6 in. long.
2. If 12-in. tiles are used, the width has nine tiles and 4 in. left over. This is too small to make a pleasing appearance, so add 12 in. to it and divide by 2. This will give 16 in. ÷ 2 or an 8-in. tile next to the opposite walls. One of the nine full tiles was divided in half and added to the small amount remaining to produce a better-looking ceiling.
3. Do the same with the length.
   10 ft 6 in. has 6 in. left over
   6 in. + 12 in. = 18 in.
   18 in. ÷ 2 = 9-in. tile on each opposite wall

   If using 16 × 16 in. tiles, change the wall distance into inches and divide by 16. This will give the number of full tiles and the amount left over. Add 16 in. to the remainder and divide by 2 to get the size of the tiles next to the opposite walls.

Installing Ceiling Tile when Ceiling Is Flat

If the ceiling is flat, such as a gypsum board ceiling, tiles can be held in place with an adhesive.

Plan the layout and cut the tiles for the one edge. Run a chalk line from one side of the room to the other. Snap a chalk mark on the ceiling, locating the edge of the tile next to two outer walls (Fig. 18-90). Place the adhesive recommended for the tile being used on the back of the tile. Adhesive placing details are shown in Fig. 18-91. Do not get the adhesive near the edges of the tile.

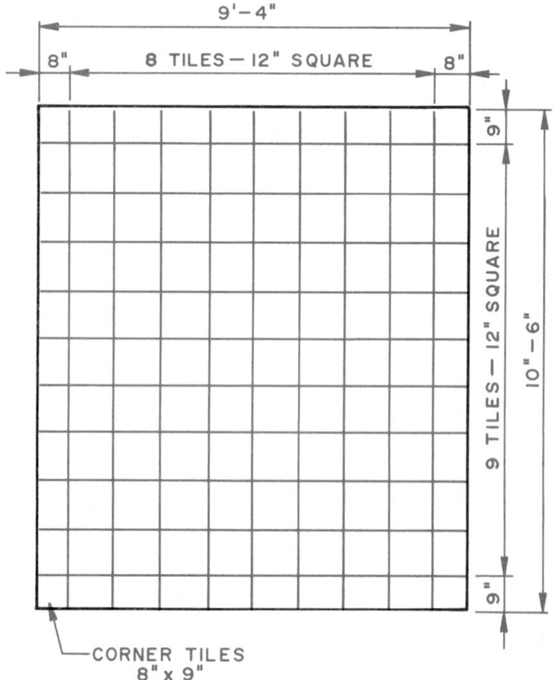

FIGURE 18-89

How to lay out a ceiling to receive tiles.

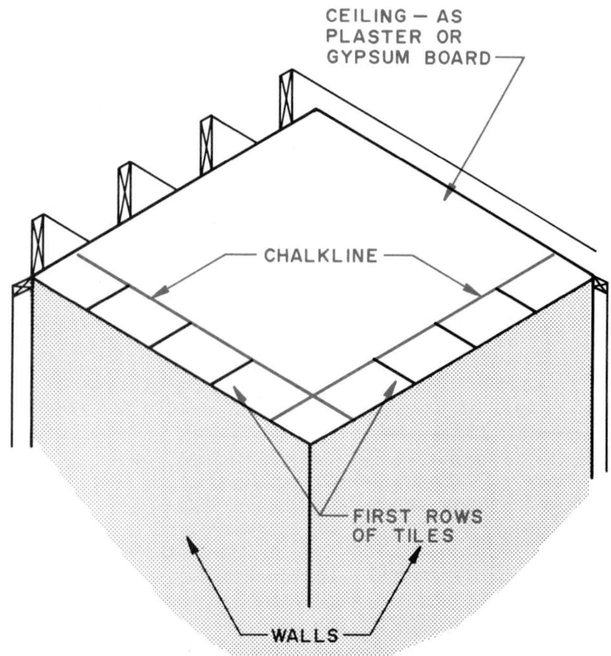

FIGURE 18-90

Locate the first row of ceiling tiles with a chalk line.

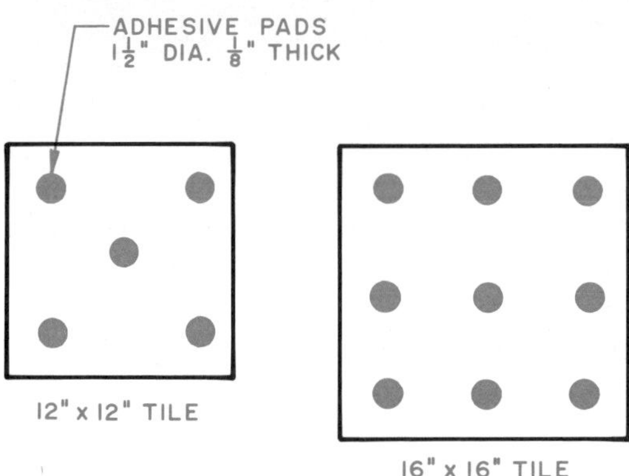

FIGURE 18-91

Proper placing of adhesives on ceiling tile.

Place the tile in position on the ceiling. Line the outer edge with the chalk line. Press it firmly to the ceiling. Sometimes two staples are driven through the flange to hold the tile while the adhesive dries. If the wall is bowed or crooked, the cut edge of the tile may have to be trimmed some more. This should be checked before putting adhesive on the tile. It is important to get this first row straight because it is the guide for all the other rows.

Installing Ceiling Tile with Furring Strips

Furring strips are nailed to the wood joists (Fig. 18-92). They are usually 1 × 2 or 1 × 3 in. material. If the joists are not all level, a wood shim can be placed between the furring strip and the joist (Fig. 18-93). If the joist extends below the others, a small amount can be cut off so that the

FIGURE 18-92

Ceiling tile can be installed on furring strips. *(Courtesy of The Celotex Corporation)*

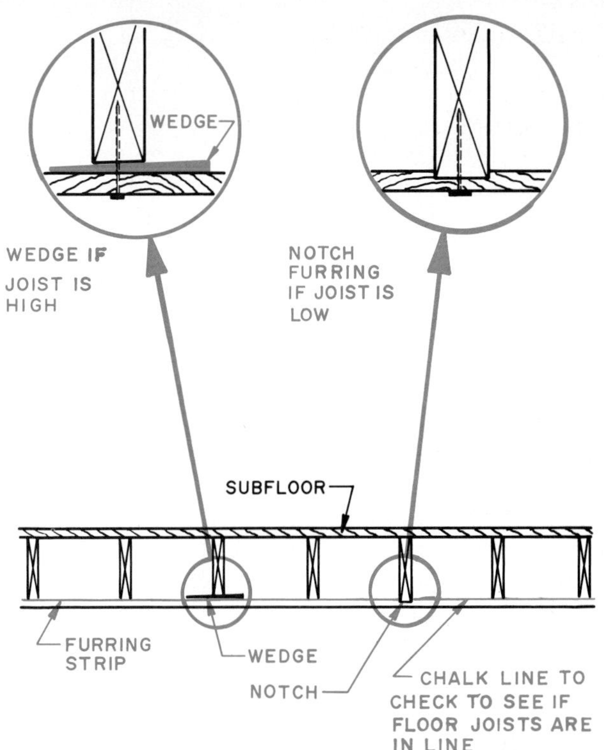

FIGURE 18-93

Furring strips can be wedged or notched to get them level.

furring strip is level. Check this with a chalk line.

The first furring strip is nailed next to the wall. The second strip is placed so that the visible edge of the first row of tile overlaps it to the center. The stapling tongue of the tile will extend ½ in. beyond the center of the wood strip. All other strips are laid 12 or 16 in. on center, depending on the size of the tile. As the opposite wall is reached, a strip is nailed next to that wall. Nail the furring with two 8d common nails at each joist. A typical layout is shown in Fig. 18-94.

Framing Projections below Joists If a water pipe or heating duct projects below the joists, the ceiling can be handled in one of several ways. If the projection is small, such as 1 in., a double layer of furring strips can be installed. Nail the first layer perpendicular to the joists and spaced 24 in. on center. Then nail a second layer over these and perpendicular to them. Space these to suit the tile layout plan (Fig. 18-95).

If the projection is more than you want to lower the entire ceiling, box it in with 2 × 2 in. framing. Consider planning the size so that it fits the tile layout plan. Wood molding can be used to conceal the inside and outside corners (Fig. 18-96).

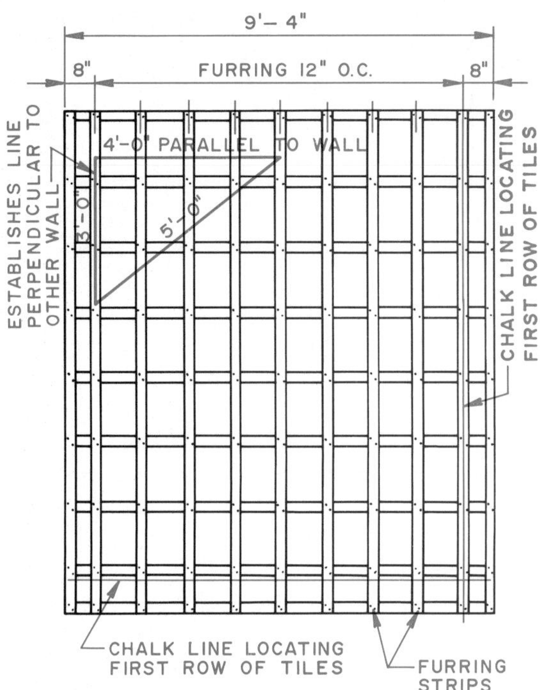

FIGURE 18-94

A typical furring layout for ceiling tile.

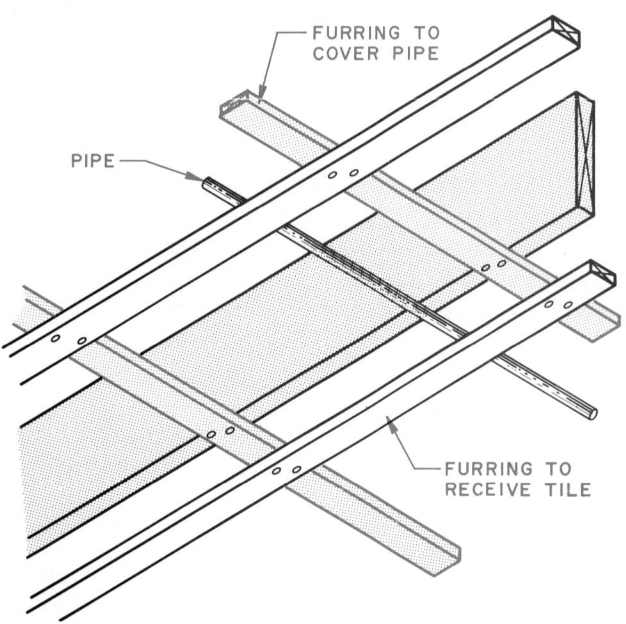

FIGURE 18-95

Items projecting below the ceiling joists can be concealed by furring below them.

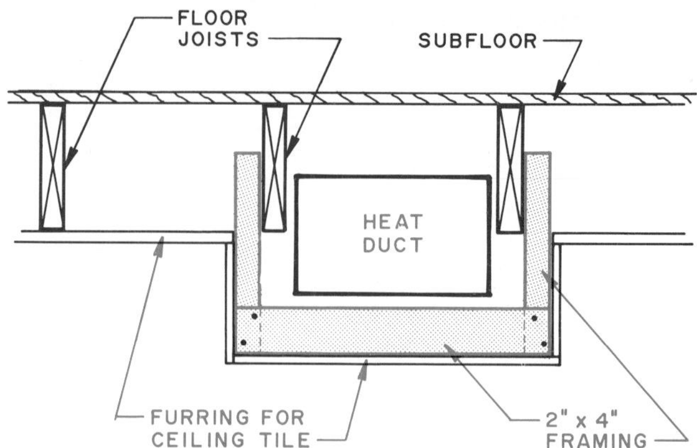

FIGURE 18-96

2″ × 4″ members are used to frame around large projections in the ceiling.

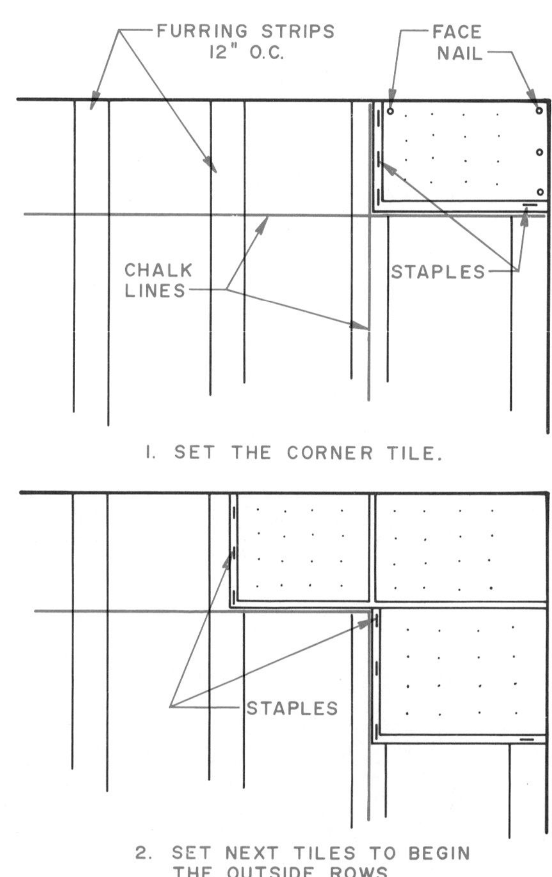

FIGURE 18-97

To install ceiling tile, set the corner tile first.

Installing the Tile on Furring Strips Using the ceiling tile layout, install chalk lines for the first row of tile on two joining walls (Fig. 18-94). In the example layout this was 8 in. on the short wall and 9 in. on the long wall. Make certain that the lines are perpendicular to each other. This can be done by measuring 4 ft on one and 3 ft on the other. The diagonal between the ends of these should measure 5 ft. Move the lines until the 5-ft measure is achieved. Refer again to Fig. 18-94. Snap the chalk lines to mark the furring.

Now cut the first tile. It will be in the corner and be cut on two sides. The tile in this example will be 8×9 in., including the flange. Cut off the sides with the groove. Leave the sides with the stapling flange. Staple and face-nail it in place (Fig. 18-97). Tile can be cut with a coping saw or very sharp utility knife. A saber saw or bandsaw with a sharp fine blade can be used. Cut with the finished surface facing up (Fig. 18-98).

Cut a second tile so that one edge fits into the tongue of the tile. Slide it into the tongue and staple (Fig. 18-97). Cut a third tile to go on the other wall. Cut it so that it fits into the corner tile. Staple to the furring.

Now staple two or three tiles along each wall and fill in the ceiling area with full-size tile. When the opposite wall is reached, cut each tile to fit. Often walls are not straight, so individual measuring and cutting is usually necessary (Fig. 18-99).

Stapling is done with 9/16- or 5/8-in. staples. Follow the tile manufacturer's recommendations. A 12 × 12 in. tile requires three staples in flange on the furring strip parallel with the wall and one in the flange perpendicular to the wall. A 16 × 16 in. tile will have four staples and a 16 × 24 in. will have five staples in the long flange. (Fig. 18-100).

Cover the space between the tile and the wall with a molding (Fig. 18-101).

SUSPENDED CEILINGS

A *suspended ceiling* consists of a grid of metal runners hung from the ceiling on wires. Removable panels are used within the grid. The

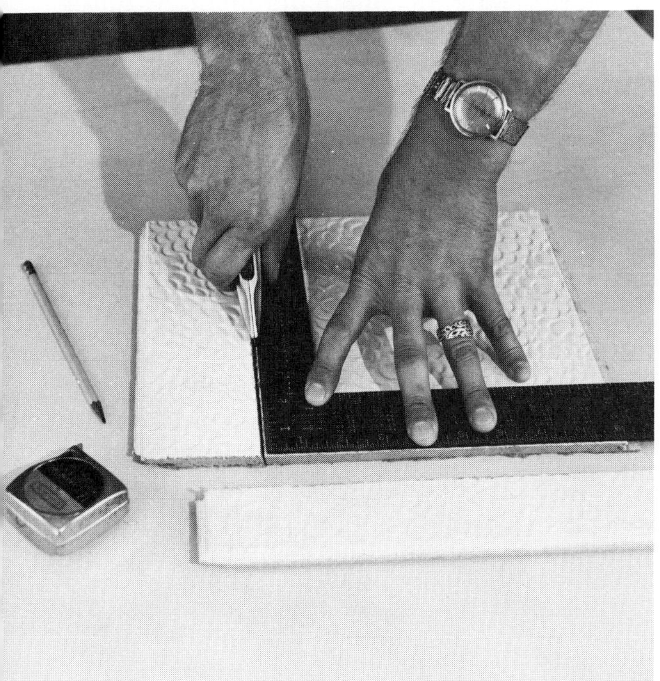

FIGURE 18-98

Ceiling tile can be cut with a utility knife. *(Courtesy of The Celotex Corporation)*

FIGURE 18-99

Staple the tiles to the furring. *(Courtesy of The Celotex Corporation)*

FIGURE 18-100

Where to locate staples in a ceiling tile. *(Courtesy of The Celotex Corporation)*

FIGURE 18-101

Cover the crack between the ceiling and wall with molding. *(Courtesy of The Celotex Corporation)*

grid has main runners which are usually in 12-ft lengths. They run the entire length of the room. Between these are cross-tees that are usually 2 ft long. This forms a grid 2 by 4 ft. The panels set in the grid are usually fiberglass. They have a finished surface which forms the ceiling. A variety of panel textures are available.

The ceiling must be dropped 2 to 3 in. below the joists so that there is room to insert the panels. Rooms with high ceilings can have the ceiling lowered by hanging the suspended ceiling on longer wires.

A suspended ceiling enables workers quickly to remove panels to get to any wiring or plumbing hidden by the ceiling. It also permits firefighters to get to fires above the ceiling by blowing the panels off with the fire hoses.

Developing the Panel Layout Generally, the panel layout is planned with the main runners parallel to the long side of the room. They are spaced 48 in. apart in most systems. This places the 4-ft panel dimension in this direction.

To locate the first main runner, measure in inches the short wall of the room. The long panel dimension parallels the short wall. If 48-in. panels are to be used, divide this by 48 in. Add the amount left over to 48 in. and divide by 2. This gives the width of the outside panel and the distance to locate the first runner (Fig. 18-102). The purpose for this division is to have the panels at the walls the same size.

For a room 13 ft 10 in. × 14 ft 6 in., the first runner would be as follows:

short wall: 13 ft 10 in. = 166 in.

$$\frac{166}{48} = 3 \text{ with 22 in. left over}$$

$$\frac{22 + 48}{2} = 35 \text{ in.}$$ (this is the distance of the main runner from the long walls)

Now figure the spacing along the long wall. Figure the distance in inches and divide by the panel width, usually 24 in. This locates the end cross-tees (Fig. 18-102). Following is the procedure.

long wall: 14 ft 6 in. = 174 in.

$$\frac{174}{24} = 7 \text{ with 6 in. left over}$$

$$\frac{6 + 24}{2} = 15 \text{ in.}$$ (this is the distance of the first cross-tee from the end wall)

This gives the grid pattern for hanging the main runners and cross-tees.

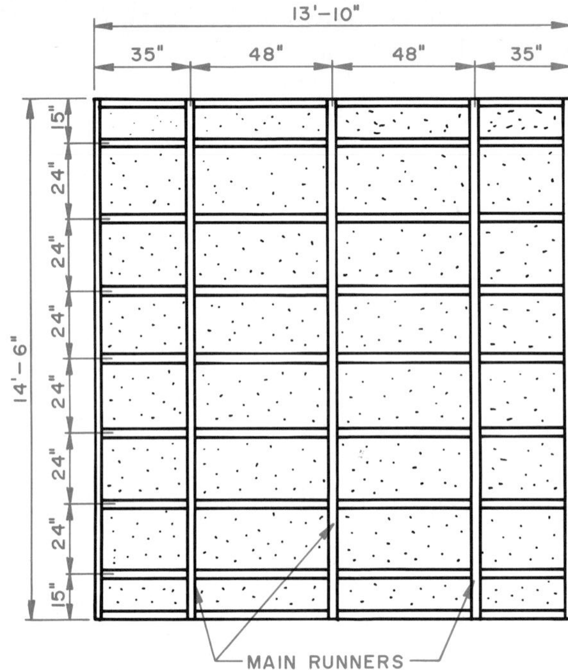

FIGURE 18-102

How to lay out ceiling panels and suspended metal runners.

Installing the Wall Molding Decide upon the height of the ceiling. Run a chalk line on each wall at this height plus the height of the main runner. Use a line level to make certain that the line is level. Pull the line tight. Snap a chalk mark on each wall (Fig. 18-103).

Place the top edge of the wall angle molding on the chalk line and nail it in place. Check

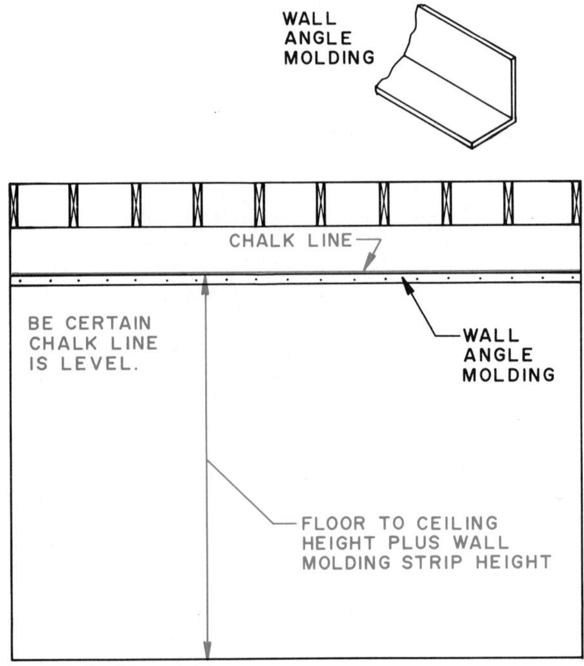

FIGURE 18-103

Use a chalk line to locate the metal runners on the walls.

with a level to be certain that it remains level (Fig. 18-104). Inside corners are formed by lapping the molding. Outside corners are mitered or overlapped (Fig. 18-105).

Installing the Main Runners The end of the main runner that meets the wall must be cut to hold the end panel. The calculated distance is measured from the point where the cross-tee connects to the main runner. In the example this was 15 in. (Fig. 18-106).

Hold the main runner in place along the chalk line that locates it from the wall. Mark the joists that are nearest the hanger holes. Install screw eyes in these joists. Wires are usually placed every 4 ft. Fasten a wire long enough to hang the ceiling in each screw eye. Now place the wire in the hanger holes on the main runner.

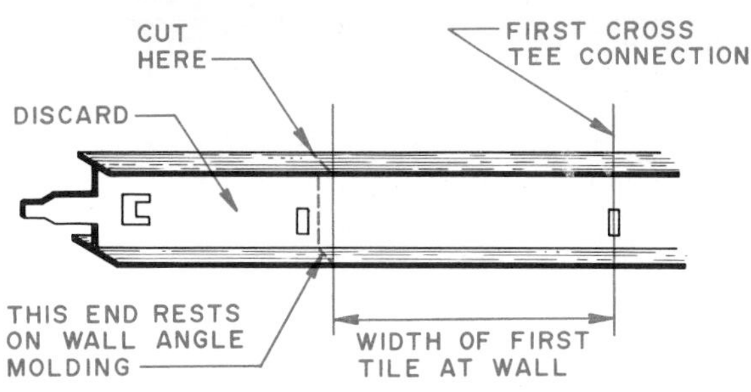

FIGURE 18-106

Cut the metal runner so that it rests on the wall runner.

Bend the wire up to hold the runner temporarily (Fig. 18-107). Do not twist the wire tight yet. If the runner is short, cut a piece for the remaining distance and join with the splicing tab provided (Fig. 18-108).

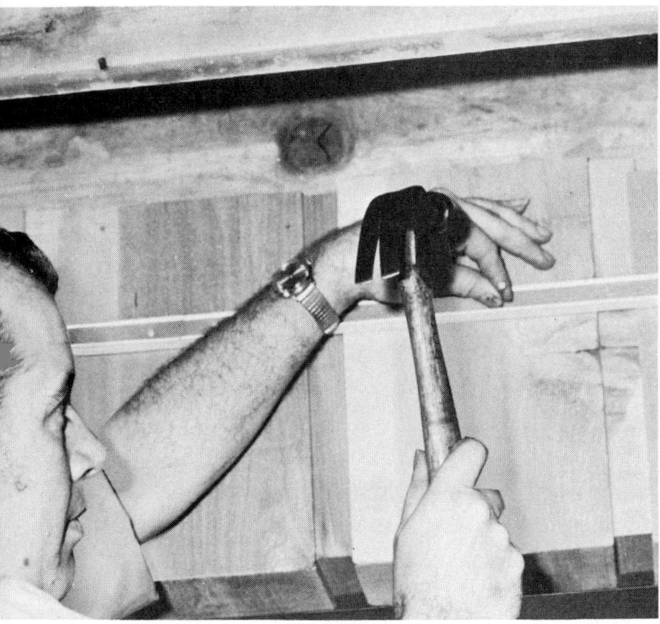

FIGURE 18-104

When the wall runners are level, nail to the wall. *(Courtesy of Chicago Metallic Corporation)*

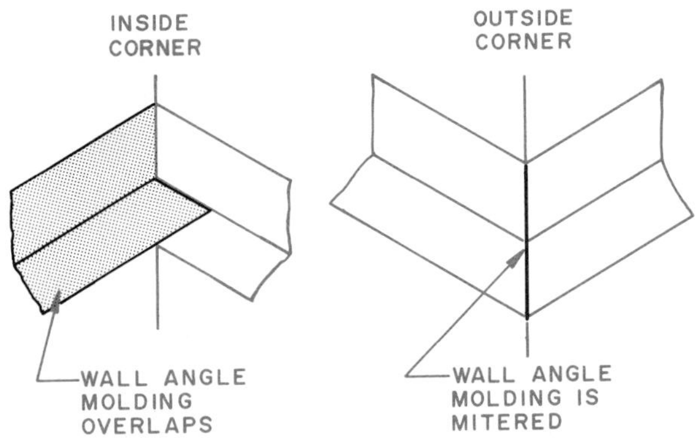

FIGURE 18-105

How to form corners with metal wall runners.

FIGURE 18-107

The runners are hung by wires from the joists. *(Courtesy of The Celotex Corporation)*

FIGURE 18-108

Runners can be spliced to extend across a long room. *(Courtesy of The Celotex Corporation)*

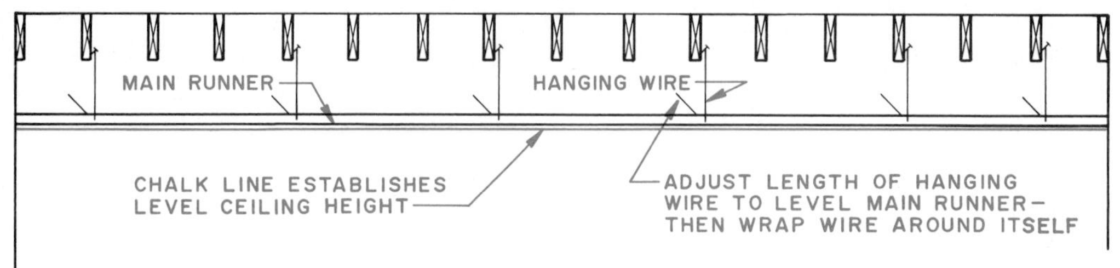

FIGURE 18-109

Use a chalk line to establish the level ceiling height for the metal runners.

Stretch a chalk line the desired ceiling height from one end of the runner to the other. Pull it tight. Adjust the position of the runner by changing the length of each wire. When the runner is level, wrap the wire around itself (Fig. 18-109). Repeat these steps to install the other main runners.

Installing the Cross-Tees **The cross-tees run perpendicular to the main runners. The main runners have openings to receive a specially prepared end on the cross-tee. Insert the cross-tees and join as directed by the manufacturer (Fig. 18-110).**

Installing the Ceiling Panels The panels will be cut to 24 × 48 in. size. They can be placed above the grid and lowered into place (Fig. 18-111). They are held by the edges of the grid members. The panels for the outer row have to be cut to fit the narrower space. Cut with a sharp utility knife, coping saw, or fine-tooth saber saw. Panels are cut to fit around pipes or other obstructions (Fig. 18-112).

FIGURE 18-110

Cross-tee runners are perpendicular to the main runners. They are joined by a special end on the cross-tee. *(Courtesy of The Celotex Corporation)*

FIGURE 18-111

Lower the panels onto the lips of the metal runners. *(Courtesy of The Celotex Corporation)*

FIGURE 18-112

Cut holes in panels for pipes, wires, and other obstructions. *(Courtesy of The Celotex Corporation)*

Estimating Materials To figure the number of ceiling tiles, refer to the tile planning drawing. It will show the total number of tiles needed. Order full tiles for those that have to be cut.

Ceiling Lights Fluorescent light fixtures are designed to replace a panel in a dropped ceiling grid. Fixtures for use on ceilings that are glued or stapled in place are mounted flush with the ceiling (Fig. 18-113).

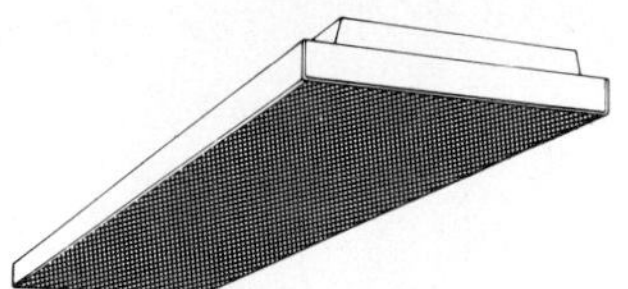

A SURFACE MOUNTED FIXTURE OFTEN USED WITH TILES GLUED OR STAPLED IN PLACE.

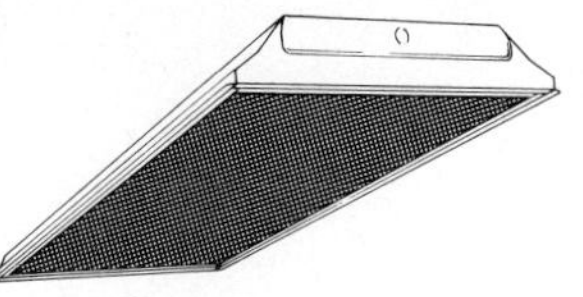

A RECESSED FIXTURE USED IN THE GRID OF A SUSPENDED CEILING.

FIGURE 18-113

Typical light fixtures used with a metal-tee grid ceiling system. *(Courtesy of The Celotex Corporation)*

STUDY QUESTIONS

After reading the chapter, answer each of the following questions. If you do not know the answer, review the chapter.

1. When is the interior finishing work started?
2. What materials are used to finish interior walls?
3. What materials are used as a base for plaster?
4. Why are plaster grounds needed?
5. How many coats of plaster are required?
6. What materials are classified as drywall finish?
7. What are the standard thicknesses of gypsum board?
8. What types of fasteners are used to install gypsum board?
9. How are inside and outside corners on a gypsum board finished?
10. If plywood paneling is applied horizontally, how are the horizontal joints treated?
11. Why is a hairline crack left between the edges of plywood panels?
12. How is plywood paneling applied over masonry walls?
13. How is the wall prepared to receive vertically applied solid wood paneling?
14. What types of water-resistant wall coverings are in use?
15. What materials are used for finished ceilings?
16. What parts make up a suspended ceiling system?

IMPORTANT TECHNICAL TERMS

Following are technical terms that you should be able to use as part of your working vocabulary. Write a brief description of the meaning of each term.

Plaster
Lath
Plaster grounds
Gypsum board
Joint compounds
Joint tape
Plywood paneling
Wainscoting
Chair rail
Hardboard paneling
Fiberboard paneling
Solid wood paneling
Suspended ceiling

19 INSTALLING INTERIOR TRIM

Interior trim is cut and nailed in place after the interior walls are finished and the finished floor is laid. Generally, doors and windows are trimmed first. Built-in cabinets, bookcases, room dividers, and wood mantels can then be installed. Finally, the base and shoe molding is cut and nailed in place.

TYPES OF MOLDINGS

There are many types and sizes of moldings available. Examples of these are in Fig. 19-1. The *crown*, *cove*, and *bed moldings* are often used in the corner of the ceiling and the wall. *Quarter rounds* are used in 90° corners. *Base shoe molding* is used next to the baseboard at the floor. *Casing* and *base mouldings* are used to trim door and window openings and are on the wall next to the floor. Chair rail is placed around the room usually about 30 to 36 in. from the floor. *Stops* are used inside door frames to keep a door from swinging through the frame. *Mullion* casings cover the space between adjoining windows. *Stools* are used on the bottom of window frames and trim where it meets the window sill (Fig. 19-2). Installation of these is explained in the following pages.

INSTALLING TRIM AND MOLDINGS

When installing trim and moldings, the carpenter must work with great care. The material must be kept clean and free from dents and abrasions. Often, the trim and moldings are stained or finished natural. Every minor blemish will be visible. The joints between sec-

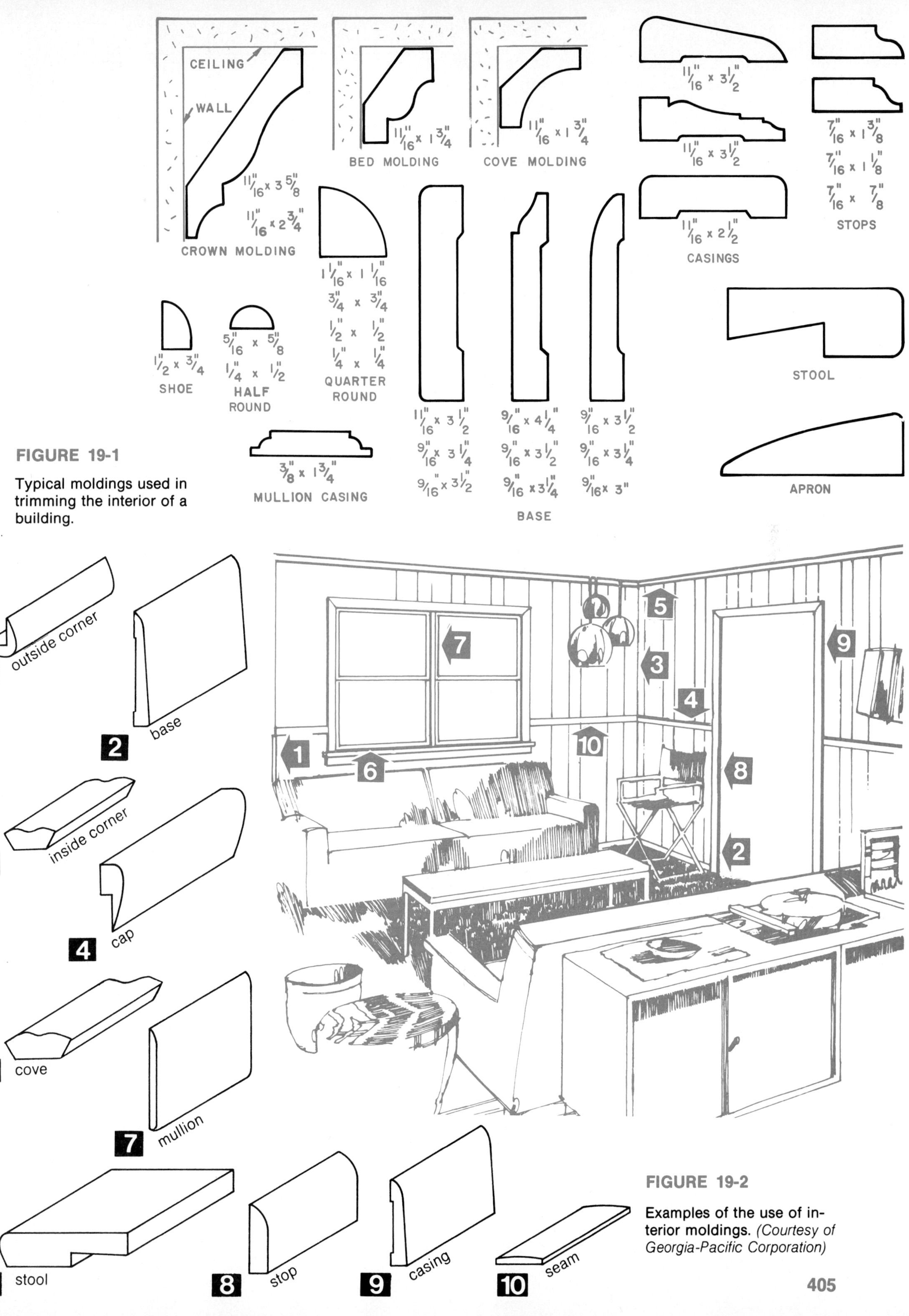

FIGURE 19-1

Typical moldings used in trimming the interior of a building.

FIGURE 19-2

Examples of the use of interior moldings. *(Courtesy of Georgia-Pacific Corporation)*

tions of trim and moldings must fit perfectly. Poorly fitting joints spoil the overall appearance of the job. All nails must be set with a nail set. Hammer marks in the wood surface are not acceptable.

TRIMMING DOOR OPENINGS

The door casing is set in from the edge of the door frame about 3/16 to 1/4 in. (Fig. 19-3). The thinner edge of the trim is nailed to the frame with 4d finishing nails. The thicker side of the casing is nailed to the stud with 6d finishing nails. The nails are placed in pairs and spaced about 16 in. on center along the edge of the casing. If the casing is a hardwood, it may split when nailed. Drill small holes slightly smaller than the nail diameter to prevent this.

Casing that has a curved or shaped surface must be mitered at the corners. Cut it with a miter box. The corners are nailed to keep the miter closed. Square-cut casing can be mitered or butted. The butted corner is toenailed (Fig. 19-4).

When measuring the casing to be mitered, be certain that the bottom end is square. Measure the length along the door frame and add 3/16 to 1/4 in. for the set back from the door frame (Fig. 19-5). Cut each side to this length. Nail temporarily to the frame with four nails. Measure and cut the top piece. Fit it in place. If it is a little long or short, the side pieces can be moved to close the miter. If the miters do not fit tightly, they can be trimmed a little with a sharp block plane. When all parts fit, nail the pieces in place. Toe-nail the miters and set the nail heads.

It is possible to buy casing precut to fit standard door frames.

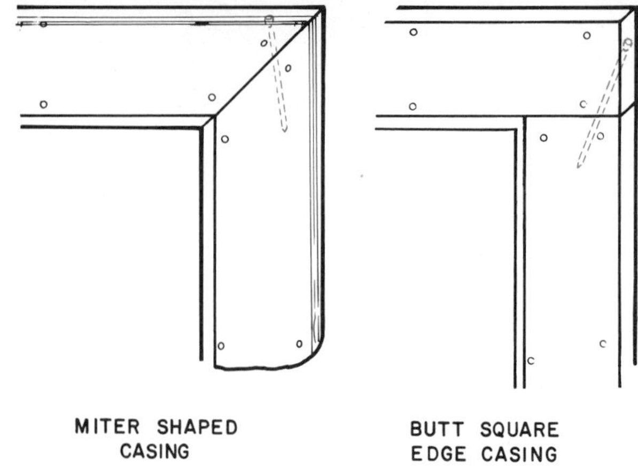

FIGURE 19-4

Door and window casings can be mitered or butted.

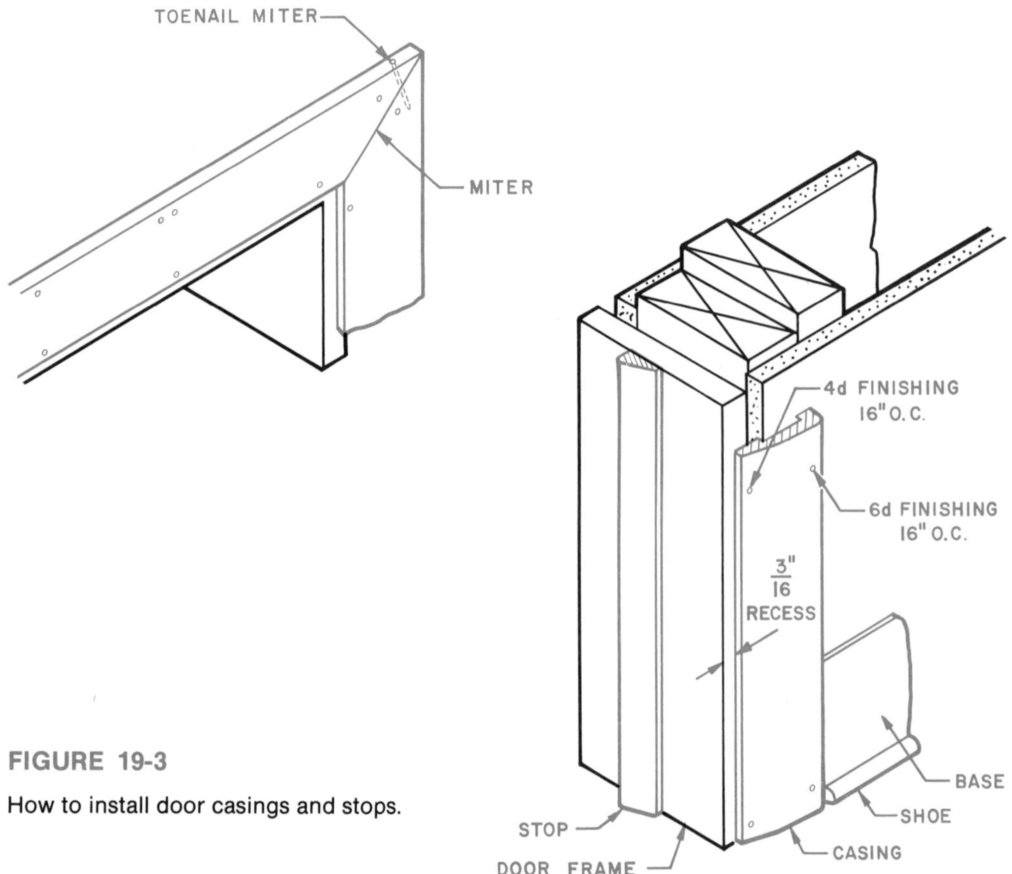

FIGURE 19-3

How to install door casings and stops.

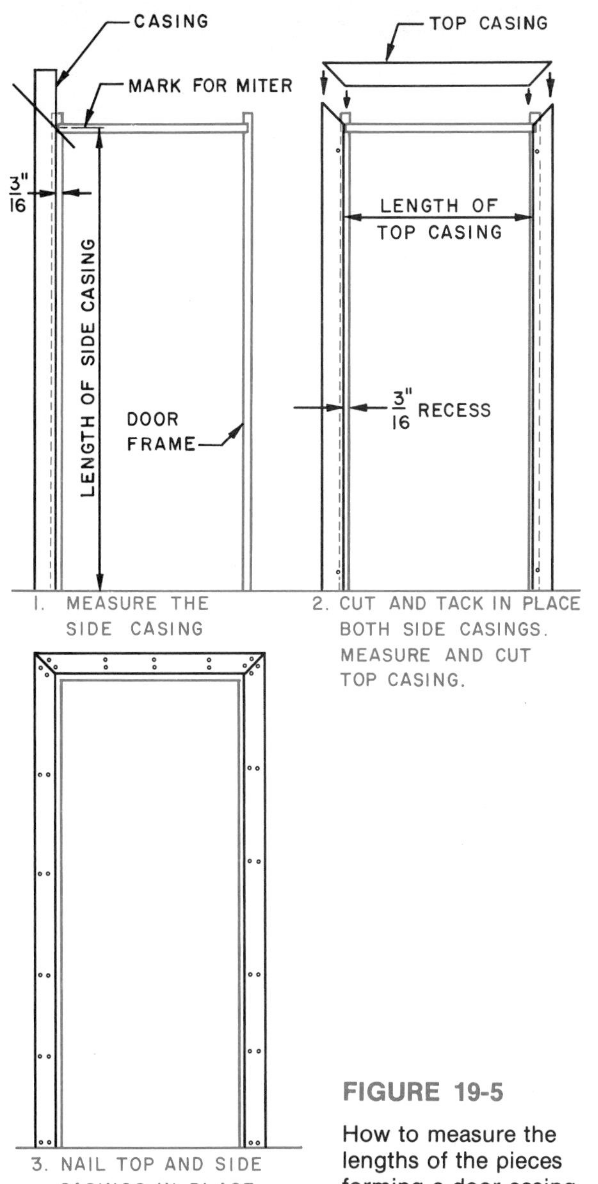

FIGURE 19-5

How to measure the lengths of the pieces forming a door casing.

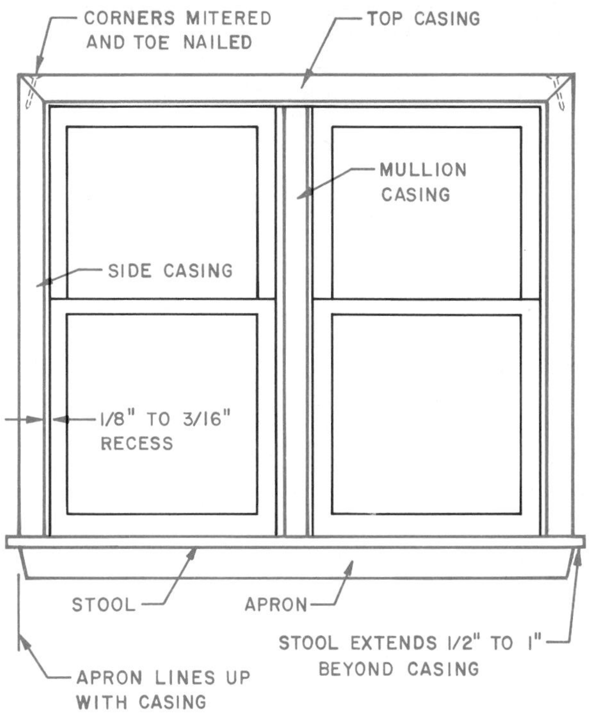

FIGURE 19-6

How to trim a window.

TRIMMING WINDOW OPENINGS

Windows require side and head casings, a stool, and an apron. If several windows are grouped together, a mullion trim is needed (Fig. 19-6).

Companies manufacturing windows supply a package of casing for each window. The window stool is cut to length and installed first. The stool must be cut to fit inside the window frame with the end notched to extend over the finished interior wall (Fig. 19-7). It usually extends ½ to 1 in. beyond the side casing (Fig. 19-8). The stool should extend over the sill to within $^1/_{16}$ in. of the sash. The stool will have a notch cut in the bottom to fit over the edge of the sloped sill (Fig. 19-9).

After cutting the stool, sand the edges so that they are smooth and the corners slightly

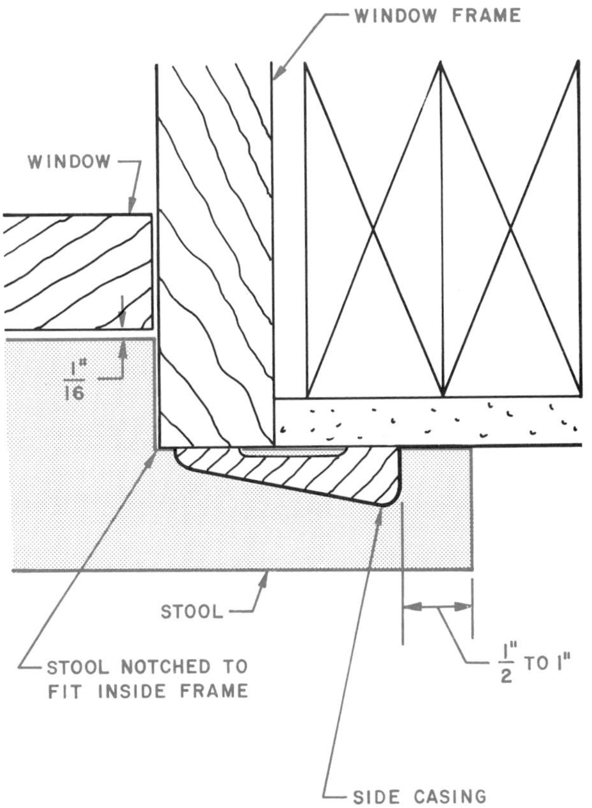

FIGURE 19-7

The window stool is cut to fit inside the window frame.

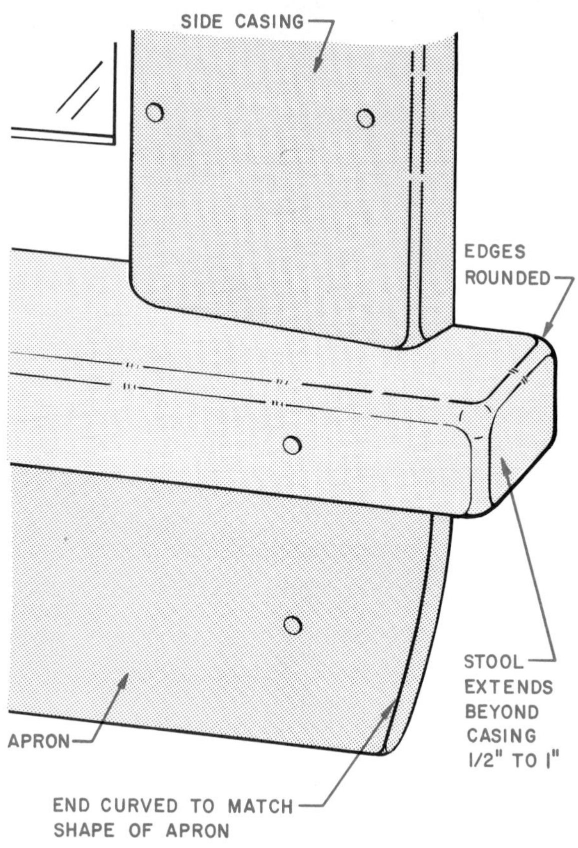

FIGURE 19-8

The window stool extends beyond the side casing.

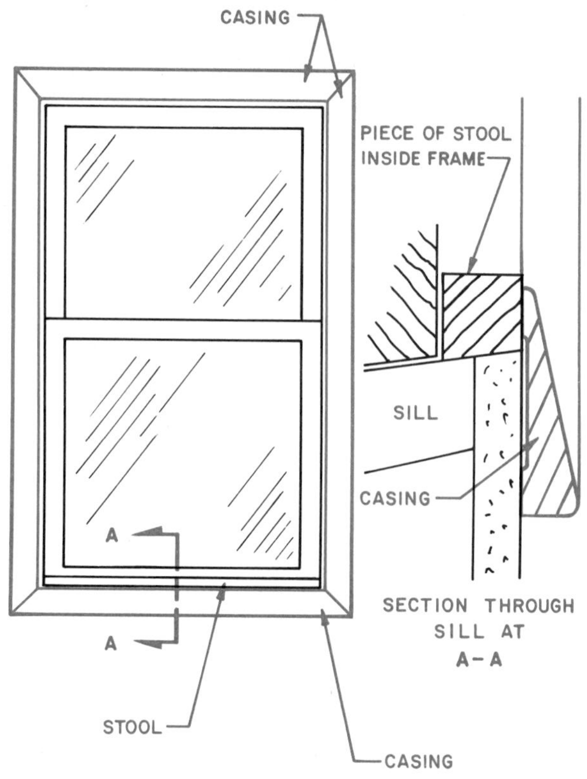

FIGURE 19-10

If the window has casing on all four sides, a small stool is installed between the sash and the casing.

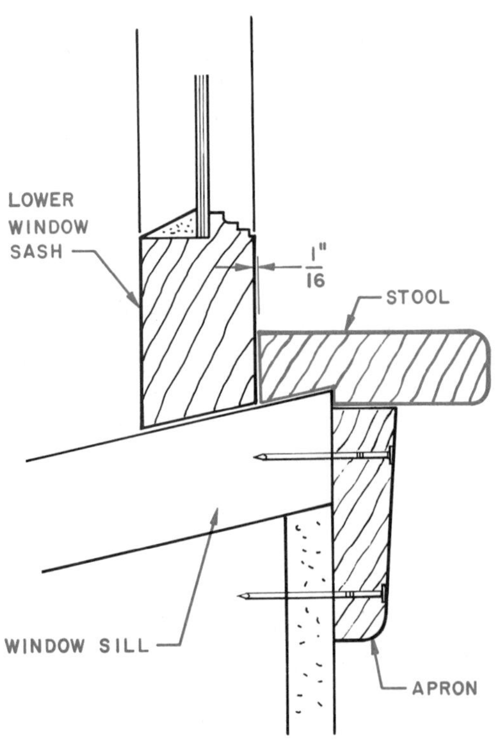

FIGURE 19-9

The window stool extends to within 1/16 in. of the lower window sash.

rounded. Set it on the sill and nail into the sill. Usually, 8d finishing nails are used. Some persons caulk the seam between the sill and stool.

Now the side casing is applied. It is marked cut and installed the same as described for door casing. It is usually set back from the face of the window frame 1/8 to 3/16 in. (Fig. 19-5).

The *apron* is cut to length last. It should extend from the outside edges of the side casing. The ends can be cut square or slightly rounded. Usually, the end is curved to match the curve of the surface of the apron (Fig. 19-8). Sand the edges so that they are smooth and free of saw marks. Nail the apron to the framing below the window (Fig. 19-9).

Sometimes a piece of stool is cut to fit inside the frame. The casing is placed on all four sides of the window (Fig. 19-10).

INSTALLING BASE AND SHOE

The base covers the joint where the wall meets the finished flooring (Fig. 19-11). It is usually installed last because it fits against the door casing and cabinets and other built-in units. The shoe is used to cover the joint between the base and the finished floor. If a room is to have wall-to-wall carpeting, the shoe is often omitted.

It should not be installed until the tile or other finished floor is in place.

The base joins the door casing or cabinets with a vertical end. It is mitered at outside corners (Fig. 19-12). Inside corners are mitered or coped. *Coping* refers to cutting the end to the shape of the base. The first piece of base is cut square on both ends and butts square against the wall. The adjoining piece has the end coped to fit the shape of the base it is to meet (Fig. 19-13).

To cope the base, first cut an inside miter on the end. The shape of the cut provides the guide for coping the end (Fig. 19-14). Cut the

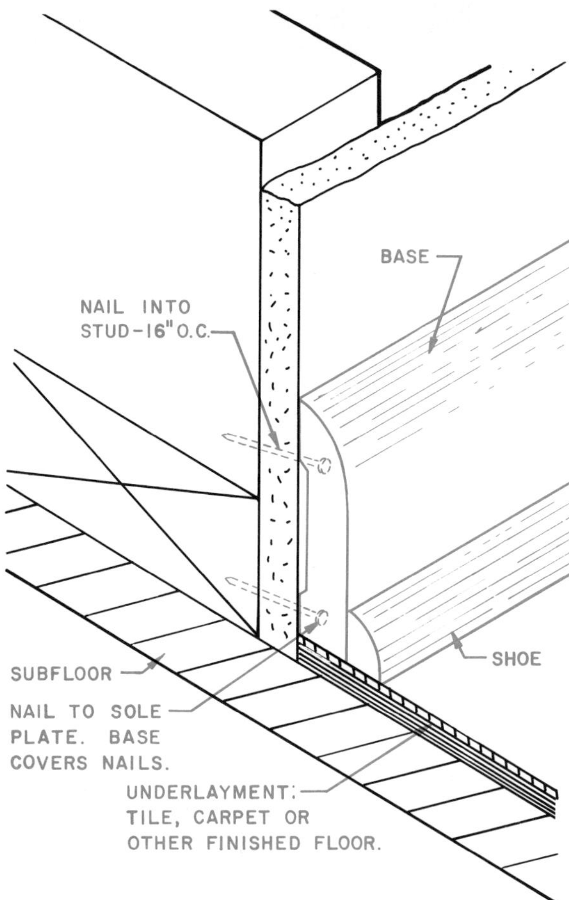

FIGURE 19-11

The base and shoe moldings are used to cover the joint between the wall and the floor.

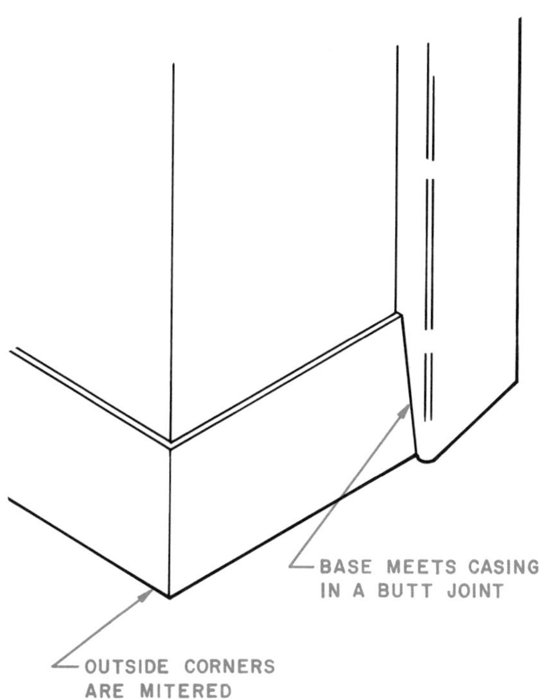

FIGURE 19-12

The base meets a casing using a butt joint.

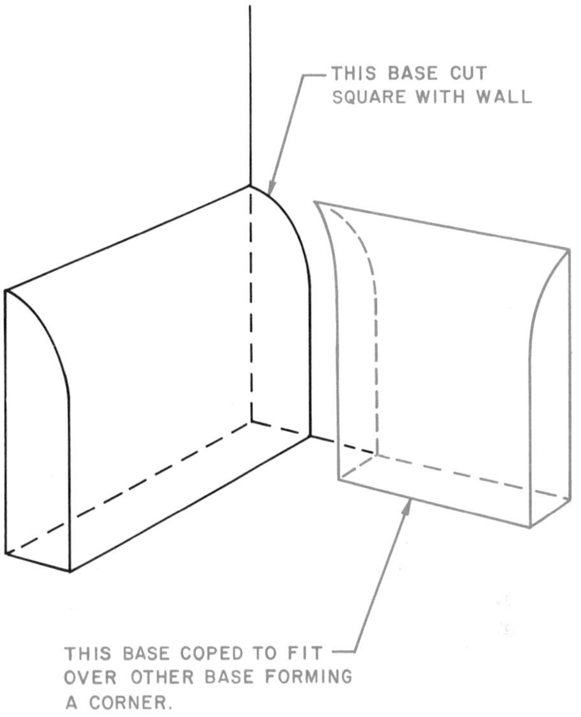

FIGURE 19-13

The base is coped to form interior corners.

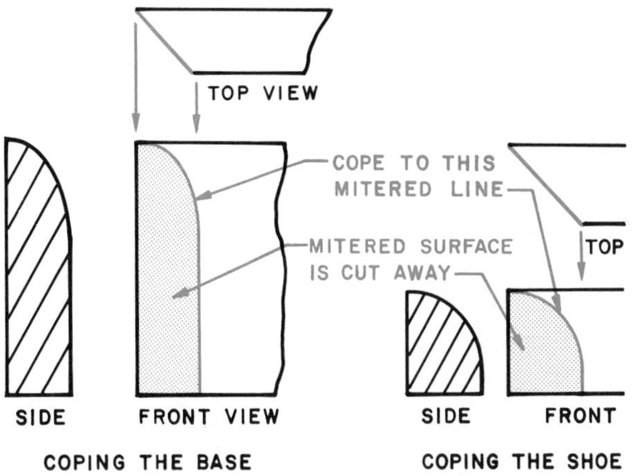

FIGURE 19-14

How to cope base and shoe molding.

mitered end to the curved shape. Use a coping saw or saber saw. This cut should be sloped slightly to the back so that the joint fits tightly. The cut can be smoothed with a half-round file or sandpaper wrapped around a round object.

Nail the base into the sole and each stud. An 8d finishing nail is often used.

If it is necessary to join two pieces of base on a long wall, use a 45° scarf joint (Fig. 19-15). Locate the joint at a stud so that both sides can be nailed to the stud.

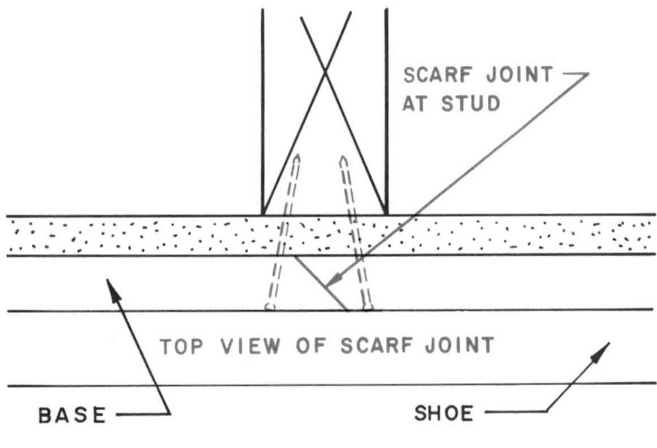

FIGURE 19-15

Base is joined end to end using a scarf joint at a stud location.

The shoe molding is laid out and cut the same as the base. Where it meets the door casing, it is usually cut on a 45° angle. This produces a neat ending (Fig. 19-16).

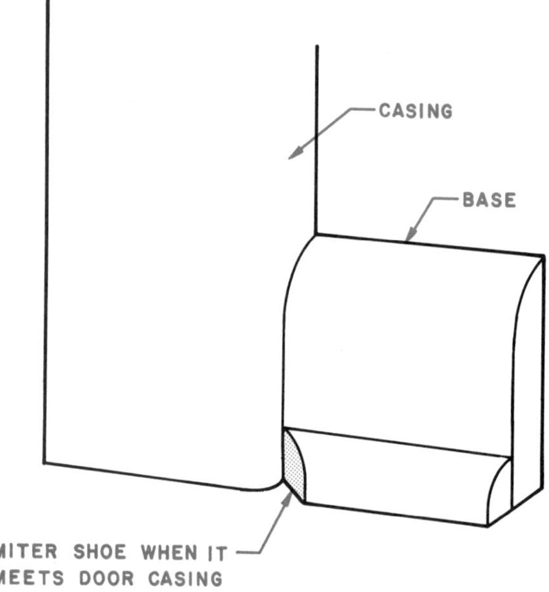

FIGURE 19-16

Shoe molding is mitered when it meets a door casing.

STUDY QUESTIONS

After reading the chapter, answer each of the following questions. If you do not know the answer, review the chapter.

1. When is the interior trim installed?
2. What are the common moldings, and for what purpose are they used?
3. When applying casing around a window, what joint is generally used?
4. What pieces of trim are required to finish a window?
5. What is done with base and shoe molding?

IMPORTANT TECHNICAL TERMS

Following are technical terms that you should be able to use as part of your working vocabulary. Write a brief description of the meaning of each term.

Crown molding
Cove molding
Bed molding
Quarter-round molding
Shoe molding
Base molding
Casing molding
Stops
Mullion molding
Stools
Apron
Cope the base

20 FINISH FLOORING

One of the last jobs to do to finish the interior of a house is to install the finish flooring. The subfloor and underlayment for the finish floor were installed earlier in the construction process. Review Chapter 10 for this information.

The most commonly used finish flooring is either resilient flooring; *natural materials* such as ceramic tile, slate, and terrazo; or wood flooring. *Resilient flooring* includes carpeting, vinyl, and vinyl asbestos material (Fig. 20-1).

RESILIENT FLOORING

Resilient flooring materials are usually laid by persons specializing in this work. *Carpeting* is sold in rolls 12 ft wide. It is also available in 12 × 12 in. squares which are joined to the subfloor with an adhesive. It takes special training to lay carpeting. The handling of large heavy sections of carpeting requires strength and know-how.

Vinyl and *vinyl asbestos* floor covering material are available in the tile and sheet form. The tiles are usually 12 × 12 in. The sheets are 12 ft wide and room length.

INSTALLING VINYL AND VINYL ASBESTOS TILE

Be certain that the underlayment is securely nailed in place. Any rough edges should be planed so that they are flat and smooth. Any ridges or nail heads not flush with the surface of the underlayment will show through finished floor. Vacuum the subfloor so that it is free of all dust and debris.

FIGURE 20-1

A resilient floor covering of vinyl asbestos tile. *(Courtesy of Azrock Floor Products)*

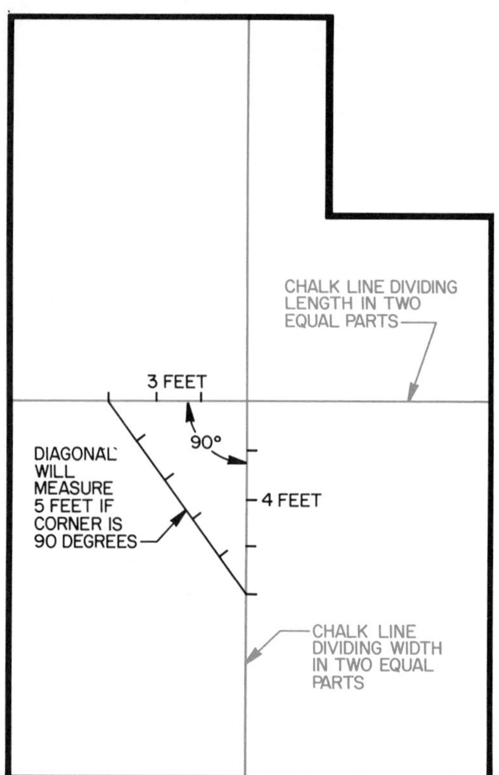

FIGURE 20-2

To prepare to lay resilient floor tile, first divide the room into four equal sections.

It is important to read carefully the directions of the company making the tile. There are available a wide variety of materials and adhesives, and the manufacturer's instructions should be carefully followed.

Following is a general procedure for laying vinyl and vinyl asbestos tile.

1. Locate the center of the room. Run a chalk line parallel with a wall from one side of the room to the other and snap a mark on the floor (Fig. 20-2). Ignore any irregularities in the shape of the room. To get an accurate mark, do not snap the entire line at one time. Hold it against the floor at some midpoint and snap only short sections.
2. Lay out the centerline of the room from the other direction. Snap a chalk line across this distance at right angles to the other line. Determine the right angle by using the 3-4-5 technique (Fig. 20-2).
3. Now make a trial layout of tile. Lay them dry along the centerline to the side wall (Fig. 20-3). If the space left to the wall is less than 2 in., consider moving the center-

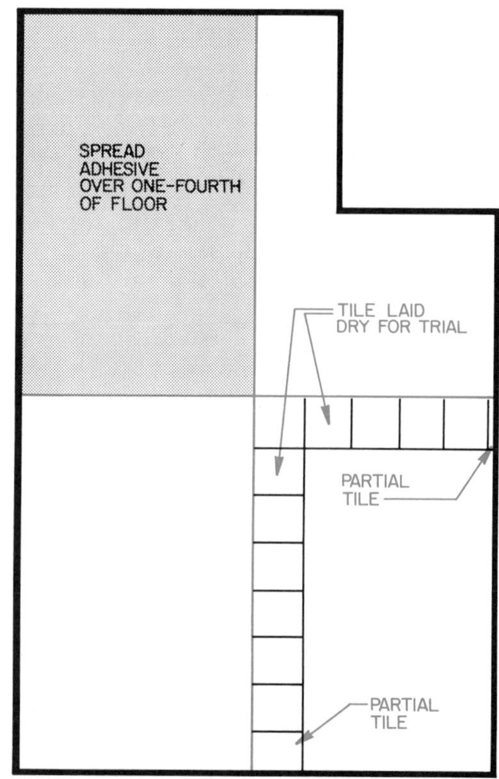

FIGURE 20-3

Make a trial layout of tile before the adhesive is applied. When the centerlines are correct, apply adhesive to one-fourth of the floor.

line 6 in. so that the end tiles are not small pieces. This will improve the appearance of the job. The larger tiles will be easier to lay and less likely to be damaged.

4. Now spread the adhesive over one-fourth of the floor area. Spread up to the chalk line but do not go over it (Fig. 20-4). Use the trowel or brush or roller as recommended by the tile manufacturer. It is important to get the proper thickness of adhesive on the underlayment. If it is too thick, the adhesive will come up between the joints. If it is too thin, the tiles will not adhere completely. If "self-stick" tiles are used, this step is not necessary.

   Generally, the adhesive must be allowed to set before laying tile. This can be 15 minutes or longer. Follow the instructions on the adhesive container. The adhesive, when ready, should feel tacky but should not stick to your fingers.

5. Now start laying the tile. Begin in the center of the room. Lay them following the chalk-line mark (Fig. 20-5). Some prefer to use a straight strip of wood to align this first row of tiles. If it is not straight, all following rows will be crooked.

   Butt the tiles squarely together. Leave no cracks or gaps. Lay each tile in place.

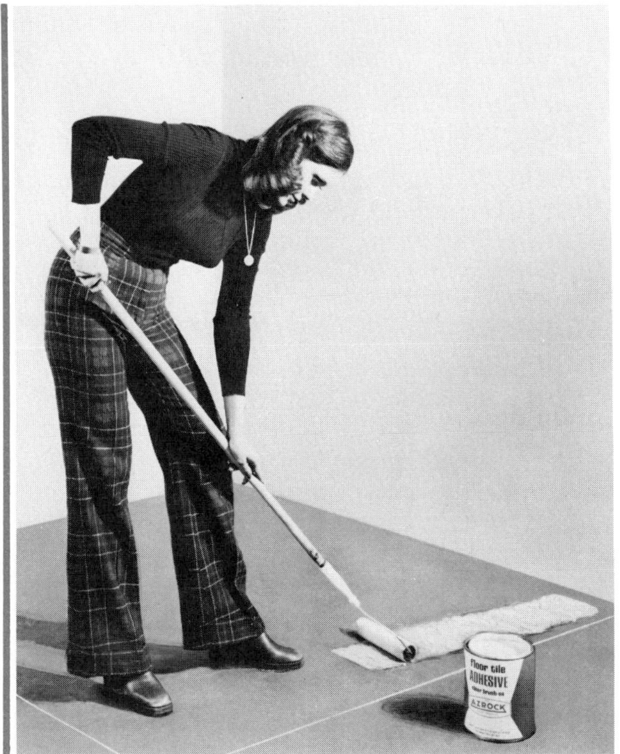

FIGURE 20-4

A thin layer of adhesive can be applied to the floor with a roller. *(Courtesy of Azrock Floor Products)*

FIGURE 20-5

Start laying the tile in the adhesive beginning in the center of the room. *(Courtesy of Azrock Floor Products)*

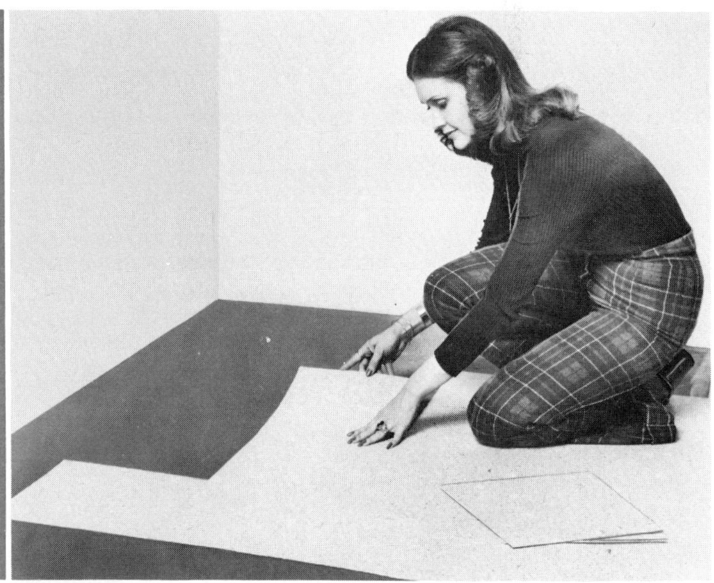

FIGURE 20-6

Lay the tile in one quarter of the room. Then apply adhesive and tile another quarter. *(Courtesy of Azrock Floor Products)*

Do not try to slide them in the adhesive. Since the adhesive does not dry for many hours, there is no hurry. Work carefully. Lay one quarter of the room at one time. Move from the chalk line to the walls (Fig. 20-6).

6. Manufacturers recommend that vinyl tile be rolled after they are laid. A weighted roller is available for this purpose. Vinyl asbestos tiles need not be rolled. Follow the recommendation of the manufacturer.

7. Repeat this process for the other three quarters of the floor.
8. Finish the job by marking and cutting the tiles to fit the space along the wall (Fig. 20-7). To mark a tile, lay the one to be cut on top of the last tile (A). Lay another tile against the wall (B) and mark its edge on the tile below. Mark it with a soft lead pencil. Cut the tile with a heavy scissors or tin snips. Cut the tile about 1/16 in. short of meeting the wall. If it touches the wall, changing temperatures may cause it to expand and it will then buckle.

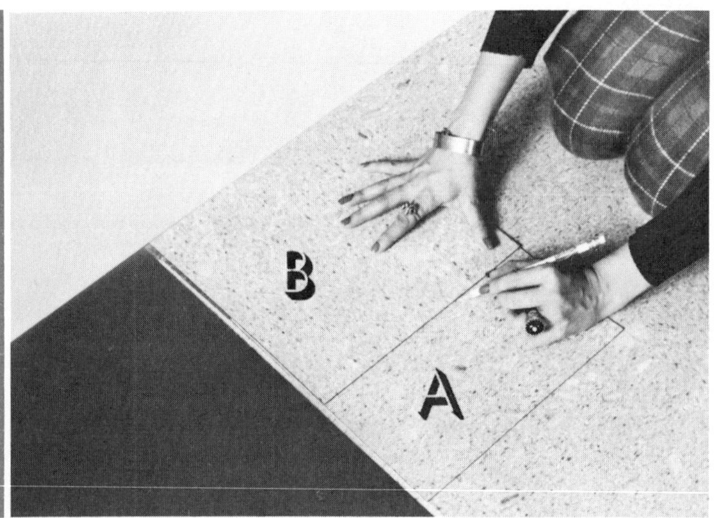

FIGURE 20-7

To cut a tile to fit in the space at the wall, place a loose tile (A) on top of the last tile cemented in place. Then place another tile (B) against the wall and mark its edge on tile A. Cut tile A on this line and glue it in the space at the wall. *(Courtesy of Azrock Floor Products)*

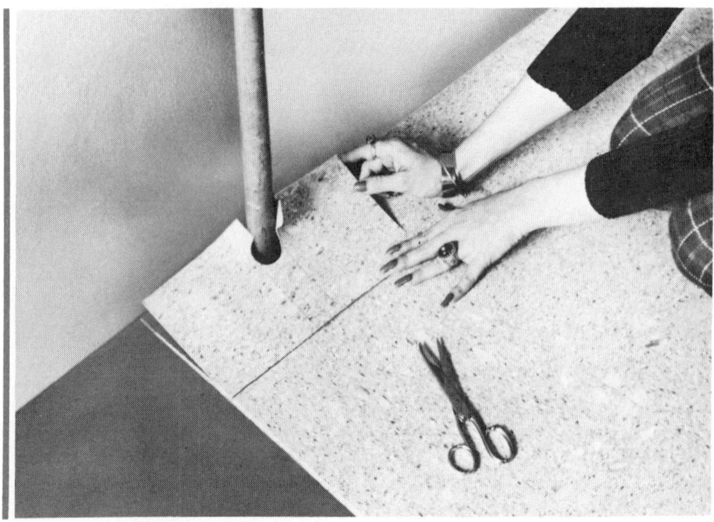

FIGURE 20-8

To fit tile around pipes, make a paper pattern to fit the space. Trace onto a tile and cut to size. Cut through the tile to the hole to permit it to slip around the pipe. *(Courtesy of Azrock Floor Products)*

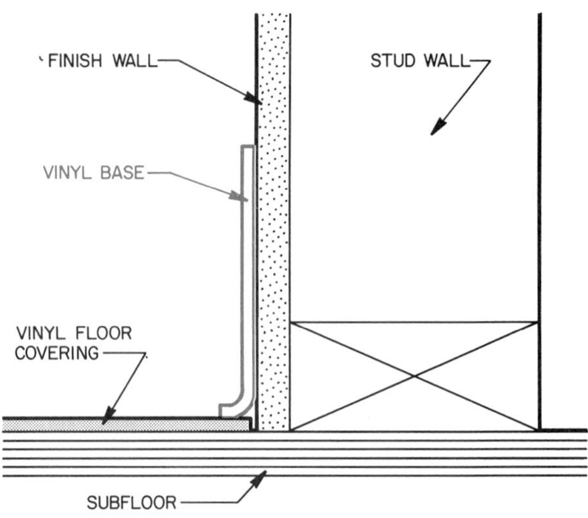

FIGURE 20-9

The joint between the tile and the wall can be covered with a vinyl cove base.

9. To fit around pipes make a paper pattern to fit the space. Trace the outline on the tile and cut with a shears. Cut on a line to the center of the hole (Fig. 20-8).
10. Cover the joint between the wall and the tile with a vinyl base of wood quarter-round molding (Fig. 20-9).
11. Now examine the entire floor carefully. Remove any spots of adhesive or stains using the cleaner recommended by the tile manufacturer.

Tiles with adhesive on their back sides are available. To adhere, peel off the protective paper covering and press in place.

Estimating Vinyl Tile Flooring Most tiles in use are 12 × 12 in., which equals 1 square foot. To figure the number of tiles to order, figure the total square feet to be covered and add a percentage for waste. Recommended waste figures are listed below.

TILE WASTE ALLOWANCES

1 to 50 sq ft	14%
50 to 100 sq ft	10%
100 to 200 sq ft	8%
200 to 300 sq ft	7%
300 to 1000 sq ft	5%
over 1000 sq ft	3%

For example, a house requires 1000 square feet to be covered. The tiles are 12 × 12 in.

$$\begin{aligned} \text{total floor area} &= 1000\,\text{ft}^2 \\ \%\ \text{waste } 5\%\ \text{of } 1000\,\text{ft}^2 &= \underline{50\,\text{ft}^2} \\ & \quad 1050\,\text{ft}^2 \end{aligned}$$

Each tile equals 1 square foot, so 1050 tiles must be ordered.

Laying Sheet-Type Vinyl Measure the room and plan the installation so that there are as few seams as possible. Seams should be located in places where they will be least noticed. Avoid seams in areas of heavy traffic.

Usually, rooms are rectangular and the sheet is run parallel with the side wall.

Measure the length of the room. Cut a length from the large roll of vinyl flooring. It should be 3 in. longer than the actual length so that there is material for trimming and fitting. One end should be cut square. This is done by placing a framing square on the edge of the vinyl (Fig. 20-10). A straightedge and a utility knife with a straight blade or a linoleum blade is used to cut the material (Fig. 20-11).

After cutting a piece to length, reroll it to prevent it from contracting.

When cutting pieces that meet at a seam, remember to fit them so that the pattern is matched.

After the length is cut, place the square end against the end wall and the edge against the side wall. Let each end curl up on the end walls. If the end and side walls are straight, the sheet can be adhered to the floor. If there is a projection, the sheet must be cut to fit around it. This is usually done by scribing or with a pattern. If the wall is crooked, the sheet will have to be scribed to it.

To scribe the projection, set a divider on the amount to be removed. Let one leg run along the wall. The other leg marks the sheet. The length can be marked by projecting from the wall (Fig. 20-12). Cut out the area of the projection.

If a pattern is used, the shape to be removed is cut from cardboard until it fits around the projection. This pattern is placed on the sheet and marked. The area to be removed is cut away (Fig. 20-13).

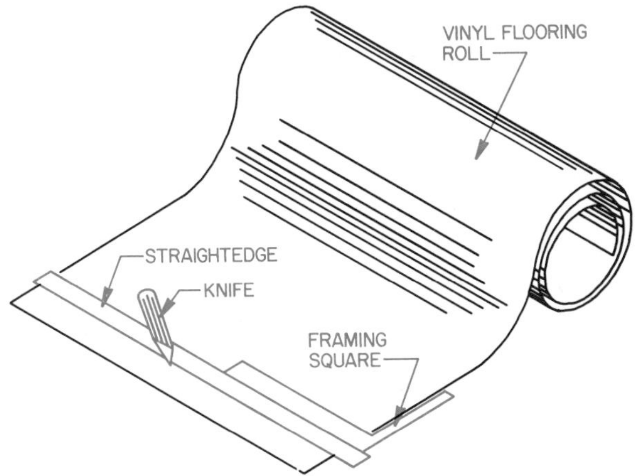

FIGURE 20-10

Sheet vinyl floor covering is cut to length with a straightedge and a knife.

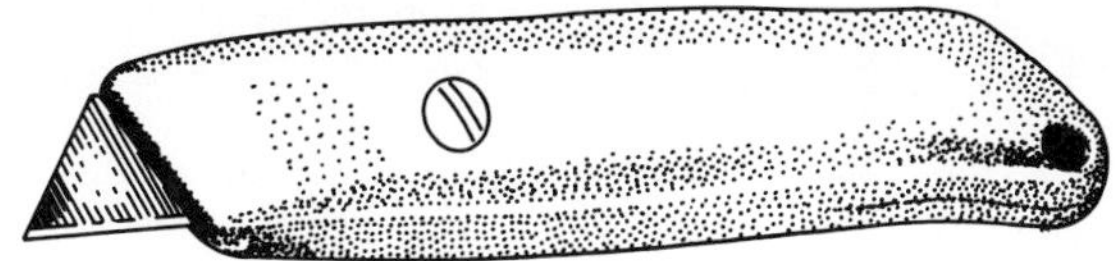

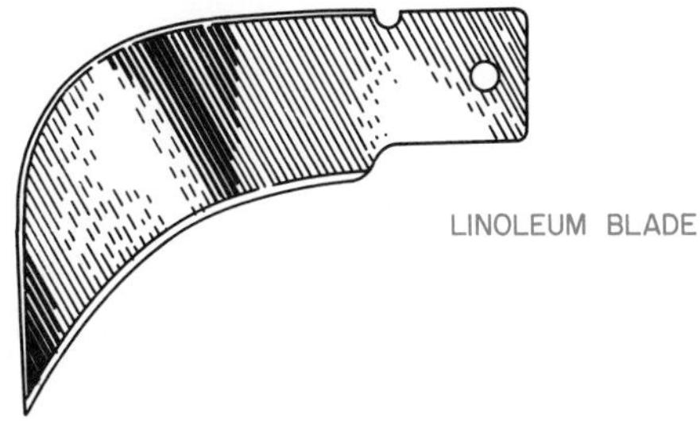

FIGURE 20-11

Knives commonly used to cut vinyl sheet floor covering.

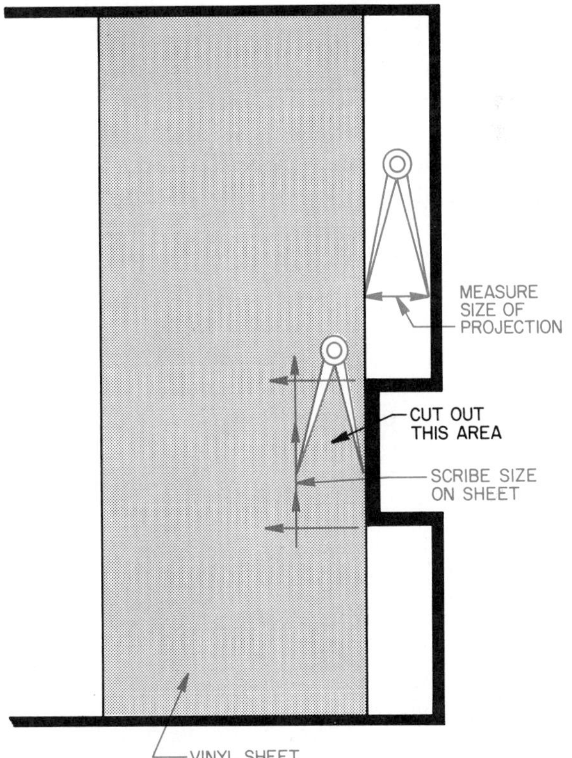

FIGURE 20-12

Cuts for projections are marked with a scriber.

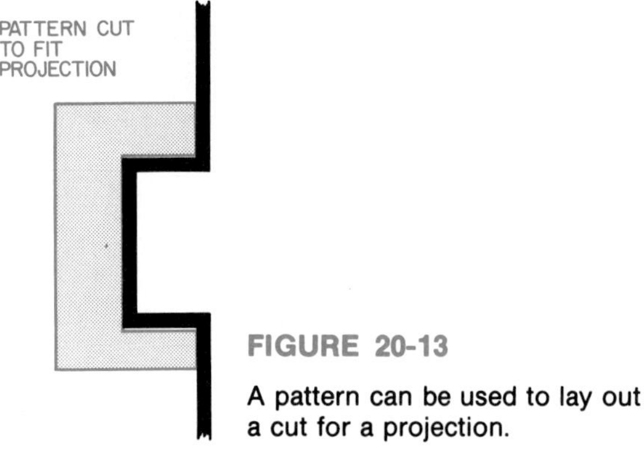

FIGURE 20-13

A pattern can be used to lay out a cut for a projection.

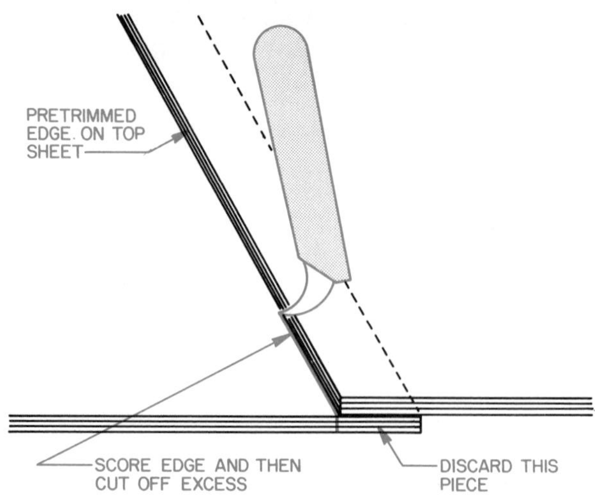

FIGURE 20-15

A single-cut seam method for cutting edge seams of sheets.

Cutting Seams Seams between sheets are made by underscribing, single cutting, or double cutting. Before seams are cut, the two sheets should be partially bonded to the subfloor. Keep the adhesive about 4 in. from the seam until it is cut. Place a piece of scrap vinyl under the seam as it is being cut. This keeps the back of the sheet from fraying as it is cut.

Underscribing uses a *recess scriber*. It is set so that the scribing pin is directly above the edge of the sheet below. Slide it along the edge of the sheet. The pin leaves a score mark on the surface. Place a straightedge on the score mark and cut the sheet with a linoleum or utility knife (Fig. 20-14).

A *single-cut seam* is used when it is important to match up a pattern. The edge of the top overlapping piece is pretrimmed to the size wanted. It overlaps the lower piece until the pattern matches. The lower sheet has a score mark cut into it with a knife. The top sheet is laid back and the lower sheet is cut with a straightedge and knife (Fig. 20-15).

A *double-cut seam* is used when neither edge is pretrimmed. Overlap the two sheets. Place a straightedge on the line to be cut. Using a knife, cut through both sheets at the same time (Fig. 20-16).

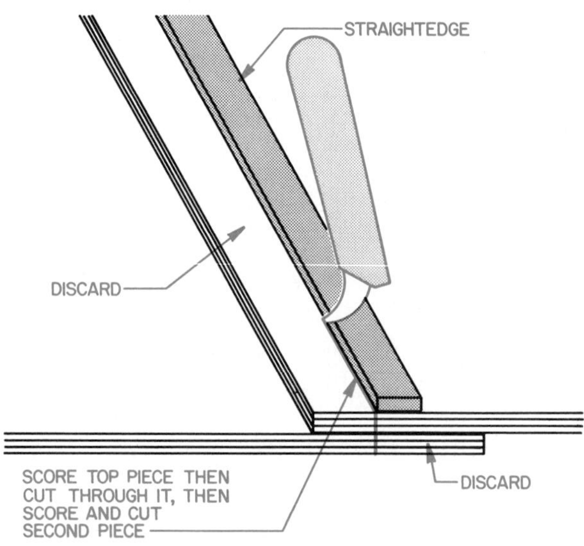

FIGURE 20-16

A double seam cut scores and cuts the two sheets when they overlap.

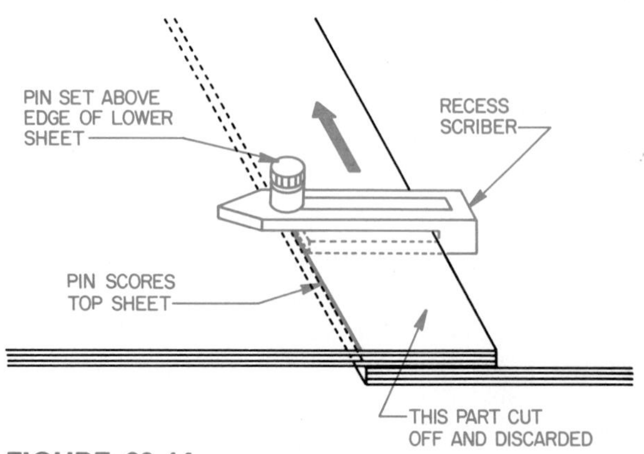

FIGURE 20-14

A recess scriber is used to underscribe overlapping vinyl sheets.

End Scribing a Sheet to Length This method is used with the lapping method of sheet installation. Adhere the center half of the sheet to the subfloor. Lay the two end sections on the dry floor up against the end walls. Since the sheet is longer than the room, the sheet will curl up on the wall. The following steps are used to scribe it to length.

1. With the sheet curled up the wall, mark a line on the sheet 12 in. from the end. Use a soft pencil. Also put this mark on the subfloor (Fig. 20-17).

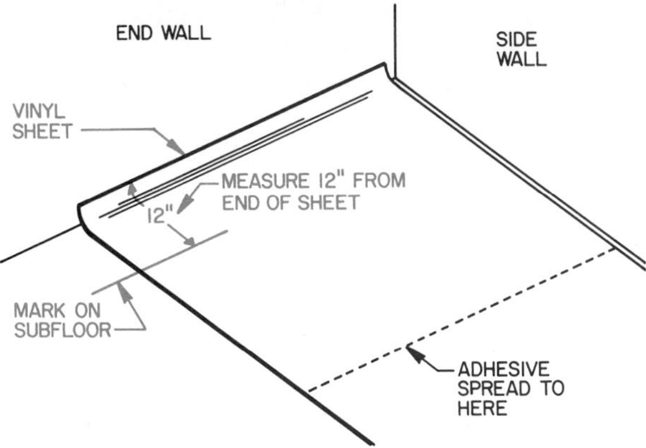

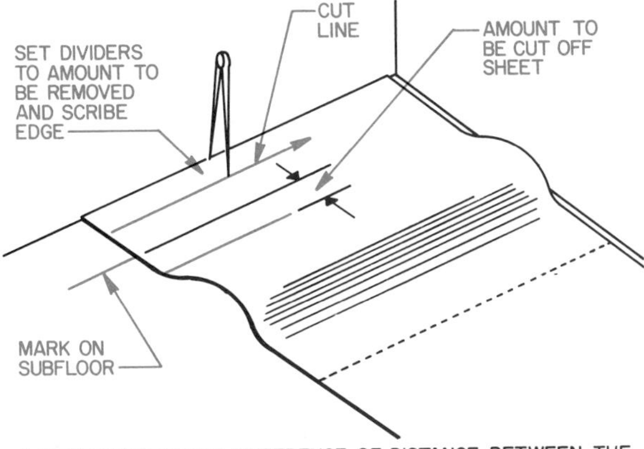

FIGURE 20-17

How to mark and scribe a sheet to length.

2. Slide the end of the sheet against the wall. This will cause the sheet to buckle (Fig. 20-17).
3. Measure the distance between the 12-in. mark on the floor and on the sheet. Set a divider on this distance.
4. Using the divider, scribe that distance on the end of the sheet. Cut on that line.

Methods for Installing Sheets Some types of vinyl flooring sheets do not have to be bonded to the subfloor. They can be cut to size and laid dry. The edges are held by the quarter round. If there is a seam, however, it would have to be cemented in place.

The two methods for installing sheet vinyl with an adhesive are the tubing method and the lapped method. The *lapped method* has the sheet folded back along its width. It is the one most generally used. The *tubing method* has the sheet folded along its length. This method is preferred for laying heavy-gauge vinyl sheets (Fig. 20-18).

The sequence of cutting, fitting, spreading the adhesive, and placing the vinyl in the adhesive is different for the lapped and tubing methods.

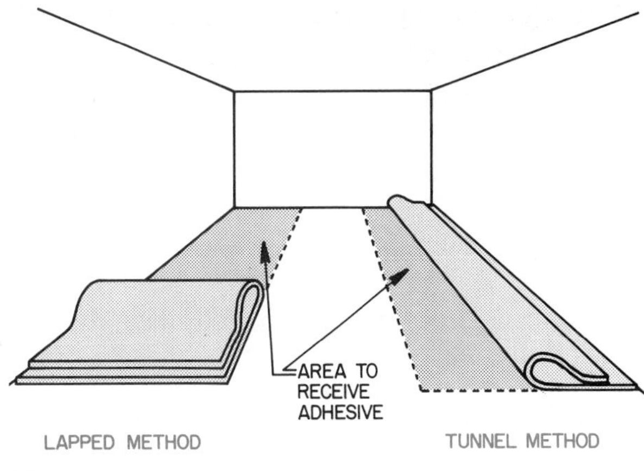

FIGURE 20-18

Two methods for folding and installing sheet vinyl flooring covering.

Tubing Sequence The following discussion is illustrated in Fig. 20-19.

1. Scribe sheet A to the side and end walls.
2. Trim edge XY so that the sheet is the desired width.
3. One-half the sheet, area 1, is bonded to the subfloor.
4. Then bond area 2 of sheet A.

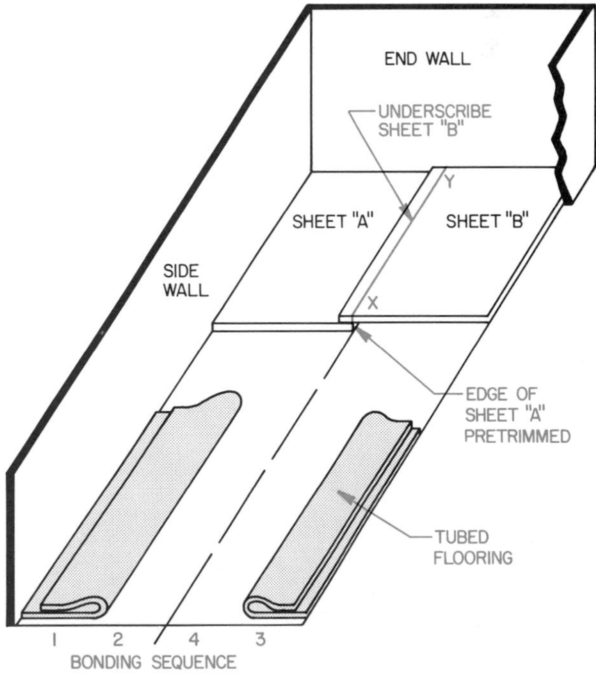

FIGURE 20-19

The steps to install vinyl sheet floor covering using the tubed method.

5. Scribe sheet B to the end and side wall. Overlap sheet A.
6. Bond one-half of sheet B, area 3.
7. Underscribe edge XY between sheets A and B and cut the edge of B to form the seam.
8. Bond other half of sheet B, area 4.

Lapping Sequence The following discussion is illustrated in Fig. 20-20.

1. Scribe sheet A to the sidewall.
2. Trim edge XY so that the sheet is the desired width.
3. Bond areas 1 and 2 to the subfloor. This joins the center section of the sheet to the subfloor.
4. Trim both ends of the sheet to length.
5. Bond areas 3 and 4.
6. Lap sheet B over sheet A so that the pattern matches. Scribe to the side wall.
7. Bond sheet B as described for sheet A.
8. Underscribe and cut sheet B at edge XY and press into the adhesive.

Applying Adhesives Follow the directions of the adhesive manufacturer as to how to apply the material. After the adhesive is ready to receive the sheet, lay it over the area. Do not try to slide the sheet around. Press the sheet into the adhesive. Each seam should be rolled with a hand roller. Then the entire sheet is rolled with a 100-lb roller.

WOOD FLOORING

Wood flooring is available in strips, planks, blocks, and parquet.

Strip flooring is cut in a variety of widths. Common widths are 1½, 2, 2¼, and 3¼ in. They are made in random lengths. In thickness they are usually ⅜, ½, and $^{25}/_{32}$ (Fig. 20-21).

Strip flooring is tongue-and-grooved on the edges and ends (Fig. 20-22). This permits a tight joint and helps hold the flooring in place.

Plank flooring is usually $^{25}/_{32}$ in. thick and from 2 to 9 in. wide. It is available with tongue-and-grooved edges and ends and square edges and ends (Fig. 20-23).

Block flooring is also $^{25}/_{32}$ in. thick. The size of the block depends upon the width of the strips used. If 2-in. strips were used, the blocks could be 6 × 6, 8 × 8, or 12 × 12 in. (Fig. 20-24). The blocks have tongue-and-grooved edges (Fig. 20-25).

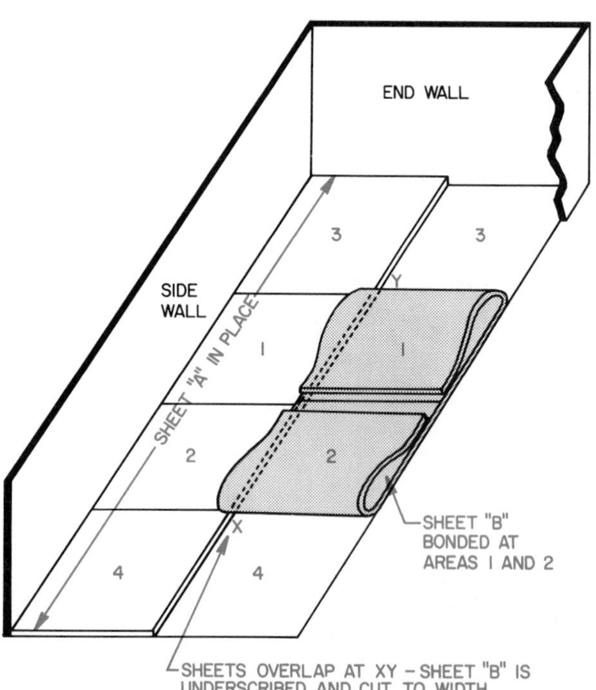

FIGURE 20-20

How to install vinyl sheet floor covering using the lapping method.

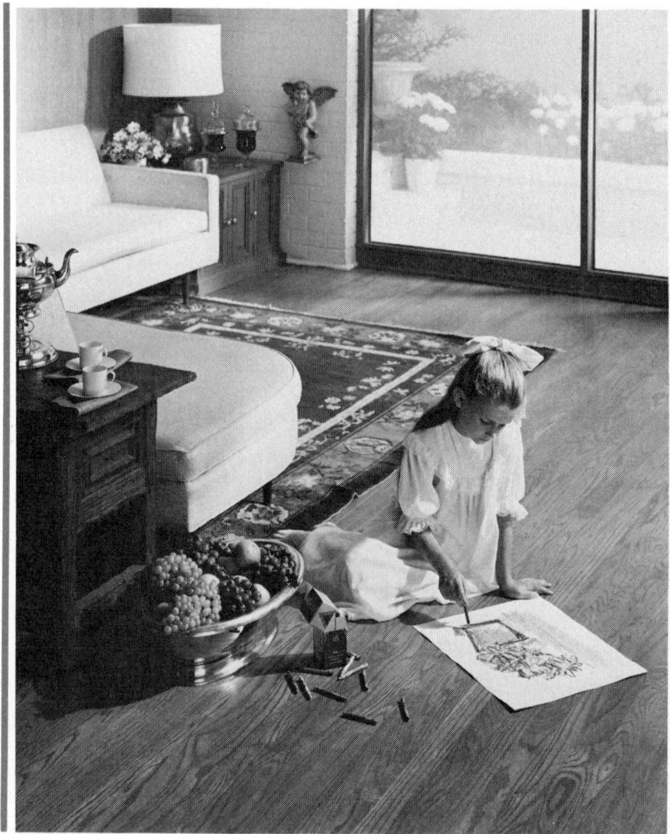

FIGURE 20-21

A finished hardwood strip floor. *(Courtesy of National Oak Flooring Manufacturers' Association)*

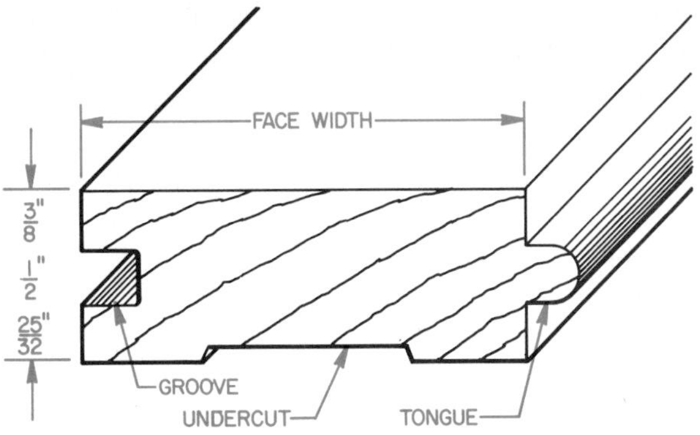

FIGURE 20-22

A typical section of tongue-and-groove hardwood strip flooring.

FIGURE 20-24

A wood block floor. *(Courtesy of National Oak Flooring Manufacturers' Association)*

FIGURE 20-23

Plank flooring is in wide widths and is pegged. *(Courtesy of Harris Manufacturing Co.)*

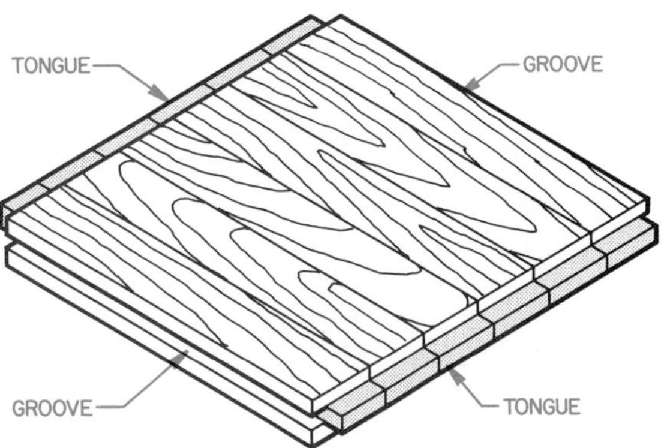

FIGURE 20-25

A typical wood block flooring unit.

Parquet flooring uses a square, rectangle, or herringbone pattern (Fig. 20-26). Many other patterns can be developed. It is manufactured in short lengths of individual pieces. The pieces have tongue-and-grooved edges and ends. Some are installed one piece at a time by nailing or with adhesives. Others are glued in blocks to heavy paper. After the block is glued to the subfloor and dried, the paper is removed (Fig. 20-27).

Wood flooring is most commonly made from oak; however, maple, beech, birch, and pecan are used. The U.S. Department of Commerce has approved the rules and regulations used to grade strip flooring. The two associations setting these standards are the National Oak Flooring Manufacturers' Association and the Maple Flooring Manufacturers' Association. There are no official grades for plank, block, and parquet flooring.

Wood flooring is available in unfinished and factory-finished condition. The unfinished is sanded and finished after it has been installed. The factory-finished needs only to be installed.

FIGURE 20-26

A parquet hardwood floor. *(Courtesy of Harris Manufacturing Co.)*

FIGURE 20-27

Typical patterns for parquet hardwood floors.

Storage of Wood Flooring Hardwood flooring is kiln-dried to a low moisture content. It should be delivered to the job and placed in the rooms in which it is to be installed 48 hours before installation. This gives it time to adjust to the moisture content of the room. It must be protected from moisture at all times. The building in which it is being installed must be heated to 70° F.

Nails and Nailing Improper nailing of the subfloor or finish flooring will produce a floor with squeaks. Tongue-and-grooved flooring is blind-nailed through the tongue. The nail is set on an angle of 45 to 50°. Square-edge flooring is face-nailed. The nails are set below the surface and covered with a proper filler (Fig. 20-28).

The three types of nails commonly used are barbed, screw, and cut steel (Fig. 20-29). Recommendations for nail sizes and spacing are in Fig. 20-30.

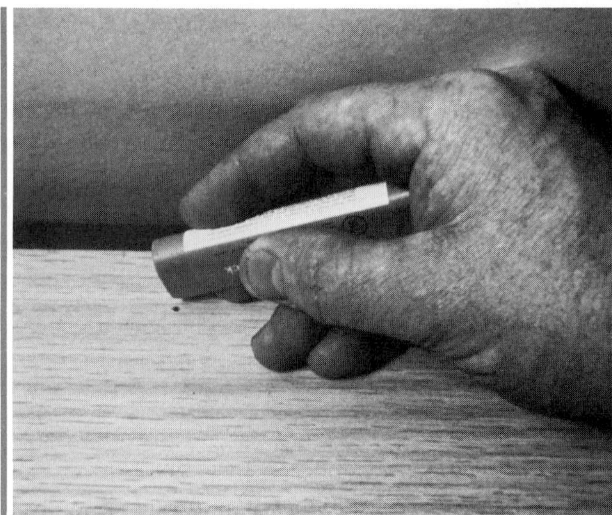

FIGURE 20-28

Nail holes are filled with a filler. *(Courtesy of National Oak Flooring Manufacturers' Association)*

FLOORING SIZE (in.)	TYPE AND SIZE OF NAIL	NAIL SPACING
Tongue-and-grooved 25/32 × 3¼ 25/32 × 2¼ 25/32 × 1½	7d or 8d screw-type or cut steel nail or machine-driven barbed fastener	10 to 12 in. apart
Tongue-and-grooved ½ × 2 ½ × 1½	5d screw-type, cut steel, or wire nail, or machine-driven barbed fastener	10 in. apart
Tongue-and-grooved[a] ⅜ × 2 ⅜ × 1½	4d bright casing, wire, screw, or cut nail, or machine-driven barbed fastener	8 in. apart
Square-edged[a] 5/16 × 2 5/16 × 1½	1-in. 15-gauge fully barbed flooring brad, cement-coated	2 nails every 7 in.

[a]Must be laid on wood subfloor.

FIGURE 20-30

Recommended nail sizes and spacing for strip hardwood flooring.

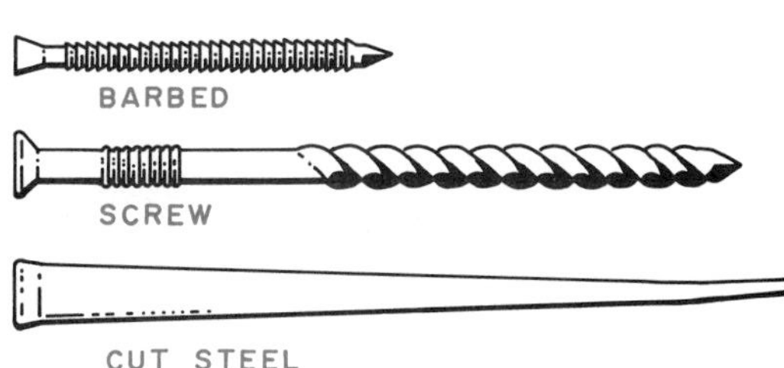

FIGURE 20-29

Nails commonly used to install strip hardwood flooring.

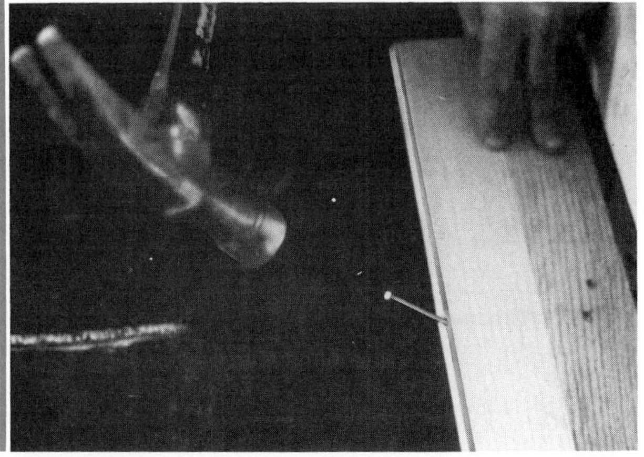

FIGURE 20-31

Installing strip hardwood flooring with a hammer. Nails are set with a nail set. *(Courtesy of National Oak Flooring Manufacturers' Association)*

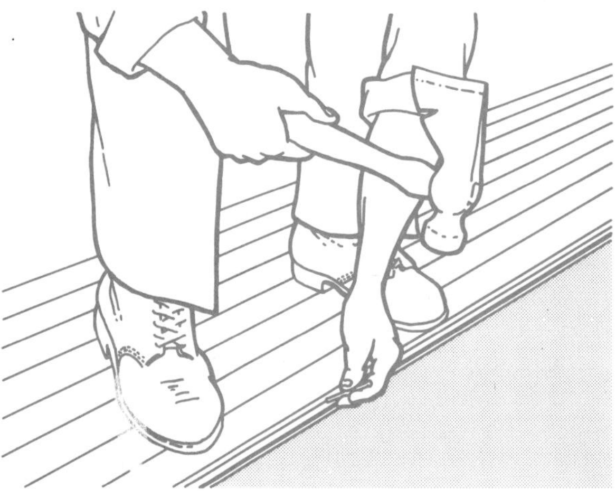

FIGURE 20-32

To hand-drive nails in strip flooring, straddle the spot to be nailed.

The nails can be driven using a hammer or hatchet (Fig. 20-31). The carpenter stands with his/her feet on each side of the nail location and leans over to drive the nail (Fig. 20-32). A nail set is used to finish driving the nail (Fig. 20-33). A nailing machine is often used. It has a lip that fits under the tongue. This positions the nail. The carpenter drives the nail by hitting the nailing block on the end of the machine (Fig. 20-34).

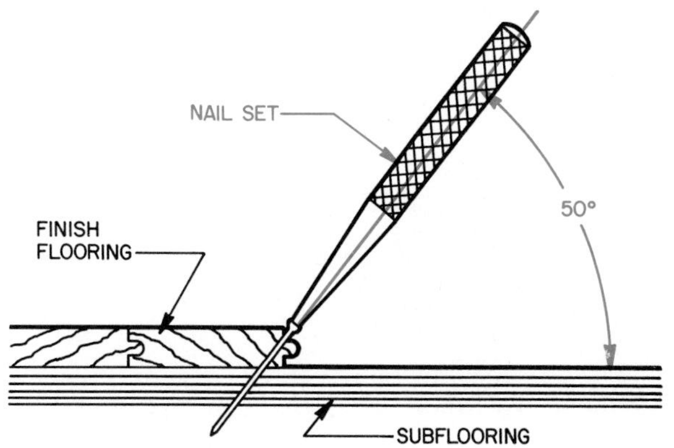

FIGURE 20-33

Set nails into the strip flooring with a nail set.

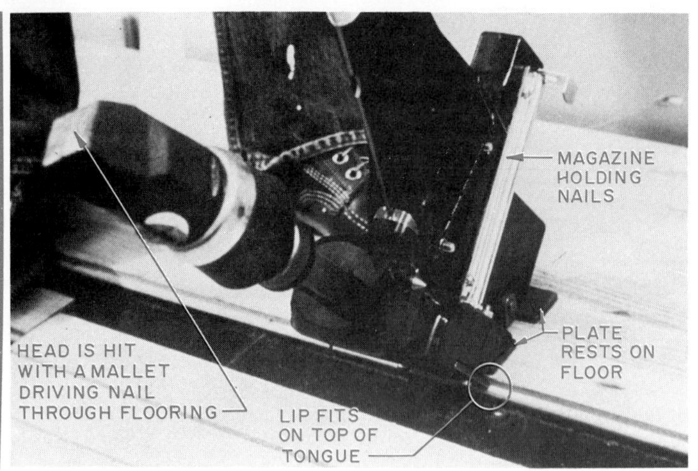

FIGURE 20-34

A hammer-activated strip wood floor nailing machine. The nails are blind-nailed on the top of the tongue.

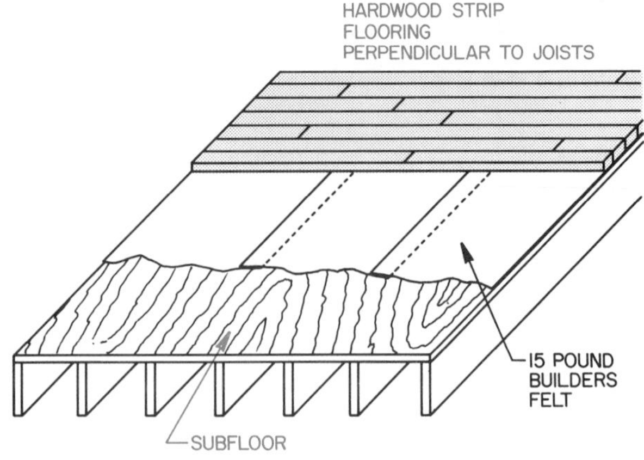

FIGURE 20-36

Strip hardwood flooring is installed over a layer of builder's felt.

Installing Strip Flooring Check the subfloor to be certain that it is firmly nailed to the joists. All raised nail heads need to be set flush. Remove all dirt and debris from the floor (Fig. 20-35). Then lay a layer of 15-lb asphalt-saturated building paper over the subfloor. Overlap the joints 4 in. (Fig. 20-36). This protects the wood floor from moisture and helps seal out cold and dust.

The building paper will cover the rows of nails in the subfloor, thus hiding the locations of the floor joists. Mark each joist on the wall before laying the paper. Chalk lines can be snapped to locate the joists on top of the building paper. Whenever possible, nail into the joists.

1. Start the first strip about ½ in. from a wall. Place the grooved side to the wall. Face-nail this strip to the subfloor (Fig. 20-37). The space allows the floor to expand. It will be covered with the base and shoe molding. Strip flooring should be installed perpendicular to the floor joists.

The first board must be carefully set in place. Use the longest piece of flooring available (Fig. 20-38). If it runs the full length of the room, this is best. If several rooms are to be floored, then a plan for the entire house must be developed. The following discussion is related to the plan in Fig. 20-39.

The beginning line was established to run the length of the living room and down the hall. The line was located by measuring out 6 in. from the wall and snapping a chalk line. The first piece was set ½ in. from the wall. It was located by measuring from the chalk line. This kept it straight. The piece was face-nailed

FIGURE 20-35

Before laying wood flooring, sweep the subfloor clear of dust and debris. *(Courtesy of National Oak Flooring Manufacturers' Association)*

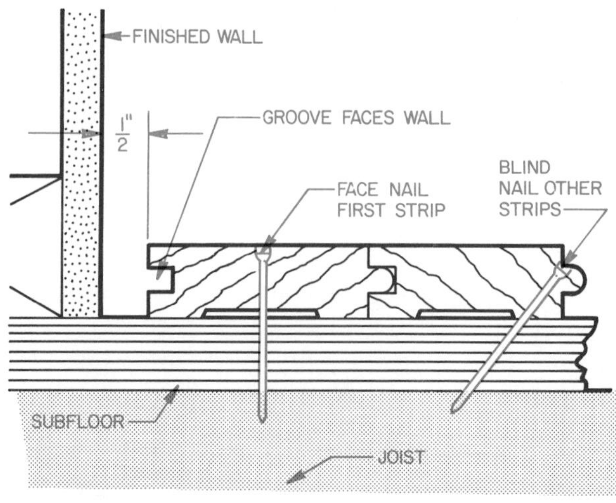

FIGURE 20-37

The first strip of flooring is placed ½ in. from the wall and is face-nailed.

FIGURE 20-38

The carpenter is face-nailing the first board parallel with the wall. *(Courtesy of National Oak Flooring Manufacturers' Association)*

through the subfloor and into the joists. Do not drive the nails completely into the flooring at this time because they may have to be removed to allow for adjustments.

Now nail the flooring across the living room and hall to the outside wall and into the front bedrooms.

Next, it is necessary to put the flooring into the dining room and the closet in bedrooms 1 and 3. The flooring has the groove facing into the area. It is necessary to get a side with a tongue facing into the area. To do this, place the next piece groove-to-groove and put a wood spline in both grooves (Fig. 20-40). Face-nail both pieces. Then lay the floor to the other wall in the normal manner.

2. All strips other than the first are *blind-nailed*. Usually, one carpenter does the nailing while another selects the pieces to be used and lays them in place (Fig. 20-41). The lengths used should vary so that the end joints are staggered. Joints in adjoining strips should never be closer than 6 in. Also, the carpenter can examine the pieces for color and grain and try to blend together pieces to produce an attractive floor. The pieces cut off on the end at one wall can be used to start the next row at the other end.

3. After laying two or three pieces, take a short length of scrap flooring, place it against the edge, and hit it with a hammer. This drives the flooring tightly together. It may be necessary to drive each piece this way if it is difficult to get a good joint (Fig. 20-42).

4. To cut the end pieces to length at the end of a course, turn it around so that the tongue end is cut off. The groove is needed to form the last end joint (Fig. 20-43).

5. As the installed floor gets to within about 3 ft of the other wall, measure the distance between the edge of the flooring and the wall. If the strip flooring is not parallel with the wall,

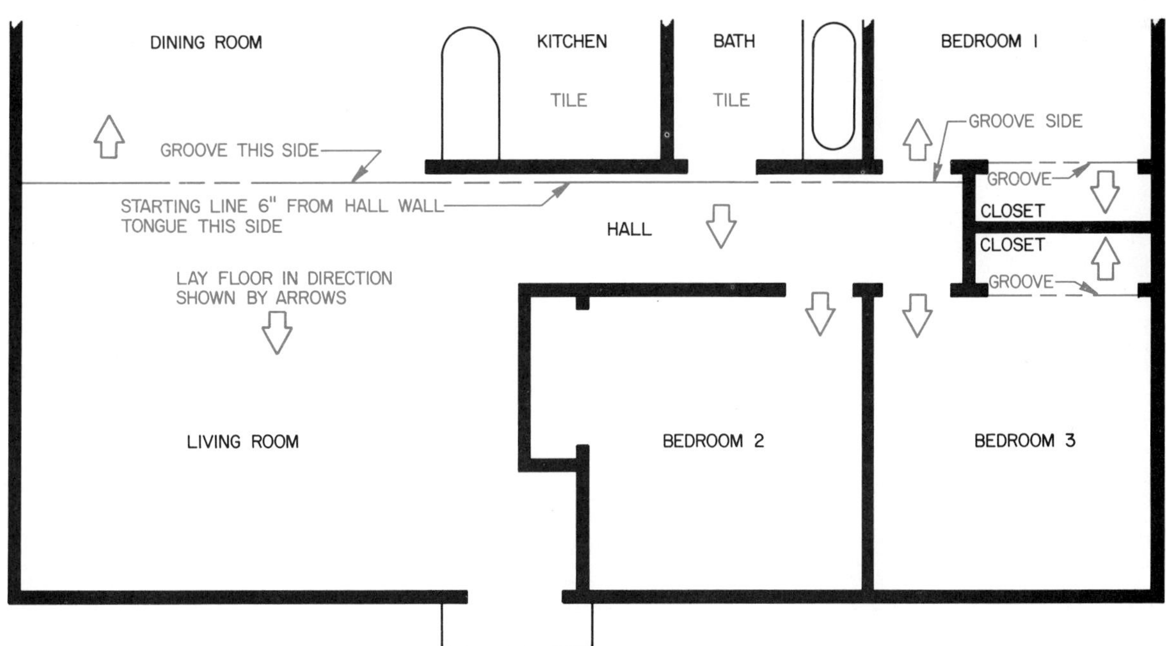

FIGURE 20-39

Develop a plan for starting and installing strip wood flooring.

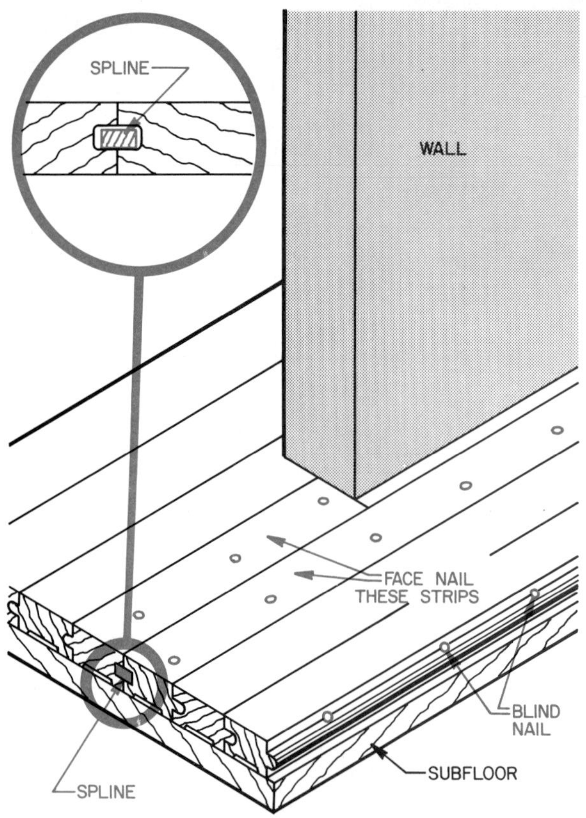

FIGURE 20-40

When going from one room to the other, a spline is used to turn the wood flooring around so that it can be blind-nailed at the tongue.

FIGURE 20-41

Lay out and match the wood strips so that they are ready for the person doing the nailing. *(Courtesy of National Oak Flooring Manufacturers' Association)*

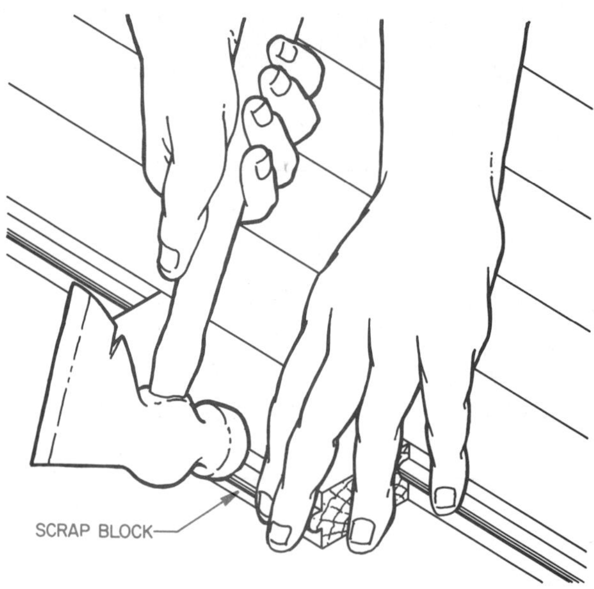

FIGURE 20-42

Hammer against a piece of scrap flooring to drive the courses together. Do this after every three or four are laid.

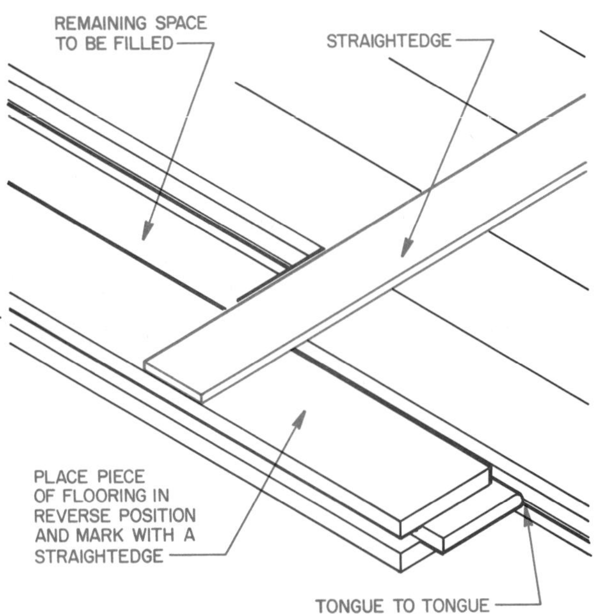

FIGURE 20-43

The end piece of flooring must be turned around and cut off so that the grooved end is left for joining with the tongue in the piece before it.

the grooved edges can be planed slightly at one end. This will taper the board a little and bring the last strip more nearly parallel with the wall.

The last few courses are too close to the wall to toenail. They must be face-nailed. They can be pulled up tight by using a crowbar to apply pressure. Be certain to protect the edge of the flooring from damage by the crowbar (Fig. 20-44).

Laying Wood Strip Flooring Over Concrete

Be certain that the slab is clean and free of debris. Treated wood 1 × 2-in. nailing strips are

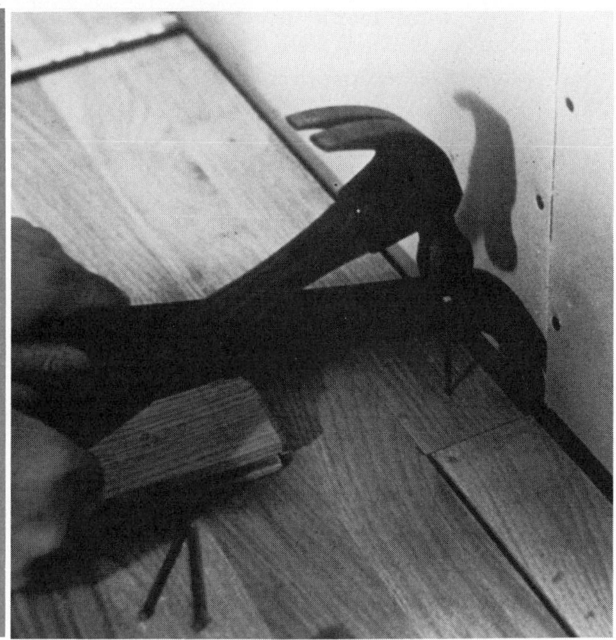

FIGURE 20-44

Use a crowbar or other pry bar to pull up the last few courses as they are nailed. Protect the edge of the flooring with cardboard or soft wood. *(Courtesy of National Oak Flooring Manufacturers' Association)*

joined to the concrete slab with adhesive and nailed to it with 1½-in. concrete nails. Space them 16 in. apart. They are run perpendicular to the direction of the flooring strips (Fig. 20-45).

Lay a sheet of polyethylene film over the strips. The edges of the sheets should be lapped over one of the nailing strips.

Now nail a second layer of 1 × 2-in. nailing strips over the polyethylene film and into the first strips. Use 4d coated nails.

The hardwood strip flooring is now installed in the same way as previously described.

Installing Plank Flooring Plank flooring is installed by blind-nailing the same as strip flooring. Two wood screws are countersunk at the end joints of each plank. The countersink is filled with a wood plug. Sometimes, blind wood plugs are put in various places to add to the decorative appearance of the flooring (Fig. 20-46).

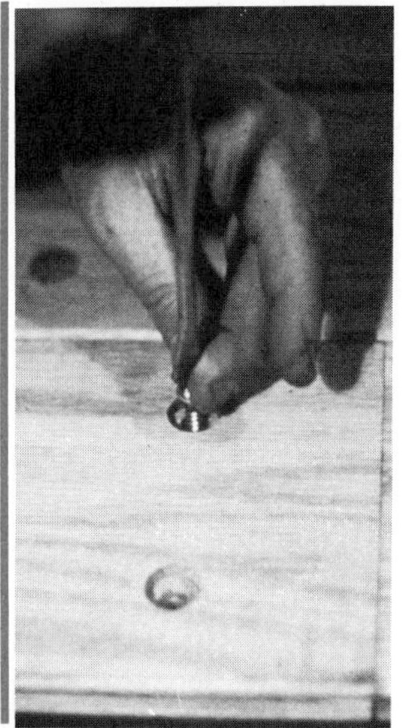

Wood screws are used to secure the ends of the planks.

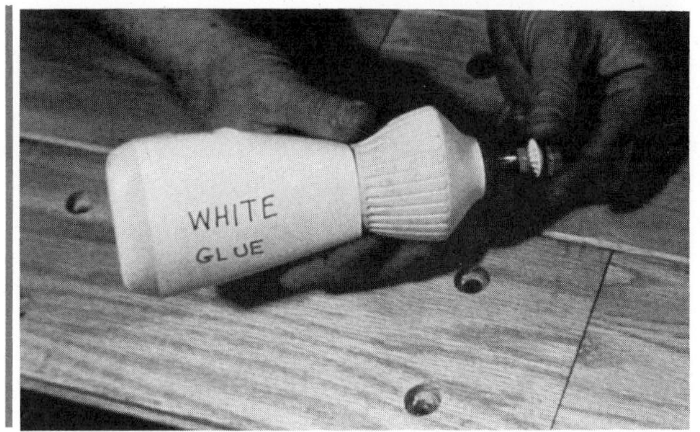

Wood plugs are glued over the screws.

FIGURE 20-46

Wood screws with plugs are used to secure the ends of plank flooring. *(Courtesy of National Oak Flooring Manufacturers' Association)*

FIGURE 20-45

Hardwood strip flooring can be laid over concrete floors by gluing treated wood strips to the concrete. Nail the flooring into these strips.

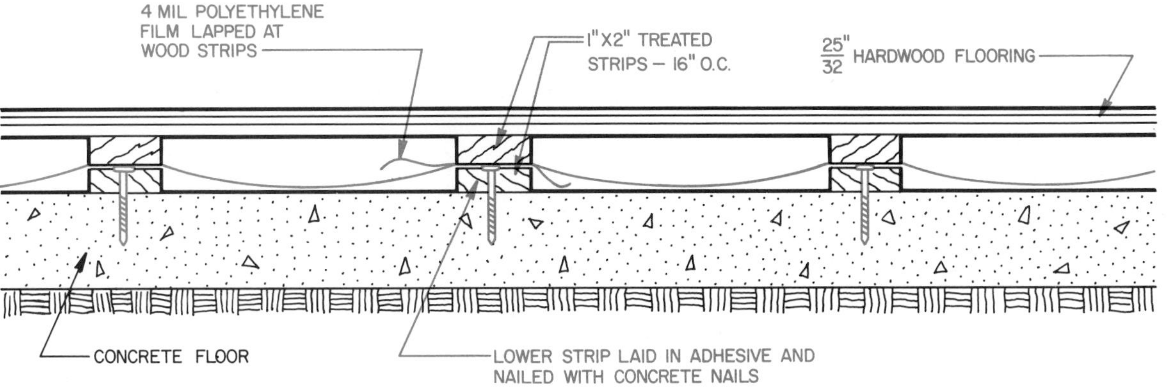

Installing Block Flooring Block flooring has tongue-and-grooved edges. They are laid using the layout procedures described for vinyl tile. They are blind-nailed at the tongue or joined to the subfloor with an adhesive. Usually, they are laid so that the grain in each block is perpendicular to the grain in the adjoining block (Fig. 20-47).

Estimating Wood Strip Flooring To figure the number of board feet of strip flooring needed, determine the number of square feet to be covered. To this add a percentage for waste and the coverage lost by the tongue-and-groove edges (Fig. 20-48).

For example, a house requires 1500 square feet to be covered. The flooring will be ½ × 2 in.

total floor area	=	1500 ft^2
% required is 30% of 1500	=	450 ft^2
total board feet to order		1950 board feet

Finishing the Floor After the floor is laid, it must be finished unless it is of the prefinished type. Wait a few days after laying before finishing it. This lets it adjust to the moisture. Be certain that all other work, such as painting, papering, or plumbing, is complete before finishing the floor. Sweep the floor clean and check for loose boards and protruding nail heads.

Unfinished wood floors are sanded with a drum sander using medium abrasive paper. Sand with the grain the entire length of the room (Fig. 20-49).

To start sanding strip flooring, locate the machine next to the right wall and about two-thirds of the room length. Start the motor and

FIGURE 20-47

Installing block floor with adhesive. *(Courtesy of National Oak Flooring Manufacturers' Association)*

PERCENT ADDED	SIZE OF STRIP FLOORING (in.)
55	25/32 × 1½
42½	25/32 × 2
38⅓	25/32 × 2¼
29	25/32 × 3¼
38⅓	⅜ × 1½
30	⅜ × 2
38⅓	½ × 1½
30	½ × 2

FIGURE 20-48

When estimating the amount of tongue and groove flooring to order, add the percentages shown to the number of square feet of floor area.

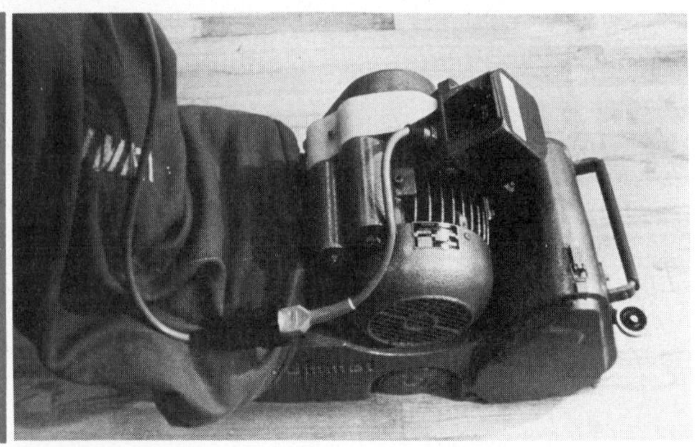

FIGURE 20-49

A drum sander being used on hardwood strip floors. *(Courtesy of National Oak Flooring Manufacturers' Association)*

lower the sanding drum carefully to the floor. Walk the sander forward to the end of the room. Ease the drum off the floor. Lower the drum and move the sander back to the starting point. Raise the drum off the floor. Move the drum to the left for the second cut. Overlapping the first cut 2 to 4 in., repeat the sanding procedure. Continue this until the other wall is reached. Then turn the sander around. Sand the final one-third of the room in the same manner. Never let the sanding machine touch the floor unless it is moving forward or backward.

Use a power edger to sand the edges that could not be sanded with the drum sander. Tight spots, such as in a corner, must be hand scraped and sanded (Fig. 20-50).

After sanding, fill all holes. The floor is ready for the finishing material to be applied.

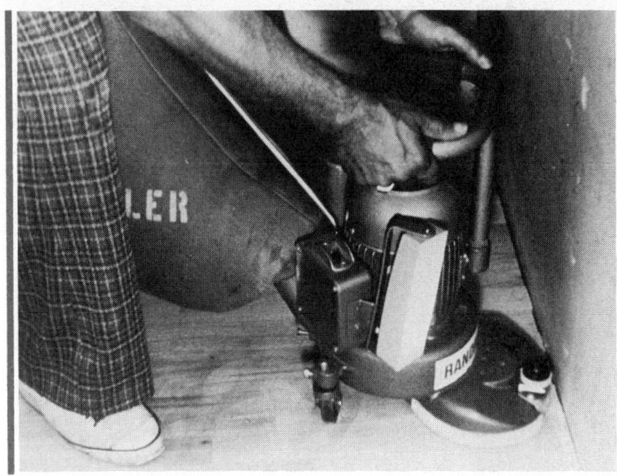

FIGURE 20-50

An edger is used in places the drum sander will not reach. *(Courtesy of National Oak Flooring Manufacturers' Association)*

STUDY QUESTIONS

After reading the chapter, answer each of the following questions. If you do not know the answer, review the chapter.

1. What types of materials are used for finished flooring?
2. In what sizes is vinyl floor covering material available?
3. Explain how you start to locate the first floor tiles.
4. How many vinyl tiles are needed to cover a floor with an area of 500 square feet?
5. How are seams between sheets of vinyl floor covering cut?
6. What are the two methods for installing sheets of vinyl floor covering?
7. In what forms is wood flooring available?
8. How should wood flooring be stored before it is installed?
9. How do you start installing strip wood flooring?
10. How is plank wood flooring fastened to the subfloor and joists?

IMPORTANT TECHNICAL TERMS

Following are technical terms that you should be able to use as part of your working vocabulary. Write a brief description of the meaning of each term.

Natural flooring
Resilient flooring
Carpet
Wood flooring
Vinyl
Vinyl asbestos
Underscribing
Scriber
Single cutting
Double cutting
Lapping method
Tubing method
Wood blocks
Parquet
Blind-nail

21 STAIRS

A stair is a series of steps used to provide a means of moving from one level of a building to another.

There are many types of stair construction. Some of those commonly used are presented in this chapter.

Generally, carpenters purchase the parts for a stair from their building supply dealer. These parts are accurately manufactured by companies specializing in this work. However, the carpenter can cut and assemble a stair on the job.

TYPES OF STAIRS

Stairs are classified into service and main stairs. Service stairs run to a basement or attic. They are built of less expensive materials. Main stairs generally form an important part of the interior design. High-quality materials such as oak, birch, or maple are used.

Stairs may be open or closed. An open stair has at least one side open to the adjoining room (Fig. 21-1). A closed stair has walls on both sides and is hidden from view.

The types of stairs in common use include straight, L-shaped, U-shaped, and winding (Fig. 21-2).

The straight stair is most commonly used. It requires a rather long section of floor area. It is the easiest to build.

The L-shaped and U-shaped stair have a platform partway to the next floor. This is used when space is rather restricted. Often, the platform is located high enough to get a door below. The area under the platform can then be used for storage.

If space is extremely tight, a stair with winders can be used. It

FIGURE 21-1

An open stair.

occupies the least space. It is more difficult to build and is dangerous to use because the tread is so narrow.

In split-level houses a short straight stair is used. The stair has a short run and has excellent ceiling height because of the stair directly above it (Fig. 21-3).

FIGURE 21-2

Commonly used types of stairs.

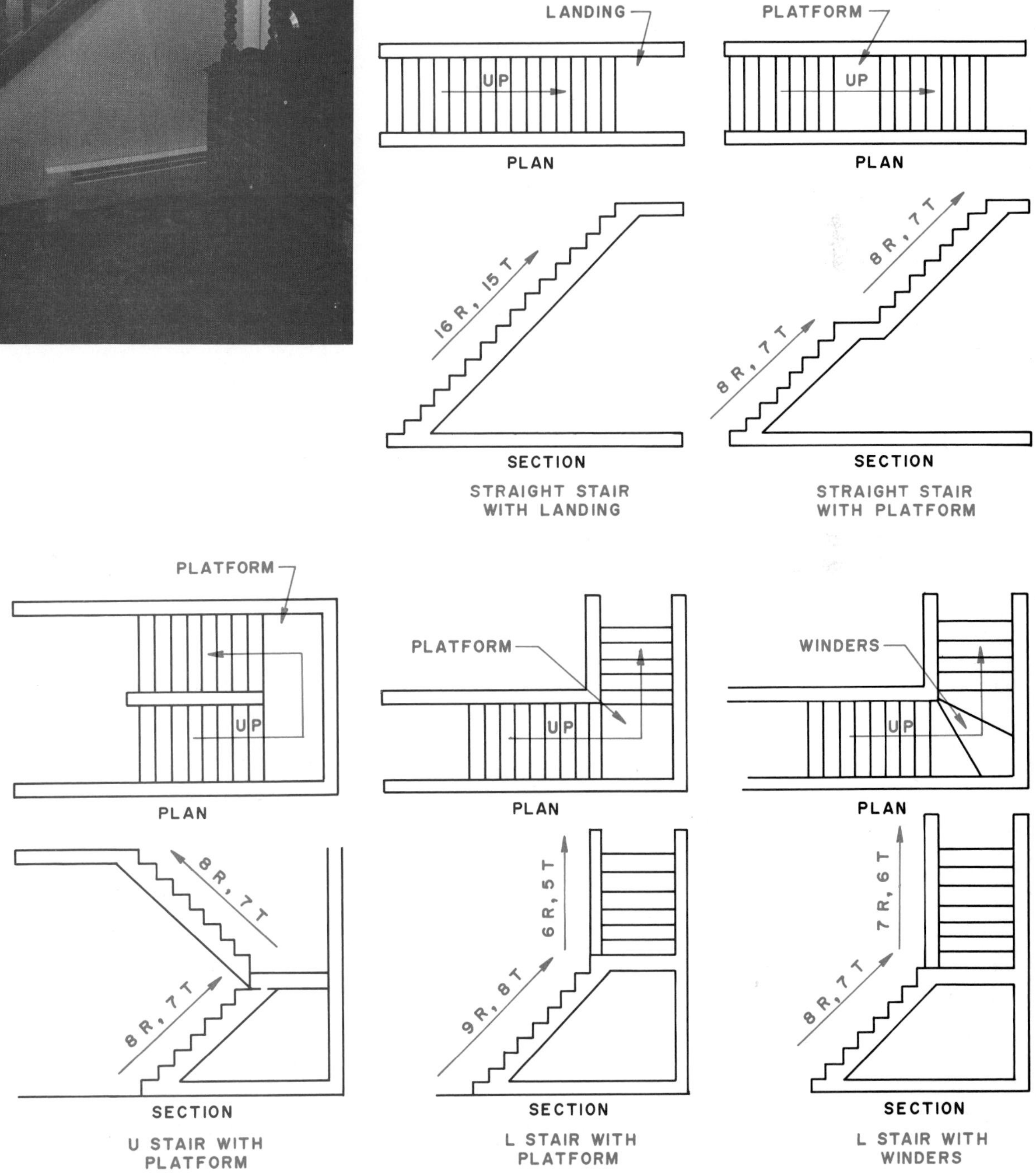

LANDING
UP
PLAN
16 R, 15 T
SECTION
STRAIGHT STAIR WITH LANDING

PLATFORM
UP
PLAN
8 R, 7 T
8 R, 7 T
SECTION
STRAIGHT STAIR WITH PLATFORM

PLATFORM
UP
PLAN
8 R, 7 T
8 R, 7 T
SECTION
U STAIR WITH PLATFORM

PLATFORM
UP
PLAN
9 R, 8 T
6 R, 5 T
SECTION
L STAIR WITH PLATFORM

WINDERS
UP
PLAN
8 R, 7 T
7 R, 6 T
SECTION
L STAIR WITH WINDERS

TERMINOLOGY

Carpenters must be familiar with the names of the parts of a stair. The structural parts are shown in Fig. 21-4. Finished stair parts are in Fig. 21-5.

Stair horse: 2-in. piece of stock with the tread and riser sections cut out. The treads and risers are nailed to it.

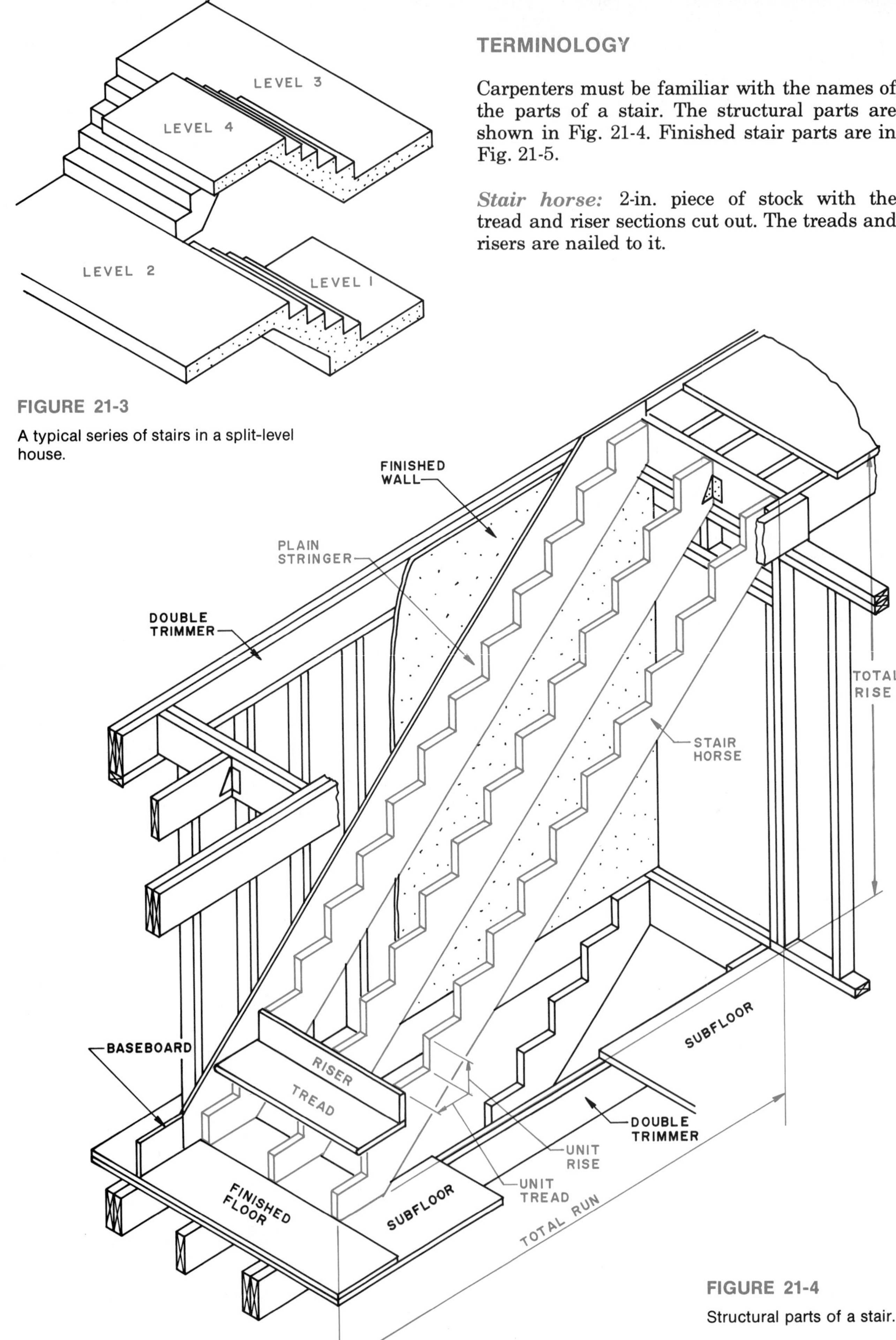

FIGURE 21-3

A typical series of stairs in a split-level house.

FIGURE 21-4

Structural parts of a stair.

FIGURE 21-5

Finished stair parts.

Square-cut stringer: member cut the same size as the stair horse. It is nailed on the exposed side of the horse to provide an attractive finished surface.

Plain stringer: finished board that is between the wall and the stair horse or is notched to match the riser and tread cuts.

Mitered stringer: just like the square-cut stringer except that the riser cuts on the side to be exposed are mitered. It is used where the stringer is to be on the open side of a stair.

Housed stringer: stringer with dados that hold the treads and risers.

Nosing: rounded projection on the edge of the tread extending beyond the riser.

Unit rise: vertical height of a step.

Unit tread: horizontal size of a step.

Riser: board enclosing the rise of a step.

Tread: board forming the part of a step upon which a person stands.

Total rise: total vertical distance from one floor to the next.

Total run: total horizontal length of the stair.

Headroom: vertical clearance between the treads and any overhead ceiling or floor.

Railing: protective barrier about open sides of a stair.

Newel: main post of the railing at the bottom of the stair. They are also used in railings on platforms.

Handrail: rail secured to the wall on one or both sides of a stair. It parallels the angle of the carriage. It is also used as the top member of a railing.

Balusters: vertical members supporting the handrail.

Landing: floor at the top or bottom of the stair where the stair begins or ends.

Platform: horizontal area in the middle of a stair.

DESIGN CONSIDERATIONS

Stair design involves tread size, riser size, headroom, and width. The minimum acceptable headroom for a main stair is 6 ft 8 in. For service stairs it is 6 ft 4 in. The minimum width is 2 ft 8 in. clear of any handrail for main and service stairs.

The minimum tread width is 9 in. plus a $1\frac{1}{8}$-in. nosing for stairs with a closed riser and $\frac{1}{2}$ in. for stairs with an open riser (Fig. 21-6).

The maximum rise is $8\frac{1}{4}$ in. except in buildings for the physically handicapped, where a $7\frac{1}{2}$-in. rise is maximum.

Winders must run from a point 12 in. from the edge of the stair horse. At this point they should not be less in width than the regular treads. Winders cannot be used in buildings for the physically handicapped (Fig. 21-7).

In a stair all treads must be the same width and risers the same height.

The width of a landing should not be less than the width of the stair. Its depth at the top of a stair must be at least 2 ft 6 in. In no case can the edge of a door swinging toward the stair overlap the top step (Fig. 21-8).

All stairs should have a continuous handrail on at least one side. It is usually placed 30 to 34 in. above the treads. It is held to the wall with brackets. They should not be placed more than 8 ft 0 in. apart (Fig. 21-9).

Open stairwells require a railing at least 36 in. high on all sides with 42 in. considered better (Fig. 21-10).

In buildings required to be accessible by the handicapped, ramps with handrails must be provided.

DETERMINING TOTAL RISE

Before the size of the stair risers can be calculated, the total rise must be found. The measurements to take will vary slightly depending upon

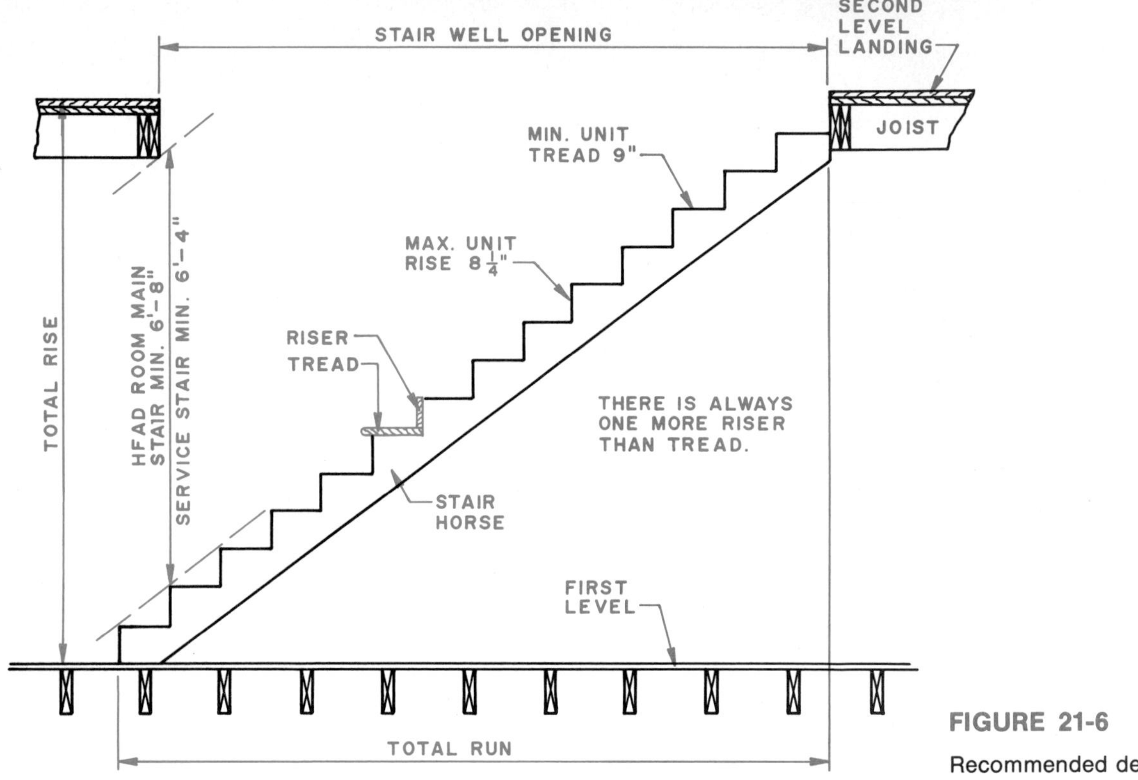

FIGURE 21-6

Recommended design sizes.

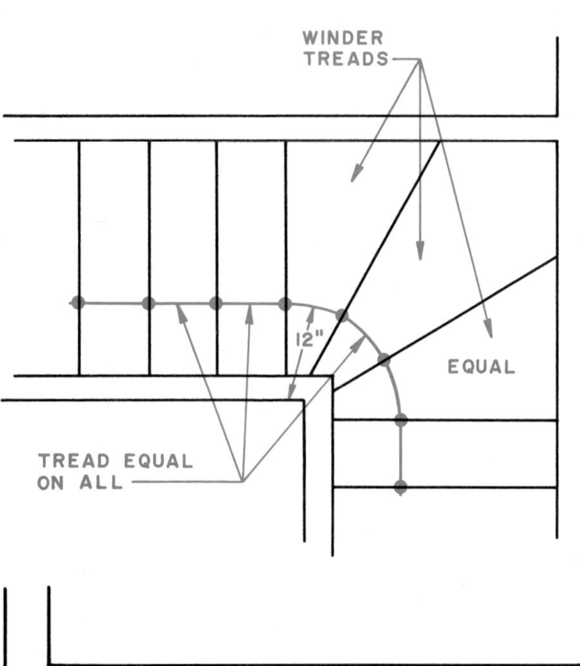

FIGURE 21-7 *(left)*

Winders should be normal tread width 12 in. in from the edge of the stair horse.

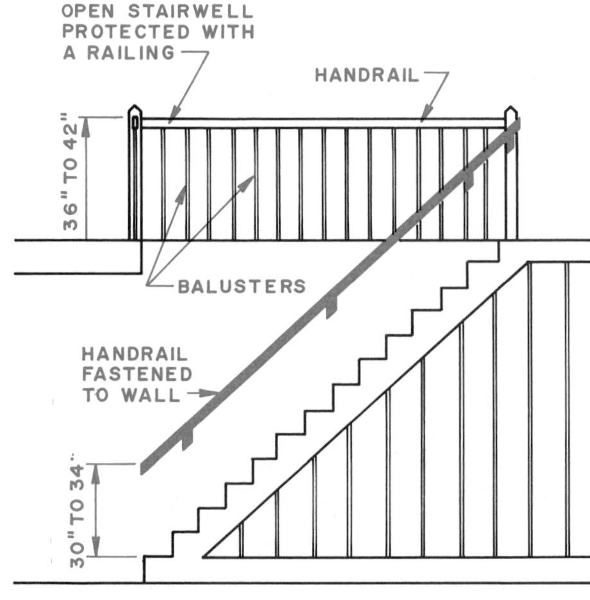

FIGURE 21-9

Handrails are used on at least one side of a stair and railings are used to enclose open stairwells.

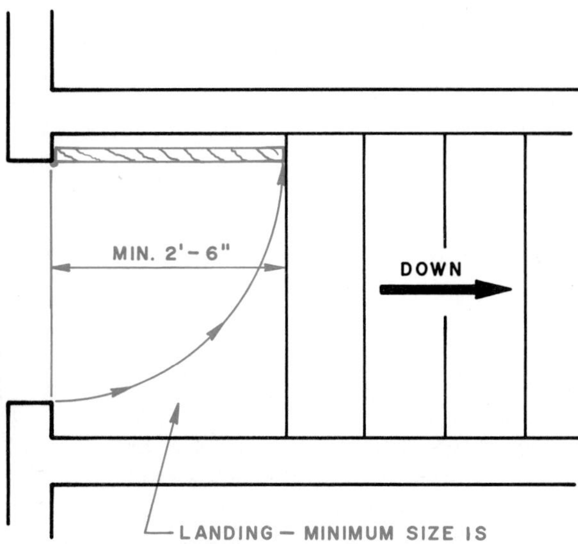

FIGURE 21-8 *(left)*

Minimum size for a landing.

FIGURE 21-10
Open stairs must have a railing.

the construction. Basically, the rise is the distance between the top of the finish flooring on the first floor to the top of the finish flooring on the second floor. Usually, when this measurement is taken only the subfloor is in place, so the thickness of the finish floor to be used must be known.

The steps to find the total rise are in Fig. 21-11. Stand a straight, smooth 1 × 2 in. board between floors. Mark the location of the finish floor at the second level. Above this, mark the location of the finish floor. On the first floor, mark the thickness of the finish floor. The distance between the top of the finish floor marks is the total rise.

FIGURE 21-11
Total rise runs from finished floor to finished floor.

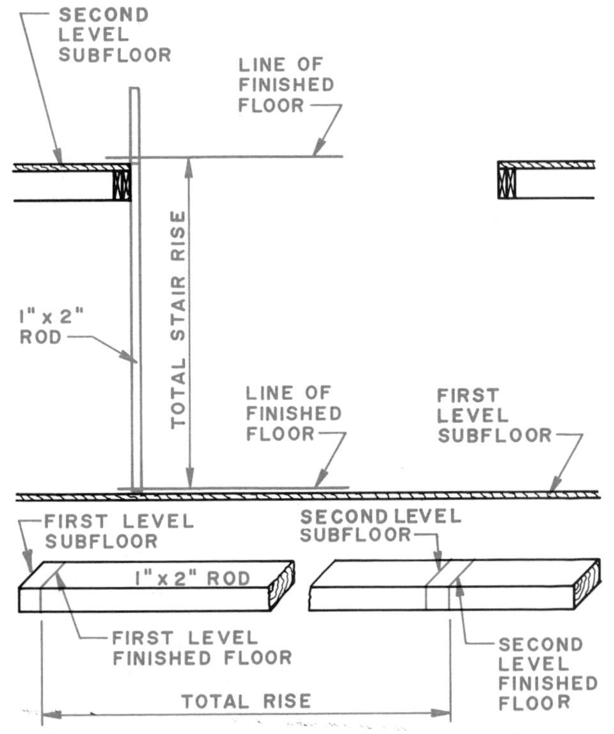

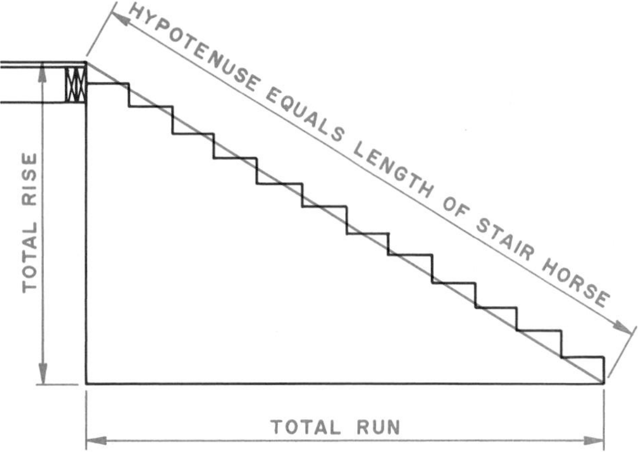

FIGURE 21-12
The approximate length of the stair horse is equal to the hypotenuse formed by the total rise and total run.

DETERMINING THE LENGTH OF THE STAIR HORSE

The approximate length of the stair horse and stringers can be found using the principles of the right triangle. The rise and run are known. They form two sides of a right triangle. The stair horse is the third side or the hypotenuse (Fig. 21-12).

To find the length of stock required for the stair horse, locate the total rise on the tongue of the framing square. Locate the total run on the blade. Measure the diagonal between these. This is the approximate length of stock needed for the stair horse and stringers (Fig. 21-13).

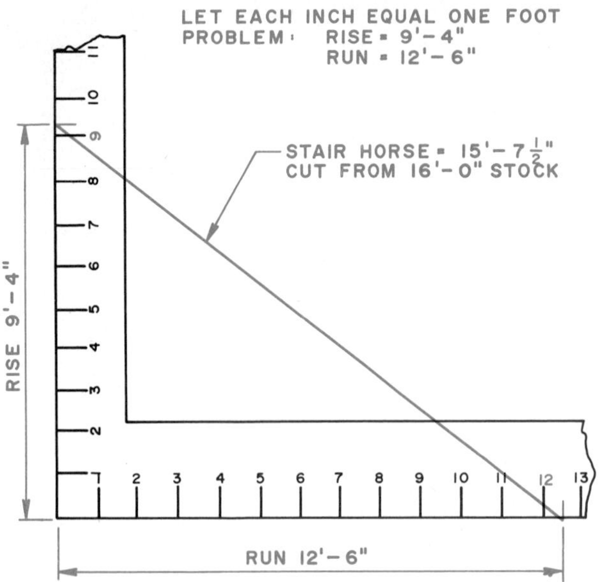

FIGURE 21-13

The approximate length of stock needed for a stair horse can be found using a framing square.

RATIO OF RISER TO TREAD

Riser and tread sizes bear a direct relationship to each other. The most satisfactory riser size is 7½ to 7¾ in. This is a comfortable height to step up. As the riser gets smaller, the tread must get wider for comfortable stair use. A rule of proportion to follow is that the *sum of the tread plus the riser equal 17 to 18 in.* For example, a riser of 7½ in. would have a tread of 10 in. Some commonly used tread and riser sizes are shown in Fig. 21-14. These riser sizes are used to calculate the number of steps between floors. This size does not include the nosing.

The total rise is divided by the desired riser size. If this does not come out to an even figure, the riser size is adjusted. For example, assume a total rise of 9 ft 0½ in. or 108½ in. (Fig. 21-15). If the desired rise is 7.75 in., this would require 14 steps. If 15 steps were desired, the riser would be 7.23 or 7 15/64 in.

The tread recommended for a rise of 7.75 in. is 10 in. (Fig. 21-14). Since there is always one less tread than riser, there are 13 treads of 10 in. or a total stair run of 130 in.

TREAD SIZE (in.)	RISER SIZE (in.)
9	8
10	7½
11	7

FIGURE 21-14

Commonly used tread and riser sizes.

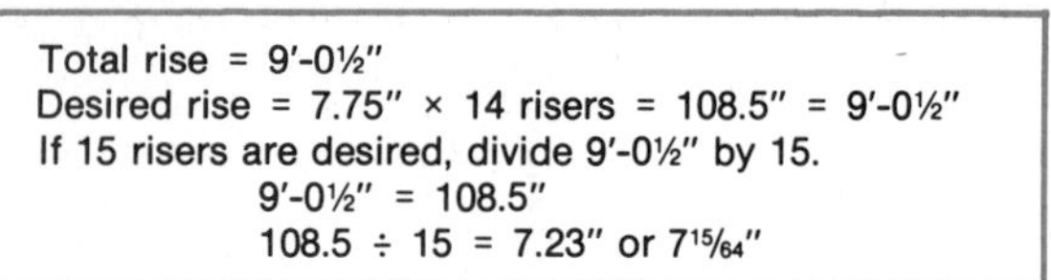
Total rise = 9'-0½"
Desired rise = 7.75" × 14 risers = 108.5" = 9'-0½"
If 15 risers are desired, divide 9'-0½" by 15.
9'-0½" = 108.5"
108.5 ÷ 15 = 7.23" or 7 15/64"

FIGURE 21-15

How to figure riser size.

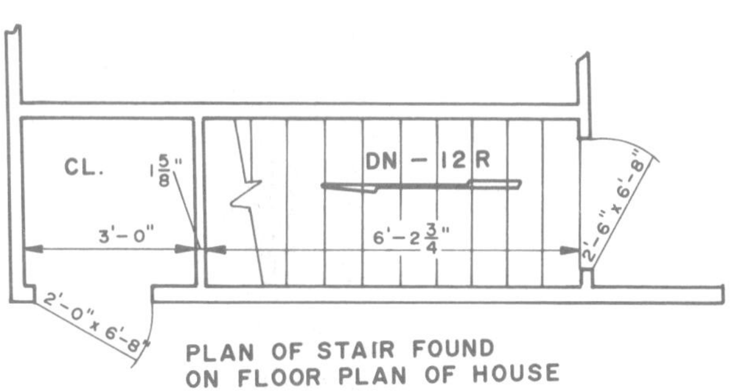

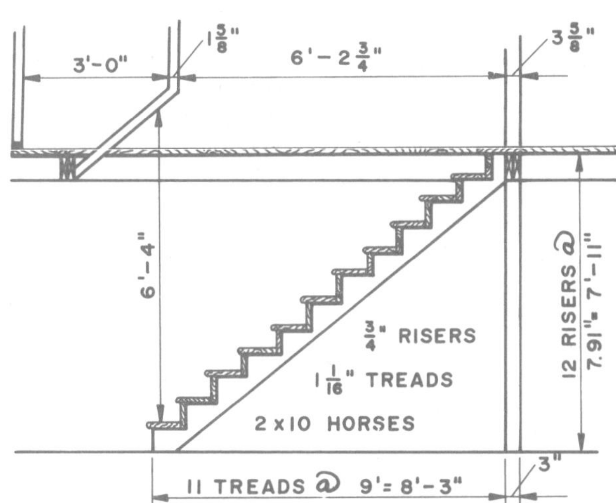

FIGURE 21-16

A typical stair detail as found on working drawings.

A stair detail as found on a typical house plan is shown in Fig. 21-16. Notice that the architect has figured the riser and tread sizes for the carpenter.

STOCK PARTS

Most parts of a stair are available from a building materials dealer. They can supply accurately cut and machined parts. Examples of these are in Fig. 21-17.

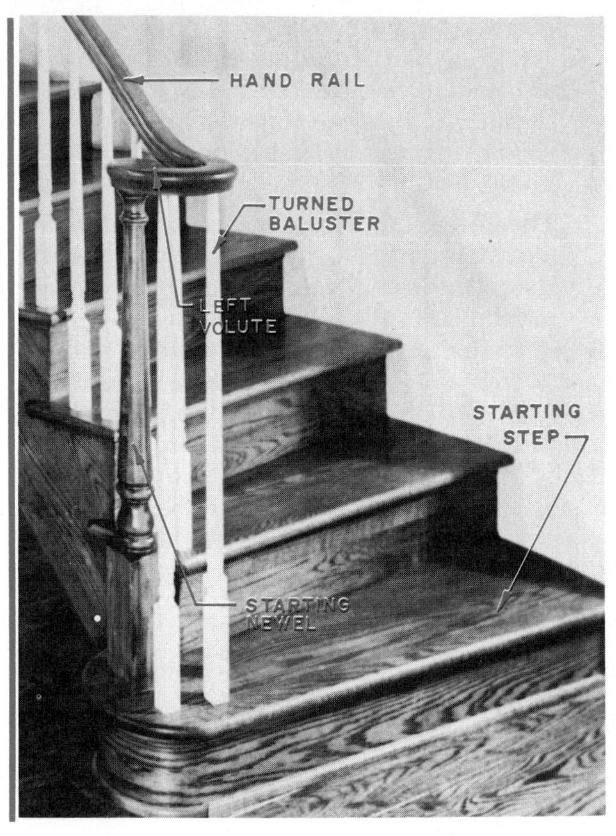

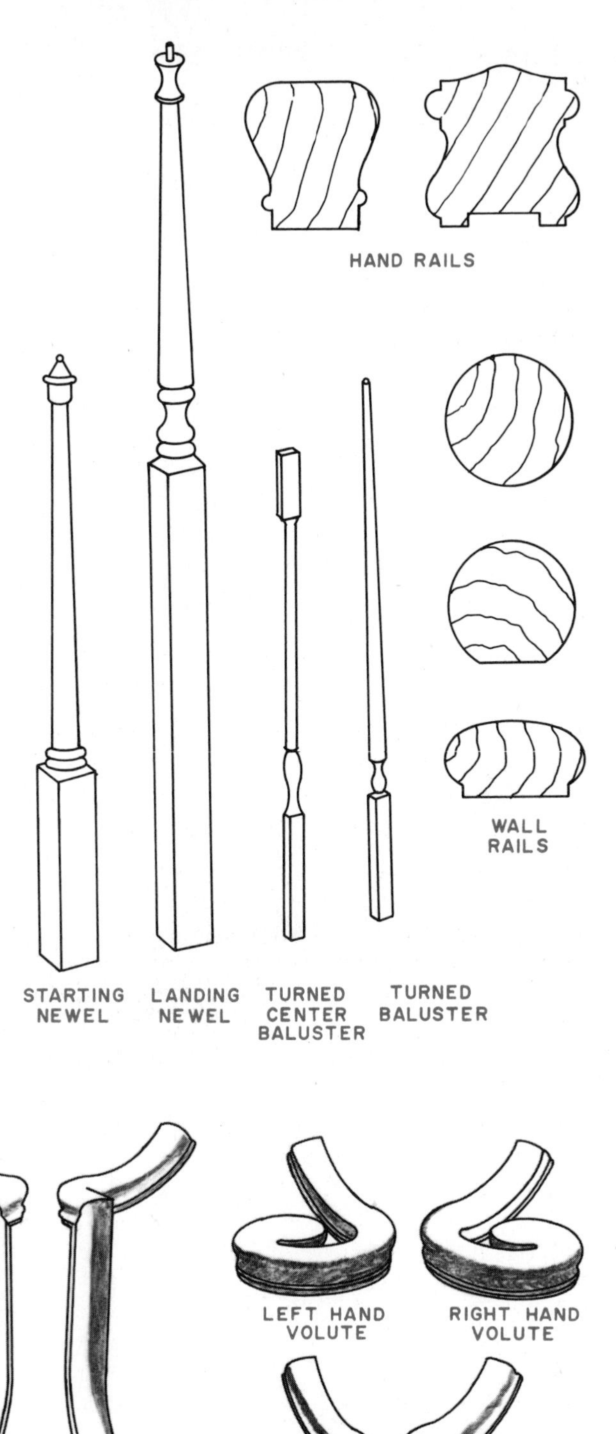

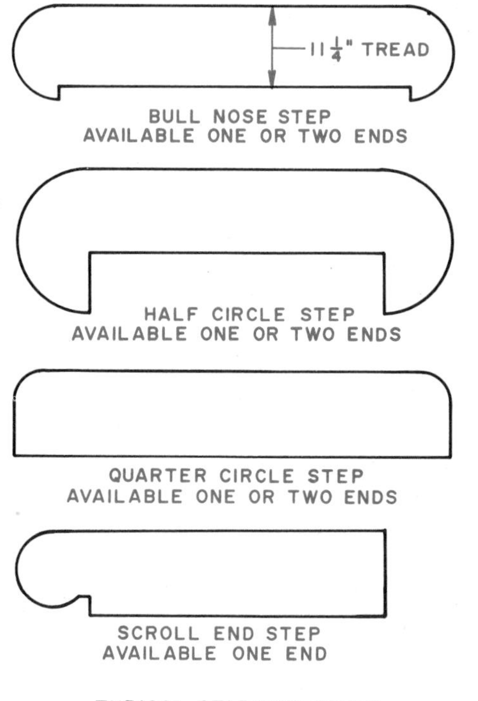

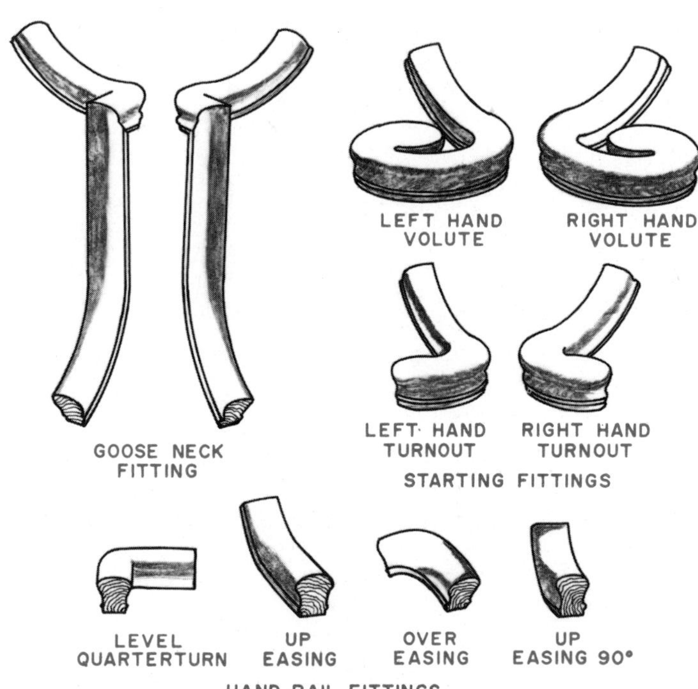

FIGURE 21-17

Examples of stock stair parts. *(Courtesy of Visador Co.)*

LAYING OUT A STAIR HORSE

Once the tread and riser size is known, the riser and tread cuts can be laid out. These must be done very carefully. Thin, accurately drawn lines are necessary. All cuts must be made accurately. If a horse is too long, the treads will slant backward. If it is too short, the treads will slant forward. Following are the steps to make the layout.

1. Position the stock so that the top edge is away from the carpenter. All measurements are taken from this top edge (Fig. 21-18).
2. Begin the layout at the left end. Place the square on the stock as shown in Fig. 21-18. Using stair gauges, set the tread measurement on the blade of the framing square and the rise on the tongue. Place the square on the left side of the stock with the stair gauges touching the top edge. Draw a line marking the rise and run.
3. Extend the tread mark on the blade until it reaches the bottom of the stock. This locates the level. Cut at the bottom of the stock.
4. Move the square to mark the next rise and the first tread. Continue marking until all are laid out. Remember to lay out one less riser than was calculated. This is because the top riser is formed by the header (Fig. 21-19).
5. The lower end of the horse must be dropped (cut off) some. This is necessary to keep all risers the same size. There are two cases in which drop is needed:
   a. If the stair horse was measured and cut from the subfloor and is to be installed on top of a finish floor, the thickness of the tread must be cut off (Fig. 21-20).
   b. If the stair horse is to be nailed to the subfloor, the amount to be cut off is equal to the difference between the thickness of the finish floor and the thickness of the tread material (Fig. 21-20).
6. Now mark and cut the top end of the stair horse to fit the header. How this is done depends on the conditions present. Examples are in Fig. 21-19.
7. After cutting one horse, use it as the pattern for the others.

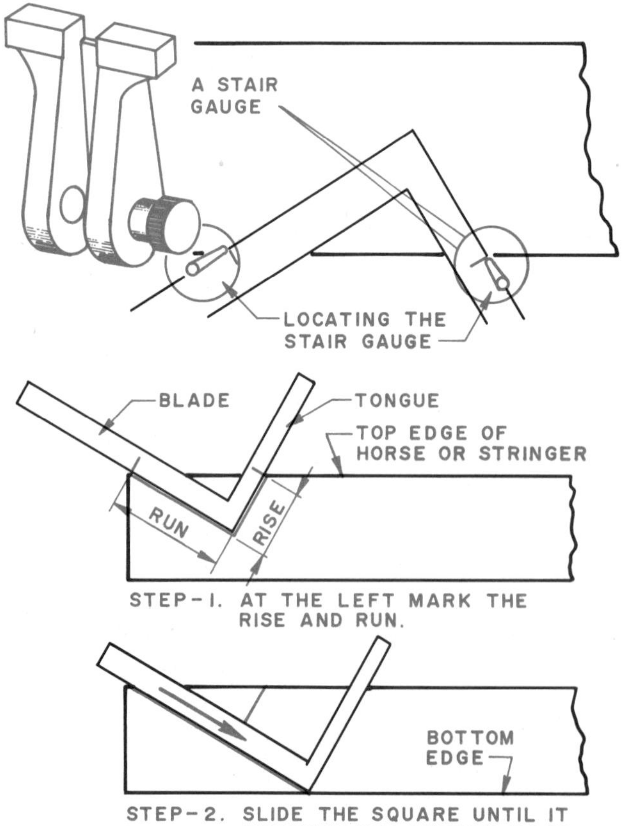

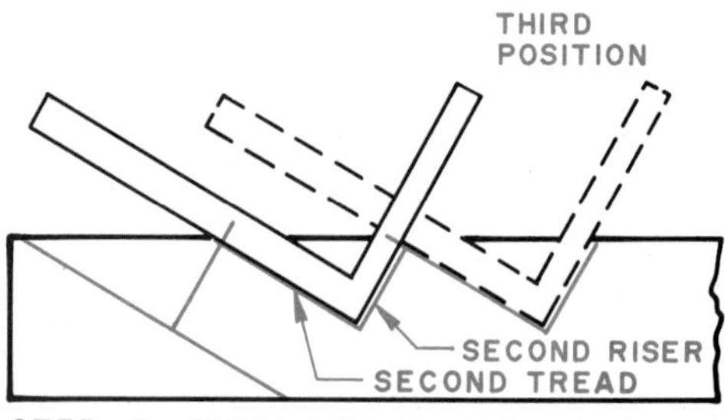

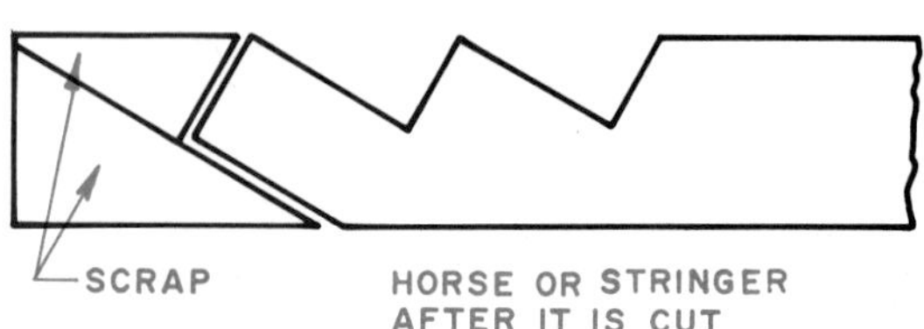

FIGURE 21-18

Steps to lay out a stair horse.

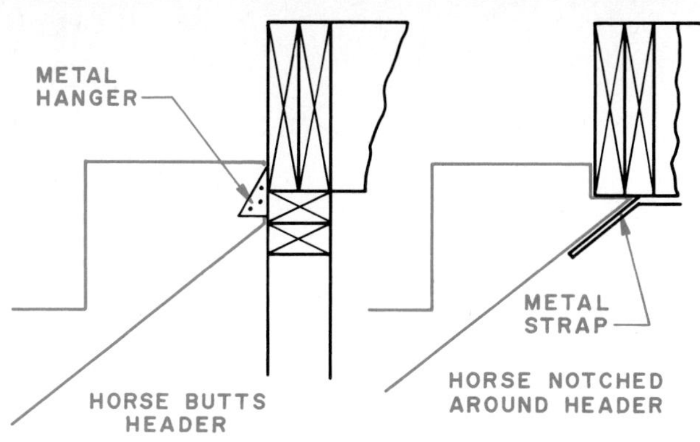

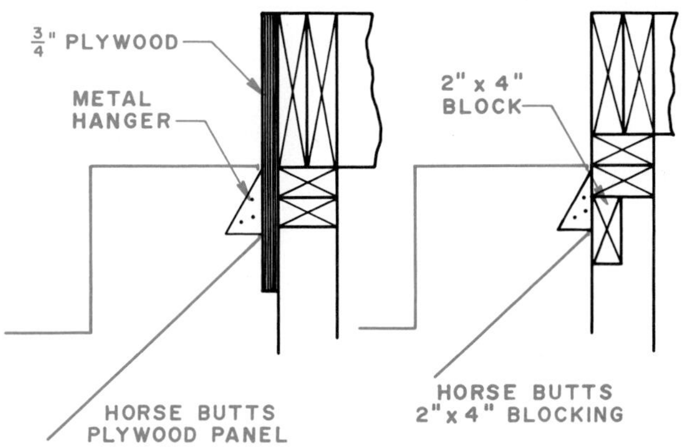

FIGURE 21-19

Typical methods of joining horse to header.

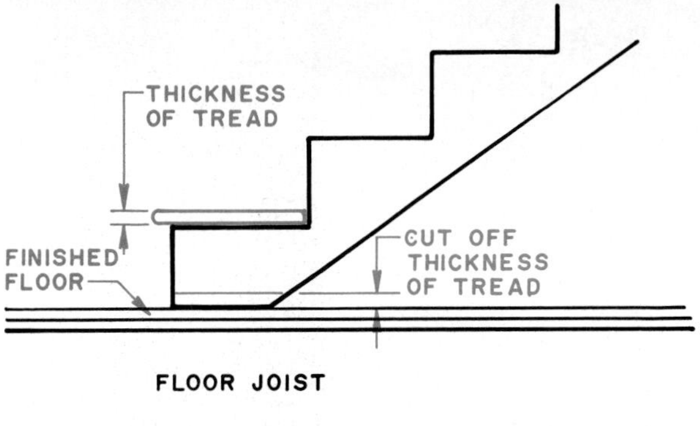

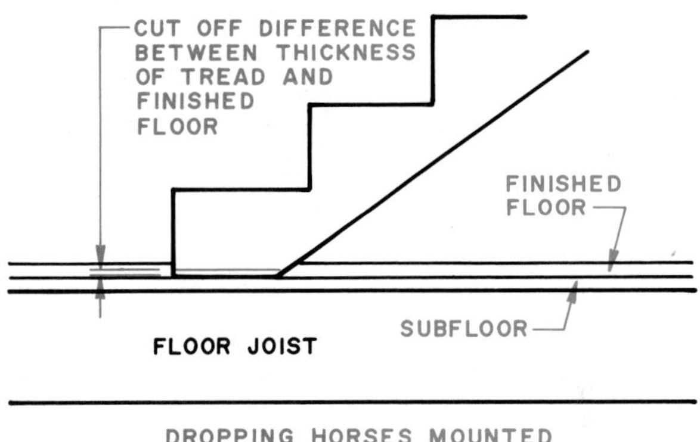

FIGURE 21-20

Two cases requiring the horse to be dropped (cut off).

LAYING OUT A PLAIN STRINGER

The plain stringer is usually made from good-quality 1 × 12 in. stock. There should be at least 2 in. of material remaining above the tread nosing. The stringer might be of the same material as the treads. To lay out a plain stringer:

1. Position the board with the top away from the carpenter.
2. Begin the layout from the left end of the board. This is the bottom end. Lay out the first-floor level cut. This is the same as the cut on the stair horse (Fig. 21-21).
3. Lay out each tread and riser. These are laid

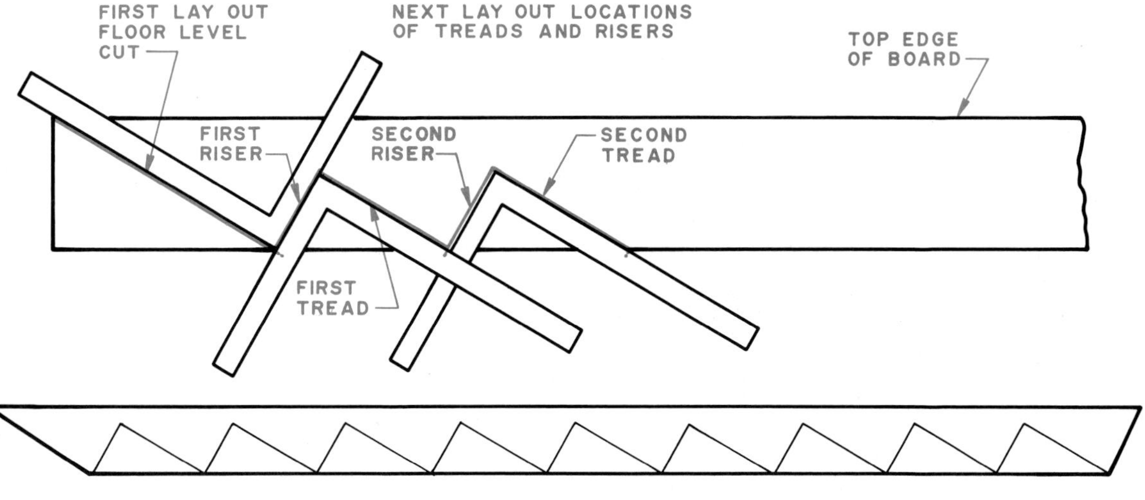

FIGURE 21-21

A finished plain stringer laid out and ends cut.

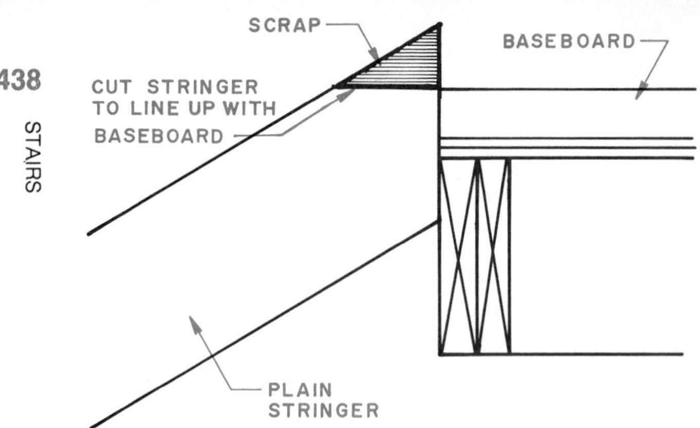

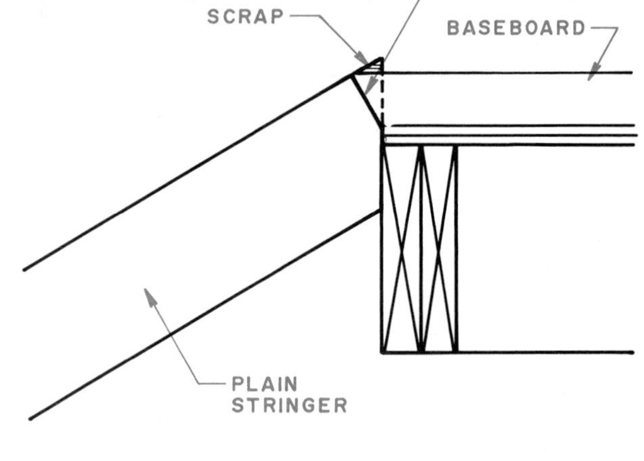

FIGURE 21-22

How to join the stringer and the baseboard.

out on the bottom of the board rather than the top as is done with the stair horse.

4. Lay out and cut the shape of the stringer at the top. The exact cut will be determined by the conditions that exist. Remember to cut the stringer to butt the baseboard. It should be cut the same height as the baseboard (Fig. 21-22).
5. Lay out the drop (amount to cut off) at the bottom of the stringer. This is usually the thickness of the finished floor. Then cut the end of the stringer to butt the baseboard (Fig. 21-23). Remember, the plain stringer can be nailed to a wall alongside the stair and the treads butt into it. It can also be cut on the tread and riser lines and fit over the top of the tread and riser (Fig. 21-24).

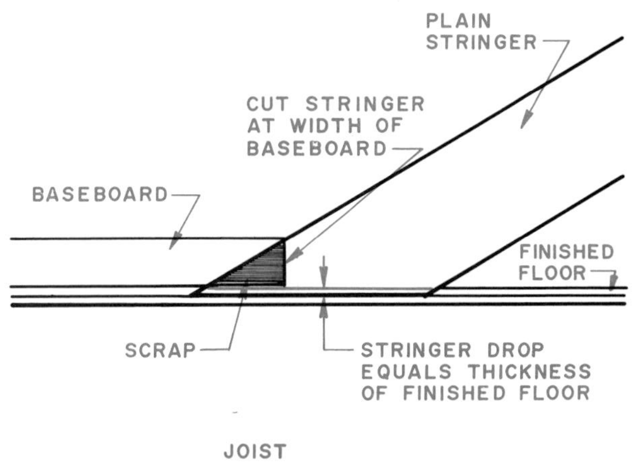

FIGURE 21-23 (*above*)

Marking the drop on a plain stringer.

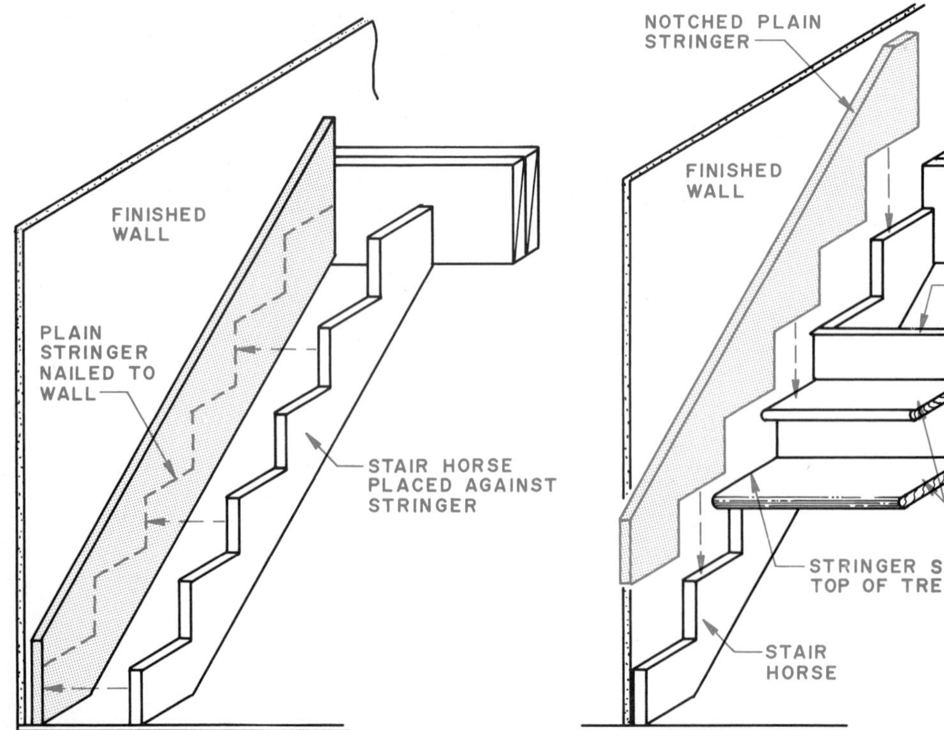

FIGURE 21-24

Two ways of cutting and installing a plain stringer.

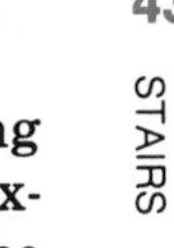

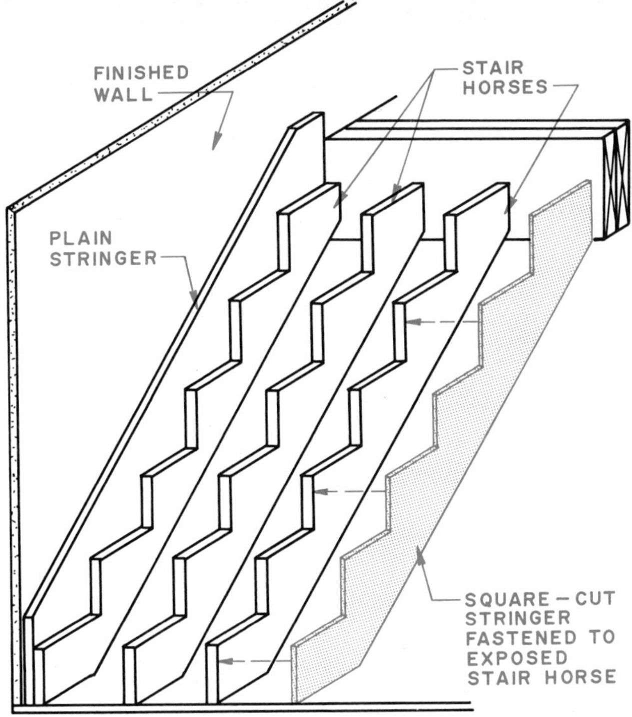

FIGURE 21-25

Installing a square-cut stringer.

LAYING OUT A SQUARE-CUT STRINGER

The square-cut stringer can be laid out using the procedure for the stair horse. Since it is exposed to view as a finished member, it must be very carefully laid out and cut to size (Fig. 21-25).

LAYING OUT A MITERED STRINGER

If a stair is open on one side, a mitered stringer is used on the exposed side. The mitered joint occurs where the riser joins the stringer. The end of the riser and the riser cut on the stringer are cut at 45°.

To lay out the stringer:

1. Lay out the stringer as described for a plain stringer.
2. Lay out a 45° line from the riser line (Fig. 21-26). The long side of the angle should be on the side to be exposed.
3. Draw this line down the face of the stringer.

LAYING OUT A HOUSED STRINGER

The housed stringer is the best method for building a stair. The dadoes are cut into the stringer. The ends of the risers and treads are hidden (Fig. 21-27).

Housed stringers are usually bought as part of a completed stair from stair manufacturing companies. If a carpenter desires to make housed stringers, the following instructions will be helpful.

1. Select a clear piece of stock for the housed stringers. Since the inside surface will be visible, it should be free of scratches and dents. Usually, 2 × 12 in. stock is used.
2. Draw a line 1½ in. from the bottom edge of the stringer. The rise and run lines are measured from this line rather than the edge of the board (Fig. 21-28).
3. Lay out the floor level cut. It is parallel with the tread.
4. Lay out the riser and tread lines using the line drawn 1½ in. from the edge of the board. The tread line does not include the nosing. This is the same procedure as described for laying out a square-cut stringer.

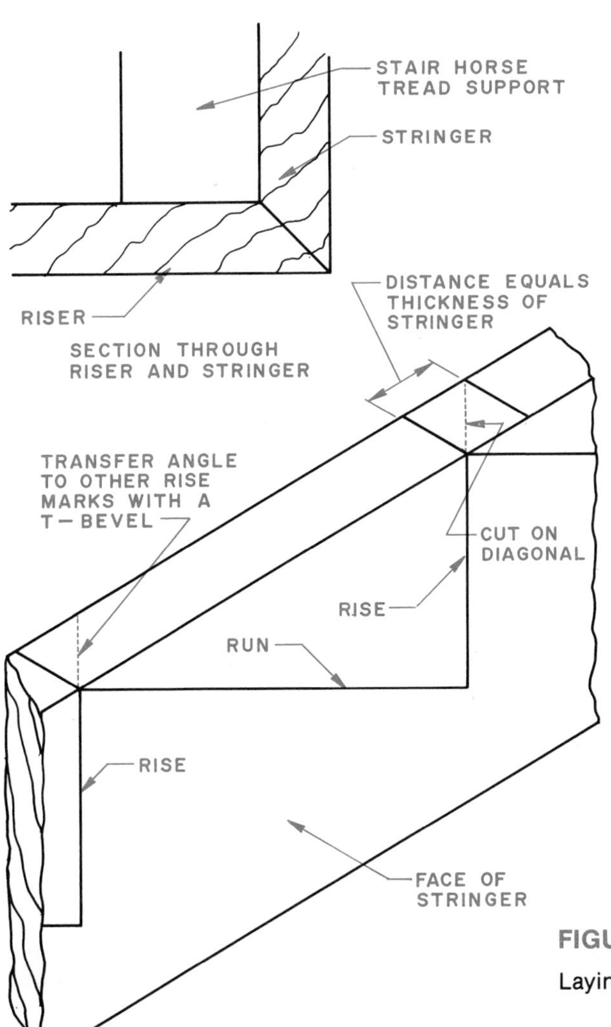

FIGURE 21-26

Laying out a mitered stringer.

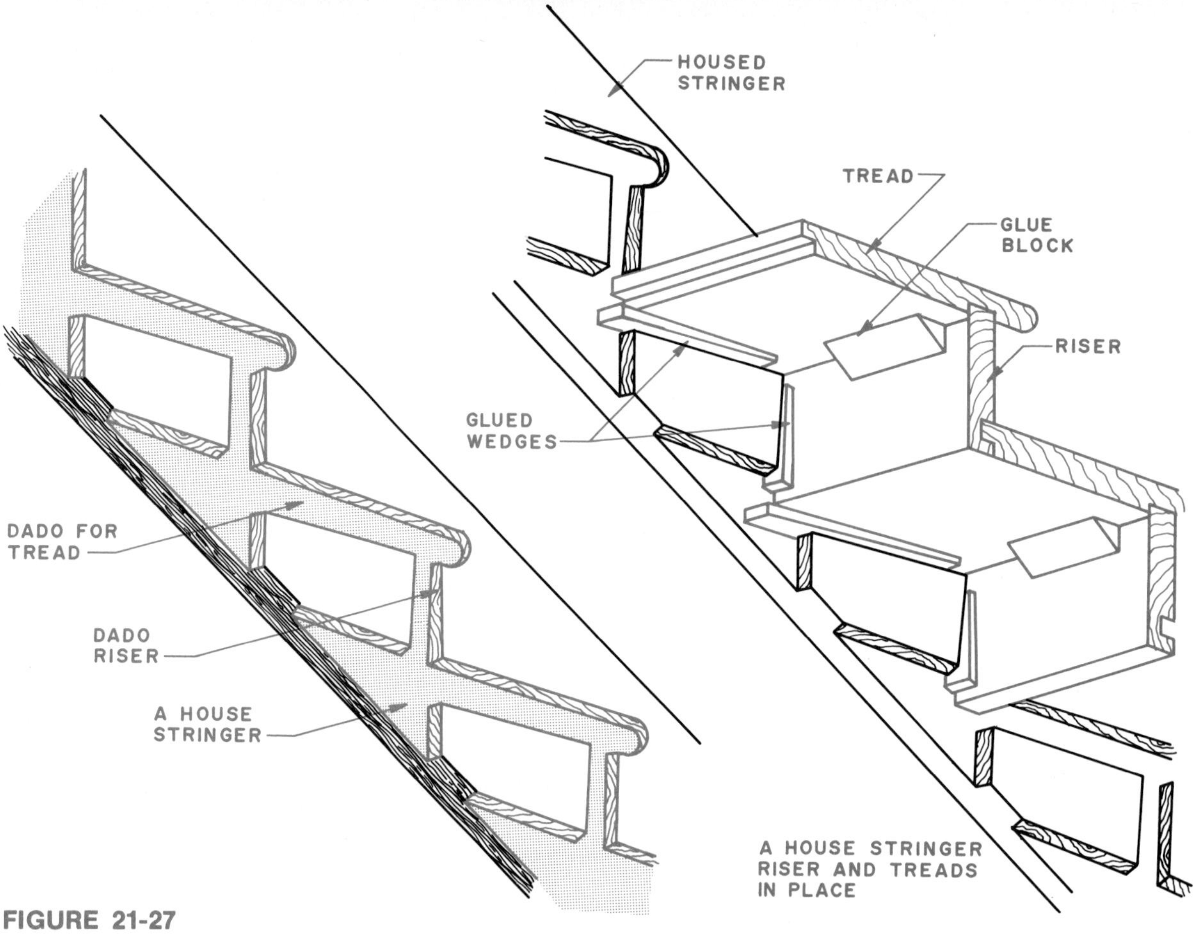

FIGURE 21-27

Treads and risers are held in dadoes in a housed stringer.

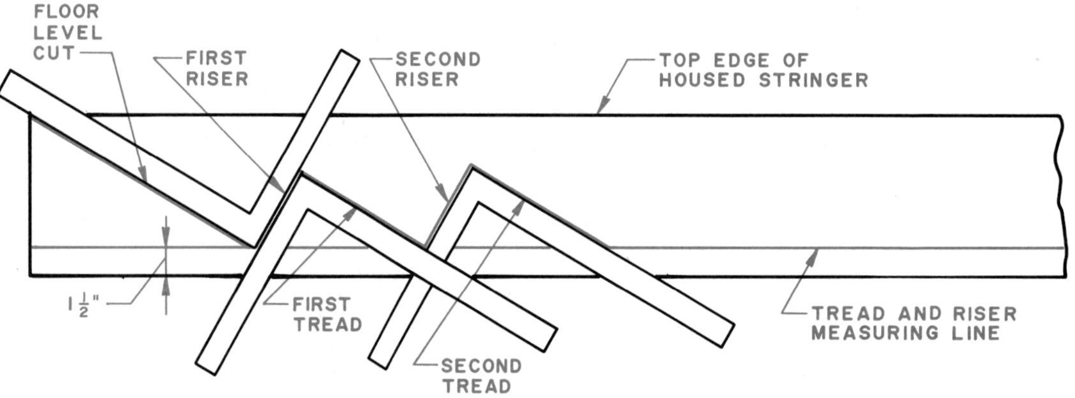

FIGURE 21-28

How to lay out riser and tread lines for a housed stringer.

5. Make templates showing the true size of the riser and tread plus the wedge to be used to hold them in place (Fig. 21-29).
6. Lay the templates on the stringer and mark their outlines. Mark the center hole of the nosing. This is used when boring the round end with a Forstner bit (Fig. 21-30).
7. Cut out the stair and tread dadoes with a router. They are usually cut ¾ in. deep.
8. Cut the top end of the stringer to meet the header. Various ways that this is done are shown in Fig. 21-19.

Generally, the stair is assembled on the floor and raised into place. Glue is placed in a tread dado and a tread is slid in place. Glue is placed on the wedge and it is driven in place. Next, a riser is installed in the same manner.

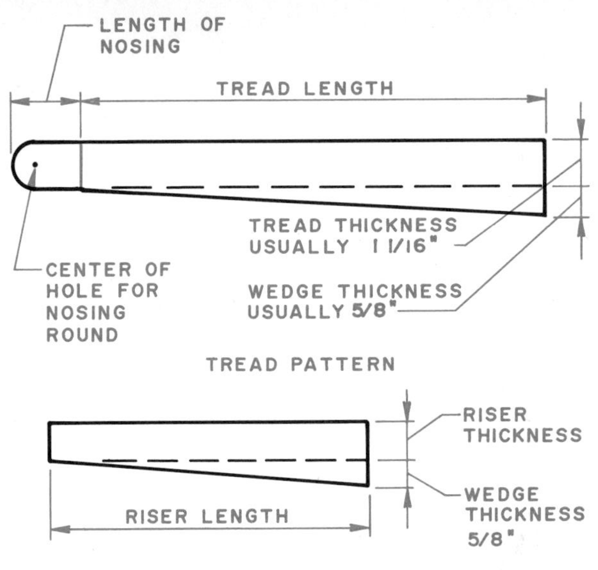

FIGURE 21-29

Templates used to mark riser and tread dadoes on a housed stringer.

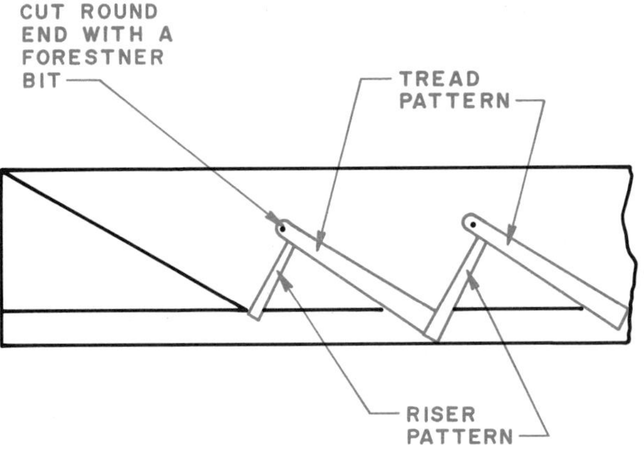

FIGURE 21-30

Marking the dadoes with tread and riser patterns.

This process is repeated until all treads and risers are in place. The lower edge of the riser is nailed to the edge of the tread. Often, the carpenter will drill small holes to help the nail penetrate the tread. The glue blocks are set in place. Sometimes they have a few small nails to hold them as the glue dries.

The assembled stair is usually ½ in. smaller in width than the stair opening. This helps slide it in place. Cove molding is placed on the edges to cover the gap.

Usually, the top step of a housed stringer stair may have a nosing that butts the finished floor (Fig. 21-31). Nosing is generally obtained cut to shape from a building material supplier. It is rabbetted so that it is the same thickness as the finished floor. The front end is rounded the same as the nosing on the treads.

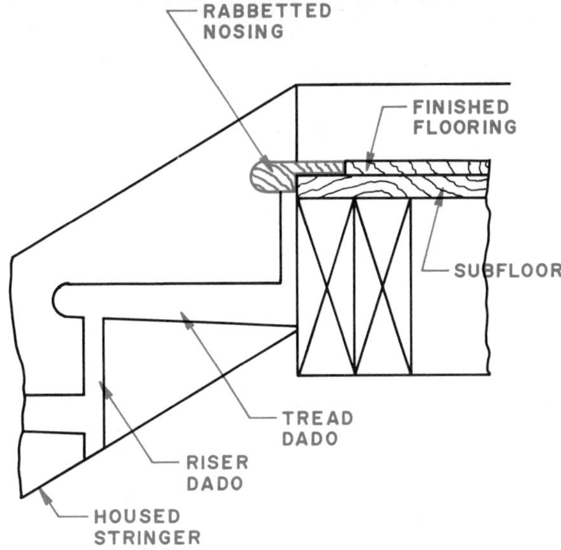

FIGURE 21-31

A rabbetted nosing is sometimes used with a housed stringer.

INSTALLATION DETAILS

The details of stair installation can vary. The following description represents typical construction details.

Straight Stair Figure 21-32 shows a series of drawings detailing the installation of a straight stair between two walls.

Most stairs require three stair horses. These are cut and nailed in place during the rough framing. It is nailed to the rough flooring. If this is done, the horse is not cut off the thickness of the finished floor. Some prefer to install the finished floor and set the horse on top of it. A 2 × 4 in. spacer is nailed on each side to hold the horse away from the wall. This provides space for the finished wall material. Check the stringers for plumb.

After the walls are finished, the plain stringer is put in place and nailed to the wall.

Finally, the risers are cut to length and nailed to the horses. Use two 6d, 8d, or 10d finishing nails, depending upon the thickness of the stock. Then the treads are cut to length and nailed to the horses. Usually, at least three 10d finishing nails are used at each bearing point. Drill holes before nailing. Set the nails.

The carpenter goes below the stair and nails the lower edge of the riser to the tread. Since this is hardwood, the nails require that holes be drilled that are slightly smaller than the diameter of the nail. Glue blocks are placed between the tread and riser at the top of the riser.

If a stair is open on one side, a mitered stringer is nailed to the outside horse (Fig. 21-33). The treads overhang the stringer. This nos-

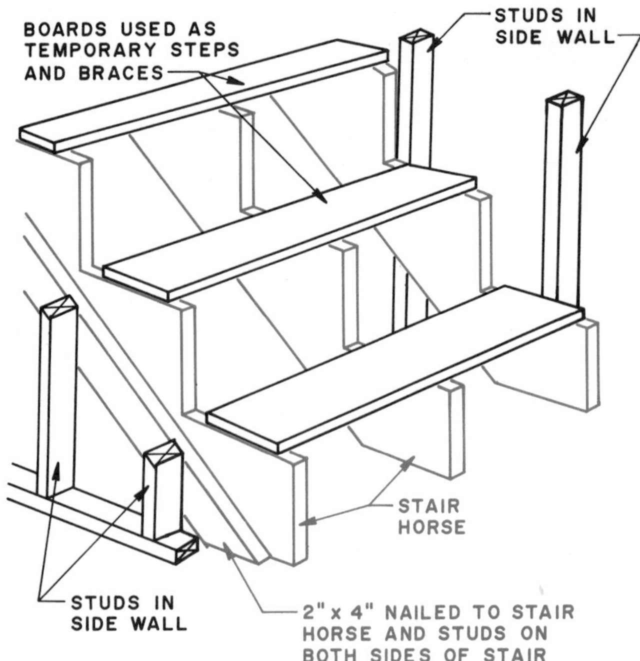
BOARDS USED AS TEMPORARY STEPS AND BRACES
STUDS IN SIDE WALL
STAIR HORSE
STUDS IN SIDE WALL
2" x 4" NAILED TO STAIR HORSE AND STUDS ON BOTH SIDES OF STAIR
1. NAIL 2" x 4" SPACERS TO OUTSIDE STAIR HORSES. NAIL STAIR HORSES TO FLOOR AND HEADER.

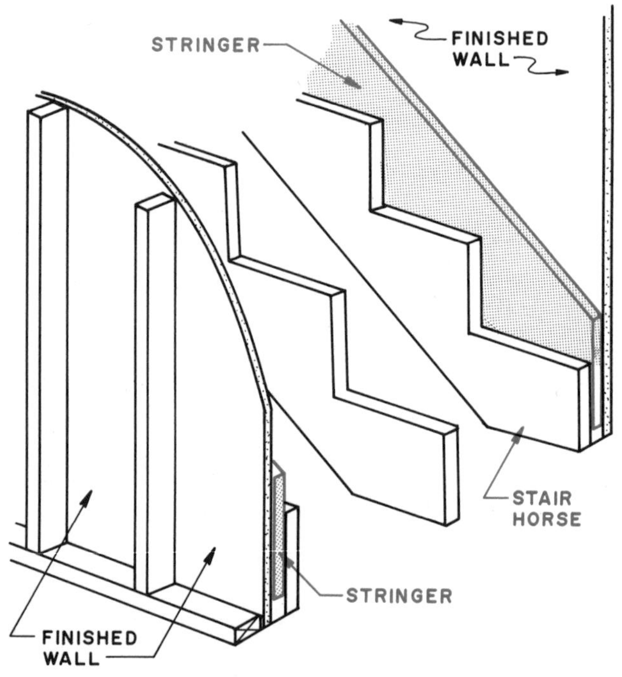
STRINGER
FINISHED WALL
STAIR HORSE
STRINGER
FINISHED WALL
2. INSTALL FINISHED WALL BESIDE STAIR HORSE. PLACE STRINGER BETWEEN STAIR HORSE AND FINISHED WALL.

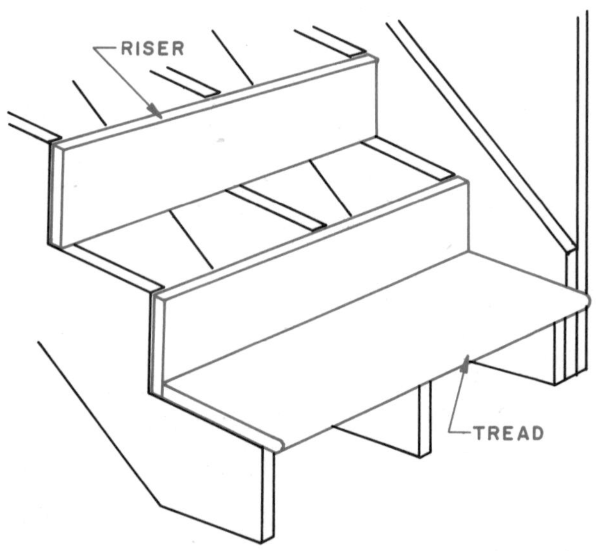
RISER
TREAD
3. INSTALL THE RISERS AND THEN THE TREADS.

FIGURE 21-32

Steps to assemble a stair between walls.

ing return is applied to the square end of the tread (Fig. 21-34).

Framing Platforms If a stair has a landing, a *platform* must be built. The stair is built as two separate units, each having the same tread and riser size. This calculation must be done before the platform is located. The framing of the platform accounts for one riser and one tread. Therefore, the height of the platform must occur at one of the tread heights.

A typical framing plan is in Fig. 21-35.

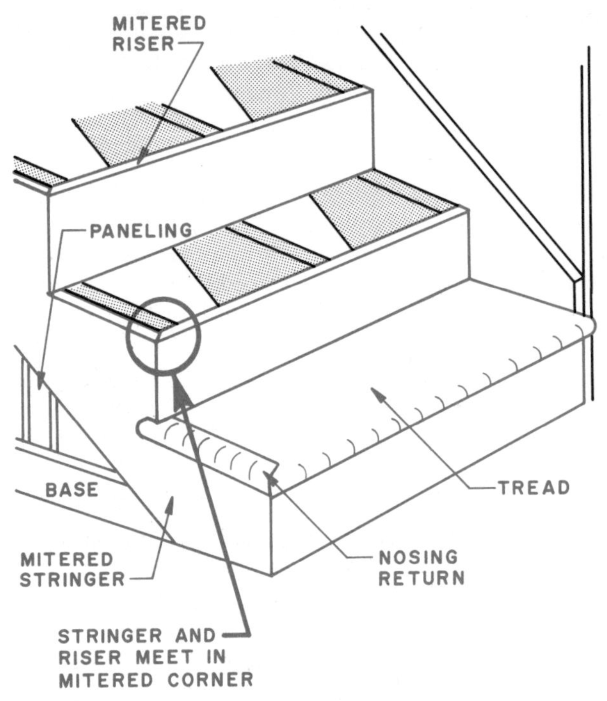

FIGURE 21-33

Open stairs used mitered risers and a nosing return on the tread.

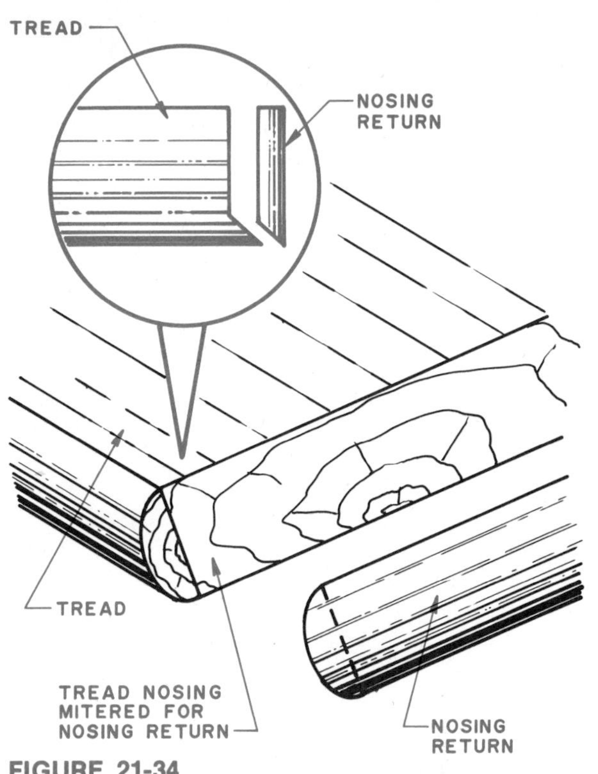

FIGURE 21-34

The nosing return is mitered to form an attractive corner.

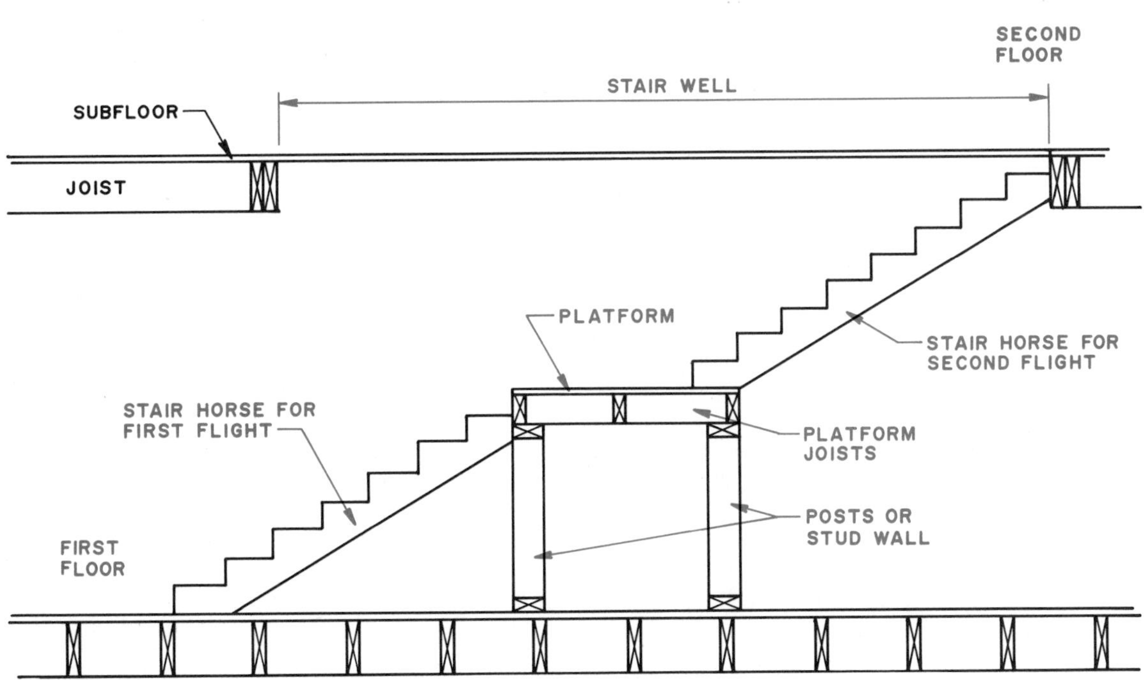

FIGURE 21-35

Typical framing for a platform.

Folding Stairs *Folding stairs* are used to provide access to attics where the area is suitable only for storage. They are not an acceptable access to living areas. They are purchased completely assembled and ready to install. The opening in the ceiling should be framed with double headers. The size of the rough opening depends upon the stair purchased. Be certain you know this size before making the opening (Fig. 21-36).

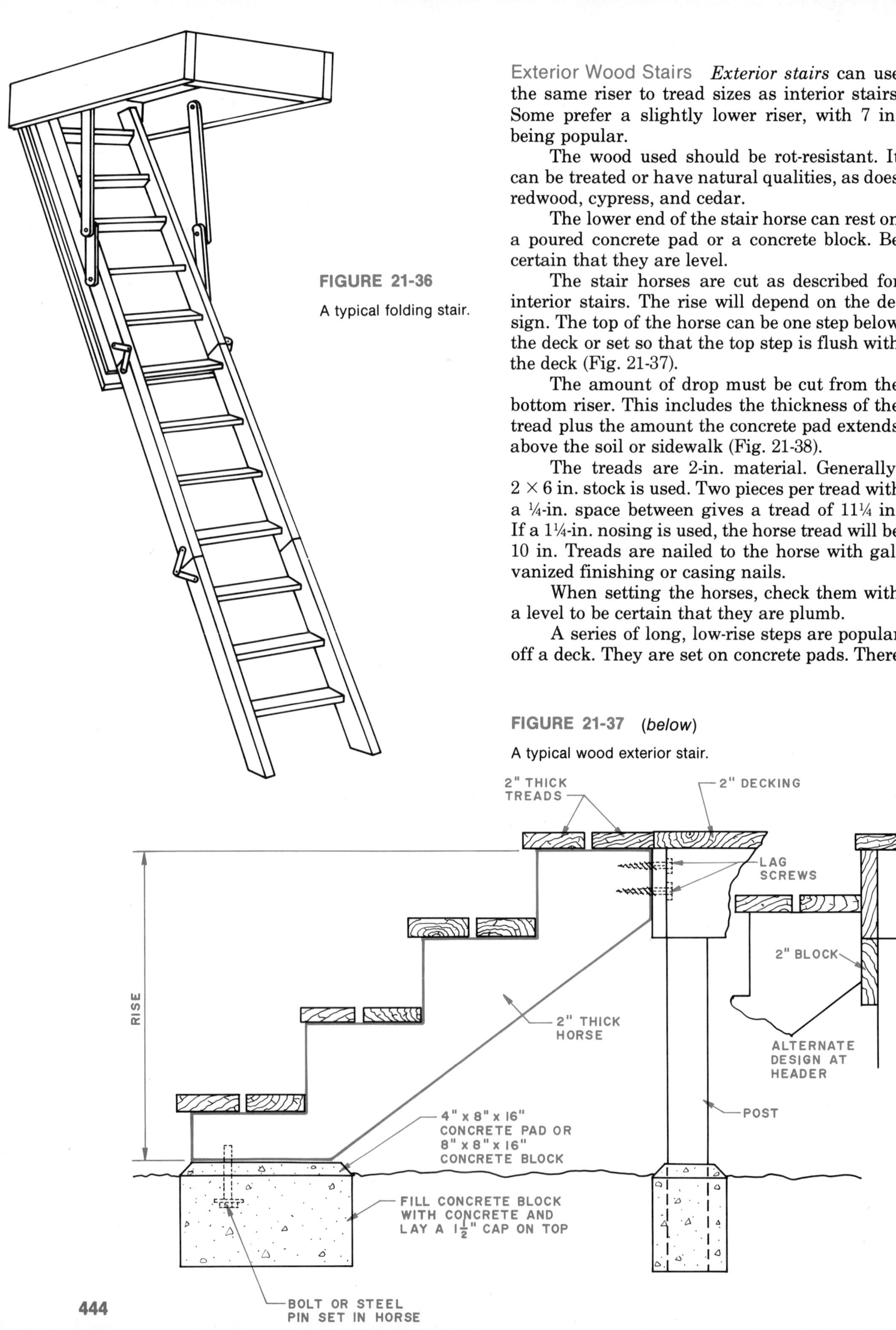

FIGURE 21-36

A typical folding stair.

Exterior Wood Stairs *Exterior stairs* can use the same riser to tread sizes as interior stairs. Some prefer a slightly lower riser, with 7 in. being popular.

The wood used should be rot-resistant. It can be treated or have natural qualities, as does redwood, cypress, and cedar.

The lower end of the stair horse can rest on a poured concrete pad or a concrete block. Be certain that they are level.

The stair horses are cut as described for interior stairs. The rise will depend on the design. The top of the horse can be one step below the deck or set so that the top step is flush with the deck (Fig. 21-37).

The amount of drop must be cut from the bottom riser. This includes the thickness of the tread plus the amount the concrete pad extends above the soil or sidewalk (Fig. 21-38).

The treads are 2-in. material. Generally, 2 × 6 in. stock is used. Two pieces per tread with a ¼-in. space between gives a tread of 11¼ in. If a 1¼-in. nosing is used, the horse tread will be 10 in. Treads are nailed to the horse with galvanized finishing or casing nails.

When setting the horses, check them with a level to be certain that they are plumb.

A series of long, low-rise steps are popular off a deck. They are set on concrete pads. There

FIGURE 21-37 *(below)*

A typical wood exterior stair.

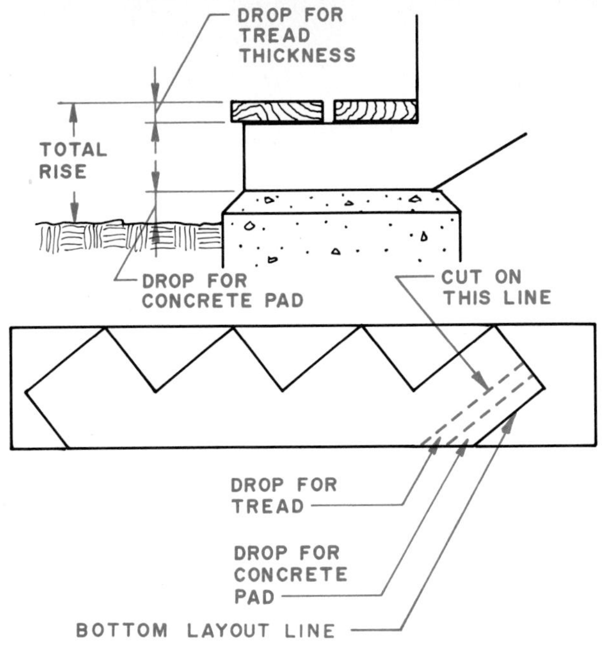

FIGURE 21-38

How to lay out a horse for an exterior stair.

are no horses. Instead, solid stock, such as 4 × 4 or 2 × 6 in. material, is used. The treads can be 2 × 4 or 2 × 6 in. material (Fig. 21-39).

Usually, exterior stairs do not have any riser boards.

Winders A stair with *winders* is not used a great deal. They are used when space is very limited.

The rise is figured the same as for a straight stair. The regular tread width is the same as for a straight stair.

To figure the size of the winders, draw them full size. Lay them out as shown in Fig. 21-40. This gives the size and shape of the treads and the angles to cut the risers on the horse.

The winder tread should be nearly the same width as the regular tread at the line of travel. The tread should be 6 in. at its smallest end. From the face of the horse in the small corner, swing a 12-in. arc which is the line travel. Divide this line into three equal parts. These divisions locate the face of the riser.

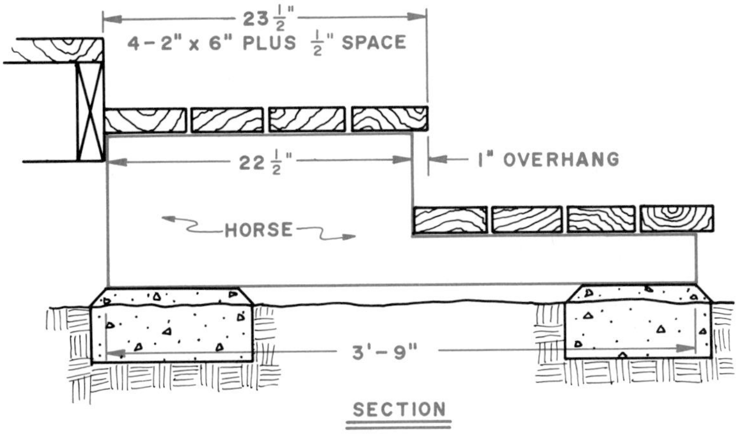

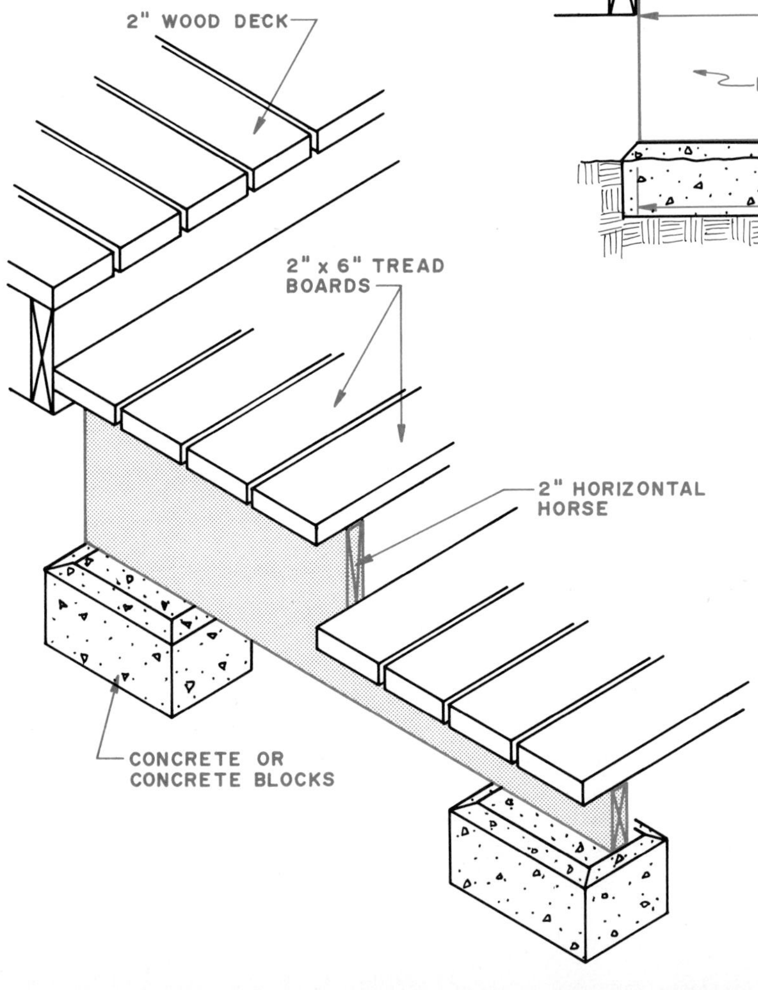

FIGURE 21-39

A low, deck-like stair leading off an exterior deck.

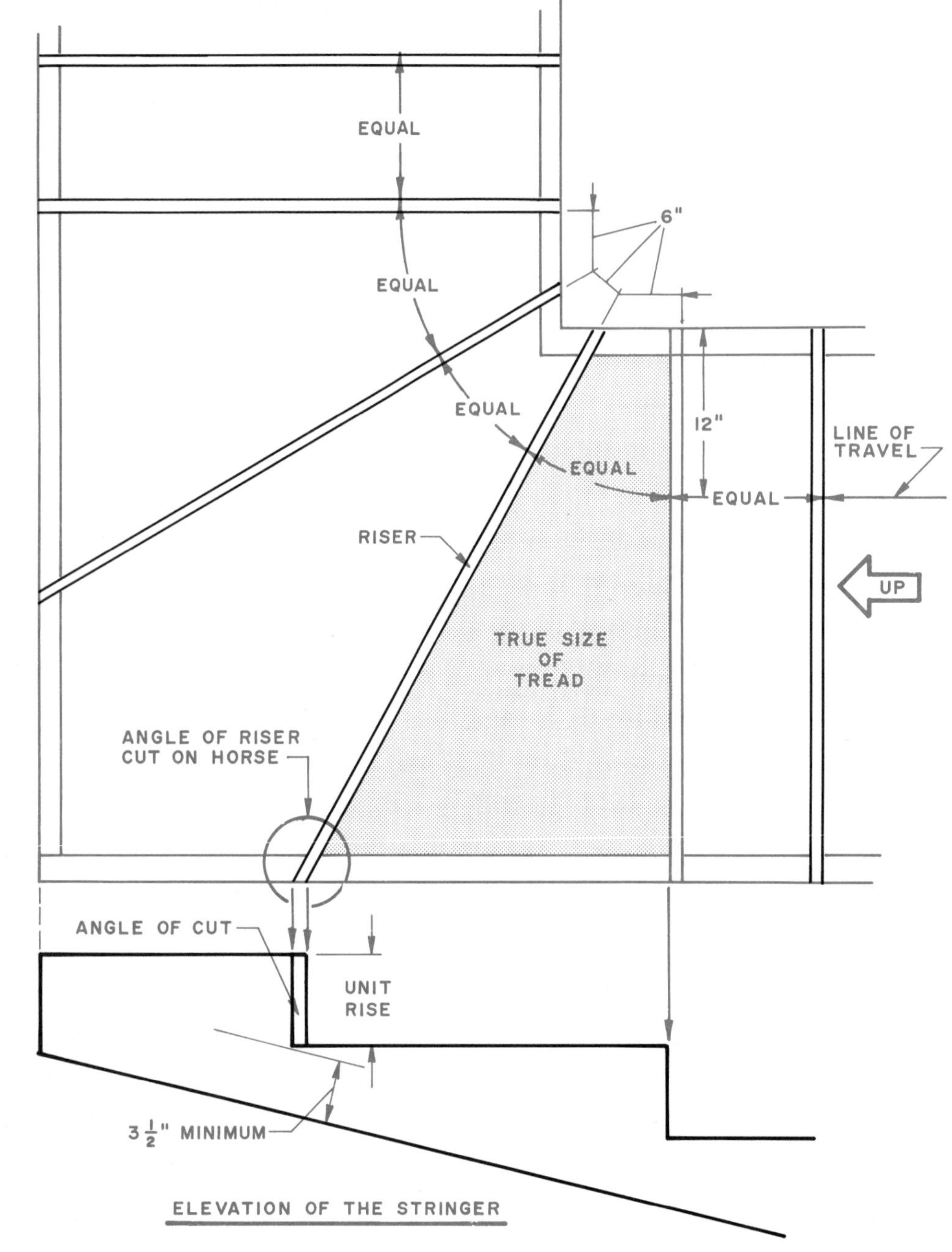

FIGURE 21-40

A layout for winders in a stair.

When laying out the horse, the regular steps are drawn as described for straight stairs. The rise and tread size for the winders are taken from the full-size drawing of the winders.

Spiral Stairs *Spiral stairs* are used where space is limited or where their decorative appearance enhances the surroundings (Fig. 21-41). They are available in diameters from 4 ft 6 in. to 8 ft 0 in. They may fit into a square area or be round (Figs. 21-42). Spiral stairs are made to turn clockwise or counterclockwise. The number of degrees of turn can also be varied.

Spiral stairs are shipped to the building site completely manufactured and ready for assembly. Handrails and protective balustrades are designed to meet building codes.

FIGURE 21-41

An all-wood full-spiral stair. *(Courtesy of Spiral Manufacturing, Inc.)*

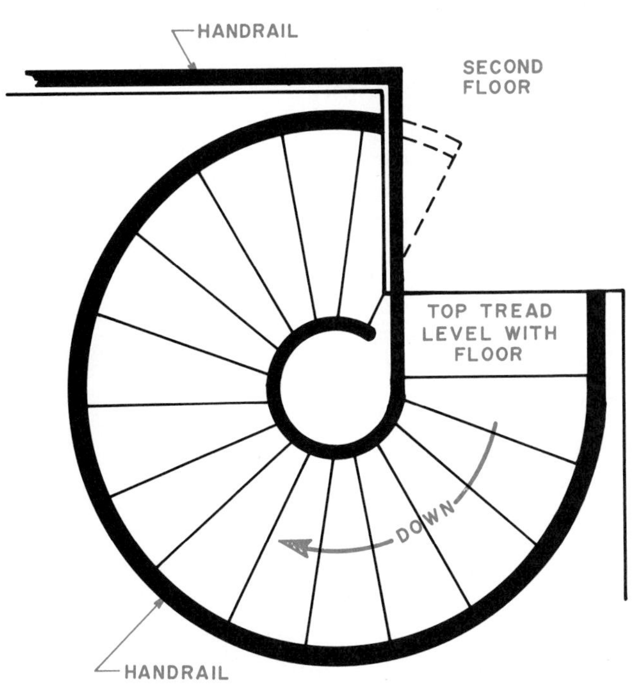

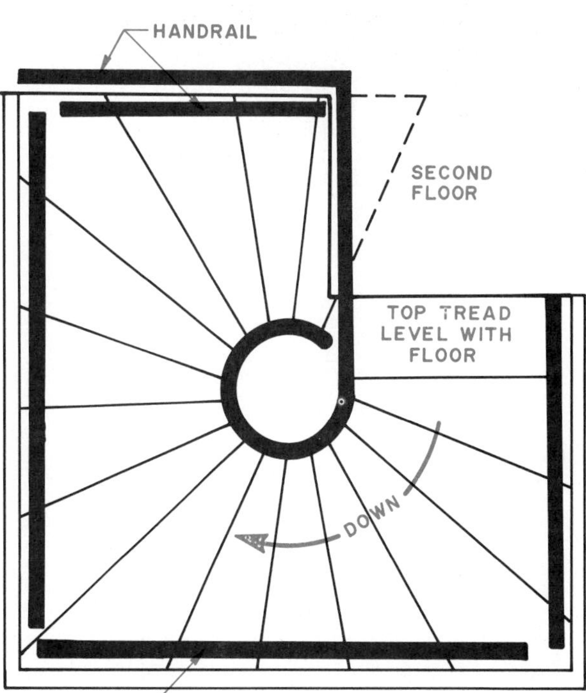

FIGURE 21-42

Typical plans for spiral stairs in circular and square forms. *(Courtesy of Spiral Manufacturing, Inc.)*

STUDY QUESTIONS

After reading the chapter, answer each of the following questions. If you do not know the answer, review the chapter.

1. What is the difference between a service stair and a main stair?
2. What are open and closed stairs?
3. What are the commonly used shapes of stairs?
4. What are the minimum headroom heights for service and main stairs?
5. What is the minimum width of a stair?
6. What is the minimum tread width and maximum riser size?
7. What is the height for a handrail?
8. How do you determine total rise and run of a stair?
9. When is a mitered stringer used?

IMPORTANT TECHNICAL TERMS

Following are technical terms that you should be able to use as part of your working vocabulary. Write a brief description of the meaning of each term.

Stair horse
Square-cut stringer
Plain stringer
Mitered stringer
Housed stringer
Nosing
Unit rise
Unit tread
Riser
Tread
Total rise
Total run
Headroom
Railing
Newel
Handrail
Baluster
Landing
Platform

22 CABINETS AND OTHER BUILT-IN UNITS

The carpenter can find information about kitchen cabinets, shelving, and other built-in furniture units on the drawings of the building. The location and plan view of the units is shown on the floor plan (Figs. 22-1 and 22-2).

Other information is given in detail drawings which show the elevation and often a section through the unit (Figs. 22-3 and 22-4).

There are three ways to secure cabinets. The carpenter can build them on the job following the details given on the plans by the architect. The carpenter must be skilled in cabinetmaking to do this. Sometimes, a cabinetmaker comes to the site to build the cabinets. A second way is to purchase ready-cut cabinets in knocked-down condition. The carpenter must assemble the cabinets and hang them. A third way is to purchase cabinets completely assembled and finished in a factory. The carpenter hangs the wall cabinets and installs the base cabinets (Fig. 22-5).

FACTORY-BUILT CABINETS

In most cases factory-built cabinets are used in residential construction. They are available in a wide range of sizes, styles, and materials. Typical sizes are shown in Figs. 22-6 and 22-7.

When factory-built units are specified, the architect will list the manufacturer's stock number for each item and indicate the design and material selected. Most manufacturers make single-door base cabinets with doors opening right or left. This must be specified as cabinets are ordered and installed. Be certain to allow the proper opening for appliances such as dishwashers and compactors.

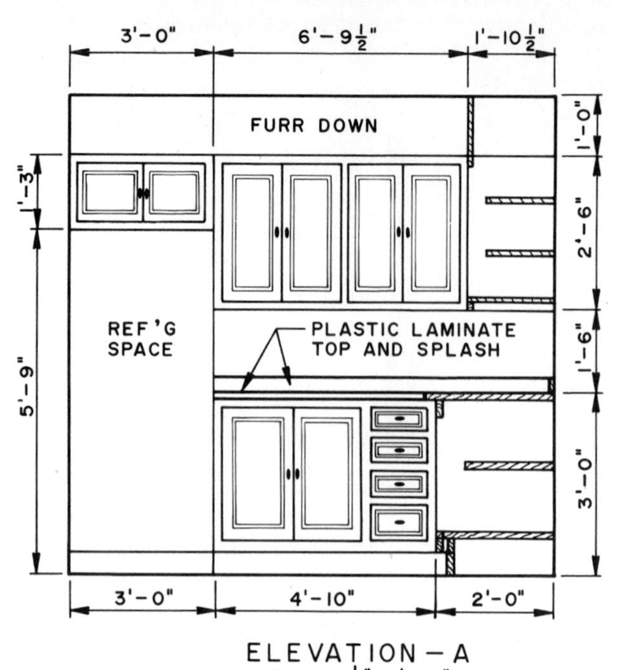

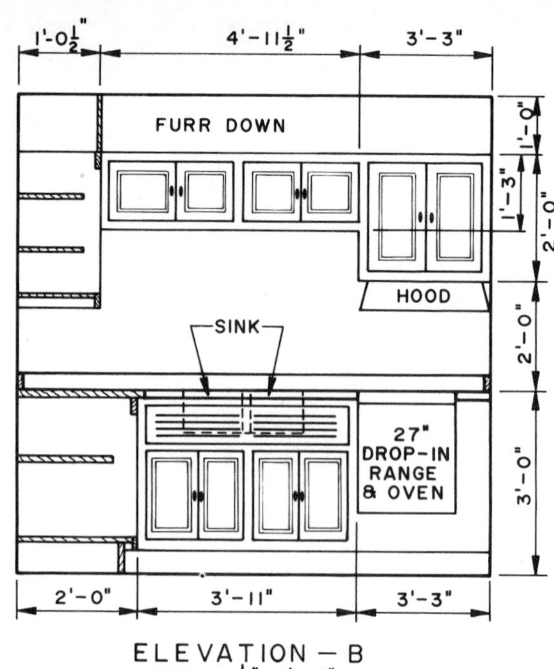

FIGURE 22-3 *(above)*

Typical elevations of kitchen cabinets as drawn on working drawings. Notice how these are identified on the plan in Fig. 22-1.

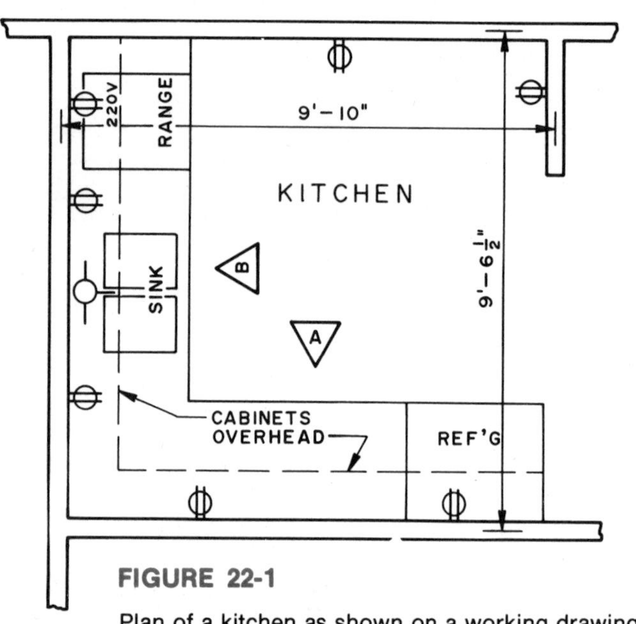

FIGURE 22-1

Plan of a kitchen as shown on a working drawing.

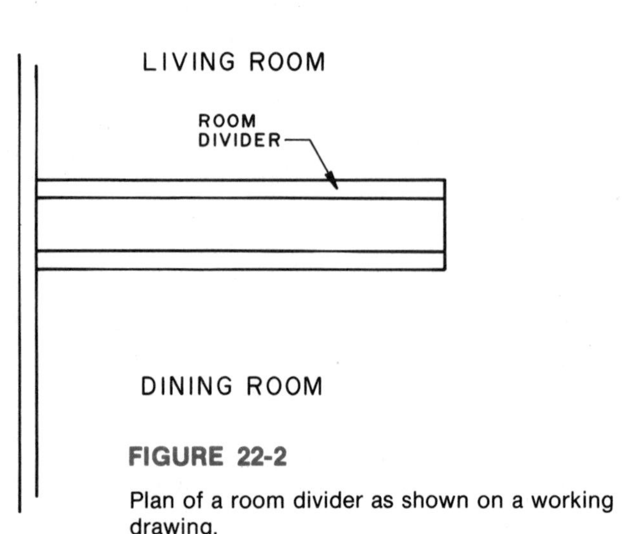

FIGURE 22-2

Plan of a room divider as shown on a working drawing.

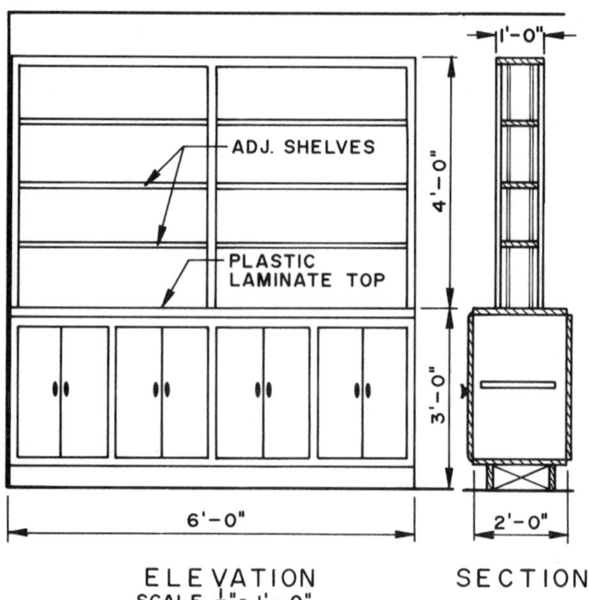

FIGURE 22-4

Elevation and section of the room divider shown in Fig. 22-2.

FIGURE 22-5

A complete kitchen installation using factory-manufactured cabinets. *(Courtesy of Coppes, Inc.)*

FIGURE 22-6

Typical sizes of kitchen wall and base cabinets.

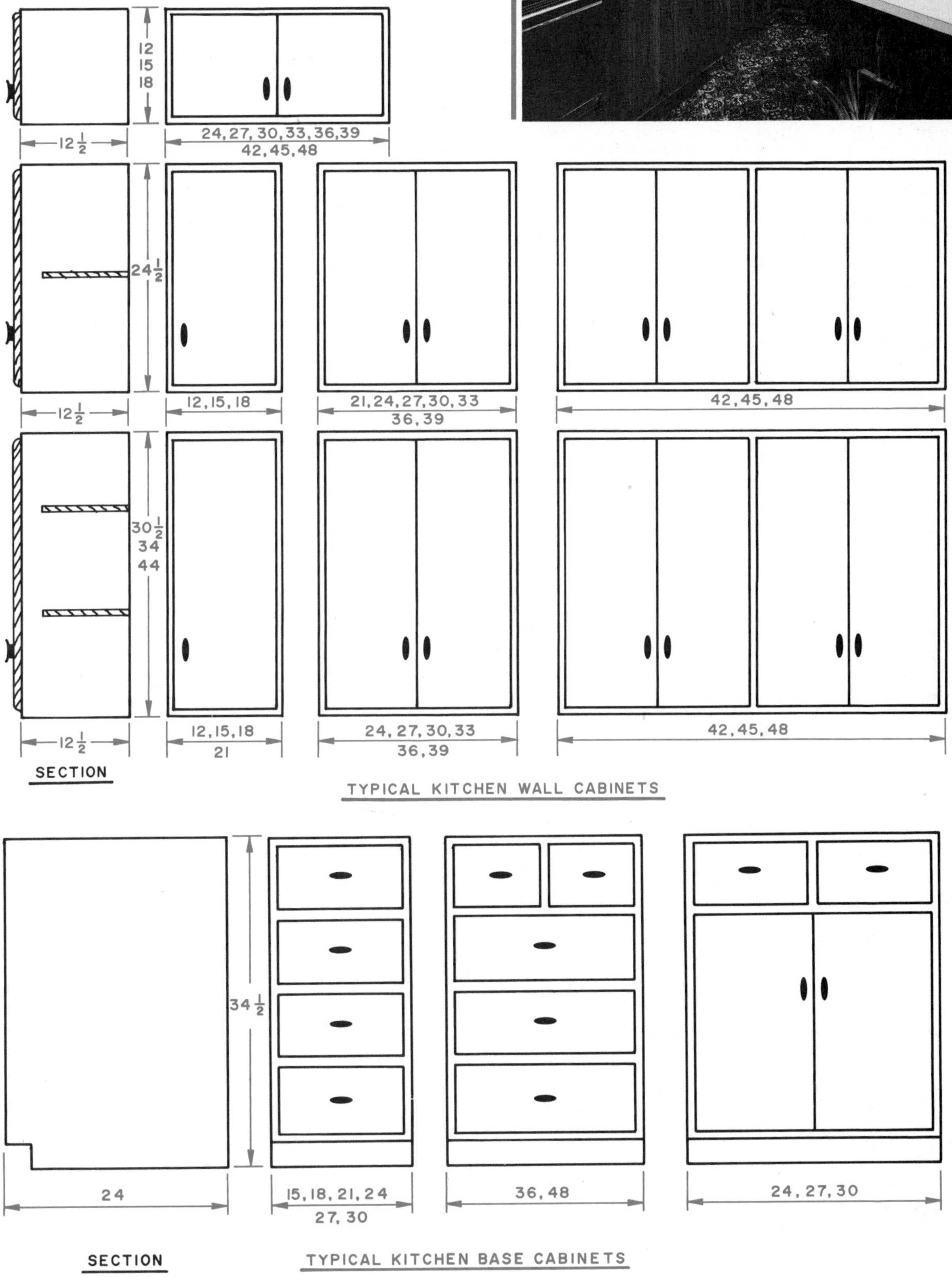

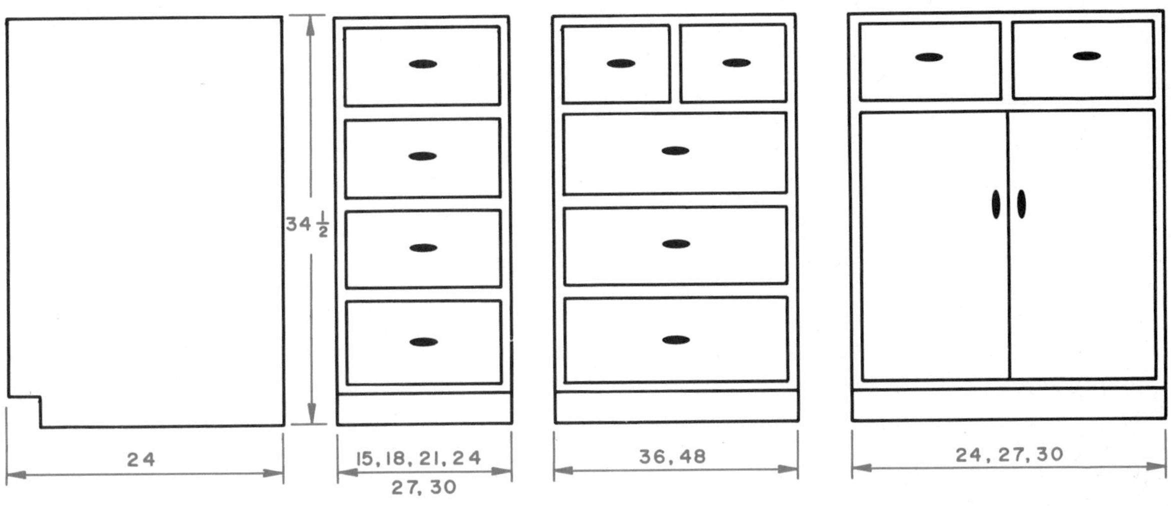

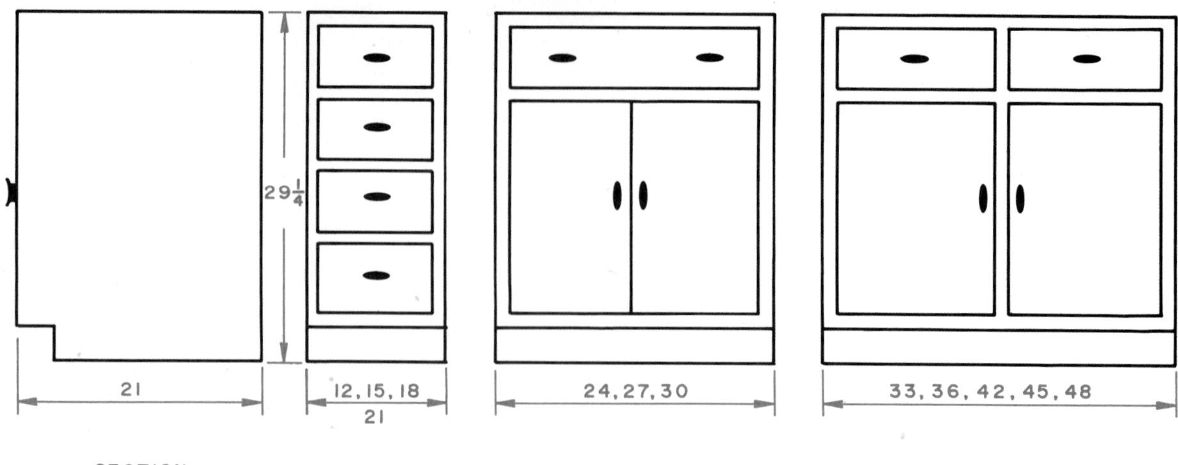

FIGURE 22-7

Typical sizes of bath lavatory cabinets.

Special attention must be given to ordering the cabinet tops. The size must be specified. Often, the carpenter must cut the hole for the sink. Some companies specialize in making cabinet tops.

Installing Factory-Built Kitchen Cabinets

Some carpenters prefer to install the *wall cabinets* first. This enables them to stand directly below the cabinet and use props to hold them at the desired level. It also reduces the possibility of damaging the base cabinets.

Installing Wall Cabinets Join the units together on the floor in the order in which they are to be on the wall. Clamp them with C-clamps. Put pieces of wood between the cabinet and the clamp to prevent marring the cabinet. Be certain that the bottom rails are in perfect alignment. Then join them together with wood screws through the front stiles. Round-head screws are preferred. At least two screws should be used to join each cabinet (Fig. 22-8).

Measure the location of any ducts or electrical service that must be put through the cabinet. Mark and cut the required openings in the cabinet.

Mark the stud locations on the wall. Measure and mark the stud locations on the back rails of the cabinet. Drill the holes for the screws through the upper and lower rear rails. Use at least four screws per cabinet. Large double-door cabinets should have six screws (Fig. 22-9).

Locate on the wall the height of the bottom of the longest cabinets from the floor. Using a chalk line, mark this height on the wall so that it is absolutely level.

Now raise the cabinets into position on

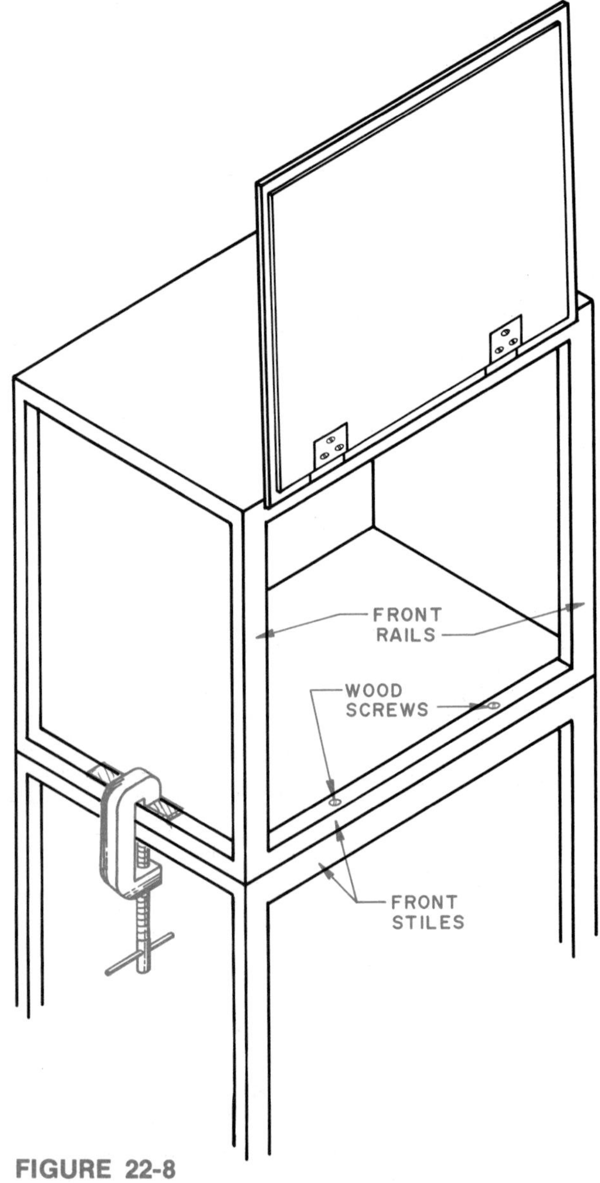

FIGURE 22-8

Wall cabinets are joined together with wood screws through the front stiles.

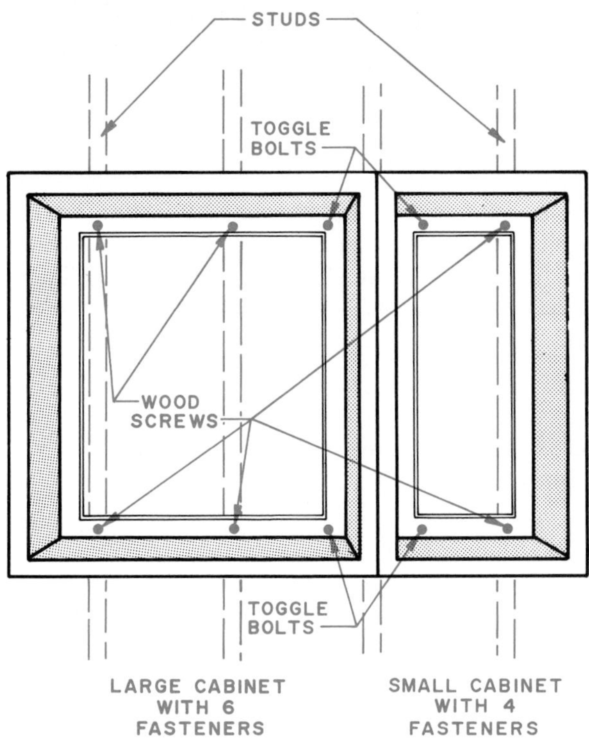

FIGURE 22-9

Wall cabinets are installed with wood screws and toggle bolts.

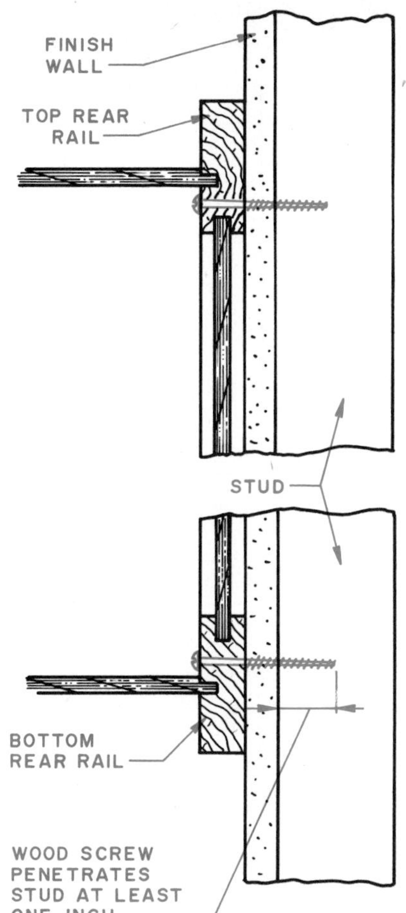

FIGURE 22-10

Wood screws holding wall cabinets to the wall should penetrate the stud at least 1 in.

the wall. Have wood members to use as props to help hold it in position. Align the bottom with the chalk line. Check for levelness with a level. When all is in order, drive several screws into the studs. The screws should penetrate the stud at least 1 in. (Fig. 22-10).

If a cabinet only crosses one stud, the second pair of fasteners can be hollow-wall anchors or toggle bolts (Fig. 22-11). If the cabinet is to carry a heavy load, it is best to nail a 2 × 4 in. hanging strip between the studs before the finish wall is applied. Then screw the cabinets to the stud or the hanging strip if no stud is available.

If the wall is irregular, it may be necessary to place wood shims behind the cabinet before giving the screws their final tightening. If the wall is crooked and the cabinets are fastened tightly to it, they may become bowed. The rails may open up. The doors may not close. The cabinet must not be under any strain that throws it out of plumb (Fig. 22-12).

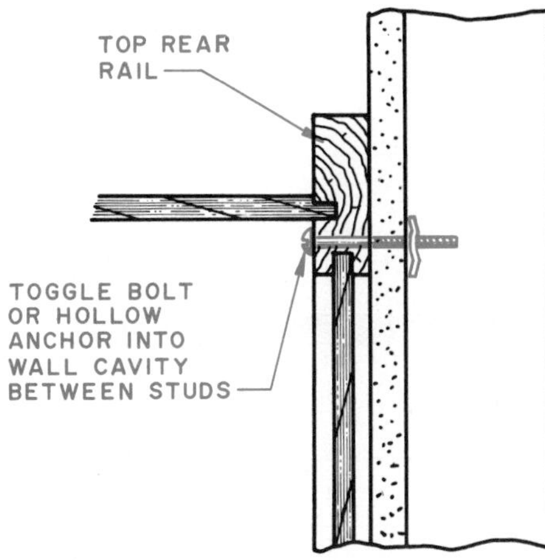

FIGURE 22-11

How to install a toggle bolt in the wall cavity.

Installing the Base Cabinet If the *base cabinet* or wall cabinet is to turn a 90° corner, the walls must be checked for squareness. If the corner is more than 90°, shim out one side the entire length of the wall. If it is less than 90°, shim the corner to form a 90° angle (Fig. 22-13). It is possible to cut into the plaster or drywall a little if it will help to get a 90° corner.

Check the floor for levelness. Find the highest spot and measure up the wall the height of the base unit. Run a level chalk line through this point. Mark the line on the wall.

Now join the base cabinets together. Hold them together with C-clamps and join with

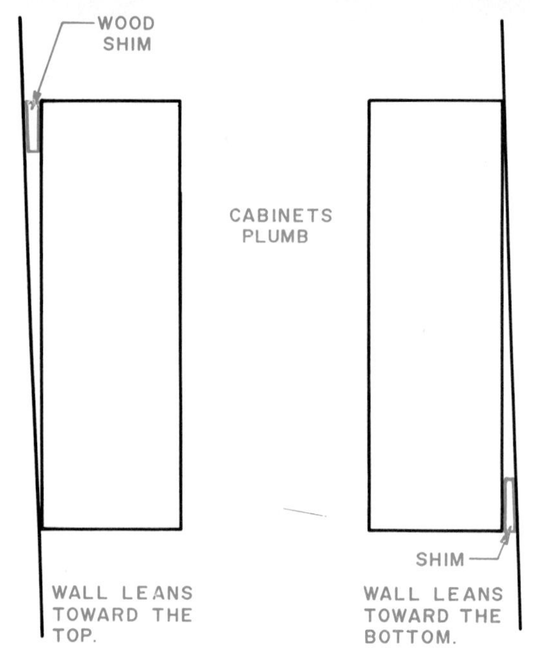

FIGURE 22-12

Use wood shims to plumb wall cabinets.

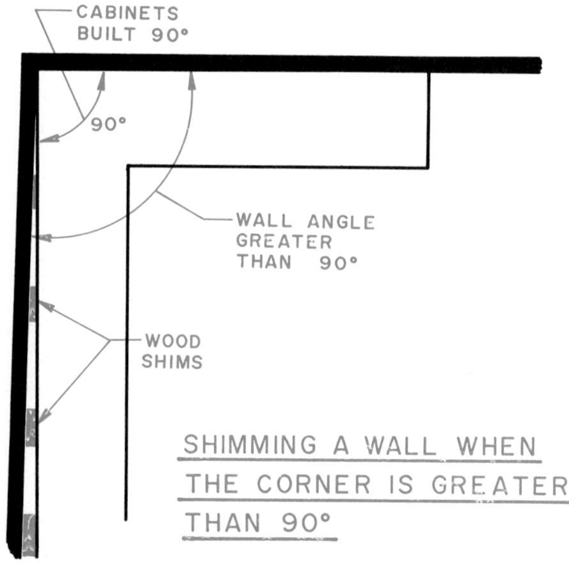

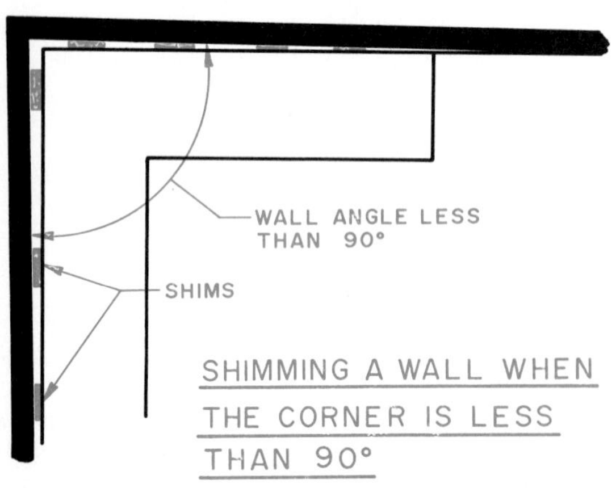

FIGURE 22-13

When a wall is out of square, use wood shims to get the cabinets in square.

round-head wood screws through the front *stiles*. Mark the location of the studs on the back *rails*. Drill the holes for the screws.

Slide them in place along the wall. If the floor is irregular, put wood shims under the cabinets and adjust until they are level (Fig. 22-14). The chalk line serves as a guide, but check with a carpenter's level. Also check their fit to the wall and shim with the wall. Then fasten to the studs with round-head screws (Fig. 22-15).

Installing the Countertop After the base cabinet is level and screwed to the wall, the top can be installed.

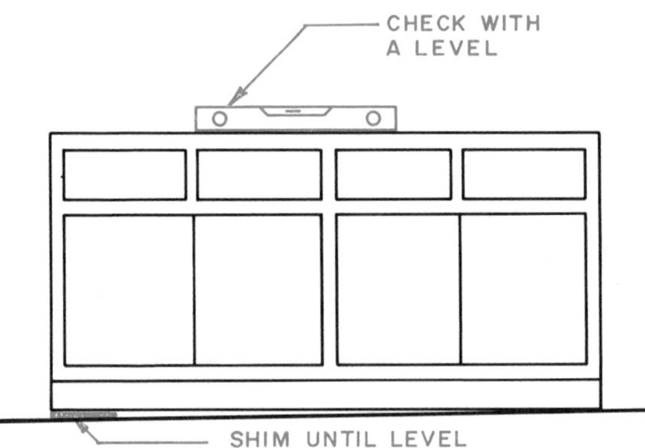

FIGURE 22-14

The base cabinet is leveled using wood shims at the floor.

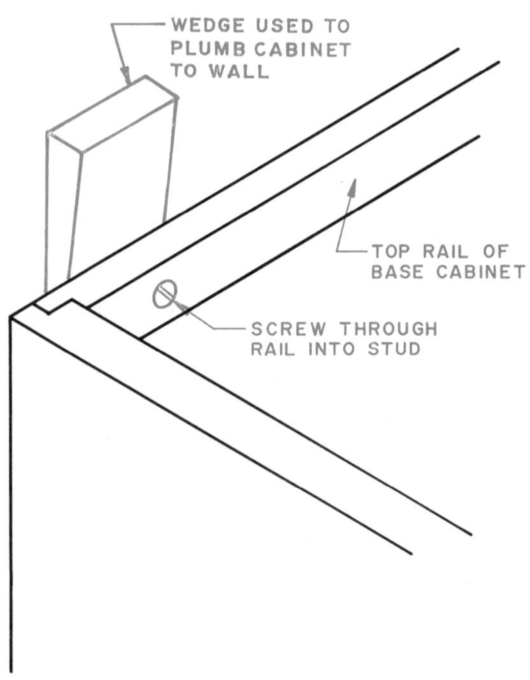

FIGURE 22-15

The base cabinets are fastened to the studs in the wall through the rear top rail of the cabinet.

Drill screw holes through the front top rails of the base cabinets. Set the top in place and carefully position it so that the overhang is uniform. Drive the screws into the countertop. Some prefer to fasten only at the front rail and not the back. Be certain that the screws do not penetrate the top.

If the plaster on the wall is irregular, it may be necessary to cut away small portions of it.

If the top is to have a sink, position the metal rim on the top in the exact location. Trace around the outside of the edge of the rim that touches the top. Be certain that there are at least 2 in. of clear space beyond the edge of the hole (Fig. 22-16). This area is needed to hold the metal clips that pull the rim, sink, and top together (Fig. 22-17).

Cut to the line with an electric saber saw. Insert the metal rim to see if it fits. A file can be used to remove high spots from the sawed opening.

Apply a thin bead of caulking to the bottom side of the metal rim. Set the sink and rim into the opening. Place the clips on the rim below the countertop and tighten. Wipe away any caulking that may be squeezed out from under the rim.

After the installation is complete, check the doors and drawers for proper operation. Adjust the door catches, if necessary.

Carefully wipe down the cabinets with a soft clean cloth. Be certain that the drawers and shelves are free of sawdust or plaster.

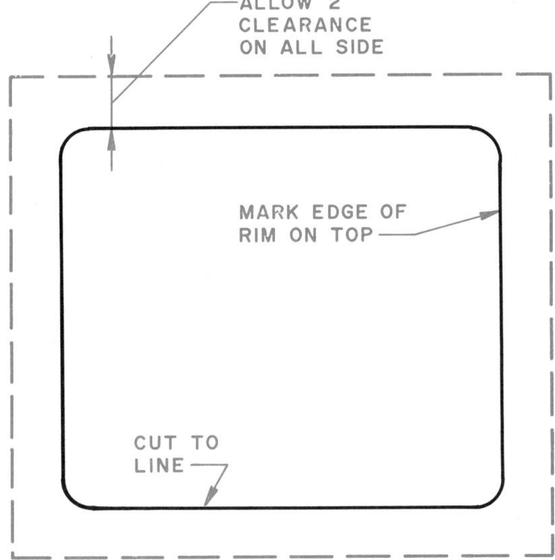

FIGURE 22-16

Be certain that there are 2 in. of clearance around the hole to be cut for the sink. This is needed for the brackets to hold the sink in the cabinet.

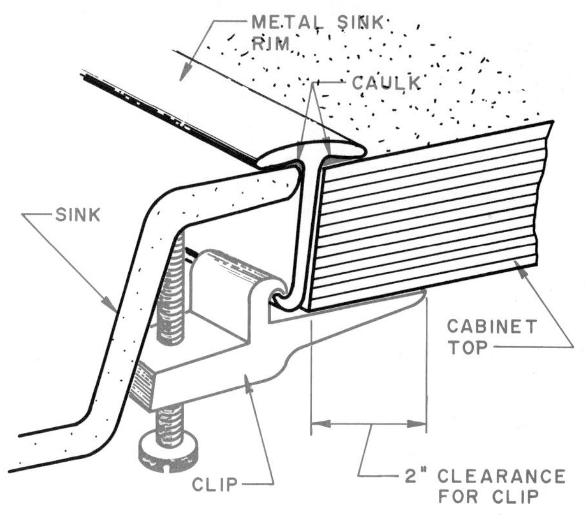

FIGURE 22-17

Metal clips clamp the sink and sink rim to the countertop.

CARPENTER-MADE CABINETS

Cabinet construction requires special skills and knowledge beyond those used in general carpentry. Those who desire to add this to their list of accomplishments should secure a book on cabinetmaking and study it in detail. The following discussion is a generalized presentation of how cabinets are often built.

The standard kitchen base cabinet is 34½ in. high without the top. It is 24 in. deep. The toe space is usually 3 in. high.

The sides can be ¾-in. plywood. The shelves in a base cabinet are often set in a dado. Shelves in a wall cabinet may be dadoed or adjustable. Shelves are usually made from ¾-in.-thick wood (Figs. 22-18 and 22-19).

The facing on the front of the cabinet is ¾-in.-thick stock usually 1⅝ to 1⅞ in. wide. The stiles and rails are joined at the corners with an open mortise and tenon joint. The facing is nailed to the frame of the unit.

Doors are often lipped to fit over the stiles and rails. A ⅜-in. lip is used. A hinge designed for a lipped door holds the doors to the stiles (Fig. 22-20).

There are many ways to make drawers. One kind often used joins the drawer front to the sides with a rabbet joint or a drawer corner joint. The tongue and groove is a stronger joint. The drawer sides and back are often ½-in. thick. The drawer bottom is usually ¼-in. plywood set in grooves cut in the drawer sides, back, and front (Fig. 22-21).

A wood *drawer guide* can be built. It is usually a wood grooved member fastened to the bottom of the drawer. A wood drawer guide is run from the front rail to the back of the cabinet (Fig. 22-22).

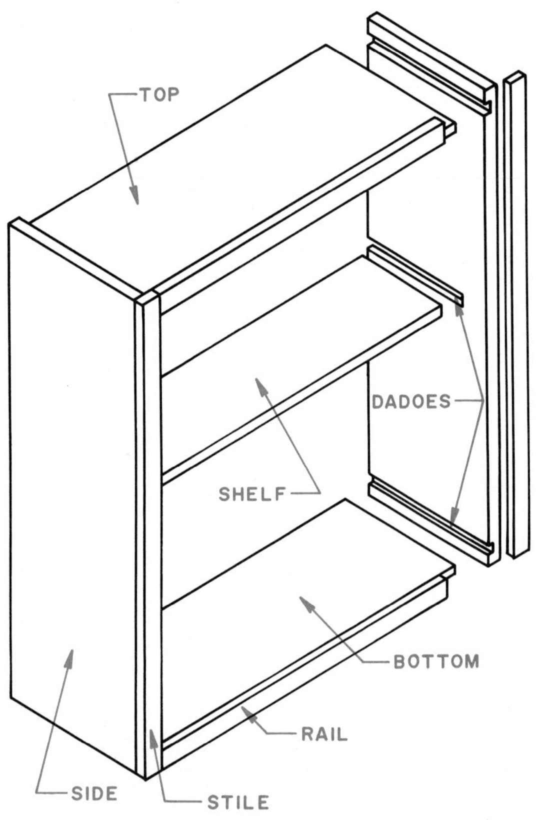

FIGURE 22-18

A typical framing plan for a carpenter-made wall cabinet.

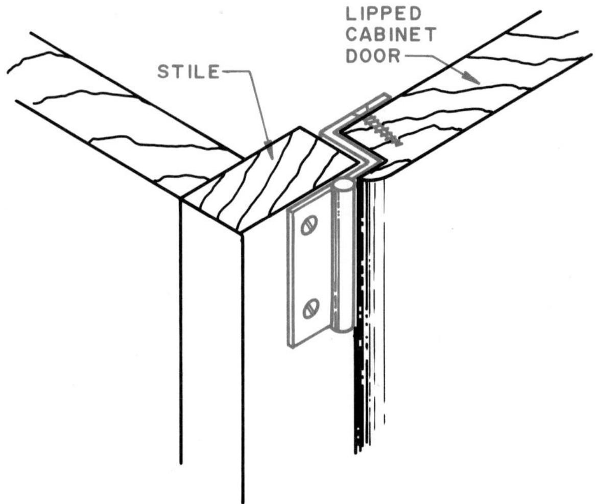

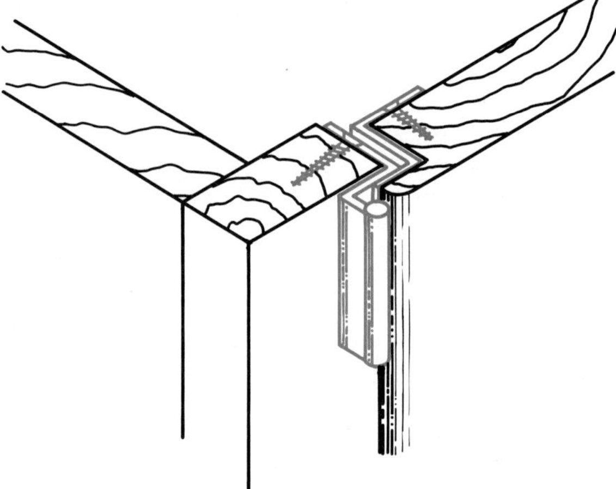

FIGURE 22-20 (*above*)

Two types of hinges for lipped cabinet doors.

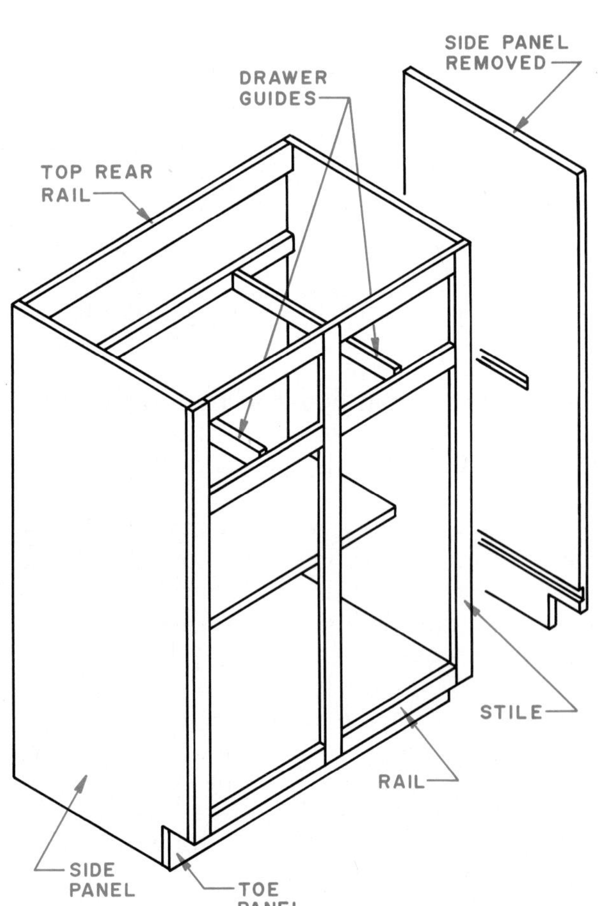

FIGURE 22-19 (*left*)

Construction details for a carpenter-made base cabinet.

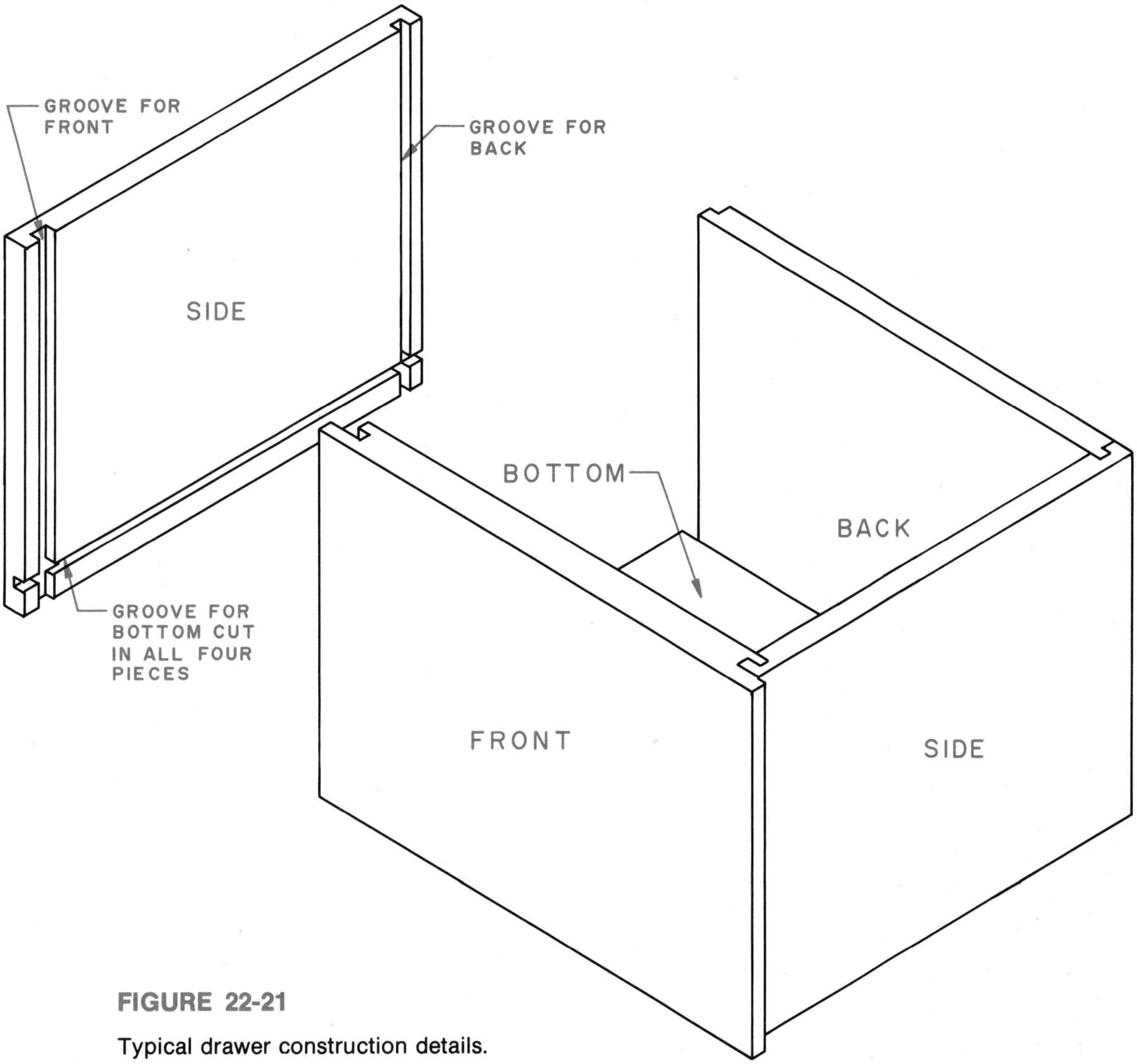

GROOVE FOR FRONT
GROOVE FOR BACK
SIDE
BOTTOM
BACK
GROOVE FOR BOTTOM CUT IN ALL FOUR PIECES
FRONT
SIDE

FIGURE 22-21

Typical drawer construction details.

BACK
FRONT
BOTTOM
DRAWER GUIDE
SECTION THROUGH LENGTH OF DRAWER
SIDES
BOTTOM
GUIDE JOINED TO FRAME
GUIDE FASTENED TO BOTTOM OF DRAWER
SECTION THROUGH WIDTH OF DRAWER

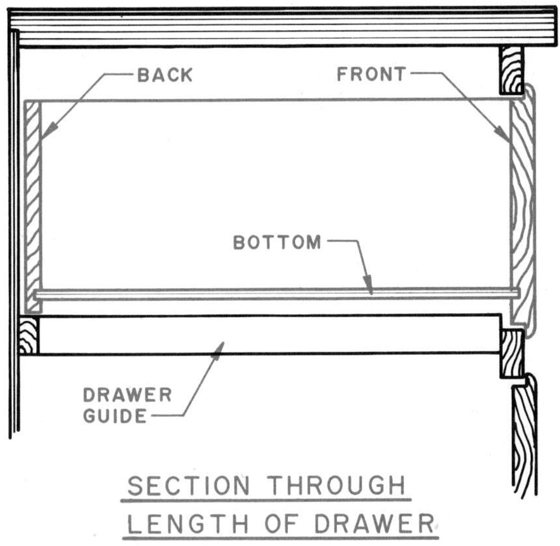

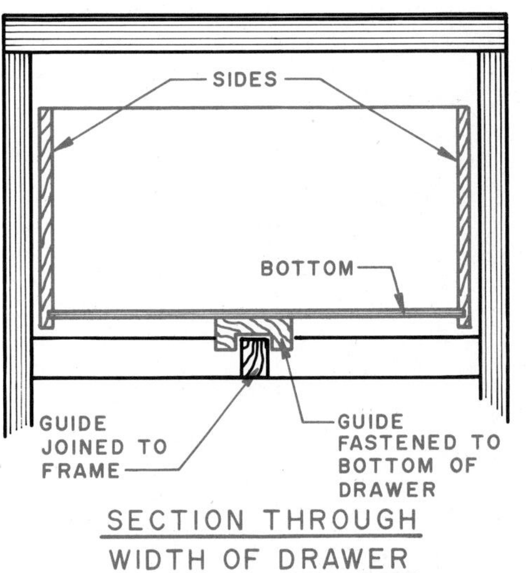

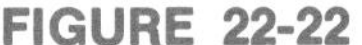

FIGURE 22-22

Wood drawer guides are used to guide the drawer in the cabinet.

Most cabinets do not have backs. They have a top and bottom rail that is used to join the cabinet to the wall. The wall surface serves as the back of the cabinet. A ¼-in.-thick plywood back can be installed if desired. It is best if it is set in rabbets on the back edges of the cabinet.

BATHROOM CABINETS

Bathroom cabinets are much the same as kitchen cabinets. The design and installation are the same. They are usually 29¼ in. high without the top and 21 in. deep (Fig. 22-23).

APPLYING PLASTIC LAMINATES

Plastic laminates are glued to plywood or particleboard countertops. Although most countertops are manufactured in a factory, sometimes a cabinetmaker or carpenter must apply a laminate on the construction site.

Countertops are usually covered with 1/16-in.-thick laminate. Vertical surfaces are covered with 1/32-in. material.

The plastic sheet is cut to size. It is usually cut with a circular saw having a blade with carbide-tipped teeth. A regular saw blade will get dull rapidly. Usually, the sheet is cut ½ in. larger than the area to be covered. It is trimmed to the finished size after it is adhered to the top.

Plastic laminates can also be cut by scoring. Place a straightedge along the line to be cut. Score the line deeply into the plastic with a scratch awl or knife. Holding the straightedge firmly against the plastic, bend the end up to break it on the line (Fig. 22-24).

If there is to be a seam, the joining edges must be straight. They can be undercut a little with a block plane, file, or sandpaper. This helps produce a tight seam.

The surface to be covered must be clean. The room temperature should be at least 70°F. Any cracks or holes in the top base material must be filled.

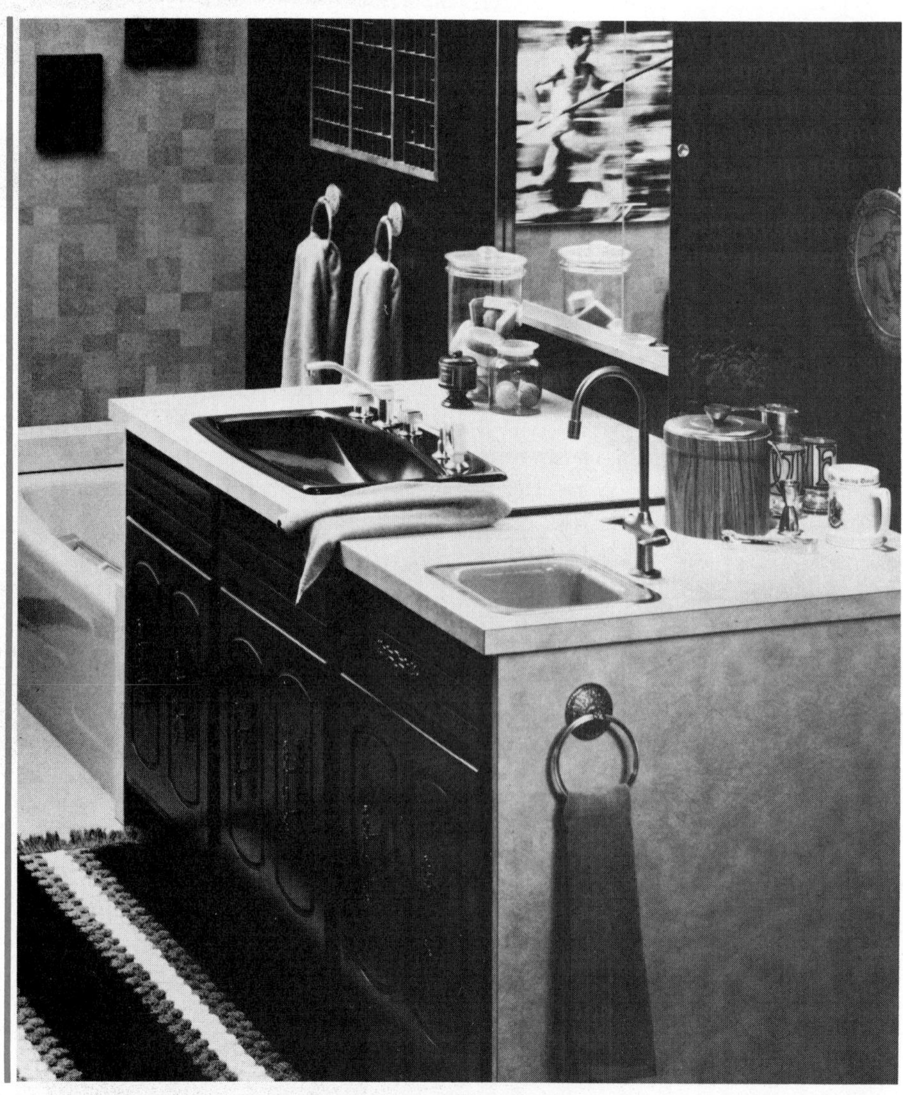

FIGURE 22-23

Lavatory base cabinets in a finished installation. *(Courtesy of Kohler Co.)*

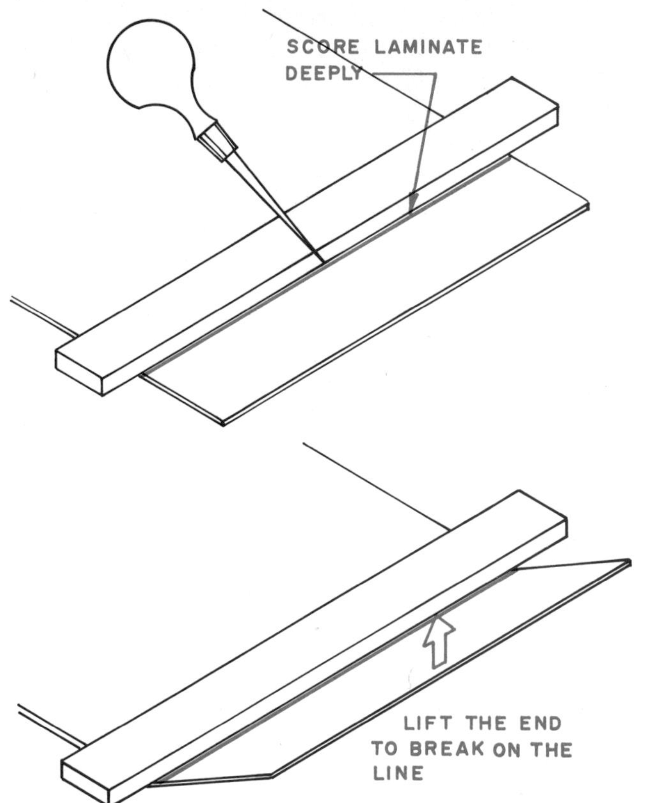

FIGURE 22-24

Plastic laminates can be cut by scoring and breaking.

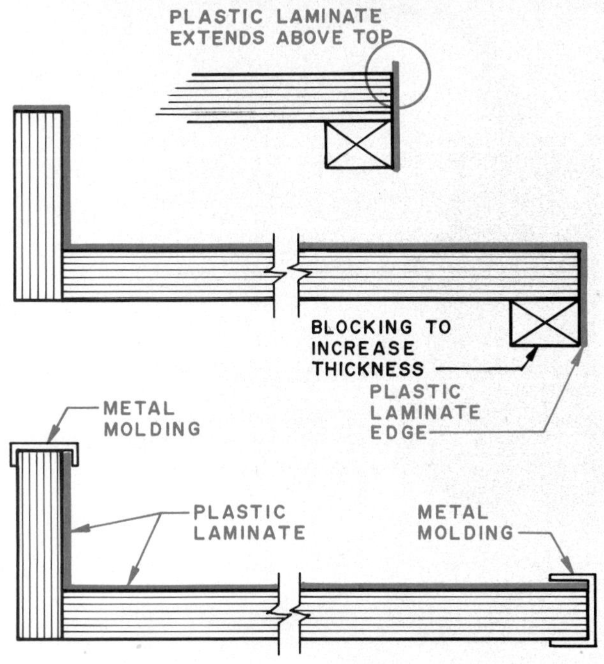

FIGURE 22-25

The plastic laminate is applied to the counter edge first. Metal molding is sometimes used instead of plastic.

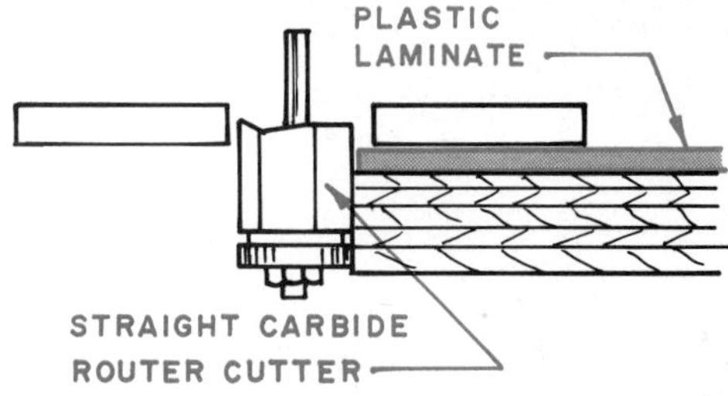

FIGURE 22-26

The excess plastic laminate is trimmed using a carbide straight cutter in an electric router.

The adhesive most frequently used is contact cement. The room must be kept ventilated when working with contact cement.

The edges of the top are covered first. This is done so that the top piece can overlap the edge. The edge to be covered is wiped clean. A rag dampened with lacquer thinner is good for cleaning the wood and the laminate. Apply the adhesive to the edge of the top and the back side of the laminate. Use a brush or small roller. Some manufacturers recommend applying a second coat after the first is dry. Then place the bottom edge of the plastic flush with the bottom edge of the wood top. The top of the plastic will stick up beyond the top of the counter (Fig. 22-25).

Press the plastic against the wood edge. Tap it in place with a wood block and hammer or a rubber mallet. Press it firmly along the entire length. After it has dried, trim it flush with the countertop. Drying usually requires 12 to 24 hours.

The best method to trim the laminate is to use an electric router with a carbide cutter. A ball-bearing roller runs along the surface of the top and controls the cut (Fig. 22-26). After trimming the edge, smooth it carefully with fine sandpaper.

A special laminate trimmer is used by those who do this work regularly. It operates much like the router but is designed especially for laminate trimming (Fig. 22-27).

Metal moldings are sometimes used to cover the edges of the top. They are applied after the plastic laminate is glued in place (Fig. 22-25).

Now apply adhesive to the wood top and the back of the laminate to cover the top (Fig. 22-28). After the adhesive is dry, place a sheet of kraft paper over the countertop. The contact cement will not stick to the paper. Then place the laminate with the adhesive down on the kraft paper. Slide the laminate around until it is in the proper position.

Then slide out about one-third of the kraft paper. This permits part of the laminate to adhere to the top. Then continue to remove the paper, permitting the laminate to adhere to the top.

FIGURE 22-27

A special laminate trimmer is used to trim the laminate flush with the surface of the countertop. *(Courtesy of Stanley Power Tools)*

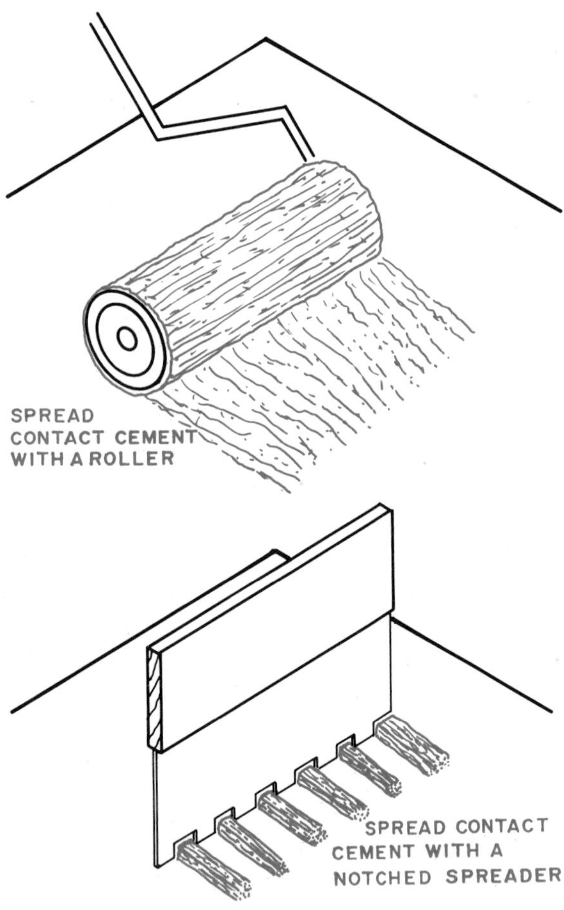

FIGURE 22-28

Apply contact cement to large surfaces with a roller or notched spreader.

Next, roll the laminate with a small roller. This helps it adhere to the top. Begin rolling in the center of the sheet and work toward the edges (Fig. 22-29).

The overhanging material on the edge is trimmed off with an electric router or laminate trimmer. Often, a slight bevel is desired on this cut (Fig. 22-30).

Preformed countertops are available. They have the plastic laminates already adhered to the top. The top is cut to length. Preshaped end caps are available to cover the ends.

These tops have the advantage of having an unbroken surface. The front edge is usually slightly raised to keep water from running over the edge of the counter (Fig. 22-31).

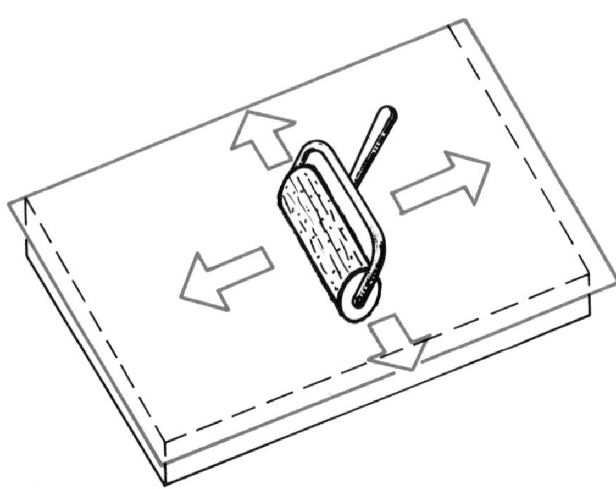

FIGURE 22-29

After the laminate adheres to the top, roll it starting from the center and working toward the edges.

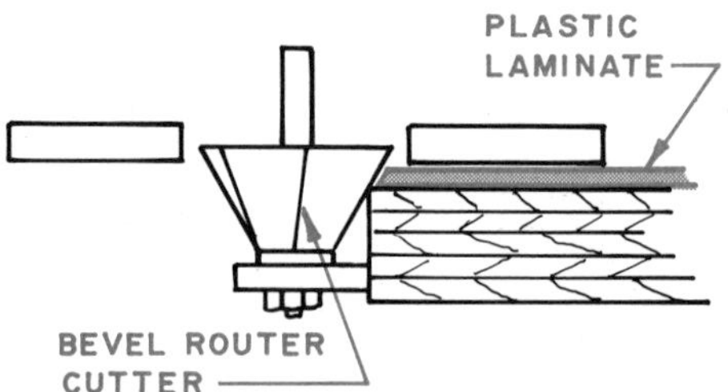

FIGURE 22-30

The top edge of the laminate can be beveled with a carbide bevel router cutter.

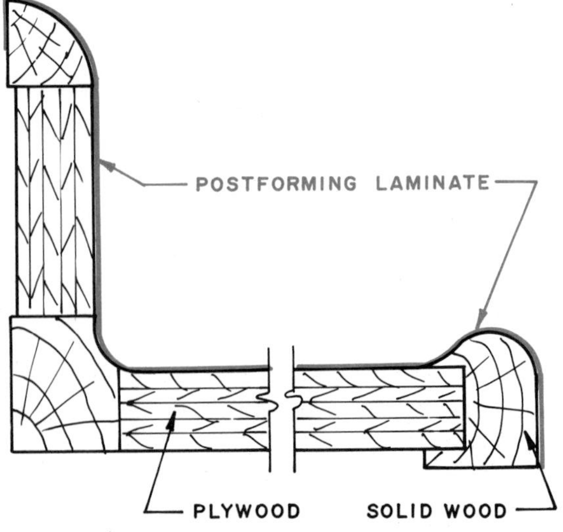

FIGURE 22-31

Preformed countertops are sold with the plastic laminate adhered to their surface.

STUDY QUESTIONS

After reading the chapter, answer each of the following questions. If you do not know the answer, review the chapter.

1. What information about cabinets is given on the working drawings?
2. What are three sources for kitchen cabinets?
3. How are factory-manufactured cabinets assembled for installation?
4. How are wall cabinets fastened to the wall?
5. How are base cabinets leveled?
6. When is the countertop installed?
7. How is the hole for a sink marked and cut?
8. What is the standard height and width of kitchen base cabinets?
9. Describe how a drawer is constructed.
10. How are plastic laminates cut?
11. Explain the steps for gluing a plastic laminate to a countertop.

IMPORTANT TECHNICAL TERMS

Following are technical terms that you should be able to use as part of your working vocabulary. Write a brief description of the meaning of each term.

Wall cabinets
Base cabinets
Stiles
Rails
Drawer guide
Plastic laminate

23 POST-AND-BEAM CONSTRUCTION

Post-and-beam construction utilizes large wood structural members for vertical support (posts), for spanning horizontal distances (beams), and for decking (planks) (Fig. 23-1).

POSTS

Posts are either solid wood or made by laminating (gluing together) 2-in. pieces of wood. For general residential work a solid 4 × 4 in. is often used. The exact size must be determined by the load it will carry and the spacing between posts. The longer a post gets, the greater it must be in cross-sectional area. Lateral bracing can reduce the needed cross-sectional area.

Building codes require posts to meet a predetermined slenderness ratio. The *slenderness ratio* is a number found by dividing the post length in inches by the actual size of the smallest dimension of the post in cross section. It is shown by the formula $1/d$. For example, a 4 × 4 in. wood post 8 ft 0 in. would be 96 in./3.5 in. = 27.5 slenderness ratio. Slenderness ratios for selected wood posts are in Fig. 23-2. The smaller the slenderness ratio, the greater the load that can be carried.

Usually, posts are spaced evenly in exterior walls. The designer usually places them on a standard module as 16, 24, or 48 in. for economy in using building materials.

BEAMS

Beams are either solid or built up. *Built-up beams* may be made of several 2-in. pieces of stock nailed together. These are used in places

FIGURE 23-1

Wood posts and beams are used for the structural framing of buildings. *(Courtesy of Timber Structures Inc.)*

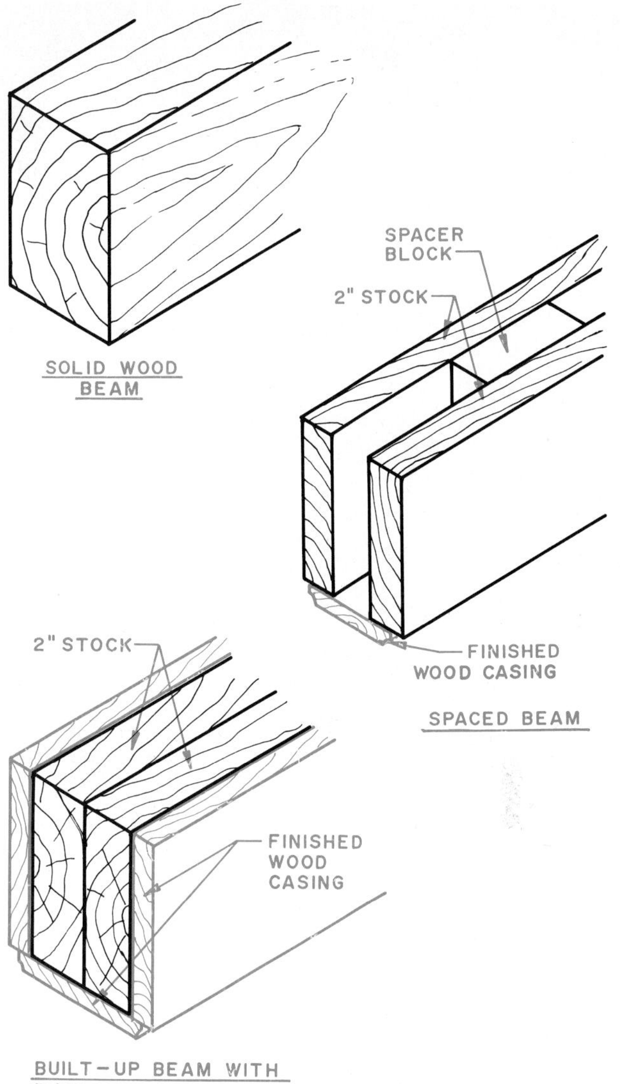

FIGURE 23-3

Wood beams can be solid, built-up, or spaced.

in which they are not seen. They may be below the floor or if exposed may be covered with finished boards and molding. Built-up beams may be made with wood spacing blocks. These are called spaced beams. They provide a cavity in which electric wires and plumbing can be run (Fig. 23-3).

Another type of beam is a *laminated beam.* It is made up of wood layers glued together under carefully controlled conditions. The thickness of the layers laminated together varies depending upon the design. Layers of ¾, 1⅜, and 1⅝ in. are often used. Laminated members are manufactured from lumber that is inspected for quality and accurately machined. They usually have a finished coating applied in the factory and are exposed as part of the interior design of the building (Fig. 23-4).

In addition to straight laminated beams, many other forms are manufactured. These include tapered beams, three-hinge arches, two-hinge arches, and trusses of various designs (Figs. 23-5 and 23-6).

NOMINAL SIZE (in.)	ACTUAL SIZE (in.)	SLENDERNESS RATIO (*l/d*)	CALIF. REDWOOD SELECT LOAD (lb)	DOUGLAS FIR SELECT STR. LOAD (lb)	SOUTHERN PINE SELECT STR. LOAD (lb)
4 × 4	3½ × 3½	27.4	6,360	7,830	8,800
4 × 6	3½ × 5½	27.4	10,000	12,300	13,830
6 × 6	5½ × 5½	17.5	31,150	30,250	37,800

Note: Data for solid wood posts 8-ft-0-in. long.

FIGURE 23-2

The slenderness ratio and safe load of posts for selected wood species.

FIGURE 23-4

Glued, laminated wood arches can span long distances.

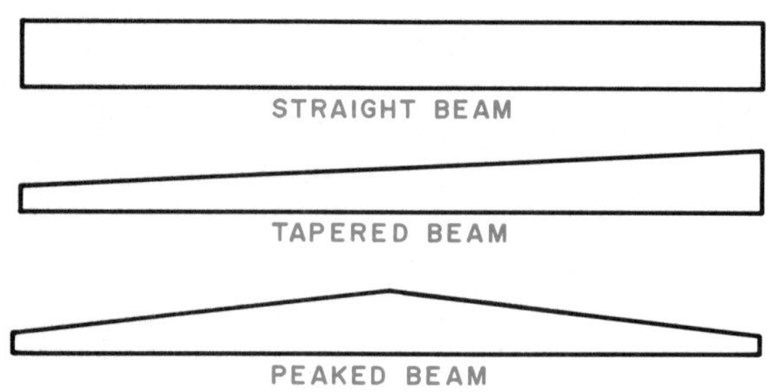

FIGURE 23-5

Typical types of laminated wood beams.

DECKING AND PLANKS

The terms "decking" and "planks" are used interchangeably to identify wood members over 1 in. thick used to cover roofs and serve as flooring. Generally, 2-in.-thick *decking* has a single tongue and groove. Three- and 4-in. decking has two tongue and grooves (Fig. 23-7). Some types join to form V-grooves at each joint. Others have a striated surface.

The distance the decking will span is influenced by its thickness, specie, grade of lumber used, installation method, and design loads. A few examples are in Fig. 23-8.

Wood decking is installed in three different ways: single-span, random length, and a combination of single-span and two-span continu-

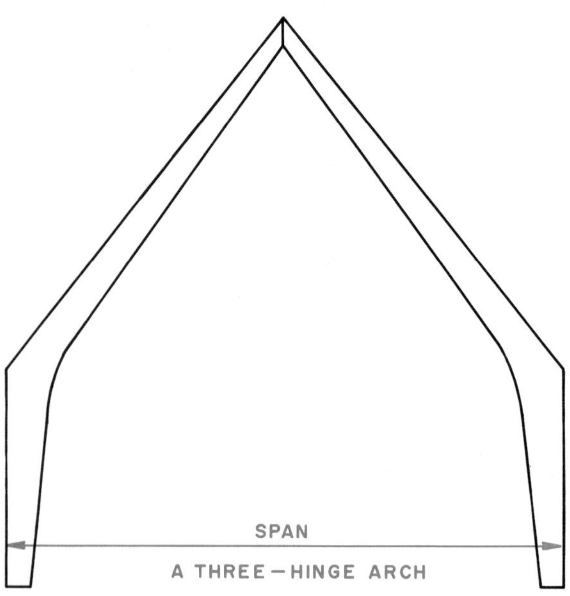

FIGURE 23-6

Types of laminated wood arches used to span long distances.

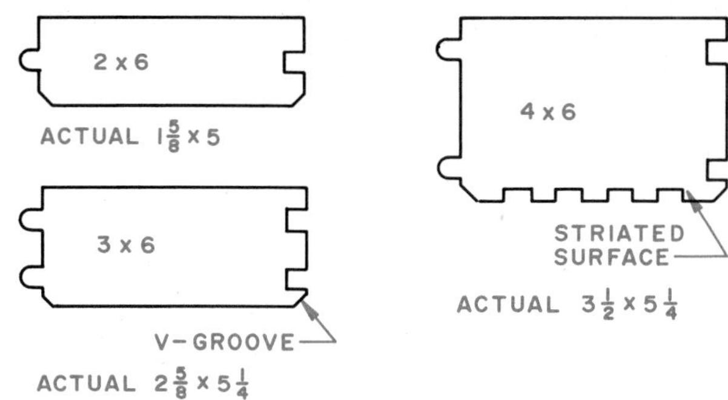

FIGURE 23-7

Heavy wood decking is used in post-and-beam construction.

ous (Fig. 23-9). The random-length decking has the ends tongue-and-grooved so that the joint need not occur over a beam.

The decking is installed by spiking each course to the preceding course at regular intervals through factory-drilled holes. Each course is then toenailed to the beam.

Since the bottom surface will form the exposed ceiling, the carpenter must be certain to avoid scoring it during installation (Fig. 23-10).

Wood decking is a good insulator. However, in cold climates it is advisable to cover it with a vapor barrier and 1-in. rigid insulation

FIGURE 23-8 (*right*)

Spans for selected grades of Douglas fir wood roof decking.

NOMINAL THICKNESS	GRADE	LIVE LOAD (lb/ft²)			
		20	30	40	50
2	Construction and Standard	10′-3″[a]	9′-0″	8′-2″	7′-7″
3	Select Deluxe and Commercial Deluxe	16′-9″	14′-6″	13′-3″	12′-3″
4	Select Deluxe and Commerical Deluxe	22′-0″	19′-3″	17′-6″	16′-3″

[a]Span in feet.

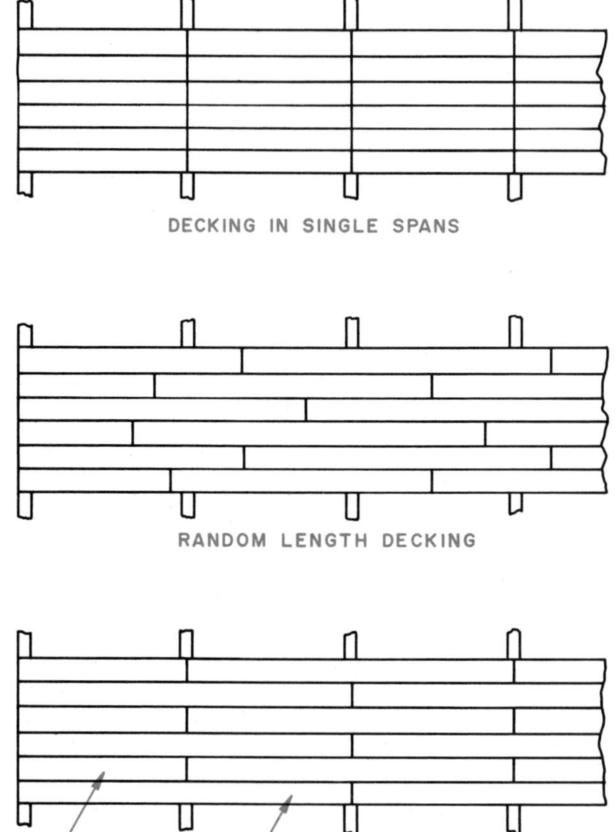

FIGURE 23-9

Ways to place wood decking.

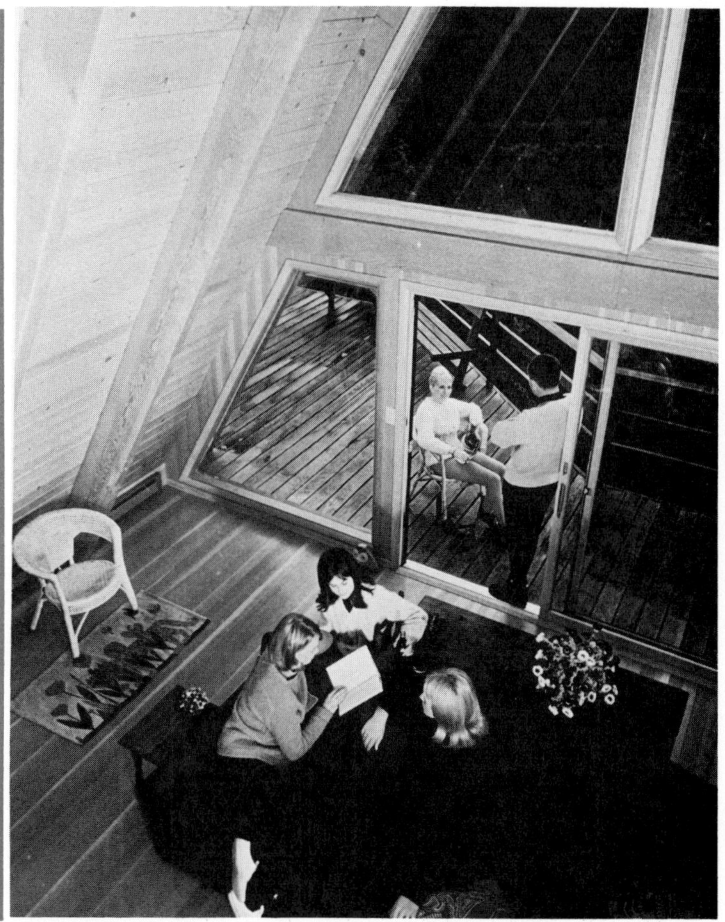

FIGURE 23-10

The bottom of wood decking is left exposed, forming the finished ceiling. *(Courtesy of Western Wood Products)*

on the outside surface. This prevents moisture from leaving the building (Fig. 23-11).

Wood fiber roof planks are made from long wood fibers bonded together with an inorganic hydraulic cement. Typical panel sizes are shown in Fig. 23-12. It is nailed to the wood beams. The surface to face the inside of the building has an attractive textured finish.

Another decking used is 2 × 4 × 1 in. plywood. It is 1⅛ in. thick, is tongue-and-grooved, and made with exterior glue. It gives the same performance as nominal 2-in. tongue-and-groove lumber decking. It is used for floor and roof construction. It is available with a textured surface which serves as a finished interior ceiling (Fig. 23-13). The end joints fall on the beams. For additional information on this system, see Chapter 10.

Stressed skin panels can be used for floors and roofs in post-and-beam construction (Fig. 23-14). The plywood skin is glued to longitudinal lumber members. Lumber blocking is used in the panel wherever there is a joint in the skin. The panels may have a single or double skin. The single-skin panels require bridging between the lumber stringers. They use a plywood splice plate below joints in the skin (Fig. 23-15).

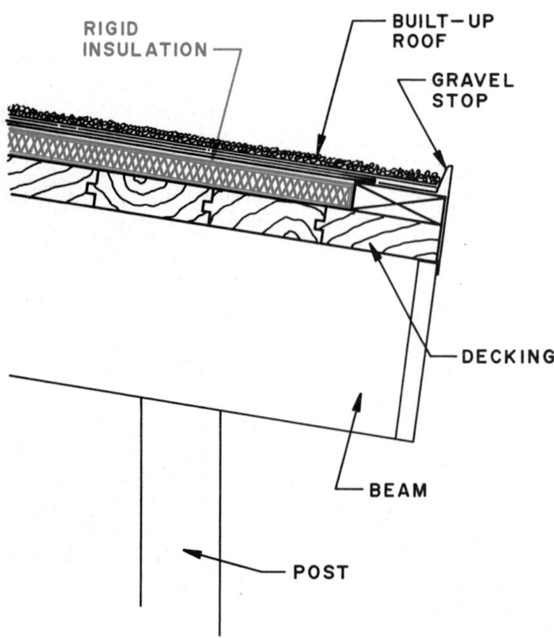

FIGURE 23-11

Wood roof decking can be covered with a rigid insulation and a vapor barrier.

Thickness (in.)	2, 2½, 3
Panel width (in.)	30¾, 46½
Panel length (in.)	
Tongue-and-grooved	48–96
Square-edged	48–144

FIGURE 23-12

Typical sizes for wood fiber roof decking.

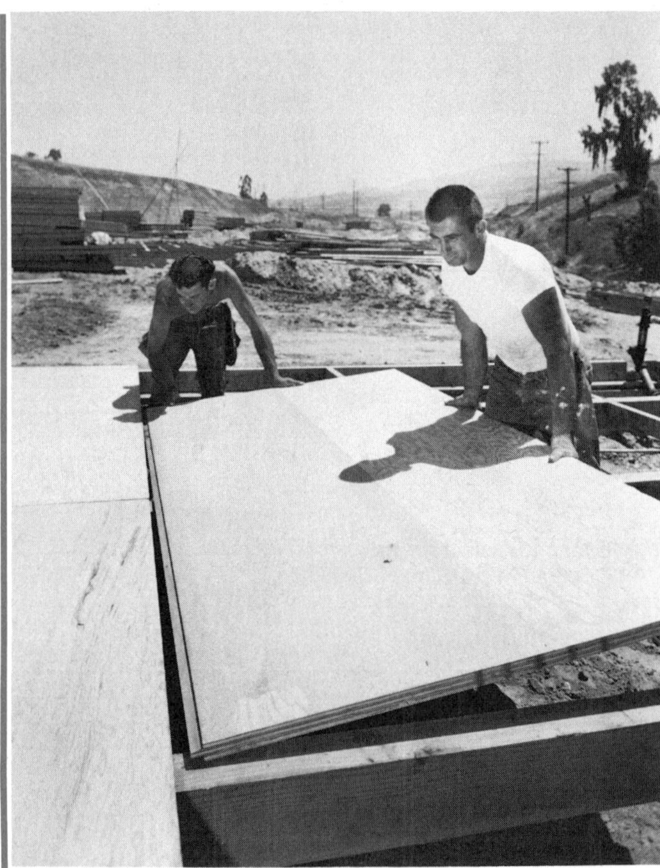

FIGURE 23-13

Plywood floor panels spanning wood floor beams. *(Courtesy of American Plywood Association)*

FIGURE 23-14

Carpenter setting a stressed skin roof panel in place.

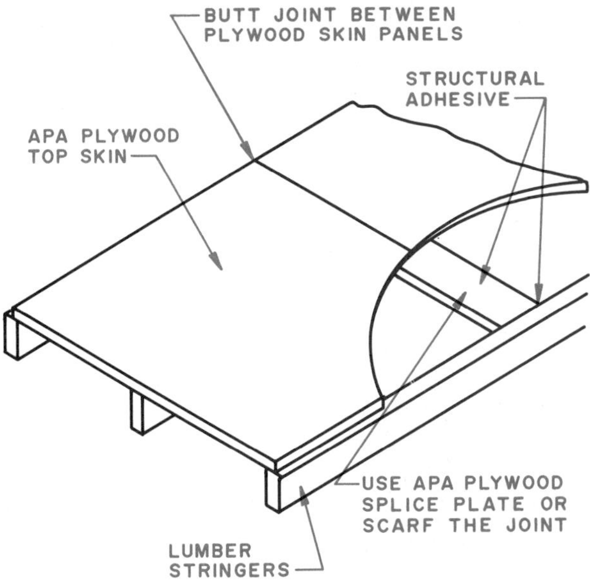

FIGURE 23-15

A typical single skin roof panel.

POST-AND-BEAM CONSTRUCTION

Post-and-beam construction permits a building to have large, open interior spaces free of supporting walls. Since the roof is supported by a row of posts, the exterior walls can have large open areas of glass or be covered with non-load-bearing wall materials (Fig. 23-16).

The system can be rapidly erected and under roof. There are several ways to frame a post-and-beam building. One commonly used method uses heavy wood beams as rafters with wood decking applied perpendicular to them (Fig. 23-17). Another runs wood beams the length of the building and the decking runs from the ridge to the outside wall (Fig. 23-18).

There are several ways the sill might be framed. The floor beam can rest on a wood sill. A header is nailed across the ends of the beam (Fig. 23-19). If it is desired to lower the appearance of the house, the beams could be set in

FIGURE 23-16

Post-and-beam construction permits the use of large glass panels in exterior walls.

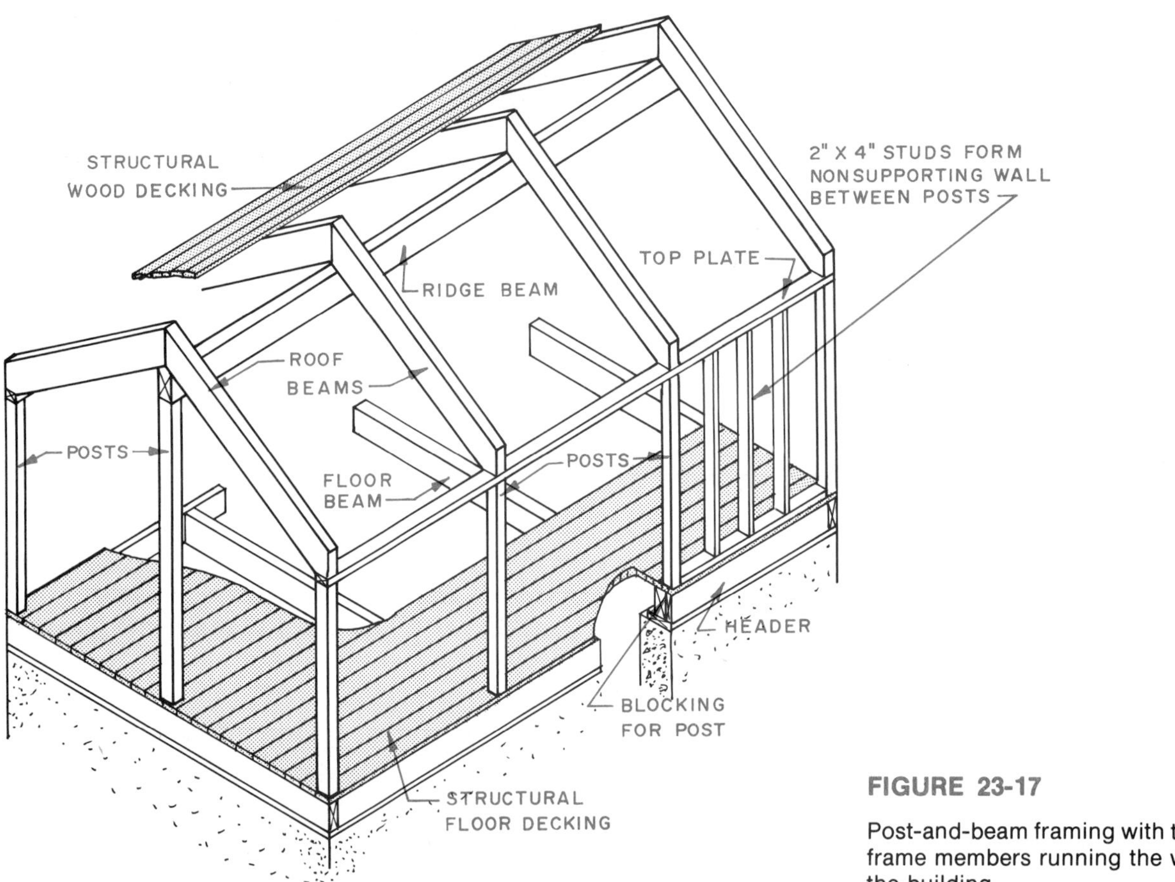

FIGURE 23-17

Post-and-beam framing with the wood frame members running the width of the building.

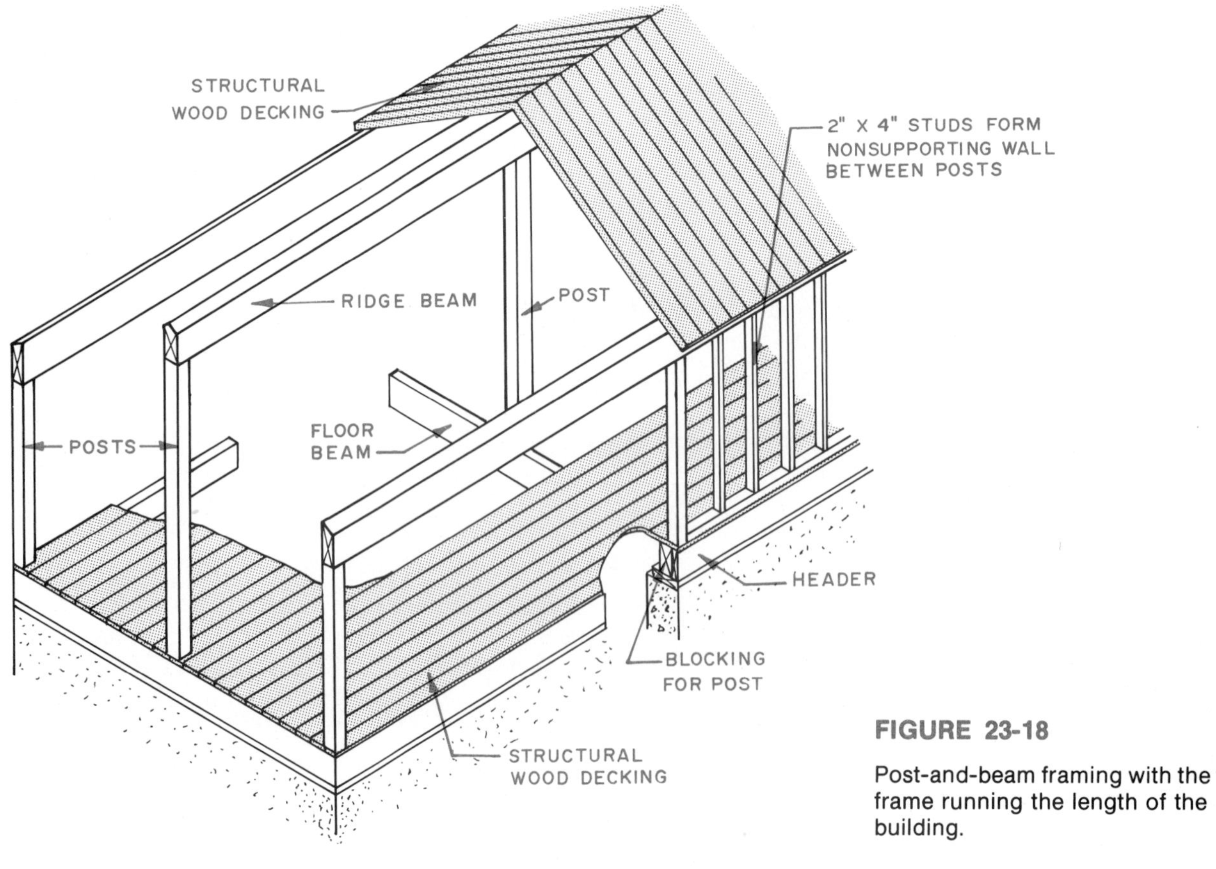

FIGURE 23-18

Post-and-beam framing with the wood frame running the length of the building.

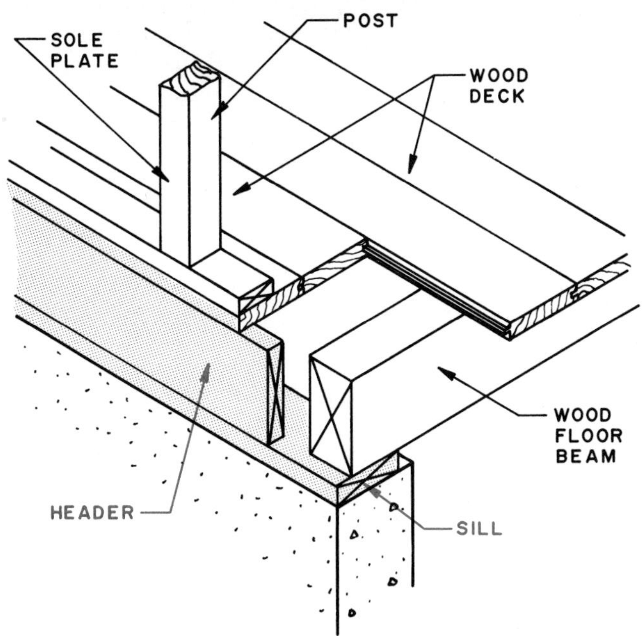

FIGURE 23-19

One way to frame the sill is to have a header run perpendicular to the floor beams.

pockets in the foundation (Fig. 23-20). The floor beams are supported in the center of the building by a column or pier (Fig. 23-21). Either of these methods enable a building to have a crawl space or a basement. If a concrete slab floor is desired, the posts are set on the foundation and held in place with a steel pin (Fig. 23-22).

The wood decking, which serves as finished flooring, is nailed to the floor beams.

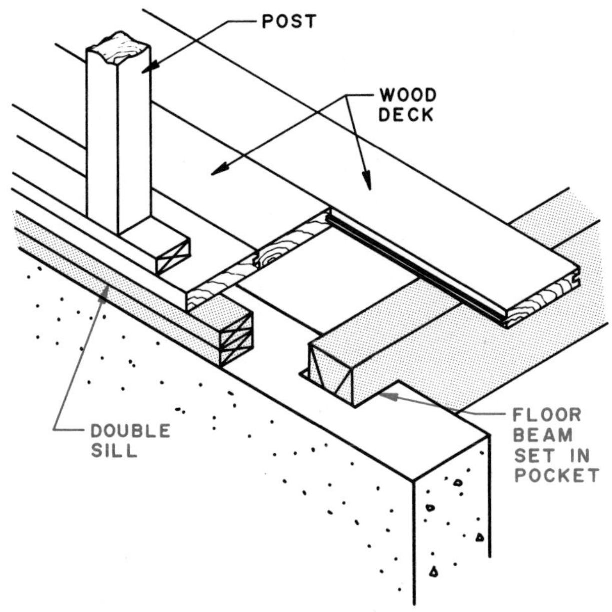

FIGURE 23-20

Floor beams can be set in pockets formed in the concrete foundation.

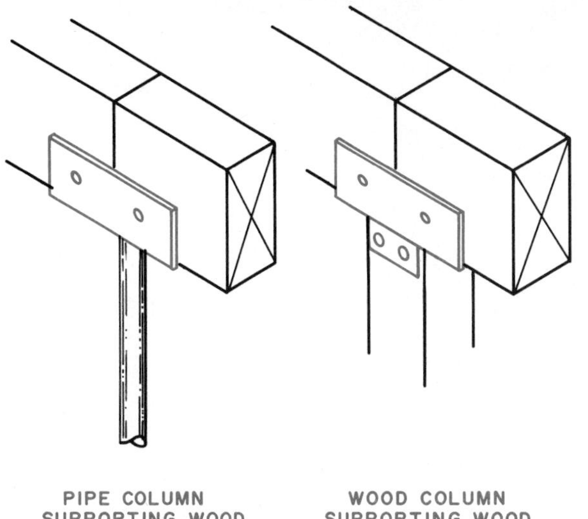

FIGURE 23-21

Floor beams are supported by a column or a pier.

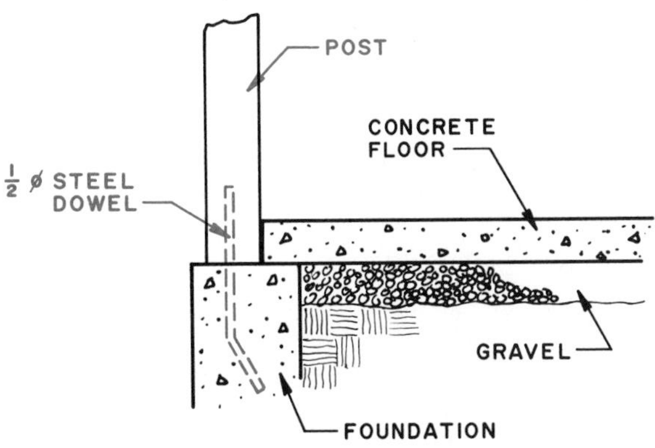

FIGURE 23-22

Sill construction for a post-and-beam framed building with a concrete slab floor.

Interior walls are installed on top of the flooring. Partitions that run perpendicular to a sloped ceiling have top plates and filler sections between the beams (Fig. 23-23). Partitions running parallel to sloped beams have a sloping top plate. Generally, the partition is framed in the normal manner. On top of this, the sloped section is built (Fig. 23-24).

When non-load-bearing partitions run perpendicular to the wood decking, no special support is needed. If they run parallel to the decking, a small beam must be added. It can be above or below the floor deck (Fig. 23-25). If it is above the deck, it can replace the sole.

Roof beams must be directly above a post. They are joined to the post with metal connectors. Sometimes a plate is also used (Fig. 23-26).

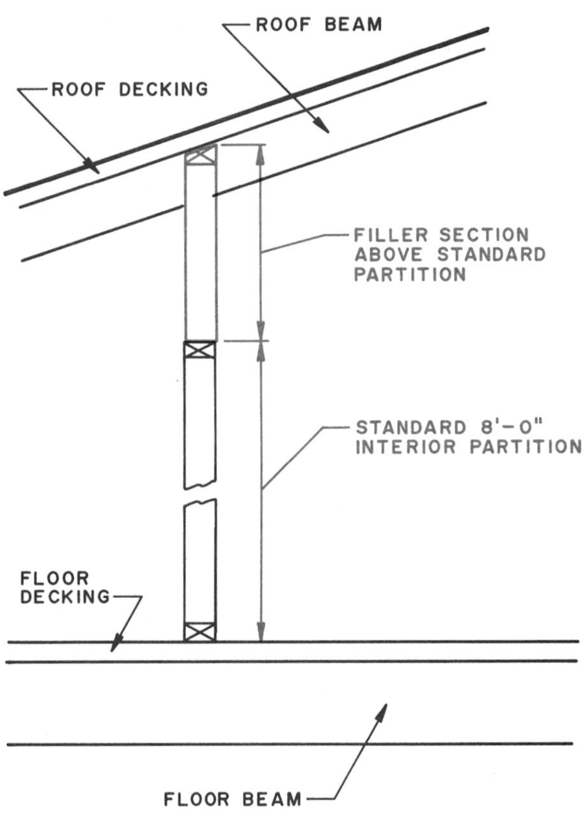

FIGURE 23-23

Filler sections are used above standard interior partitions. This partition is running perpendicular to the floor and roof beams.

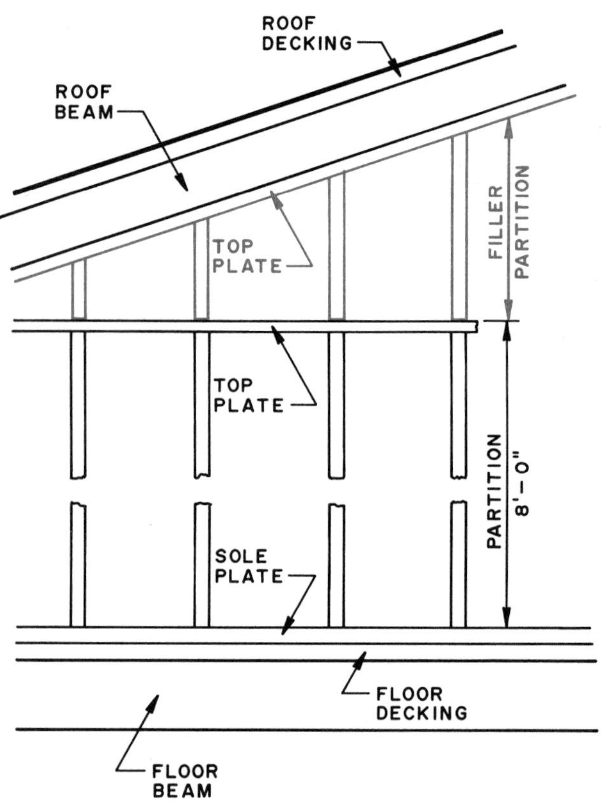

FIGURE 23-24

This filler partition is running parallel with the roof beams.

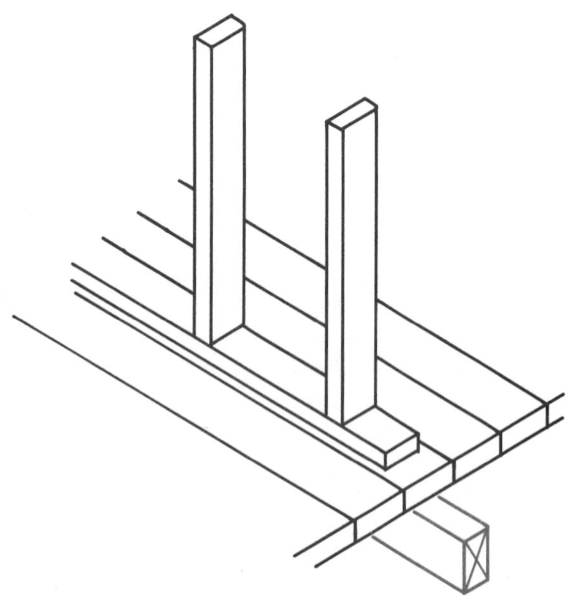

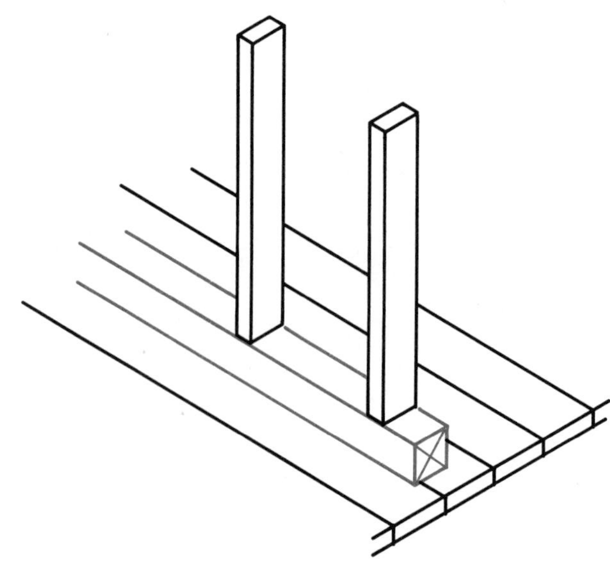

FIGURE 23-25

How to frame interior partitions that run parallel with the floor beams.

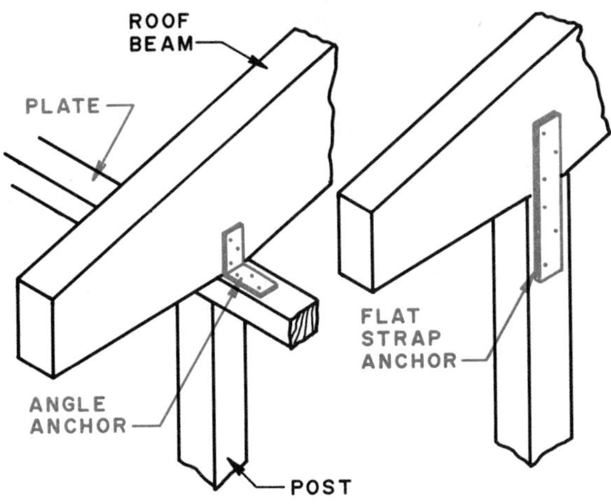

FIGURE 23-26

Roof beams must be placed directly above a post.

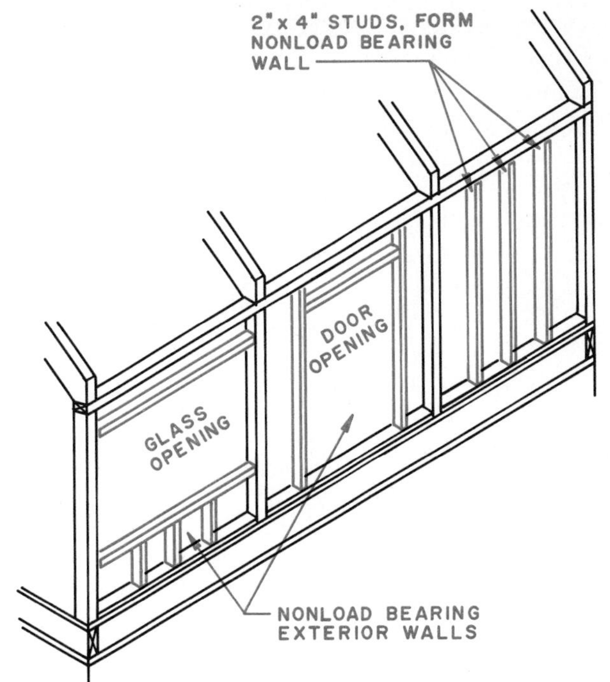

FIGURE 23-28

Typical framing of an exterior wall between posts using conventional 2 × 4-in. studs.

The ridge beam rests on posts on each end, spaced as needed to carry the load on the interior of the building. The roof beam can rest on top of the ridge or butt it. When they rest on top, the roof beam is often notched to seat on the ridge. Metal straps and plates are used to tie them together (Fig. 23-27).

Exterior walls can be framed using 2×4 in. studs in the normal manner. Since the walls are non-load-bearing, the designer has great freedom in using large windows (Fig. 23-28).

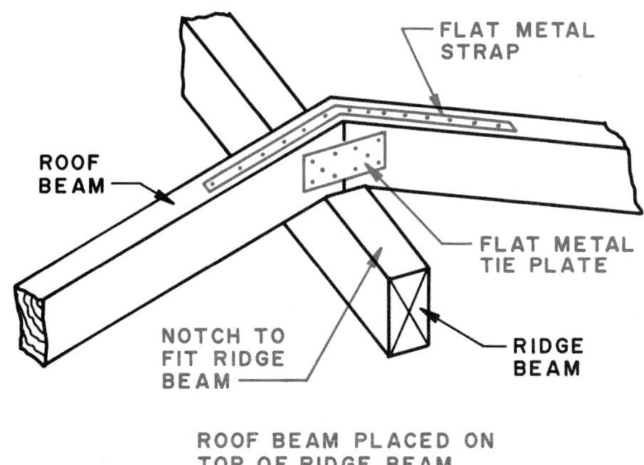

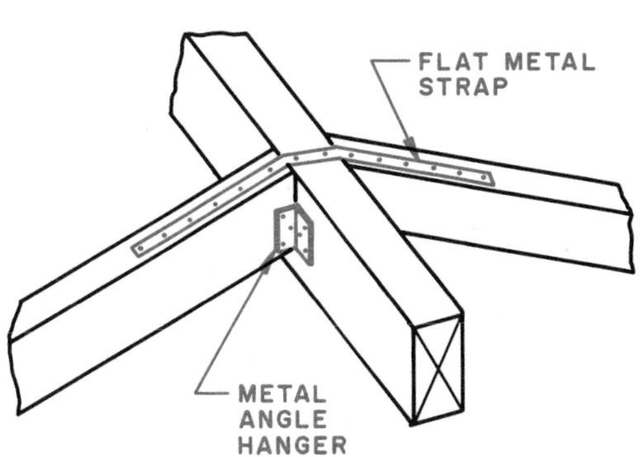

FIGURE 23-27

Rafter beam can butt against or be placed on top of the ridge beam.

STUDY QUESTIONS

After reading the chapter, answer each of the following questions. If you do not know the answer, review the chapter.

1. What types of posts are used?
2. What is the slenderness ratio?
3. What types of beams are used?
4. What material is used for the floor and roof?
5. What are the three ways in which wood decking can be installed?
6. What type of plywood can be used in place of 2-in. solid wood decking?
7. What is a stressed skin panel?
8. What are the two ways in which post-and-beam buildings are framed?
9. What must be done if a load-bearing partition runs parallel with the direction of the decking?

IMPORTANT TECHNICAL TERMS

Following are technical terms that you should be able to use as part of your working vocabulary. Write a brief description of the meaning of each term.

Posts
Slenderness ratio
Beams
Laminated beam
Decking
Wood fiber planks
Stressed skin panels

24 FACTORY-MANUFACTURED HOMES

Many single- and multiple-family dwellings are manufactured in a factory and shipped to the building site. Most of the carpentry work is performed indoors, enabling production to continue in periods of inclement weather. A lot of the wiring, plumbing, and interior finish is also installed in the factory.

These units are cut and assembled using many specially designed machines and jigs. These machines hold the various pieces and nail and glue them together. Some even can turn the unit over so that work can be performed on both sides (Fig. 24-1).

Production of this nature enables the manufacturer to purchase materials in large quantities, thus reducing the cost of the unit. The quality of material going into a house can be carefully controlled. Since the assembly of the materials is done in a controlled environment with machines designed to nail and glue the members together, a structurally sound building will result.

FACTORY-ASSEMBLED COMPONENTS

Even houses built from materials cut to size by the carpenter on the building site use many factory-made *components*. These include things such as windows, doors, door frames, cabinets of all kinds, and soffit systems. Structural units such as trusses, laminated arches, stressed-skin panels, box beams, and floor and roof panels are factory built.

FIGURE 24-1

Roof trusses are assembled in power-nailing jigs. *(Courtesy of National Homes Corporation)*

FACTORY-BUILT BUILDINGS

Factory-built buildings generally fall into several basic types. These include precut, panelized, and modular units.

Precut Buildings The parts for a *precut building* are cut to their proper length and labeled in a factory. They are shipped packaged for a particular section of the house. For example, the floor joists and subfloor are shipped as a package. They are assembled on the site by carpenters much the same as conventionally built buildings.

Panelized Buildings *Panelized buildings* are constructed with factory-manufactured wall panels, partitions, trusses, and other components (Fig. 24-2). The panels are either open or closed. An open panel has the finished exterior wall installed plus doors and windows. The wiring, plumbing, insulation, and interior wall finish are installed on the job.

A closed panel is completely finished, including interior and exterior finish, insulation, doors, windows, wiring, and plumbing.

The stock used to build panels is rapidly cut to size on saws with automatic feeding devices (Fig. 24-3). The various members are placed in jigs on assembly tables. Here they are

FIGURE 24-2

A residence constructed using factory-manufactured panels and components. *(Courtesy of National Homes Corporation)*

FIGURE 24-3

The structural members are cut to length on power-fed saws. *(Courtesy of National Homes Corporation)*

nailed and/or glued together (Fig. 24-4). The sheathing is applied to the studs (Fig. 24-5) and the finished siding is nailed in place. Doors and windows are usually installed at this time (Fig. 24-6).

The finished wall units are moved about the factory on overhead carriers (Fig. 24-7). When ready to be shipped, they are loaded on trucks for delivery to the site.

At the site the foundation has been previously built. Then the floor is assembled. The joists are cut to length at the factory and labeled. The header is notched to locate each joist. The carpenter assembles the floor without having to measure the joist locations (Fig. 24-8). After the tongue-and-grooved plywood subfloor is in place, the wall panels are set in place (Fig. 24-9).

FIGURE 24-4

The panels are assembled in jigs. *(Courtesy of National Homes Corporation)*

FIGURE 24-5

The sheathing is applied. *(Courtesy of National Homes Corporation)*

FIGURE 24-6

After the finished exterior material is in place, the windows are installed. *(Courtesy of National Homes Corporation)*

FIGURE 24-7

The finished wall panels on conveyors ready to be shipped. *(Courtesy of National Homes Corporation)*

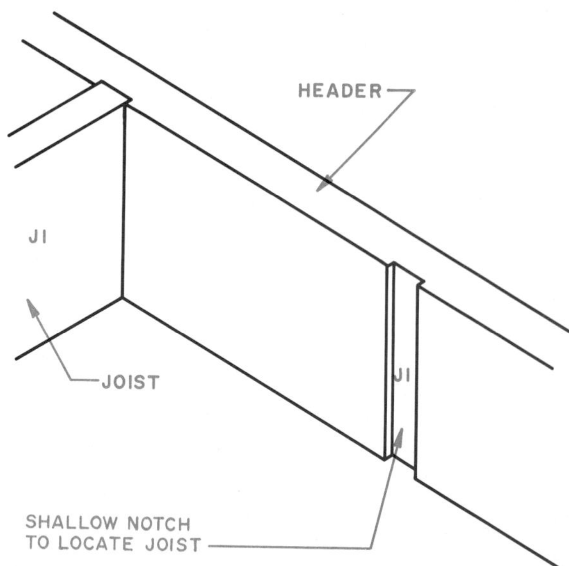

FIGURE 24-8

The header can be notched to locate each joist.

FIGURE 24-9

The tongue-and-groove subfloor is set in place. *(Courtesy of American Plywood Association)*

A corner wall panel is lifted from the truck and set in place. Next, an adjoining panel is set forming a corner (Fig. 24-10). After all the exterior panels are in place, the interior partitions are lifted into the building and nailed in place. Then the gable end is secured (Fig. 24-11) and the roof trusses are nailed to the walls. The plywood roof sheathing is lifted on to the roof and nailed to the trusses. The finished roof material is installed in the conventional manner.

Some companies manufacture a series of standard-size wall panels. These are made with doors, windows, and in solid sections of various lengths (Fig. 24-12).

As a house is designed, the length and width and door and window locations are chosen so as to utilize the panels available. These are assembled into a wall (Fig. 24-13).

The general construction and erection of these units is the same as those just discussed.

FIGURE 24-10

The exterior wall panels are set in place. *(Courtesy of National Homes Corporation)*

FIGURE 24-11

The gable ends are lifted into place. *(Courtesy of National Homes Corporation)*

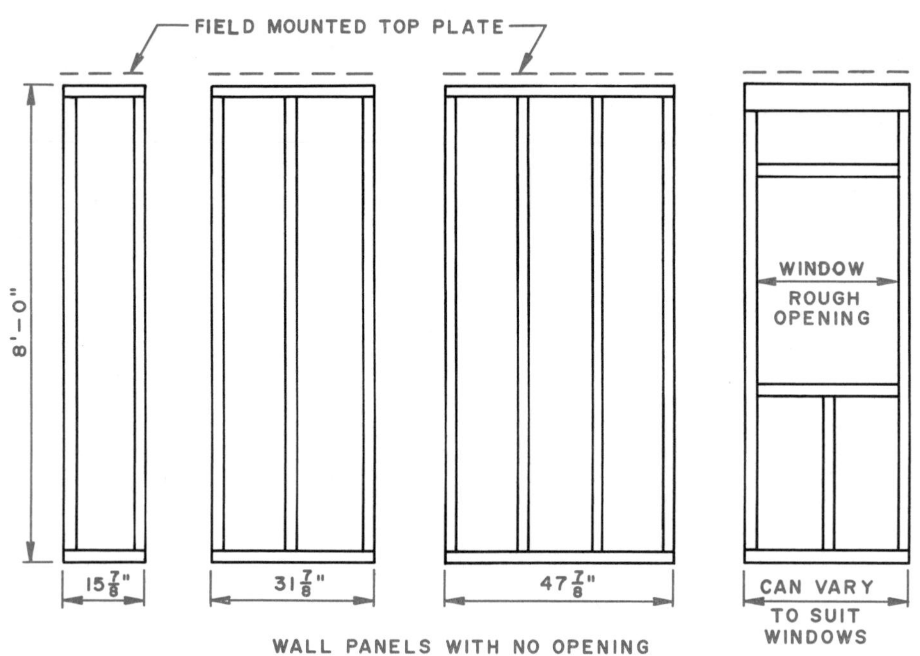

FIGURE 24-12

Typical wall panels, showing stud placement and window framing details. Sheathing is fastened over the studs in the factory.

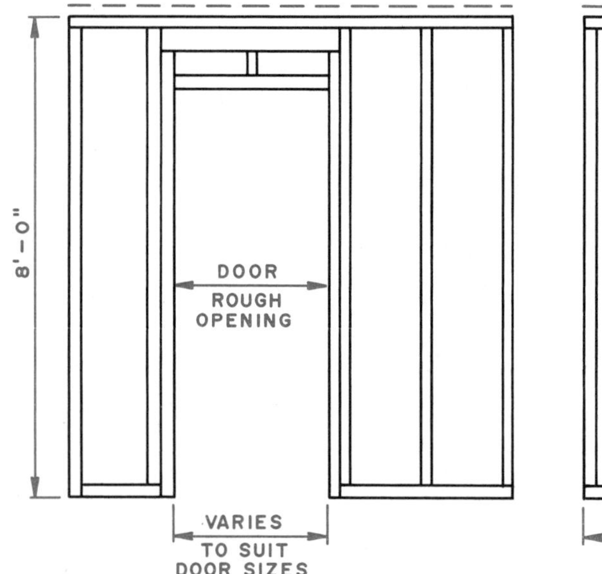

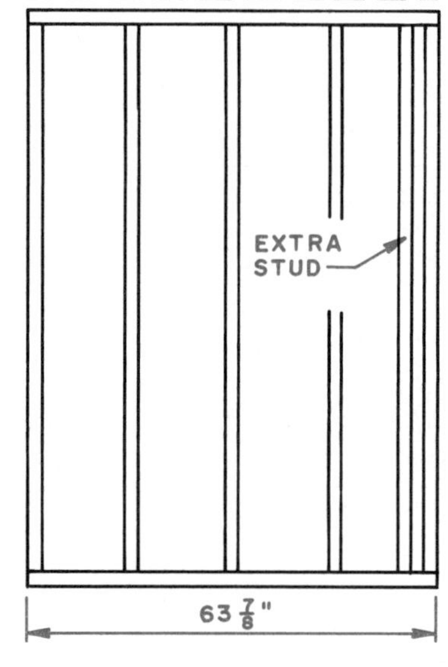

FIGURE 24-13

Standard wall panels are assembled into an exterior wall.

Modular Buildings A *modular building* is one made up of factory-finished units composed of the floor, walls, and roof. The unit is usually completely finished inside and outside. The siding, doors, windows, roof shingles, and soffit are all finished. Inside, the unit has been wired, the plumbing is in place, and interior walls are finished. The floor tile or carpet is usually installed. All cabinets and other built-in furniture are sometimes in place (Fig. 24.14).

The finished units are moved to the building site on specially designed trailers. They are set on the foundation with a crane. Here the electrical and plumbing are connected to the service brought to the foundation (Fig. 24-15).

Usually, the unit can be occupied in a few days.

Modular building units are limited in the width they can be built. Since they are moved on highways, the width is regulated by laws applying to trailers and mobile homes. The units are set on foundations (Fig. 24-16). Piers are used in the center where the two modules join together. The structural members in the floor over the piers form a beam to support the floor load (Fig. 24-17).

Construction details are similar to conventionally built buildings. Since the building is moved on a trailer-type vehicle, the floor must

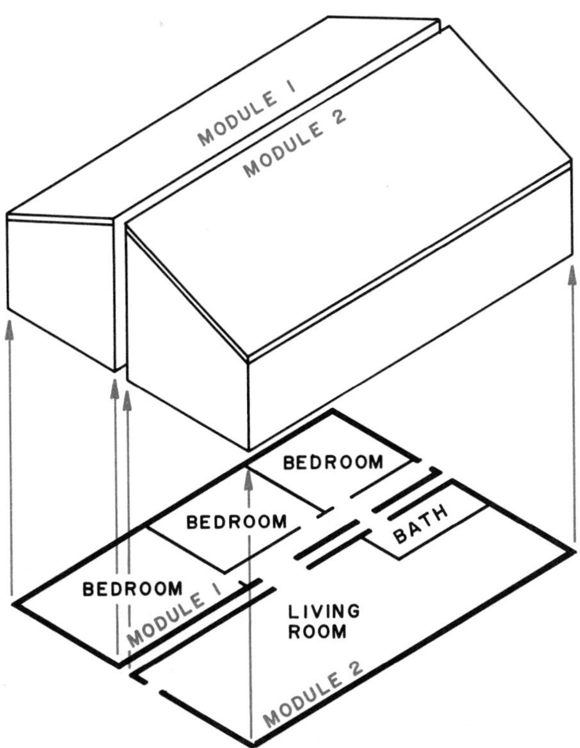

FIGURE 24-14

A typical plan for a two-module residence.

FIGURE 24-15

The modular units are moved to the site on special trailers. *(Courtesy of Insta Housing, Inc.)*

FIGURE 24-16

Cranes place the modular units on the foundation. *(Courtesy of Insta Housing, Inc.)*

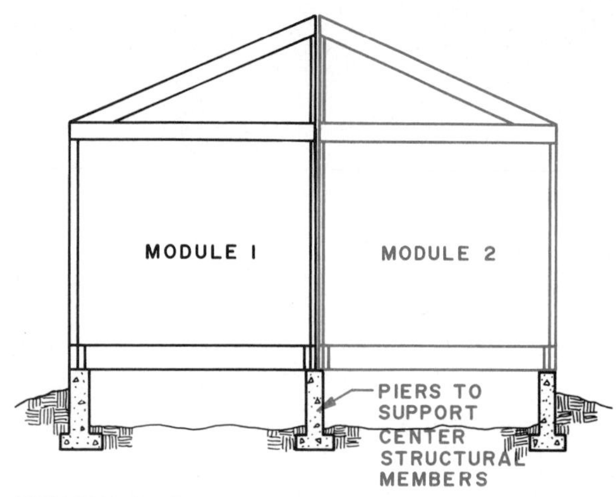

FIGURE 24-17

Piers support the modular units where they join.

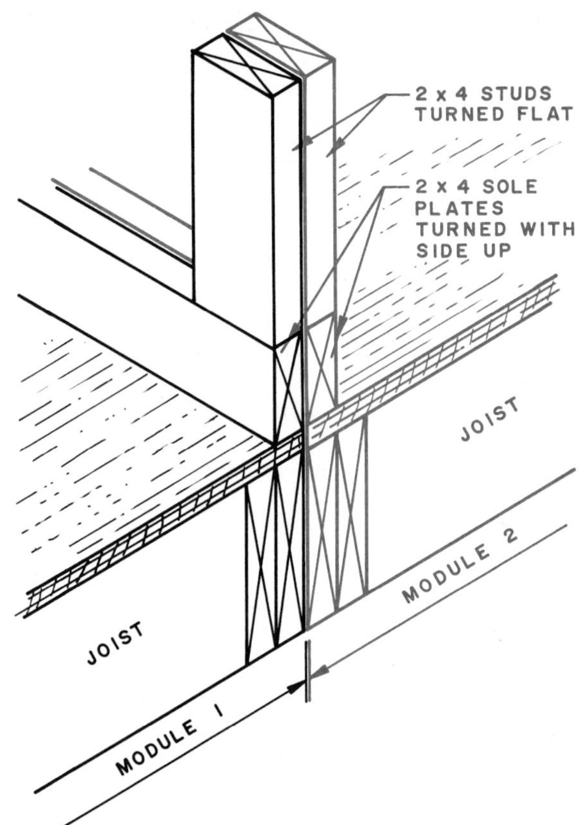

FIGURE 24-19

The two modules are bolted together in the center.

be designed for the extra strain. Usually, the headers on all sides of the unit are doubled. The end floor joists are also doubled.

One type of roof construction is a half truss (Fig. 24-18). The trusses and interior walls are joined as shown in Fig. 24-19. The studs on the joining interior walls are turned flatwise and joined together.

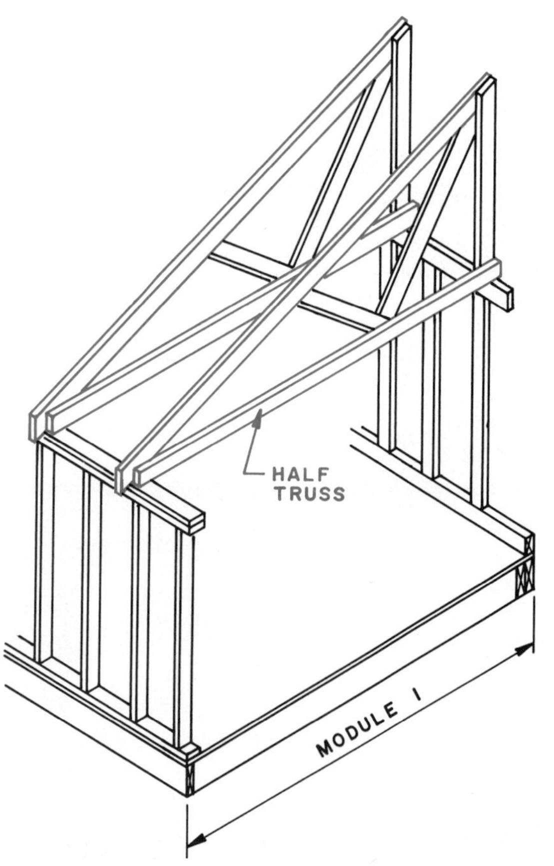

FIGURE 24-18

A module can be built with a half roof truss.

Since a wide overhang would reduce the width of the rooms, it is sometimes left off the unit and applied after they are on the foundation.

Modular units are built on an assembly line. The floor system is framed and the subfloor is applied. The walls are assembled on jigs using power nailing and gluing techniques. The finished wall is moved by conveyor to the floor, where it is set in place (Fig. 24-20). The plumbing and wiring is installed and the interior walls are finished (Fig. 24-21). The exterior siding, doors, and windows are installed and painted if required. The roof trusses are set in place and sheathed. The finished roof is stapled to the sheathing. All exterior details such as corner boards or soffits are complete. The unit is lifted onto a trailer frame for delivery.

FIGURE 24-20

Finished walls on a conveyor ready to be joined to the floor. *(Courtesy of Guerdon Industries, Inc.)*

FIGURE 24-21

Walls being attached to the floor. Notice the plumbing that has been installed. *(Courtesy of Guerdon Industries, Inc.)*

STUDY QUESTIONS

After reading the chapter, answer each of the following questions. If you do not know the answer, review the chapter.

1. What are advantages to manufacturing buildings in a factory?
2. List some factory-assembled components used in a conventionally built house.
3. What is the difference between panelized, precut, and modular home construction?

IMPORTANT TECHNICAL TERMS

Following are technical terms that you should be able to use as part of your working vocabulary. Write a brief description of the meaning of each term.

Components
Precut
Panelized
Modular

GLOSSARY

A

Acoustical Materials: Materials that will absorb sound waves. Use mainly on walls and ceilings.

Actual Size of Lumber: The size of lumber after it is dried and dressed.

Adhesive: A pliable mixture used to bond one material to another.

Air-Dried Lumber: Lumber stored out-of-doors in carefully arranged stacks that allow air to circulate between the boards and remove moisture.

Anchor Bolt: A steel bolt used to secure a wood sill to a concrete or masonry foundation.

Apprenticeship: Educational preparation for those desiring to learn a skilled trade by working on a job and going to special classes at night.

Apron: A piece of trim placed beneath the window stool on the inside of a window.

B

Backfill: The replacement of earth in the space excavated on the sides of a foundation.

Ballon Framing: A type of frame construction in which the studs extend in one piece from the sill to the roof plate. Second-floor joists are held by ledger boards.

Baluster: Small vertical members in a railing that run between the tread and the top handrail.

Balustrade: A railing made up of balusters, a top rail, and sometimes a bottom rail. It is used on the open side of a stair, balcony, or porch.

Baseboard: A finished piece of interior trim placed against the wall where it meets the floor.

Base Shoe: A small molding nailed next to the baseboard at the floor.

Batten: A narrow strip of wood

nailed over cracks between adjoining boards or panels.

Batter Boards: Horizontal boards nailed to posts driven in the ground near the corners of an excavation. They are used to locate the desired corner of the foundation and establish its level.

Bay Window: A window with three window units that project beyond the exterior wall.

Beam: A large piece of wood or steel used to support a load over a long distance.

Bearing Wall: A wall that carries a load, such as a ceiling or roof load.

Bench Mark: A reference mark on the construction site from which land measurements and elevations are measured.

Bidding: When contractors estimate the cost of a job and indicate how much they will charge if selected to do the work.

Bird's-Mouth: A notch cut in the end of a rafter to permit it to sit firmly on the top plate of the exterior wall.

Blind Nailing: Placing nails so that they are hidden as the unit is built.

Board: A sawed piece of wood up to 2 in. thick and 2 in. or more wide used for sheathing and subflooring.

Board Foot: Equals 144 cubic inches of wood.

Board Measure: A system for the measurement of the quantity of lumber. The quantity is given in terms of board feet.

Box Beam: A beam built by assembling solid wood stringers and plywood panels following engineered design sizes and nailing patterns.

Brace: A wood member nailed to some part of a structure to stiffen it. It may be permanent or tempoary.

Brick Veneer: A type of construction in which a wood-framed building has a layer of bricks laid over its exterior forming the siding.

Bridging: Wood or metal cross braces placed between floor and ceiling joists.

Building Code: A legal document specifying the regulations that must be observed when designing and building buildings.

Building Permit: A permit issued by a local government agency granting permission to construct a building.

Built-Up Beam: A beam made by nailing and/or gluing together several layers of wood.

Built-Up Roof: A roof covering composed of layers of tar and builder's felt. The top layer is usually gravel. It is used for flat and low-sloped roofs.

C

Cabinet: A unit built to store articles. It has sides, bottom, top, shelves, doors, and sometimes drawers.

Cantilever: A projection beyond the point of support, as floor joists extending beyond the foundation wall.

Casing: A wood trim placed around a door or window either inside or outside the building.

Caulk: To seal cracks and joints with a pliable material to make them watertight.

Chord: The horizontal member of a truss.

Collar Beam: A horizontal member nailed to pairs of opposite roof rafters to stiffen the roof.

Column: A vertical supporting member.

Common Rafter: A rafter that extends from the ridge to the top plate.

Coped Joint: A cut on a molding that shapes it to fit the face of a piece it will meet.

Core: The central layer of a sheet of plywood.

Corner Boards: Boards used to finish the external corners of a house. The siding butts against them.

Cornice: The exterior trim formed where the roof and exterior wall join. It usually consists of a fascia board, a soffit, and moldings.

Counterflashing: Flashing used on chimneys at the roof line which covers the shingle flashing.

Cove Molding: Molding with a concave face. It is used to finish interior corners such as between a wall and the ceiling.

Crawl Space: The space below the floor joists in houses without basements.

Cricket: A single- or double-sloped structure used to shed water away from roof surfaces which meet in a sharp V, such as a roof and a chimney.

Cripples: Short studs such as those used above and below window openings.

Cross-bands: Layers of veneer glued at right angles to each other, forming plywood sheets.

Crosscutting: Cutting wood across the grain.

D

d: A unit used to indicate nail sizes.

Dead Load: A load that is permanently acting on the structure of a building, such as the weight of the roof.

Dimension Lumber: Surfaced softwood lumber manufactured to standard sizes and used for framing buildings.

Dressed Size of Lumber: The size of lumber after it has been dried and smoothed by planing.

Drip Cap: A molding at the exterior top side of doors and windows to help shed rain.

Drywall: A system of finishing interior walls using gypsum panels with joints taped, plywood, or hardboard panels.

Door Frame: The casing surrounding the door to which the door is hinged.

Door Head: The top of the door frame.

Dormer: A projection built out from a sloping roof to hold one or more windows. It has its own roof and sides, which join the sloping roof.

Duplex nail: A double-headed nail used when the nail will eventually have to be removed, such as in concrete formwork.

E

Eaves: The lower part of a roof that projects over the exterior wall.

Edge-Matched: Lumber that has tongue-and-grooved edges.

Elevation: Drawings of buildings which show the design of the exterior walls. Also, the height of some part of a building above an established measuring point.

End-Matched: Lumber that has tongue-and-grooved ends.

Exposure: The amount a shingle or siding is exposed to the weather.

F

Face Nail: To drive nails through the face of a board so that they are visible when in place.

Fascia: The flat, vertical member of a cornice which is nailed along the outer edge of the eave.

Finish Grade: The elevation of the ground in relation to the building when land is landscaped.

Finish Lumber: High-quality softwood lumber which can be stained or finished natural.

Fire Stops: Wood members placed horizontally between studs to stop drafts up the inside of the wall.

Flashing: Sheet material, usually metal, used to prevent water from leaking into joints in a building. Common locations are between the chimney and the roof and above windows.

Floor Plan: A drawing showing the width and length of a building and the location of rooms, doors, windows, partitions, and other details.

Fly-Rafter: The end rafter that overhangs the exterior wall and is supported by roof sheathing and lookouts.

Footing: A concrete pad in the ground upon which a foundation is built.

Forms: Panels that when joined together form molds into which concrete is poured and held until it hardens.

Foundation: A concrete or wood wall extending from the footing to the first-floor joists.

Foundation Plan: A drawing showing the width and length of the foundation and the location and size of footings, piers, columns, and foundation walls.

Framing: The process of joining together the various wood structural members.

Frieze: A horizontal member joining the siding to the soffit.

Frost Line: The depth that frost penetrates the earth.

Furring: The placing of wood strips on masonry or concrete walls so that another material can be nailed over the wall.

G

Gable: That part of an exterior wall located between the slopes of the roof rafters.

Gable Roof: A roof that slopes to a point from two sides.

Gain: A notch cut to receive a hinge, lock, or other hardware.

Gambrel Roof: A roof with two surfaces, each at a different slope. The lower slope is steeper than the upper slope.

Girder: A large main structural member carrying loads between piers or columns, often called a beam.

Girder Pocket: A notch in the foundation into which the girder is placed.

Green Lumber: Lumber that has not been dried.

Grounds: Wood strips nailed around openings in walls and at the floor. Used to control the application of plaster.

Gusset: A plywood piece glued and nailed to wood members at joints to strengthen the joint.

Gypsum Board: A panel made with a core of gypsum between two layers of fibrous absorbent paper.

H

Hardboard: A panel made from wood fibers put under heat and pressure.

Hardwood: Wood from trees that shed their leaves in the winter.

Header: Structural wood members used to frame around openings in floors and walls.

Heartwood: The mature wood located in the center of the tree.

Hip Rafter: A rafter that runs from the wall top plate to the hip ridge.

Hip Roof: A roof in which all sides slope to the ridge.

I

Identification Index: A stamp placed on plywood sheets which gives information about its grade.

Impact Insulation Class: The ability of a material to resist the transmission of impact noise.

Insulation: Material used to resist the transfer of heat or sound.

Insulation Board: Panels made from wood cane or other types of cellulose fibers.

Interior Finish: Materials used to cover the interior wall and ceiling surfaces.

Interior Trim: The moldings, casings, baseboard, and other trim items used to finish the interior of a building.

J

Jack Rafter: A rafter shorter than the common rafter.

Jamb: The side frame of doors and windows.

Joist: Structural members placed in horizontal positions. Usually used to support floors and ceilings.

Joist Hanger: Metal brackets used to support joists when they join beams or other joists.

K

Kerf: The slot made by a saw when it cuts through a board.

Kiln-Dried Lumber: Lumber dried in a kiln using steam and heated air under controlled conditions.

Knocked Down: When something is delivered to the job with parts cut to size but not assembled.

L

Ladder Jack: A metal frame that is fastened to extension ladder rungs and supports scaffolding boards.

Lally Column: A steel pipe used to support girders.

Landing: The floor area at the top or bottom of a stair.

Lath: Metal or gypsum panels that are nailed to studs. They form the base upon which plaster is applied.

Ledger Strip: A wood member nailed to studs or beams to provide support for other members which join them.

Let-in Brace: A brace that is set into studs by notching the studs.

Level: A tool used to see if items are exactly horizontal or vertical.

Lintel: A structural member over a small opening such as a door or window.

Live Load: A load that is placed on the structure of a building but which may be moved or removed, such as furniture.

Lookout: Members that extend beyond the exterior wall to support a fly rafter, fascia, and soffit.

Louver: A unit with inclined slots used for ventilating attics and crawl spaces.

Lumber: Products produced at the sawmill by cutting logs into standard sizes.

M

Matched Lumber: Lumber that has its edges shaped to form a tongue-and-groove joint.

Measuring Line: An imaginary line running from the outside corner of the double plate parallel with the edge of the rafter to the centerline of the ridge.

Metric System: A system of measurement widely used throughout the world. Linear measure is based on the meter.

Mil: A unit of thickness equal to one thousandth of an inch.

Millwork: Wood products mass produced in a manufacturing plant, such as doors, windows, panels, moldings, and trim.

Miter: Cutting a piece of wood on a 45° angle.

Moisture Content: The amount of moisture remaining in wood expressed as a percentage of the weight of the wood if it were oven-dry.

Molding: Wood strips machined to various shapes to be used for decorative purposes.

Mullion: A vertical member in the frame separating doors and windows which are next to each other.

Muntin: A wood member that separates the panes of glass in doors and windows.

N

Newel: The large, major post at the foot of a staircase.

Nominal Size of Lumber: The size of lumber after it is rough cut at the sawmill.

Non-Load-Bearing Wall: A wall that does not support any weight other than its own.

Nosing: The extension of a stair tread beyond the front of the riser.

O

On Center: A term used to indicate a measurement is taken from the center of one member to the center of the next member. Also shown as o.c.

OSHA: Occupational Safety and Health Act of 1973.

Out-of-Plumb: When a member is not in proper vertical alignment, it is "out-of-plumb."

P

Particleboard: A processed wood panel made of wood fibers, flakes, and shavings bonded together with a synthetic resin.

Penny: A term used to designate the length of nails. It is shown by the symbol d.

Pier: A masonry column used to support a beam.

Pitch: The ratio between the rise of a roof and its span.

Plancier: The underside of an eave or cornice. It is also called a soffit.

Plate: A horizontal structural member. It can be placed on top of the studs in a wall as a top plate, on the bottom of the studs as a sole plate, or on the foundation as a sill plate.

Platform: A horizontal area occurring in the middle of a stair to provide a break in the stair.

Platform Framing: A framing system in which studs extend for only one floor and rest upon the subfloor of each story.

Plot Plan: A drawing showing the size and slope of the lot and location of the building.

Plumb: Means that a member is in a vertical position.

Plywood: A panel made by gluing layers of veneer together, keeping their grains at 90° angles.

Post: A timber set on end and used to support a beam or other weight.

Purlin: Horizontal members in a roof which run between rafters to help support the sheathing.

Q

Quarter Round: A small molding with a cross section of one-fourth of a cylinder.

Quarter-Sawed: Lumber cut at a 90° angle to the annular growth rings.

R

Rafter: A structural member forming the roof that supports the sheathing.

Rail: Horizontal cross members of panel doors or window sash.

Rake: The flat, inclined edge on a gable roof.

Ribbon: A narrow board nailed to studs to add support to floor or ceiling joists.

Ridge Board: A horizontal member at the very peak of the roof to which rafters are nailed.

Ripping: Cutting wood with the direction of the grain.

Rise: The total vertical height of a stair.

Riser: The vertical board on a stair running between two treads.

Roofing Bracket: A metal bracket nailed over the roof ridge which is used to support scaffolding boards.

Rough Opening: An opening in a wall, floor, or ceiling formed by the structural framing members.

Roughing In: The installation of plumbing, electrical, and mechanical systems prior to applying the interior finish.

Run: The total horizontal length of a stair.

***R*-value:** The measure of a material's resistance to the flow of heat.

S

Sapwood: The light-colored living wood located outside the heartwood near the bark.

Sash: A frame containing the glass in a window.

Scab: A short piece of wood nailed to two butting pieces of wood to splice them together.

Scaffold: A platform used to hold workers when they must work at heights above the ground.

Scarfing: Joining the ends of stock with a sloping lap-type joint.

Scribing: Marking material to be cut to fit on an irregular surface.

Sheathing: Material used to cover roof rafters or exterior wall studs.

Shed Roof: A roof with only one sloping surface.

Shingles: Units nailed over roof and wall sheathing forming the finished exterior surface.

Shoring: Materials used to temporarily brace and hold something in place.

Siding: Materials applied over sheathing on exterior walls to provide a finished surface.

Sill: The bottom member of a window or door frame.

Sill Plate: A structural member bolted to the top of the foundation.

Sleeper: A wood member fastened to a concrete floor, enabling a wood floor to be nailed over the concrete floor.

Slope: The rise of the roof in inches per foot of run.

Soffit: The undersurface of a cornice from the fascia to the exterior wall.

Softwood: Wood from trees that keep their leaves in the winter.

Sole or Sole Plate: A plate nailed to the subfloor upon which the wall studs are fastened.

Sound-Transmission Class: A measure of the resistance of a material to the passage of airborne sound.

Span: The distance between two supports.

Specifications: A typed set of instructions written by the architect governing materials, workmanship, and procedures which cannot be clearly shown on the working drawings.

Stair: A series of steps that enable people to walk from one floor to the next.

Stair Horse: An inclined member of a stair with the tread and riser section cut into it.

Stile: The vertical members forming the structure of a panel door.

Stool: A molding shaped to fit on top of the window sill between the jambs.

Story Pole: A rod used to lay out needed measurements such as window height or the location of shingle courses.

Strike Plate: A metal plate fastened on the door frame to receive the door lock plunger.

Stringer: A finished board that is placed between the wall and the stair horse.

Subcontractor: A company that specializes in one type of work, such as electrical wiring.

Subfloor: Material applied to floor joists which forms the base upon which a finished flooring material is laid.

Stud: A vertical structural member used to form the framework of walls.

T

Tail: The end of the rafter that extends beyond the exterior wall.

Threshold: The member that closes the space between the door sill and the bottom of the door.

Timber: Wood members 5 in. or more in their least dimension.

Toenail: To drive nails at a slant.

Tread: The horizontal board on a stair upon which a person steps.

Trim: The finished moldings used inside and outside a building.

Trimmer: A beam or joist into which a header is framed.

Truss: An assembly of structural members that forms a unit to support the roof and ceiling material.

U

Underlayment: Materials laid on top of the subfloor to provide a smoother surface for the finished flooring.

U.S. Product Standard PSI-74: Standard grading regulations for softwood plywood.

***U*-Value:** The coefficient of heat transmission through several materials such as a wall or ceiling.

V

Valley: The intersection of two sloping roofs forming a V-shaped depression.

Valley Rafter: A rafter that runs from a wall top plate to the ridge board in a valley.

Vapor Barrier: A material that resists the flow of water vapor.

Veneer: Thin sheets of wood cut from logs and assembled into sheets of plywood.

W

Wainscoting: Paneling applied to the lower part of an interior wall.

Waler: Wood structural members used to stiffen concrete forms.

Warp: Any variation of wood from a flat plane. The types of warp are bow, crook, cup, twist, and wind.

Web: The diagonal supports used in trusses.

Winders: Treads of steps that are used on a winding staircase and are triangular in shape.

Working Drawings: A complete set of drawings giving all details about a building.

INDEX

C

H

I

J

K

L

Q

R

S

T

U

V

W

Y